Lines

Slope of line through (x_1, y_1) and (x_2, y_2):

$$m = \frac{y_2 - y_1}{x_2 - x_1}$$

Point-slope equation of line through (x_1, y_1) with slope m:

$$y - y_1 = m(x - x_1)$$

Slope-intercept equation of line with slope m and y-intercept b:

$$y = b + mx$$

Rules of Exponents

$$a^x a^t = a^{x+t}$$
$$\frac{a^x}{a^t} = a^{x-t}$$
$$(a^x)^t = a^{xt}$$

Definition of Natural Log

$y = \ln x \quad$ means $\quad e^y = x$
ex: $\ln 1 = 0$ since $e^0 = 1$

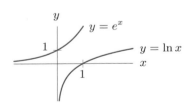

Identities

$$\ln e^x = x$$
$$e^{\ln x} = x$$

Rules of Natural Logarithms

$$\ln(AB) = \ln A + \ln B$$
$$\ln\left(\frac{A}{B}\right) = \ln A - \ln B$$
$$\ln A^p = p \ln A$$

Distance and Midpoint Formulas

Distance D between (x_1, y_1) and (x_2, y_2):

$$D = \sqrt{(x_2 - x_1)^2 + (y_2 - y_1)^2}$$

Midpoint of (x_1, y_1) and (x_2, y_2):

$$\left(\frac{x_1 + x_2}{2}, \frac{y_1 + y_2}{2}\right)$$

Quadratic Formula

If $ax^2 + bx + c = 0$, then

$$x = \frac{-b \pm \sqrt{b^2 - 4ac}}{2a}$$

Factoring Special Polynomials

$$x^2 - y^2 = (x + y)(x - y)$$
$$x^3 + y^3 = (x + y)(x^2 - xy + y^2)$$
$$x^3 - y^3 = (x - y)(x^2 + xy + y^2)$$

Circles

Center (h, k) and radius r:

$$(x - h)^2 + (y - k)^2 = r^2$$

Ellipse

$$\frac{x^2}{a^2} + \frac{y^2}{b^2} = 1$$

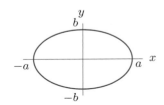

Hyperbola

$$\frac{x^2}{a^2} - \frac{y^2}{b^2} = 1$$

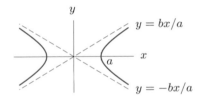

Geometric Formulas

Conversion Between Radians and Degrees: π radians $= 180°$

Triangle

$A = \frac{1}{2}bh$

$\quad = \frac{1}{2}ab\sin\theta$

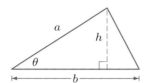

Circle

$A = \pi r^2$

$C = 2\pi r$

Sector of Circle

$A = \frac{1}{2}r^2\theta \quad$ (θ in radians)

$s = r\theta \quad$ (θ in radians)

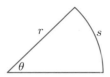

Sphere

$V = \frac{4}{3}\pi r^3 \quad A = 4\pi r^2$

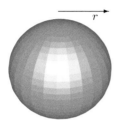

Cylinder

$V = \pi r^2 h$

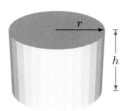

Cone

$V = \frac{1}{3}\pi r^2 h$

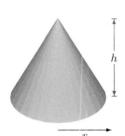

Trigonometric Functions

$\sin\theta = \dfrac{y}{r}$

$\cos\theta = \dfrac{x}{r}$

$\tan\theta = \dfrac{y}{x}$

$\tan\theta = \dfrac{\sin\theta}{\cos\theta}$

$\cos^2\theta + \sin^2\theta = 1$

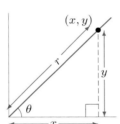

$\sin(A \pm B) = \sin A \cos B \pm \cos A \sin B$

$\cos(A \pm B) = \cos A \cos B \mp \sin A \sin B$

$\sin(2A) = 2\sin A \cos A$

$\cos(2A) = 2\cos^2 A - 1 = 1 - 2\sin^2 A$

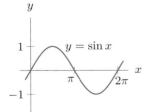

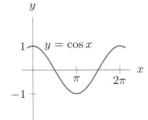

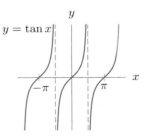

The Binomial Theorem

$(x+y)^n = x^n + nx^{n-1}y + \dfrac{n(n-1)}{1\cdot 2}x^{n-2}y^2 + \dfrac{n(n-1)(n-2)}{1\cdot 2\cdot 3}x^{n-3}y^3 + \cdots + nxy^{n-1} + y^n$

$(x-y)^n = x^n - nx^{n-1}y + \dfrac{n(n-1)}{1\cdot 2}x^{n-2}y^2 - \dfrac{n(n-1)(n-2)}{1\cdot 2\cdot 3}x^{n-3}y^3 + \cdots \pm nxy^{n-1} \mp y^n$

MULTIVARIABLE CALCULUS

Sixth Edition: International Student Version

MULTIVARIABLE CALCULUS

Sixth Edition: International Student Version

Produced by the Calculus Consortium and initially funded by a National Science Foundation Grant.

William G. McCallum
University of Arizona

Deborah Hughes-Hallett
University of Arizona

Daniel Flath
Macalester College

David Mumford
Brown University

Andrew M. Gleason
Harvard University

Brad G. Osgood
Stanford University

Selin Kalaycıoğlu
New York University

Cody L. Patterson
University of Arizona

Brigitte Lahme
Sonoma State University

Douglas Quinney
University of Keele

Patti Frazer Lock
St. Lawrence University

Adam H. Spiegler
Loyola University Chicago

Guadalupe I. Lozano
University of Arizona

Jeff Tecosky-Feldman
Haverford College

Jerry Morris
Sonoma State University

Thomas W. Tucker
Colgate University

with the assistance of
Adrian Iovita
University of Washington

Coordinated by
Elliot J. Marks

WILEY

This material is based upon work supported by the National Science Foundation under Grant No. DUE-9352905. Opinions expressed are those of the authors and not necessarily those of the Foundation.

ISBN-13 978-1118-57221-4

Printed in Asia

10 9 8 7 6 5 4 3 2

PREFACE

Calculus is one of the greatest achievements of the human intellect. Inspired by problems in astronomy, Newton and Leibniz developed the ideas of calculus 300 years ago. Since then, each century has demonstrated the power of calculus to illuminate questions in mathematics, the physical sciences, engineering, and the social and biological sciences.

Calculus has been so successful both because its central theme—change—is pivotal to an analysis of the natural world and because of its extraordinary power to reduce complicated problems to simple procedures. Therein lies the danger in teaching calculus: it is possible to teach the subject as nothing but procedures—thereby losing sight of both the mathematics and of its practical value. This edition of *Calculus* continues our effort to promote courses in which understanding and computation reinforce each other.

Mathematical Thinking Supported by Theory and Modeling

The first stage in the development of mathematical thinking is the acquisition of a clear intuitive picture of the central ideas. In the next stage, the student learns to reason with the intuitive ideas in plain English. After this foundation has been laid, there is a choice of direction. All students benefit from both theory and modeling, but the balance may differ for different groups. Some students, such as mathematics majors, may prefer more theory, while others may prefer more modeling. For instructors wishing to emphasize the connection between calculus and other fields, the text includes:

- A variety of problems from the **physical sciences** and **engineering**.
- Examples from the **biological sciences** and **economics**.
- Models from the **health sciences** and of **population growth**.

Origin of the Text

From the beginning, this textbook grew out of a community of mathematics instructors eager to find effective ways for students to learn calculus. This Sixth Edition of *Calculus* reflects the many voices of users at research universities, four-year colleges, community colleges, and secondary schools. Their input and that of our partner disciplines, engineering and the natural and social sciences, continue to shape our work.

Active Learning: Good Problems

As instructors ourselves, we know that interactive classrooms and well-crafted problems promote student learning. Since its inception, the hallmark of our text has been its innovative and engaging problems. These problems probe student understanding in ways often taken for granted. Praised for their creativity and variety, the influence of these problems has extended far beyond the users of our textbook.

The Sixth Edition continues this tradition. Under our approach, which we called the "Rule of Four," ideas are presented graphically, numerically, symbolically, and verbally, thereby encouraging students with a variety of learning styles to expand their knowledge. This edition expands the types of problems available:

- New **Strengthen Your Understanding** problems at the end of every section. These problems ask students to reflect on what they have learned by deciding "What is wrong?" with a statement and to "Give an example" of an idea.
- **ConcepTests** promote active learning in the classroom.
  Available in a book or on the web at www.wiley.com/college/hughes-hallett.
- **Class Worksheets** allow instructors to engage students in individual or group class-work. Samples are available in the Instructor's Manual, and all are on the web at www.wiley.com/college/hughes-hallett.
- Updated **Data and Models**. For example, the new project on the relation between noise levels and flight patterns a tLondon's Heathrow Airport in Chapter 12 underscores how mathematics can influence social and political decisions.

- **Projects** at the end of each chapter provide opportunities for a sustained investigation, often using skills from different parts of the course.
- **Drill Exercises** build student skill and confidence.
- **Online Problems** available in WileyPLUS or WeBWorK, for example. Many problems are randomized, providing students with expanded opportunities for practice with immediate feedback.

Symbolic Manipulation and Technology

To use calculus effectively, students need skill in both symbolic manipulation and the use of technology. The balance between the two may vary, depending on the needs of the students and the wishes of the instructor. The book is adaptable to many different combinations.

The book does not require any specific software or technology. It has been used with graphing calculators, graphing software, and computer algebra systems. Any technology with the ability to graph functions and perform numerical integration will suffice. Students are expected to use their own judgment to determine where technology is useful.

Content

This content represents our vision of how calculus can be taught. It is flexible enough to accommodate individual course needs and requirements. Topics can easily be added or deleted, or the order changed.

Changes to the text in the Sixth Edition are in italics. In all chapters, many new problems were added and others were updated.

Chapter 12: Functions of Several Variables

This chapter introduces functions of many variables from several points of view, using surface graphs, contour diagrams, and tables. We assume throughout that functions of two or more variables are defined on regions with piecewise smooth boundaries. We conclude with a section on continuity.

Chapter 13: A Fundamental Tool: Vectors

This chapter introduces vectors geometrically and algebraically and discusses the dot and cross product.

Chapter 14: Differentiating Functions of Several Variables

Partial derivatives, directional derivatives, gradients, and local linearity are introduced. The chapter also discusses higher order partial derivatives, quadratic Taylor approximations, and differentiability.

Chapter 15: Optimization

The ideas of the previous chapter are applied to optimization problems, both constrained and unconstrained.

Chapter 16: Integrating Functions of Several Variables

This chapter discusses double and triple integrals in Cartesian, polar, cylindrical, and spherical coordinates.
The former Section 16.7 has been moved to the new Chapter 21.

Chapter 17: Parameterization and Vector Fields

This chapter discusses parameterized curves and motion, vector fields and flowlines.
The former Section 17.5 has been moved to the new Chapter 21.

Chapter 18: Line Integrals

This chapter introduces line integrals and shows how to calculate them using parameterizations. Conservative fields, gradient fields, the Fundamental Theorem of Calculus for Line Integrals, and Green's Theorem are discussed.

Chapter 19: Flux Integrals and Divergence

This chapter introduces flux integrals and shows how to calculate them over surface graphs, portions of cylinders, and portions of spheres. The divergence is introduced and its relationship to flux integrals discussed in the Divergence Theorem.

This new chapter combines Sections 19.1 and 19.2 with Sections 20.1 and 20.2 from the fifth edition

Chapter 20: The Curl and Stokes' Theorem

The purpose of this chapter is to give students a practical understanding of the curl and of Stokes' Theorem and to lay out the relationship between the theorems of vector calculus.

This chapter consists of Sections 20.3–20.5 from the fifth edition.

Chapter 21: Parameters, Coordinates, and Integrals

This new chapter covers parameterized surfaces, the change of variable formula in a double or triple integral, and flux though a parameterized surface.

Appendices

There are appendices on roots, accuracy, and bounds; complex numbers; Newton's Method; and determinants.

Projects

There are new projects in Chapter 12: "Heathrow"; Chapter 19: "Solid Angle"; and Chapter 20: "Magnetic field generated by a current in a wire".

Supplementary Materials and Additional Resources

Supplements for the instructor can be obtained online at the book companion site or by contacting your Wiley representative. The following supplementary materials are available for this edition:

- **Instructor's Manual** containing teaching tips, calculator programs, overhead transparency masters, sample worksheets, and sample syllabi.
- **Computerized Test Bank**, comprised of nearly 7,000 questions, mostly algorithmically-generated, which allows for multiple versions of a single test or quiz.
- **Instructor's Solution Manual** with complete solutions to all problems.
- **Student Solution Manual** with complete solutions to half the odd-numbered problems.
- **Additional Material**, elaborating specially marked points in the text and password-protected electronic versions of the instructor ancillaries, can be found on the web at www.wiley.com/college/hughes-hallett.

ConcepTests

ConcepTests, modeled on the pioneering work of Harvard physicist Eric Mazur, are questions designed to promote active learning during class, particularly (but not exclusively) in large lectures. Our evaluation data show students taught with ConcepTests outperformed students taught by traditional lecture methods 73% versus 17% on conceptual questions, and 63% versus 54% on computational problems.

Faculty Resource Network

A peer-to-peer network of academic faculty dedicated to the effective use of technology in the classroom, this group can help you apply innovative classroom techniques and implement specific software packages. Visit www.facultyresourcenetwork.com or speak to your Wiley representative.

WileyPLUS

WileyPLUS, Wiley's digital learning environment, is loaded with all of the supplements above, and also features:

- Online version of the text, featuring hyperlinks to referenced content, applets, and supplements.

- Homework management tools, which easily enable the instructor to assign and automatically graded questions, using a rich set of options and controls.

- QuickStart pre-designed reading and homework assignments. Use them as-is or customize them to fit the needs of your classroom.

- Guided Online (GO) Exercises, which prompt students to build solutions step-by-step. Rather than simply grading an exercise answer as wrong, GO problems show students precisely where they are making a mistake.

- Animated applets, which can be used in class to present and explore key ideas graphically and dynamically—especially useful for display of three-dimensional graphs in multivariable calculus.

- Graphing Calculator Manual, to help students get the most out of their graphing calculator, and to show how they can apply the numerical and graphing functions of their calculators to their study of calculus.

Acknowledgements

First and foremost, we want to express our appreciation to the National Science Foundation for their faith in our ability to produce a revitalized calculus curriculum and, in particular, to our program officers, Louise Raphael, John Kenelly, John Bradley, and James Lightbourne. We also want to thank the members of our Advisory Board, Benita Albert, Lida Barrett, Simon Bernau, Robert Davis, M. Lavinia DeConge-Watson, John Dossey, Ron Douglas, Eli Fromm, William Haver, Seymour Parter, John Prados, and Stephen Rodi.

In addition, a host of other people around the country and abroad deserve our thanks for their contributions to shaping this edition. They include: Huriye Arikan, Ruth Baruth, Paul Blanchard, Lewis Blake, David Bressoud, Stephen Boyd, Lucille Buonocore, Jo Cannon, Ray Cannon, Phil Cheifetz, Scott Clark, Jailing Dai, Ann Davidian, Tom Dick, Srdjan Divac, Tevian Dray, Steven Dunbar, David Durlach, John Eggers, Wade Ellis, Johann Engelbrecht, Brad Ernst, Sunny Fawcett, Paul Feehan, Sol Friedberg, Melanie Fulton, Tom Gearhart, David Glickenstein, Chris Goff, Sheldon P. Gordon, Salim Haïdar, Elizabeth Hentges, Rob Indik, Adrian Iovita, David Jackson, Sue Jensen, Alex Kasman, Matthias Kawski, Mike Klucznik, Donna Krawczyk, Stephane Lafortune, Andrew Lawrence, Carl Leinert, Andrew Looms, Bin Lu, Alex Mallozzi, Corinne Manogue, Jay Martin, Eric Mazur, Abby McCallum, Dan McGee, Ansie Meiring, Lang Moore, Jerry Morris, Hideo Nagahashi, Kartikeya Nagendra, Alan Newell, Steve Olson, John Orr, Arnie Ostebee, Andrew Pasquale, Wayne Raskind, Maria Robinson, Laurie Rosatone, Ayse Sahin, Nataliya Sandler, Ken Santor, Anne Scanlan-Rohrer, Ellen Schmierer, Michael Sherman, Pat Shure, Scott Pilzer, David Smith, Ernie Solheid, Misha Stepanov, Steve Strogatz, Peter Taylor, Dinesh Thakur, Sally Thomas, Joe Thrash, Alan Tucker, Doug Ulmer, Ignatios Vakalis, Bill Vélez, Joe Vignolini, Stan Wagon, Hannah Winkler, Debra Wood, Aaron Wootton, Deane Yang, Bruce Yoshiwara, Kathy Yoshiwara, and Paul Zorn.

Reports from the following reviewers were most helpful for the fifth edition:

Lewis Blake, Patrice Conrath, Christopher Ennis, John Eggers, Paul DeLand, Dana Fine, Dave Folk, Elizabeth Hodes, Richard Jenson, Emelie Kenney, Michael Kinter, Douglas Lapp, Glenn Ledder, Eric Marland, Cindy Moss, Michael Naylor, Genevra Neumann, Dennis Piontkowski, Robert Reed, Laurence Small, Ed Soares, Diana Staats, Kurt Verdeber, Elizabeth Wilcox, and Deborah Yoklic.

Reports from the following reviewers were most helpful for the sixth edition:
Barbara Armenta, James Baglama, Jon Clauss, Ann Darke, Marcel Finan, Dana Fine, Michael Huber, Greg Marks, Wes Ostertag, Ben Smith, Mark Turner, Aaron Weinberg, and Jianying Zhang.

William G. McCallum	Patti Frazer Lock	Douglas Quinney
Deborah Hughes-Hallett	Guadalupe I. Lozano	Adam Spiegler
Daniel E. Flath	Jerry Morris	Jeff Tecosky-Feldman
Andrew M. Gleason	David Mumford	Thomas W. Tucker
Selin Kalaycıoğlu	Brad G. Osgood	
Brigitte Lahme	Cody L. Patterson	

To Students: How to Learn from this Book

- This book may be different from other math textbooks that you have used, so it may be helpful to know about some of the differences in advance. This book emphasizes at every stage the *meaning* (in practical, graphical or numerical terms) of the symbols you are using. There is much less emphasis on "plug-and-chug" and using formulas, and much more emphasis on the interpretation of these formulas than you may expect. You will often be asked to explain your ideas in words or to explain an answer using graphs.

- The book contains the main ideas of multivariable calculus in plain English. Your success in using this book will depend on your reading, questioning, and thinking hard about the ideas presented. Although you may not have done this with other books, you should plan on reading the text in detail, not just the worked examples.

- There are very few examples in the text that are exactly like the homework problems. This means that you can't just look at a homework problem and search for a similar–looking "worked out" example. Success with the homework will come by grappling with the ideas of calculus.

- Many of the problems that we have included in the book are open-ended. This means that there may be more than one approach and more than one solution, depending on your analysis. Many times, solving a problem relies on common sense ideas that are not stated in the problem but which you will know from everyday life.

- Some problems in this book assume that you have access to a graphing calculator or computer; preferably one that can draw surface graphs, contour diagrams, and vector fields, and can compute multivariable integrals and line integrals numerically. There are many situations where you may not be able to find an exact solution to a problem, but you can use a calculator or computer to get a reasonable approximation.

- This book attempts to give equal weight to three methods for describing functions: graphical (a picture), numerical (a table of values) and algebraic (a formula). Sometimes you may find it easier to translate a problem given in one form into another. For example, if you have to find the maximum of a function, you might use a contour diagram to estimate its approximate position, use its formula to find equations that give the exact position, then use a numerical method to solve the equations. The best idea is to be flexible about your approach: if one way of looking at a problem doesn't work, try another.

- Students using this book have found discussing these problems in small groups very helpful. There are a great many problems which are not cut-and-dried; it can help to attack them with the other perspectives your colleagues can provide. If group work is not feasible, see if your instructor can organize a discussion session in which additional problems can be worked on.

- You are probably wondering what you'll get from the book. The answer is, if you put in a solid effort, you will get a real understanding of one of the most important accomplishments of the millennium – calculus – as well as a real sense of the power of mathematics in the age of technology.

CONTENTS

Chapter Twelve

FUNCTIONS OF SEVERAL VARIABLES

Contents

12.1 FUNCTIONS OF TWO VARIABLES

Function Notation

Suppose you want to calculate your monthly payment on a five-year car loan; this depends on both the amount of money you borrow and the interest rate. These quantities can vary separately: the loan amount can change while the interest rate remains the same, or the interest rate can change while the loan amount remains the same. To calculate your monthly payment you need to know both. If the monthly payment is $\$m$, the loan amount is $\$L$, and the interest rate is $r\%$, then we express the fact that m is a function of L and r by writing:

$$m = f(L, r).$$

This is just like the function notation of one-variable calculus. The variable m is called the dependent variable, and the variables L and r are called the independent variables. The letter f stands for the *function* or rule that gives the value of m corresponding to given values of L and r.

A function of two variables can be represented graphically, numerically by a table of values, or algebraically by a formula. In this section, we give examples of each.

Graphical Example: A Weather Map

Figure 12.1 shows a weather map from a newspaper. What information does it convey? It displays the predicted high temperature, T, in degrees Celsius ($°C$), throughout the US on that day. The curves on the map, called *isotherms*, separate the country into zones, according to whether T is in the 40s, 30s, 20s, 10s, or 0s. (*Iso* means same and *therm* means heat.) Notice that the isotherm separating the 20s and 30s zones connects all the points where the temperature is exactly $30°C$.

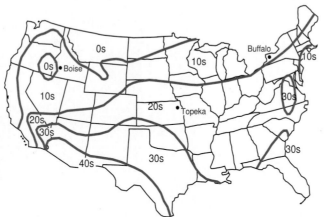

Figure 12.1: Weather map showing predicted high temperatures, T, on a particular day

Example 1   Estimate the predicted value of T in Boise, Idaho; Topeka, Kansas; and Buffalo, New York.

Solution   Boise and Buffalo are in the 10s region, and Topeka is in the 20s region. Thus, the predicted temperature in Boise and Buffalo is between 10 and 20 while the predicted temperature in Topeka is between 20 and 30. In fact, we can say more. Although both Boise and Buffalo are in the 10s region, Boise is quite close to the $T = 10$ isotherm, whereas Buffalo is quite close to the $T = 20$ isotherm. So we estimate the temperature to be in the low 10s in Boise and in the high 10s in Buffalo. Topeka is about halfway between the $T = 20$ isotherm and the $T = 30$ isotherm. Thus, we guess the temperature in Topeka to be in the mid-20s. In fact, the actual high temperatures for that day were 11°C for Boise, 19°C for Buffalo, and 26°C for Topeka.

The predicted high temperature, T, illustrated by the weather map is a function of (that is, depends on) two variables, often longitude and latitude, or kilometres east-west and kilometres north-south of a fixed point, say, Topeka. The weather map in Figure 12.1 is called a *contour map* or *contour diagram* of that function. Section 12.2 shows another way of visualising functions of two variables using surfaces; Section 12.3 looks at contour maps in detail.

Numerical Example: Beef Consumption

Suppose you are a beef producer and you want to know how much beef people will buy. This depends on how much money people have and on the price of beef. The consumption of beef, C (in kilograms per week per household) is a function of household income, I (in thousands of dollars per year), and the price of beef, p (in dollars per kilogram). In function notation, we write:

$$C = f(I, p).$$

Table 12.1 contains values of this function. Values of p are shown across the top, values of I are down the left side, and corresponding values of $f(I, p)$ are given in the table.[1] For example, to find the value of $f(40, 7.00)$, we look in the row corresponding to $I = 40$ under $p = 7.00$, where we find the number 2.03. Thus,

$$f(40, 7.00) = 2.03.$$

This means that, on average, if a household's income is $40,000 a year and the price of beef is $7.00/kg, the family will buy 2.03 kg of beef per week.

Table 12.1   *Quantity of beef bought (kilograms/household/week)*

		Price of beef ($/kg)			
		6.00	7.00	8.00	9.00
Household	20	1.32	1.30	1.25	1.22
income	40	2.07	2.03	1.97	1.94
per year,	60	2.56	2.50	2.48	2.42
I	80	2.67	2.65	2.60	2.54
(1000)	100	2.90	2.86	2.80	2.76

Notice how this differs from the table of values of a one-variable function, where one row or one column is enough to list the values of the function. Here many rows and columns are needed because the function has a value for every *pair* of values of the independent variables.

Algebraic Examples: Formulas

In both the weather map and beef consumption examples, there is no formula for the underlying function. That is usually the case for functions representing real-life data. On the other hand, for many idealised models in physics, engineering, or economics, there are exact formulas.

---

[1] Adapted from Richard G. Lipsey, *An Introduction to Positive Economics*, 3rd ed, (London: Weidenfeld and Nicolson, 1971).

Example 2   Give a formula for the function $M = f(B, t)$ where M is the amount of money in a bank account t years after an initial investment of B dollars, if interest is accrued at a rate of 1.2% per year compounded annually.

Solution   Annual compounding means that M increases by a factor of 1.012 every year, so
$$M = f(B, t) = B(1.012)^t.$$

Example 3   A cylinder with closed ends has radius r and height h. If its volume is V and its surface area is A, find formulas for the functions $V = f(r, h)$ and $A = g(r, h)$.

Solution   Since the area of the circular base is πr^2, we have
$$V = f(r, h) = \text{Area of base} \cdot \text{Height} = \pi r^2 h.$$

The surface area of the side is the circumference of the bottom, $2\pi r$, times the height h, giving $2\pi rh$. Thus,
$$A = g(r, h) = 2 \cdot \text{Area of base} + \text{Area of side} = 2\pi r^2 + 2\pi rh.$$

A Tour of 3-Space

In Section 12.2 we see how to visualise a function of two variables as a surface in space. Now we see how to locate points in three-dimensional space (3-space).

Imagine three coordinate axes meeting at the *origin*: a vertical axis, and two horizontal axes at right angles to each other. (See Figure 12.2.) Think of the xy-plane as being horizontal, while the z-axis extends vertically above and below the plane. The labels x, y, and z show which part of each axis is positive; the other side is negative. We generally use *right-handed axes* in which looking down the positive z-axis gives the usual view of the xy-plane. We specify a point in 3-space by giving its coordinates (x, y, z) with respect to these axes. Think of the coordinates as instructions telling you how to get to the point: start at the origin, go x units along the x-axis, then y units in the direction parallel to the y-axis, and finally z units in the direction parallel to the z-axis. The coordinates can be positive, zero or negative; a zero coordinate means "don't move in this direction," and a negative coordinate means "go in the negative direction parallel to this axis." For example, the origin has coordinates $(0, 0, 0)$, since we get there from the origin by doing nothing at all.

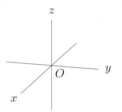

Figure 12.2: Coordinate axes in three-dimensional space

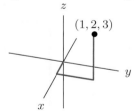

Figure 12.3: The point $(1, 2, 3)$ in 3-space

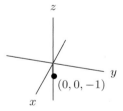

Figure 12.4: The point $(0, 0, -1)$ in 3-space

Example 4   Describe the position of the points with coordinates $(1, 2, 3)$ and $(0, 0, -1)$.

Solution   We get to the point $(1, 2, 3)$ by starting at the origin, going 1 unit along the x-axis, 2 units in the direction parallel to the y-axis, and 3 units up in the direction parallel to the z-axis. (See Figure 12.3.)

To get to $(0, 0, -1)$, we don't move at all in the x- and the y-direction, but move 1 unit in the negative z-direction. So the point is on the negative z-axis. (See Figure 12.4.) You can check that the position of the point is independent of the order of the x, y, and z displacements.

Example 5   You start at the origin, go along the y-axis a distance of 2 units in the positive direction, and then move vertically upward a distance of 1 unit. What are the coordinates of your final position?

Solution   You started at the point $(0, 0, 0)$. When you went along the y-axis, your y-coordinate increased to 2. Moving vertically increased your z-coordinate to 1; your x-coordinate did not change because you did not move in the x-direction. So your final coordinates are $(0, 2, 1)$. (See Figure 12.5.)

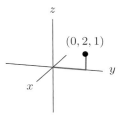

Figure 12.5: The point $(0, 2, 1)$ is reached by moving 2 along the y-axis and 1 upward

It is often helpful to picture a three-dimensional coordinate system in terms of a room. The origin is a corner at floor level where two walls meet the floor. The z-axis is the vertical intersection of the two walls; the x- and the y-axis are the intersections of each wall with the floor. Points with negative coordinates lie behind a wall in the next room or below the floor.

Graphing Equations in 3-Space

We can graph an equation involving the variables x, y, and z in 3-space; such a graph is a picture of all points (x, y, z) that satisfy the equation.

Example 6   What do the graphs of the equations $z = 0$, $z = 3$, and $z = -1$ look like?

Solution   To graph $z = 0$, we visualise the set of points whose z-coordinate is zero. If the z-coordinate is 0, then we must be at the same vertical level as the origin, that is, we are in the horizontal plane containing the origin. So the graph of $z = 0$ is the middle plane in Figure 12.6. The graph of $z = 3$ is a plane parallel to the graph of $z = 0$, but three units above it. The graph of $z = -1$ is a plane parallel to the graph of $z = 0$, but one unit below it.

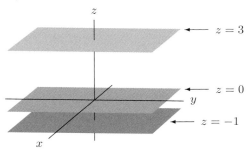

Figure 12.6: The planes $z = -1$, $z = 0$, and $z = 3$

The plane $z = 0$ contains the x- and the y-coordinate axis, and is called the xy-plane. There are two other coordinate planes. The yz-plane contains both the y- and the z-axis, and the xz-plane contains the x- and the z-axis. (See Figure 12.7.)

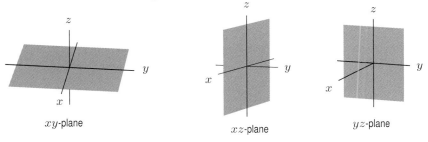

Figure 12.7: The three coordinate planes

Example 7   Which of the points $A = (1, -1, 0)$, $B = (0, 3, 4)$, $C = (2, 2, 1)$, and $D = (0, -4, 0)$ lies closest to the xz-plane? Which point lies on the y-axis?

Solution   The magnitude of the y-coordinate gives the distance to the xz-plane. The point A lies closest to that plane, because it has the smallest y-coordinate in magnitude. To get to a point on the y-axis, we move along the y-axis, but we don't move at all in the x- or the z-direction. Thus, a point on the y-axis has both its x- and z-coordinate equal to zero. The only point of the four that satisfies this is D. (See Figure 12.8.)

In general, if a point has one of its coordinates equal to zero, it lies in one of the coordinate planes. If a point has two of its coordinates equal to zero, it lies on one of the coordinate axes.

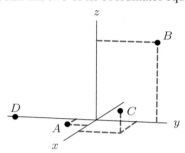

Figure 12.8: Which point lies closest to the xz-plane? Which point lies on the y-axis?

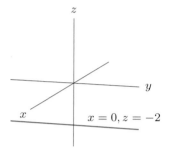

Figure 12.9: The line $x = 0$, $z = -2$

Example 8   You are 2 units below the xy-plane and in the yz-plane. What are your coordinates?

Solution   Since you are 2 units below the xy-plane, your z-coordinate is -2. Since you are in the yz-plane, your x-coordinate is 0; your y-coordinate can be anything. Thus, you are at the point $(0, y, -2)$. The set of all such points forms a line parallel to the y-axis, 2 units below the xy-plane, and in the yz-plane. (See Figure 12.9.)

Example 9   You are standing at the point $(4, 5, 2)$, looking at the point $(0.5, 0, 3)$. Are you looking up or down?

Solution   The point you are standing at has z-coordinate 2, whereas the point you are looking at has z-coordinate 3; hence you are looking up.

Example 10   Imagine that the yz-plane in Figure 12.7 is a page of this book. Describe the region behind the page algebraically.

Solution   The positive part of the x-axis pokes out of the page; moving in the positive x-direction brings you out in front of the page. The region behind the page corresponds to negative values of x, so it is the set of all points in 3-space satisfying the inequality $x < 0$.

Distance Between Two Points

In 2-space, the formula for the distance between two points (x, y) and (a, b) is given by

$$\text{Distance} = \sqrt{(x - a)^2 + (y - b)^2}.$$

The distance between two points (x, y, z) and (a, b, c) in 3-space is represented by PG in Figure 12.10. The side PE is parallel to the x-axis, EF is parallel to the y-axis, and FG is parallel to the z-axis.

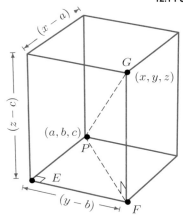

Figure 12.10: The diagonal PG gives the distance between the points (x, y, z) and (a, b, c)

Using Pythagoras' theorem twice gives

$$(PG)^2 = (PF)^2 + (FG)^2 = (PE)^2 + (EF)^2 + (FG)^2 = (x - a)^2 + (y - b)^2 + (z - c)^2.$$

Thus, a formula for the distance between the points (x, y, z) and (a, b, c) in 3-space is

$$\text{Distance} = \sqrt{(x - a)^2 + (y - b)^2 + (z - c)^2}.$$

Example 11   Find the distance between $(1, 2, 1)$ and $(-3, 1, 2)$.

Solution $\text{Distance} = \sqrt{(-3 - 1)^2 + (1 - 2)^2 + (2 - 1)^2} = \sqrt{18} = 4.24.$

Example 12   Find an expression for the distance from the origin to the point (x, y, z).

Solution   The origin has coordinates $(0, 0, 0)$, so the distance from the origin to (x, y, z) is given by

$$\text{Distance} = \sqrt{(x - 0)^2 + (y - 0)^2 + (z - 0)^2} = \sqrt{x^2 + y^2 + z^2}.$$

Example 13   Find an equation for a sphere of radius 1 with centre at the origin.

Solution   The sphere consists of all points (x, y, z) whose distance from the origin is 1, that is, which satisfy the equation

$$\sqrt{x^2 + y^2 + z^2} = 1.$$

This is an equation for the sphere. If we square both sides we get the equation in the form

$$x^2 + y^2 + z^2 = 1.$$

Note that this equation represents the *surface* of the sphere. The solid ball enclosed by the sphere is represented by the inequality $x^2 + y^2 + z^2 \le 1$.

Exercises and Problems for Section 12.1

Exercises

1. Which of the points $A = (23, 92, 48)$, $B = (-60, 0, 0)$, $C = (60, 1, -92)$ is closest to the yz-plane? Which lies on the xz-plane? Which is farthest from the xy-plane?

2. Which of the points $A = (1.3, -2.7, 0)$, $B = (0.9, 0, 3.2)$, $C = (2.5, 0.1, -0.3)$ is closest to the yz-plane? Which one lies on the xz-plane? Which one is farthest from the xy-plane?

3. Which of the points $P = (1, 2, 1)$ and $Q = (2, 0, 0)$ is closest to the origin?

4. Which two of the three points $P_1 = (1, 2, 3)$, $P_2 = (3, 2, 1)$ and $P_3 = (1, 1, 0)$ are closest to each other?

5. You are at the point $(3, 1, 1)$, standing upright and facing the yz-plane. You walk 2 units forward, turn left, and walk another 2 units. What is your final position? From the point of view of an observer looking at the coordinate system in Figure 12.2 on page 686, are you in front of or behind the yz-plane? To the left or to the right of the xz-plane? Above or below the xy-plane?

6. You are at the point $(-1, -3, -3)$, standing upright and facing the yz-plane. You walk 2 units forward, turn left, and walk for another 2 units. What is your final position? From the point of view of an observer looking at the coordinate system in Figure 12.2 on page 686, are you in front of or behind the yz-plane? To the left or to the right of the xz-plane? Above or below the xy-plane?

7. What is the midpoint of the line segment joining the points $(-1, 3, 9)$ and $(5, 6, -3)$?

8. On January 1st 2001, a mysterious black steel monolith (a rectangular solid) appeared at Seattle's Magnuson Park, USA, only to vanish a couple of days later. Place a system of coordinate axes with origin at one of the monolith's vertices and axes parallel to its sides (of lengths 1, 4 and 9, respectively), in a way that no vertices have any negative coordinates. Sketch the monolith with this coordinate system and give the coordinates of its eight vertices.

In Exercises 9–12 sketch graphs of the equations in 3-space.

9. $z = 4$

10. $x = -3$

11. $y = 1$

12. $z = 2$ and $y = 4$

13. With the z-axis vertical, a sphere has center $(2, 3, 7)$ and lowest point $(2, 3, -1)$. What is the highest point on the sphere?

14. Find an equation of the sphere with radius 5 centered at the origin.

15. Find the equation of the sphere with radius 2 and centered at $(1, 0, 0)$.

16. Find the equation of the vertical plane perpendicular to the y-axis and through the point $(2, 3, 4)$.

Exercises 17–19 refer to the map in Figure 12.1 on page 684.

17. Give the range of daily high temperatures for:

(a) Pennsylvania   (b) North Dakota

(c) California

18. Sketch a possible graph of the predicted high temperature T on a line north-south through Topeka.

19. Sketch possible graphs of the predicted high temperature on a north-south line and an east-west line through Boise.

For Exercises 20–22, refer to Table 12.1 on page 685, where p is the price of beef and I is annual household income.

20. Give tables for beef consumption as a function of p, with I fixed at $I = 20$ and $I = 100$. Give tables for beef consumption as a function of I, with p fixed at $p = 6.00$ and $p = 8.00$. Comment on what you see in the tables.

21. Make a table of the proportion, P, of household income spent on beef per week as a function of price and income. (Note that P is the fraction of income spent on beef.)

22. How does beef consumption vary as a function of household income if the price of beef is held constant?

Problems

23. The temperature adjusted for wind chill is a temperature which tells you how cold it feels, as a result of the combination of wind and temperature.[2] See Table 12.2.

Table 12.2 *Temperature adjusted for wind chill (°C) as a function of wind speed and temperature*

		Temperature (°C)							
		2	−1	−4	−7	−10	−13	−16	−19
	8	−1	−4	−7	−11	−14	−17	−21	−24
Wind	16	−3	−6	−9	−13	−16	−20	−23	−27
Speed	24	−4	−7	−11	−14	−18	−22	−25	−28
(km/hr)	32	−4	−8	−12	−15	−19	−23	−26	−30
	40	−5	−9	−13	−16	−20	−24	−27	−31

(a) If the temperature is $-19°C$ and the wind speed is 24 km/hr, how cold does it feel?

(b) If the temperature is $2°C$, what wind speed makes it feel like $-3°C$?

(c) If the temperature is $-7°C$, what wind speed makes it feel like $-12°C$?

(d) If the wind is blowing at 24 km/hr, what temperature feels like $-23°C$?

In Problems 24–25, use Table 12.2 to make tables with the given properties.

24. The temperature adjusted for wind chill as a function of wind speed for temperatures of $-7°C$ and $-19°C$.

25. The temperature adjusted for wind chill as a function of temperature for wind speeds of 8 km/hr and 32 km/hr.

---

[2]Data from www.nws.noaa.gov/om/windchill, accessed on May 30, 2004.

26. A car rental company charges $40 a day and 15 cents a mile for its cars.

  (a) Write a formula for the cost, C, of renting a car as a function, f, of the number of days, d, and the number of miles driven, m.

  (b) If $C = f(d, m)$, find $f(5, 300)$ and interpret it.

27. The gravitational force, F newtons, exerted on an object by the earth depends on the object's mass, m kilograms, and its distance, r meters, from the center of the earth, so $F = f(m, r)$. Interpret the following statement in terms of gravitation: $f(100, 7000000) \approx 820$.

28. Consider the acceleration due to gravity, g, at a distance h from the center of a planet of mass m.

  (a) If m is held constant, is g an increasing or decreasing function of h? Why?

  (b) If h is held constant, is g an increasing or decreasing function of m? Why?

29. A cube is located such that its top four corners have the coordinates $(-1, -2, 2)$, $(-1, 3, 2)$, $(4, -2, 2)$ and $(4, 3, 2)$. Give the coordinates of the center of the cube.

30. Describe the set of points whose distance from the x-axis is 2.

31. Describe the set of points whose distance from the x-axis equals the distance from the yz-plane.

32. Find a formula for the shortest distance between a point (a, b, c) and the y-axis.

33. Find the equations of planes that just touch the sphere $(x - 2)^2 + (y - 3)^2 + (z - 3)^2 = 16$ and are parallel to

  (a) The xy-plane   **(b)** The yz-plane

  (c) The xz-plane

34. Find an equation of the largest sphere contained in the cube determined by the planes $x = 2, x = 6; y = 5, y = 9$; and $z = -1, z = 3$.

35. A cube has edges parallel to the axes. One corner is at $A = (5, 1, 2)$ and the corner at the other end of the longest diagonal through A is $B = (12, 7, 4)$.

  (a) What are the coordinates of the other three vertices on the bottom face?

  (b) What are the coordinates of the other three vertices on the top face?

36. Which of the points $P_1 = (-3, 2, 15)$, $P_2 = (0, -10, 0)$, $P_3 = (-6, 5, 3)$ and $P_4 = (-4, 2, 7)$ is closest to $P = (6, 0, 4)$?

37. **(a)** Find the equations of the circles (if any) where the sphere $(x - 1)^2 + (y + 3)^2 + (z - 2)^2 = 4$ intersects each coordinate plane.

  (b) Find the points (if any) where this sphere intersects each coordinate axis.

38. A rectangular solid lies with its length parallel to the y-axis, and its top and bottom faces parallel to the plane $z = 0$. If the center of the object is at $(1, 1, -2)$ and it has a length of 13, a height of 5 and a width of 6, give the coordinates of all eight corners and draw the figure labeling the eight corners.

39. An equilateral triangle is standing vertically with a vertex above the xy-plane and its two other vertices at $(7, 0, 0)$ and $(9, 0, 0)$. What is its highest point?

40. **(a)** Find the midpoint of the line segment joining $A = (1, 5, 7)$ to $B = (5, 13, 19)$.

  (b) Find the point one quarter of the way along the line segment from A to B.

  (c) Find the point one quarter of the way along the line segment from B to A.

Strengthen Your Understanding

In Problems 41–43, explain what is wrong with the statement.

41. In 3-space, $y = 1$ is a line parallel to the x-axis.

42. The xy-plane has equation $xy = 0$.

43. The distance from $(2, 3, 4)$ to the x-axis is 2.

In Problems 44–45, give an example of:

44. A formula for a function $f(x, y)$ that is increasing in x and decreasing in y.

45. A point in 3-space with all its coordinates negative and farther from the xz-plane than from the plane $z = -5$.

Are the statements in Problems 46–59 true or false? Give reasons for your answer.

46. If $f(x, y)$ is a function of two variables defined for all x and y, then $f(10, y)$ is a function of one variable.

47. The volume V of a box of height h and square base of side length s is a function of h and s.

48. If $H = f(t, d)$ is the function giving the water temperature $H°C$ of a lake at time t hours after midnight and depth d meters, then t is a function of d and H.

49. A table for a function $f(x, y)$ cannot have any values of f appearing twice.

50. If $f(x)$ and $g(y)$ are both functions of a single variable, then the product $f(x) \cdot g(y)$ is a function of two variables.

51. The point $(1, 2, 3)$ lies above the plane $z = 2$.

52. The graph of the equation $z = 2$ is a plane parallel to the xz-plane.

53. The points $(1, 0, 1)$ and $(0, -1, 1)$ are the same distance from the origin.

54. The point $(2, -1, 3)$ lies on the graph of the sphere $(x - 2)^2 + (y + 1)^2 + (z - 3)^2 = 25$.

55. There is only one point in the yz-plane that is a distance 3 from the point $(3, 0, 0)$.

56. There is only one point in the yz-plane that is distance 5 from the point $(3, 0, 0)$.

57. If the point $(0, b, 0)$ has distance 4 from the plane $y = 0$, then b must be 4.

58. A line parallel to the z-axis can intersect the graph of $f(x, y)$ at most once.

59. A line parallel to the y-axis can intersect the graph of $f(x, y)$ at most once.

12.2   GRAPHS AND SURFACES

The weather map on page 684 is one way of visualising a function of two variables. In this section we see how to visualise a function of two variables in another way, using a surface in 3-space.

Visualizing a Function of Two Variables Using a Graph

For a function of one variable, $y = f(x)$, the graph of f is the set of all points (x, y) in 2-space such that $y = f(x)$. In general, these points lie on a curve in the plane. When a computer or calculator graphs f, it approximates by plotting points in the xy-plane and joining consecutive points by line segments. The more points, the better the approximation.

Now consider a function of two variables.

> The **graph** of a function of two variables, f, is the set of all points (x, y, z) such that $z = f(x, y)$. In general, the graph of a function of two variables is a surface in 3-space.

Plotting the Graph of the Function $f(x, y) = x^2 + y^2$

To sketch the graph of f we connect points as for a function of one variable. We first make a table of values of f, such as in Table 12.3.

Table 12.3   *Table of values of* $f(x, y) = x^2 + y^2$

		y						
		-3	-2	-1	0	1	2	3
	-3	18	13	10	9	10	13	18
	-2	13	8	5	4	5	8	13
	-1	10	5	2	1	2	5	10
x	0	9	4	1	0	1	4	9
	1	10	5	2	1	2	5	10
	2	13	8	5	4	5	8	13
	3	18	13	10	9	10	13	18

Now we plot points. For example, we plot $(1, 2, 5)$ because $f(1, 2) = 5$ and we plot $(0, 2, 4)$ because $f(0, 2) = 4$. Then, we connect the points corresponding to the rows and columns in the table. The result is called a *wire-frame* picture of the graph. Filling in between the wires gives a surface. That is the way a computer drew the graphs in Figure 12.11 and 12.12. As more points are plotted, we get the surface in Figure 12.13, called a *paraboloid*.

You should check to see if the sketches make sense. Notice that the graph goes through the origin since $(x, y, z) = (0, 0, 0)$ satisfies $z = x^2 + y^2$. Observe that if x is held fixed and y is allowed to vary, the graph dips down and then goes back up, just like the entries in the rows of Table 12.3. Similarly, if y is held fixed and x is allowed to vary, the graph dips down and then goes back up, just like the columns of Table 12.3.

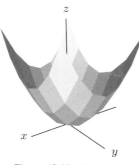

Figure 12.11: Wire frame picture of $f(x, y) = x^2 + y^2$ for $-3 \leq x \leq 3, -3 \leq y \leq 3$

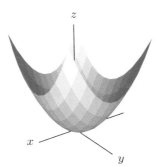

Figure 12.12: Wire frame picture of $f(x, y) = x^2 + y^2$ with more points plotted

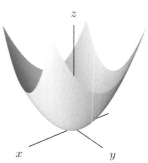

Figure 12.13: Graph of $f(x, y) = x^2 + y^2$ for $-3 \leq x \leq 3, -3 \leq y \leq 3$

New Graphs from Old

We can use the graph of a function to visualise the graphs of related functions.

Example 1   Let $f(x, y) = x^2 + y^2$. Describe in words the graphs of the following functions:
(a) $g(x, y) = x^2 + y^2 + 3$,   (b) $h(x, y) = 5 - x^2 - y^2$,   (c) $k(x, y) = x^2 + (y - 1)^2$.

Solution   We know from Figure 12.13 that the graph of f is a paraboloid, or bowl with its vertex at the origin. From this we can work out what the graphs of g, h, and k will look like.

(a) The function $g(x, y) = x^2 + y^2 + 3 = f(x, y) + 3$, so the graph of g is the graph of f, but raised by 3 units. See Figure 12.14.

(b) Since $-x^2 - y^2$ is the negative of $x^2 + y^2$, the graph of $-x^2 - y^2$ is a paraboloid opening downward. Thus, the graph of $h(x, y) = 5 - x^2 - y^2 = 5 - f(x, y)$ looks like a downward-opening paraboloid with vertex at $(0, 0, 5)$, as in Figure 12.15.

(c) The graph of $k(x, y) = x^2 + (y - 1)^2 = f(x, y - 1)$ is a paraboloid with vertex at $x = 0, y = 1$, since that is where $k(x, y) = 0$, as in Figure 12.16.

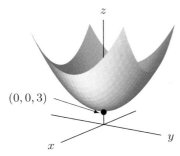

Figure 12.14: Graph of $g(x, y) = x^2 + y^2 + 3$

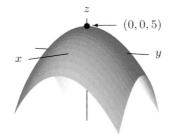

Figure 12.15: Graph of $h(x, y) = 5 - x^2 - y^2$

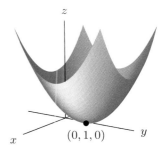

Figure 12.16: Graph of $k(x, y) = x^2 + (y - 1)^2$

Example 2   Describe the graph of $G(x, y) = e^{-(x^2 + y^2)}$. What symmetry does it have?

Solution   Since the exponential function is always positive, the graph lies entirely above the xy-plane. From the graph of $x^2 + y^2$ we see that $x^2 + y^2$ is zero at the origin and gets larger as we move farther from the origin in any direction. Thus, $e^{-(x^2 + y^2)}$ is 1 at the origin, and gets smaller as we move away from the origin in any direction. It can't go below the xy-plane; instead it flattens out, getting closer and closer to the plane. We say the surface is *asymptotic* to the xy-plane. (See Figure 12.17.)

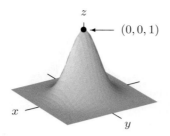

Figure 12.17: Graph of $G(x, y) = e^{-(x^2+y^2)}$

Now consider a point (x, y) on the circle $x^2 + y^2 = r^2$. Since

$$G(x, y) = e^{-(x^2+y^2)} = e^{-r^2},$$

the value of the function G is the same at all points on this circle. Thus, we say the graph of G has *circular symmetry*.

Cross-Sections and the Graph of a Function

We have seen that a good way to analyse a function of two variables is to let one variable vary while the other is kept fixed.

For a function $f(x, y)$, the function we get by holding x fixed and letting y vary is called a **cross-section** of f with x fixed. The graph of the cross-section of $f(x, y)$ with $x = c$ is the curve, or cross-section, we get by intersecting the graph of f with the plane $x = c$. We define a cross-section of f with y fixed similarly.

For example, the cross-section of $f(x, y) = x^2 + y^2$ with $x = 2$ is $f(2, y) = 4 + y^2$. The graph of this cross-section is the curve we get by intersecting the graph of f with the plane perpendicular to the x-axis at $x = 2$. (See Figure 12.18.)

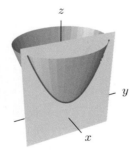

Figure 12.18: Cross-section of the surface $z = f(x, y)$ by the plane $x = 2$

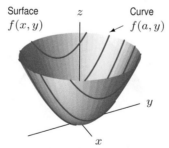

Figure 12.19: The curves $z = f(a, y)$ with a constant: cross-sections with x fixed

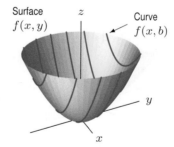

Figure 12.20: The curves $z = f(x, b)$ with b constant: cross-sections with y fixed

Figure 12.19 shows graphs of other cross-sections of f with x fixed; Figure 12.20 shows graphs of cross-sections with y fixed.

Example 3   Describe the cross-sections of the function $g(x, y) = x^2 - y^2$ with y fixed and then with x fixed. Use these cross-sections to describe the shape of the graph of g.

Solution   The cross-sections with y fixed at $y = b$ are given by

$$z = g(x, b) = x^2 - b^2.$$

Thus, each cross-section with y fixed gives a parabola opening upward, with minimum $z = -b^2$. The cross-sections with x fixed are of the form

$$z = g(a, y) = a^2 - y^2,$$

which are parabolas opening downward with a maximum of $z = a^2$. (See Figures 12.21 and 12.22.) The graph of g is shown in Figure 12.23. Notice the upward-opening parabolas in the x-direction and the downward-opening parabolas in the y-direction. We say that the surface is *saddle-shaped*.

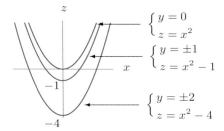

Figure 12.21: Cross-sections of $g(x, y) = x^2 - y^2$ with y fixed

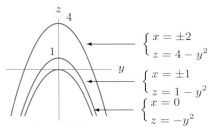

Figure 12.22: Cross-sections of $g(x, y) = x^2 - y^2$ with x fixed

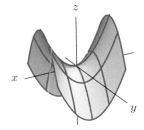

Figure 12.23: Graph of $g(x, y) = x^2 - y^2$ showing cross sections

Linear Functions

Linear functions are central to single-variable calculus; they are equally important in multivariable calculus. You may be able to guess the shape of the graph of a linear function of two variables. (It's a plane.) Let's look at an example.

Example 4   Describe the graph of $f(x, y) = 1 + x - y$.

Solution   The plane $x = a$ is vertical and parallel to the yz-plane. Thus, the cross-section with $x = a$ is the line $z = 1 + a - y$ which slopes downward in the y-direction. Similarly, the plane $y = b$ is parallel to the xz-plane. Thus, the cross-section with $y = b$ is the line $z = 1 + x - b$ which slopes upward in the x-direction. Since all the cross-sections are lines, you might expect the graph to be a flat plane, sloping down in the y-direction and up in the x-direction. This is indeed the case. (See Figure 12.24.)

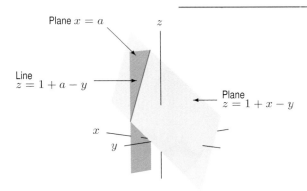

Figure 12.24: Graph of the plane $z = 1 + x - y$ showing cross-section with $x = a$

When One Variable is Missing: Cylinders

Suppose we graph an equation like $z = x^2$ which has one variable missing. What does the surface look like? Since y is missing from the equation, the cross-sections with y fixed are all the same parabola, $z = x^2$. Letting y vary up and down the y-axis, this parabola sweeps out the trough-shaped surface shown in Figure 12.25. The cross-sections with x fixed are horizontal lines obtained by cutting the surface by a plane perpendicular to the x-axis. This surface is called a *parabolic cylinder*, because it is formed from a parabola in the same way that an ordinary cylinder is formed from a circle; it has a parabolic cross-section instead of a circular one.

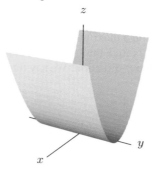

Figure 12.25: A parabolic cylinder $z = x^2$

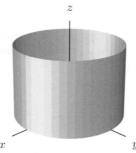

Figure 12.26: Circular cylinder $x^2 + y^2 = 1$

Example 5   Graph the equation $x^2 + y^2 = 1$ in 3-space.

Solution   Although the equation $x^2 + y^2 = 1$ does not represent a function, the surface representing it can be graphed by the method used for $z = x^2$. The graph of $x^2 + y^2 = 1$ in the xy-plane is a circle. Since z does not appear in the equation, the intersection of the surface with any horizontal plane will be the same circle $x^2 + y^2 = 1$. Thus, the surface is the cylinder shown in Figure 12.26.

Exercises and Problems for Section 12.2

Exercises

1. Without a calculator or computer, match the functions with their graphs in Figure 12.27.

  (a) $z = 2 + x^2 + y^2$ **(b)** $z = 2 - x^2 - y^2$

  (c) $z = 2(x^2 + y^2)$ **(d)** $z = 2 + 2x - y$

  (e) $z = 2$

2. Without a calculator or computer, match the functions with their graphs in Figure 12.28.

  (a) $z = \dfrac{1}{x^2 + y^2}$ **(b)** $z = -e^{-x^2 - y^2}$

  (c) $z = x + 2y + 3$ **(d)** $z = -y^2$

  (e) $z = x^3 - \sin y.$

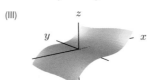

Figure 12.27

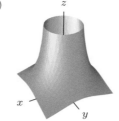

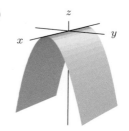

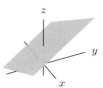

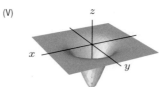

Figure 12.28

In Exercises 3–10, sketch a graph of the surface and briefly describe it in words.

3. $z = 3$

4. $x^2 + y^2 + z^2 = 9$

5. $z = x^2 + y^2 + 4$

6. $z = 5 - x^2 - y^2$

7. $z = y^2$

8. $2x + 4y + 3z = 12$

9. $x^2 + y^2 = 4$

10. $x^2 + z^2 = 4$

In Exercises 11–13, find the equation of the surface.

11. A cylinder of radius $\sqrt{7}$ with its axis along the y-axis.

12. A sphere of radius 3 centered at $(0, \sqrt{7}, 0)$.

13. The paraboloid obtained by moving the surface $z = x^2 + y^2$ so that its vertex is at $(1, 3, 5)$, its axis is parallel to the x-axis, and the surface opens towards negative x values.

Problems

14. Suppose you know the graph of a function $z = f(x, y)$. Describe in words the relationship between this graph and the graph of:

   (a) $z = f(x, y) + 5$
   (b) $z = f(x, y + 5)$
   (c) $z = |f(x, y)|$
   (d) $z = f(|x|, |y|)$

Problems 15–17 concern the concentration, C, in mg per liter, of a drug in the blood as a function of x, the amount, in mg, of the drug given and t, the time in hours since the injection. For $0 \leq x \leq 4$ and $t \geq 0$, we have $C = f(x, t) = te^{-t(5-x)}$.

15. Find $f(3, 2)$. Give units and interpret in terms of drug concentration.

16. Graph the following single-variable functions and explain their significance in terms of drug concentration.

   (a) $f(4, t)$
   (b) $f(x, 1)$

17. Graph $f(a, t)$ for $a = 1, 2, 3, 4$ on the same axes. Describe how the graph changes as a increases and explain what this means in terms of drug concentration.

18. Consider the function f given by $f(x, y) = y^3 + xy$. Draw graphs of cross-sections with:

   (a) x fixed at $x = -1$, $x = 0$, and $x = 1$.
   (b) y fixed at $y = -1$, $y = 0$, and $y = 1$.

19. Without a computer or calculator, match the equations (a)–(i) with the graphs (I)–(IX).

   (a) $z = xye^{-(x^2+y^2)}$ **(b)** $z = \cos(\sqrt{x^2 + y^2})$

   (c) $z = \sin y$ **(d)** $z = -\dfrac{1}{x^2 + y^2}$

   (e) $z = \cos^2 x \cos^2 y$ **(f)** $z = \dfrac{\sin(x^2 + y^2)}{x^2 + y^2}$

   (g) $z = \cos(xy)$ **(h)** $z = |x||y|$

   (i) $z = (2x^2 + y^2)e^{1 - x^2 - y^2}$

(I)
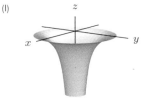

(II)
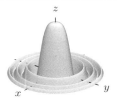

(III)

(IV)
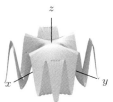

(V)
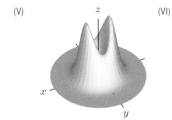

(VI)
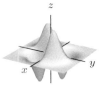

(VII)
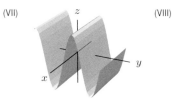

(VIII)
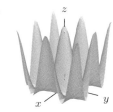

(IX)

20. Decide whether the graph of each of the following equations is the shape of a bowl, a plate, or neither. Consider a plate to be any flat surface and a bowl to be anything that could hold water, assuming the positive z-axis is up.

   (a) $z = x^2 + y^2$ **(b)** $z = 1 - x^2 - y^2$
   (c) $x + y + z = 1$ **(d)** $z = -\sqrt{5 - x^2 - y^2}$
   (e) $z = 3$

21. Sketch cross-sections for each function in Problem 20.

For Problems 22–25, give a formula for a function whose graph is described. Sketch it using a computer or calculator.

22. A bowl which opens upward and has its vertex at 5 on the z-axis.

23. A plane which has its x-, y-, and z-intercepts all positive.

24. A parabolic cylinder opening upward from along the line $y = x$ in the xy-plane.

25. A cone of circular cross-section opening downward and with its vertex at the origin.

26. You like pizza and you like chocolate. Which of the graphs in Figure 12.29 represents your happiness as a function of how many pizzas and how much chocolate you have if

   (a) There is no such thing as too many pizzas and too much chocolate?
   (b) There is such a thing as too many pizzas or too much chocolate?
   (c) There is such a thing as too much chocolate but no such thing as too many pizzas?

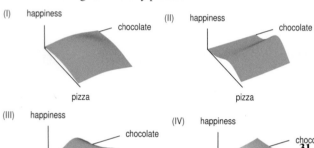

Figure 12.29

27. By setting one variable constant, find a plane that intersects the graph of $z = 4x^2 - y^2 + 1$ in a:

   (a) Parabola opening upward
   (b) Parabola opening downward
   (c) Pair of intersecting straight lines

28. By setting one variable constant, find a plane that intersects the graph of $z = (x^2 + 1) \sin y + xy^2$ in a:

   (a) Parabola
   (b) Straight line
   (c) Sine curve

29. For each of the graphs I-IV in Problem 26 draw:

   (a) Two cross-sections with pizza fixed
   (b) Two cross-sections with chocolate fixed.

30. A wave travels along a canal. Let x be the distance along the canal, t be the time, and z be the height of the water above the equilibrium level. The graph of z as a function of x and t is in Figure 12.30.

   (a) Draw the profile of the wave for $t = -1, 0, 1, 2$. (Put the x-axis to the right and the z-axis vertical.)
   (b) Is the wave traveling in the direction of increasing or decreasing x?
   (c) Sketch a surface representing a wave traveling in the opposite direction.

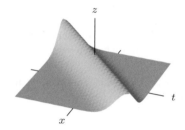

Figure 12.30

31. At time t, the displacement of a point on a vibrating guitar string stretched between $x = 0$ and $x = \pi$ is given by

$$f(x, t) = \cos t \sin x, \quad 0 \le x \le \pi, \quad 0 \le t \le 2\pi.$$

   (a) Sketch the cross-sections of this function with t fixed at $t = 0, \pi/4$ and the cross-sections with x fixed at $x = \pi/4, \pi/2$.
   (b) What is the value of f if $x = 0$ or $x = \pi$? Explain why this is to be expected.
   (c) Explain the relation of the cross-sections to the surface representing f.

Strengthen Your Understanding

In Problems 32–33, explain what is wrong with the statement.

32. The graph of the function $f(x, y) = x^2 + y^2$ is a circle.

33. Cross-sections of the function $f(x, y) = x^2$ with x fixed are parabolas.

In Problems 34–36, give an example of:

34. A function whose graph lies above the xy-plane and intersects the plane $z = 2$ in a single point.

35. A function which intersects the xz-plane in a parabola and the yz-plane in a line.

36. A function which intersects the xy-plane in a circle.

Are the statements in Problems 37–50 true or false? Give reasons for your answer.

37. The function given by the formula $f(v, w) = e^v/w$ is an increasing function of v when w is a nonzero constant.

38. A function $f(x, y)$ can be an increasing function of x with y held fixed, and be a decreasing function of y with x held fixed.

39. A function $f(x, y)$ can have the property that $g(x) = f(x, 5)$ is increasing, whereas $h(x) = f(x, 10)$ is decreasing.

40. The plane $x + 2y - 3z = 1$ passes through the origin.

41. The plane $x + y + z = 3$ intersects the x-axis when $x = 3$.

42. The sphere $x^2 + y^2 + z^2 = 10$ intersects the plane $x = 10$.

43. The cross-section of the function $f(x, y) = x + y^2$ with $y = 1$ is a line.

44. The function $g(x, y) = 1 - y^2$ has identical parabolas for all cross-sections with x constant.

45. The function $g(x, y) = 1 - y^2$ has lines for all cross-sections with y constant.

46. The graphs of $f(x, y) = \sin(xy)$ and $g(x, y) = \sin(xy) + 2$ never intersect.

47. The graphs of $f(x, y) = x^2 + y^2$ and $g(x, y) = 1 - x^2 - y^2$ intersect in a circle.

48. If all the cross-sections of the graph of $f(x, y)$ with x constant are lines, then the graph of f is a plane.

49. The only point of intersection of the graphs of $f(x, y)$ and $-f(x, y)$ is the origin.

50. The point $(0, 0, 10)$ is the highest point on the graph of the function $f(x, y) = 10 - x^2 - y^2$.

51. The object in 3-space described by $x = 2$ is

   (a) A point   **(b)** A line

   (c) A plane   **(d)** Undefined.

12.3 CONTOUR DIAGRAMS

The surface which represents a function of two variables often gives a good idea of the function's general behaviour—for example, whether it is increasing or decreasing as one of the variables increases. However it is difficult to read numerical values off a surface and it can be hard to see all of the function's behaviour from a surface. Thus, functions of two variables are often represented by contour diagrams like the weather map on page 684. Contour diagrams have the additional advantage that they can be extended to functions of three variables.

Topographical Maps

One of the most common examples of a contour diagram is a topographical map like that shown in Figure 12.31. It gives the elevation in the region and is a good way of getting an overall picture of the terrain: where the mountains are, where the flat areas are. Such topographical maps are frequently coloured green at the lower elevations and brown, red, or white at the higher elevations.

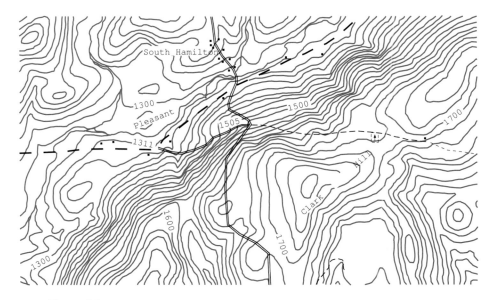

Figure 12.31: A topographical map showing the region around South Hamilton, NY

The curves on a topographical map that separate lower elevations from higher elevations are called *contour lines* because they outline the contour or shape of the land.[3] Because every point along the same contour has the same elevation, contour lines are also called *level curves* or *level sets*. The more closely spaced the contours, the steeper the terrain; the more widely spaced the contours, the flatter the terrain (provided, of course, that the elevation between contours varies by a constant amount). Certain features have distinctive characteristics. A mountain peak is typically surrounded by contour lines like those in Figure 12.32. A pass in a range of mountains may have contours that look like Figure 12.33. A long valley has parallel contour lines indicating the rising elevations on both sides of the valley (see Figure 12.34); a long ridge of mountains has the same type of contour lines, only the elevations decrease on both sides of the ridge. Notice that the elevation numbers on the contour lines are as important as the curves themselves. We usually draw contours for equally spaced values of z.

Figure 12.32: Mountain peak

Figure 12.33: Pass between two mountains

Figure 12.34: Long valley

Figure 12.35: Impossible contour lines

Notice that two contours corresponding to different elevations cannot cross each other as shown in Figure 12.35. If they did, the point of intersection of the two curves would have two different elevations, which is impossible (assuming the terrain has no overhangs).

Corn Production

Contour maps can display information about a function of two variables without reference to a surface. Consider the effect of weather conditions on US corn production. Figure 12.36 gives corn production $C = f(R, T)$ as a function of the total rainfall, R, in centimetres, and average temperature, T, in degrees Celsius, during the growing season.[4] At the present time, $R = 37.5$ cm and $T = 25°$C. Production is measured as a percentage of the present production; thus, the contour through $R = 37.5$, $T = 25$, has value 100, that is, $C = f(37.5, 25) = 100$.

Example 1   Use Figure 12.36 to estimate $f(45, 26)$ and $f(30, 25)$ and interpret in terms of corn production.

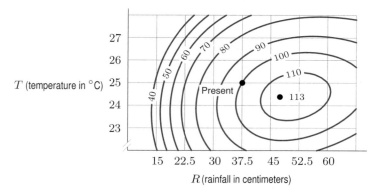

Figure 12.36: Corn production, C, as a function of rainfall and temperature

---

[3]In fact they are usually not straight lines, but curves. They may also be in disconnected pieces.

[4]Adapted from S. Beaty and R. Healy, "The Future of American Agriculture," *Scientific American* 248, No. 2, February 1983.

Solution   The point with R-coordinate 45 and T-coordinate 26 is on the contour $C = 100$, so $f(45, 26) = 100$. This means that if the annual rainfall were 45 centimetres and the temperature were 26°C, the country would produce about the same amount of corn as at present, although it would be wetter and warmer than it is now.

The point with R-coordinate 30 and T-coordinate 25 is about halfway between the $C = 80$ and the $C = 90$ contours, so $f(30, 25) \approx 85$. This means that if the rainfall fell to 30 centimetres and the temperature stayed at 25°, then corn production would drop to about 85% of what it is now.

Example 2   Use Figure 12.36 to describe in words the cross-sections with T and R constant through the point representing present conditions. Give a common-sense explanation of your answer.

Solution   To see what happens to corn production if the temperature stays fixed at 25°C but the rainfall changes, look along the horizontal line $T = 25$. Starting from the present and moving left along the line $T = 25$, the values on the contours decrease. In other words, if there is a drought, corn production decreases. Conversely, as rainfall increases, that is, as we move from the present to the right along the line $T = 25$, corn production increases, reaching a maximum of more than 110% when $R = 52$, and then decreases (too much rainfall floods the fields).

If, instead, rainfall remains at the present value and temperature increases, we move up the vertical line $R = 37.5$. Under these circumstances corn production decreases; a 1° increase causes a 10% drop in production. This makes sense since hotter temperatures lead to greater evaporation and hence drier conditions, even with rainfall constant at $R = 37.5$ centimetres. Similarly, a decrease in temperature leads to a very slight increase in production, reaching a maximum of around 102% when $T = 24$, followed by a decrease (the corn won't grow if it is too cold).

Contour Diagrams and Graphs

Contour diagrams and graphs are two different ways of representing a function of two variables. How do we go from one to the other? In the case of the topographical map, the contour diagram was created by joining all the points at the same height on the surface and dropping the curve into the xy-plane.

How do we go the other way? Suppose we wanted to plot the surface representing the corn production function $C = f(R, T)$ given by the contour diagram in Figure 12.36. Along each contour the function has a constant value; if we take each contour and lift it above the plane to a height equal to this value, we get the surface in Figure 12.37.

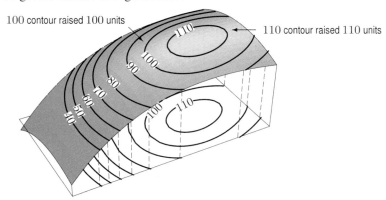

Figure 12.37: Getting the graph of the corn yield function from the contour diagram

Notice that the raised contours are the curves we get by slicing the surface horizontally. In general, we have the following result:

> Contour lines, or level curves, are obtained from a surface by slicing it with horizontal planes. A contour diagram is a collection of level curves labeled with function values.

Finding Contours Algebraically

Algebraic equations for the contours of a function f are easy to find if we have a formula for $f(x, y)$. Suppose the surface has equation

$$z = f(x, y).$$

A contour is obtained by slicing the surface with a horizontal plane with equation $z = c$. Thus, the equation for the contour at height c is given by:

$$f(x, y) = c.$$

Example 3   Find equations for the contours of $f(x, y) = x^2 + y^2$ and draw a contour diagram for f. Relate the contour diagram to the graph of f.

Solution   The contour at height c is given by

$$f(x, y) = x^2 + y^2 = c.$$

This is a contour only for $c \geq 0$, For $c > 0$ it is a circle of radius $\sqrt{c}$. For $c = 0$, it is a single point (the origin). Thus, the contours at an elevation of $c = 1, 2, 3, 4, \ldots$ are all circles centred at the origin of radius $1, \sqrt{2}, \sqrt{3}, 2, \ldots$. The contour diagram is shown in Figure 12.38. The bowl–shaped graph of f is shown in Figure 12.39. Notice that the graph of f gets steeper as we move further away from the origin. This is reflected in the fact that the contours become more closely packed as we move further from the origin; for example, the contours for $c = 6$ and $c = 8$ are closer together than the contours for $c = 2$ and $c = 4$.

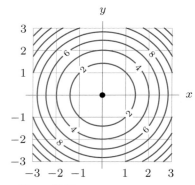

Figure 12.38: Contour diagram for $f(x, y) = x^2 + y^2$ (even values of c only)

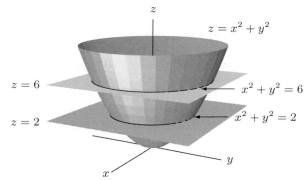

Figure 12.39: The graph of $f(x, y) = x^2 + y^2$

Example 4   Draw a contour diagram for $f(x, y) = \sqrt{x^2 + y^2}$ and relate it to the graph of f.

Solution   The contour at level c is given by

$$f(x, y) = \sqrt{x^2 + y^2} = c.$$

For $c > 0$ this is a circle, just as in the previous example, but here the radius is c instead of $\sqrt{c}$. For $c = 0$, it is the origin. Thus, if the level c increases by 1, the radius of the contour increases by 1. This means the contours are equally spaced concentric circles (see Figure 12.40) which do not become more closely packed further from the origin. Thus, the graph of f has the same constant slope as we move away from the origin (see Figure 12.41), making it a cone rather than a bowl.

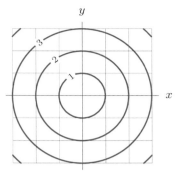

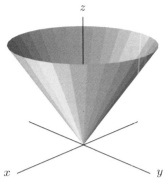

Figure 12.40: A contour diagram for
$f(x, y) = \sqrt{x^2 + y^2}$

Figure 12.41: The graph of
$f(x, y) = \sqrt{x^2 + y^2}$

In both of the previous examples the level curves are concentric circles because the surfaces have circular symmetry. Any function of two variables which depends only on the quantity $(x^2 + y^2)$ has such symmetry: for example, $G(x, y) = e^{-(x^2+y^2)}$ or $H(x, y) = \sin(\sqrt{x^2 + y^2})$.

Example 5   Draw a contour diagram for $f(x, y) = 2x + 3y + 1$.

Solution   The contour at level c has equation $2x + 3y + 1 = c$. Rewriting this as $y = -(2/3)x + (c - 1)/3$, we see that the contours are parallel lines with slope $-2/3$. The y-intercept for the contour at level c is $(c - 1)/3$; each time c increases by 3, the y-intercept moves up by 1. The contour diagram is shown in Figure 12.42.

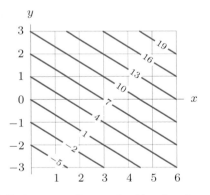

Figure 12.42: A contour diagram for $f(x, y) = 2x + 3y + 1$

Contour Diagrams and Tables

Sometimes we can get an idea of what the contour diagram of a function looks like from its table.

Example 6   Relate the values of $f(x, y) = x^2 - y^2$ in Table 12.4 to its contour diagram in Figure 12.43.

Table 12.4   *Table of values of $f(x, y) = x^2 - y^2$*

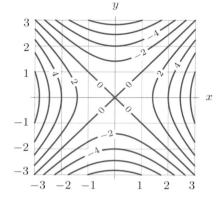

	3	0	−5	−8	−9	−8	−5	0
	2	5	0	−3	−4	−3	0	5
	1	8	3	0	−1	0	3	8
y	0	9	4	1	0	1	4	9
	−1	8	3	0	−1	0	3	8
	−2	5	0	−3	−4	−3	0	5
	−3	0	−5	−8	−9	−8	−5	0
		−3	−2	−1	0	1	2	3
					x			

Figure 12.43: Contour map of $f(x, y) = x^2 - y^2$

Solution   One striking feature of the values in Table 12.4 is the zeros along the diagonals. This occurs because $x^2 - y^2 = 0$ along the lines $y = x$ and $y = -x$. So the $z = 0$ contour consists of these two lines. In the triangular region of the table that lies to the right of both diagonals, the entries are positive. To the left of both diagonals, the entries are also positive. Thus, in the contour diagram, the positive contours lie in the triangular regions to the right and left of the lines $y = x$ and $y = -x$. Further, the table shows that the numbers on the left are the same as the numbers on the right; thus, each contour has two pieces, one on the left and one on the right. See Figure 12.43. As we move away from the origin along the x-axis, we cross contours corresponding to successively larger values. On the saddle-shaped graph of $f(x, y) = x^2 - y^2$ shown in Figure 12.44, this corresponds to climbing out of the saddle along one of the ridges. Similarly, the negative contours occur in pairs in the top and bottom triangular regions; the values get more and more negative as we go out along the y-axis. This corresponds to descending from the saddle along the valleys that are submerged below the xy-plane in Figure 12.44. Notice that we could also get the contour diagram by graphing the family of hyperbolas $x^2 - y^2 = 0, \pm 2, \pm 4, \ldots$.

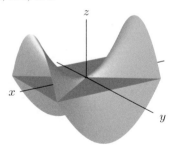

Figure 12.44: Graph of $f(x, y) = x^2 - y^2$ showing plane $z = 0$

Using Contour Diagrams: The Cobb-Douglas Production Function

Suppose you decide to expand your small printing business. Should you start a night shift and hire more workers? Should you buy more expensive but faster computers which will enable the current staff to keep up with the work? Or should you do some combination of the two?

Obviously, the way such a decision is made in practice involves many other considerations—such as whether you could get a suitably trained night shift, or whether there are any faster computers available. Nevertheless, you might model the quantity, P, of work produced by your business as a function of two variables: your total number, N, of workers, and the total value, V, of your equipment. What might the contour diagram of the production function look like?

Example 7   Explain why the contour diagram in Figure 12.45 does not model the behaviour expected of the production function, whereas the contour diagram in Figure 12.46 does.

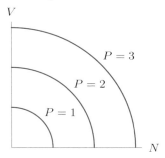

Figure 12.45: Incorrect contours for printing production

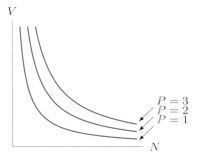

Figure 12.46: Correct contours for printing production

Solution   Look at Figure 12.45. Notice that the contour $P = 1$ intersects the N- and the V- axis, suggesting that it is possible to produce work with no workers or with no equipment; this is unreasonable. However, no contours in Figure 12.46 intersect either the N- or the V-axis.

In Figure 12.46, fixing V and letting N increase corresponds to moving to the right, crossing contours less and less frequently. Production increases more and more slowly because hiring additional workers does little to boost production if the machines are already used to capacity.

Similarly, if we fix N and let V increase, Figure 12.46 shows production increasing, but at a decreasing rate. Buying machines without enough people to use them does not increase production much. Thus Figure 12.46 fits the expected behaviour of the production function best.

Formula for a Production Function

Production functions are often approximated by formulas of the form

$$P = f(N, V) = cN^{\alpha}V^{\beta}$$

where P is the quantity produced and c, α, and β are positive constants, $0 < \alpha < 1$ and $0 < \beta < 1$.

Example 8   Show that the contours of the function $P = cN^{\alpha}V^{\beta}$ have approximately the shape of the contours in Figure 12.46.

Solution   The contours are the curves where P is equal to a constant value, say P_0, that is, where

$$cN^{\alpha}V^{\beta} = P_0.$$

Solving for V we get

$$V = \left(\frac{P_0}{c}\right)^{1/\beta} N^{-\alpha/\beta}.$$

Thus, V is a power function of N with a negative exponent, so its graph has the shape shown in Figure 12.46.

The Cobb-Douglas Production Model

In 1928, Cobb and Douglas used a similar function to model the production of the entire US economy in the first quarter of this century. Using government estimates of P, the total yearly production between 1899 and 1922, of K, the total capital investment over the same period, and of L, the total labour force, they found that P was well approximated by the *Cobb-Douglas production function*

$$P = 1.01L^{0.75}K^{0.25}.$$

This function turned out to model the US economy surprisingly well, both for the period on which it was based, and for some time afterward.

Exercises and Problems for Section 12.3

Exercises

1. Match the surfaces (a)–(e) in Figure 12.47 with the contour diagrams (I)–(V) in Figure 12.48.

2. Match Tables 12.5–12.8 with contour diagrams (I)–(IV) in Figure 12.49.

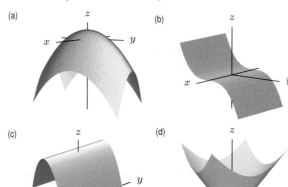

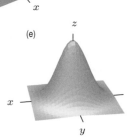

Figure 12.47

Table 12.5

$y \backslash x$	-1	0	1
-1	2	1	2
0	1	0	1
1	2	1	2

Table 12.6

$y \backslash x$	-1	0	1
-1	0	1	0
0	1	2	1
1	0	1	0

Table 12.7

$y \backslash x$	-1	0	1
-1	2	0	2
0	2	0	2
1	2	0	2

Table 12.8

$y \backslash x$	-1	0	1
-1	2	2	2
0	0	0	0
1	2	2	2

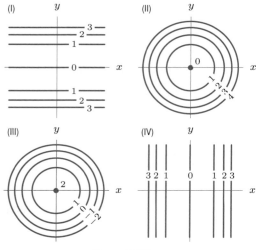

Figure 12.49

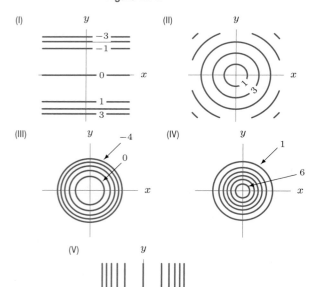

Figure 12.48

In Exercises 3–11, sketch a contour diagram for the function with at least four labeled contours. Describe in words the contours and how they are spaced.

3. $f(x, y) = x + y$

4. $f(x, y) = 3x + 3y$

5. $f(x, y) = x^2 + y^2$

6. $f(x, y) = -x^2 - y^2 + 1$

7. $f(x, y) = xy$

8. $f(x, y) = y - x^2$

9. $f(x, y) = x^2 + 2y^2$

10. $f(x, y) = \sqrt{x^2 + 2y^2}$

11. $f(x, y) = \cos \sqrt{x^2 + y^2}$

In Exercises 12–15, sketch a possible contour diagram for each surface, marked with reasonable z-values. (Note: There are many possible answers.)

12.

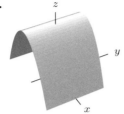

13.

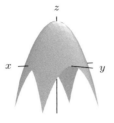

14.

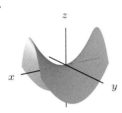

15.

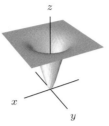

16. Find an equation for the contour of $f(x, y) = 7x^2 y^2 - 2x + 4y^2 - 10$ that goes through the point $(2, 4)$.

17. Let $f(x, y) = 3x^2 y + 7x + 20$. Find an equation for the contour that goes through the point $(5, 10)$.

18. (a) For $z = f(x, y) = xy$, sketch and label the level curves $z = \pm 1$, $z = \pm 2$.
   (b) Sketch and label cross-sections of f with $x = \pm 1$, $x = \pm 2$.
   (c) The surface $z = xy$ is cut by a vertical plane containing the line $y = x$. Sketch the cross-section.

19. Figure 12.50 shows a graph of $f(x, y) = (\sin x)(\cos y)$ for $-2\pi \le x \le 2\pi$, $-2\pi \le y \le 2\pi$. Use the surface $z = 1/2$ to sketch the contour $f(x, y) = 1/2$.

Figure 12.50

Problems

20. Total sales, Q, of a product are a function of its price and the amount spent on advertising. Figure 12.51 shows a contour diagram for total sales. Which axis corresponds to the price of the product and which to the amount spent on advertising? Explain.

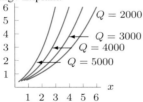

Figure 12.51

21. Figure 12.52 shows contours of $f(x, y) = 100e^x - 50y^2$. Find the values of f on the contours. They are equally spaced multiples of 10.

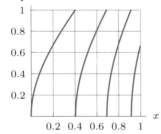

Figure 12.52

22. Figure 12.53 shows contour diagrams of $f(x, y)$ and $g(x, y)$. Sketch the smooth curve with equation $f(x, y) = g(x, y)$.

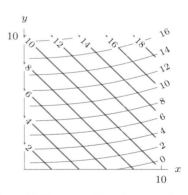

Figure 12.53: Black: $f(x, y)$. Blue: $g(x, y)$

23. Figure 12.54 shows the level curves of the temperature H in a room near a recently opened window. Label the three level curves with reasonable values of H if the house is in the following locations.

   (a) Siberia in winter (where winters are harsh).

(b) Palermo in winter (where winters are mild).
(c) Houston in summer (where summers are hot).
(d) Oregon in summer (where summers are mild).

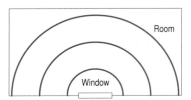

Figure 12.54

24. Match the functions (a)–(f) with the level curves (I)–(VI):

(a) $f(x,y) = x^2 - y^2 - 2x + 4y - 3$
(b) $g(x,y) = x^2 + y^2 - 2x - 4y + 15$
(c) $h(x,y) = -x^2 - y^2 + 2x + 4y - 8$
(d) $j(x,y) = -x^2 + y^2 + 2x - 4y + 3$
(e) $k(x,y) = \sqrt{(x-1)^2 + (y-2)^2}$
(f) $l(x,y) = -\sqrt{(x-1)^2 + (y-2)^2}$

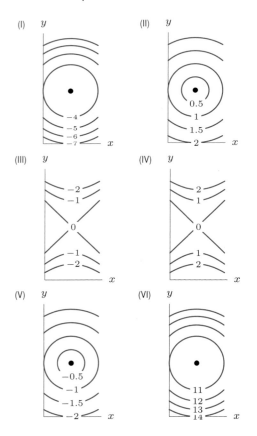

25. Figure 12.55 shows a contour map of a hill with two paths, A and B.

(a) On which path, A or B, will you have to climb more steeply?

(b) On which path, A or B, will you probably have a better view of the surrounding countryside? (Assume trees do not block your view.)
(c) Alongside which path is there more likely to be a stream?

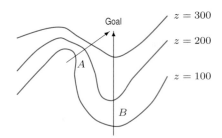

Figure 12.55

26. Figure 12.56 is a contour diagram of the monthly payment on a 5-year car loan as a function of the interest rate and the amount you borrow. The interest rate is 13% and you borrow $6000 for a used car.

(a) What is your monthly payment?
(b) If interest rates drop to 11%, how much more can you borrow without increasing your monthly payment?
(c) Make a table of how much you can borrow, without increasing your monthly payment, as a function of the interest rate.

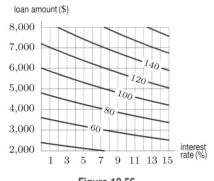

Figure 12.56

27. Hiking on a level trail going due east, you decide to leave the trail and climb toward the mountain on your left. The farther you go along the trail before turning off, the gentler the climb. Sketch a possible topographical map showing the elevation contours.

28. Describe in words the level surfaces of the function $g(x,y,z) = \cos(x + y + z)$.

29. Match the functions (a)–(d) with the shapes of their level curves (I)–(IV). Sketch each contour diagram.

(a) $f(x, y) = x^2$ (b) $f(x, y) = x^2 + 2y^2$

(c) $f(x, y) = y - x^2$ (d) $f(x, y) = x^2 - y^2$

I. Lines II. Parabolas

III. Hyperbolas IV. Ellipses

30. Figure 12.57 shows the density of the fox population P (in foxes per square kilometer) for southern England. Draw two different cross-sections along a north-south line and two different cross-sections along an east-west line of the population density P.

kilometers north

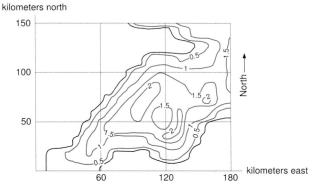

Figure 12.57

31. A manufacturer sells two goods, one at a price of $3000 a unit and the other at a price of $12,000 a unit. A quantity q_1 of the first good and q_2 of the second good are sold at a total cost of $4000 to the manufacturer.

(a) Express the manufacturer's profit, π, as a function of q_1 and q_2.
(b) Sketch curves of constant profit in the q_1q_2-plane for $\pi = 10,000$, $\pi = 20,000$, and $\pi = 30,000$ and the break-even curve $\pi = 0$.

32. Match each Cobb-Douglas production function (a)–(c) with a graph in Figure 12.58 and a statement (D)–(G).

(a) $F(L, K) = L^{0.25} K^{0.25}$
(b) $F(L, K) = L^{0.5} K^{0.5}$
(c) $F(L, K) = L^{0.75} K^{0.75}$
(D) Tripling each input triples output.
(E) Quadrupling each input doubles output.
(G) Doubling each input almost triples output.

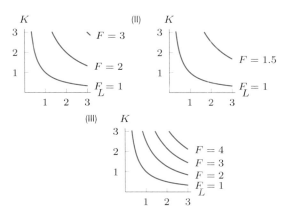

Figure 12.58

33. A Cobb-Douglas production function has the form

$$P = cL^\alpha K^\beta \quad \text{with } \alpha, \beta > 0.$$

What happens to production if labor and capital are both scaled up? For example, does production double if both labor and capital are doubled? Economists talk about

- *increasing returns to scale* if doubling L and K more than doubles P,
- *constant returns to scale* if doubling L and K exactly doubles P,
- *decreasing returns to scale* if doubling L and K less than doubles P.

What conditions on α and β lead to increasing, constant, or decreasing returns to scale?

34. (a) Match $f(x, y) = x^{0.2}y^{0.8}$ and $g(x, y) = x^{0.8}y^{0.2}$ with the level curves in Figures (I) and (II). All scales on the axes are the same.
(b) Figure (III) shows the level curves of $h(x, y) = x^\alpha y^{1-\alpha}$ for $0 < \alpha < 1$. Find the range of possible values for α. Again, the scales are the same on both axes.

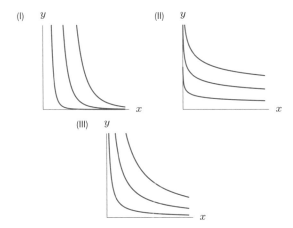

35. Match the functions (a)–(d) with the contour diagrams in Figures I–IV.

   (a) $f(x,y) = 0.7 \ln x + 0.3 \ln y$
   (b) $g(x,y) = 0.3 \ln x + 0.7 \ln y$
   (c) $h(x,y) = 0.3x^2 + 0.7y^2$
   (d) $j(x,y) = 0.7x^2 + 0.3y^2$

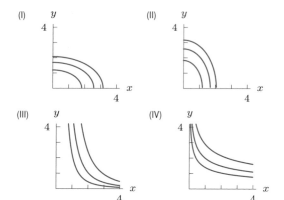

36. Figure 12.59 is the contour diagram of $f(x,y)$. Sketch the contour diagram of each of the following functions.

   (a) $3f(x,y)$ **(b)** $f(x,y) - 10$
   (c) $f(x-2, y-2)$ **(d)** $f(-x,y)$

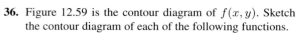

Figure 12.59

37. Figure 12.60 shows part of the contour diagram of $f(x,y)$. Complete the diagram for $x < 0$ if

   (a) $f(-x,y) = f(x,y)$ **(b)** $f(-x,y) = -f(x,y)$

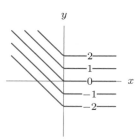

Figure 12.60

38. Values of $f(x,y) = \frac{1}{2}(x+y-2)(x+y-1) + y$ are in Table 12.9.

   (a) Find a pattern in the table. Make a conjecture and use it to complete Table 12.9 without computation. Check by using the formula for f.
   (b) Using the formula, check that the pattern holds for all $x \geq 1$ and $y \geq 1$.

Table 12.9

		\multicolumn{6}{c}{y}
		1	2	3	4	5	6
x	1	1	3	6	10	15	21
	2	2	5	9	14	20	
	3	4	8	13	19		
	4	7	12	18			
	5	11	17				
	6	16					

39. Let $f(x,y) = x^2 - y^2 = (x-y)(x+y)$. Use the factored form to sketch the contour $f(x,y) = 0$ and to find the regions in the xy-plane where $f(x,y) > 0$ and the regions where $f(x,y) < 0$. Explain how this sketch shows that the graph of $f(x,y)$ is saddle-shaped at the origin.

40. Use Problem 39 to find a formula for a "monkey saddle" surface $z = g(x,y)$ which has three regions with $g(x,y) > 0$ and three with $g(x,y) < 0$.

41. Use the contour diagram for $f(x,t) = \cos t \sin x$ in Figure 12.61 to describe in words the cross-sections of f with t fixed and the cross-sections of f with x fixed. Explain what you see in terms of the behavior of the string.

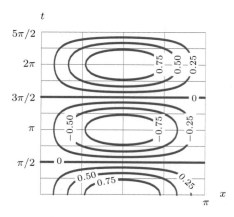

Figure 12.61

42. The power P produced by a windmill is proportional to the square of the diameter d of the windmill and to the cube of the speed v of the wind.[5]

(a) Write a formula for P as a function of d and v.
(b) A windmill generates 100 kW of power at a cer-

tain wind speed. If a second windmill is built having twice the diameter of the original, what fraction of the original wind speed is needed by the second windmill to produce 100 kW?
(c) Sketch a contour diagram for P.

Strengthen Your Understanding

In Problems 43–44, explain what is wrong with the statement.

43. A contour diagram for $z = f(x, y)$ is a surface in xyz-space.

44. The functions $f(x, y) = \sqrt{x^2 + y^2}$ and $g(x, y) = x^2 + y^2$ have the same contour diagram.

In Problems 45–46, give an example of:

45. A function $f(x, y)$ whose $z = 10$ contour consists of two or more parallel lines.

46. A function whose contours are all parabolas.

Decide if the statements in Problems 47–51 must be true, might be true, or could not be true. The function $z = f(x, y)$ is defined everywhere.

47. The level curves corresponding to $z = 1$ and $z = -1$ cross at the origin.

48. The level curve $z = 1$ consists of the circle $x^2 + y^2 = 2$ and the circle $x^2 + y^2 = 3$, but no other points.

49. The level curve $z = 1$ consists of two lines which intersect at the origin.

50. If $z = e^{-(x^2 + y^2)}$, there is a level curve for every value of z.

51. If $z = e^{-(x^2 + y^2)}$, there is a level curve through every point (x, y).

Are the statements in Problems 52–59 true or false? Give reasons for your answer.

52. Two isotherms representing distinct temperatures on a weather map cannot intersect.

53. A weather map can have two isotherms representing the same temperature that do not intersect.

54. The contours of the function $f(x, y) = y^2 + (x - 2)^2$ are either circles or a single point.

55. If the contours of $g(x, y)$ are concentric circles, then the graph of g is a cone.

56. If the contours for $f(x, y)$ get closer together in a certain direction, then f is increasing in that direction.

57. If all of the contours of $f(x, y)$ are parallel lines, then the graph of f is a plane.

58. If the $f = 10$ contour of the function $f(x, y)$ is identical to the $g = 10$ contour of the function $g(x, y)$, then $f(x, y) = g(x, y)$ for all (x, y).

59. The $f = 5$ contour of the function $f(x, y)$ is identical to the $g = 0$ contour of the function $g(x, y) = f(x, y) - 5$.

12.4 LINEAR FUNCTIONS

What is a Linear Function of Two Variables?

Linear functions played a central role in one-variable calculus because many one-variable functions have graphs that look like a line when we zoom in. In two-variable calculus, a *linear function* is one whose graph is a plane. In Chapter 14, we see that many two-variable functions have graphs which look like planes when we zoom in.

What Makes a Plane Flat?

What makes the graph of the function $z = f(x, y)$ a plane? Linear functions of *one* variable have straight line graphs because they have constant slope. On a plane, the situation is a bit more complicated. If we walk around on a tilted plane, the slope is not always the same: it depends on the direction in which we walk. However, at every point on the plane, the slope is the same as long as we choose the same direction. If we walk parallel to the x-axis, we always find ourselves walking up or down with the same slope;[6] the same is true if we walk parallel to the y-axis. In other words, the slope ratios $\Delta z / \Delta x$ (with y fixed) and $\Delta z / \Delta y$ (with x fixed) are each constant.

---

[5] From www.ecolo.org/documents/documents_in_english/WindmillFormula.htm, accessed on October 9, 2011.
[6] To be precise, walking in a vertical plane parallel to the x-axis while rising or falling with the plane you are on.

Example 1   A plane cuts the z-axis at $z = 5$ and has slope 2 in the x-direction and slope -1 in the y-direction. What is the equation of the plane?

Solution   Finding the equation of the plane means constructing a formula for the z-coordinate of the point on the plane directly above the point (x, y) in the xy-plane. To get to that point start from the point above the origin, where $z = 5$. Then walk x units in the x-direction. Since the slope in the x-direction is 2, the height increases by $2x$. Then walk y units in the y-direction; since the slope in the y-direction is -1, the height decreases by y units. Since the height has changed by $2x - y$ units, the z-coordinate is $5 + 2x - y$. Thus, the equation for the plane is

$$z = 5 + 2x - y.$$

For any linear function, if we know its value at a point (x_0, y_0), its slope in the x-direction, and its slope in the y-direction, then we can write the equation of the function. This is just like the equation of a line in the one-variable case, except that there are two slopes instead of one.

If a **plane** has slope m in the x-direction, has slope n in the y-direction, and passes through the point (x_0, y_0, z_0), then its equation is

$$z = z_0 + m(x - x_0) + n(y - y_0).$$

This plane is the graph of the **linear function**

$$f(x, y) = z_0 + m(x - x_0) + n(y - y_0).$$

If we write $c = z_0 - mx_0 - ny_0$, then we can write $f(x, y)$ in the equivalent form

$$f(x, y) = c + mx + ny.$$

Just as in 2-space a line is determined by two points, so in 3-space a plane is determined by three points, provided they do not lie on a line.

Example 2   Find the equation of the plane passing through the points $(1, 0, 1)$, $(1, -1, 3)$, and $(3, 0, -1)$.

Solution   The first two points have the same x-coordinate, so we use them to find the slope of the plane in the y-direction. As the y-coordinate changes from 0 to -1, the z-coordinate changes from 1 to 3, so the slope in the y-direction is $n = \Delta z / \Delta y = (3 - 1)/(-1 - 0) = -2$. The first and third points have the same y-coordinate, so we use them to find the slope in the x-direction; it is $m = \Delta z / \Delta x = (-1 - 1)/(3 - 1) = -1$. Because the plane passes through $(1, 0, 1)$, its equation is

$$z = 1 - (x - 1) - 2(y - 0) \quad \text{or} \quad z = 2 - x - 2y.$$

You should check that this equation is also satisfied by the points $(1, -1, 3)$ and $(3, 0, -1)$.

Example 2 was made easier by the fact that two of the points had the same x-coordinate and two had the same y-coordinate. An alternative method, which works for any three points, is to substitute the x, y, and z-values of each of the three points into the equation $z = c + mx + ny$. The resulting three equations in c, m, n are then solved simultaneously.

Linear Functions from a Numerical Point of View

To avoid flying planes with empty seats, airlines sell some tickets at full price and some at a discount. Table 12.10 shows an airline's revenue in dollars from tickets sold on a particular route, as a function of the number of full-price tickets sold, f, and the number of discount tickets sold, d.

Table 12.10 *Revenue from ticket sales (dollars)*

		Full-price tickets (f)		
	100	200	300	400
200	39,700	63,600	87,500	111,400
400	55,500	79,400	103,300	127,200
Discount tickets (d) 600	71,300	95,200	119,100	143,000
800	87,100	111,000	134,900	158,800
1000	102,900	126,800	150,700	174,600

In every column, the revenue jumps by $15,800 for each extra 200 discount tickets. Thus, each column is a linear function of the number of discount tickets sold. In addition, every column has the same slope, $15,800/200 = 79$ dollars/ticket. This is the price of a discount ticket. Similarly, each row is a linear function and all the rows have the same slope, 239, which is the price in dollars of a full-fare ticket. Thus, R is a linear function of f and d, given by:

$$R = 239f + 79d.$$

We have the following general result:

> A **linear function** can be recognised from its table by the following features:
> - Each row and each column is linear.
> - All the rows have the same slope.
> - All the columns have the same slope (although the slope of the rows and the slope of the columns are generally different).

Example 3   The table contains values of a linear function. Fill in the blank and give a formula for the function.

$x\backslash y$	1.5	2.0
2	0.5	1.5
3	−0.5	?

Solution   In the first column the function decreases by 1 (from 0.5 to −0.5) as x goes from 2 to 3. Since the function is linear, it must decrease by the same amount in the second column. So the missing entry must be $1.5 - 1 = 0.5$. The slope of the function in the x-direction is −1. The slope in the y-direction is 2, since in each row the function increases by 1 when y increases by 0.5. From the table we get $f(2, 1.5) = 0.5$. Therefore, the formula is

$$f(x, y) = 0.5 - (x - 2) + 2(y - 1.5) = -0.5 - x + 2y.$$

What Does the Contour Diagram of a Linear Function Look Like?

The formula for the airline revenue function in Table 12.10 is $R = 239f + 79d$, where f is the number of full-fares and d is the number of discount fares sold.

Notice that the contours of this function in Figure 12.62 are parallel straight lines. What is the practical significance of the slope of these contour lines? Consider the contour $R = 100,000$; that means we are looking at combinations of ticket sales that yield $100,000 in revenue. If we move down and to the right on the contour, the f-coordinate increases and the d-coordinate decreases, so we sell more full-fares and fewer discount fares. This is because to receive a fixed revenue of $100,000, we must sell more full-fares if we sell fewer discount fares. The exact trade-off depends on the slope of the contour; the diagram shows that each contour has a slope of about -3. This means that for a fixed revenue, we must sell three discount fares to replace one full-fare. This can also be seen by comparing prices. Each full fare brings in $239; to earn the same amount in discount fares we need to sell $239/79 \approx 3.03 \approx 3$ fares. Since the price ratio is independent of how many of each type of fare we sell, this slope remains constant over the whole contour map; thus, the contours are all parallel straight lines.

Notice also that the contours are evenly spaced. Thus, no matter which contour we are on, a fixed increase in one of the variables causes the same increase in the value of the function. In terms of revenue, no matter how many fares we have sold, an extra fare, whether full or discount, brings the same revenue as before.

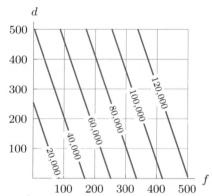

Figure 12.62: Revenue as a function of full and discount fares, $R = 239f + 79d$

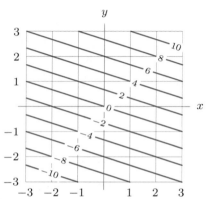

Figure 12.63: Contour map of linear function $f(x, y)$

Example 4   Find the equation of the linear function whose contour diagram is in Figure 12.63.

Solution   Suppose we start at the origin on the $z = 0$ contour. Moving 2 units in the y-direction takes us to the $z = 6$ contour; so the slope in the y-direction is $\Delta z/\Delta y = 6/2 = 3$. Similarly, a move of 2 in the x-direction from the origin takes us to the $z = 2$ contour, so the slope in the x-direction is $\Delta z/\Delta x = 2/2 = 1$. Since $f(0, 0) = 0$, we have $f(x, y) = x + 3y$.

Exercises and Problems for Section 12.4

Exercises

In Exercises 1–2, could the contour diagram represent a linear function?

1.

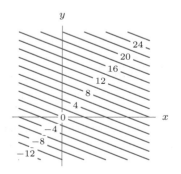

2.

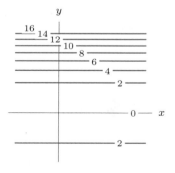

In Exercises 3–6, could the tables of values represent a linear function?

3.

x \ y	0	1	2
0	10	13	16
1	6	9	12
2	2	5	8

4.

x \ y	0	1	2
0	0	1	4
1	1	0	1
2	4	1	0

5.

x \ y	0	1	2
0	0	5	10
1	2	7	12
2	4	9	14

6.

x \ y	0	1	2
0	5	7	9
1	6	9	12
2	7	11	15

Exercises 7–8 each contain a partial table of values for a linear function. Fill in the blanks.

7.

$x\backslash y$	0.0	1.0
0.0		1.0
2.0	3.0	5.0

8.

$x\backslash y$	−1.0	0.0	1.0
2.0	4.0		
3.0		3.0	5.0

Problems

15. A store sells CDs at one price and DVDs at another price. Figure 12.64 shows the revenue (in dollars) of the music store as a function of the number, c, of CDs and the number, d, of DVDs that it sells. What is the price of a CD? What is the price of a DVD?

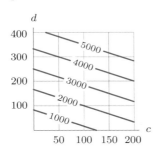

Figure 12.64

16. The charge, C, in dollars, to use an Internet service is a function of m, the number of months of use, and t, the total number of minutes on-line:

$$C = f(m, t) = 35 + 15m + 0.05t.$$

(a) Is f a linear function?
(b) Give units for the coefficients of m and t, and interpret them as charges.
(c) Interpret the intercept 35 as a charge.
(d) Find $f(3, 800)$ and interpret your answer.

9. Find the equation of the linear function $z = c + mx + ny$ whose graph contains the points $(0, 0, 0)$, $(0, 2, -1)$, and $(-3, 0, -4)$.

10. Find the linear function whose graph is the plane through the points $(4, 0, 0)$, $(0, 3, 0)$ and $(0, 0, 2)$.

11. Find an equation for the plane containing the line in the xy-plane where $y = 1$, and the line in the xz-plane where $z = 2$.

12. Find the equation of the linear function $z = c + mx + ny$ whose graph intersects the xz-plane in the line $z = 3x + 4$ and intersects the yz-plane in the line $z = y + 4$.

13. Suppose that z is a linear function of x and y with slope 2 in the x-direction and slope 3 in the y-direction.

(a) A change of 0.5 in x and -0.2 in y produces what change in z?
(b) If $z = 2$ when $x = 5$ and $y = 7$, what is the value of z when $x = 4.9$ and $y = 7.2$?

14. (a) Find a formula for the linear function whose graph is a plane passing through point $(4, 3, -2)$ with slope 5 in the x-direction and slope -3 in the y-direction.
(b) Sketch the contour diagram for this function.

17. For the contour diagrams (I)–(IV) on $-2 \le x, y \le 2$, pick the corresponding function.

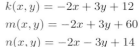

$f(x, y) = 2x + 3y + 10$
$g(x, y) = 2x + 3y + 60$
$h(x, y) = 2x - 3y + 12$
$j(x, y) = 2x - 3y + 60$

$k(x, y) = -2x + 3y + 12$
$m(x, y) = -2x + 3y + 60$
$n(x, y) = -2x - 3y + 14$
$p(x, y) = -2x - 3y + 60$

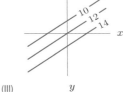

(I)

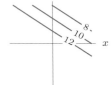

(II)

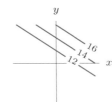

(III)

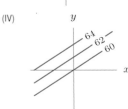

(IV)

18. A college admissions office uses the following equation to predict the grade point average of an incoming student:

$$z = 0.003x + 0.8y - 4,$$

where z is the predicted college GPA on a scale of 0 to 4.3, and x is the sum of the student's SAT Math and SAT

Verbal on a scale of 400 to 1600, and y is the student's high school GPA on a scale of 0 to 4.3. The college admits students whose predicted GPA is at least 2.3.

(a) Will a student with SATs of 1050 and high school GPA of 3.0 be admitted?
(b) Will every student with SATs of 1600 be admitted?
(c) Will every student with a high school GPA of 4.3 be admitted?
(d) Draw a contour diagram for the predicted GPA z with $400 \leq x \leq 1600$ and $0 \leq y \leq 4.3$. Shade the points corresponding to students who will be admitted.
(e) Which is more important, an extra 100 points on the SAT or an extra 0.5 of high school GPA?

19. A manufacturer makes two products out of two raw materials. Let q_1, q_2 be the quantities sold of the two products, p_1, p_2 their prices, and m_1, m_2 the quantities purchased of the two raw materials. Which of the following functions do you expect to be linear, and why? In each case, assume that all variables except the ones mentioned are held fixed.

(a) Expenditure on raw materials as a function of m_1 and m_2.
(b) Revenue as a function of q_1 and q_2.
(c) Revenue as a function of p_1 and q_1.

Problems 20–22 concern Table 12.11, which gives the number of calories burned per minute for someone roller-blading, as a function of the person's weight and speed.[7]

Table 12.11

	Calories burned per minute			
Weight	14 km/h	15 km/h	16 km/h	17 km/h
55 kg	4.2	5.8	7.4	8.9
65 kg	5.1	6.7	8.3	9.9
75 kg	6.1	7.7	9.2	10.8
85kg	7.0	8.6	10.2	11.7
95 kg	7.9	9.5	11.1	12.6

20. Does the data in Table 12.11 look approximately linear? Give a formula for B, the number of calories burned per minute in terms of the weight, w, and the speed, s. Does the formula make sense for all weights or speeds?

21. Who burns more total calories to go 10 kilometers: A 55 kg person going 16 km/h or a 85 kg person going 14 km/h? Which of these two people burns more calories per kilogram for the 10-kilometer trip?

22. Use Problem 20 to give a formula for P, the number of calories burned per kilogram, in terms of w and s, for a person weighing w kg roller-blading 10 kilometers at s km/h.

For Problems 23–24, find an equation for the linear function with the given values.

23.

$x \backslash y$	−1	0	1	2
0	1.5	1	0.5	0
1	3.5	3	2.5	2
2	5.5	5	4.5	4
3	7.5	7	6.5	6

24.

$x \backslash y$	10	20	30	40
100	3	6	9	12
200	2	5	8	11
300	1	4	7	10
400	0	3	6	9

For Problems 25–26, find possible equations for linear functions with the given contour diagrams.

25.

26.

It is difficult to graph a linear function by hand. One method that works if the x, y, and z-intercepts are positive is to plot the intercepts and join them by a triangle as shown in Figure 12.65; this shows the part of the plane in the octant where $x \geq 0$, $y \geq 0$, $z \geq 0$. If the intercepts are not all positive, the same method works if the x, y, and z-axes are drawn from a different perspective. Use this method to graph the linear functions in Problems 27–30.

Figure 12.65

27. $z = 2 - 2x + y$

28. $z = 2 - x - 2y$

29. $z = 4 + x - 2y$

30. $z = 6 - 2x - 3y$

---

[7]From the August 28, 1994, issue of *Parade Magazine*.

31. Let f be the linear function $f(x, y) = c + mx + ny$, where c, m, n are constants and $n \neq 0$.

    (a) Show that all the contours of f are lines of slope $-m/n$.

    (b) For all x and y, show $f(x + n, y - m) = f(x, y)$.

    (c) Explain the relation between parts (a) and (b).

Problems 32–33 refer to the linear function $z = f(x, y)$ whose values are in Table 12.12.

Table 12.12

	y				
	4	6	8	10	12
5	3	6	9	12	15
10	7	10	13	16	19
x 15	11	14	17	20	23
20	15	18	21	24	27
25	19	22	25	28	31

32. Each column of Table 12.12 is linear with the same slope, $m = \Delta z / \Delta x = 4/5$. Each row is linear with the same slope, $n = \Delta z / \Delta y = 3/2$. We now investigate the slope obtained by moving through the table along lines that are neither rows nor columns.

Strengthen Your Understanding

In Problems 34–35, explain what is wrong with the statement.

34. If the contours of f are all parallel lines, then f is linear.

35. A function $f(x, y)$ with linear cross-sections for x fixed and linear cross-sections for y fixed is a linear function.

In Problems 36–37, give an example of:

36. A table of values, with three rows and three columns, for a nonlinear function that is linear in each row and in each column.

37. A linear function whose contours are lines with slope 2.

Are the statements in Problems 38–49 true or false? Give reasons for your answer.

38. The planes $z = 3 + 2x + 4y$ and $z = 5 + 2x + 4y$ intersect.

39. The function represented in Table 12.13 is linear.

Table 12.13

$u \backslash v$	1.1	1.2	1.3	1.4
3.2	11.06	12.06	13.06	14.06
3.4	11.75	12.82	13.89	14.96
3.6	12.44	13.58	14.72	15.86
3.8	13.13	14.34	15.55	16.76
4.0	13.82	15.10	16.38	17.66

(a) Move down the diagonal of the table from the upper left corner ($z = 3$) to the lower right corner ($z = 31$). What do you notice about the changes in z? Now move diagonally from $z = 6$ to $z = 27$. What do you notice about the changes in z now?

(b) Move in the table along a line right one step, up two steps from $z = 19$ to $z = 9$. Then move in the same direction from $z = 22$ to $z = 12$. What do you notice about the changes in z?

(c) Show that $\Delta z = m\Delta x + n\Delta y$. Use this to explain what you observed in parts (a) and (b).

33. If we hold y fixed, that is we keep $\Delta y = 0$, and step in the positive x-direction, we get the x-slope, m. If instead we keep $\Delta x = 0$ and step in the positive y-direction, we get the y-slope, n. Fix a step in which neither $\Delta x = 0$ nor $\Delta y = 0$. The slope in the $\Delta x, \Delta y$ direction is

$$\text{Slope} = \frac{\text{Rise}}{\text{Run}} = \frac{\Delta z}{\text{Length of step}}$$
$$= \frac{\Delta z}{\sqrt{(\Delta x)^2 + (\Delta y)^2}}.$$

(a) Compute the slopes for the linear function in Table 12.12 in the direction of $\Delta x = 5, \Delta y = 2$.

(b) Compute the slopes for the linear function in Table 12.12 in the direction of $\Delta x = -10, \Delta y = 2$.

40. Contours of $f(x, y) = 3x + 2y$ are lines with slope 3.

41. If f is linear, then the contours of f are parallel lines.

42. If $f(0, 0) = 1, f(0, 1) = 4, f(0, 3) = 5$, then f cannot be linear.

43. The graph of a linear function is always a plane.

44. The cross-section $x = c$ of a linear function $f(x, y)$ is always a line.

45. There is no linear function $f(x, y)$ with a graph parallel to the xy-plane.

46. There is no linear function $f(x, y)$ with a graph parallel to the xz-plane.

47. A linear function $f(x, y) = 2x + 3y - 5$, has exactly one point (a, b) satisfying $f(a, b) = 0$.

48. In a table of values of a linear function, the columns have the same slope as the rows.

49. There is exactly one linear function $f(x, y)$ whose $f = 0$ contour is $y = 2x + 1$.

12.5 FUNCTIONS OF THREE VARIABLES

In applications of calculus, functions of any number of variables can arise. The density of matter in the universe is a function of three variables, since it takes three numbers to specify a point in space. Models of the US economy often use functions of ten or more variables. We need to be able to apply calculus to functions of arbitrarily many variables.

One difficulty with functions of more than two variables is that it is hard to visualise them. The graph of a function of one variable is a curve in 2-space, the graph of a function of two variables is a surface in 3-space, so the graph of a function of three variables would be a solid in 4-space. Since we can't easily visualise 4-space, we won't use the graphs of functions of three variables. On the other hand, it is possible to draw contour diagrams for functions of three variables, only now the contours are surfaces in 3-space.

Representing a Function of Three Variables Using a Family of Level Surfaces

A function of two variables, $f(x, y)$, can be represented by a family of level curves of the form $f(x, y) = c$ for various values of the constant, c.

> A **level surface**, or **level set** of a function of three variables, $f(x, y, z)$, is a surface of the form $f(x, y, z) = c$, where c is a constant. The function f can be represented by the family of level surfaces obtained by allowing c to vary.

The value of the function, f, is constant on each level surface.

Example 1   The temperature, in °C, at a point (x, y, z) is given by $T = f(x, y, z) = x^2 + y^2 + z^2$. What do the level surfaces of the function f look like and what do they mean in terms of temperature?

Solution   The level surface corresponding to $T = 100$ is the set of all points where the temperature is $100°C$. That is, where $f(x, y, z) = 100$, so

$$x^2 + y^2 + z^2 = 100.$$

This is the equation of a sphere of radius 10, with centre at the origin. Similarly, the level surface corresponding to $T = 200$ is the sphere with radius $\sqrt{200}$. The other level surfaces are concentric spheres. The temperature is constant on each sphere. We may view the temperature distribution as a set of nested spheres, like concentric layers of an onion, each one labelled with a different temperature, starting from low temperatures in the middle and getting hotter as we go out from the centre. (See Figure 12.66.) The level surfaces become more closely spaced as we move farther from the origin because the temperature increases more rapidly the farther we get from the origin.

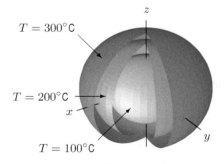

Figure 12.66: Level surfaces of $T = f(x, y, z) = x^2 + y^2 + z^2$, each one having a constant temperature

Example 2   What do the level surfaces of $f(x, y, z) = x^2 + y^2$ and $g(x, y, z) = z - y$ look like?

Solution   The level surface of f corresponding to the constant c is the surface consisting of all points satisfying the equation

$$x^2 + y^2 = c.$$

Since there is no z-coordinate in the equation, z can take any value. For $c > 0$, this is a circular cylinder of radius $\sqrt{c}$ around the z-axis. The level surfaces are concentric cylinders; on the narrow ones near the z-axis, f has small values; on the wider ones, f has larger values. See Figure 12.67.

The level surface of g corresponding to the constant c is the plane

$$z - y = c.$$

Since there is no x variable in the equation, these plane are parallel to the x-axis and cut the yz-plane in the line $z - y = c$. See Figure 12.68.

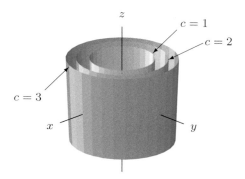

Figure 12.67: Level surfaces of $f(x, y, z) = x^2 + y^2$

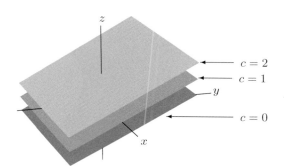

Figure 12.68: Level surfaces of $g(x, y, z) = z - y$

Example 3   What do the level surfaces of $f(x, y, z) = x^2 + y^2 - z^2$ look like?

Solution   In Section 12.3, we saw that the two-variable quadratic function $g(x, y) = x^2 - y^2$ has a saddle-shaped graph and three types of contours. The contour equation $x^2 - y^2 = c$ gives a hyperbola opening right-left when $c > 0$, a hyperbola opening up-down when $c < 0$, and a pair of intersecting lines when $c = 0$. Similarly, the three-variable quadratic function $f(x, y, z) = x^2 + y^2 - z^2$ has three types of level surfaces depending on the value of c in the equation $x^2 + y^2 - z^2 = c$.

Suppose that $c > 0$, say $c = 1$. Rewrite the equation as $x^2 + y^2 = z^2 + 1$ and think of what happens as we cut the surface perpendicular to the z-axis by holding z fixed. The result is a circle, $x^2 + y^2 = \text{constant}$, of radius at least 1 (since the constant $z^2 + 1 \geq 1$). The circles get larger as z gets larger. If we take the $x = 0$ cross-section instead, we get the hyperbola $y^2 - z^2 = 1$. The result is shown in Figure 12.72, with $a = b = c = 1$.

Suppose instead $c < 0$, say $c = -1$. Then the horizontal cross-sections of $x^2 + y^2 = z^2 - 1$ are again circles except that the radii shrink to 0 at $z = \pm 1$ and between $z = -1$ and $z = 1$ there are no cross-sections at all. The result is shown in Figure 12.73 with $a = b = c = 1$.

When $c = 0$, we get the equation $x^2 + y^2 = z^2$. Again the horizontal cross-sections are circles, this time with the radius shrinking down to exactly 0 when $z = 0$. The resulting surface, shown in Figure 12.74 with $a = b = c = 1$, is the cone $z = \sqrt{x^2 + y^2}$ studied in Section 12.3, together with the lower cone $z = -\sqrt{x^2 + y^2}$.

A Catalogue of Surfaces

For later reference, here is a small catalogue of the surfaces we have encountered.

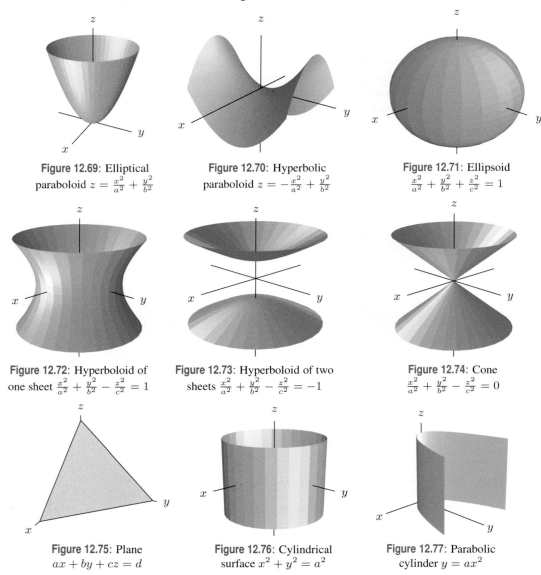

Figure 12.69: Elliptical paraboloid $z = \frac{x^2}{a^2} + \frac{y^2}{b^2}$

Figure 12.70: Hyperbolic paraboloid $z = -\frac{x^2}{a^2} + \frac{y^2}{b^2}$

Figure 12.71: Ellipsoid $\frac{x^2}{a^2} + \frac{y^2}{b^2} + \frac{z^2}{c^2} = 1$

Figure 12.72: Hyperboloid of one sheet $\frac{x^2}{a^2} + \frac{y^2}{b^2} - \frac{z^2}{c^2} = 1$

Figure 12.73: Hyperboloid of two sheets $\frac{x^2}{a^2} + \frac{y^2}{b^2} - \frac{z^2}{c^2} = -1$

Figure 12.74: Cone $\frac{x^2}{a^2} + \frac{y^2}{b^2} - \frac{z^2}{c^2} = 0$

Figure 12.75: Plane $ax + by + cz = d$

Figure 12.76: Cylindrical surface $x^2 + y^2 = a^2$

Figure 12.77: Parabolic cylinder $y = ax^2$

(These are viewed as equations in three variables x, y, and z)

How Surfaces Can Represent Functions of Two Variables and Functions of Three Variables

You may have noticed that we have used surfaces to represent functions in two different ways. First, we used a *single* surface to represent a two-variable function $f(x, y)$. Second, we used a *family* of level surfaces to represent a three-variable function $g(x, y, z)$. These level surfaces have equation $g(x, y, z) = c$.

What is the relation between these two uses of surfaces? For example, consider the function

$$f(x, y) = x^2 + y^2 + 3.$$

Define

$$g(x, y, z) = x^2 + y^2 + 3 - z$$

The points on the graph of f satisfy $z = x^2 + y^2 + 3$, so they also satisfy $x^2 + y^2 + 3 - z = 0$. Thus the graph of f is the same as the level surface

$$g(x, y, z) = x^2 + y^2 + 3 - z = 0.$$

In general, we have the following result:

A single surface that is the graph of a two-variable function $f(x, y)$ can be thought of as one member of the family of level surfaces representing the three-variable function

$$g(x, y, z) = f(x, y) - z.$$

The graph of f is the level surface $g = 0$.

Conversely, a single level surface $g(x, y, z) = c$ can be regarded as the graph of a function $f(x, y)$ if it is possible to solve for z. Sometimes the level surface is pieced together from the graphs of two or more two-variable functions. For example, if $g(x, y, z) = x^2 + y^2 + z^2$, then one member of the family of level surfaces is the sphere

$$x^2 + y^2 + z^2 = 1.$$

This equation defines z implicitly as a function of x and y. Solving it gives two functions

$$z = \sqrt{1 - x^2 - y^2} \qquad \text{and} \qquad z = -\sqrt{1 - x^2 - y^2}.$$

The graph of the first function is the top half of the sphere and the graph of the second function is the bottom half.

Exercises and Problems for Section 12.5

Exercises

1. Match the following functions with the level surfaces in Figure 12.78.

 (a) $f(x, y, z) = y^2 + z^2$ **(b)** $h(x, y, z) = x^2 + z^2$.

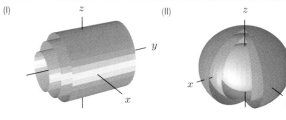

Figure 12.79

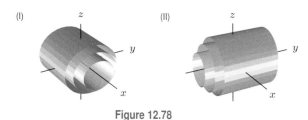

Figure 12.78

2. Match the functions with the level surfaces in Figure 12.79.

 (a) $f(x, y, z) = x^2 + y^2 + z^2$
 (b) $g(x, y, z) = x^2 + z^2$.

3. Write the level surface $x + 2y + 3z = 5$ as the graph of a function $f(x, y)$.

4. Find a formula for a function $f(x, y, z)$ whose level surface $f = 4$ is a sphere of radius 2, centered at the origin.

5. Write the level surface $x^2 + y + \sqrt{z} = 1$ as the graph of a function $f(x, y)$.

6. Find a formula for a function $f(x, y, z)$ whose level surfaces are spheres centered at the point (a, b, c).

7. Which of the graphs in the catalog of surfaces on page 720 is the graph of a function of x and y?

Use the catalog on page 720 to identify the surfaces in Exercises 8–11.

8. $x^2 + y^2 - z = 0$

9. $-x^2 - y^2 + z^2 = 1$

10. $x + y = 1$

11. $x^2 + y^2/4 + z^2 = 1$

In Exercises 12–15, decide if the given level surface can be expressed as the graph of a function, $f(x, y)$.

12. $z - x^2 - 3y^2 = 0$

13. $2x + 3y - 5z - 10 = 0$

14. $x^2 + y^2 + z^2 - 1 = 0$

15. $z^2 = x^2 + 3y^2$

Problems

In Problems 16–18, represent the surface whose equation is given as the graph of a two-variable function, $f(x, y)$, and as the level surface of a three-variable function, $g(x, y, z) = c$. There are many possible answers.

16. The plane $4x - y - 2z = 6$

17. The top half of the sphere $x^2 + y^2 + z^2 - 10 = 0$

18. The bottom half of the ellipsoid $x^2 + y^2 + z^2/2 = 1$

19. Suppose the function $f(x, y, z) = 2x - 3y + z - 20$ gives the temperature, in degrees Celsius, at a point (x, y, z).

   (a) Describe the isothermal surfaces of f.
   (b) Calculate and interpret $f_z(0, 0, 0)$.
   (c) If you are standing at the point $(0, 0, 0)$, in what direction should you move to increase your temperature the fastest?
   (d) Is $f(x, y) = -2x + 3y + 17$ an isothermal surface of f? If so, what is the temperature on this isotherm?

20. Find a function $f(x, y, z)$ whose level surface $f = 1$ is the graph of the function $g(x, y) = x + 2y$.

21. Find two functions $f(x, y)$ and $g(x, y)$ so that the graphs of both together form the ellipsoid $x^2 + y^2/4 + z^2/9 = 1$.

22. Find a formula for a function $g(x, y, z)$ whose level surfaces are planes parallel to the plane $z = 2x + 3y - 5$.

23. The surface S is the graph of $f(x, y) = \sqrt{1 - x^2 - y^2}$.

   (a) Explain why S is the upper hemisphere of radius 1, with equator in the xy-plane, centered at the origin.
   (b) Find a level surface $g(x, y, z) = c$ representing S.

24. The surface S is the graph of $f(x, y) = \sqrt{1 - y^2}$.

   (a) Explain why S is the upper half of a circular cylinder of radius 1, centered along the x-axis.
   (b) Find a level surface $g(x, y, z) = c$ representing S.

25. A cone C, with height 1 and radius 1, has its base in the xz-plane and its vertex on the positive y-axis. Find a function $g(x, y, z)$ such that C is part of the level surface $g(x, y, z) = 0$. [Hint: The graph of $f(x, y) = \sqrt{x^2 + y^2}$ is a cone which opens up and has vertex at the origin.]

26. Describe the level surface $f(x, y, z) = x^2/4 + z^2 = 1$ in words.

27. Describe the level surface $g(x, y, z) = x^2 + y^2/4 + z^2 = 1$ in words. [Hint: Look at cross-sections with constant x, y, and z values.]

28. Describe in words the level surfaces of the function $g(x, y, z) = x + y + z$.

29. Describe in words the level surfaces of $f(x, y, z) = \sin(x + y + z)$.

30. Describe the surface $x^2 + y^2 = (2 + \sin z)^2$. In general, if $f(z) \geq 0$ for all z, describe the surface $x^2 + y^2 = (f(z))^2$.

31. What do the level surfaces of $f(x, y, z) = x^2 - y^2 + z^2$ look like? [Hint: Use cross-sections with y constant instead of cross-sections with z constant.]

32. Describe in words the level surfaces of $g(x, y, z) = e^{-(x^2 + y^2 + z^2)}$.

33. Sketch and label level surfaces of $h(x, y, z) = e^{z - y}$ for $h = 1, e, e^2$.

34. Sketch and label level surfaces of $f(x, y, z) = 4 - x^2 - y^2 - z^2$ for $f = 0, 1, 2$.

35. Sketch and label level surfaces of $g(x, y, z) = 1 - x^2 - y^2$ for $g = 0, -1, -2$.

Strengthen Your Understanding

In Problems 36–38, explain what is wrong with the statement.

36. The graph of a function $f(x, y, z)$ is a surface in 3-space.

37. The level surfaces of $f(x, y, z) = x^2 - y^2$ are all saddle-shaped.

38. The level surfaces of $f(x, y, z) = x^2 + y^2$ are paraboloids.

In Problems 39–42, give an example of:

39. A function $f(x, y, z)$ whose level surfaces are equally spaced planes perpendicular to the yz-plane.

40. A function $f(x, y, z)$ whose level sets are concentric cylinders centered on the y-axis.

41. A nonlinear function $f(x, y, z)$ whose level sets are parallel planes.

42. A function $f(x, y, z)$ whose level sets are paraboloids.

Are the statements in Problems 43–52 true or false? Give reasons for your answer.

43. The graph of the function $f(x, y) = x^2 + y^2$ is the same as the level surface $g(x, y, z) = x^2 + y^2 - z = 0$.

44. The graph of $f(x, y) = \sqrt{1 - x^2 - y^2}$ is the same as the level surface $g(x, y, z) = x^2 + y^2 + z^2 = 1$.

45. Any surface which is the graph of a two-variable function $f(x, y)$ can also be represented as the level surface of a three-variable function $g(x, y, z)$.

46. Any surface which is the level surface of a three-variable function $g(x, y, z)$ can also be represented as the graph of a two-variable function $f(x, y)$.

47. The level surfaces of the function $g(x, y, z) = x + 2y + z$ are parallel planes.

48. The level surfaces of $g(x, y, z) = x^2 + y + z^2$ are cylinders with axis along the y-axis.

49. A level surface of a function $g(x, y, z)$ cannot be a single point.

50. If $g(x, y, z) = ax + by + cz + d$, where a, b, c, d are nonzero constants, then the level surfaces of g are planes.

51. If the level surfaces of g are planes, then $g(x, y, z) = ax + by + cz + d$, where a, b, c, d are constants.

52. If the level surfaces $g(x, y, z) = k_1$ and $g(x, y, z) = k_2$ are the same surface, then $k_1 = k_2$.

12.6 LIMITS AND CONTINUITY

The sheer face of Half Dome, in Yosemite National Park in California, was caused by glacial activity during the Ice Age. (See Figure 12.80.) As we scale the rock from the west, the height of the terrain rises abruptly by over 1500 metres from the valley floor, 600 metres of it vertical.

If we consider the function h giving the height of the terrain above sea level in terms of longitude and latitude, then h has a *discontinuity* along the path at the base of the cliff of Half Dome. Looking at the contour map of the region in Figure 12.81, we see that in most places a small change in position results in a small change in height, except near the cliff. There, no matter how small a step we take, we get a large change in height. (You can see how crowded the contours get near the cliff; some end abruptly along the discontinuity.)

This geological feature illustrates the ideas of continuity and discontinuity. Roughly speaking, a function is said to be *continuous* at a point if its values at places near the point are close to the value at the point. If this is not the case, the function is said to be *discontinuous*.

The property of continuity is one that, practically speaking, we usually assume of the functions we are studying. Informally, we expect (except under special circumstances) that values of a function do not change drastically when making small changes to the input variables. Whenever we model a one-variable function by an unbroken curve, we are making this assumption. Even when functions come to us as tables of data, we usually make the assumption that the missing function values between data points are close to the measured ones.

In this section we study limits and continuity a bit more formally in the context of functions

Figure 12.80: Half Dome in Yosemite National Park

Figure 12.81: A contour map of Half Dome

of several variables. For simplicity we study these concepts for functions of two variables, but our discussion can be adapted to functions of three or more variables.

One can show that sums, products, and compositions of continuous functions are continuous, while the quotient of two continuous functions is continuous everywhere the denominator function is nonzero. Thus, each of the functions

$$\cos(x^2 y), \qquad \ln(x^2 + y^2), \qquad \frac{e^{x+y}}{x+y}, \qquad \ln(\sin(x^2 + y^2))$$

is continuous at all points (x, y) where it is defined. As for functions of one variable, the graph of a continuous function over an unbroken domain is unbroken—that is, the surface has no holes or rips in it.

Example 1   From Figures 12.82–12.85, which of the following functions appear to be continuous at $(0, 0)$?

(a) $f(x, y) = \begin{cases} \dfrac{x^2 y}{x^2 + y^2}, & (x, y) \neq (0, 0), \\ 0, & (x, y) = (0, 0). \end{cases}$ (b) $g(x, y) = \begin{cases} \dfrac{x^2}{x^2 + y^2}, & (x, y) \neq (0, 0), \\ 0, & (x, y) = (0, 0). \end{cases}$

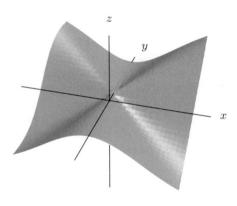

Figure 12.82: Graph of $z = x^2 y/(x^2 + y^2)$

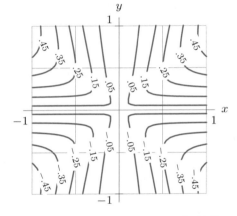

Figure 12.83: Contour diagram of $z = x^2 y/(x^2 + y^2)$

Figure 12.84: Graph of $z = x^2/(x^2 + y^2)$

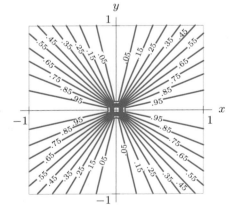

Figure 12.85: Contour diagram of $z = x^2/(x^2 + y^2)$

Solution   (a) The graph and contour diagram of f in Figures 12.82 and 12.83 suggest that f is close to 0 when (x, y) is close to $(0, 0)$. That is, the figures suggest that f is continuous at the point $(0, 0)$; the graph appears to have no rips or holes there.

However, the figures cannot tell us for sure whether f is continuous. To be certain we must investigate the limit analytically, as is done in Example 2(a) on page 725.

(b) The graph of g and its contours near $(0,0)$ in Figure 12.84 and 12.85 suggest that g behaves differently from f: The contours of g seem to "crash" at the origin and the graph rises rapidly from 0 to 1 near $(0,0)$. Small changes in (x,y) near $(0,0)$ can yield large changes in g, so we expect that g is not continuous at the point $(0,0)$. Again, a more precise analysis is given in Example 2(b) on page 725.

The previous example suggests that continuity *at* a point depends on a function's behaviour *near* the point. To study behaviour near a point more carefully we need the idea of a limit of a function of two variables. Suppose that $f(x,y)$ is a function defined on a set in 2-space, not necessarily containing the point (a,b), but containing points (x,y) arbitrarily close to (a,b); suppose that L is a number.

The function f has a **limit** L at the point (a,b), written

$$\lim_{(x,y)\to(a,b)} f(x,y) = L,$$

if $f(x,y)$ is as close to L as we please whenever the distance from the point (x,y) to the point (a,b) is sufficiently small, but not zero.

We define continuity for functions of two variables in the same way as for functions of one variable:

A function f is **continuous at the point** (a,b) if

$$\lim_{(x,y)\to(a,b)} f(x,y) = f(a,b).$$

A function is **continuous on a region** R in the xy-plane if it is continuous at each point in R.

Thus, if f is continuous at the point (a,b), then f must be defined at (a,b) and the limit, $\lim_{(x,y)\to(a,b)} f(x,y)$, must exist and be equal to the value $f(a,b)$. If a function is defined at a point (a,b) but is not continuous there, then we say that f is *discontinuous* at (a,b).

We now apply the definition of continuity to the functions in Example 1, showing that f is continuous at $(0,0)$ and that g is discontinuous at $(0,0)$.

Example 2   Let f and g be the functions in Example 1. Use the definition of the limit to show that:
(a) $\lim\limits_{(x,y)\to(0,0)} f(x,y) = 0$ (b) $\lim\limits_{(x,y)\to(0,0)} g(x,y)$ does not exist.

Solution   To investigate these limits of f and g, we consider values of these functions near, but not at, the origin, where they are given by the formulas

$$f(x,y) = \frac{x^2 y}{x^2 + y^2} \qquad g(x,y) = \frac{x^2}{x^2 + y^2}.$$

(a) The graph and contour diagram of f both suggest that $\lim_{(x,y)\to(0,0)} f(x,y) = 0$. To use the definition of the limit, we estimate $|f(x,y) - L|$ with $L = 0$:

$$|f(x,y) - L| = \left| \frac{x^2 y}{x^2 + y^2} - 0 \right| = \left| \frac{x^2}{x^2 + y^2} \right| |y| \le |y| \le \sqrt{x^2 + y^2}.$$

Now $\sqrt{x^2 + y^2}$ is the distance from (x, y) to $(0, 0)$. Thus, to make $|f(x, y) - 0| < 0.001$, for example, we need only require (x, y) be within 0.001 of $(0, 0)$. More generally, for any positive number u, no matter how small, we are sure that $|f(x, y) - 0| < u$ whenever (x, y) is no farther than u from $(0, 0)$. This is what we mean by saying that the difference $|f(x, y) - 0|$ can be made as small as we wish by choosing the distance to be sufficiently small. Thus, we conclude that

$$\lim_{(x,y)\to(0,0)} f(x, y) = \lim_{(x,y)\to(0,0)} \frac{x^2 y}{x^2 + y^2} = 0.$$

Notice that since this limit equals $f(0, 0)$, the function f is continuous at $(0, 0)$.

(b) Although the formula defining the function g looks similar to that of f, we saw in Example 1 that g's behaviour near the origin is quite different. If we consider points $(x, 0)$ lying along the x-axis near $(0, 0)$, then the values $g(x, 0)$ are equal to 1, while if we consider points $(0, y)$ lying along the y-axis near $(0, 0)$, then the values $g(0, y)$ are equal to 0. Thus, within any distance (no matter how small) from the origin, there are points where $g = 0$ and points where $g = 1$. Therefore the limit $\lim_{(x,y)\to(0,0)} g(x, y)$ does not exist, and thus g is not continuous at $(0, 0)$.

While the notions of limit and continuity look formally the same for one- and two-variable functions, they are somewhat more subtle in the multivariable case. The reason for this is that on the line (1-space), we can approach a point from just two directions (left or right) but in 2-space there are an infinite number of ways to approach a given point.

Exercises and Problems for Section 12.6

Exercises

In Exercises 1–6, is the function continuous at all points in the given region?

1. $\dfrac{1}{x^2 + y^2}$ on the square $-1 \le x \le 1, -1 \le y \le 1$

2. $\dfrac{1}{x^2 + y^2}$ on the square $1 \le x \le 2, 1 \le y \le 2$

3. $\dfrac{y}{x^2 + 2}$ on the disk $x^2 + y^2 \le 1$

4. $\dfrac{e^{\sin x}}{\cos y}$ on the rectangle $-\frac{\pi}{2} \le x \le \frac{\pi}{2}, 0 \le y \le \frac{\pi}{4}$

5. $\tan(xy)$ on the square $-2 \le x \le 2, -2 \le y \le 2$

6. $\sqrt{2x - y}$ on the disk $x^2 + y^2 \le 4$

In Exercises 7–11, find the limit as $(x, y) \to (0, 0)$ of $f(x, y)$. Assume that polynomials, exponentials, logarithmic, and trigonometric functions are continuous.

7. $f(x, y) = e^{-x-y}$

8. $f(x, y) = x^2 + y^2$

9. $f(x, y) = \dfrac{x}{x^2 + 1}$

10. $f(x, y) = \dfrac{x + y}{(\sin y) + 2}$

11. $f(x, y) = \dfrac{\sin(x^2 + y^2)}{x^2 + y^2}$ [Hint: $\lim\limits_{t \to 0} \dfrac{\sin t}{t} = 1$.]

Problems

In Problems 12–13, show that the function $f(x, y)$ does not have a limit as $(x, y) \to (0, 0)$. [Hint: Use the line $y = mx$.]

12. $f(x, y) = \dfrac{x + y}{x - y}, \qquad x \ne y$

13. $f(x, y) = \dfrac{x^2 - y^2}{x^2 + y^2}$

14. Show that $f(x, y)$ has no limit as $(x, y) \to (0, 0)$ if

$$f(x, y) = \frac{xy}{|xy|}, \qquad x \ne 0 \text{ and } y \ne 0.$$

15. Show that the function f does not have a limit at $(0, 0)$ by examining the limits of f as $(x, y) \to (0, 0)$ along the line $y = x$ and along the parabola $y = x^2$:

$$f(x, y) = \frac{x^2 y}{x^4 + y^2}, \qquad (x, y) \ne (0, 0).$$

16. Show that the function f does not have a limit at $(0, 0)$ by examining the limits of f as $(x, y) \to (0, 0)$ along the curve $y = kx^2$ for different values of k:

$$f(x, y) = \frac{x^2}{x^2 + y}, \qquad x^2 + y \ne 0.$$

17. Explain why the following function is not continuous along the line $y = 0$:

$$f(x, y) = \begin{cases} 1 - x, & y \geq 0, \\ -2, & y < 0. \end{cases}$$

In Problems 18–19, determine whether there is a value for c making the function continuous everywhere. If so, find it. If not, explain why not.

18. $f(x, y) = \begin{cases} c + y, & x \leq 3, \\ 5 - x, & x > 3. \end{cases}$

19. $f(x, y) = \begin{cases} c + y, & x \leq 3, \\ 5 - y, & x > 3. \end{cases}$

20. Is the following function continuous at $(0, 0)$?

$$f(x, y) = \begin{cases} x^2 + y^2 & \text{if } (x, y) \neq (0, 0) \\ 2 & \text{if } (x, y) = (0, 0) \end{cases}$$

21. What value of c makes the following function continuous at $(0, 0)$?

$$f(x, y) = \begin{cases} x^2 + y^2 + 1 & \text{if } (x, y) \neq (0, 0) \\ c & \text{if } (x, y) = (0, 0) \end{cases}$$

22. (a) Use a computer to draw the graph and the contour diagram of the following function:

$$f(x, y) = \begin{cases} \dfrac{xy(x^2 - y^2)}{x^2 + y^2}, & (x, y) \neq (0, 0), \\ 0, & (x, y) = (0, 0). \end{cases}$$

(b) Do your answers to part (a) suggest that f is continuous at $(0, 0)$? Explain your answer.

23. The function f, whose graph and contour diagram are in Figures 12.86 and 12.87, is given by

$$f(x, y) = \begin{cases} \dfrac{xy}{x^2 + y^2}, & (x, y) \neq (0, 0), \\ 0, & (x, y) = (0, 0). \end{cases}$$

(a) Show that $f(0, y)$ and $f(x, 0)$ are each continuous functions of one variable.

(b) Show that rays emanating from the origin are contained in contours of f.

(c) Is f continuous at $(0, 0)$?

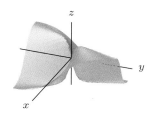

Figure 12.86: Graph of $z = xy/(x^2 + y^2)$

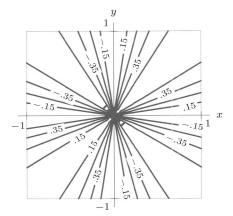

Figure 12.87: Contour diagram of $z = xy/(x^2 + y^2)$

Strengthen Your Understanding

In Problems 24–25, explain what is wrong with the statement.

24. If a function $f(x, y)$ has a limit as (x, y) approaches (a, b), then it is continuous at (a, b).

25. If both f and g are continuous at (a, b), then so are $f + g, fg$ and f/g.

In Problems 26–27, give an example of:

26. A function $f(x, y)$ which is continuous everywhere except at $(0, 0)$ and $(1, 2)$.

27. A function $f(x, y)$ that approaches 1 as (x, y) approaches $(0, 0)$ along the x-axis and approaches 2 as (x, y) approaches $(0, 0)$ along the y-axis.

In Problems 28–30, construct a function $f(x, y)$ with the given property.

28. Not continuous along the line $x = 2$; continuous everywhere else.

29. Not continuous at the point $(2, 0)$; continuous everywhere else.

30. Not continuous along the curve $x^2 + y^2 = 1$; continuous everywhere else.

CHAPTER SUMMARY (see also Ready Reference at the end of the book)

- **3-Space**
  Cartesian coordinates, x-, y- and z-axes, xy-, xz- and yz-planes, distance formula.
- **Functions of Two Variables**
  Represented by: tables, graphs, formulas, cross-sections (one variable fixed), contours (function value fixed); cylinders (one variable missing).
- **Linear Functions**

Recognizing linear functions from tables, graphs, contour diagrams, formulas. Converting from one representation to another.

- **Functions of Three Variables**
  Sketching level surfaces (function value fixed) in 3-space; graph of $z = f(x, y)$ is same as level surface $g(x, y, z) = f(x, y) - z = 0$.
- **Continuity**

REVIEW EXERCISES AND PROBLEMS FOR CHAPTER TWELVE

Exercises

1. On a set of x, y, and z axes oriented as in Figure 12.5 on page 687, draw a straight line through the origin, lying in the xz-plane and such that if you move along the line with your x-coordinate increasing, your z-coordinate is decreasing.

2. On a set of x, y and z axes oriented as in Figure 12.5 on page 687, draw a straight line through the origin, lying in the yz-plane and such that if you move along the line with your y-coordinate increasing, your z-coordinate is increasing.

In Exercises 3–5, determine if z is a function of x and y. If so, find a formula for the function.

3. $6x - 4y + 2z = 10$

4. $x^2 + y^2 + z^2 = 100$

5. $3x^2 - 5y^2 + 5z = 10 + x + y$

6. Figure 12.88 shows the parabolas $z = f(x, b)$ for $b = -2, -1, 0, 1, 2$. Which of the graphs of $z = f(x, y)$ in Figure 12.89 best fits this information?

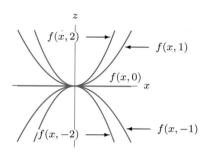

Figure 12.88

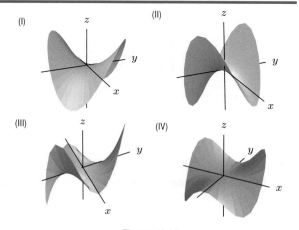

Figure 12.89

7. Match the pairs of functions (a)–(d) with the contour diagrams (I)–(IV). In each case, which contours represent f and which represent g? (The x- and y-scales are equal.)

(a) $f(x, y) = x + y$, $g(x, y) = x - y$
(b) $f(x, y) = 2x + 3y$, $g(x, y) = 2x - 3y$
(c) $f(x, y) = x^2 - y$, $g(x, y) = 2y + \ln|x|$
(d) $f(x, y) = x^2 - y^2$, $g(x, y) = xy$

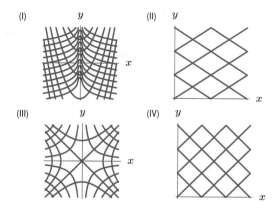

8. Match the contour diagrams (a)–(d) with the surfaces (I)–(IV). Give reasons for your choice.

(a)

(b)

(c)

(d)

(I)

(II)

(III)

(IV)

In Exercises 9–12, make a contour plot for the function in the region $-2 < x < 2$ and $-2 < y < 2$. What is the equation and the shape of the contours?

9. $z = 3x - 5y + 1$

10. $z = \sin y$

11. $z = 2x^2 + y^2$

12. $z = e^{-2x^2 - y^2}$

13. Describe the set of points whose x coordinate is 2 and whose y coordinate is 1.

14. Find the equation of the sphere of radius 5 centered at $(1, 2, 3)$.

15. Find the equation of the plane through the points $(0, 0, 2), (0, 3, 0), (5, 0, 0)$.

16. Find the center and radius of the sphere with equation $x^2 + 4x + y^2 - 6y + z^2 + 12z = 0$.

Which of the contour diagrams in Exercises 17–18 could represent linear functions?

17.

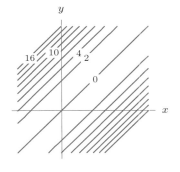

18.

19. **(a)** Complete the table with values of a linear function $f(x, y)$.
   (b) Find a formula for $f(x, y)$.

		y	
	2.5	3.0	3.50
-1	6		8
x 1		1	2
3	-6		

20. Find a formula for a function $f(x, y, z)$ whose level surfaces look like those in Figure 12.90.

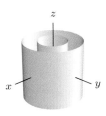

Figure 12.90

In Exercises 21–24, represent the surface as the graph of a function, $f(x, y)$, and by level surfaces of the form $g(x, y, z) = c$. (There are many possible answers.)

21. Paraboloid obtained by shifting $z = x^2 + y^2$ vertically 5 units

22. Plane with intercepts $x = 2$, $y = 3$, $z = 4$.

23. Upper half of unit sphere centered at the origin.

24. Lower half of sphere of radius 2 centered at $(3, 0, 0)$.

Use the catalog on page 720 to identify the surfaces in Exercises 25–26.

25. $x^2 + z^2 = 1$ **26.** $-x^2 + y^2 - z^2 = 0$

27. (a) What features of the contour diagram of $g(x, y)$ in Figure 12.91 suggest that g is linear?
(b) Assuming g is linear, find a formula for $g(x, y)$.

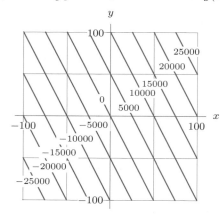

Figure 12.91

28. Figure 12.92 shows the graph of $z = f(x, y)$.

(a) Suppose y is fixed and positive. Does z increase or decrease as x increases? Graph z against x.

(b) Suppose x is fixed and positive. Does z increase or decrease as y increases? Graph z against y.

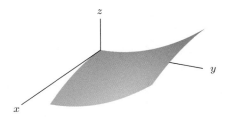

Figure 12.92

Problems

29. Use a computer or calculator to draw the graph of the vibrating guitar string function:

$$g(x, t) = \cos t \sin 2x, \quad 0 \le x \le \pi, \quad 0 \le t \le 2\pi.$$

Relate the shape of the graph to the cross-sections with t fixed and those with x fixed.

30. Consider the Cobb-Douglas production function $P = f(L, K) = 1.01 L^{0.75} K^{0.25}$. What is the effect on production of doubling both labor and capital?

31. (a) Sketch level curves of $f(x, y) = \sqrt{x^2 + y^2} + x$ for $f = 1, 2, 3$.
(b) For what values of c can level curves $f = c$ be drawn?

32. By approaching the origin along the positive x-axis and the positive y-axis, show that the following limit does not exist:

$$\lim_{(x,y) \to (0,0)} \frac{2x - y^2}{2x + y^2}.$$

33. By approaching the origin along the positive x-axis and the positive y-axis, show that the following limit does not exist:

$$\lim_{(x,y) \to (0,0)} \frac{x + y^2}{2x + y}.$$

34. You are in a room 12 meters long with a heater at one end. In the morning the room is $18°C$. You turn on the heater, which quickly warms up to $26°C$. Let $H(x, t)$ be the temperature x meters from the heater, t minutes after the heater is turned on. Figure 12.93 shows the contour diagram for H. How warm is it 4 meters from the heater 5 minutes after it was turned on? 10 minutes after it was turned on?

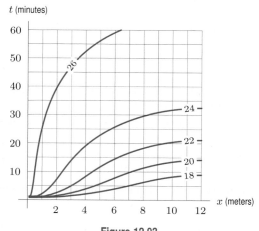

Figure 12.93

35. Using the contour diagram in Figure 12.93, sketch the graphs of the one-variable functions $H(x, 5)$ and $H(x, 20)$. Interpret the two graphs in practical terms, and explain the difference between them.

36. Let $f(x, y) = \begin{cases} \dfrac{|x|}{x} y & \text{for } x \neq 0 \\ 0 & \text{for } x = 0. \end{cases}$

    Is $f(x, y)$ continuous

    (a) On the x-axis? **(b)** On the y-axis?

    (c) At $(0, 0)$?

37. Sketch cross-sections of $f(r, h) = \pi r^2 h$, first keeping h fixed, then keeping r fixed.

38. The temperature T (in °C) at any point in the region $-10 \leq x \leq 10, -10 \leq y \leq 10$ is given by the function

    $$T(x, y) = 100 - x^2 - y^2.$$

    (a) Sketch isothermal curves (curves of constant temperature) for $T = 100°C$, $T = 75°C$, $T = 50°C$, $T = 25°C$, and $T = 0°C$.

    (b) A heat-seeking bug is put down at a point on the xy-plane. In which direction should it move to increase its temperature fastest? How is that direction related to the level curve through that point?

39. Find a linear function whose graph is the plane that intersects the xy-plane along the line $y = 2x + 2$ and contains the point $(1, 2, 2)$.

40. **(a)** Sketch the level curves of $z = \cos \sqrt{x^2 + y^2}$.

    (b) Sketch a cross-section through the surface $z = \cos \sqrt{x^2 + y^2}$ in the plane containing the x- and z-axes. Put units on your axes.

    (c) Sketch the cross-section through the surface $z = \cos \sqrt{x^2 + y^2}$ in the plane containing the z-axis and the line $y = x$ in the xy-plane.

Problems 41–44 concern a vibrating guitar string. Snapshots of the guitar string at millisecond intervals are in Figure 12.94.

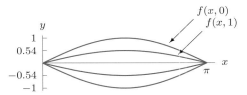

Figure 12.94: A vibrating guitar string: $f(x, t) = \cos t \sin x$ for four t values.

The guitar string is stretched tight along the x-axis from $x = 0$ to $x = \pi$. Each point on the string has an x-value, $0 \leq x \leq \pi$. As the string vibrates, each point on the string moves back and forth on either side of the x-axis. Let $y = f(x, t)$ be the displacement at time t of the point on the string located x units from the left end. A possible formula is

$$y = f(x, t) = \cos t \sin x, \quad 0 \leq x \leq \pi, \quad t \text{ in milliseconds.}$$

41. Explain what the functions $f(x, 0)$ and $f(x, 1)$ represent in terms of the vibrating string.

42. Explain what the functions $f(0, t)$ and $f(1, t)$ represent in terms of the vibrating string.

43. **(a)** Sketch graphs of y versus x for fixed t values, $t = 0$, $\pi/4, \pi/2, 3\pi/4, \pi$.

    (b) Use your graphs to explain why this function could represent a vibrating guitar string.

44. Describe the motion of the guitar strings whose displacements are given by the following:

    (a) $y = g(x, t) = \cos 2t \sin x$

    (b) $y = h(x, t) = \cos t \sin 2x$

CAS Challenge Problems

45. Let $A = (0, 0, 0)$ and $B = (2, 0, 0)$.

    (a) Find a point C in the xy-plane that is a distance 2 from both A and B.

    (b) Find a point D in 3-space that is a distance 2 from each of A, B, and C.

    (c) Describe the figure obtained by joining A, B, C, and D with straight lines.

46. Let $f(x, y) = 3 + x + 2y$.

    (a) Find formulas for $f(x, f(x, y)), f(x, f(x, f(x, y)))$ by hand.

    (b) Consider $f(x, f(x, f(x, f(x, f(x, f(x, y))))))$. Conjecture a formula for this function and check your answer with a computer algebra system.

47. A function $f(x, y, z)$ has the property that $f(1, 0, 1) = 20$, $f(1, 1, 1) = 16$, and $f(1, 1, 2) = 21$.

    (a) Estimate $f(1, 1, 3)$ and $f(1, 2, 1)$, assuming f is a linear function of each variable with the other variables held fixed.

    (b) Suppose in fact that $f(x, y, z) = ax^2 + byz + czx^3 + d2^{x-y}$, for constants a, b, c and d. Which of your estimates in part (a) do you expect to be exact?

    (c) Suppose in addition that $f(0, 0, 1) = 6$. Find an exact formula for f by solving for a, b, c, and d.

    (d) Use the formula in part (c) to evaluate $f(1, 1, 3)$ and $f(1, 2, 1)$ exactly. Do the values confirm your answer to part (b)?

PROJECTS FOR CHAPTER TWELVE

1. Noise Levels at London's Heathrow Airport

The measure used by the UK government to monitor aircraft noise is called the Loudness Equivalent or Leq.[8] This takes the sound energy from each aircraft in decibels or dB and averages it over a 16-hour day for several months. Based on the official noise output of each aircraft type and historical data on aircraft movements and flight paths, the Leq is calculated for a wide area around Heathrow Airport, London. The results are shown as noise contour maps such as Figure 12.95, where contours are labeled in dB. The 57 dB contour is significant since a major study indicated that community annoyance generally becomes significant at this noise level.

(a) Identify the position of the main runways at Heathrow and estimate the noise level in dB on the runways.

(b) Planes prefer to take off and land facing into the wind. Using the fact that the wind is generally from the west, explain the shape of the contours at the western and eastern ends of the runways.

(c) In which direction does the sound level fall most rapidly? Explain.

(d) A sound can be measured by the physical intensity of the sound, L. The decibel measure of the sound, B, is obtained from the ratio of the sound intensity, L, to a base intensity, L_0:

$$B = 10 \log_{10}\left(\frac{L}{L_0}\right).$$

The contour levels are show at 3 dB intervals. How do sound intensity levels (L) compare on adjacent contours?

(e) If the noise level of the next generation of aircraft is 50% of current values, how will the contour map change?

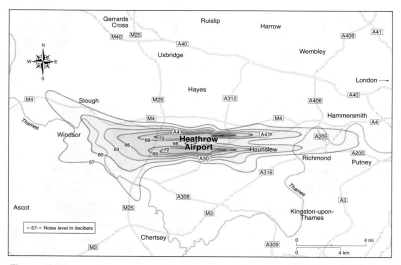

Figure 12.95: Contours, in decibels (dB), showing noise levels around Heathrow Airport, London

2. A Heater in a Room

Figure 12.96 shows the contours of the temperature along one wall of a heated room through one winter day, with time indicated as on a 24-hour clock. The room has a heater located at the leftmost corner of the wall and one window in the wall. The heater is controlled by a thermostat about 60 cm from the window.

(a) Where is the window?

---

[8]ERCD Report 1001, Environmental Research and Consultancy Department, CAA report March 2010.

(b) When is the window open?

(c) When is the heat on?

(d) Draw graphs of the temperature along the wall of the room at 6 am, at 11 am, at 3 pm (15 hours) and at 5 pm (17 hours).

(e) Draw a graph of the temperature as a function of time at the heater, at the window and midway between them.

(f) The temperature at the window at 5 pm (17 hours) is less than at 11 am. Why do you think this might be?

(g) To what temperature do you think the thermostat is set? How do you know?

(h) Where is the thermostat?

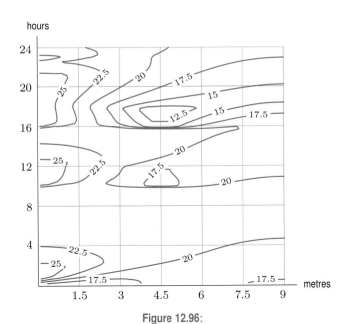

Figure 12.96:

3. Light in a Wave-guide

Figure 12.97 shows the contours of light intensity as a function of location and time in a microscopic wave-guide.

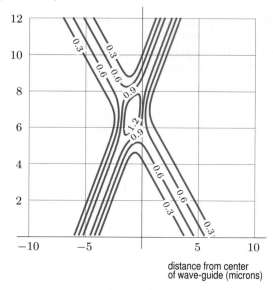

time (nanoseconds)

distance from center
of wave-guide (microns)

Figure 12.97

(a) Draw graphs showing intensity as a function of location at times 0, 2, 4, 6, 8, and 10 nanoseconds.

(b) If you could create an animation showing how the graph of intensity as a function of location varies as time passes, what would it look like?

(c) Draw a graph of intensity as a function of time at locations -5, 0, and 5 microns from center of wave-guide.

(d) Describe what the light beams are doing in this wave-guide.

Chapter Thirteen

A FUNDAMENTAL TOOL: VECTORS

Contents

13.1   DISPLACEMENT VECTORS

Suppose you are a pilot planning a flight from Dallas to Pittsburgh. There are two things you must know: the distance to be travelled (so you have enough fuel to make it) and in what direction to go (so you don't miss Pittsburgh). Both these quantities together specify the displacement or *displacement vector* between the two cities.

> The **displacement vector** from one point to another is an arrow with its tail at the first point and its tip at the second. The **magnitude** (or length) of the displacement vector is the distance between the points and is represented by the length of the arrow. The **direction** of the displacement vector is the direction of the arrow.

Figure 13.1 shows a map with the displacement vectors from Dallas to Pittsburgh, from Albuquerque to Oshkosh, and from Los Angeles to Buffalo, SD. These displacement vectors have the same length and the same direction. We say that the displacement vectors between the corresponding cities are the same, even though they do not coincide. In other words

> Displacement vectors which point in the same direction and have the same magnitude are considered to be the same, even if they do not coincide.

Figure 13.1: Displacement vectors between cities

Notation and Terminology

The displacement vector is our first example of a vector. Vectors have both magnitude and direction; in comparison, a quantity specified only by a number, but no direction, is called a *scalar*.[1] For instance, the time taken by the flight from Dallas to Pittsburgh is a scalar quantity. Displacement is a vector since it requires both distance and direction to specify it.

In this book, vectors are written with an arrow over them, $\vec{v}$, to distinguish them from scalars. Other books use a bold **v** to denote a vector. We use the notation $\overrightarrow{PQ}$ to denote the displacement vector from a point P to a point Q. The magnitude, or length, of a vector $\vec{v}$ is written $\|\vec{v}\|$.

Addition and Subtraction of Displacement Vectors

Suppose NASA commands a robot on Mars to move 75 metres in one direction and then 50 metres in another direction. (See Figure 13.2.) Where does the robot end up? Suppose the displacements are represented by the vectors $\vec{v}$ and $\vec{w}$, respectively. Then the sum $\vec{v} + \vec{w}$ gives the final position.

---

[1]So named by W. R. Hamilton because they are merely numbers on the *scale* from $-\infty$ to ∞.

The **sum**, $\vec{v} + \vec{w}$, of two vectors $\vec{v}$ and $\vec{w}$ is the combined displacement resulting from first applying $\vec{v}$ and then $\vec{w}$. (See Figure 13.3.) The sum $\vec{w} + \vec{v}$ gives the same displacement.

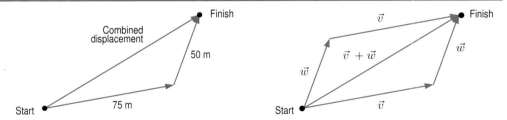

Figure 13.2: Sum of displacements of robots on Mars   Figure 13.3: The sum $\vec{v} + \vec{w} = \vec{w} + \vec{v}$

Suppose two different robots start from the same location. One moves along a displacement vector $\vec{v}$ and the second along a displacement vector $\vec{w}$. What is the displacement vector, $\vec{x}$, from the first robot to the second? (See Figure 13.4.) Since $\vec{v} + \vec{x} = \vec{w}$, we define $\vec{x}$ to be the difference $\vec{x} = \vec{w} - \vec{v}$. In other words, $\vec{w} - \vec{v}$ gets you from the first robot to the second.

The **difference**, $\vec{w} - \vec{v}$, is the displacement vector that, when added to $\vec{v}$, gives $\vec{w}$. That is, $\vec{w} = \vec{v} + (\vec{w} - \vec{v})$. (See Figure 13.4.)

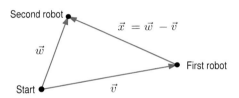

Figure 13.4: The difference $\vec{w} - \vec{v}$

If the robot ends up where it started, then its total displacement vector is the *zero vector*, $\vec{0}$. The zero vector has no direction.

The **zero vector**, $\vec{0}$, is a displacement vector with zero length.

Scalar Multiplication of Displacement Vectors

If $\vec{v}$ represents a displacement vector, the vector $2\vec{v}$ represents a displacement of twice the magnitude in the same direction as $\vec{v}$. Similarly, $-2\vec{v}$ represents a displacement of twice the magnitude in the opposite direction. (See Figure 13.5.)

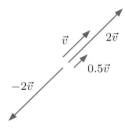

Figure 13.5: Scalar multiples of the vector $\vec{v}$

If λ is a scalar and $\vec{v}$ is a displacement vector, the **scalar multiple of $\vec{v}$ by** λ, written $\lambda\vec{v}$, is the displacement vector with the following properties:

- The displacement vector $\lambda\vec{v}$ is parallel to $\vec{v}$, pointing in the same direction if $\lambda > 0$ and in the opposite direction if $\lambda < 0$.
- The magnitude of $\lambda\vec{v}$ is $|\lambda|$ times the magnitude of $\vec{v}$, that is, $\|\lambda\vec{v}\| = |\lambda| \|\vec{v}\|$.

Note that $|\lambda|$ represents the absolute value of the scalar λ while $\|\lambda\vec{v}\|$ represents the magnitude of the vector $\lambda\vec{v}$.

Example 1   Explain why $\vec{w} - \vec{v} = \vec{w} + (-1)\vec{v}$.

Solution   The vector $(-1)\vec{v}$ has the same magnitude as $\vec{v}$, but points in the opposite direction. Figure 13.6 shows that the combined displacement $\vec{w} + (-1)\vec{v}$ is the same as the displacement $\vec{w} - \vec{v}$.

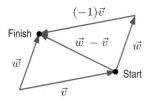

Figure 13.6: Explanation for why $\vec{w} - \vec{v} = \vec{w} + (-1)\vec{v}$

Parallel Vectors

Two vectors $\vec{v}$ and $\vec{w}$ are *parallel* if one is a scalar multiple of the other, that is, if $\vec{w} = \lambda\vec{v}$, for some scalar λ.

Components of Displacement Vectors: The Vectors $\vec{i}$, $\vec{j}$, and $\vec{k}$

Suppose that you live in a city with equally spaced streets running east-west and north-south and that you want to tell someone how to get from one place to another. You'd be likely to tell them how many blocks east-west and how many blocks north-south to go. For example, to get from P to Q in Figure 13.7, we go 4 blocks east and 1 block south. If $\vec{i}$ and $\vec{j}$ are as shown in Figure 13.7, then the displacement vector from P to Q is $4\vec{i} - \vec{j}$.

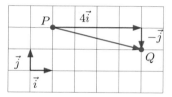

Figure 13.7: The displacement vector from P to Q is $4\vec{i} - \vec{j}$

We extend the same idea to 3 dimensions. First we choose a Cartesian system of coordinate axes. The three vectors of length 1 shown in Figure 13.8 are the vector $\vec{i}$, which points along the positive x-axis, the vector $\vec{j}$, along the positive y-axis, and the vector $\vec{k}$, along the positive z-axis.

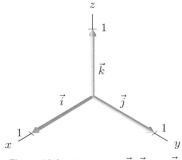

Figure 13.8: The vectors $\vec{i}$, $\vec{j}$ and $\vec{k}$ in 3-space

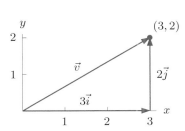

Figure 13.9: We resolve $\vec{v}$ into components by writing $\vec{v} = 3\vec{i} + 2\vec{j}$

Writing Displacement Vectors Using $\vec{i}$, $\vec{j}$, $\vec{k}$

Any displacement in 3-space or the plane can be expressed as a combination of displacements in the coordinate directions. For example, Figure 13.9 shows that the displacement vector $\vec{v}$ from the origin to the point $(3, 2)$ can be written as a sum of displacement vectors along the x- and y-axes:

$$\vec{v} = 3\vec{i} + 2\vec{j}.$$

This is called *resolving $\vec{v}$ into components*. In general:

> We **resolve** $\vec{v}$ into components by writing $\vec{v}$ in the form
>
> $$\vec{v} = v_1\vec{i} + v_2\vec{j} + v_3\vec{k},$$
>
> where v_1, v_2, v_3 are scalars. We call $v_1\vec{i}$, $v_2\vec{j}$, and $v_3\vec{k}$ the **components** of $\vec{v}$.

An Alternative Notation for Vectors

Many people write a vector in three dimensions as a string of three numbers, that is, as

$$\vec{v} = (v_1, v_2, v_3) \quad \text{instead of} \quad \vec{v} = v_1\vec{i} + v_2\vec{j} + v_3\vec{k}.$$

Since the first notation can be confused with a point and the second cannot, we usually use the second form.

Example 2 Resolve the displacement vector, $\vec{v}$, from the point $P_1 = (2, 4, 10)$ to the point $P_2 = (3, 7, 6)$ into components.

Solution To get from P_1 to P_2, we move 1 unit in the positive x-direction, 3 units in the positive y-direction, and 4 units in the negative z-direction. Hence $\vec{v} = \vec{i} + 3\vec{j} - 4\vec{k}$.

Example 3 Decide whether the vector $\vec{v} = 2\vec{i} + 3\vec{j} + 5\vec{k}$ is parallel to each of the following vectors:
$$\vec{w} = 4\vec{i} + 6\vec{j} + 10\vec{k}, \quad \vec{a} = -\vec{i} - 1.5\vec{j} - 2.5\vec{k}, \quad \vec{b} = 4\vec{i} + 6\vec{j} + 9\vec{k}.$$

Solution Since $\vec{w} = 2\vec{v}$ and $\vec{a} = -0.5\vec{v}$, the vectors $\vec{v}$, $\vec{w}$, and $\vec{a}$ are parallel. However, $\vec{b}$ is not a multiple of $\vec{v}$ (since, for example, $4/2 \neq 9/5$), so $\vec{v}$ and $\vec{b}$ are not parallel.

In general, Figure 13.10 shows us how to express the displacement vector between two points in components:

Components of Displacement Vectors

The displacement vector from the point $P_1 = (x_1, y_1, z_1)$ to the point $P_2 = (x_2, y_2, z_2)$ is given in components by

$$\overrightarrow{P_1 P_2} = (x_2 - x_1)\vec{i} + (y_2 - y_1)\vec{j} + (z_2 - z_1)\vec{k}.$$

Position Vectors: Displacement of a Point from the Origin

A displacement vector whose tail is at the origin is called a *position vector*. Thus, any point (x_0, y_0, z_0) in space has associated with it the position vector $\vec{r}_0 = x_0\vec{i} + y_0\vec{j} + z_0\vec{k}$. (See Figure 13.11.) In general, a position vector gives the displacement of a point from the origin.

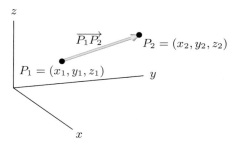

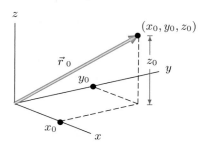

Figure 13.10: The displacement vector $\overrightarrow{P_1 P_2} = (x_2 - x_1)\vec{i} + (y_2 - y_1)\vec{j} + (z_2 - z_1)\vec{k}$

Figure 13.11: The position vector $\vec{r}_0 = x_0\vec{i} + y_0\vec{j} + z_0\vec{k}$

The Components of the Zero Vector

The zero displacement vector has magnitude equal to zero and is written $\vec{0}$. So $\vec{0} = 0\vec{i} + 0\vec{j} + 0\vec{k}$.

The Magnitude of a Vector in Components

For a vector, $\vec{v} = v_1\vec{i} + v_2\vec{j}$, the Pythagorean theorem is used to find its magnitude, $\|\vec{v}\|$. (See Figure 13.12.) The angle θ gives the direction of $\vec{v}$.

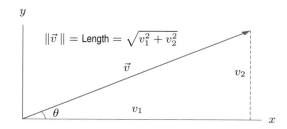

Figure 13.12: Magnitude, $\|\vec{v}\|$, of a 2-dimensional vector, $\vec{v}$

In three dimensions, for a vector $\vec{v} = v_1\vec{i} + v_2\vec{j} + v_3\vec{k}$, we have

---

Magnitude of $\vec{v}$ $= \|\vec{v}\| =$ Length of the arrow $= \sqrt{v_1^2 + v_2^2 + v_3^2}.$

For instance, if $\vec{v} = 3\vec{i} - 4\vec{j} + 5\vec{k}$, then $\|\vec{v}\| = \sqrt{3^2 + (-4)^2 + 5^2} = \sqrt{50}$.

Addition and Scalar Multiplication of Vectors in Components

Suppose the vectors $\vec{v}$ and $\vec{w}$ are given in components:

$$\vec{v} = v_1\vec{i} + v_2\vec{j} + v_3\vec{k} \quad \text{and} \quad \vec{w} = w_1\vec{i} + w_2\vec{j} + w_3\vec{k}.$$

Then

$$\vec{v} + \vec{w} = (v_1 + w_1)\vec{i} + (v_2 + w_2)\vec{j} + (v_3 + w_3)\vec{k},$$

and

$$\lambda\vec{v} = \lambda v_1\vec{i} + \lambda v_2\vec{j} + \lambda v_3\vec{k}.$$

Figures 13.13 and 13.14 illustrate these properties in two dimensions. Finally, $\vec{v} - \vec{w} = \vec{v} + (-1)\vec{w}$, so we can write $\vec{v} - \vec{w} = (v_1 - w_1)\vec{i} + (v_2 - w_2)\vec{j} + (v_3 - w_3)\vec{k}$.

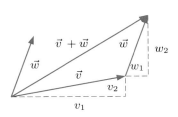

Figure 13.13: Sum $\vec{v} + \vec{w}$ in components

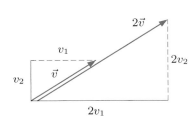
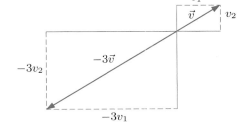

Figure 13.14: Scalar multiples of vectors showing $\vec{v}$, $2\vec{v}$, and $-3\vec{v}$

How to Resolve a Vector into Components

You may wonder how we find the components of a 2-dimensional vector, given its length and direction. Suppose the vector $\vec{v}$ has length v and makes an angle of θ with the x-axis, measured counterclockwise, as in Figure 13.15. If $\vec{v} = v_1\vec{i} + v_2\vec{j}$, Figure 13.15 shows that

$$v_1 = v\cos\theta \quad \text{and} \quad v_2 = v\sin\theta.$$

Thus, we resolve $\vec{v}$ into components by writing

$$\vec{v} = (v\cos\theta)\vec{i} + (v\sin\theta)\vec{j}.$$

Vectors in 3-space are resolved using direction cosines; see Problem 56 on page 772.

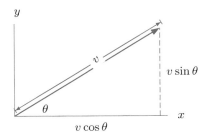

Figure 13.15: Resolving a vector: $\vec{v} = (v\cos\theta)\vec{i} + (v\sin\theta)\vec{j}$

Example 4   Resolve $\vec{v}$ into components if $\|\vec{v}\| = 2$ and $\theta = \pi/6$.

Solution   We have $\vec{v} = 2\cos(\pi/6)\vec{i} + 2\sin(\pi/6)\vec{j} = 2(\sqrt{3}/2)\,\vec{i} + 2(1/2)\vec{j} = \sqrt{3}\vec{i} + \vec{j}$.

Unit Vectors

A *unit vector* is a vector whose magnitude is 1. The vectors $\vec{i}$, $\vec{j}$, and $\vec{k}$ are unit vectors in the directions of the coordinate axes. It is often helpful to find a unit vector in the same direction as a given vector $\vec{v}$. Suppose that $\|\vec{v}\| = 10$; a unit vector in the same direction as $\vec{v}$ is $\vec{v}/10$. In general, a unit vector in the direction of any nonzero vector $\vec{v}$ is

$$\vec{u} = \frac{\vec{v}}{\|\vec{v}\|}.$$

Example 5   Find a unit vector, $\vec{u}$, in the direction of the vector $\vec{v} = \vec{i} + 3\vec{j}$.

Solution   If $\vec{v} = \vec{i} + 3\vec{j}$, then $\|\vec{v}\| = \sqrt{1^2 + 3^2} = \sqrt{10}$. Thus, a unit vector in the same direction is given by

$$\vec{u} = \frac{\vec{v}}{\sqrt{10}} = \frac{1}{\sqrt{10}}(\vec{i} + 3\vec{j}) = \frac{1}{\sqrt{10}}\vec{i} + \frac{3}{\sqrt{10}}\vec{j} \approx 0.32\vec{i} + 0.95\vec{j}.$$

Example 6   Find a unit vector at the point (x, y, z) that points radially outward away from the origin.

Solution   The vector from the origin to (x, y, z) is the position vector

$$\vec{r} = x\vec{i} + y\vec{j} + z\vec{k}.$$

Thus, if we put its tail at (x, y, z) it will point away from the origin. Its magnitude is

$$\|\vec{r}\| = \sqrt{x^2 + y^2 + z^2},$$

so a unit vector pointing in the same direction is

$$\frac{\vec{r}}{\|\vec{r}\|} = \frac{x\vec{i} + y\vec{j} + z\vec{k}}{\sqrt{x^2 + y^2 + z^2}} = \frac{x}{\sqrt{x^2 + y^2 + z^2}}\vec{i} + \frac{y}{\sqrt{x^2 + y^2 + z^2}}\vec{j} + \frac{z}{\sqrt{x^2 + y^2 + z^2}}\vec{k}.$$

Exercises and Problems for Section 13.1

Exercises

In Exercises 1–6, resolve the vectors into components.

1.

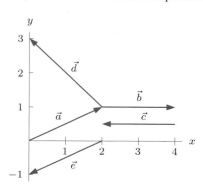

2.

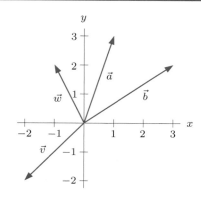

3. A vector starting at the point $P = (1, 2)$ and ending at the point $Q = (4, 6)$.

4. A vector starting at the point $Q = (4, 6)$ and ending at the point $P = (1, 2)$.

5.

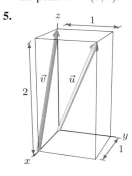

6.

18. $\vec{v} = \vec{i} - \vec{j} + 3\vec{k}$

19. $\vec{v} = 7.2\vec{i} - 1.5\vec{j} + 2.1\vec{k}$

20. $\vec{v} = 1.2\vec{i} - 3.6\vec{j} + 4.1\vec{k}$

For Exercises 21–26, perform the indicated operations on the following vectors:

$$\vec{a} = \vec{j} + 2\vec{k}, \quad \vec{b} = 2\vec{i} - 3\vec{j} + \vec{k}, \quad \vec{c} = 2\vec{i} - 2\vec{j},$$

$$\vec{x} = -3\vec{i} + 7\vec{j}, \quad \vec{y} = 2\vec{i} + 5\vec{j}, \quad \vec{z} = 2\vec{i} + \vec{j} - \vec{k}.$$

21. $3\vec{z}$ **22.** $3\vec{a} - 4\vec{b}$

23. $\vec{a} + 2\vec{z}$ **24.** $7\vec{c} + 2\vec{x}$

For Exercises 7–15, perform the indicated computation.

7. $(2\vec{i} - \vec{j}) + (-3\vec{i} + 2\vec{j})$

8. $(4\vec{i} + 2\vec{j}) - (3\vec{i} - \vec{j})$

9. $(\vec{i} + 2\vec{j}) + (-3)(2\vec{i} + \vec{j})$

10. $-4(\vec{i} - 2\vec{j}) - 0.5(\vec{i} - \vec{k})$

11. $2(0.45\vec{i} - 0.9\vec{j} - 0.01\vec{k}) - 0.5(1.2\vec{i} - 0.1\vec{k})$

12. $(3\vec{i} - 4\vec{j} + 2\vec{k}) - (6\vec{i} + 8\vec{j} - \vec{k})$

13. $(4\vec{i} - 3\vec{j} + 7\vec{k}) - 2(5\vec{i} + \vec{j} - 2\vec{k})$

14. $(0.6\vec{i} + 0.2\vec{j} - \vec{k}) + (0.3\vec{i} + 0.3\vec{k})$

15. $\frac{1}{2}(2\vec{i} - \vec{j} + 3\vec{k}) + 3(\vec{i} - \frac{1}{6}\vec{j} + \frac{1}{2}\vec{k})$

In Exercises 16–20, find the length of the vectors.

16. $\vec{v} = \vec{i} - \vec{j} + 2\vec{k}$ **17.** $\vec{z} = \vec{i} - 3\vec{j} - \vec{k}$

25. $3\vec{a} - 2\vec{b} - 3\vec{z}$ **26.** $\|2\vec{y} - \vec{x}\|$

27. (a) Draw the position vector for $\vec{v} = 5\vec{i} - 7\vec{j}$.
(b) What is $\|\vec{v}\|$?
(c) Find the angle between $\vec{v}$ and the positive x-axis.

28. Find the unit vector in the direction of $0.06\vec{i} - 0.08\vec{k}$.

29. Find the unit vector in the opposite direction to $\vec{i} - \vec{j} + \vec{k}$.

30. Find a unit vector in the opposite direction to $2\vec{i} - \vec{j} - \sqrt{11}\vec{k}$.

31. Find a vector with length 2 that points in the same direction as $\vec{i} - \vec{j} + 2\vec{k}$.

Problems

32. Find the value(s) of a making $\vec{v} = 5a\vec{i} - 3\vec{j}$ parallel to $\vec{w} = a^2\vec{i} + 6\vec{j}$.

33. (a) Find a unit vector from the point $P = (1, 2)$ and toward the point $Q = (4, 6)$.
(b) Find a vector of length 10 pointing in the same direction.

34. If north is the direction of the positive y-axis and east is the direction of the positive x-axis, give the unit vector pointing northwest.

35. Resolve the following vectors into components:

(a) The vector in 2-space of length 2 pointing up and to the right at an angle of $\pi/4$ with the x-axis.

(b) The vector in 3-space of length 1 lying in the xz-plane pointing upward at an angle of $\pi/6$ with the positive x-axis.

36. (a) From Figure 13.16, read off the coordinates of the five points, A, B, C, D, E, and thus resolve into components the following two vectors:

$$\vec{u} = (2.5)\overrightarrow{AB} + (-0.8)\overrightarrow{CD}, \quad \vec{v} = (2.5)\overrightarrow{BA} - (-0.8)\overrightarrow{CD}$$

(b) What is the relation between $\vec{u}$ and $\vec{v}$? Why was this to be expected?

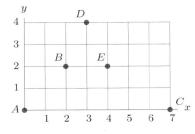

Figure 13.16

37. Find the components of a vector $\vec{p}$ that has the same direction as $\overrightarrow{EA}$ in Figure 13.16 and whose length equals two units.

38. For each of the four statements below, answer the following questions: Does the statement make sense? If yes, is it true for all possible choices of $\vec{a}$ and $\vec{b}$? If no, why not?

(a) $\vec{a} + \vec{b} = \vec{b} + \vec{a}$
(b) $\vec{a} + \|\vec{b}\| = \|\vec{a} + \vec{b}\|$
(c) $\|\vec{b} + \vec{a}\| = \|\vec{a} + \vec{b}\|$
(d) $\|\vec{a} + \vec{b}\| = \|\vec{a}\| + \|\vec{b}\|$.

39. Two adjacent sides of a regular hexagon are given as the vectors $\vec{u}$ and $\vec{v}$ in Figure 13.17. Label the remaining sides in terms of $\vec{u}$ and $\vec{v}$.

Figure 13.17

40. For what values of t are the following pairs of vectors parallel?

(a) $2\vec{i} + (t^2 + \frac{2}{3}t + 1)\vec{j} + t\vec{k}$, $6\vec{i} + 8\vec{j} + 3\vec{k}$

(b) $t\vec{i} + \vec{j} + (t-1)\vec{k}$, $2\vec{i} - 4\vec{j} + \vec{k}$,
(c) $2t\vec{i} + t\vec{j} + t\vec{k}$, $6\vec{i} + 3\vec{j} + 3\vec{k}$.

41. Find all vectors $\vec{v}$ in 2 dimensions having $\|\vec{v}\| = 5$ such that the $\vec{i}$-component of $\vec{v}$ is $3\vec{i}$.

42. Find all vectors $\vec{v}$ in the plane such that $\|\vec{v}\| = 1$ and $\|\vec{v} + \vec{i}\| = 1$.

43. Figure 13.18 shows a molecule with four atoms at O, A, B and C. Check that every atom in the molecule is 2 units away from every other atom.

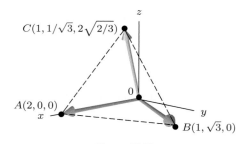

Figure 13.18

44. Show that the medians of a triangle intersect at a point $\frac{1}{3}$ of the way along each median from the side it bisects.

Strengthen Your Understanding

In Problems 45–48, explain what is wrong with the statement.

45. If $\|\vec{u}\| = 1$ and $\|\vec{v}\| > 0$, then $\|\vec{u} + \vec{v}\| \geq 1$.

46. The vector $c\vec{u}$ has the same direction as $\vec{u}$.

47. $\|\vec{v} - \vec{u}\|$ is the length of the shorter of the two diagonals of the parallelogram determined by $\vec{u}$ and $\vec{v}$.

48. Given three vectors $\vec{u}, \vec{v}$, and $\vec{w}$, if $\vec{u} + \vec{w} = \vec{u}$ then it is possible for $\vec{v} + \vec{w} \neq \vec{v}$.

In Problems 49–51, give an example of:

49. A vector $\vec{v}$ of length 2 with a positive $\vec{k}$-component and lying on a plane parallel to the yz-plane.

50. Two unit vectors $\vec{u}$ and $\vec{v}$ for which $\vec{v} - \vec{u}$ is also a unit vector.

51. Two vectors $\vec{u}$ and $\vec{v}$ that have difference vector $\vec{w} = 2\vec{i} + 3\vec{j}$.

Are the statements in Problems 52–61 true or false? Give reasons for your answer.

52. There is exactly one unit vector parallel to a given nonzero vector $\vec{v}$.

53. The vector $\frac{1}{\sqrt{3}}\vec{i} + \frac{-1}{\sqrt{3}}\vec{j} + \frac{2}{\sqrt{3}}\vec{k}$ is a unit vector.

54. The length of the vector $2\vec{v}$ is twice the length of the vector $\vec{v}$.

55. If $\vec{v}$ and $\vec{w}$ are any two vectors, then $\|\vec{v} + \vec{w}\| = \|\vec{v}\| + \|\vec{w}\|$.

56. If $\vec{v}$ and $\vec{w}$ are any two vectors, then $\|\vec{v} - \vec{w}\| = \|\vec{v}\| - \|\vec{w}\|$.

57. The vectors $2\vec{i} - \vec{j} + \vec{k}$ and $\vec{i} - 2\vec{j} + \vec{k}$ are parallel.

58. The vector $\vec{u} + \vec{v}$ is always larger in magnitude than both $\vec{u}$ and $\vec{v}$.

59. For any scalar c and vector $\vec{v}$ we have $\|c\vec{v}\| = c\|\vec{v}\|$.

60. The displacement vector from $(1, 1, 1)$ to $(1, 2, 3)$ is $-\vec{j} - 2\vec{k}$.

61. The displacement vector from (a, b) to (c, d) is the same as the displacement vector from (c, d) to (a, b).

13.2 VECTORS IN GENERAL

Besides displacement, there are many quantities that have both magnitude and direction and are added and multiplied by scalars in the same way as displacements. Any such quantity is called a *vector* and is represented by an arrow in the same manner we represent displacements. The length of the arrow is the *magnitude* of the vector, and the direction of the arrow is the direction of the vector.

Velocity Versus Speed

The speed of a moving body tells us how fast it is moving, say 80 km/hr. The speed is just a number; it is therefore a scalar. The velocity, on the other hand, tells us both how fast the body is moving and the direction of motion; it is a vector. For instance, if a car is heading northeast at 80 km/hr, then its velocity is a vector of length 80 pointing northeast.

> The **velocity vector** of a moving object is a vector whose magnitude is the speed of the object and whose direction is the direction of its motion.

The velocity vector is the displacement vector if the object moves at constant velocity for one unit of time.

Example 1   A car is travelling north at a speed of 100 km/hr, while a plane above is flying horizontally southwest at a speed of 500 km/hr. Draw the velocity vectors of the car and the plane.

Solution   Figure 13.19 shows the velocity vectors. The plane's velocity vector is five times as long as the car's, because its speed is five times as great.

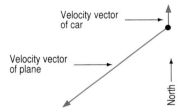

Figure 13.19: Velocity vector of the car is 100 km/hr north and of the plane is 500 km/hr southwest

---

The next example illustrates that the velocity vectors for two motions add to give the velocity vector for the combined motion, just as displacements do.

Example 2   A riverboat is moving with velocity $\vec{v}$ and a speed of 8 km/hr relative to the water. In addition, the river has a current $\vec{c}$ and a speed of 1 km/hr. (See Figure 13.20.) What is the physical significance of the vector $\vec{v} + \vec{c}$?

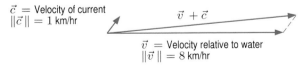

Figure 13.20: Boat's velocity relative to the river bed is the sum, $\vec{v} + \vec{c}$

Solution   The vector $\vec{v}$ shows how the boat is moving relative to the water, while $\vec{c}$ shows how the water is moving relative to the riverbed. During an hour, imagine that the boat first moves 8 km relative to the water, which remains still; this displacement is represented by $\vec{v}$. Then imagine the water moving 1 km while the boat remains stationary relative to the water; this displacement is represented by $\vec{c}$. The combined displacement is represented by $\vec{v} + \vec{c}$. Thus, the vector $\vec{v} + \vec{c}$ is the velocity of the boat relative to the riverbed.

Note that the effective speed of the boat is not necessarily 9 km/hr unless the boat is moving in the direction of the current. Although we add the velocity vectors, we do not necessarily add their lengths.

Scalar multiplication also makes sense for velocity vectors. For example, if $\vec{v}$ is a velocity vector, then $-2\vec{v}$ represents a velocity of twice the magnitude in the opposite direction.

Example 3   A ball is moving with velocity $\vec{v}$ when it hits a wall at a right angle and bounces straight back, with its speed reduced by 20%. Express its new velocity in terms of the old one.

Solution   The new velocity is $-0.8\vec{v}$, where the negative sign expresses the fact that the new velocity is in the direction opposite to the old.

We can represent velocity vectors in components in the same way we did on page 741.

Example 4   Represent the velocity vectors of the car and the plane in Example 1 using components. Take north to be the positive y-axis, east to be the positive x-axis, and upward to be the positive z-axis.

Solution   The car is travelling north at 100 km/hr, so the y-component of its velocity is $100\vec{j}$ and the x-component is $0\vec{i}$. Since it is travelling horizontally, the z-component is $0\vec{k}$. So we have

$$\text{Velocity of car} = 0\vec{i} + 100\vec{j} + 0\vec{k} = 100\vec{j}.$$

The plane's velocity vector also has $\vec{k}$ component equal to zero. Since it is travelling southwest, its $\vec{i}$ and $\vec{j}$ components have negative coefficients (north and east are positive). Since the plane is travelling at 500 km/hr, in one hour it is displaced $500/\sqrt{2} \approx 354$ km to the west and 354 km to the south. (See Figure 13.21.) Thus,
$$\text{Velocity of plane} = -(500\cos 45°)\vec{i} - (500\sin 45°)\vec{j} \approx -354\vec{i} - 354\vec{j}.$$

Of course, if the car were climbing a hill or if the plane were descending for a landing, then the $\vec{k}$ component would not be zero.

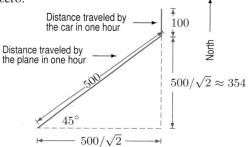

Figure 13.21: Distance travelled by the plane and car in one hour

---

Acceleration

Another example of a vector quantity is acceleration. Acceleration, like velocity, is specified by both a magnitude and a direction — for example, the acceleration due to gravity is 9.81 m/sec² vertically downward.

Force

Force is another example of a vector quantity. Suppose you push on an open door. The result depends both on how hard you push and in what direction. Thus, to specify a force we must give its magnitude (or strength) and the direction in which it is acting. For example, the gravitational force exerted on an object by the earth is a vector pointing from the object toward the centre of the earth; its magnitude is the strength of the gravitational force.

Example 5   The earth travels around the sun in an ellipse. The gravitational force on the earth and the velocity of the earth are governed by the following laws:

Newton's Law of Gravitation: The gravitational attraction, $\vec{F}$, of a mass m_1 on a mass m_2 at a distance r has magnitude $||\vec{F}|| = Gm_1m_2/r^2$, where G is a constant, and is directed from m_2 toward m_1.

Kepler's Second Law: The line joining a planet to the sun sweeps out equal areas in equal times.

(a) Sketch vectors representing the gravitational force of the sun on the earth at two different positions in the earth's orbit.

(b) Sketch the velocity vector of the earth at two points in its orbit.

Solution   (a) Figure 13.22 shows the earth orbiting the sun. Note that the gravitational force vector always points toward the sun and is larger when the earth is closer to the sun because of the r^2 term in the denominator. (In fact, the real orbit looks much more like a circle than we have shown here.)

(b) The velocity vector points in the direction of motion of the earth. Thus, the velocity vector is tangent to the ellipse. See Figure 13.23. Furthermore, the velocity vector is longer at points of the orbit where the planet is moving quickly, because the magnitude of the velocity vector is the speed. Kepler's Second Law enables us to determine when the earth is moving quickly and when it is moving slowly. Over a fixed period of time, say one month, the line joining the earth to the sun sweeps out a sector having a certain area. Figure 13.23 shows two sectors swept out in two different one-month time-intervals. Kepler's law says that the areas of the two sectors are the same. Thus, the earth must move farther in a month when it is close to the sun than when it is far from the sun. Therefore, the earth moves faster when it is closer to the sun and slower when it is farther away.

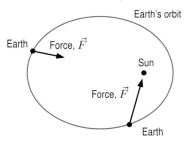

Figure 13.22: Gravitational force, $\vec{F}$, exerted by the sun on the earth: Greater magnitude closer to sun

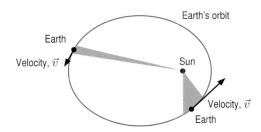

Figure 13.23: The velocity vector, $\vec{v}$, of the earth: Greater magnitude closer to the sun

Properties of Addition and Scalar Multiplication

In general, vectors add, subtract, and are multiplied by scalars in the same way as displacement vectors. Thus, for any vectors $\vec{u}$, $\vec{v}$, and $\vec{w}$ and any scalars α and β, we have the following properties:

Commutativity
1. $\vec{v} + \vec{w} = \vec{w} + \vec{v}$

Associativity
2. $(\vec{u} + \vec{v}) + \vec{w} = \vec{u} + (\vec{v} + \vec{w})$
3. $\alpha(\beta\vec{v}) = (\alpha\beta)\vec{v}$

Distributivity
4. $(\alpha + \beta)\vec{v} = \alpha\vec{v} + \beta\vec{v}$
5. $\alpha(\vec{v} + \vec{w}) = \alpha\vec{v} + \alpha\vec{w}$

Identity
6. $1\vec{v} = \vec{v}$
7. $0\vec{v} = \vec{0}$
8. $\vec{v} + \vec{0} = \vec{v}$
9. $\vec{w} + (-1)\vec{v} = \vec{w} - \vec{v}$

Problems 29–36 at the end of this section ask for a justification of these results in terms of displacement vectors.

Using Components

Example 6   A plane, heading due east at an airspeed of 600 km/hr, experiences a wind of 50 km/hr blowing toward the northeast. Find the plane's direction and ground speed.

Solution   We choose a coordinate system with the x-axis pointing east and the y-axis pointing north. See Figure 13.24.

The airspeed tells us the speed of the plane relative to still air. Thus, the plane is moving due east with velocity $\vec{v} = 600\vec{i}$ relative to still air. In addition, the air is moving with a velocity $\vec{w}$. Writing $\vec{w}$ in components, we have

$$\vec{w} = (50\cos45°)\vec{i} + (50\sin45°)\vec{j} = 35.4\vec{i} + 35.4\vec{j}.$$

The vector $\vec{v} + \vec{w}$ represents the displacement of the plane in one hour relative to the ground. Therefore, $\vec{v} + \vec{w}$ is the velocity of the plane relative to the ground. In components, we have

$$\vec{v} + \vec{w} = 600\vec{i} + \left(35.4\vec{i} + 35.4\vec{j}\right) = 635.4\vec{i} + 35.4\vec{j}.$$

The direction of the plane's motion relative to the ground is given by the angle θ in Figure 13.24, where

$$\tan\theta = \frac{35.4}{635.4}$$

so

$$\theta = \arctan\left(\frac{35.4}{635.4}\right) = 3.2°.$$

The ground speed is the speed of the plane relative to the ground, so

$$\text{Groundspeed} = \sqrt{635.4^2 + 35.4^2} = 636.4 \text{ km/hr}.$$

Thus, the speed of the plane relative to the ground has been increased slightly by the wind. (This is as we would expect, as the wind has a positive component in the direction in which the plane is travelling.) The angle θ shows how far the plane is blown off course by the wind.

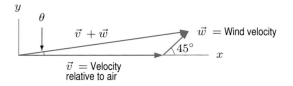

Figure 13.24: Plane's velocity relative to the ground is the sum $\vec{v} + \vec{w}$

Vectors in n Dimensions

Using the alternative notation $\vec{v} = (v_1, v_2, v_3)$ for a vector in 3-space, we can define a vector in n dimensions as a string of n numbers. Thus, a vector in n dimensions can be written as

$$\vec{c} = (c_1, c_2, \ldots, c_n).$$

Addition and scalar multiplication are defined by the formulas

$$\vec{v} + \vec{w} = (v_1, v_2, \ldots, v_n) + (w_1, w_2, \ldots, w_n) = (v_1 + w_1, v_2 + w_2, \ldots, v_n + w_n)$$

and

$$\lambda \vec{v} = \lambda(v_1, v_2, \ldots, v_n) = (\lambda v_1, \lambda v_2, \ldots, \lambda v_n).$$

Why Do We Want Vectors in n Dimensions?

Vectors in two and three dimensions can be used to model displacement, velocities, or forces. But what about vectors in n dimensions? There is another interpretation of 3-dimensional vectors (or 3-vectors) that is useful: they can be thought of as listing three different quantities — for example, the displacements parallel to the x-, y-, and z-axes. Similarly, the n-vector

$$\vec{c} = (c_1, c_2, \ldots, c_n)$$

can be thought of as a way of keeping n different quantities organised. For example, a *population* vector $\vec{N}$ shows the number of children and adults in a population:

$$\vec{N} = (\text{Number of children, Number of adults}),$$

or, if we are interested in a more detailed breakdown of ages, we might give the number in each ten-year age bracket in the population (up to age 110) in the form

$$\vec{N} = (N_1, N_2, N_3, N_4, \ldots, N_{10}, N_{11}),$$

where N_1 is the population aged 0–9, and N_2 is the population aged 10–19, and so on.

A *consumption* vector,

$$\vec{q} = (q_1, q_2, \ldots, q_n)$$

shows the quantities q_1, q_2, ..., q_n consumed of each of n different goods. A *price* vector

$$\vec{p} = (p_1, p_2, \ldots, p_n)$$

contains the prices of n different items.

In 1907, Hermann Minkowski used vectors with four components when he introduced *space-time coordinates*, whereby each event is assigned a vector position $\vec{v}$ with four coordinates, three for its position in space and one for time:

$$\vec{v} = (x, y, z, t).$$

Example 7   Suppose the vector $\vec{I}$ represents the number of copies, in thousands, made by each of four copy centres in the month of December and $\vec{J}$ represents the number of copies made at the same four copy centres during the previous eleven months (the "year-to-date"). If $\vec{I} = (25, 211, 818, 642)$, and $\vec{J} = (331, 3227, 1377, 2570)$, compute $\vec{I} + \vec{J}$. What does this sum represent?

Solution   The sum is

$$\vec{I} + \vec{J} = (25 + 331, 211 + 3227, 818 + 1377, 642 + 2570) = (356, 3438, 2195, 3212).$$

Each term in $\vec{I} + \vec{J}$ represents the sum of the number of copies made in December plus those in the previous eleven months, that is, the total number of copies made during the entire year at that particular copy centre.

Example 8   The price vector $\vec{p} = (p_1, p_2, p_3)$ represents the prices in dollars of three goods. Write a vector that gives the prices of the same goods in cents.

Solution   The prices in cents are $100 p_1$, $100 p_2$, and $100 p_3$ respectively, so the new price vector is

$$(100 p_1, 100 p_2, 100 p_3) = 100 \vec{p}.$$

Exercises and Problems for Section 13.2

Exercises

In Exercises 1–6, say whether the given quantity is a vector or a scalar.

(a) East   **(b)** South

(c) Southeast   **(d)** Northwest.

1. Your bank balance.

2. The populations of each of the 13 South American countries.

3. The magnetic field at a point on the earth's surface.

4. The population of the US.

5. The distance from Vienna to Venice.

6. The temperature at a point on the earth's surface.

7. A car is traveling at a speed of 50 km/hr. The positive y-axis is north and the positive x-axis is east. Resolve the car's velocity vector (in 2-space) into components if the car is traveling in each of the following directions:

8. Give the components of the velocity vector for wind blowing at 10 km/hr toward the southeast. (Assume north is in the positive y-direction.)

9. Give the components of the velocity vector of a boat that is moving at 40 km/hr in a direction $20°$ south of west. (Assume north is in the positive y-direction.)

10. Which is traveling faster, a car whose velocity vector is $21\vec{i} + 35\vec{j}$, or a car whose velocity vector is $40\vec{i}$, assuming that the units are the same for both directions?

11. What angle does a force of $\vec{F} = 15\vec{i} + 18\vec{j}$ make with the x-axis?

Problems

12. The velocity of the current in a river is $\vec{c} = 0.6\vec{i} + 0.8\vec{j}$ km/hr. A boat moves relative to the water with velocity $\vec{v} = 8\vec{i}$ km/hr.

    (a) What is the speed of the boat relative to the riverbed?

    (b) What angle does the velocity of the boat relative to the riverbed make with the vector $\vec{v}$? What does this angle tell us in practical terms?

13. Suppose the current in Problem 12 is twice as fast and in the opposite direction. What is the speed of the boat with respect to the riverbed?

14. An airplane is flying at an airspeed of 500 km/hr in a wind blowing at 60 km/hr toward the southeast. In what direction should the plane head to end up going due east? What is the airplane's speed relative to the ground?

15. A plane is heading due east and climbing at the rate of 80 km/hr. If its airspeed is 480 km/hr and there is a wind blowing 100 km/hr to the northeast, what is the ground speed of the plane?

16. An airplane is flying at an airspeed of 600 km/hr in a cross-wind that is blowing from the northeast at a speed of 50 km/hr. In what direction should the plane head to end up going due east?

17. The current in a river is pushing a boat in direction $25°$ north of east with a speed of 12 km/hr. The wind is pushing the same boat in a direction $80°$ south of east with a speed of 7 km/hr. Find the velocity vector of the boat's engine (relative to the water) if the boat actually moves due east at a speed of 40 km/hr relative to the ground.

18. A large ship is being towed by two tugs. The larger tug exerts a force which is 25% greater than the smaller tug and at an angle of 30 degrees north of east. Which direction must the smaller tug pull to ensure that the ship travels due east?

19. A particle moving with speed v hits a barrier at an angle of $60°$ and bounces off at an angle of $60°$ in the opposite direction with speed reduced by 20 percent. See Figure 13.25. Find the velocity vector of the object after impact.

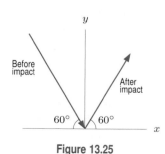

Figure 13.25

20. There are five students in a class. Their scores on the midterm (out of 100) are given by the vector $\vec{v} = (73, 80, 91, 65, 84)$. Their scores on the final (out of 100) are given by $\vec{w} = (82, 79, 88, 70, 92)$. If the final counts twice as much as the midterm, find a vector giving the total scores (as a percentage) of the students.

21. The price vector of beans, rice, and tofu is $(0.60, 0.40, 1.00)$ in dollars per kilogram. Express it in dollars per gram.

22. Two forces, represented by the vectors $\vec{F}_1 = 8\vec{i} - 6\vec{j}$ and $\vec{F}_2 = 3\vec{i} + 2\vec{j}$, are acting on an object. Give a vector representing the force that must be applied to the object if it is to remain stationary.

23. One force is pushing an object in a direction 50° south of east with a force of 25 newtons. A second force is simultaneously pushing the object in a direction 70° north of west with a force of 60 newtons. If the object is to remain stationary, give the direction and magnitude of the third force that must be applied to the object to counterbalance the first two.

24. An object P is pulled by a force $\vec{F_1}$ of magnitude 15 lb at an angle of 20 degrees north of east. In what direction must a force $\vec{F_2}$ of magnitude 20 lb pull to ensure that P moves due east?

25. An airplane heads northeast at an airspeed of 700 km/hr, but there is a wind blowing from the west at 60 km/hr. In what direction does the plane end up flying? What is its speed relative to the ground?

26. A man wishes to row the shortest possible distance from north to south across a river that is flowing at 4 km/hr from the east. He can row at 5 km/hr.

    (a) In which direction should he steer?
    (b) If there is a wind of 10 km/hr from the southwest, in which direction should he steer to try and go directly across the river? What happens?

27. An object is moving counterclockwise at a constant speed around the circle $x^2 + y^2 = 1$, where x and y are measured in meters. It completes one revolution every minute.

    (a) What is its speed?
    (b) What is its velocity vector 30 seconds after it passes the point $(1, 0)$? Does your answer change if the object is moving clockwise? Explain.

28. An object is attached by a string to a fixed point and rotates 30 times per minute in a horizontal plane. Show that the speed of the object is constant but the velocity is not. What does this imply about the acceleration?

In Problems 29–36, use the geometric definition of addition and scalar multiplication to explain each of the properties.

29. $\vec{w} + \vec{v} = \vec{v} + \vec{w}$

30. $(\alpha + \beta)\vec{v} = \alpha\vec{v} + \beta\vec{v}$

31. $\alpha(\vec{v} + \vec{w}) = \alpha\vec{v} + \alpha\vec{w}$

32. $\alpha(\beta\vec{v}) = (\alpha\beta)\vec{v}$

33. $\vec{v} + \vec{0} = \vec{v}$

34. $1\vec{v} = \vec{v}$

35. $\vec{v} + (-1)\vec{w} = \vec{v} - \vec{w}$

36. $(\vec{u} + \vec{v}) + \vec{w} = \vec{u} + (\vec{v} + \vec{w})$

37. The earth is at the origin, the moon is at the point $(384, 0)$, and a spaceship is at $(280, 90)$, where distance is in thousands of kilometers.

    (a) What is the displacement vector of the moon relative to the earth? Of the spaceship relative to the earth? Of the spaceship relative to the moon?
    (b) How far is the spaceship from the earth? From the moon?
    (c) The gravitational force on the spaceship from the earth is 461 newtons and from the moon is 26 newtons. What is the resulting force?

38. In the game of laser tag, you shoot a harmless laser gun and try to hit a target worn at the waist by other players. Suppose you are standing at the origin of a three-dimensional coordinate system and that the xy-plane is the floor. Suppose that waist-high is 1 metre above floor level and that eye level is 1.75 metres above the floor. Three of your friends are your opponents. One is standing so that his target is 10 metres along the x-axis, another lying down so that his target is at the point $x = 6$, $y = 5$, and the third lying in ambush so that his target is at a point 3 metres above the point $x = 5$, $y = 10$.

    (a) If you aim with your gun at eye level, find the vector from your gun to each of the three targets.
    (b) If you shoot from waist height, with your gun 30 cm to the right of the centre of your body as you face along the x-axis, find the vector from your gun to each of the three targets.

Strengthen Your Understanding

In Problems 39–40, explain what is wrong with the statement.

39. Two vectors in 3-space that have equal $\vec{k}$-components and the same magnitude must be the same vector.

40. A vector $\vec{v}$ in the plane whose $\vec{i}$-component is 0.5 has smaller magnitude than the vector $\vec{w} = 2\vec{i}$.

In Problems 41–42, give an example of:

41. A non-zero vector $\vec{F}$ on the plane that when combined with the force vector $\vec{G} = \vec{i} + \vec{j}$ results in a combined force vector $\vec{R}$ with a positive $\vec{i}$-component and a negative $\vec{j}$-component.

42. Non-zero vectors $\vec{u}$ and $\vec{v}$ such that $\|\vec{u} + \vec{v}\| = \|\vec{u}\| + \|\vec{v}\|$.

In Problems 43–48, is the quantity a vector? Give a reason for your answer.

43. Velocity

44. Speed

45. Force

46. Area

47. Acceleration

48. Volume

13.3 THE DOT PRODUCT

We have seen how to add vectors; can we multiply two vectors together? In the next two sections we will see two different ways of doing so: the *scalar product* (or *dot product*), which produces a scalar, and the *vector product* (or *cross product*), which produces a vector.

Definition of the Dot Product

The dot product links geometry and algebra. We already know how to calculate the length of a vector from its components; the dot product gives us a way of computing the angle between two vectors. For any two vectors $\vec{v} = v_1\vec{i} + v_2\vec{j} + v_3\vec{k}$ and $\vec{w} = w_1\vec{i} + w_2\vec{j} + w_3\vec{k}$, shown in Figure 13.26, we define a scalar as follows:

The following two definitions of the **dot product**, or **scalar product**, $\vec{v} \cdot \vec{w}$, are equivalent:
- **Geometric definition**

  $\vec{v} \cdot \vec{w} = \|\vec{v}\|\|\vec{w}\| \cos\theta$ where θ is the angle between $\vec{v}$ and $\vec{w}$ and $0 \le \theta \le \pi$.
- **Algebraic definition**

  $\vec{v} \cdot \vec{w} = v_1w_1 + v_2w_2 + v_3w_3.$

Notice that the dot product of two vectors is a *number*, not a vector.

Why don't we give just one definition of $\vec{v} \cdot \vec{w}$? The reason is that both definitions are equally important; the geometric definition gives us a picture of what the dot product means and the algebraic definition gives us a way of calculating it.

How do we know the two definitions are equivalent—that is, they really do define the same thing? First, we observe that the two definitions give the same result in a particular example. Then we show why they are equivalent in general.

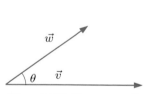

Figure 13.26: The vectors $\vec{v}$ and $\vec{w}$

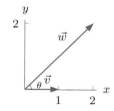

Figure 13.27: Calculating the dot product of the vectors $v = \vec{i}$ and $\vec{w} = 2\vec{i} + 2\vec{j}$ geometrically and algebraically gives the same result

---

Example 1   Suppose $\vec{v} = \vec{i}$ and $\vec{w} = 2\vec{i} + 2\vec{j}$. Compute $\vec{v} \cdot \vec{w}$ both geometrically and algebraically.

Solution   To use the geometric definition, see Figure 13.27. The angle between the vectors is $\pi/4$, or $45°$, and the lengths of the vectors are given by

$$\|\vec{v}\| = 1 \quad \text{and} \quad \|\vec{w}\| = 2\sqrt{2}.$$

Thus,

$$\vec{v} \cdot \vec{w} = \|\vec{v}\|\|\vec{w}\| \cos\theta = 1 \cdot 2\sqrt{2} \cos\left(\frac{\pi}{4}\right) = 2.$$

Using the algebraic definition, we get the same result:

$$\vec{v} \cdot \vec{w} = 1 \cdot 2 + 0 \cdot 2 = 2.$$

Why the Two Definitions of the Dot Product Give the Same Result

In the previous example, the two definitions give the same value for the dot product. To show that the geometric and algebraic definitions of the dot product always give the same result, we must show that, for any vectors $\vec{v} = v_1\vec{i} + v_2\vec{j} + v_3\vec{k}$ and $\vec{w} = w_1\vec{i} + w_2\vec{j} + w_3\vec{k}$ with an angle θ between them:

$$\|\vec{v}\|\|\vec{w}\|\cos\theta = v_1w_1 + v_2w_2 + v_3w_3.$$

One method follows; a method that does not use trigonometry is given in Problem 70 on page 761.

Using the Law of Cosines. Suppose that $0 < \theta < \pi$, so that the vectors $\vec{v}$ and $\vec{w}$ form a triangle. (See Figure 13.28.) By the Law of Cosines, we have

$$\|\vec{v} - \vec{w}\|^2 = \|\vec{v}\|^2 + \|\vec{w}\|^2 - 2\|\vec{v}\|\|\vec{w}\|\cos\theta.$$

This result is also true for $\theta = 0$ and $\theta = \pi$. We calculate the lengths using components:

$$\|\vec{v}\|^2 = v_1^2 + v_2^2 + v_3^2$$
$$\|\vec{w}\|^2 = w_1^2 + w_2^2 + w_3^2$$
$$\|\vec{v} - \vec{w}\|^2 = (v_1 - w_1)^2 + (v_2 - w_2)^2 + (v_3 - w_3)^2$$
$$= v_1^2 - 2v_1w_1 + w_1^2 + v_2^2 - 2v_2w_2 + w_2^2 + v_3^2 - 2v_3w_3 + w_3^2.$$

Substituting into the Law of Cosines and cancelling, we see that

$$-2v_1w_1 - 2v_2w_2 - 2v_3w_3 = -2\|\vec{v}\|\|\vec{w}\|\cos\theta.$$

Therefore we have the result we wanted, namely that:

$$v_1w_1 + v_2w_2 + v_3w_3 = \|\vec{v}\|\|\vec{w}\|\cos\theta.$$

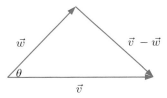

Figure 13.28: Triangle used in the justification of $\|\vec{v}\|\|\vec{w}\|\cos\theta = v_1w_1 + v_2w_2 + v_3w_3$

Properties of the Dot Product

The following properties of the dot product can be justified using the algebraic definition; see Problem 64 on page 760. For a geometric interpretation of Property 3, see Problem 67.

Properties of the Dot Product. For any vectors $\vec{u}$, $\vec{v}$, and $\vec{w}$ and any scalar λ,

1. $\vec{v} \cdot \vec{w} = \vec{w} \cdot \vec{v}$
2. $\vec{v} \cdot (\lambda\vec{w}) = \lambda(\vec{v} \cdot \vec{w}) = (\lambda\vec{v}) \cdot \vec{w}$
3. $(\vec{v} + \vec{w}) \cdot \vec{u} = \vec{v} \cdot \vec{u} + \vec{w} \cdot \vec{u}$

Perpendicularity, Magnitude, and Dot Products

Two vectors are perpendicular if the angle between them is $\pi/2$ or $90°$. Since $\cos(\pi/2) = 0$, if $\vec{v}$ and $\vec{w}$ are perpendicular, then $\vec{v} \cdot \vec{w} = 0$. Conversely, provided that $\vec{v} \cdot \vec{w} = 0$, then $\cos\theta = 0$, so $\theta = \pi/2$ and the vectors are perpendicular. Thus, we have the following result:

> Two non-zero vectors $\vec{v}$ and $\vec{w}$ are **perpendicular**, or **orthogonal**, if and only if
>
> $$\vec{v} \cdot \vec{w} = 0.$$

For example: $\vec{i} \cdot \vec{j} = 0, \vec{j} \cdot \vec{k} = 0, \vec{i} \cdot \vec{k} = 0.$

If we take the dot product of a vector with itself, then $\theta = 0$ and $\cos\theta = 1$. For any vector $\vec{v}$:

> Magnitude and dot product are related as follows:
>
> $$\vec{v} \cdot \vec{v} = \|\vec{v}\|^2.$$

For example: $\vec{i} \cdot \vec{i} = 1, \vec{j} \cdot \vec{j} = 1, \vec{k} \cdot \vec{k} = 1.$

Using the Dot Product

Depending on the situation, one definition of the dot product may be more convenient to use than the other. In Example 2, the geometric definition is the only one that can be used because we are not given components. In Example 3, the algebraic definition is used.

Example 2   Suppose the vector $\vec{b}$ is fixed and has length 2; the vector $\vec{a}$ is free to rotate and has length 3. What are the maximum and minimum values of the dot product $\vec{a} \cdot \vec{b}$ as the vector $\vec{a}$ rotates through all possible positions? What positions of $\vec{a}$ and $\vec{b}$ lead to these values?

Solution   The geometric definition gives $\vec{a} \cdot \vec{b} = \|\vec{a}\|\|\vec{b}\|\cos\theta = 3 \cdot 2\cos\theta = 6\cos\theta$. Thus, the maximum value of $\vec{a} \cdot \vec{b}$ is 6, and it occurs when $\cos\theta = 1$ so $\theta = 0$, that is, when $\vec{a}$ and $\vec{b}$ point in the same direction. The minimum value of $\vec{a} \cdot \vec{b}$ is -6, and it occurs when $\cos\theta = -1$ so $\theta = \pi$, that is, when $\vec{a}$ and $\vec{b}$ point in opposite directions. (See Figure 13.29.)

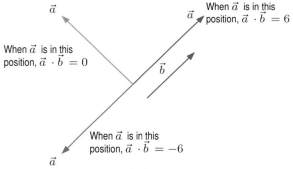

Figure 13.29: Maximum and minimum values of $\vec{a} \cdot \vec{b}$ obtained from a fixed vector $\vec{b}$ of length 2 and rotating vector $\vec{a}$ of length 3

Example 3   Which pairs from the following list of 3-dimensional vectors are perpendicular to one another?

$$\vec{u} = \vec{i} + \sqrt{3}\,\vec{k}, \quad \vec{v} = \vec{i} + \sqrt{3}\,\vec{j}, \quad \vec{w} = \sqrt{3}\,\vec{i} + \vec{j} - \vec{k}.$$

Solution   The geometric definition tells us that two vectors are perpendicular if and only if their dot product is zero. Since the vectors are given in components, we calculate dot products using the algebraic definition:

$$\vec{v} \cdot \vec{u} = (\vec{i} + \sqrt{3}\,\vec{j} + 0\vec{k}) \cdot (\vec{i} + 0\vec{j} + \sqrt{3}\,\vec{k}) = 1 \cdot 1 + \sqrt{3} \cdot 0 + 0 \cdot \sqrt{3} = 1,$$

$$\vec{v} \cdot \vec{w} = (\vec{i} + \sqrt{3}\,\vec{j} + 0\vec{k}) \cdot (\sqrt{3}\,\vec{i} + \vec{j} - \vec{k}) = 1 \cdot \sqrt{3} + \sqrt{3} \cdot 1 + 0(-1) = 2\sqrt{3},$$

$$\vec{w} \cdot \vec{u} = (\sqrt{3}\,\vec{i} + \vec{j} - \vec{k}) \cdot (\vec{i} + 0\vec{j} + \sqrt{3}\,\vec{k}) = \sqrt{3} \cdot 1 + 1 \cdot 0 + (-1) \cdot \sqrt{3} = 0.$$

So the only two vectors that are perpendicular are $\vec{w}$ and $\vec{u}$.

Example 4   Compute the angle between the vectors $\vec{v}$ and $\vec{w}$ from Example 3.

Solution   We know that $\vec{v} \cdot \vec{w} = \|\vec{v}\|\|\vec{w}\| \cos\theta$, so $\cos\theta = \dfrac{\vec{v} \cdot \vec{w}}{\|\vec{v}\|\|\vec{w}\|}$. From Example 3, we know that $\vec{v} \cdot \vec{w} = 2\sqrt{3}$. This gives:

$$\cos\theta = \frac{2\sqrt{3}}{\|\vec{v}\|\|\vec{w}\|} = \frac{2\sqrt{3}}{\sqrt{1^2 + \left(\sqrt{3}\right)^2 + 0^2}\sqrt{\left(\sqrt{3}\right)^2 + 1^2 + (-1)^2}} = \frac{\sqrt{3}}{\sqrt{5}}$$

so $\theta = \arccos\left(\dfrac{\sqrt{3}}{\sqrt{5}}\right) = 39.2315°.$

Normal Vectors and the Equation of a Plane

In Section 12.4 we wrote the equation of a plane given its x-slope, y-slope and z-intercept. Now we write the equation of a plane using a vector $\vec{n}$ and a point P_0. The key idea is that all the displacement vectors from P_0 that are perpendicular to $\vec{n}$ form a plane. To picture this, imagine a pencil balanced on a table, with other pencils fanned out on the table in different directions. The upright pencil is $\vec{n}$, its base is P_0, the other pencils are perpendicular displacement vectors, and the table is the plane.

More formally, a *normal vector* to a plane is a vector that is perpendicular to the plane, that is, it is perpendicular to every displacement vector between any two points in the plane. Let $\vec{n} = a\vec{i} + b\vec{j} + c\vec{k}$ be a normal vector to the plane, let $P_0 = (x_0, y_0, z_0)$ be a fixed point in the plane, and let $P = (x, y, z)$ be any other point in the plane. Then $\overrightarrow{P_0P} = (x - x_0)\vec{i} + (y - y_0)\vec{j} + (z - z_0)\vec{k}$ is a vector whose head and tail both lie in the plane. (See Figure 13.30.) Thus, the vectors $\vec{n}$ and $\overrightarrow{P_0P}$ are perpendicular, so $\vec{n} \cdot \overrightarrow{P_0P} = 0$. The algebraic definition of the dot product gives $\vec{n} \cdot \overrightarrow{P_0P} = a(x - x_0) + b(y - y_0) + c(z - z_0)$, so we obtain the following result:

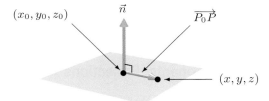

Figure 13.30: Plane with normal $\vec{n}$ and containing a fixed point (x_0, y_0, z_0)

The **equation of the plane** with normal vector $\vec{n} = a\vec{i} + b\vec{j} + c\vec{k}$ and containing the point $P_0 = (x_0, y_0, z_0)$ is

$$a(x - x_0) + b(y - y_0) + c(z - z_0) = 0.$$

Letting $d = ax_0 + by_0 + cz_0$ (a constant), we can write the equation of the plane in the form

$$ax + by + cz = d.$$

Example 5   (a) Find the equation of the plane perpendicular to $\vec{n} = -\vec{i} + 3\vec{j} + 2\vec{k}$ and passing through the point $(1, 0, 4)$.
(b) Find a vector parallel to the plane.

Solution   (a) The equation of the plane is

$$-(x - 1) + 3(y - 0) + 2(z - 4) = 0,$$

which can be written as

$$-x + 3y + 2z = 7.$$

(b) Any vector $\vec{v}$ that is perpendicular to n is also parallel to the plane, so we look for any vector satisfying $\vec{v} \cdot \vec{n} = 0$; for example, $\vec{v} = 3\vec{i} + \vec{j}$. There are many other possible vectors.

Example 6   Find a normal vector to the plane with equation   (a) $x - y + 2z = 5$ (b) $z = 0.5x + 1.2y.$

Solution   (a) Since the coefficients of $\vec{i}$, $\vec{j}$, and $\vec{k}$ in a normal vector are the coefficients of x, y, and z in the equation of the plane, a normal vector is $\vec{n} = \vec{i} - \vec{j} + 2\vec{k}$.

(b) Before we can find a normal vector, we rewrite the equation of the plane in the form

$$0.5x + 1.2y - z = 0.$$

Thus, a normal vector is $\vec{n} = 0.5\vec{i} + 1.2\vec{j} - \vec{k}$.

The Dot Product in n Dimensions

The algebraic definition of the dot product can be extended to vectors in higher dimensions.

> If $\vec{u} = (u_1, \ldots, u_n)$ and $\vec{v} = (v_1, \ldots, v_n)$ then the dot product of $\vec{u}$ and $\vec{v}$ is the **scalar**
>
> $$\vec{u} \cdot \vec{v} = u_1 v_1 + \cdots + u_n v_n.$$

Example 7   A video store sells videos, tapes, CDs, and computer games. We define the quantity vector $\vec{q} = (q_1, q_2, q_3, q_4)$, where q_1, q_2, q_3, q_4 denote the quantities sold of each of the items, and the price vector $\vec{p} = (p_1, p_2, p_3, p_4)$, where p_1, p_2, p_3, p_4 denote the price per unit of each item. What does the dot product $\vec{p} \cdot \vec{q}$ represent?

Solution   The dot product is $\vec{p} \cdot \vec{q} = p_1 q_1 + p_2 q_2 + p_3 q_3 + p_4 q_4$. The quantity $p_1 q_1$ represents the revenue received by the store for the videos, $p_2 q_2$ represents the revenue for the tapes, and so on. The dot product represents the total revenue received by the store for the sale of these four items.

Resolving a Vector into Components: Projections

In Section 13.1, we resolved a vector into components parallel to the axes. Now we see how to resolve a vector, $\vec{v}$, into components, called $\vec{v}_{\text{parallel}}$ and $\vec{v}_{\text{perp}}$, which are parallel and perpendicular, respectively, to a given non-zero vector, $\vec{u}$. (See Figure 13.31.)

Figure 13.31: Resolving $\vec{v}$ into components parallel and perpendicular to $\vec{u}$
(a) $0 < \theta < \pi/2$ (b) $\pi/2 < \theta < \pi$

The projection of $\vec{v}$ on $\vec{u}$, written $\vec{v}_{\text{parallel}}$, measures (in some sense) how much the vector $\vec{v}$ is aligned with the vector $\vec{u}$. The length of $\vec{v}_{\text{parallel}}$ is the length of the shadow cast by $\vec{v}$ on a line in the direction of $\vec{u}$.

To compute $\vec{v}_{\text{parallel}}$, we assume $\vec{u}$ is a unit vector. (If not, create one by dividing by its length.) Then Figure 13.31(a) shows that, if $0 \leq \theta \leq \pi/2$:

$$\|\vec{v}_{\text{parallel}}\| = \|\vec{v}\| \cos\theta = \vec{v} \cdot \vec{u} \qquad (\text{since } \|\vec{u}\| = 1).$$

Now $\vec{v}_{\text{parallel}}$ is a scalar multiple of $\vec{u}$, and since $\vec{u}$ is a unit vector,

$$\vec{v}_{\text{parallel}} = (\|\vec{v}\| \cos\theta)\vec{u} = (\vec{v} \cdot \vec{u})\vec{u}.$$

A similar argument shows that if $\pi/2 < \theta \le \pi$, as in Figure 13.31(b), this formula for $\vec{v}_{\text{parallel}}$ still holds. The vector $\vec{v}_{\text{perp}}$ is specified by

$$\vec{v}_{\text{perp}} = \vec{v} - \vec{v}_{\text{parallel}}.$$

Thus, we have the following results:

Projection of $\vec{v}$ on the Line in the Direction of the Unit Vector $\vec{u}$

If $\vec{v}_{\text{parallel}}$ and $\vec{v}_{\text{perp}}$ are components of $\vec{v}$ that are parallel and perpendicular, respectively, to $\vec{u}$, then

$$\text{Projection of } \vec{v} \text{ onto } \vec{u} = \vec{v}_{\text{parallel}} = (\vec{v} \cdot \vec{u})\vec{u} \qquad \text{provided } \|\vec{u}\| = 1$$

and $\qquad \vec{v} = \vec{v}_{\text{parallel}} + \vec{v}_{\text{perp}} \qquad$ so $\qquad \vec{v}_{\text{perp}} = \vec{v} - \vec{v}_{\text{parallel}}.$

Example 8   Figure 13.32 shows the force the wind exerts on the sail of a sailboat. Find the component of the force in the direction in which the sailboat is travelling.

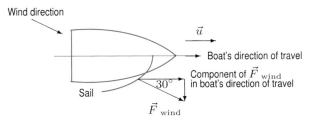

Figure 13.32: Wind moving a sailboat

Solution   Let $\vec{u}$ be a unit vector in the direction of travel. The force of the wind on the sail makes an angle of $30°$ with $\vec{u}$. Thus, the component of this force in the direction of $\vec{u}$ is

$$\vec{F}_{\text{parallel}} = (\vec{F} \cdot \vec{u})\vec{u} = \|\vec{F}\|(\cos 30°)\vec{u} = 0.87\|\vec{F}\|\vec{u}.$$

Thus, the boat is being pushed forward with about 87% of the total force due to the wind. (In fact, the interaction of wind and sail is much more complex than this model suggests.)

A Physical Interpretation of the Dot Product: Work

In physics, the word "work" has a different meaning from its everyday meaning. In physics, when a force of magnitude F acts on an object through a distance d, we say the *work*, W, done by the force is

$$W = Fd,$$

provided the force and the displacement are in the same direction. For example, if a 1 kg body falls 10 metres under the force of gravity, which is 9.8 newtons, then the work done by gravity is

$$W = (9.8 \text{ newtons}) \cdot (10 \text{ metres}) = 98 \text{ joules}.$$

What if the force and the displacement are not in the same direction? Suppose a force $\vec{F}$ acts on an object as it moves along a displacement vector $\vec{d}$. Let θ be the angle between $\vec{F}$ and $\vec{d}$. First, we assume $0 \le \theta \le \pi/2$. Figure 13.33 shows how we can resolve $\vec{F}$ into components that are parallel and perpendicular to $\vec{d}$:

$$\vec{F} = \vec{F}_{\text{parallel}} + \vec{F}_{\text{perp}}.$$

Then the work done by $\vec{F}$ is defined to be

$$W = \|\vec{F}_{\text{parallel}}\|\,\|\vec{d}\|.$$

We see from Figure 13.33 that $\vec{F}_{\text{parallel}}$ has magnitude $\|\vec{F}\|\cos\theta$. So the work is given by the dot product:

$$W = (\|\vec{F}\|\cos\theta)\|\vec{d}\| = \|\vec{F}\|\|\vec{d}\|\cos\theta = \vec{F}\cdot\vec{d}.$$

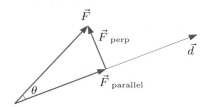

Figure 13.33: Resolving the force $\vec{F}$ into two forces, one parallel to $\vec{d}$, one perpendicular to $\vec{d}$

The formula $W = \vec{F}\cdot\vec{d}$ holds when $\pi/2 < \theta \leq \pi$ also. In that case, the work done by the force is negative and the object is moving against the force. Thus, we have the following definition:

> The **work**, W, done by a force $\vec{F}$ acting on an object through a displacement $\vec{d}$ is given by
>
> $$W = \vec{F}\cdot\vec{d}.$$

Example 9   How much work does the wind do on the sailboat from Example 8 if the boat moves 20 m and the wind's force is 120 newtons?

Solution   From Example 8, we know that the force of the wind $\vec{F}$ makes a 30°angle with the boat's displacement $\vec{d}$. Since $\|\vec{F}\| = 120$ and $\|\vec{d}\| = 20$, the work done by the wind on the boat is

$$W = \vec{F}\cdot\vec{d} = \|\vec{F}\|\|\vec{d}\|\cos 30° = 2078.461 \text{ joules.}$$

Notice that if the vectors $\vec{F}$ and $\vec{d}$ are parallel and in the same direction, with magnitudes F and d, then $\cos\theta = \cos 0 = 1$, so $W = \|\vec{F}\|\|\vec{d}\| = Fd$, which is the original definition. When the vectors are perpendicular, $\cos\theta = \cos(\pi/2) = 0$, so $W = 0$ and no work is done in the technical definition of the word. For example, if you carry a heavy box across the room at the same horizontal height, no work is done by gravity because the force of gravity is vertical but the motion is horizontal.

Exercises and Problems for Section 13.3

Exercises

For Exercises 1–10, perform the following operations on the given 3-dimensional vectors.

$\vec{a} = 2\vec{j} + \vec{k}$ $\vec{b} = -3\vec{i} + 5\vec{j} + 4\vec{k}$ $\vec{c} = \vec{i} + 6\vec{j}$

$\vec{y} = 4\vec{i} - 7\vec{j}$ $\vec{z} = \vec{i} - 3\vec{j} - \vec{k}$

1. $\vec{a}\cdot\vec{c}$

2. $\vec{a}\cdot\vec{y}$

3. $\vec{c}\cdot\vec{y}$

4. $\vec{a}\cdot\vec{b}$

5. $\vec{a}\cdot\vec{z}$

6. $\vec{c}\cdot\vec{a} + \vec{a}\cdot\vec{y}$

7. $\vec{a} \cdot (\vec{c} + \vec{y})$

8. $(\vec{a} \cdot \vec{b})\vec{a}$

9. $(\vec{a} \cdot \vec{y})(\vec{c} \cdot \vec{z})$

10. $((\vec{c} \cdot \vec{c})\vec{a}) \cdot \vec{a}$

19. $\vec{i} + \vec{j}$ and $\vec{i} + 2\vec{j} - \vec{k}$.

20. $\vec{i}$ and $2\vec{i} + 3\vec{j} - \vec{k}$.

In Exercises 21–27, find an equation of a plane that satisfies the given conditions.

In Exercises 11–15, find a normal vector to the plane.

11. $2x + y - z = 23$

12. $1.5x + 3.2y + z = 0$

13. $z = 3x + 4y - 7$

14. $z - 5(x - 2) = 3(5 - y)$

15. $\pi(x - 1) = (1 - \pi)(y - z) + \pi$

21. Through $(1, 5, 2)$ perpendicular to $3\vec{i} - \vec{j} + 4\vec{k}$

22. Through $(2, -1, 3)$ perpendicular to $5\vec{i} + 4\vec{j} - \vec{k}$.

23. Through $(1, 3, 5)$ and normal to $\vec{i} - \vec{j} + \vec{k}$.

24. Perpendicular to $5\vec{i} + \vec{j} - 2\vec{k}$ and passing through $(0, 1, -1)$.

25. Parallel to $2x + 4y - 3z = 1$ and through $(1, 0, -1)$.

In Exercises 16–20, compute the angle between the vectors.

16. $\vec{i} + \vec{j} + \vec{k}$ and $\vec{i} - \vec{j} - \vec{k}$.

17. $\vec{i} + \vec{k}$ and $\vec{j} - \vec{k}$.

18. $\vec{i} + \vec{j} - \vec{k}$ and $2\vec{i} + 3\vec{j} + \vec{k}$.

26. Through $(-2, 3, 2)$ and parallel to $3x + y + z = 4$.

27. Perpendicular to $\vec{v} = 2\vec{i} - 3\vec{j} + 5\vec{k}$ and through $(4, 5, -2)$.

Problems

28. Give a unit vector

    (a) In the same direction as $\vec{v} = 2\vec{i} + 3\vec{j}$.

    (b) Perpendicular to $\vec{v}$.

29. A plane has equation $z = 5x - 2y + 7$.

    (a) Find a value of λ making the vector $\lambda\vec{i} + \vec{j} + 0.5\vec{k}$ normal to the plane.

    (b) Find a value of a so that the point $(a + 1, a, a - 1)$ lies on the plane.

30. Consider the plane $5x - y + 7z = 21$.

    (a) Find a point on the x-axis on this plane.

    (b) Find two other points on the plane.

    (c) Find a vector perpendicular to the plane.

    (d) Find a vector parallel to the plane.

31. **(a)** Find a vector perpendicular to the plane $z = 2 + 3x - y$.

    (b) Find a vector parallel to the plane.

32. **(a)** Find a vector perpendicular to the plane $z = 2x + 3y$.

    (b) Find a vector parallel to the plane.

33. Match the planes in (a)–(d) with one or more of the descriptions in (I)–(IV). No reasons needed.

    (a) $3x - y + z = 0$ **(b)** $4x + y + 2z - 5 = 0$

    (c) $x + y = 5$ **(d)** $x = 5$

      I Goes through the origin.

      II Has a normal vector parallel to the xy-plane.

      III Goes through the point $(0, 5, 0)$.

      IV Has a normal vector whose dot products with $\vec{i}, \vec{j}, \vec{k}$ are all positive.

34. Which pairs (if any) of vectors from the following list

    (a) Are perpendicular?

    (b) Are parallel?

    (c) Have an angle less than $\pi/2$ between them?

    (d) Have an angle of more than $\pi/2$ between them?

$$\vec{a} = \vec{i} - 3\vec{j} - \vec{k}, \qquad \vec{b} = \vec{i} + \vec{j} + 2\vec{k},$$
$$\vec{c} = -2\vec{i} - \vec{j} + \vec{k}, \qquad \vec{d} = -\vec{i} - \vec{j} + \vec{k}.$$

35. List any vectors that are parallel to each other and any vectors that are perpendicular to each other:

$$\vec{v}_1 = \vec{i} - 2\vec{j} \qquad\qquad \vec{v}_2 = 2\vec{i} + 4\vec{j}$$
$$\vec{v}_3 = 3\vec{i} + 1.5\vec{j} \qquad\qquad \vec{v}_4 = -1.2\vec{i} + 2.4\vec{j}$$
$$\vec{v}_5 = -5\vec{i} - 2.5\vec{j} \qquad\qquad \vec{v}_6 = 12\vec{i} - 12\vec{j}$$
$$\vec{v}_7 = 4\vec{i} + 2\vec{j} \qquad\qquad \vec{v}_8 = 3\vec{i} - 6\vec{j}$$
$$\vec{v}_9 = 0.70\vec{i} - 0.35\vec{j}$$

36. **(a)** Give a vector that is parallel to, but not equal to, $\vec{v} = 4\vec{i} + 3\vec{j}$.

    (b) Give a vector that is perpendicular to $\vec{v}$.

37. For what values of t are $\vec{u} = t\vec{i} - \vec{j} + \vec{k}$ and $\vec{v} = t\vec{i} + t\vec{j} - 2\vec{k}$ perpendicular? Are there values of t for which $\vec{u}$ and $\vec{v}$ are parallel?

38. Let θ be the angle between $\vec{v}$ and $\vec{w}$, with $0 < \theta < \pi/2$. What is the effect on $\vec{v} \cdot \vec{w}$ of increasing each of the following quantities? Does $\vec{v} \cdot \vec{w}$ increase or decrease?

    (a) $||\vec{v}||$ **(b)** θ

39. Write $\vec{a} = 3\vec{i} + 2\vec{j} - 6\vec{k}$ as the sum of two vectors, one parallel, and one perpendicular, to $\vec{d} = 2\vec{i} - 4\vec{j} + \vec{k}$.

40. Find angle BAC if $A = (2, 2, 2)$, $B = (4, 2, 1)$, and $C = (2, 3, 1)$.

41. The points $(5, 0, 0)$, $(0, -3, 0)$, and $(0, 0, 2)$ form a triangle. Find the lengths of the sides of the triangle and each of its angles.

42. Let S be the triangle with vertices $A = (2,2,2)$, $B = (4,2,1)$, and $C = (2,3,1)$.

   (a) Find the length of the shortest side of S.
   (b) Find the cosine of the angle BAC at vertex A.

In Problems 43–48, given $\vec{v} = 3\vec{i} + 4\vec{j}$ and force vector $\vec{F}$, find:

(a) The component of $\vec{F}$ parallel to $\vec{v}$.
(b) The component of $\vec{F}$ perpendicular to $\vec{v}$.
(c) The work, W, done by force $\vec{F}$ through displacement $\vec{v}$.

43. $\vec{F} = 0.2\vec{i} - 0.5\vec{j}$ **44.** $\vec{F} = 4\vec{i} + \vec{j}$

45. $\vec{F} = -0.4\vec{i} + 0.3\vec{j}$ **46.** $\vec{F} = 9\vec{i} + 12\vec{j}$

47. $\vec{F} = -3\vec{i} - 5\vec{j}$ **48.** $\vec{F} = -6\vec{i} - 8\vec{j}$

In Problems 49–52, the force on an object is $\vec{F} = -20\vec{j}$. For vector $\vec{v}$, find:

(a) The component of $\vec{F}$ parallel to $\vec{v}$.
(b) The component of $\vec{F}$ perpendicular to $\vec{v}$.
(c) The work, W, done by force $\vec{F}$ through displacement $\vec{v}$.

49. $\vec{v} = 5\vec{i} - \vec{j}$ **50.** $\vec{v} = 2\vec{i} + 3\vec{j}$

51. $\vec{v} = 5\vec{i}$ **52.** $\vec{v} = 3\vec{j}$

53. A basketball gymnasium is 25 meters high, 80 meters wide and 200 meters long. For a half-time stunt, the cheerleaders want to run two strings, one from each of the two corners above one basket to the diagonally opposite corners of the gym floor. What is the cosine of the angle made by the strings as they cross?

54. A 100-meter dash is run on a track in the direction of the vector $\vec{v} = 2\vec{i} + 6\vec{j}$. The wind velocity $\vec{w}$ is $5\vec{i} + \vec{j}$ km/hr. The rules say that a legal wind speed measured in the direction of the dash must not exceed 5 km/hr. Will the race results be disqualified due to an illegal wind? Justify your answer.

55. An airplane is flying toward the southeast. Which of the following wind velocity vectors increases the plane's speed the most? Which slows down the plane the most?

$$\vec{w}_1 = -4\vec{i} - \vec{j} \qquad \vec{w}_2 = \vec{i} - 2\vec{j} \qquad \vec{w}_3 = -\vec{i} + 8\vec{j}$$
$$\vec{w}_4 = 10\vec{i} + 2\vec{j} \qquad \vec{w}_5 = 5\vec{i} - 2\vec{j}$$

56. A canoe is moving with velocity $\vec{v} = 5\vec{i} + 3\vec{j}$ m/sec relative to the water. The velocity of the current in the water is $\vec{c} = \vec{i} + 2\vec{j}$ m/sec.

   (a) What is the speed of the current?
   (b) What is the speed of the current in the direction of the canoe's motion?

57. Find a vector that bisects the smaller of the two angles formed by $3\vec{i} + 4\vec{j}$ and $5\vec{i} - 12\vec{j}$.

58. Find the shortest distance between the planes $2x - 5y + z = 10$ and $z = 5y - 2x$.

59. A street vendor sells six items, with prices p_1 dollars per unit, p_2 dollars per unit, and so on. The vendor's price vector is $\vec{p} = (p_1, p_2, p_3, p_4, p_5, p_6) = (1.00, 3.50, 4.00, 2.75, 5.00, 3.00)$. The vendor sells q_1 units of the first item, q_2 units of the second item, and so on. The vendor's quantity vector is $\vec{q} = (q_1, q_2, q_3, q_4, q_5, q_6) = (43, 57, 12, 78, 20, 35)$. Find $\vec{p} \cdot \vec{q}$, give its units, and explain its significance to the vendor.

60. A course has four exams, weighted 10%, 15%, 25%, 50%, respectively. The class average on each of these exams is 75%, 91%, 84%, 87%, respectively. What do the vectors $\vec{a} = (0.75, 0.91, 0.84, 0.87)$ and $\vec{w} = (0.1, 0.15, 0.25, 0.5)$ represent, in terms of the course? Calculate the dot product $\vec{w} \cdot \vec{a}$. What does it represent, in terms of the course?

61. A consumption vector of three goods is defined by $\vec{x} = (x_1, x_2, x_3)$, where x_1, x_2 and x_3 are the quantities consumed of the three goods. A budget constraint is represented by the equation $\vec{p} \cdot \vec{x} = k$, where $\vec{p}$ is the price vector of the three goods and k is a constant. Show that the difference between two consumption vectors corresponding to points satisfying the same budget constraint is perpendicular to the price vector $\vec{p}$.

62. What does Property 2 of the dot product in the box on page 753 say geometrically?

63. Show that the vectors $(\vec{b} \cdot \vec{c})\vec{a} - (\vec{a} \cdot \vec{c})\vec{b}$ and $\vec{c}$ are perpendicular.

64. Show why each of the properties of the dot product in the box on page 753 follows from the algebraic definition of the dot product:

$$\vec{v} \cdot \vec{w} = v_1 w_1 + v_2 w_2 + v_3 w_3.$$

65. Show that if $\vec{u}$ and $\vec{v}$ are two vectors such that

$$\vec{u} \cdot \vec{w} = \vec{v} \cdot \vec{w}$$

for every vector $\vec{w}$, then

$$\vec{u} = \vec{v}.$$

66. Show that

$$\frac{\vec{u}}{\|\vec{u}\|^2} - \frac{\vec{v}}{\|\vec{v}\|^2} \quad \text{and} \quad \frac{\vec{u}}{\|\vec{u}\|\|\vec{v}\|} - \frac{\vec{v}}{\|\vec{u}\|\|\vec{v}\|}$$

have the same magnitude where $\vec{u}$ and $\vec{v}$ are nonzero vectors.

67. Figure 13.34 shows that, given three vectors $\vec{u}$, $\vec{v}$, and $\vec{w}$, the sum of the components of $\vec{v}$ and $\vec{w}$ in the direction of $\vec{u}$ is the component of $\vec{v} + \vec{w}$ in the direction of $\vec{u}$. (Although the figure is drawn in two dimensions, this result is also true in three dimensions.) Use this figure to explain why the geometric definition of the dot product satisfies $(\vec{v} + \vec{w}) \cdot \vec{u} = \vec{v} \cdot \vec{u} + \vec{w} \cdot \vec{u}$.

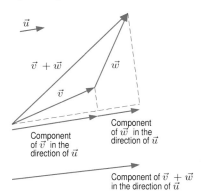

Figure 13.34: Component of $\vec{v} + \vec{w}$ in the direction of $\vec{u}$ is the sum of the components of $\vec{v}$ and $\vec{w}$ in that direction

68. (a) Using the geometric definition of the dot product, show that

$$\vec{u} \cdot (-\vec{v}) = -(\vec{u} \cdot \vec{v}).$$

[Hint: What happens to the angle when you multiply $\vec{v}$ by -1?]

(b) Using the geometric definition of the dot product, show that for any negative scalar λ

$$\vec{u} \cdot (\lambda\vec{v}) = \lambda(\vec{u} \cdot \vec{v})$$
$$(\lambda\vec{u}) \cdot \vec{v} = \lambda(\vec{u} \cdot \vec{v}).$$

69. The Law of Cosines for a triangle with side lengths a, b, and c, and with angle C opposite side c, says

$$c^2 = a^2 + b^2 - 2ab\cos C.$$

On page 753, we used the Law of Cosines to show that the two definitions of the dot product are equivalent. In this problem, use the geometric definition of the dot product and its properties in the box on page 753 to prove the Law of Cosines. [Hint: Let $\vec{u}$ and $\vec{v}$ be the displacement vectors from C to the other two vertices, and express c^2 in terms of $\vec{u}$ and $\vec{v}$.]

70. Use Problems 67 and 68 and the following steps to show (without trigonometry) that the geometric and algebraic definitions of the dot product are equivalent. Let $\vec{u} = u_1\vec{i} + u_2\vec{j} + u_3\vec{k}$ and $\vec{v} = v_1\vec{i} + v_2\vec{j} + v_3\vec{k}$ be any vectors. Write $(\vec{u} \cdot \vec{v})_{\text{geom}}$ for the result of the dot product computed geometrically. Substitute $\vec{u} = u_1\vec{i} + u_2\vec{j} + u_3\vec{k}$ and use Problems 67–68 to expand $(\vec{u} \cdot \vec{v})_{\text{geom}}$. Substitute for $\vec{v}$ and expand. Then calculate the dot products $\vec{i} \cdot \vec{i}$, $\vec{i} \cdot \vec{j}$, etc. geometrically.

71. For any vectors $\vec{v}$ and $\vec{w}$, consider the following function of t:

$$q(t) = (\vec{v} + t\vec{w}) \cdot (\vec{v} + t\vec{w}).$$

(a) Explain why $q(t) \geq 0$ for all real t.
(b) Expand $q(t)$ as a quadratic polynomial in t using the properties on page 753.
(c) Using the discriminant of the quadratic, show that

$$|\vec{v} \cdot \vec{w}| \leq \|\vec{v}\|\|\vec{w}\|.$$

Strengthen Your Understanding

In Problems 72–74, explain what is wrong with the statement.

72. For any 3-dimensional vectors $\vec{u}$, $\vec{v}$, $\vec{w}$, we have $(\vec{u} \cdot \vec{v}) \cdot \vec{w} = \vec{u} \cdot (\vec{v} \cdot \vec{w})$.

73. If $\vec{u} = \vec{i} + \vec{j}$ and $\vec{v} = 2\vec{i} + \vec{j}$, then the component of $\vec{v}$ parallel to $\vec{u}$ is $\vec{v}_{\text{parallel}} = (\vec{v} \cdot \vec{u})\vec{u} = 3\vec{i} + 3\vec{j}$.

74. A normal vector for the plane $z = 2x + 3y$ is $2\vec{i} + 3\vec{j}$.

In Problems 75–76, give an example of:

75. A point (a, b) such that the displacement vector from $(1, 1)$ to (a, b) is perpendicular to $\vec{i} + 2\vec{j}$.

76. A linear function $f(x, y) = mx + ny + c$ whose graph is perpendicular to $\vec{i} + 2\vec{j} + 3\vec{k}$.

Are the statements in Problems 77–86 true or false? Give reasons for your answer.

77. The quantity $\vec{u} \cdot \vec{v}$ is a vector.

78. The plane $x + 2y - 3z = 5$ has normal vector $\vec{i} + 2\vec{j} - 3\vec{k}$.

79. If $\vec{u} \cdot \vec{v} < 0$ then the angle between $\vec{u}$ and $\vec{v}$ is greater than $\pi/2$.

80. An equation of the plane with normal vector $\vec{i} + \vec{j} + \vec{k}$ containing the point $(1, 2, 3)$ is $z = x + y$.

81. The triangle in 3-space with vertices $(1, 1, 0), (0, 1, 0)$ and $(0, 1, 1)$ has a right angle.

82. The dot product $\vec{v} \cdot \vec{v}$ is never negative.

83. If $\vec{u} \cdot \vec{v} = 0$ then either $\vec{u} = 0$ or $\vec{v} = 0$.

84. If $\vec{u}$, $\vec{v}$ and $\vec{w}$ are all nonzero, and $\vec{u} \cdot \vec{v} = \vec{u} \cdot \vec{w}$, then $\vec{v} = \vec{w}$.

85. For any vectors $\vec{u}$ and $\vec{v}$: $(\vec{u} + \vec{v}) \cdot (\vec{u} - \vec{v}) = \|\vec{u}\|^2 - \|\vec{v}\|^2$.

86. If $\|\vec{u}\| = 1$, then the vector $\vec{v} - (\vec{v} \cdot \vec{u})\vec{u}$ is perpendicular to $\vec{u}$.

13.4   THE CROSS PRODUCT

In the previous section we combined two vectors to get a number, the dot product. In this section we see another way of combining two vectors, this time to get a vector, the *cross product*. Any two vectors in 3-space form a parallelogram. We define the cross product using this parallelogram.

The Area of a Parallelogram

Consider the parallelogram formed by the vectors $\vec{v}$ and $\vec{w}$ with an angle of θ between them. Then Figure 13.35 shows

$$\text{Area of parallelogram} = \text{Base} \cdot \text{Height} = \|\vec{v}\|\|\vec{w}\| \sin\theta.$$

How would we compute the area of the parallelogram if we were given $\vec{v}$ and $\vec{w}$ in components, $\vec{v} = v_1\vec{i} + v_2\vec{j} + v_3\vec{k}$ and $\vec{w} = w_1\vec{i} + w_2\vec{j} + w_3\vec{k}$? Project 1 on page 773 shows that if $\vec{v}$ and $\vec{w}$ are in the xy-plane so that $v_3 = w_3 = 0$, then

$$\text{Area of parallelogram} = |v_1 w_2 - v_2 w_1|.$$

What if $\vec{v}$ and $\vec{w}$ do not lie in the xy-plane? The cross product will enable us to compute the area of the parallelogram formed by any two vectors.

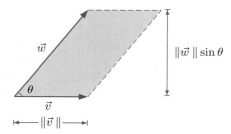

Figure 13.35: Parallelogram formed by $\vec{v}$ and $\vec{w}$ has
Area $= \|\vec{v}\|\|\vec{w}\| \sin\theta$

Definition of the Cross Product

We define the cross product of the vectors $\vec{v}$ and $\vec{w}$, written $\vec{v} \times \vec{w}$, to be a vector perpendicular to both $\vec{v}$ and $\vec{w}$. The magnitude of this vector is the area of the parallelogram formed by the two vectors. The direction of $\vec{v} \times \vec{w}$ is given by the normal vector, $\vec{n}$, to the plane defined by $\vec{v}$ and $\vec{w}$. If we require that $\vec{n}$ be a unit vector, there are two choices for $\vec{n}$, pointing out of the plane in opposite directions. We pick one by the following rule (see Figure 13.36):

The right-hand rule: Place $\vec{v}$ and $\vec{w}$ so that their tails coincide and curl the fingers of your right hand through the smaller of the two angles from $\vec{v}$ to $\vec{w}$; your thumb points in the direction of the normal vector, $\vec{n}$.

Like the dot product, there are two equivalent definitions of the cross product:

The following two definitions of the **cross product** or **vector product** $\vec{v} \times \vec{w}$ are equivalent:

- **Geometric definition**

  If $\vec{v}$ and $\vec{w}$ are not parallel, then

  $$\vec{v} \times \vec{w} = \begin{pmatrix} \text{Area of parallelogram} \\ \text{with edges } \vec{v} \text{ and } \vec{w} \end{pmatrix} \vec{n} = (\|\vec{v}\|\|\vec{w}\| \sin\theta)\vec{n},$$

  where $0 \leq \theta \leq \pi$ is the angle between $\vec{v}$ and $\vec{w}$ and $\vec{n}$ is the unit vector perpendicular to $\vec{v}$ and $\vec{w}$ pointing in the direction given by the right-hand rule. If $\vec{v}$ and $\vec{w}$ are parallel, then $\vec{v} \times \vec{w} = \vec{0}$.

- **Algebraic definition**

  $$\vec{v} \times \vec{w} = (v_2 w_3 - v_3 w_2)\vec{i} + (v_3 w_1 - v_1 w_3)\vec{j} + (v_1 w_2 - v_2 w_1)\vec{k}$$

  where $\vec{v} = v_1\vec{i} + v_2\vec{j} + v_3\vec{k}$ and $\vec{w} = w_1\vec{i} + w_2\vec{j} + w_3\vec{k}$.

Problems 47 and 50 at the end of this section show that the geometric and algebraic definitions of the cross product give the same result.

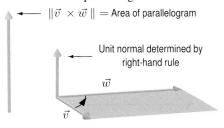

Figure 13.36: Area of parallelogram $= \|\vec{v} \times \vec{w}\|$

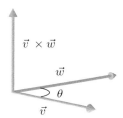

Figure 13.37: The cross product $\vec{v} \times \vec{w}$

The geometric definition shows us that the cross product is *rotation invariant*. Imagine the two vectors $\vec{v}$ and $\vec{w}$ as two metal rods welded together. Attach a third rod whose direction and length correspond to $\vec{v} \times \vec{w}$. (See Figure 13.37.) Then, no matter how we turn this set of rods, the third will still be the cross product of the first two.

The algebraic definition is more easily remembered by writing it as a 3×3 determinant. (See Appendix E.)

$$\vec{v} \times \vec{w} = \begin{vmatrix} \vec{i} & \vec{j} & \vec{k} \\ v_1 & v_2 & v_3 \\ w_1 & w_2 & w_3 \end{vmatrix} = (v_2 w_3 - v_3 w_2)\vec{i} + (v_3 w_1 - v_1 w_3)\vec{j} + (v_1 w_2 - v_2 w_1)\vec{k}.$$

Example 1   Find $\vec{i} \times \vec{j}$ and $\vec{j} \times \vec{i}$.

Solution   The vectors $\vec{i}$ and $\vec{j}$ both have magnitude 1 and the angle between them is $\pi/2$. By the right-hand rule, the vector $\vec{i} \times \vec{j}$ is in the direction of $\vec{k}$, so $\vec{n} = \vec{k}$ and we have

$$\vec{i} \times \vec{j} = \left(\|\vec{i}\|\|\vec{j}\| \sin\frac{\pi}{2} \right)\vec{k} = \vec{k}.$$

Similarly, the right-hand rule says that the direction of $\vec{j} \times \vec{i}$ is $-\vec{k}$, so

$$\vec{j} \times \vec{i} = (\|\vec{j}\|\|\vec{i}\| \sin\frac{\pi}{2})(-\vec{k}) = -\vec{k}.$$

Similar calculations show that $\vec{j} \times \vec{k} = \vec{i}$ and $\vec{k} \times \vec{i} = \vec{j}$.

Example 2 For any vector $\vec{v}$, find $\vec{v} \times \vec{v}$.

Solution Since $\vec{v}$ is parallel to itself, $\vec{v} \times \vec{v} = \vec{0}$.

Example 3 Find the cross product of $\vec{v} = 2\vec{i} + \vec{j} - 2\vec{k}$ and $\vec{w} = 3\vec{i} + \vec{k}$ and check that the cross product is perpendicular to both $\vec{v}$ and $\vec{w}$.

Solution Writing $\vec{v} \times \vec{w}$ as a determinant and expanding it into three two-by-two determinants, we have

$$\vec{v} \times \vec{w} = \begin{vmatrix} \vec{i} & \vec{j} & \vec{k} \\ 2 & 1 & -2 \\ 3 & 0 & 1 \end{vmatrix} = \vec{i} \begin{vmatrix} 1 & -2 \\ 0 & 1 \end{vmatrix} - \vec{j} \begin{vmatrix} 2 & -2 \\ 3 & 1 \end{vmatrix} + \vec{k} \begin{vmatrix} 2 & 1 \\ 3 & 0 \end{vmatrix}$$

$$= \vec{i} \, (1(1) - 0(-2)) - \vec{j} \, (2(1) - 3(-2)) + \vec{k} \, (2(0) - 3(1))$$

$$= \vec{i} - 8\vec{j} - 3\vec{k}.$$

To check that $\vec{v} \times \vec{w}$ is perpendicular to $\vec{v}$, we compute the dot product:

$$\vec{v} \cdot (\vec{v} \times \vec{w}) = (2\vec{i} + \vec{j} - 2\vec{k}) \cdot (\vec{i} - 8\vec{j} - 3\vec{k}) = 2 - 8 + 6 = 0.$$

Similarly,

$$\vec{w} \cdot (\vec{v} \times \vec{w}) = (3\vec{i} + 0\vec{j} + \vec{k}) \cdot (\vec{i} - 8\vec{j} - 3\vec{k}) = 3 + 0 - 3 = 0.$$

Thus, $\vec{v} \times \vec{w}$ is perpendicular to both $\vec{v}$ and $\vec{w}$.

Properties of the Cross Product

The right-hand rule tells us that $\vec{v} \times \vec{w}$ and $\vec{w} \times \vec{v}$ point in opposite directions. The magnitudes of $\vec{v} \times \vec{w}$ and $\vec{w} \times \vec{v}$ are the same, so $\vec{w} \times \vec{v} = -(\vec{v} \times \vec{w})$. (See Figure 13.38.)

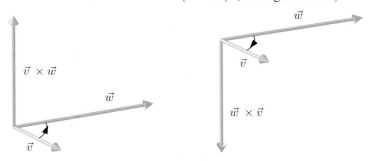

Figure 13.38: Diagram showing $\vec{v} \times \vec{w} = -(\vec{w} \times \vec{v})$

This explains the first of the following properties. The other two are derived in Problems 40, 41, and 50 at the end of this section.

Properties of the Cross Product

For vectors $\vec{u}$, $\vec{v}$, $\vec{w}$ and scalar λ

1. $\vec{w} \times \vec{v} = -(\vec{v} \times \vec{w})$
2. $(\lambda \vec{v}) \times \vec{w} = \lambda(\vec{v} \times \vec{w}) = \vec{v} \times (\lambda \vec{w})$
3. $\vec{u} \times (\vec{v} + \vec{w}) = \vec{u} \times \vec{v} + \vec{u} \times \vec{w}$.

The Equation of a Plane Through Three Points

As we saw on page 755, the equation of a plane is determined by a point $P_0 = (x_0, y_0, z_0)$ on the plane, and a normal vector, $\vec{n} = a\vec{i} + b\vec{j} + c\vec{k}$:

$$a(x - x_0) + b(y - y_0) + c(z - z_0) = 0.$$

However, a plane can also be determined by three points on it (provided they do not lie on the same line). In that case we can find an equation of the plane by first determining two vectors in the plane and then finding a normal vector using the cross product, as in the following example.

Example 4   Find an equation of the plane containing the points $P = (1, 3, 0)$, $Q = (3, 4, -3)$, and $R = (3, 6, 2)$.

Solution   Since the points P and Q are in the plane, the displacement vector between them, $\overrightarrow{PQ}$, is in the plane, where
$$\overrightarrow{PQ} = (3 - 1)\vec{i} + (4 - 3)\vec{j} + (-3 - 0)\vec{k} = 2\vec{i} + \vec{j} - 3\vec{k}.$$

The displacement vector $\overrightarrow{PR}$ is also in the plane, where

$$\overrightarrow{PR} = (3 - 1)\vec{i} + (6 - 3)\vec{j} + (2 - 0)\vec{k} = 2\vec{i} + 3\vec{j} + 2\vec{k}.$$

Thus, a normal vector, $\vec{n}$, to the plane is given by

$$\vec{n} = \overrightarrow{PQ} \times \overrightarrow{PR} = \begin{vmatrix} \vec{i} & \vec{j} & \vec{k} \\ 2 & 1 & -3 \\ 2 & 3 & 2 \end{vmatrix} = 11\vec{i} - 10\vec{j} + 4\vec{k}.$$

Since the point $(1, 3, 0)$ is on the plane, the equation of the plane is

$$11(x - 1) - 10(y - 3) + 4(z - 0) = 0,$$

which simplifies to
$$11x - 10y + 4z = -19.$$

You should check that P, Q, and R satisfy this equation, since they lie on the plane.

Areas and Volumes Using the Cross Product and Determinants

We can use the cross product to calculate the area of the parallelogram with sides $\vec{v}$ and $\vec{w}$. We say that $\vec{v} \times \vec{w}$ is the *area vector* of the parallelogram. The geometric definition of the cross product tells us that $\vec{v} \times \vec{w}$ is normal to the parallelogram and gives us the following result:

Area of a parallelogram with edges $\vec{v} = v_1\vec{i} + v_2\vec{j} + v_3\vec{k}$ and $\vec{w} = w_1\vec{i} + w_2\vec{j} + w_3\vec{k}$ is given by

$$\text{Area} = \|\vec{v} \times \vec{w}\|, \qquad \text{where} \quad \vec{v} \times \vec{w} = \begin{vmatrix} \vec{i} & \vec{j} & \vec{k} \\ v_1 & v_2 & v_3 \\ w_1 & w_2 & w_3 \end{vmatrix}.$$

Example 5   Find the area of the parallelogram with edges $\vec{v} = 2\vec{i} + \vec{j} - 3\vec{k}$ and $\vec{w} = \vec{i} + 3\vec{j} + 2\vec{k}$.

Solution   We calculate the cross product:

$$\vec{v} \times \vec{w} = \begin{vmatrix} \vec{i} & \vec{j} & \vec{k} \\ 2 & 1 & -3 \\ 1 & 3 & 2 \end{vmatrix} = (2+9)\vec{i} - (4+3)\vec{j} + (6-1)\vec{k} = 11\vec{i} - 7\vec{j} + 5\vec{k}.$$

The area of the parallelogram with edges $\vec{v}$ and $\vec{w}$ is the magnitude of the vector $\vec{v} \times \vec{w}$:

$$\text{Area} = \|\vec{v} \times \vec{w}\| = \sqrt{11^2 + (-7)^2 + 5^2} = \sqrt{195}.$$

Volume of a Parallelepiped

Consider the parallelepiped with sides formed by $\vec{a}, \vec{b}$, and $\vec{c}$. (See Figure 13.39.) Since the base is formed by the vectors $\vec{b}$ and $\vec{c}$, we have

$$\text{Area of base of parallelepiped} = \|\vec{b} \times \vec{c}\|.$$

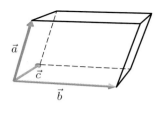
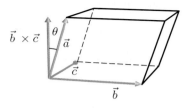
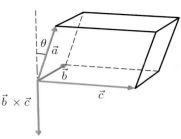

Figure 13.39: Volume of a parallelepiped

Figure 13.40: The vectors $\vec{a}, \vec{b}, \vec{c}$ are called a right-handed set

Figure 13.41: The vectors $\vec{a}, \vec{b}, \vec{c}$ are called a left-handed set

The vectors $\vec{a}$, $\vec{b}$, and $\vec{c}$ can be arranged either as in Figure 13.40 or as in Figure 13.41. In either case,

$$\text{Height of parallelepiped} = \|\vec{a}\| \cos\theta,$$

where θ is the angle shown in the figures. In Figure 13.40 the angle θ is less than $\pi/2$, so the product, $(\vec{b} \times \vec{c}) \cdot \vec{a}$, called the *triple product*, is positive. Thus, in this case

$$\text{Volume of parallelepiped} = \text{Base} \cdot \text{Height} = \|\vec{b} \times \vec{c}\| \cdot \|\vec{a}\| \cos\theta = (\vec{b} \times \vec{c}) \cdot \vec{a}.$$

In Figure 13.41, the angle, $\pi - \theta$, between $\vec{a}$ and $\vec{b} \times \vec{c}$ is more than $\pi/2$, so the product $(\vec{b} \times \vec{c}) \cdot \vec{a}$ is negative. Thus, in this case we have

$$\text{Volume} = \text{Base} \cdot \text{Height} = \|\vec{b} \times \vec{c}\| \cdot \|\vec{a}\| \cos\theta = -\|\vec{b} \times \vec{c}\| \cdot \|\vec{a}\| \cos(\pi - \theta)$$
$$= -(\vec{b} \times \vec{c}) \cdot \vec{a} = \left|(\vec{b} \times \vec{c}) \cdot \vec{a}\right|.$$

Therefore, in both cases the volume is given by $\left|(\vec{b} \times \vec{c}) \cdot \vec{a}\right|$. Using determinants, we can write

---

Volume of a parallelepiped with edges $\vec{a}, \vec{b}, \vec{c}$ is given by

$$\text{Volume} = \left|(\vec{b} \times \vec{c}) \cdot \vec{a}\right| = \text{Absolute value of the determinant} \begin{vmatrix} a_1 & a_2 & a_3 \\ b_1 & b_2 & b_3 \\ c_1 & c_2 & c_3 \end{vmatrix}.$$

Exercises and Problems for Section 13.4

Exercises

In Exercises 1–8, use the algebraic definition to find $\vec{v} \times \vec{w}$.

1. $\vec{v} = -\vec{j}, \vec{w} = \vec{i}$

2. $\vec{v} = \vec{k}, \vec{w} = \vec{j}$

3. $\vec{v} = -\vec{i}, \vec{w} = \vec{j} + \vec{k}$

4. $\vec{v} = \vec{i} + \vec{k}, \vec{w} = \vec{i} + \vec{j}$

5. $\vec{v} = \vec{i} + \vec{j} + \vec{k}, \vec{w} = \vec{i} + \vec{j} + -\vec{k}$

6. $\vec{v} = 2\vec{i} - 3\vec{j} + \vec{k}, \vec{w} = \vec{i} + 2\vec{j} - \vec{k}$

7. $\vec{v} = 2\vec{i} - \vec{j} - \vec{k}, \vec{w} = -6\vec{i} + 3\vec{j} + 3\vec{k}$

8. $\vec{v} = -3\vec{i} + 5\vec{j} + 4\vec{k}, \vec{w} = \vec{i} - 3\vec{j} - \vec{k}$

Use the geometric definition in Exercises 9–10 to find:

9. $2\vec{i} \times (\vec{i} + \vec{j})$

10. $(\vec{i} + \vec{j}) \times (\vec{i} - \vec{j})$

In Exercises 11–12, use the properties on page 764 to find:

11. $\left((\vec{i} + \vec{j}) \times \vec{i}\right) \times \vec{j}$

12. $(\vec{i} + \vec{j}) \times (\vec{i} \times \vec{j})$

In Exercises 13–14, find an equation for the plane through the points.

13. $(1, 0, 0), (0, 1, 0), (0, 0, 1)$.

14. $(3, 4, 2), (-2, 1, 0), (0, 2, 1)$.

In Exercises 15–18, find the volume of the parallelogram with edges $\vec{a}, \vec{b}, \vec{c}$.

15. $\vec{a} = 3\vec{i} + 4\vec{j} + 5\vec{k}, \vec{b} = 5\vec{i} + 4\vec{j} + 3\vec{k}, \vec{c} = \vec{i} + \vec{j} + \vec{k}$.

16. $\vec{a} = -\vec{i} + \vec{j} + \vec{k}, \vec{b} = \vec{i} - \vec{j} + \vec{k}, \vec{c} = \vec{i} + \vec{j} - \vec{k}$.

17. $\vec{a} = -\vec{i} + 8\vec{j} + 7\vec{k}, \vec{b} = 2\vec{j} + 9\vec{k}, \vec{c} = 3\vec{k}$.

18. $\vec{a} = \vec{i} + \vec{j} + 2\vec{k}, \vec{b} = \vec{i} + \vec{k}, \vec{c} = \vec{j} + \vec{k}$.

19. For $\vec{a} = 3\vec{i} + \vec{j} - \vec{k}$ and $\vec{b} = \vec{i} - 4\vec{j} + 2\vec{k}$, find $\vec{a} \times \vec{b}$ and check that it is perpendicular to both $\vec{a}$ and $\vec{b}$.

20. If $\vec{v} = 3\vec{i} - 2\vec{j} + 4\vec{k}$ and $\vec{w} = \vec{i} + 2\vec{j} - \vec{k}$, find $\vec{v} \times \vec{w}$ and $\vec{w} \times \vec{v}$. What is the relation between the two answers?

Problems

21. Find a vector parallel to the line of intersection of the planes given by the equations $2x - 3y + 5z = 2$ and $4x + y - 3z = 7$.

22. Find the equation of the plane through the origin that is perpendicular to the line of intersection of the planes in Problem 21.

23. Find the equation of the plane through the point $(4, 5, 6)$ and perpendicular to the line of intersection of the planes in Problem 21.

24. Find an equation for the plane through the origin containing the points $(1, 3, 0)$ and $(2, 4, 1)$.

25. Find a vector parallel to the line of intersection of the two planes $4x - 3y + 2z = 12$ and $x + 5y - z = 25$.

26. Find a vector parallel to the intersection of the planes $2x - 3y + 5z = 2$ and $4x + y - 3z = 7$.

27. Find the equation of the plane through the origin that is perpendicular to the line of intersection of the planes in Problem 26.

28. Find the equation of the plane through the point $(4, 5, 6)$ that is perpendicular to the line of intersection of the planes in Problem 26.

29. Find the equation of a plane through the origin and perpendicular to $x - y + z = 5$ and $2x + y - 2z = 7$.

30. Let $P = (0, 1, 0), Q = (-1, 1, 2), R = (2, 1, -1)$. Find

   (a) The area of the triangle PQR.

   (b) The equation for a plane that contains P, Q, and R.

31. Let $A = (-1, 3, 0), B = (3, 2, 4)$, and $C = (1, -1, 5)$.

   (a) Find an equation for the plane that passes through these three points.

   (b) Find the area of the triangle determined by these three points.

32. If $\vec{v}$ and $\vec{w}$ are both parallel to the xy-plane, what can you conclude about $\vec{v} \times \vec{w}$? Explain.

33. Suppose $\vec{v} \cdot \vec{w} = 5$ and $\|\vec{v} \times \vec{w}\| = 3$, and the angle between $\vec{v}$ and $\vec{w}$ is θ. Find

   (a) $\tan \theta$ **(b)** θ.

34. If $\vec{v} \times \vec{w} = 2\vec{i} - 3\vec{j} + 5\vec{k}$, and $\vec{v} \cdot \vec{w} = 3$, find $\tan \theta$ where θ is the angle between $\vec{v}$ and $\vec{w}$.

35. Suppose $\vec{v} \cdot \vec{w} = 8$ and $\vec{v} \times \vec{w} = 12\vec{i} - 3\vec{j} + 4\vec{k}$ and that the angle between $\vec{v}$ and $\vec{w}$ is θ. Find

   (a) $\tan \theta$ **(b)** θ

36. Why does a baseball curve? The baseball in Figure 13.42 has velocity $\vec{v}$ meters/sec and is spinning at ω radians per second about an axis in the direction of the unit vector $\vec{n}$. The ball experiences a force, called the Magnus force,[2] $\vec{F}_M$, that is proportional to $\omega \vec{n} \times \vec{v}$.

---

[2]Named after German physicist Heinrich Magnus, who first described it in 1853.

(a) What is the effect on $\vec{F}_M$ of increasing ω?

(b) The ball in Figure 13.42 is moving away from you. What is the direction of the Magnus force?

Figure 13.42: Spinning baseball

37. The point P in Figure 13.43 has position vector $\vec{v}$ obtained by rotating the position vector $\vec{r}$ of the point (x, y) by $90°$ counterclockwise about the origin.

(a) Use the geometric definition of the cross product to explain why $\vec{v} = \vec{k} \times \vec{r}$.
(b) Find the coordinates of P.

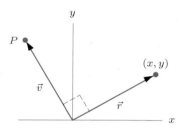

Figure 13.43

38. The points $P_1 = (0, 0, 0)$, $P_2 = (2, 4, 2)$, $P_3 = (3, 0, 0)$, and $P_4 = (5, 4, 2)$ are vertices of a parallelogram.

(a) Find the displacement vectors along each of the four sides. Check that these are equal in pairs.
(b) Find the area of the parallelogram.

39. Using the parallelogram in Problem 38 as a base, create a parallelopiped with side $\overrightarrow{P_1P_5}$ where $P_5 = (1, 0, 4)$. Find the volume of this parallelepiped.

40. Use the algebraic definition to check that

$$\vec{a} \times (\vec{b} + \vec{c}) = (\vec{a} \times \vec{b}) + (\vec{a} \times \vec{c}).$$

41. If $\vec{v}$ and $\vec{w}$ are non-zero vectors, use the geometric definition of the cross product to explain why

$$(\lambda \vec{v}) \times \vec{w} = \lambda(\vec{v} \times \vec{w}) = \vec{v} \times (\lambda \vec{w}).$$

Consider the cases $\lambda > 0$, and $\lambda = 0$, and $\lambda < 0$ separately.

42. Use a parallelepiped to show that $\vec{a} \cdot (\vec{b} \times \vec{c}) = (\vec{a} \times \vec{b}) \cdot \vec{c}$ for any vectors $\vec{a}$, $\vec{b}$, and $\vec{c}$.

43. Show that $\|\vec{a} \times \vec{b}\|^2 = \|\vec{a}\|^2 \|\vec{b}\|^2 - (\vec{a} \cdot \vec{b})^2$.

44. If $\vec{a} + \vec{b} + \vec{c} = \vec{0}$, show that

$$\vec{a} \times \vec{b} = \vec{b} \times \vec{c} = \vec{c} \times \vec{a}.$$

Geometrically, what does this imply about $\vec{a}$, $\vec{b}$, and $\vec{c}$?

45. Consider the triangle ABC in Figure 13.44.

(a) Use the cross-product to show that the area of ABC is $hc/2$.
(b) If A, B, and C all lie on the unit circle, what must be true about A, B, and C so that the area of ABC is as large as possible?

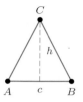

Figure 13.44

46. If $\vec{a} = a_1\vec{i} + a_2\vec{j} + a_3\vec{k}$, $\vec{b} = b_1\vec{i} + b_2\vec{j} + b_3\vec{k}$ and $\vec{c} = c_1\vec{i} + c_2\vec{j} + c_3\vec{k}$ are any three vectors in space, show that

$$\vec{a} \cdot (\vec{b} \times \vec{c}) = \begin{vmatrix} a_1 & a_2 & a_3 \\ b_1 & b_2 & b_3 \\ c_1 & c_2 & c_3 \end{vmatrix}.$$

47. Use the fact that $\vec{i} \times \vec{i} = \vec{0}$, $\vec{i} \times \vec{j} = \vec{k}$, $\vec{i} \times \vec{k} = -\vec{j}$, and so on, together with the properties on page 764 to derive the algebraic definition for the cross product.

48. In this problem, we arrive at the algebraic definition for the cross product by a different route. Let $\vec{a} = a_1\vec{i} + a_2\vec{j} + a_3\vec{k}$ and $\vec{b} = b_1\vec{i} + b_2\vec{j} + b_3\vec{k}$. We seek a vector $\vec{v} = x\vec{i} + y\vec{j} + z\vec{k}$ that is perpendicular to both $\vec{a}$ and $\vec{b}$. Use this requirement to construct two equations for x, y, and z. Eliminate x and solve for y in terms of z. Then eliminate y and solve for x in terms of z. Since z can be any value whatsoever (the direction of $\vec{v}$ is unaffected), select the value for z which eliminates the denominator in the equation you obtained. How does the resulting expression for $\vec{v}$ compare to the formula we derived on page 763?

49. For vectors $\vec{a}$ and $\vec{b}$, let $\vec{c} = \vec{a} \times (\vec{b} \times \vec{a})$.

(a) Show that $\vec{c}$ lies in the plane containing $\vec{a}$ and $\vec{b}$.
(b) Use Problems 42 and 43 to show that $\vec{a} \cdot \vec{c} = 0$ and $\vec{b} \cdot \vec{c} = \|\vec{a}\|^2 \|\vec{b}\|^2 - (\vec{a} \cdot \vec{b})^2$.
(c) Show that

$$\vec{a} \times (\vec{b} \times \vec{a}) = \|\vec{a}\|^2 \vec{b} - (\vec{a} \cdot \vec{b})\vec{a}.$$

50. Use the result of Problem 42 to show that the cross product distributes over addition. First, use distributivity for the dot product to show that for any vector $\vec{d}$,

$$[(\vec{a} + \vec{b}) \times \vec{c}] \cdot \vec{d} = [(\vec{a} \times \vec{c}) + (\vec{b} \times \vec{c})] \cdot \vec{d}.$$

Next, show that for any vector $\vec{d}$,

$$[((\vec{a} + \vec{b}) \times \vec{c}) - (\vec{a} \times \vec{c}) - (\vec{b} \times \vec{c})] \cdot \vec{d} = 0.$$

Finally, explain why you can conclude that

$$(\vec{a} + \vec{b}) \times \vec{c} = (\vec{a} \times \vec{c}) + (\vec{b} \times \vec{c}).$$

51. Figure 13.45 shows the tetrahedron determined by three vectors $\vec{a}, \vec{b}, \vec{c}$. The *area vector* of a face is a vector perpendicular to the face, pointing outward, whose magnitude is the area of the face. Show that the sum of the four outward pointing area vectors of the faces equals the zero vector.

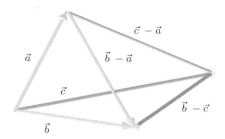

Figure 13.45

In Problems 52–54, find the vector representing the area of a surface. The magnitude of the vector equals the magnitude of the area; the direction is perpendicular to the surface. Since there are two perpendicular directions, we pick one by giving an orientation for the surface.

52. The rectangle with vertices $(0,0,0)$, $(0,1,0)$, $(2,1,0)$, and $(2,0,0)$, oriented so that it faces downward.

53. The circle of radius 2 in the yz-plane, facing in the direction of the positive x-axis.

54. The triangle ABC, oriented upward, where $A = (1,2,3)$, $B = (3,1,2)$, and $C = (2,1,3)$.

55. This problem relates the area of a parallelogram S lying in the plane $z = mx + ny + c$ to the area of its projection R in the xy-plane. Let S be determined by the vectors $\vec{u} = u_1\vec{i} + u_2\vec{j} + u_3\vec{k}$ and $\vec{v} = v_1\vec{i} + v_2\vec{j} + v_3\vec{k}$. See Figure 13.46.

  (a) Find the area of S.
  (b) Find the area of R.
  (c) Find m and n in terms of the components of $\vec{u}$ and $\vec{v}$.
  (d) Show that

$$\text{Area of } S = \sqrt{1 + m^2 + n^2} \cdot \text{Area of } R.$$

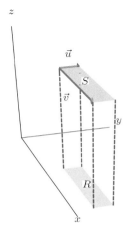

Figure 13.46

Strengthen Your Understanding

In Problems 56–57, explain what is wrong with the statement.

56. There is only one unit vector perpendicular to two nonparallel vectors in 3-space.

57. $\vec{u} \times \vec{v} = \vec{0}$ when $\vec{u}$ and $\vec{v}$ are perpendicular.

In Problems 58–59, give an example of:

58. A vector $\vec{u}$ whose cross product with $\vec{v} = \vec{i} + \vec{j}$ is parallel to $\vec{k}$.

59. A vector $\vec{v}$ such that $\|\vec{u} \times \vec{v}\| = 10$, where $\vec{u} = 3\vec{i} + 4\vec{j}$.

Are the statements in Problems 60–69 true or false? Give reasons for your answer.

60. $\vec{u} \times \vec{v}$ is a vector.

61. $\vec{u} \times \vec{v}$ has direction parallel to both $\vec{u}$ and $\vec{v}$.

62. $\|\vec{u} \times \vec{v}\| = \|\vec{u}\|\|\vec{v}\|$.

63. $(\vec{i} \times \vec{j}) \cdot \vec{k} = \vec{i} \cdot (\vec{j} \times \vec{k})$.

64. If $\vec{v}$ is a non-zero vector and $\vec{v} \times \vec{u} = \vec{v} \times \vec{w}$, then $\vec{u} = \vec{w}$.

65. The value of $\vec{v} \cdot (\vec{v} \times \vec{w})$ is always 0.

66. The value of $\vec{v} \times \vec{w}$ is never the same as $\vec{v} \cdot \vec{w}$.

67. The area of the triangle with two sides given by $\vec{i} + \vec{j}$ and $\vec{j} + 2\vec{k}$ is 3/2.

68. Given a non-zero vector $\vec{v}$ in 3-space, there is a non-zero vector $\vec{w}$ such that $\vec{v} \times \vec{w} = \vec{0}$.

69. It is never true that $\vec{v} \times \vec{w} = \vec{w} \times \vec{v}$.

CHAPTER SUMMARY (see also Ready Reference at the end of the book)

- **Vectors**
  Geometric definition of vector addition, subtraction and scalar multiplication, resolving into $\vec{i}, \vec{j}$, and $\vec{k}$ components, magnitude of a vector, algebraic properties of addition and scalar multiplication.

- **Dot Product**
  Geometric and algebraic definition, algebraic properties, using dot products to find angles and determine perpen-

dicularity, the equation of a plane with given normal vector passing through a given point, projection of a vector in a direction given by a unit vector.

- **Cross Product**
  Geometric and algebraic definition, algebraic properties, cross product and volume, finding the equation of a plane through three points.

REVIEW EXERCISES AND PROBLEMS FOR CHAPTER THIRTEEN

Exercises

In Exercises 1–2, is the quantity a vector or a scalar? Compute it.

1. $\vec{u} \cdot \vec{v}$, where $\vec{u} = 2\vec{i} - 3\vec{j} - 4\vec{k}$ and $\vec{v} = \vec{k} - \vec{j}$

2. $\vec{u} \times \vec{v}$, where $\vec{u} = 2\vec{i} - 3\vec{j} - 4\vec{k}$ and $\vec{v} = 3\vec{i} - \vec{j} + \vec{k}$.

In Exercises 3–4, calculate the quantity.

3. $(2\vec{i} - 3\vec{j} + 4\vec{k}) \cdot (2\vec{i} + 3\vec{j} + \vec{k})$

4. $\vec{i} \cdot (\vec{k} \times \vec{j})$

5. Resolve the vectors in Figure 13.47 into components.

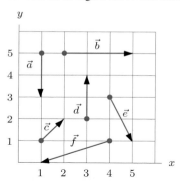

Figure 13.47

6. Resolve vector $\vec{v}$ into components if $\|\vec{v}\| = 8$ and the direction of $\vec{v}$ is shown in Figure 13.48.

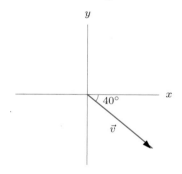

Figure 13.48

For Exercises 7–9, perform the indicated operations on the following vectors:

$$\vec{c} = \vec{i} + 6\vec{j}, \quad \vec{x} = -2\vec{i} + 9\vec{j}, \quad \vec{y} = 4\vec{i} - 7\vec{j}.$$

7. $5\vec{c}$ **8.** $\vec{c} + \vec{x} + \vec{y}$ **9.** $\|\vec{x} - \vec{c}\|$

In Exercises 10–19, use $\vec{v} = 2\vec{i} + 3\vec{j} - \vec{k}$ and $\vec{w} = \vec{i} - \vec{j} + 2\vec{k}$ to calculate the given quantities.

10. $\vec{v} + 2\vec{w}$ **11.** $3\vec{v} - \vec{w} - \vec{v}$

12. $\|\vec{v} + \vec{w}\|$ **13.** $\vec{v} \cdot \vec{w}$

14. $\vec{v} \times \vec{w}$ **15.** $\vec{v} \times \vec{v}$

16. $(\vec{v} \cdot \vec{w})\vec{v}$ **17.** $(\vec{v} \times \vec{w}) \cdot \vec{w}$

18. $(\vec{v} \times \vec{w}) \times \vec{w}$ **19.** $(\vec{v} \times \vec{w}) \times (\vec{v} \times \vec{w})$

20. Find the equation of the plane through the origin which is parallel to $z = 4x - 3y + 8$.

21. Let $\vec{v} = 3\vec{i} + 2\vec{j} - 2\vec{k}$ and $\vec{w} = 4\vec{i} - 3\vec{j} + \vec{k}$. Find each of the following:

    (a) $\vec{v} \cdot \vec{w}$
    (b) $\vec{v} \times \vec{w}$
    (c) A vector of length 5 parallel to vector $\vec{v}$
    (d) The angle between vectors $\vec{v}$ and $\vec{w}$
    (e) The component of $\vec{v}$ in the direction of $\vec{w}$
    (f) A vector perpendicular to vector $\vec{v}$
    (g) A vector perpendicular to both vectors $\vec{v}$ and $\vec{w}$

In Exercises 22–28, find a vector with the given property.

22. Length 10, parallel to $2\vec{i} + 3\vec{j} - \vec{k}$.

23. Unit vector perpendicular to $\vec{i} + \vec{j}$ and $\vec{i} - \vec{j} - \vec{k}$

24. Unit vector in the xy-plane perpendicular to $3\vec{i} - 2\vec{j}$.

25. Normal to $4(x - 1) + 6(z + 3) = 12$.

26. Perpendicular to $x - y = 1 + z$.

27. The vector obtained from $4\vec{i} + 3\vec{j}$ by rotating it $90°$ counterclockwise.

28. A non-zero vector perpendicular to $\vec{v} = 3\vec{i} - \vec{j} + \vec{k}$ and $\vec{w} = \vec{i} - 2\vec{j} + \vec{k}$.

29. Which of the following vectors are parallel?

$$\vec{u} = 2\vec{i} + 4\vec{j} - 2\vec{k}, \quad \vec{p} = \vec{i} + \vec{j} + \vec{k},$$
$$\vec{v} = \vec{i} - \vec{j} + 3\vec{k}, \quad \vec{q} = 4\vec{i} - 4\vec{j} + 12\vec{k},$$
$$\vec{w} = -\vec{i} - 2\vec{j} + \vec{k}, \quad \vec{r} = \vec{i} - \vec{j} + \vec{k}.$$

In Exercises 30–35, find the parallel and perpendicular components of the force vector $\vec{F}$ in the direction of the displacement vector $\vec{d}$. Then find the work W done by $\vec{F}$ though the displacement $\vec{d}$.

30. $\vec{F} = 2\vec{i} + 4\vec{j}, \quad \vec{d} = \vec{i} + 2\vec{j}$

31. $\vec{F} = -2\vec{i} - 4\vec{j}, \quad \vec{d} = \vec{i} + 2\vec{j}$

32. $\vec{F} = 2\vec{i} + 4\vec{j}, \quad \vec{d} = 2\vec{i} - 1\vec{j}$

33. $\vec{F} = 2\vec{i} + 4\vec{j}, \quad \vec{d} = 3\vec{i} - 4\vec{j}$

34. $\vec{F} = 2\vec{i}, \quad \vec{d} = \vec{i} + \vec{j}$

35. $\vec{F} = 5\vec{i} + 2\vec{j}, \quad \vec{d} = 3\vec{j}$

36. Find the area of the triangle with vectors $\vec{a} = \vec{i} + 2\vec{j} - \vec{k}$ and $\vec{b} = 4\vec{i} - 2\vec{j} + \vec{k}$ as sides.

Problems

37. Figure 13.49 shows a rectangular box containing several vectors. Are the following statements true or false? Explain.

(a) $\vec{c} = \vec{f}$ (b) $\vec{a} = \vec{d}$ (c) $\vec{a} = -\vec{b}$

(d) $\vec{g} = \vec{f} + \vec{a}$ (e) $\vec{e} = \vec{a} - \vec{b}$ (f) $\vec{d} = \vec{g} - \vec{c}$

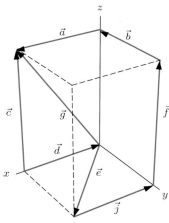

Figure 13.49

38. Shortly after takeoff, a plane is climbing northwest through still air at an airspeed of 200 km/hr, and rising at a rate of 300 m/min. Resolve its velocity vector into components. The x-axis points east, the y-axis points north, and the z-axis points up.

39. A boat is heading due east at 25 km/hr (relative to the water). The current is moving toward the southwest at 10 km/hr.

(a) Give the vector representing the actual movement of the boat.

(b) How fast is the boat going, relative to the ground?

(c) By what angle does the current push the boat off of its due east course?

40. A model rocket is shot into the air at an angle with the earth of about $60°$. The rocket is going fast initially but slows down as it reaches its highest point. It picks up speed again as it falls to earth.

(a) Sketch a graph showing the path of the rocket. Draw several velocity vectors on your graph.

(b) A second rocket has a parachute that deploys as it begins its descent. How do the velocity vectors from part (a) change for this rocket?

41. A car drives clockwise around the track in Figure 13.50, slowing down at the curves and speeding up along the straight portions. Sketch velocity vectors at the points P, Q, and R.

Figure 13.50

42. A racing car drives clockwise around the track shown in Figure 13.50 at a constant speed. At what point on the track does the car have the longest acceleration vector, and in roughly what direction is it pointing? (Recall that acceleration is the rate of change of velocity.)

43. Which pairs of the vectors $\sqrt{3}\vec{i} + \vec{j}, 3\vec{i} + \sqrt{3}\vec{j}, \vec{i} - \sqrt{3}\vec{j}$ are parallel and which are perpendicular?

44. An object is to be moved vertically upward by a crane. As the crane cannot get directly above the object, three ropes are attached to guide the object. One rope is pulled parallel to the ground with a force of 100 newtons in a direction $30°$ north of east. The second rope is pulled parallel to the ground with a force of 70 newtons in a direction $80°$ south of east. If the crane is attached to the third rope and can pull with a total force of 3000 newtons, find the force vector for the crane. What is the resulting (total) force on the object? (Assume vector $\vec{i}$ points east, vector $\vec{j}$ points north, and vector $\vec{k}$ points vertically up.)

45. What values of a make $\vec{v} = 2a\vec{i} - a\vec{j} + 16\vec{k}$ perpendicular to $\vec{w} = 5\vec{i} + a\vec{j} - \vec{k}$?

In Problems 46–47, find an equation of a plane that satisfies the given conditions.

46. Perpendicular to the vector $-\vec{i} + 2\vec{j} + \vec{k}$ and passing through the point $(1, 0, 2)$.

47. Perpendicular to the vector $2\vec{i} - 3\vec{j} + 7\vec{k}$ and passing through the point $(1, -1, 2)$.

48. Let $A = (0, 4)$, $B = (-1, -3)$, and $C = (-5, 1)$. Draw triangle ABC and find each of its interior angles.

49. Find the area of the triangle with vertices $P = (-2, 2, 0)$, $Q = (1, 3, -1)$, and $R = (-4, 2, 1)$.

50. A plane is drawn through the points $A = (2, 1, 0)$, $B = (0, 1, 3)$ and $C = (1, 0, 1)$. Find

(a) Two vectors lying in the plane.
(b) A vector perpendicular to the plane.
(c) The equation of the plane.

51. Given the points $P = (1, 2, 3)$, $Q = (3, 5, 7)$, and $R = (2, 5, 3)$, find:

(a) A unit vector perpendicular to a plane containing P, Q, R.
(b) The angle between PQ and PR.
(c) The area of the triangle PQR.
(d) The distance from R to the line through P and Q.

52. Find the distance from the point $P = (2, -1, 3)$ to the plane $2x + 4y - z = -1$.

53. Find an equation of the plane passing through the three points $(1, 1, 1)$, $(1, 4, 5)$, $(-3, -2, 0)$. Find the distance from the origin to the plane.

54. An airport is at the point $(200, 10, 0)$ and an approaching plane is at the point $(550, 60, 4)$. Assume that the xy-plane is horizontal, with the x-axis pointing eastward and the y-axis pointing northward. Also assume that the z-axis is upward and that all distances are measured in kilometers. The plane flies due west at a constant altitude at a speed of 500 km/hr for half an hour. It then descends at 200 km/hr, heading straight for the airport.

(a) Find the velocity vector of the plane while it is flying at constant altitude.
(b) Find the coordinates of the point at which the plane starts to descend.
(c) Find a vector representing the velocity of the plane when it is descending.

55. Find the vector $\vec{v}$ with all of the following properties:

• Magnitude 10
• Angle of $45°$ with positive x-axis
• Angle of $75°$ with positive y-axis
• Positive $\vec{k}$-component.

56. (a) A vector $\vec{v}$ of magnitude v makes an angle α with the positive x-axis, angle β with the positive y-axis, and angle γ with the positive z-axis. Show that

$$\vec{v} = v \cos \alpha \vec{i} + v \cos \beta \vec{j} + v \cos \gamma \vec{k}.$$

(b) Cos α, cos β, and cos γ are called *direction cosines*. Show that

$$\cos^2 \alpha + \cos^2 \beta + \cos^2 \gamma = 1.$$

57. Three people are trying to hold a ferocious lion still for the veterinarian. The lion, in the center, is wearing a collar with three ropes attached to it and each person has hold of a rope. Charlie is pulling in the direction $62°$ west of north with a force of 175 newtons and Sam is pulling in the direction $43°$ east of north with a force of 200 newtons. What is the direction and magnitude of the force that must be exerted by Alice on the third rope to counterbalance Sam and Charlie?

CAS Challenge Problems

58. Let $\vec{a} = x\vec{i} + y\vec{j} + z\vec{k}$, $\vec{b} = u\vec{i} + v\vec{j} + w\vec{k}$, and $\vec{c} = m\vec{a} + n\vec{b}$. Compute $(\vec{a} \times \vec{b}) \cdot \vec{c}$ and $(\vec{a} \times \vec{b}) \times (\vec{a} \times \vec{c})$, and explain the geometric meaning of your answers.

59. Let $\vec{a} = x\vec{i} + y\vec{j} + z\vec{k}$, $\vec{b} = u\vec{i} + v\vec{j} + w\vec{k}$ and $\vec{c} = r\vec{i} + s\vec{j} + t\vec{k}$. Show that the parallelepiped with edges $\vec{a}$, $\vec{b}$, $\vec{c}$ has the same volume as the parallelepiped with edges $\vec{a}$, $\vec{b}$, $2\vec{a} - \vec{b} + \vec{c}$. Explain this result geometrically.

60. Let $\vec{a} = \vec{i} + 2\vec{j} + 3\vec{k}$ and $\vec{b} = 2\vec{i} + \vec{j} + 2\vec{k}$, and let θ be the angle between $\vec{a}$ and $\vec{b}$.

(a) For $\vec{c} = x\vec{i} + y\vec{j} + z\vec{k}$, write the following condi-

tions as equations in x, y, z and solve them:

$$\vec{a} \cdot \vec{c} = 0, \quad \vec{b} \cdot \vec{c} = 0, \quad \|\vec{c}\|^2 = \|\vec{a}\|^2\|\vec{b}\|^2 \sin^2 \theta.$$

[Hint: Use the dot product to find $\sin^2 \theta$.]

(b) Compute the cross product $\vec{a} \times \vec{b}$ and compare with your answer in part (a). What do you notice? Explain.

61. Let $A = (0, 0, 0)$, $B = (2, 0, 0)$, $C = (1, \sqrt{3}, 0)$ and $D = (1, 1/\sqrt{3}, 2\sqrt{2/3})$.

(a) Show that A, B, C, D are all the same distance from each other.

(b) Find the point $P = (x, y, z)$ that is equidistant from A, B, C and D by setting up and solving three equations in x, y, and z.

(c) Use the dot product to find the angle APB. (In chemistry, this angle is often approximated by $109.5°$. A methane molecule can be represented by four hydrogen atoms at points A, B, C and D, and a carbon atom at P.)

62. Let $P = (x, y, z)$, $Q = (u, v, w)$ and $R = (r, s, t)$ be

points on the plane $ax + by + cz = d$.

(a) What is the relation between $\overrightarrow{PQ} \times \overrightarrow{PR}$ and the normal vector to the plane, $a\vec{i} + b\vec{j} + c\vec{k}$?

(b) Express $\overrightarrow{PQ} \times \overrightarrow{PR}$ in terms of $x, y, z, u, v, w, r, s, t$.

(c) Use the equation for the plane to eliminate z, w, and t from the expression you obtained in part (b), and simplify. Does your answer agree with what you said in part (a)?

PROJECTS FOR CHAPTER THIRTEEN

1. Cross Product of Vectors in the Plane

Let $\vec{a} = a_1\vec{i} + a_2\vec{j}$ and $\vec{b} = b_1\vec{i} + b_2\vec{j}$ be two nonparallel vectors in 2-space, as in Figure 13.51.

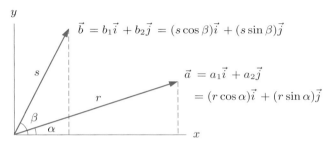

Figure 13.51

(a) Use the identity $\sin(\beta - \alpha) = (\sin \beta \cos \alpha - \cos \beta \sin \alpha)$ to derive the formula for the area of the parallelogram formed by $\vec{a}$ and $\vec{b}$:

$$\text{Area of parallelogram} = |a_1 b_2 - a_2 b_1|.$$

(b) Show that $a_1 b_2 - a_2 b_1$ is positive when the rotation from $\vec{a}$ to $\vec{b}$ is counterclockwise, and negative when it is clockwise.

(c) Use parts (a) and (b) to show that the geometric and algebraic definitions of $\vec{a} \times \vec{b}$ give the same result.

2. The Dot Product in Genetics[3]

We define[4] the angle between two n-dimensional vectors, $\vec{v}$ and $\vec{w}$, using the dot product:

$$\cos \theta = \frac{\vec{v} \cdot \vec{w}}{\|\vec{v}\|\|\vec{w}\|} = \frac{v_1 w_1 + v_2 w_2 + \cdots + v_n w_n}{\|\vec{v}\|\|\vec{w}\|}, \qquad \text{provided } \|\vec{v}\|, \|\vec{w}\| \neq 0.$$

We use this idea of angle to measure how close two populations are to one another genetically. The table shows the relative frequencies of four alleles (variants of a gene) in four populations.

Allele	Eskimo	Bantu	English	Korean
A_1	0.29	0.10	0.21	0.22
A_2	0.00	0.09	0.07	0.00
B	0.03	0.12	0.06	0.21
O	0.68	0.69	0.66	0.57

---

[3] Adapted from L. L. Cavalli-Sforza and A. W. F. Edwards, "Models and Estimation Procedures," Am. J. Hum. Genet., Vol. 19 (1967), pp. 223-57.

[4] The result of Problem 71 on page 761 shows that the quantity on the right-hand side of this equation is between -1 and 1, so this definition makes sense.

Let $\vec{a}_1$ be the 4-vector showing the square roots of the relative frequencies of the alleles in the Eskimo population. Let $\vec{a}_2, \vec{a}_3, \vec{a}_4$ be the corresponding vectors for the Bantu, English, and Korean populations, respectively. The genetic distance between two populations is defined as the angle between the corresponding vectors.

(a) Using this definition, is the English population closer genetically to the Bantus or to the Koreans? Explain.

(b) Is the English population closer to a half Eskimo, half Bantu population than to the Bantu population alone?

(c) Among all possible populations that are a mix of Eskimo and Bantu, find the mix that is closest to the English population.

3. A Warren Truss

A Warren truss is a structure for bearing a weight such as a roof or a bridge with two supports at either end of a gap. The truss in Figure 13.52 is loaded by weights at points D and E and is supported by vertical forces at points A and C. The horizontal bars in the truss are 10 ft long and the diagonal bars are 12 ft. Angles A and C are 65.38°.

Each bar exerts a force at the two joints at its ends. The two force vectors are parallel to the bar, equal in magnitude, and opposite in direction. If the bar pushes on the joints at its ends, then the bar is under compression, and if it pulls it is under tension, and the magnitude of the force is called the magnitude of the tension or compression.

Engineers need to know the magnitude of the compression or tension in each of the bars of the truss to prevent them from bending or breaking. To determine these magnitudes, we use the fact that at each joint the sum of the external forces from the weights and supports and the pushing and pulling forces exerted by the bars is zero. Find the magnitudes for all seven bars in this order:

(a) Joint A; Bars AB, AE

(b) Joint C; Bars BC, CD

(c) Joint D; Bars BD, DE

(d) Joint E; Bar BE

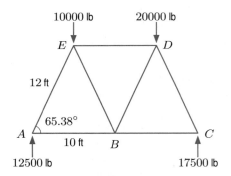

Figure 13.52

Chapter Fourteen

DIFFERENTIATING FUNCTIONS OF SEVERAL VARIABLES

Contents

14.1 THE PARTIAL DERIVATIVE

The derivative of a one-variable function measures its rate of change. In this section we see how a two-variable function has two rates of change: one as x changes (with y held constant) and one as y changes (with x held constant).

Rate of Change of Temperature in a Metal Rod: a One-Variable Problem

Imagine an unevenly heated metal rod lying along the x-axis, with its left end at the origin and x measured in metres. (See Figure 14.1.) Let $u(x)$ be the temperature (in °C) of the rod at the point x. Table 14.1 gives values of $u(x)$. We see that the temperature increases as we move along the rod, reaching its maximum at $x = 4$, after which it starts to decrease.

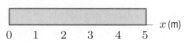

Figure 14.1: Unevenly heated metal rod

Table 14.1   *Temperature $u(x)$ of the rod*

x (m)	0	1	2	3	4	5
$u(x)$ (°C)	125	128	135	160	175	160

Example 1   Estimate the derivative $u'(2)$ using Table 14.1 and explain what the answer means in terms of temperature.

Solution   The derivative $u'(2)$ is defined as a limit of difference quotients:

$$u'(2) = \lim_{h \to 0} \frac{u(2+h) - u(2)}{h}.$$

Choosing $h = 1$ so that we can use the data in Table 14.1, we get

$$u'(2) \approx \frac{u(2+1) - u(2)}{1} = \frac{160 - 135}{1} = 25.$$

This means that the temperature increases at a rate of approximately 25°C per metre as we go from left to right, past $x = 2$.

Rate of Change of Temperature in a Metal Plate

Imagine an unevenly heated thin rectangular metal plate lying in the xy-plane with its lower left corner at the origin and x and y measured in metres. The temperature (in °C) at the point (x, y) is $T(x, y)$. See Figure 14.2 and Table 14.2. How does T vary near the point $(2, 1)$? We consider the horizontal line $y = 1$ containing the point $(2, 1)$. The temperature along this line is the cross section, $T(x, 1)$, of the function $T(x, y)$ with $y = 1$. Suppose we write $u(x) = T(x, 1)$.

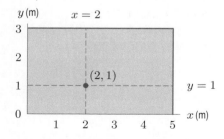

Figure 14.2: Unevenly heated metal plate

Table 14.2   *Temperature (°C) of a metal plate*

y (m)						
3	85	90	**110**	135	155	180
2	100	110	**120**	145	190	170
1	**125**	**128**	135	**160**	**175**	**160**
0	120	135	**155**	160	160	150
	0	1	2	3	4	5

x (m)

What is the meaning of the derivative $u'(2)$? It is the rate of change of temperature T *in the x-direction* at the point $(2, 1)$, keeping y fixed. Denote this rate of change by $T_x(2, 1)$, so that

$$T_x(2, 1) = u'(2) = \lim_{h \to 0} \frac{u(2+h) - u(2)}{h} = \lim_{h \to 0} \frac{T(2+h, 1) - T(2, 1)}{h}.$$

We call $T_x(2, 1)$ the *partial derivative of T with respect to x at the point* $(2, 1)$. Taking $h = 1$, we can read values of T from the row with $y = 1$ in Table 14.2, giving

$$T_x(2, 1) \approx \frac{T(3, 1) - T(2, 1)}{1} = \frac{160 - 135}{1} = 25°\text{C/m}.$$

The fact that $T_x(2, 1)$ is positive means that the temperature of the plate is increasing as we move past the point $(2, 1)$ in the direction of increasing x (that is, horizontally from left to right in Figure 14.2).

Example 2   Estimate the rate of change of T in the y-direction at the point $(2, 1)$.

Solution   The temperature along the line $x = 2$ is the cross-section of T with $x = 2$, that is, the function $v(y) = T(2, y)$. If we denote the rate of change of T in the y-direction at $(2, 1)$ by $T_y(2, 1)$, then

$$T_y(2, 1) = v'(1) = \lim_{h \to 0} \frac{v(1 + h) - v(1)}{h} = \lim_{h \to 0} \frac{T(2, 1 + h) - T(2, 1)}{h}.$$

We call $T_y(2, 1)$ the *partial derivative of T with respect to y at the point* $(2, 1)$. Taking $h = 1$ so that we can use the column with $x = 2$ in Table 14.2, we get

$$T_y(2, 1) \approx \frac{T(2, 1 + 1) - T(2, 1)}{1} = \frac{120 - 135}{1} = -15°\text{C/m}.$$

The fact that $T_y(2, 1)$ is negative means that the temperature decreases as y increases.

Definition of the Partial Derivative

We study the influence of x and y separately on the value of the function $f(x, y)$ by holding one fixed and letting the other vary. This leads to the following definitions.

Partial Derivatives of f With Respect to x and y

For all points at which the limits exist, we define the **partial derivatives at the point** (a, b) by

$$f_x(a, b) = \begin{array}{c} \text{Rate of change of } f \text{ with respect to } x \\ \text{at the point } (a, b) \end{array} = \lim_{h \to 0} \frac{f(a + h, b) - f(a, b)}{h},$$

$$f_y(a, b) = \begin{array}{c} \text{Rate of change of } f \text{ with respect to } y \\ \text{at the point } (a, b) \end{array} = \lim_{h \to 0} \frac{f(a, b + h) - f(a, b)}{h}.$$

If we let a and b vary, we have the **partial derivative functions** $f_x(x, y)$ and $f_y(x, y)$.

Just as with ordinary derivatives, there is an alternative notation:

Alternative Notation for Partial Derivatives

If $z = f(x, y)$, we can write

$$f_x(x, y) = \frac{\partial z}{\partial x} \quad \text{and} \quad f_y(x, y) = \frac{\partial z}{\partial y},$$

$$f_x(a, b) = \frac{\partial z}{\partial x}\bigg|_{(a,b)} \quad \text{and} \quad f_y(a, b) = \frac{\partial z}{\partial y}\bigg|_{(a,b)}.$$

We use the symbol ∂ to distinguish partial derivatives from ordinary derivatives. In cases where the independent variables have names different from x and y, we adjust the notation accordingly. For example, the partial derivatives of $f(u, v)$ are denoted by f_u and f_v.

Visualising Partial Derivatives on a Graph

The ordinary derivative of a one-variable function is the slope of its graph. How do we visualise the partial derivative $f_x(a, b)$? The graph of the one-variable function $f(x, b)$ is the curve where the vertical plane $y = b$ cuts the graph of $f(x, y)$. (See Figure 14.3.) Thus, $f_x(a, b)$ is the slope of the tangent line to this curve at $x = a$.

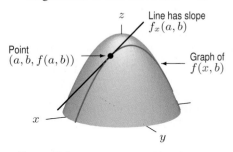

Figure 14.3: The curve $z = f(x, b)$ on the graph of f has slope $f_x(a, b)$ at $x = a$

Figure 14.4: The curve $z = f(a, y)$ on the graph of f has slope $f_y(a, b)$ at $y = b$

Similarly, the graph of the function $f(a, y)$ is the curve where the vertical plane $x = a$ cuts the graph of f, and the partial derivative $f_y(a, b)$ is the slope of this curve at $y = b$. (See Figure 14.4.)

Example 3 At each point labelled on the graph of the surface $z = f(x, y)$ in Figure 14.5, say whether each partial derivative is positive or negative.

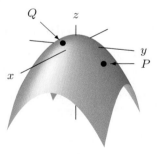

Figure 14.5: Decide the signs of f_x and f_y at P and Q

Solution The positive x-axis points out of the page. Imagine heading off in this direction from the point marked P; we descend steeply. So the partial derivative with respect to x is negative at P, with quite a large absolute value. The same is true for the partial derivative with respect to y at P, since there is also a steep descent in the positive y-direction.

At the point marked Q, heading in the positive x-direction results in a gentle descent, whereas heading in the positive y-direction results in a gentle ascent. Thus, the partial derivative f_x at Q is negative but small (that is, near zero), and the partial derivative f_y is positive but small.

Estimating Partial Derivatives from a Contour Diagram

The graph of a function $f(x, y)$ often makes clear the sign of the partial derivatives. However, numerical estimates of these derivatives are more easily made from a contour diagram than a surface graph. If we move parallel to one of the axes on a contour diagram, the partial derivative is the rate of change of the value of the function on the contours. For example, if the values on the contours are increasing as we move in the positive direction, then the partial derivative must be positive.

Example 4   Figure 14.6 shows the contour diagram for the temperature $H(x, t)$ (in °C) in a room as a function of distance x (in metres) from a heater and time t (in minutes) after the heater has been turned on. What are the signs of $H_x(10, 20)$ and $H_t(10, 20)$? Estimate these partial derivatives and explain the answers in practical terms.

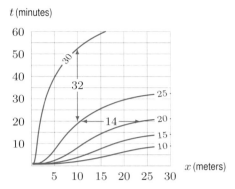

Figure 14.6: Temperature in a heated room: Heater at $x = 0$ is turned on at $t = 0$

Solution   The point $(10, 20)$ is nearly on the $H = 25$ contour. As x increases, we move toward the $H = 20$ contour, so H is decreasing and $H_x(10, 20)$ is negative. This makes sense because the $H = 30$ contour is to the left: As we move further from the heater, the temperature drops. On the other hand, as t increases, we move toward the $H = 30$ contour, so H is increasing; as t decreases H decreases. Thus, $H_t(10, 20)$ is positive. This says that as time passes, the room warms up.

To estimate the partial derivatives, use a difference quotient. Looking at the contour diagram, we see there is a point on the $H = 20$ contour about 14 units to the right of the point $(10, 20)$. Hence, H decreases by 5 when x increases by 14, so we find

$$\text{Rate of change of } H \text{ with respect to } x = H_x(10, 20) \approx \frac{-5}{14} \approx -0.36°\text{C/metre.}$$

This means that near the point 10 m from the heater, after 20 minutes the temperature drops about 0.36, or one third, of a degree, for each metre we move away from the heater.

To estimate $H_t(10, 20)$, we notice that the $H = 30$ contour is about 32 units directly above the point $(10, 20)$. So H increases by 5 when t increases by 32. Hence,

$$\text{Rate of change of } H \text{ with respect to } t = H_t(10, 20) \approx \frac{5}{32} = 0.16°\text{C/minute.}$$

This means that after 20 minutes the temperature is going up about 0.16, or 1/6, of a degree each minute at the point 10 m from the heater.

Using Units to Interpret Partial Derivatives

The meaning of a partial derivative can often be explained using units.

Example 5   Suppose that your weight w in kilograms is a function $f(c, n)$ of the number c of calories you consume daily and the number n of minutes you exercise daily. Using the units for w, c and n, interpret in everyday terms the statements

$$\frac{\partial w}{\partial c}(2000, 15) = 0.01 \quad \text{and} \quad \frac{\partial w}{\partial n}(2000, 15) = -0.012.$$

Solution   The units of $\partial w / \partial c$ are kilograms per calorie. The statement

$$\frac{\partial w}{\partial c}(2000, 15) = 0.01$$

means that if you are presently consuming 2000 calories daily and exercising 15 minutes daily, you will weigh 0.01 kilograms more for each extra calorie you consume daily, or about 1 kilogram for each extra 100 calories per day. The units of $\partial w/\partial n$ are kilograms per minute. The statement

$$\frac{\partial w}{\partial n}(2000, 15) = -0.012$$

means that for the same calorie consumption and number of minutes of exercise, you will weigh 0.012 kilograms less for each extra minute you exercise daily, or about 1 kilogram less for each extra 80 minutes per day. So if you eat an extra 100 calories each day and exercise about 80 minutes more each day, your weight should remain roughly steady.

Exercises and Problems for Section 14.1

Exercises

1. Given the following table of values for $z = f(x, y)$, estimate $f_x(3, 2)$ and $f_y(3, 2)$. Assume that f is differentiable.

$x \backslash y$	0	2	5
1	1	2	4
3	−1	1	2
6	−3	0	0

2. Using difference quotients, estimate $f_x(3, 2)$ and $f_y(3, 2)$ for the function given by

$$f(x, y) = \frac{x^2}{y + 1}.$$

[Recall: A difference quotient is an expression of the form $(f(a + h, b) - f(a, b))/h$.]

3. Use difference quotients with $\Delta x = 0.1$ and $\Delta y = 0.1$ to estimate $f_x(1, 3)$ and $f_y(1, 3)$ where

$$f(x, y) = e^{-x}y^2.$$

Then give better estimates by using $\Delta x = 0.01$ and $\Delta y = 0.01$.

4. Use difference quotients with $\Delta x = 0.1$ and $\Delta y = 0.1$ to estimate $f_x(1, 3)$ and $f_y(1, 3)$ where

$$f(x, y) = e^{-x} \sin y.$$

Then give better estimates by using $\Delta x = 0.01$ and $\Delta y = 0.01$.

5. The price P in dollars to purchase a used car is a function of its original cost, C, in dollars, and its age, A, in years.

  (a) What are the units of $\partial P/\partial A$?
  (b) What is the sign of $\partial P/\partial A$ and why?
  (c) What are the units of $\partial P/\partial C$?
  (d) What is the sign of $\partial P/\partial C$ and why?

6. You borrow $\$A$ at an interest rate of $r\%$ (per month) and pay it off over t months by making monthly payments of $P = g(A, r, t)$ dollars. In financial terms, what do the following statements tell you?

  (a) $g(8000, 1, 24) = 376.59$

(b) $\left. \dfrac{\partial g}{\partial A} \right|_{(8000,1,24)} = 0.047$

(c) $\left. \dfrac{\partial g}{\partial r} \right|_{(8000,1,24)} = 44.83$

7. Your monthly car payment in dollars is $P = f(P_0, t, r)$, where $\$P_0$ is the amount you borrowed, t is the number of months it takes to pay off the loan, and $r\%$ is the interest rate. What are the units, the financial meaning, and the signs of $\partial P/\partial t$ and $\partial P/\partial r$?

8. A drug is injected into a patient's blood vessel. The function $c = f(x, t)$ represents the concentration of the drug at a distance x mm in the direction of the blood flow measured from the point of injection and at time t seconds since the injection. What are the units of the following partial derivatives? What are their practical interpretations? What do you expect their signs to be?

  (a) $\partial c/\partial x$ **(b)** $\partial c/\partial t$

9. The sales of a product, $S = f(p, a)$, are a function of the price, p, of the product (in dollars per unit) and the amount, a, spent on advertising (in thousands of dollars).

  (a) Do you expect f_p to be positive or negative? Why?
  (b) Explain the meaning of the statement $f_a(8, 12) = 150$ in terms of sales.

10. The quantity, Q, of beef purchased at a store, in kilograms per week, is a function of the price of beef, b, and the price of chicken, c, both in dollars per kilogram.

  (a) Do you expect $\partial Q/\partial b$ to be positive or negative? Explain.
  (b) Do you expect $\partial Q/\partial c$ to be positive or negative? Explain.
  (c) Interpret the statement $\partial Q/\partial b = -213$ in terms of quantity of beef purchased.

In Exercises 11–14, determine the sign of f_x and f_y at the point using the contour diagram of f in Figure 14.7.

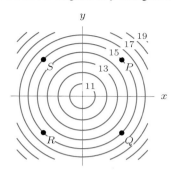

Figure 14.7

11. P **12.** Q **13.** R **14.** S

For Exercises 15–17, refer to Table 12.2 on page 690 giving the temperature adjusted for wind chill, C, in $°C$, as a function $f(w, T)$ of the wind speed, w, in km/hr, and the temperature, T, in $°C$. The temperature adjusted for wind chill tells you how cold it feels, as a result of the combination of wind and temperature.

15. Estimate $f_w(16, -4)$. What does your answer mean in practical terms?

16. Estimate $f_T(8, -7)$. What does your answer mean in practical terms?

17. From Table 12.2 you can see that when the temperature is $-10°C$, the temperature adjusted for wind-chill drops by an average of about $0.25°C$ with every 1 km/hr increase in wind speed from 8 km/hr to 16 km/hr. Which partial derivative is this telling you about?

Problems

18. When riding your bike in winter, the windchill temperature is a measure of how cold you feel as a result of the induced breeze caused by your travel. If W represents windchill temperature (in $°C$) that you experience, then $W = f(T, v)$, where T is the actual air temperature (in $°C$) and v is your speed, in meters per second. Match each of the practical interpretations below with a mathematical statement that most accurately describes it below. For the remaining mathematical statement, give a practical interpretation.

   (i) "The faster you ride, the colder you'll feel."
   (ii) "The warmer the day, the warmer you'll feel."

   (a) $f_T(T, v) > 0$
   (b) $f(0, v) \leq 0$
   (c) $f_v(T, v) < 0$

19. Approximate $f_x(3, 5)$ using the contour diagram of $f(x, y)$ in Figure 14.8.

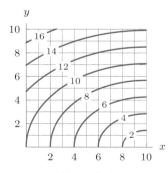

Figure 14.8

20. Figure 14.9 is a contour diagram for $z = f(x, y)$. Is f_x positive or negative? Is f_y positive or negative? Estimate $f(2, 1)$, $f_x(2, 1)$, and $f_y(2, 1)$.

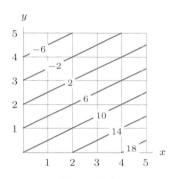

Figure 14.9

21. The quantity Q (in kg) of beef that a certain community buys during a week is a function $Q = f(b, c)$ of the prices of beef, b, and chicken, c, during the week. Do you expect $\partial Q / \partial b$ to be positive or negative? What about $\partial Q / \partial c$?

22. The average price of new cars is x and the average price of a litre of petrol is y. The number, q_1, of new cars bought in a year, depends on both x and y, so $q_1 = f(x, y)$. Similarly, if q_2 is the number of litres of petrol bought to fill new cars in a year, then $q_2 = g(x, y)$.

   (a) What do you expect the signs of $\partial q_1 / \partial x$ and $\partial q_2 / \partial y$ to be? Explain.
   (b) What do you expect the signs of $\partial q_1 / \partial y$ and $\partial q_2 / \partial x$ to be? Explain.

23. An experiment to measure the toxicity of formaldehyde yielded the data in Table 14.3. The values show the percent, $P = f(t, c)$, of rats surviving an exposure to formaldehyde at a concentration of c (in parts per million, ppm) after t months. Estimate $f_t(18, 6)$ and $f_c(18, 6)$. Interpret your answers in terms of formaldehyde toxicity.

Table 14.3

	Time t (months)					
		14	16	18	20	22	24
	0	100	100	100	99	97	95
Conc. c (ppm)	2	100	99	98	97	95	92
	6	96	95	93	90	86	80
	15	96	93	82	70	58	36

24. Figure 14.10 shows contours of $f(x, y)$ with values of f on the contours omitted. If $f_x(P) > 0$, find the sign of

(a) $f_y(P)$ **(b)** $f_y(Q)$ **(c)** $f_x(Q)$

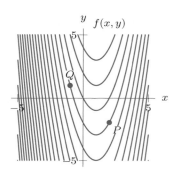

Figure 14.10

25. Figure 14.11 shows the contour diagram of $g(x, y)$. Mark the points on the contours where

(a) $g_x = 0$ **(b)** $g_y = 0$

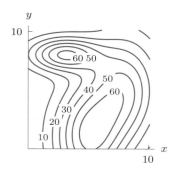

Figure 14.11

26. The surface $z = f(x, y)$ is shown in Figure 14.12. The points A and B are in the xy-plane.

(a) What is the sign of
   (i) $f_x(A)$?   (ii) $f_y(A)$?

(b) The point P in the xy-plane moves along a straight line from A to B. How does the sign of $f_x(P)$ change? How does the sign of $f_y(P)$ change?

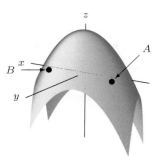

Figure 14.12

27. Figure 14.13 shows the saddle-shaped surface $z = f(x, y)$.

(a) What is the sign of $f_x(0, 5)$?
(b) What is the sign of $f_y(0, 5)$?

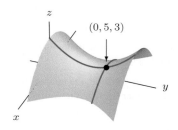

Figure 14.13

28. Figure 14.14 shows the graph of the function $f(x, y)$ on the domain $0 \le x \le 4$ and $0 \le y \le 4$. Use the graph to rank the following quantities in order from smallest to largest: $f_x(3, 2)$, $f_x(1, 2)$, $f_y(3, 2)$, $f_y(1, 2)$, 0.

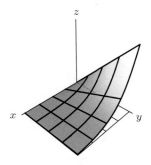

Figure 14.14

29. Figure 14.15 shows a contour diagram for the monthly payment P as a function of the interest rate, $r\%$, and the amount, L, of a 5-year loan. Estimate $\partial P/\partial r$ and $\partial P/\partial L$ at the following points. In each case, give the units and the everyday meaning of your answer.

(a) $r = 8, L = 4000$ **(b)** $r = 8, L = 6000$
(c) $r = 13, L = 7000$

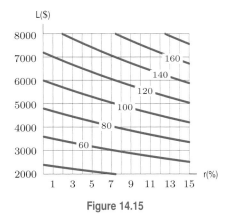

Figure 14.15

30. People commuting to a city can choose to go either by bus or by train. The number of people who choose either method depends in part upon the price of each. Let $f(P_1, P_2)$ be the number of people who take the bus when P_1 is the price of a bus ride and P_2 is the price of a train ride. What can you say about the signs of $\partial f/\partial P_1$ and $\partial f/\partial P_2$? Explain your answers.

31. Figure 14.16 shows a contour diagram for the temperature T (in °C) along a wall in a heated room as a function of distance x along the wall and time t in minutes. Estimate $\partial T/\partial x$ and $\partial T/\partial t$ at the given points. Give units and interpret your answers.

(a) $x = 15, t = 20$ **(b)** $x = 5, t = 12$

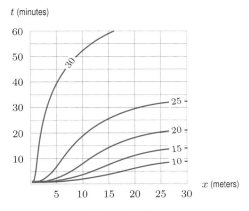

Figure 14.16

An airport can be cleared of fog by heating the air. The amount of heat required depends on the air temperature and the wetness of the fog. Problems 32–34 involve Figure 14.17, which shows the heat $H(T, w)$ required (in calories per cubic meter of fog) as a function of the temperature T (in degrees Celsius) and the water content w (in grams per cubic meter of fog). Note that Figure 14.17 is not a contour diagram, but shows cross-sections of H with w fixed at 0.1, 0.2, 0.3, and 0.4.

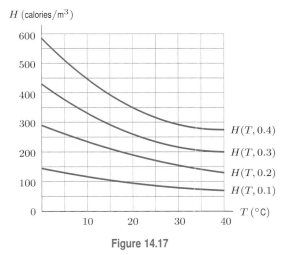

Figure 14.17

32. Use Figure 14.17 to estimate $H_T(10, 0.1)$. Interpret the partial derivative in practical terms.

33. Make a table of values for $H(T, w)$ from Figure 14.17, and use it to estimate $H_T(T, w)$ for $T = 10, 20,$ and 30 and $w = 0.1, 0.2,$ and 0.3.

34. Repeat Problem 33 for $H_w(T, w)$ at $T = 10, 20,$ and 30 and $w = 0.1, 0.2,$ and 0.3. What is the practical meaning of these partial derivatives?

35. The cardiac output, represented by c, is the volume of blood flowing through a person's heart per unit time. The systemic vascular resistance (SVR), represented by s, is the resistance to blood flowing through veins and arteries. Let p be a person's blood pressure. Then p is a function of c and s, so $p = f(c, s)$.

(a) What does $\partial p/\partial c$ represent?

Suppose now that $p = kcs$, where k is a constant.

(b) Sketch the level curves of p. What do they represent? Label your axes.
(c) For a person with a weak heart, it is desirable to have the heart pumping against less resistance, while maintaining the same blood pressure. Such a person may be given the drug nitroglycerine to decrease the SVR and the drug Dopamine to increase the cardiac output. Represent this on a graph showing level curves. Put a point A on the graph representing the person's state before drugs are given and a point B for after.

(d) Right after a heart attack, a patient's cardiac output drops, thereby causing the blood pressure to drop. A common mistake made by medical residents is to get the patient's blood pressure back to normal by using drugs to increase the SVR, rather than by increasing the cardiac output. On a graph of the level curves of p, put a point D representing the patient before the heart attack, a point E representing the patient right after the heart attack, and a third point F represent-

ing the patient after the resident has given the drugs to increase the SVR.

36. In each case, give a possible contour diagram for the function $f(x, y)$ if

(a) $f_x > 0$ and $f_y > 0$ **(b)** $f_x > 0$ and $f_y < 0$
(c) $f_x < 0$ and $f_y > 0$ **(d)** $f_x < 0$ and $f_y < 0$

Strengthen Your Understanding

In Problems 37–38, explain what is wrong with the statement.

37. For $f(x, y)$, $\partial f / \partial x$ has the same units as $\partial f / \partial y$.

38. The partial derivative with respect to y is not defined for functions such as $f(x, y) = x^2 + 5$ that have a formula that does not contain y explicitly.

In Problems 39–40, give an example of:

39. A table of values with three rows and three columns of a linear function $f(x, y)$ with $f_x < 0$ and $f_y > 0$.

40. A function $f(x, y)$ with $f_x > 0$ and $f_y < 0$ everywhere.

Are the statements in Problems 41–50 true or false? Give reasons for your answer.

41. If $f(x, y)$ is a function of two variables and $f_x(10, 20)$ is defined, then $f_x(10, 20)$ is a scalar.

42. If $f(x, y) = x^2 + y^2$, then $f_y(1, 1) < 0$.

43. If the graph of $f(x, y)$ is a hemisphere centered at the origin, then $f_x(0, 0) = f_y(0, 0) = 0$.

44. If $P = f(T, V)$ is a function expressing the pressure P (in grams/cm^3) of gas in a piston in terms of the temperature T (in degrees $°$C) and volume V (in cm^3), then $\partial P / \partial V$ has units of grams.

45. If $f_x(a, b) > 0$, then the values of f decrease as we move in the negative x-direction near (a, b).

46. If $g(r, s) = r^2 + s$, then for fixed s, the partial derivative g_r increases as r increases.

47. If $g(u, v) = (u + v)^u$, then $2.3 \leq g_u(1, 1) \leq 2.4$.

48. Let $P = f(m, d)$ be the purchase price (in dollars) of a used car that has m kilometers on its engine and originally cost d dollars when new. Then $\partial P / \partial m$ and $\partial P / \partial d$ have the same sign.

49. If $f(x, y)$ is a function with the property that $f_x(x, y)$ and $f_y(x, y)$ are both constant, then f is linear.

50. If $f(x, y)$ has $f_x(a, b) = f_y(a, b) = 0$ at the point (a, b), then f is constant everywhere.

14.2 COMPUTING PARTIAL DERIVATIVES ALGEBRAICALLY

Since the partial derivative $f_x(x, y)$ is the ordinary derivative of the function $f(x, y)$ with y held constant and $f_y(x, y)$ is the ordinary derivative of $f(x, y)$ with x held constant, we can use all the differentiation formulas from one-variable calculus to find partial derivatives.

Example 1   Let $f(x, y) = \dfrac{x^2}{y + 1}$. Find $f_x(3, 2)$ algebraically.

Solution   We use the fact that $f_x(3, 2)$ equals the derivative of $f(x, 2)$ at $x = 3$. Since

$$f(x, 2) = \frac{x^2}{2 + 1} = \frac{x^2}{3},$$

differentiating with respect to x, we have

$$f_x(x, 2) = \frac{\partial}{\partial x}\left(\frac{x^2}{3}\right) = \frac{2x}{3}, \qquad \text{and so} \qquad f_x(3, 2) = 2.$$

Example 2   Compute the partial derivatives with respect to x and with respect to y for the following functions.
(a) $f(x, y) = y^2 e^{3x}$ **(b)** $z = (3xy + 2x)^5$ **(c)** $g(x, y) = e^{x+3y} \sin(xy)$

Solution   (a) This is the product of a function of x (namely e^{3x}) and a function of y (namely y^2). When we differentiate with respect to x, we think of the function of y as a constant, and vice versa. Thus,

$$f_x(x, y) = y^2 \frac{\partial}{\partial x}\left(e^{3x}\right) = 3y^2 e^{3x},$$

$$f_y(x, y) = e^{3x} \frac{\partial}{\partial y}(y^2) = 2ye^{3x}.$$

(b) Here we use the chain rule:

$$\frac{\partial z}{\partial x} = 5(3xy + 2x)^4 \frac{\partial}{\partial x}(3xy + 2x) = 5(3xy + 2x)^4(3y + 2),$$

$$\frac{\partial z}{\partial y} = 5(3xy + 2x)^4 \frac{\partial}{\partial y}(3xy + 2x) = 5(3xy + 2x)^4 3x = 15x(3xy + 2x)^4.$$

(c) Since each function in the product is a function of both x and y, we need to use the product rule for each partial derivative:

$$g_x(x, y) = \left(\frac{\partial}{\partial x}(e^{x+3y})\right) \sin(xy) + e^{x+3y} \frac{\partial}{\partial x}(\sin(xy)) = e^{x+3y} \sin(xy) + e^{x+3y} y \cos(xy),$$

$$g_y(x, y) = \left(\frac{\partial}{\partial y}(e^{x+3y})\right) \sin(xy) + e^{x+3y} \frac{\partial}{\partial y}(\sin(xy)) = 3e^{x+3y} \sin(xy) + e^{x+3y} x \cos(xy).$$

For functions of three or more variables, we find partial derivatives by the same method: Differentiate with respect to one variable, regarding the other variables as constants. For a function $f(x, y, z)$, the partial derivative $f_x(a, b, c)$ gives the rate of change of f with respect to x along the line $y = b, z = c$.

Example 3   Find all the partial derivatives of $f(x, y, z) = \dfrac{x^2 y^3}{z}$.

Solution   To find $f_x(x, y, z)$, for example, we consider y and z as fixed, giving

$$f_x(x, y, z) = \frac{2xy^3}{z}, \quad \text{and} \quad f_y(x, y, z) = \frac{3x^2 y^2}{z}, \quad \text{and} \quad f_z(x, y, z) = -\frac{x^2 y^3}{z^2}.$$

Interpretation of Partial Derivatives

Example 4   A vibrating guitar string, originally at rest along the x-axis, is shown in Figure 14.18. Let x be the distance in metres from the left end of the string. At time t seconds the point x has been displaced $y = f(x, t)$ metres vertically from its rest position, where

$$y = f(x, t) = 0.003 \sin(\pi x) \sin(2765t).$$

Evaluate $f_x(0.3, 1)$ and $f_t(0.3, 1)$ and explain what each means in practical terms.

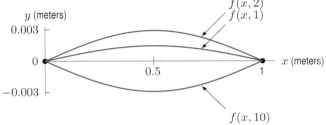

Figure 14.18: The position of a vibrating guitar string at several different times: Graph of $f(x, t)$ for $t = 1, 2, 10$.

Solution   Differentiating $f(x,t) = 0.003\sin(\pi x)\sin(2765t)$ with respect to x, we have

$$f_x(x,t) = 0.003\pi\cos(\pi x)\sin(2765t).$$

In particular, substituting $x = 0.3$ and $t = 1$ gives

$$f_x(0.3,1) = 0.003\pi\cos(\pi(0.3))\sin(2765) \approx 0.002.$$

To see what $f_x(0.3,1)$ means, think about the function $f(x,1)$. The graph of $f(x,1)$ in Figure 14.19 is a snapshot of the string at the time $t = 1$. Thus, the derivative $f_x(0.3,1)$ is the slope of the string at the point $x = 0.3$ at the instant when $t = 1$.

Similarly, taking the derivative of $f(x,t) = 0.003\sin(\pi x)\sin(2765t)$ with respect to t, we get

$$f_t(x,t) = (0.003)(2765)\sin(\pi x)\cos(2765t) = 8.3\sin(\pi x)\cos(2765t).$$

Since $f(x,t)$ is in metres and t is in seconds, the derivative $f_t(0.3,1)$ is in m/sec. Thus, substituting $x = 0.3$ and $t = 1$,

$$f_t(0.3,1) = 8.3\sin(\pi(0.3))\cos(2765(1)) \approx 6 \text{ m/sec}.$$

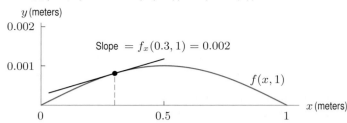

Figure 14.19: Graph of $f(x,1)$: Snapshot of the shape of the string at $t = 1$ sec

To see what $f_t(0.3,1)$ means, think about the function $f(0.3,t)$. The graph of $f(0.3,t)$ is a position versus time graph that tracks the up-and-down movement of the point on the string where $x = 0.3$. (See Figure 14.20.) The derivative $f_t(0.3,1) = 6$ m/sec is the velocity of that point on the string at time $t = 1$. The fact that $f_t(0.3,1)$ is positive indicates that the point is moving upward when $t = 1$.

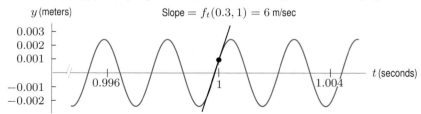

Figure 14.20: Graph of $f(0.3,t)$: Position versus time graph of the point $x = 0.3$ m from the end of the guitar string

Exercises and Problems for Section 14.2

Exercises

1. **(a)** If $f(x,y) = 2x^2 + xy + y^2$, approximate $f_y(3,2)$ using $\Delta y = 0.01$.
   (b) Find the exact value of $f_y(3,2)$.

2. **(a)** If $f(x,y) = x^3 + 2xy + e^x$, approximate $f_x(2,1)$ using $\Delta y = 0.01$.
   (b) Find the exact value of $f_y(1,2)$.

Find the partial derivatives in Exercises 3–42. Assume the variables are restricted to a domain on which the function is defined.

3. f_x and f_y if $f(x,y) = x^2y + x^3 - 7xy^6$

4. $f_x(1,2)$ and $f_y(1,2)$ if $f(x,y) = x^3 + 3x^2y - 2y^2$

5. $\dfrac{\partial z}{\partial x}$ and $\dfrac{\partial z}{\partial y}$ if $z = (x^2 + x - y)^7$

6. $\dfrac{\partial}{\partial y}(3x^5y^7 - 32x^4y^3 + 5xy)$

7. f_x and f_y if $f(x,y) = A^\alpha x^{\alpha+\beta}y^{1-\alpha-\beta}$

8. f_x and f_y if $f(x,y) = \ln(x^{0.6}y^{0.4})$

9. z_x if $z = \dfrac{1}{2x^2ay} + \dfrac{3x^5abc}{y}$

10. z_x if $z = x^2y + 2x^5y$ **11.** V_r if $V = \frac{1}{3}\pi r^2 h$

12. z_y if $z = \dfrac{3x^2y^7 - y^2}{15xy - 8}$ **13.** $\dfrac{\partial}{\partial T}\left(\dfrac{2\pi r}{T}\right)$

14. $\dfrac{\partial}{\partial x}(a\sqrt{x})$ **15.** $\dfrac{\partial}{\partial x}(xe^{\sqrt{xy}})$

16. $\dfrac{\partial}{\partial t}e^{\sin(x+ct)}$ **17.** F_m if $F = mg$

18. a_v if $a = \dfrac{v^2}{r}$ **19.** $\dfrac{\partial A}{\partial h}$ if $A = \frac{1}{2}(a+b)h$

20. $\dfrac{\partial}{\partial m}\left(\frac{1}{2}mv^2\right)$ **21.** $\dfrac{\partial}{\partial B}\left(\dfrac{1}{u_0}B^2\right)$

22. $\dfrac{\partial}{\partial r}\left(\dfrac{2\pi r}{v}\right)$ **23.** F_v if $F = \dfrac{mv^2}{r}$

24. $\dfrac{\partial}{\partial v_0}(v_0 + at)$

25. $\dfrac{\partial F}{\partial m_2}$ if $F = \dfrac{Gm_1m_2}{r^2}$

26. $\dfrac{\partial}{\partial x}\left(\dfrac{1}{a}e^{-x^2/a^2}\right)$

27. $\dfrac{\partial}{\partial a}\left(\dfrac{1}{a}e^{-x^2/a^2}\right)$

28. f_x if $f(x,y) = e^{xy}(\ln y)$

29. $\dfrac{\partial}{\partial t}\left(v_0 t + \frac{1}{2}at^2\right)$

30. $\dfrac{\partial}{\partial \theta}(\sin(\pi\theta\phi) + \ln(\theta^2 + \phi))$

31. $\dfrac{\partial}{\partial M}\left(\dfrac{2\pi r^{3/2}}{\sqrt{GM}}\right)$

32. f_a if $f(a,b) = e^a \sin(a+b)$

33. z_x if $z = \sin(5x^3y - 3xy^2)$

34. g_x if $g(x,y) = \ln(ye^{xy})$

35. F_L if $F(L,K) = 3\sqrt{LK}$

36. $\dfrac{\partial V}{\partial r}$ and $\dfrac{\partial V}{\partial h}$ if $V = \frac{4}{3}\pi r^2 h$

37. u_E if $u = \dfrac{1}{2}\epsilon_0 E^2 + \dfrac{1}{2\mu_0}B^2$

38. $\dfrac{\partial}{\partial x}\left(\dfrac{1}{\sqrt{2\pi}\sigma}e^{-(x-\mu)^2/(2\sigma^2)}\right)$

39. $\dfrac{\partial Q}{\partial K}$ if $Q = c(a_1 K^{b_1} + a_2 L^{b_2})^\gamma$

40. z_x and z_y for $z = x^7 + 2^y + x^y$

41. $\dfrac{\partial z}{\partial y}\bigg|_{(1,0.5)}$ if $z = e^{x+2y}\sin y$

42. $\dfrac{\partial f}{\partial x}\bigg|_{(\pi/3,1)}$ if $f(x,y) = x\ln(y\cos x)$

Problems

43. **(a)** Let $f(x,y) = x^2 + y^2$. Estimate $f_x(2,1)$ and $f_y(2,1)$ using the contour diagram for f in Figure 14.21.
   (b) Estimate $f_x(2,1)$ and $f_y(2,1)$ from a table of values for f with $x = 1.9, 2, 2.1$ and $y = 0.9, 1, 1.1$.
   (c) Compare your estimates in parts (a) and (b) with the exact values of $f_x(2,1)$ and $f_y(2,1)$ found algebraically.

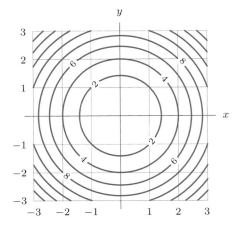

Figure 14.21

44. **(a)** Let $f(w,z) = e^{w \ln z}$. Use difference quotients with $h = 0.01$ to approximate $f_w(2,2)$ and $f_z(2,2)$.
   (b) Now evaluate $f_w(2,2)$ and $f_z(2,2)$ exactly.

45. **(a)** The surface S is given, for some constant a, by
$$z = 3x^2 + 4y^2 - axy$$
   Find the values of a which ensure that S is sloping upward when we move in the positive x-direction from the point $(1,2)$.
   (b) With the values of a from part (a), if you move in the positive y-direction from the point $(1,2)$, does the surface slope up or down? Explain.

46. Money in a bank account earns interest at a continuous rate, r. The amount of money, $\$B$, in the account depends on the amount deposited, $\$P$, and the time, t, it has been in the bank according to the formula
$$B = Pe^{rt}.$$
Find $\partial B/\partial t$ and $\partial B/\partial P$ and interpret each in financial terms.

47. The acceleration g due to gravity, at a distance r from the center of a planet of mass m, is given by

$$g = \frac{Gm}{r^2},$$

where G is the universal gravitational constant.

(a) Find $\partial g/\partial m$ and $\partial g/\partial r$.

(b) Interpret each of the partial derivatives you found in part (a) as the slope of a graph in the plane and sketch the graph.

48. The Dubois formula relates a person's surface area, s, in m^2, to weight, w, in kg, and height, h, in cm, by

$$s = f(w, h) = 0.01w^{0.25}h^{0.75}.$$

Find $f(65, 160)$, $f_w(65, 160)$, and $f_h(65, 160)$. Interpret your answers in terms of surface area, height, and weight.

49. The energy, E, of a body of mass m moving with speed v is given by the formula

$$E = mc^2 \left(\frac{1}{\sqrt{1 - v^2/c^2}} - 1 \right).$$

The speed, v, is nonnegative and less than the speed of light, c, which is a constant.

(a) Find $\partial E/\partial m$. What would you expect the sign of $\partial E/\partial m$ to be? Explain.

(b) Find $\partial E/\partial v$. Explain what you would expect the sign of $\partial E/\partial v$ to be and why.

50. Let $h(x, t) = 5 + \cos(0.5x - t)$ describe a wave. The value of $h(x, t)$ gives the depth of the water in cm at a distance x meters from a fixed point and at time t seconds. Evaluate $h_x(2, 5)$ and $h_t(2, 5)$ and interpret each in terms of the wave.

51. A one-meter long bar is heated unevenly, with temperature in °C at a distance x meters from one end at time t given by

$$H(x, t) = 100e^{-0.1t} \sin(\pi x) \qquad 0 \le x \le 1.$$

(a) Sketch a graph of H against x for $t = 0$ and $t = 1$.

(b) Calculate $H_x(0.2, t)$ and $H_x(0.8, t)$. What is the practical interpretation (in terms of temperature) of these two partial derivatives? Explain why each one has the sign it does.

(c) Calculate $H_t(x, t)$. What is its sign? What is its interpretation in terms of temperature?

52. Show that the Cobb-Douglas function

$$Q = bK^\alpha L^{1-\alpha} \quad \text{where} \quad 0 < \alpha < 1$$

satisfies the equation

$$K \frac{\partial Q}{\partial K} + L \frac{\partial Q}{\partial L} = Q.$$

53. Is there a function f which has the following partial derivatives? If so what is it? Are there any others?

$$f_x(x, y) = 4x^3y^2 - 3y^4,$$
$$f_y(x, y) = 2x^4y - 12xy^3.$$

Strengthen Your Understanding

In Problems 54–55, explain what is wrong with the statement.

54. The partial derivative of $f(x, y) = x^2y^3$ is $2xy^3 + 3y^2x^2$.

55. For $f(x, y)$, if $\dfrac{f(0.01, 0) - f(0, 0)}{0.01} > 0$, then $f_x(0, 0) > 0$.

In Problems 56–58, give an example of:

56. A nonlinear function $f(x, y)$ such that $f_x(0, 0) = 2$ and $f_y(0, 0) = 3$.

57. Functions $f(x, y)$ and $g(x, y)$ such that $f_x = g_x$ but $f_y \ne g_y$.

58. A non-constant function $f(x, y)$ such that $f_x = 0$ everywhere.

Are the statements in Problems 59–66 true or false? Give reasons for your answer.

59. There is a function $f(x, y)$ with $f_x(x, y) = y$ and $f_y(x, y) = x$.

60. The function $z(u, v) = u \cos v$ satisfies the equation

$$\cos v \frac{\partial z}{\partial u} - \frac{\sin v}{u} \frac{\partial z}{\partial v} = 1.$$

61. If $f(x, y)$ is a function of two variables and $g(x)$ is a function of a single variable, then

$$\frac{\partial}{\partial y} (g(x)f(x, y)) = g(x)f_y(x, y).$$

62. The function $k(r, s) = rse^s$ is increasing in the s-direction at the point $(r, s) = (-1, 2)$.

63. There is a function $f(x, y)$ with $f_x(x, y) = y^2$ and $f_y(x, y) = x^2$.

64. If $f(x, y)$ has $f_y(x, y) = 0$ then f must be a constant.

65. If $f(x, y) = ye^{g(x)}$ then $f_x = f$.

66. If f is a symmetric two-variable function, that is $f(x, y) = f(y, x)$, then $f_x(x, y) = f_y(x, y)$.

67. Which of the following functions satisfy the following equation (called Euler's Equation):

(a) x^2y^3 **(b)** $x+y+1$ **(c)** x^2+y^2 **(d)** $x^{0.4}y^{0.6}$

$$xf_x + yf_y = f?$$

14.3 LOCAL LINEARITY AND THE DIFFERENTIAL

In Sections 14.1 and 14.2 we studied a function of two variables by allowing one variable at a time to change. We now let both variables change at once to develop a linear approximation for functions of two variables.

Zooming In to See Local Linearity

For a function of one variable, local linearity means that as we zoom in on the graph, it looks like a straight line. As we zoom in on the graph of a two-variable function, the graph usually looks like a plane, which is the graph of a linear function of two variables. (See Figure 14.22.)

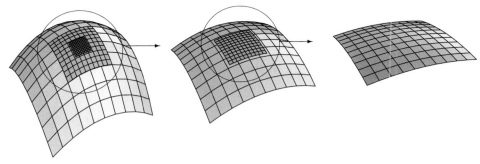

Figure 14.22: Zooming in on the graph of a function of two variables until the graph looks like a plane

Similarly, Figure 14.23 shows three successive views of the contours near a point. As we zoom in, the contours look more like equally spaced parallel lines, which are the contours of a linear function. (As we zoom in, we have to add more contours.)

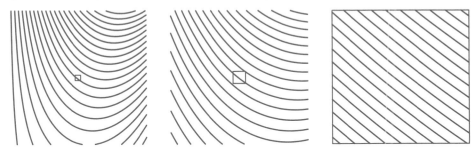

Figure 14.23: Zooming in on a contour diagram until the lines look parallel and equally spaced

This effect can also be seen numerically by zooming in with tables of values. Table 14.4 shows three tables of values for $f(x, y) = x^2 + y^3$ near $x = 2$, $y = 1$, each one a closer view than the previous one. Notice how each table looks more like the table of a linear function.

Table 14.4 *Zooming in on values of $f(x, y) = x^2 + y^3$ near $(2, 1)$ until the table looks linear*

		y	
	0	1	2
1	1	2	9
2	4	5	12
3	9	10	17

		y	
	0.9	1.0	1.1
1.9	4.34	4.61	4.94
2.0	4.73	5.00	5.33
2.1	5.14	5.41	5.74

		y	
	0.99	1.00	1.01
1.99	4.93	4.96	4.99
2.00	4.97	5.00	5.03
2.01	5.01	5.04	5.07

Zooming in Algebraically: Differentiability

Seeing a plane when we zoom in at a point tells us (provided the plane is not vertical) that $f(x, y)$ is closely approximated near that point by a linear function, $L(x, y)$:

$$f(x, y) \approx L(x, y).$$

The graph of the function $z = L(x, y)$ is the tangent plane at that point. Provided the approximation is sufficiently good, we say that $f(x, y)$ is *differentiable* at the point. Section 14.8 on page 834 defines precisely what is meant by the approximation being sufficiently good. The functions we encounter are differentiable at most points in their domain.

The Tangent Plane

The plane that we see when we zoom in on a surface is called the *tangent plane* to the surface at the point. Figure 14.24 shows the graph of a function with a tangent plane.

What is the equation of the tangent plane? At the point (a, b), the x-slope of the graph of f is the partial derivative $f_x(a, b)$ and the y-slope is $f_y(a, b)$. Thus, using the equation for a plane on page 712 of Chapter 12, we have the following result:

Tangent Plane to the Surface $z = f(x, y)$ at the Point (a, b)

Assuming f is differentiable at (a, b), the equation of the tangent plane is

$$z = f(a, b) + f_x(a, b)(x - a) + f_y(a, b)(y - b).$$

Here we are thinking of a and b as fixed, so $f(a, b)$, and $f_x(a, b)$, and $f_y(a, b)$ are constants. Thus, the right side of the equation is a linear function of x and y.

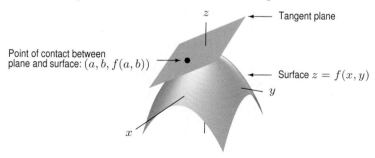

Point of contact between plane and surface: $(a, b, f(a, b))$ — Tangent plane — z — Surface $z = f(x, y)$ — y — x

Figure 14.24: The tangent plane to the surface $z = f(x, y)$ at the point (a, b)

Example 1   Find the equation for the tangent plane to the surface $z = x^2 + y^2$ at the point $(3, 4)$.

Solution   We have $f_x(x, y) = 2x$, so $f_x(3, 4) = 6$, and $f_y(x, y) = 2y$, so $f_y(3, 4) = 8$. Also, $f(3, 4) = 3^2 + 4^2 = 25$. Thus, the equation for the tangent plane at $(3, 4)$ is

$$z = 25 + 6(x - 3) + 8(y - 4).$$

Local Linearisation

Since the tangent plane lies close to the surface near the point at which they meet, z-values on the tangent plane are close to values of $f(x, y)$ for points near (a, b). Thus, replacing z by $f(x, y)$ in the equation of the tangent plane, we get the following approximation:

Tangent Plane Approximation to $f(x, y)$ for (x, y) Near the Point (a, b)

Provided f is differentiable at (a, b), we can approximate $f(x, y)$:

$$f(x, y) \approx f(a, b) + f_x(a, b)(x - a) + f_y(a, b)(y - b).$$

We are thinking of a and b as fixed, so the expression on the right side is linear in x and y. The right side of this approximation gives the **local linearisation** of f near $x = a$, $y = b$.

Figure 14.25 shows the tangent plane approximation graphically.

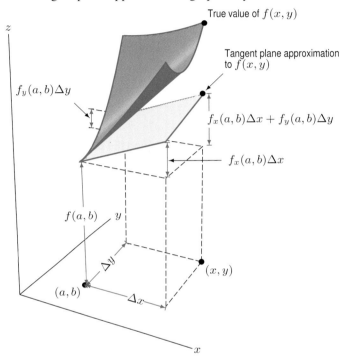

Figure 14.25: Local linearisation: Approximating $f(x, y)$ by the z-value from the tangent plane

Example 2   Find the local linearisation of $f(x, y) = x^2 + y^2$ at the point $(3, 4)$. Estimate $f(2.9, 4.2)$ and $f(2, 2)$ using the linearisation and compare your answers to the true values.

Solution   Let $z = f(x, y) = x^2 + y^2$. In Example 1 on page 790, we found the equation of the tangent plane at $(3, 4)$ to be

$$z = 25 + 6(x - 3) + 8(y - 4).$$

Therefore, for (x, y) near $(3, 4)$, we have the local linearisation

$$f(x, y) \approx 25 + 6(x - 3) + 8(y - 4).$$

Substituting $x = 2.9, y = 4.2$ gives

$$f(2.9, 4.2) \approx 25 + 6(-0.1) + 8(0.2) = 26.$$

This compares favourably with the true value $f(2.9, 4.2) = (2.9)^2 + (4.2)^2 = 26.05$.

However, the local linearisation does not give a good approximation at points far away from $(3, 4)$. For example, if $x = 2, y = 2$, the local linearisation gives

$$f(2, 2) \approx 25 + 6(-1) + 8(-2) = 3,$$

whereas the true value of the function is $f(2, 2) = 2^2 + 2^2 = 8$.

Example 3   Designing safe boilers depends on knowing how steam behaves under changes in temperature and pressure. Steam tables, such as Table 14.5, are published giving values of the function $V = f(T, P)$ where V is the volume (in m^3) of one kilogram of steam at a temperature T (in °C) and pressure P (in kp, where 1 kp = 1000 newton/m^2).

(a) Give a linear function approximating $V = f(T, P)$ for T near 260°C and P near 170 kp.

(b) Estimate the volume of a pound of steam at a temperature of 262°C and a pressure of 171 kp.

Table 14.5   *Volume (in cubic metres) of one kilogram of steam at various temperatures and pressures*

		Pressure P (kp)			
		140	155	170	185
Temperature T (°C)	250	0.7868	0.7151	0.6552	0.6043
	260	0.8041	0.7306	0.6693	0.6176
	270	0.8210	0.7461	0.6887	0.6309
	280	0.8380	0.7614	0.6978	0.6438

Solution   (a) We want the local linearisation around the point $T = 260$, $P = 170$, which is

$$f(T, P) \approx f(260, 170) + f_T(260, 170)(T - 260) + f_P(260, 170)(P - 170).$$

We read the value $f(260, 170) = 0.6693$ from the table.

Next we approximate $f_T(260, 170)$ by a difference quotient. From the $P = 170$ column, we compute the average rate of change between $T = 260$ and $T = 270$:

$$f_T(260, 170) \approx \frac{f(270, 170) - f(260, 170)}{270 - 260} = \frac{0.6887 - 0.6693}{10} = 0.00194.$$

Note that $f_T(260, 170)$ is positive, because steam expands when heated.

Next we approximate $f_P(260, 170)$ by looking at the $T = 260$ row and computing the average rate of change between $P = 170$ and $P = 185$:

$$f_P(260, 170) \approx \frac{f(260, 185) - f(260, 170)}{185 - 170} = \frac{0.6176 - 0.6693}{15} = -0.003447.$$

Note that $f_P(260, 170)$ is negative, because increasing the pressure on steam decreases its volume. Using these approximations for the partial derivatives, we obtain the local linearisation:

$$V = f(T, P) \approx 0.6693 + 0.00194(T - 260) - 0.003447(P - 170) \, m^3 \quad \begin{matrix} \text{for } T \text{ near } 260 \,°C \\ \text{and } P \text{ near } 170 \text{ kp.} \end{matrix}$$

(b) We are interested in the volume at $T = 262$°C and $P = 171$ kp. Since these values are close to $T = 260$°C and $P = 170$ kp, we use the linear relation obtained in part (a).

$$V \approx 0.6693 + 0.00194(262 - 260) - 0.003447(171 - 170) = 0.6697 \, m^3.$$

Local Linearity with Three or More Variables

Local linear approximations for functions of three or more variables follow the same pattern as for functions of two variables. The local linearisation of $f(x, y, z)$ at (a, b, c) is given by

$$f(x, y, z) \approx f(a, b, c) + f_x(a, b, c)(x - a) + f_y(a, b, c)(y - b) + f_z(a, b, c)(z - c).$$

The Differential

We are often interested in the change in the value of the function as we move from the point (a, b) to a nearby point (x, y). Then we use the notation

$$\Delta f = f(x, y) - f(a, b) \quad \text{and} \quad \Delta x = x - a \quad \text{and} \quad \Delta y = y - b$$

to rewrite the tangent plane approximation

$$f(x, y) \approx f(a, b) + f_x(a, b)(x - a) + f_y(a, b)(y - b)$$

in the form

$$\Delta f \approx f_x(a, b)\Delta x + f_y(a, b)\Delta y.$$

For fixed a and b, the right side of this is a linear function of Δx and Δy that can be used to estimate Δf. We call this linear function the *differential*. To define the differential in general, we introduce new variables dx and dy to represent changes in x and y.

The Differential of a Function $z = f(x, y)$

The **differential**, df (or dz), at a point (a, b) is the linear function of dx and dy given by the formula

$$df = f_x(a, b)\, dx + f_y(a, b)\, dy.$$

The differential at a general point is often written $df = f_x\, dx + f_y\, dy$.

Note that the differential, df, is a function of four variables a, b, and dx, dy.

Example 4   Compute the differentials of the following functions.
  (a) $f(x, y) = x^2 e^{5y}$ (b) $z = x \sin(xy)$ (c) $f(x, y) = x \cos(2x)$

Solution   (a) Since $f_x(x, y) = 2xe^{5y}$ and $f_y(x, y) = 5x^2 e^{5y}$, we have

$$df = 2xe^{5y}\, dx + 5x^2 e^{5y}\, dy.$$

  (b) Since $\partial z / \partial x = \sin(xy) + xy \cos(xy)$ and $\partial z / \partial y = x^2 \cos(xy)$, we have

$$dz = (\sin(xy) + xy \cos(xy))\, dx + x^2 \cos(xy)\, dy.$$

  (c) Since $f_x(x, y) = \cos(2x) - 2x \sin(2x)$ and $f_y(x, y) = 0$, we have

$$df = (\cos(2x) - 2x \sin(2x))\, dx + 0\, dy = (\cos(2x) - 2x \sin(2x))\, dx.$$

Example 5   The density ρ (in g/cm^3) of carbon dioxide gas CO_2 depends upon its temperature T (in °C) and pressure P (in atmospheres). The ideal gas model for CO_2 gives what is called the state equation:

$$\rho = \frac{0.5363P}{T + 273.15}.$$

Compute the differential $d\rho$. Explain the signs of the coefficients of dT and dP.

Solution   The differential for $\rho = f(T, P)$ is

$$d\rho = f_T(T, P)\, dT + f_P(T, P)dP = \frac{-0.5363P}{(T + 273.15)^2}\, dT + \frac{0.5363}{T + 273.15}\, dP.$$

The coefficient of dT is negative because increasing the temperature expands the gas (if the pressure is kept constant) and therefore decreases its density. The coefficient of dP is positive because increasing the pressure compresses the gas (if the temperature is kept constant) and therefore increases its density.

Where Does the Notation for the Differential Come From?

We write the differential as a linear function of the new variables dx and dy. You may wonder why we chose these names for our variables. The reason is historical: The people who invented calculus thought of dx and dy as "infinitesimal" changes in x and y. The equation

$$df = f_x dx + f_y dy$$

was regarded as an infinitesimal version of the local linear approximation

$$\Delta f \approx f_x \Delta x + f_y \Delta y.$$

In spite of the problems with defining exactly what "infinitesimal" means, some mathematicians, scientists, and engineers think of the differential in terms of infinitesimals.

Figure 14.26 illustrates a way of thinking about differentials that combines the definition with this informal point of view. It shows the graph of f along with a view of the graph around the point $(a, b, f(a, b))$ under a microscope. Since f is locally linear at the point, the magnified view looks like the tangent plane. Under the microscope, we use a magnified coordinate system with its origin at the point $(a, b, f(a, b))$ and with coordinates dx, dy, and dz along the three axes. The graph of the differential df is the tangent plane, which has equation $dz = f_x(a, b)\, dx + f_y(a, b)\, dy$ in the magnified coordinates.

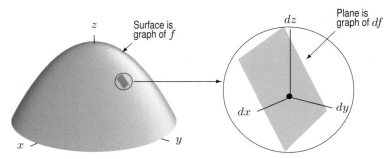

Figure 14.26: The graph of f, with a view through a microscope showing the tangent plane in the magnified coordinate system

Exercises and Problems for Section 14.3

Exercises

In Exercises 1–8, find the equation of the tangent plane at the given point.

1. $z = \frac{1}{2}(x^2 + 4y^2)$ at the point $(2, 1, 4)$

2. $z = ye^{x/y}$ at the point $(1, 1, e)$

3. $z = e^y + x + x^2 + 6$ at the point $(1, 0, 9)$

4. $z = \ln(x^2 + 1) + y^2$ at the point $(0, 3, 9)$

5. $z = \sin(xy)$ at $x = 2$, $y = 3\pi/4$

6. $x^2 + y^2 - z = 1$ at the point $(1, 3, 9)$

7. $x^2 y^2 + z - 40 = 0$ at $x = 2$, $y = 3$

8. $x^2 y + \ln(xy) + z = 6$ at the point $(4, 0.25, 2)$

In Exercises 9–13, find the differential of the function.

9. $g(u, v) = u^2 + uv$

10. $f(x, y) = \sin(xy)$

11. $z = \cos(xy) + x^2$

12. $z = e^{-x} \cos y$

13. $h(x, t) = e^{-3t} \sin(x + 5t)$

In Exercises 14–17, find the differential of the function at the given point.

14. $f(x, y) = xe^{-y}$ at $(1, 0)$

15. $g(x, t) = x^2 \sin(2t)$ at $(2, \pi/4)$

16. $F(m, r) = Gm/r^2$ at $(100, 10)$

17. $P(L, K) = 1.01 L^{0.25} K^{0.75}$ at $(100, 1)$

Problems

18. At a distance of x meters from the beach, the price in dollars of a plot of land of area a square meters is $f(a, x)$.

   (a) What are the units of $f_a(a, x)$?
   (b) What does $f_a(1000, 300) = 47$ mean in practical terms?
   (c) What are the units of $f_x(a, x)$?
   (d) What does $f_x(1000, 300) = -6$ mean in practical terms?
   (e) Which is cheaper: 1005 square meters that are 305 meters from the beach or 998 square meters that are 295 meters from the beach? Justify your answer.

19. A student was asked to find the equation of the tangent plane to the surface $z = x^3 - y^2$ at the point $(x, y) = (2, 3)$. The student's answer was

$$z = 3x^2(x - 2) - 2y(y - 3) - 1.$$

   (a) At a glance, how do you know this is wrong?
   (b) What mistake did the student make?
   (c) Answer the question correctly.

20. **(a)** Check the local linearity of $f(x, y) = e^{-x} \sin y$ near $x = 1$, $y = 2$ by making a table of values of f for $x = 0.9$, 1.0, 1.1 and $y = 1.9$, 2.0, 2.1. Express values of f with 4 digits after the decimal point. Then make a table of values for $x = 0.99$, 1.00, 1.01 and $y = 1.99$, 2.00. 2.01, again showing 4 digits after the decimal point. Do both tables look nearly linear? Does the second table look more linear than the first?

   (b) Give the local linearization of $f(x, y) = e^{-x} \sin y$ at $(1, 2)$, first using your tables, and second using the fact that $f_x(x, y) = -e^{-x} \sin y$ and $f_y(x, y) = e^{-x} \cos y$.

21. Find the local linearization of the function $f(x, y) = x^2 y$ at the point $(3, 1)$.

22. For the differentiable function $h(x, y)$, we are told that $h(600, 100) = 300$ and $h_x(600, 100) = 12$ and $h_y(600, 100) = -8$. Estimate $h(605, 98)$.

23. **(a)** Find the equation of the plane tangent to the graph of $f(x, y) = x^2 e^{xy}$ at $(1, 0)$.
   (b) Find the linear approximation of $f(x, y)$ for (x, y) near $(1, 0)$.
   (c) Find the differential of f at the point $(1, 0)$.

24. Find the differential of $f(x, y) = \sqrt{x^2 + y^3}$ at the point $(1, 2)$. Use it to estimate $f(1.04, 1.98)$.

25. **(a)** Find the differential of $g(u, v) = u^2 + uv$.
   (b) Use your answer to part (a) to estimate the change in g as you move from $(1, 2)$ to $(1.2, 2.1)$.

26. An unevenly heated plate has temperature $T(x, y)$ in $°C$ at the point (x, y). If $T(2, 1) = 135$, and $T_x(2, 1) = 16$, and $T_y(2, 1) = -15$, estimate the temperature at the point $(2.04, 0.97)$.

27. A right circular cylinder has a radius of 50 cm and a height of 100 cm. Use differentials to estimate the change in volume of the cylinder if its height and radius are both increased by 1 cm.

28. Give the local linearization for the monthly car-loan payment function at each of the points investigated in Problem 29 on page 783.

29. In Example 3 on page 792 we found a linear approximation for $V = f(T, P)$ near $(260, 170)$. Now find a linear approximation near $(250, 140)$.

30. In Example 3 on page 792 we found a linear approximation for $V = f(T, p)$ near $(260, 170)$.

   (a) Test the accuracy of this approximation by comparing its predicted value with the four neighbouring values in the table. What do you notice? Which predicted values are accurate? Which are not? Explain your answer.
   (b) Suggest a linear approximation for $f(T, p)$ near $(260, 170)$ that does not have the property you noticed in part (a). [Hint: Estimate the partial derivatives in a different way.]

31. In a room, the temperature is given by $T = f(x, t)$ degrees Celsius, where x is the distance from a heater (in meters) and t is the elapsed time (in minutes) since the heat has been turned on. A person standing 3 meters from the heater 5 minutes after it has been turned on observes the following: (1) The temperature is increasing by $1.2°C$ per minute, and (2) Walking away from the heater, the temperature decreases by $2°C$ per meter as time is held constant. Estimate how much cooler or warmer it would be 2.5 meters from the heater after 6 minutes.

32. Van der Waal's equation relates the pressure, P, and the volume, V, of a fixed quantity of a gas at constant temperature T. For a, b, n, R constants, the equation is

$$\left(P + \frac{n^2 a}{V^2} \right) (V - nb) = nRT.$$

   (a) Express P as a function of T and V.
   (b) Write a linear approximation for the change in pressure, $\Delta P = P - P_0$, resulting from a change in temperature $\Delta T = T - T_0$ and a change in pressure, $\Delta V = V - V_0$.

33. The gas equation for one mole of oxygen relates its pressure, P (in atmospheres), its temperature, T (in K), and its volume, V (in cubic decimeters, dm^3):

$$T = 16.574 \frac{1}{V} - 0.52754 \frac{1}{V^2} - 0.3879P + 12.187VP.$$

   (a) Find the temperature T and differential dT if the volume is 25 dm^3 and the pressure is 1 atmosphere.
   (b) Use your answer to part (a) to estimate how much the volume would have to change if the pressure increased by 0.1 atmosphere and the temperature remained constant.

34. The coefficient, β, of thermal expansion of a liquid relates the change in the volume V (in m^3) of a fixed quantity of a liquid to an increase in its temperature T (in $^\circ$C):

$$dV = \beta V \, dT.$$

(a) Let ρ be the density (in kg/m^3) of water as a function of temperature. (For a mass m of liquid, we have $\rho = m/V$.) Write an expression for $d\rho$ in terms of ρ and dT.
(b) The graph in Figure 14.27 shows density of water as a function of temperature. Use it to estimate β when $T = 20^\circ$C and when $T = 80^\circ$C.

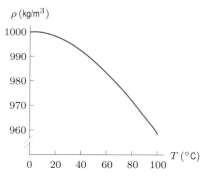

Figure 14.27

35. A fluid moves through a tube of length 1 meter and radius $r = 0.005 \pm 0.00025$ meters under a pressure $p = 10^5 \pm 1000$ pascals, at a rate $v = 0.625 \cdot 10^{-9}$ m^3 per unit time. Use differentials to estimate the maximum error in the viscosity η given by

$$\eta = \frac{\pi}{8} \frac{pr^4}{v}.$$

36. The period, T, of oscillation in seconds of a pendulum clock is given by $T = 2\pi\sqrt{l/g}$, where g is the acceleration due to gravity. The length of the pendulum, l, depends on the temperature, t, according to the formula $l = l_0(1 + \alpha(t - t_0))$ where l_0 is the length of the pendulum at temperature t_0 and α is a constant which characterizes the clock. The clock is set to the correct period at the temperature t_0. How many seconds a day does the clock gain or lose when the temperature is $t_0 + \Delta t$? Show that this gain or loss is independent of l_0.

37. Two functions that have the same local linearization at a point have contours that are tangent at this point.

(a) If $f_x(a, b)$ or $f_y(a, b)$ is nonzero, use the local linearization to show that an equation of the line tangent at (a, b) to the contour of f through (a, b) is $f_x(a, b)(x - a) + f_y(a, b)(y - b) = 0$.
(b) Find the slope of the tangent line if $f_y(a, b) \neq 0$.
(c) Find an equation for the line tangent to the contour of $f(x, y) = x^2 + xy$ at $(3, 4)$.

Strengthen Your Understanding

In Problems 38–40, explain what is wrong with the statement.

38. An equation for the tangent plane to the surface $z = f(x, y)$ at the point $(3, 4)$ is

$$z = f(3, 4) + f_x(3, 4)x + f_y(3, 4)y.$$

39. If $f_x(0, 0) = g_x(0, 0)$ and $f_y(0, 0) = g_y(0, 0)$, then the surfaces $z = f(x, y)$ and $z = g(x, y)$ have the same tangent planes at the point $(0, 0)$.

40. The tangent plane to the surface $z = x^2y$ at the point $(1, 2)$ has equation

$$z = 2 + 2xy(x - 1) + x^2(y - 2).$$

In Problems 41–42, give an example of:

41. Two different functions with the same differential.

42. A surface in three space whose tangent plane at $(0, 0, 3)$ is the plane $z = 3$.

Are the statements in Problems 43–50 true or false? Give reasons for your answer.

43. The tangent plane approximation of $f(x, y) = ye^{x^2}$ at the point $(0, 1)$ is $f(x, y) \approx y$.

44. If f is a function with differential $df = 2y \, dx + \sin(xy) \, dy$, then f changes by about -0.4 between the points $(1, 2)$ and $(0.9, 2.0002)$.

45. The local linearization of $f(x, y) = x^2 + y^2$ at $(1,1)$ gives an overestimate of the value of $f(x, y)$ at the point $(1.04, 0.95)$.

46. If two functions f and g have the same differential at the point $(1, 1)$, then $f = g$.

47. If two functions f and g have the same tangent plane at a point $(1, 1)$, then $f = g$.

48. If $f(x, y)$ is a constant function, then $df = 0$.

49. If $f(x, y)$ is a linear function, then df is a linear function of dx and dy.

50. If you zoom close enough near a point (a, b) on the contour diagram of any differentiable function, the contours will be *precisely* parallel and *exactly* equally spaced.

14.4 GRADIENTS AND DIRECTIONAL DERIVATIVES IN THE PLANE

The Rate of Change in an Arbitrary Direction: The Directional Derivative

The partial derivatives of a function f tell us the rate of change of f in the directions parallel to the coordinate axes. In this section we see how to compute the rate of change of f in an arbitrary direction.

Example 1   Figure 14.28 shows the temperature, in °C, at the point (x, y). Estimate the average rate of change of temperature as we walk from point A to point B.

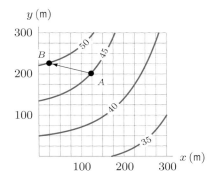

Figure 14.28: Estimating rate of change on a temperature map

Solution   At the point A we are on the $H = 45°C$ contour. At B we are on the $H = 50°C$ contour. The displacement vector from A to B has x component approximately $-100\vec{i}$ and y component approximately $25\vec{j}$, so its length is $\sqrt{(-100)^2 + 25^2} \approx 103$. Thus, the temperature rises by 5°C as we move 103 metres, so the average rate of change of the temperature in that direction is about $5/103 \approx 0.05°C/m$.

Suppose we want to compute the rate of change of a function $f(x, y)$ at the point $P = (a, b)$ in the direction of the unit vector $\vec{u} = u_1\vec{i} + u_2\vec{j}$. For $h > 0$, consider the point $Q = (a + hu_1, b + hu_2)$ whose displacement from P is $h\vec{u}$. (See Figure 14.29.) Since $\|\vec{u}\| = 1$, the distance from P to Q is h. Thus,

$$\begin{array}{l} \text{Average rate of change} \\ \text{in } f \text{ from } P \text{ to } Q \end{array} = \frac{\text{Change in } f}{\text{Distance from } P \text{ to } Q} = \frac{f(a + hu_1, b + hu_2) - f(a, b)}{h}.$$

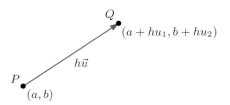

Figure 14.29: Displacement of $h\vec{u}$ from the point (a, b)

Taking the limit as $h \to 0$ gives the instantaneous rate of change and the following definition:

Directional Derivative of f at (a, b) in the Direction of a Unit Vector $\vec{u}$

If $\vec{u} = u_1\vec{i} + u_2\vec{j}$ is a unit vector, we define the directional derivative, $f_{\vec{u}}$, by

$$f_{\vec{u}}(a, b) = \begin{array}{c} \text{Rate of change} \\ \text{of } f \text{ in direction} \\ \text{of } \vec{u} \text{ at } (a, b) \end{array} = \lim_{h \to 0} \frac{f(a + hu_1, b + hu_2) - f(a, b)}{h},$$

provided the limit exists.

Notice that if $\vec{u} = \vec{i}$, so $u_1 = 1$, $u_2 = 0$, then the directional derivative is f_x, since

$$f_{\vec{i}}(a, b) = \lim_{h \to 0} \frac{f(a + h, b) - f(a, b)}{h} = f_x(a, b).$$

Similarly, if $\vec{u} = \vec{j}$ then the directional derivative $f_{\vec{j}} = f_y$.

What If We Do Not Have a Unit Vector?

We defined $f_{\vec{u}}$ for $\vec{u}$ a unit vector. If $\vec{v}$ is not a unit vector, $\vec{v} \neq \vec{0}$, we construct a unit vector $\vec{u} = \vec{v}/\|\vec{v}\|$ in the same direction as $\vec{v}$ and define the rate of change of f in the direction of $\vec{v}$ as $f_{\vec{u}}$.

Example 2 For each of the functions f, g, and h in Figure 14.30, decide whether the directional derivative at the indicated point is positive, negative, or zero, in the direction of the vector $\vec{v} = \vec{i} + 2\vec{j}$, and in the direction of the vector $\vec{w} = 2\vec{i} + \vec{j}$.

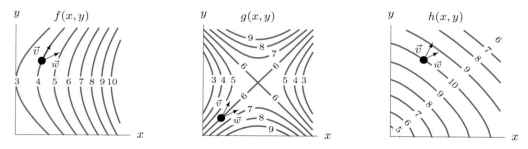

Figure 14.30: Contour diagrams of three functions with direction vectors $\vec{v} = \vec{i} + 2\vec{j}$ and $\vec{w} = 2\vec{i} + \vec{j}$ marked on each

Solution On the contour diagram for f, the vector $\vec{v} = \vec{i} + 2\vec{j}$ appears to be tangent to the contour. Thus, in this direction, the value of the function is not changing, so the directional derivative in the direction of $\vec{v}$ is zero. The vector $\vec{w} = 2\vec{i} + \vec{j}$ points from the contour marked 4 toward the contour marked 5. Thus, the values of the function are increasing and the directional derivative in the direction of $\vec{w}$ is positive.

On the contour diagram for g, the vector $\vec{v} = \vec{i} + 2\vec{j}$ points from the contour marked 6 toward the contour marked 5, so the function is decreasing in that direction. Thus, the rate of change is negative. On the other hand, the vector $\vec{w} = 2\vec{i} + \vec{j}$ points from the contour marked 6 toward the contour marked 7, and hence the directional derivative in the direction of $\vec{w}$ is positive.

Finally, on the contour diagram for h, both vectors point from the $h = 10$ contour to the $h = 9$ contour, so both directional derivatives are negative.

Example 3   Calculate the directional derivative of $f(x, y) = x^2 + y^2$ at $(1, 0)$ in the direction of the vector $\vec{i} + \vec{j}$.

Solution   First we have to find the unit vector in the same direction as the vector $\vec{i} + \vec{j}$. Since this vector has magnitude $\sqrt{2}$, the unit vector is

$$\vec{u} = \frac{1}{\sqrt{2}}(\vec{i} + \vec{j}) = \frac{1}{\sqrt{2}}\vec{i} + \frac{1}{\sqrt{2}}\vec{j}.$$

Thus,

$$f_{\vec{u}}(1, 0) = \lim_{h \to 0} \frac{f(1 + h/\sqrt{2}, h/\sqrt{2}) - f(1, 0)}{h} = \lim_{h \to 0} \frac{(1 + h/\sqrt{2})^2 + (h/\sqrt{2})^2 - 1}{h}$$

$$= \lim_{h \to 0} \frac{\sqrt{2}h + h^2}{h} = \lim_{h \to 0}(\sqrt{2} + h) = \sqrt{2}.$$

Computing Directional Derivatives from Partial Derivatives

If f is differentiable, we will now see how to use local linearity to find a formula for the directional derivative which does not involve a limit. If $\vec{u}$ is a unit vector, the definition of $f_{\vec{u}}$ says

$$f_{\vec{u}}(a, b) = \lim_{h \to 0} \frac{f(a + hu_1, b + hu_2) - f(a, b)}{h} = \lim_{h \to 0} \frac{\Delta f}{h},$$

where $\Delta f = f(a + hu_1, b + hu_2) - f(a, b)$ is the change in f. We write Δx for the change in x, so $\Delta x = (a + hu_1) - a = hu_1$; similarly $\Delta y = hu_2$. Using local linearity, we have

$$\Delta f \approx f_x(a, b)\Delta x + f_y(a, b)\Delta y = f_x(a, b)hu_1 + f_y(a, b)hu_2.$$

Thus, dividing by h gives

$$\frac{\Delta f}{h} \approx \frac{f_x(a, b)hu_1 + f_y(a, b)hu_2}{h} = f_x(a, b)u_1 + f_y(a, b)u_2.$$

This approximation becomes exact as $h \to 0$, so we have the following formula:

$$f_{\vec{u}}(a, b) = f_x(a, b)u_1 + f_y(a, b)u_2.$$

Example 4   Use the preceding formula to compute the directional derivative in Example 3. Check that we get the same answer as before.

Solution   We calculate $f_{\vec{u}}(1, 0)$, where $f(x, y) = x^2 + y^2$ and $\vec{u} = \frac{1}{\sqrt{2}}\vec{i} + \frac{1}{\sqrt{2}}\vec{j}$.

The partial derivatives are $f_x(x, y) = 2x$ and $f_y(x, y) = 2y$. So, as before,

$$f_{\vec{u}}(1, 0) = f_x(1, 0)u_1 + f_y(1, 0)u_2 = (2)\left(\frac{1}{\sqrt{2}}\right) + (0)\left(\frac{1}{\sqrt{2}}\right) = \sqrt{2}.$$

The Gradient Vector

Notice that the expression for $f_{\vec{u}}(a, b)$ can be written as a dot product of $\vec{u}$ and a new vector:

$$f_{\vec{u}}(a, b) = f_x(a, b)u_1 + f_y(a, b)u_2 = (f_x(a, b)\vec{i} + f_y(a, b)\vec{j}) \cdot (u_1\vec{i} + u_2\vec{j}).$$

The new vector, $f_x(a, b)\vec{i} + f_y(a, b)\vec{j}$, turns out to be important. Thus, we make the following definition:

> **The Gradient Vector** of a differentiable function f at the point (a, b) is
>
> $$\text{grad } f(a, b) = f_x(a, b)\vec{i} + f_y(a, b)\vec{j}$$

The formula for the directional derivative can be written in terms of the gradient as follows:

> ### The Directional Derivative and the Gradient
>
> If f is differentiable at (a, b) and $\vec{u} = u_1\vec{i} + u_2\vec{j}$ is a unit vector, then
>
> $$f_{\vec{u}}(a, b) = f_x(a, b)u_1 + f_y(a, b)u_2 = \text{grad } f(a, b) \cdot \vec{u}.$$

Example 5   Find the gradient vector of $f(x, y) = x + e^y$ at the point $(1, 1)$.

Solution   Using the definition, we have

$$\text{grad } f = f_x\vec{i} + f_y\vec{j} = \vec{i} + e^y\vec{j},$$

so at the point $(1, 1)$

$$\text{grad } f(1, 1) = \vec{i} + e\vec{j}.$$

Alternative Notation for the Gradient

You can think of $\dfrac{\partial f}{\partial x}\vec{i} + \dfrac{\partial f}{\partial y}\vec{j}$ as the result of applying the vector operator (pronounced "del")

$$\nabla = \frac{\partial}{\partial x}\vec{i} + \frac{\partial}{\partial y}\vec{j}$$

to the function f. Thus, we get the alternative notation

$$\text{grad } f = \nabla f.$$

If $z = f(x, y)$, we can write $\text{grad } z$ or ∇z for $\text{grad } f$ or for ∇f.

What Does the Gradient Tell Us?

The fact that $f_{\vec{u}} = \text{grad } f \cdot \vec{u}$ enables us to see what the gradient vector represents. Suppose θ is the angle between the vectors $\text{grad} f$ and $\vec{u}$. At the point (a, b), we have

$$f_{\vec{u}} = \text{grad } f \cdot \vec{u} = \| \text{grad } f\| \underbrace{\|\vec{u}\|}_{1} \cos \theta = \| \text{grad } f\| \cos\theta.$$

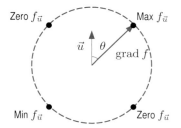

Figure 14.31: Values of the directional derivative at different angles to the gradient

Imagine that grad f is fixed and that $\vec{u}$ can rotate. (See Figure 14.31.) The maximum value of $f_{\vec{u}}$ occurs when $\cos \theta = 1$, so $\theta = 0$ and $\vec{u}$ is pointing in the direction of grad f. Then

$$\text{Maximum } f_{\vec{u}} = \| \operatorname{grad} f \| \cos 0 = \| \operatorname{grad} f \|.$$

The minimum value of $f_{\vec{u}}$ occurs when $\cos \theta = -1$, so $\theta = \pi$ and $\vec{u}$ is pointing in the direction opposite to grad f. Then

$$\text{Minimum } f_{\vec{u}} = \| \operatorname{grad} f \| \cos \pi = -\| \operatorname{grad} f \|.$$

When $\theta = \pi/2$ or $3\pi/2$, so $\cos \theta = 0$, the directional derivative is zero.

Properties of the Gradient Vector

We have seen that the gradient vector points in the direction of the greatest rate of change at a point and the magnitude of the gradient vector is that rate of change.

Figure 14.32 shows that the gradient vector at a point is perpendicular to the contour through that point. If the contours represent equally spaced f-values and f is differentiable, local linearity tells us that the contours of f around a point appear straight, parallel, and equally spaced. The greatest rate of change is obtained by moving in the direction that takes us to the next contour in the shortest possible distance; that is, perpendicular to the contour. Thus, we have the following:

Geometric Properties of the Gradient Vector in the Plane

If f is a differentiable function at the point (a, b) and grad $f(a, b) \neq \vec{0}$, then:
- The direction of grad $f(a, b)$ is
  - Perpendicular[1] to the contour of f through (a, b);
  - In the direction of the maximum rate of increase of f.
- The magnitude of the gradient vector, $\| \operatorname{grad} f \|$, is
  - The maximum rate of change of f at that point;
  - Large when the contours are close together and small when they are far apart.

---

[1]This assumes that the same scale is used on both axes.

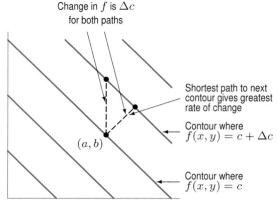

Change in f is Δc
for both paths

Shortest path to next
contour gives greatest
rate of change

Contour where
$f(x, y) = c + \Delta c$

(a, b)

Contour where
$f(x, y) = c$

Figure 14.32: Close-up view of the contours around (a, b),
showing the gradient is perpendicular to the contours

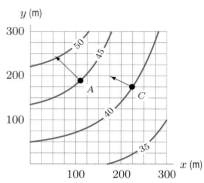

Figure 14.33: A temperature map showing
directions and relative magnitudes of two
gradient vectors

Examples of Directional Derivatives and Gradient Vectors

Example 6 Explain why the gradient vectors at points A and C in Figure 14.33 have the direction and the relative magnitudes they do.

Solution The gradient vector points in the direction of greatest increase of the function. This means that in Figure 14.33, the gradient points directly toward warmer temperatures. The magnitude of the gradient vector measures the rate of change. The gradient vector at A is longer than the gradient vector at C because the contours are closer together at A, so the rate of change is larger.

Example 2 on page 798 shows how the contour diagram can tell us the sign of the directional derivative. In the next example we compute the directional derivative in three directions, two that are close to that of the gradient vector and one that is not.

Example 7 Use the gradient to find the directional derivative of $f(x, y) = x + e^y$ at the point $(1, 1)$ in the direction of the vectors $\vec{i} - \vec{j}, \vec{i} + 2\vec{j}, \vec{i} + 3\vec{j}$.

Solution In Example 5 we found
$$\text{grad } f(1, 1) = \vec{i} + e\vec{j}.$$
A unit vector in the direction of $\vec{i} - \vec{j}$ is $\vec{s} = (\vec{i} - \vec{j})/\sqrt{2}$, so
$$f_{\vec{s}}(1, 1) = \text{grad } f(1, 1) \cdot \vec{s} = (\vec{i} + e\vec{j}) \cdot \left(\frac{\vec{i} - \vec{j}}{\sqrt{2}}\right) = \frac{1 - e}{\sqrt{2}} \approx -1.215.$$

A unit vector in the direction of $\vec{i} + 2\vec{j}$ is $\vec{v} = (\vec{i} + 2\vec{j})/\sqrt{5}$, so
$$f_{\vec{v}}(1, 1) = \text{grad } f(1, 1) \cdot \vec{v} = (\vec{i} + e\vec{j}) \cdot \left(\frac{\vec{i} + 2\vec{j}}{\sqrt{5}}\right) = \frac{1 + 2e}{\sqrt{5}} \approx 2.879.$$

A unit vector in the direction of $\vec{i} + 3\vec{j}$ is $\vec{w} = (\vec{i} + 3\vec{j})/\sqrt{10}$, so
$$f_{\vec{w}}(1, 1) = \text{grad } f(1, 1) \cdot \vec{w} = (\vec{i} + e\vec{j}) \cdot \left(\frac{\vec{i} + 3\vec{j}}{\sqrt{10}}\right) = \frac{1 + 3e}{\sqrt{10}} \approx 2.895.$$

Now look back at the answers and compare with the value of $\| \operatorname{grad} f \| = \sqrt{1 + e^2} \approx 2.896$. One answer is not close to this value; the other two, $f_{\vec{v}} = 2.879$ and $f_{\vec{w}} = 2.895$, are close but slightly smaller than $\| \operatorname{grad} f \|$. Since $\| \operatorname{grad} f \|$ is the maximum rate of change of f at the point, we have for *any* unit vector $\vec{u}$:

$$f_{\vec{u}}(1, 1) \leq \| \operatorname{grad} f \|.$$

with equality when $\vec{u}$ is in the direction of $\operatorname{grad} f$. Since $e \approx 2.718$, the vectors $\vec{i} + 2\vec{j}$ and $\vec{i} + 3\vec{j}$ both point roughly, but not exactly, in the direction of the gradient vector $\operatorname{grad} f(1, 1) = \vec{i} + e\vec{j}$. Thus, the values of $f_{\vec{v}}$ and $f_{\vec{w}}$ are both close to the value of $\| \operatorname{grad} f \|$. The direction of the vector $\vec{i} - \vec{j}$ is not close to the direction of $\operatorname{grad} f$ and the value of $f_{\vec{s}}$ is not close to the value of $\| \operatorname{grad} f \|$.

Exercises and Problems for Section 14.4

Exercises

In Exercises 1–14, find the gradient of the function. Assume the variables are restricted to a domain on which the function is defined.

1. $f(x, y) = \frac{3}{2}x^5 - \frac{4}{7}y^6$
2. $f(m, n) = m^2 + n^2$
3. $Q = 50K + 100L$
4. $z = xe^y$
5. $z = (x + y)e^y$
6. $f(K, L) = K^{0.3}L^{0.7}$
7. $f(x, y) = \sqrt{x^2 + y^2}$
8. $f(r, h) = \pi r^2 h$
9. $f(r, \theta) = r \sin \theta$
10. $z = \sin(x^2 + y^2)$
11. $z = \sin(x/y)$
12. $z = \tan^{-1}(x/y)$
13. $f(\alpha, \beta) = \dfrac{2\alpha + 3\beta}{2\alpha - 3\beta}$
14. $z = x\dfrac{e^y}{x + y}$

In Exercises 15–22, find the gradient at the point.

15. $f(x, y) = x^2 y + 7xy^3$, at $(1, 2)$
16. $f(m, n) = 5m^2 + 3n^4$, at $(5, 2)$
17. $f(r, h) = 2\pi rh + \pi r^2$, at $(2, 3)$
18. $f(x, y) = \sin(x^2) + \cos y$, at $(\frac{\sqrt{\pi}}{2}, 0)$
19. $f(x, y) = 1/(x^2 + y^2)$, at $(-1, 3)$
20. $f(x, y) = e^{\sin y}$, at $(0, \pi)$
21. $f(x, y) = \ln(x^2 + xy)$, at $(4, 1)$
22. $f(x, y) = \sqrt{\tan x + y}$, at $(0, 1)$

In Exercises 23–26, find the directional derivative $f_{\vec{u}}(1, 2)$ for the function f with $\vec{u} = (3\vec{i} - 4\vec{j})/5$.

23. $f(x, y) = 3x - 4y$
24. $f(x, y) = x^2 - y^2$
25. $f(x, y) = xy + y^3$
26. $f(x, y) = \sin(2x - y)$

27. If $f(x, y) = x^2 y$ and $\vec{v} = 4\vec{i} - 3\vec{j}$, find the directional derivative at the point $(2, 6)$ in the direction of $\vec{v}$.

In Exercises 28–30, find the differential df from the gradient.

28. $\operatorname{grad} f = 12\vec{i} + 7\vec{j}$
29. $\operatorname{grad} f = y\vec{i} + x\vec{j}$
30. $\operatorname{grad} f = (2x + 3e^y)\vec{i} + 3xe^y\vec{j}$

In Exercises 31–32, find $\operatorname{grad} f$ from the differential.

31. $df = 2xdx + 10ydy$
32. $df = (x + 1)ye^x dx + xe^x dy$

In Exercises 33–38, use the contour diagram of f in Figure 14.34 to decide if the specified directional derivative is positive, negative, or approximately zero.

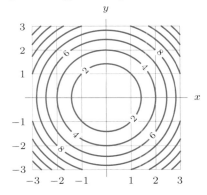

Figure 14.34

33. At point $(-2, 2)$, in direction $\vec{i}$.
34. At point $(0, -2)$, in direction $\vec{j}$.
35. At point $(-1, 1)$, in direction $\vec{i} + \vec{j}$.
36. At point $(-1, 1)$, in direction $-\vec{i} + \vec{j}$.
37. At point $(0, -2)$, in direction $\vec{i} + 2\vec{j}$.
38. At point $(0, -2)$, in direction $\vec{i} - 2\vec{j}$.

In Exercises 39–46, use the contour diagram of f in Figure 14.34 to find the approximate direction of the gradient vector at the given point.

39. $(2, 0)$
40. $(0, 2)$
41. $(-2, 0)$
42. $(0, -2)$
43. $(2, 2)$
44. $(2, -2)$
45. $(-2, 2)$
46. $(-2, -2)$

Problems

47. A student was asked to find the directional derivative of $f(x, y) = x^2 e^y$ at the point $(1, 0)$ in the direction of $\vec{v} = 4\vec{i} + 3\vec{j}$. The student's answer was

$$f_{\vec{u}}(1, 0) = \text{grad } f(1, 0) \cdot \vec{u} = \frac{8}{5}\vec{i} + \frac{3}{5}\vec{j}.$$

(a) At a glance, how do you know this is wrong?
(b) What is the correct answer?

48. Let $f(P) = 15$ and $f(Q) = 20$ where $P = (3, 4)$ and $Q = (3.03, 3.96)$. Approximate the directional derivative of f at P in the direction of Q.

49. **(a)** Give Q, the point at a distance of 0.1 from $P = (4, 5)$ in the direction of $\vec{v} = -\vec{i} + 3\vec{j}$. Give five decimal places in your answer.
(b) Use P and Q to approximate the directional derivative of $f(x, y) = \sqrt{x + y}$ in the direction of $\vec{v}$.
(c) Give the exact value for the directional derivative you estimated in part (b).

50. Find the directional derivative of $f(x, y) = e^x \tan(y) + 2x^2 y$ at the point $(0, \pi/4)$ in the following directions

(a) $\vec{i} - \vec{j}$ **(b)** $\vec{i} + \sqrt{3}\vec{j}$

51. Find the rate of change of $f(x, y) = x^2 + y^2$ at the point $(1, 2)$ in the direction of the vector $\vec{u} = 0.6\vec{i} + 0.8\vec{j}$.

In Problems 52–55, do the level curves of $f(x, y)$ cross the level curves of $g(x, y)$ at right angles? Sketch contour diagrams.

52. $f(x, y) = x + y, g(x, y) = x - y$

53. $f(x, y) = 2x + 3y, g(x, y) = 2x - 3y$

54. $f(x, y) = x^2 - y, g(x, y) = 2y + \ln|x|$

55. $f(x, y) = x^2 - y^2, g(x, y) = xy$

For Problems 56–60 use Figure 14.35, showing level curves of $f(x, y)$, to estimate the directional derivatives.

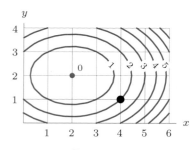

Figure 14.35

56. $f_{\vec{i}}(4, 1)$ **57.** $f_{\vec{j}}(4, 1)$

58. $f_{\vec{u}}(4, 1)$ where $\vec{u} = (\vec{i} - \vec{j})/\sqrt{2}$

59. $f_{\vec{u}}(4, 1)$ where $\vec{u} = (-\vec{i} + \vec{j})/\sqrt{2}$

60. $f_{\vec{u}}(4, 1)$ with $\vec{u} = (-2\vec{i} + \vec{j})/\sqrt{5}$

In Problems 61–64, check that the point $(2, 3)$ lies on the curve. Then, viewing the curve as a contour of $f(x, y)$, use grad $f(2, 3)$ to find a vector normal to the curve at $(2, 3)$ and an equation for the tangent line to the curve at $(2, 3)$.

61. $x^2 + y^2 = 13$ **62.** $xy = 6$

63. $y = x^2 - 1$ **64.** $(y - x)^2 + 2 = xy - 3$

65. Consider the functions $F(x, y) = y^2 + 4xy - x^4$, $G(x, y) = x^2 + 2y^3 - 4xy$ and $R(x, y) = x^2 + y^2$.

(a) Find all points $P(a, 1)$ such that the gradients of G and R at the point P are perpendicular.
(b) Find all points $Q(1, b)$ such that the gradients of F and R at the point Q are parallel.
(c) Figure 14.36 shows the contour line $F(x, y) = 11$. If you consider all the points on it, which one is the closest to the origin? [Hint: imagine a growing circle centered at the origin..]

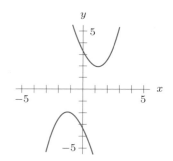

Figure 14.36

66. The surface $z = g(x, y)$ is in Figure 14.37. What is the sign of each of the following directional derivatives?

(a) $g_{\vec{u}}(2, 5)$ where $\vec{u} = (\vec{i} - \vec{j})/\sqrt{2}$.
(b) $g_{\vec{u}}(2, 5)$ where $\vec{u} = (\vec{i} + \vec{j})/\sqrt{2}$.

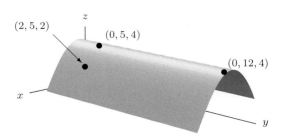

Figure 14.37

67. The table gives values of a differentiable function $f(x, y)$. At the point $(1.2, 0)$, into which quadrant does the gradient vector of f point? Justify your answer.

	y		
	-1	0	1
x			
1.0	0.7	0.1	-0.5
1.2	4.8	4.2	3.6
1.4	8.9	8.3	7.7

68. Figure 14.38 represents the level curves $f(x, y) = c$; the values of f on each curve are marked. In each of the following parts, decide whether the given quantity is positive, negative or zero. Explain your answer.

(a) The value of $\nabla f \cdot \vec{i}$ at P.
(b) The value of $\nabla f \cdot \vec{j}$ at P.
(c) $\partial f/\partial x$ at Q.
(d) $\partial f/\partial y$ at Q.

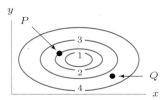

Figure 14.38

69. In Figure 14.38, which is larger: $\|\nabla f\|$ at P or $\|\nabla f\|$ at Q? Explain how you know.

70. (a) Let $f(x, y) = (x+y)/(1+x^2)$. Find the directional derivative of f at $P = (1, -2)$ in the direction of:
   (i) $\vec{v} = 3\vec{i} - 2\vec{j}$ (ii) $\vec{v} = -\vec{i} + 4\vec{j}$
   (b) What is the direction of greatest increase of f at P?

71. Let $f(x, y) = x^2 y^3$. At the point $(-1, 2)$, find a vector
   (a) In the direction of maximum rate of change.
   (b) In the direction of minimum rate of change.
   (c) In a direction in which the rate of change is zero.

72. You are at the point $(\pi/4, 1)$ and start to move in the direction of the point $(1 + \pi/4, 2)$. At what rate does the value of $f(x, y) = \sin(xy)$ change as you leave $(\pi/4, 1)$? Give your answer in units of f per unit distance.

73. (a) Let $f(x, y) = x^2 + \ln y$. Find the average rate of change of f as you go from $(3, 1)$ to $(1, 2)$.
   (b) Find the instantaneous rate of change of f as you leave the point $(3, 1)$ heading toward $(1, 2)$.

74. (a) What is the rate of change of $f(x, y) = 3xy + y^2$ at the point $(2, 3)$ in the direction $\vec{v} = 3\vec{i} - \vec{j}$?
   (b) What is the direction of maximum rate of change of f at $(2, 3)$?
   (c) What is the maximum rate of change?

75. (a) Sketch the surface $z = f(x, y) = y^2$ in three dimensions.
   (b) Sketch the level curves of f in the xy-plane.
   (c) If you are standing on the surface $z = y^2$ at the point $(2, 3, 9)$, in which direction should you move to climb the fastest? (Give your answer as a 2-vector.)

76. You are standing above the point $(1, 3)$ on the surface $z = 20 - (2x^2 + y^2)$.
   (a) In which direction should you walk to descend fastest? (Give your answer as a 2-vector.)
   (b) If you start to move in this direction, what is the slope of your path?

77. Let P be a fixed point in the plane and let $f(x, y)$ be the distance from P to (x, y). Answer the following questions using geometric interpretations, not formulas.
   (a) What are the level curves of f?
   (b) In what direction does $\operatorname{grad} f(x, y)$ point?
   (c) What is the magnitude $\| \operatorname{grad} f(x, y) \|$?

78. The directional derivative of $z = f(x, y)$ at $(2, 1)$ in the direction toward the point $(1, 3)$ is $-2/\sqrt{5}$, and the directional derivative in the direction toward the point $(5, 5)$ is 1. Compute $\partial z/\partial x$ and $\partial z/\partial y$ at $(2, 1)$.

79. Consider the function $f(x, y)$. If you start at the point $(4, 5)$ and move toward the point $(5, 6)$, the directional derivative is 2. Starting at the point $(4, 5)$ and moving toward the point $(6, 6)$ gives a directional derivative of 3. Find ∇f at the point $(4, 5)$.

80. (a) For $g(x, y) = \sqrt{x^2 + 3y + 3}$, find $\operatorname{grad} g(1, 4)$.
   (b) Find the best linear approximation of $g(x, y)$ for (x, y) near $(1, 4)$.
   (c) Use the approximation in part (b) to estimate $g(1.01, 3.98)$.

81. Find the directional derivative of $z = x^2 - y^2$ at the point $(3, -1)$ in the direction making an angle $\theta = \pi/4$ with the x-axis. In which direction is the directional derivative the largest?

82. The temperature H in °Celsius y kilometers north of the Canadian border t hours after 4 P.M. is given by $H = 30 - 0.05y - 5t$. A moose trots north at a speed of 20 km/h. At what rate does the moose perceive the temperature to be changing?

83. At a certain point on a heated plate, the greatest rate of temperature increase, $5°$ C per meter, is toward the northeast. If an object at this point moves directly north, at what rate is the temperature increasing?

84. You are climbing a mountain by the steepest route at a slope of $20°$ when you come upon a trail branching off at a $30°$ angle from yours. What is the angle of ascent of the branch trail?

85. Figure 14.40 is a graph of the directional derivative, $f_{\vec{u}}$, at the point (a, b) versus θ, the angle in Figure 14.39.

  (a) Which points on the graph in Figure 14.40 correspond to the greatest rate of increase of f? The greatest rate of decrease?

  (b) Mark points on the circle in Figure 14.39 corresponding to the points P, Q, R, S.

  (c) What is the amplitude of the function graphed in Figure 14.40? What is its formula?

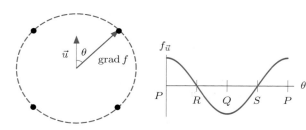

Figure 14.39   **Figure 14.40**

86. You are standing at the point $(1, 1, 3)$ on the hill whose equation is given by $z = 5y - x^2 - y^2$.

  (a) If you choose to climb in the direction of steepest ascent, what is your initial rate of ascent relative to the horizontal distance?

  (b) If you decide to go straight northwest, will you be ascending or descending? At what rate?

  (c) If you decide to maintain your altitude, in what directions can you go?

87. In this problem we see another way of obtaining the formula $f_{\vec{u}}(a, b) = \operatorname{grad} f(a, b) \cdot \vec{u}$. Imagine zooming in on a function $f(x, y)$ at a point (a, b). By local linearity, the contours around (a, b) look like the contours of a linear function. See Figure 14.41. Suppose you want to find the directional derivative $f_{\vec{u}}(a, b)$ in the direction of a unit vector $\vec{u}$. If you move from P to Q, a small distance h in the direction of $\vec{u}$, then the directional derivative is approximated by the difference quotient

$$\frac{\text{Change in } f \text{ between } P \text{ and } Q}{h}.$$

  (a) Use the gradient to show that

$$\text{Change in } f \approx \|\operatorname{grad} f\|(h \cos \theta).$$

  (b) Use part (a) to obtain $f_{\vec{u}}(a, b) = \operatorname{grad} f(a, b) \cdot \vec{u}$.

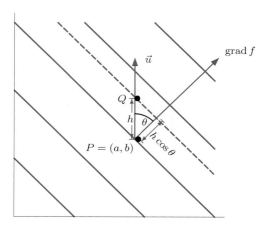

Figure 14.41

88. Let L be a line tangent to the ellipse $x^2/2 + y^2 = 1$ at the point (a, b). See Figure 14.42.

  (a) Find a vector perpendicular to L.

  (b) Find the distance p from $P = (-1, 0)$ to L as a function of a.

  (c) Find the distance q from $Q = (1, 0)$ to L as a function of a.

  (d) Show that $pq = 1$.

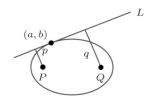

Figure 14.42

89. Let C be the contour C of $f(x, y)$ through (a, b) and $\operatorname{grad} f(a, b) \neq \vec{0}$. Show that

  (a) The vector $-f_y(a, b)\vec{i} + f_x(a, b)\vec{j}$ is tangent to C at (a, b).

  (b) The slope of the line tangent to C at the point (a, b) is $-f_x(a, b)/f_y(a, b)$ if the tangent line is not vertical.

90. Let $\|\operatorname{grad} f(x, y)\| = \|\operatorname{grad} g(x, y)\|$ at a point P where these gradients are not the zero vector. Show that at P, the direction of the most rapid increase of $f + g$

  (a) Increases f and g at equal rates.

  (b) Bisects the angle between the contours of f and g that pass through P.

Strengthen Your Understanding

In Problems 91–93, explain what is wrong with the statement.

91. A function f has a directional derivative given by $f_{\vec{u}}(0,0) = 3\vec{i} + 4\vec{j}$.

92. A function f has gradient $\operatorname{grad} f(0,0) = 7$.

93. The gradient vector $\operatorname{grad} f(x,y)$ is perpendicular to the contours of f, and the closer together the contours for equally spaced values of f, the shorter the gradient vector.

In Problems 94–95, give an example of:

94. A unit vector $\vec{u}$ such that $f_{\vec{u}}(0,0) < 0$, given that $f_x(0,0) = 2$ and $f_y(0,0) = 3$.

95. A contour diagram of a function with two points in the domain where the gradients are parallel but different lengths.

96. For the gradient $\nabla f(P)$ of f at a point P, describe the geometric interpretation of its

   (a) Direction
   (b) Magnitude
   (c) Dot product with a unit vector $\vec{u}$

Are the statements in Problems 97–108 true or false? Give reasons for your answer.

97. If the point (a,b) is on the contour $f(x,y) = k$, then the slope of the line tangent to this contour at (a,b) is $f_y(a,b)/f_x(a,b)$.

98. The gradient vector $\operatorname{grad} f(a,b)$ is a vector in 3-space.

99. $\operatorname{grad}(fg) = (\operatorname{grad} f) \cdot (\operatorname{grad} g)$

100. The gradient vector $\operatorname{grad} f(a,b)$ is tangent to the contour of f at (a,b).

101. If you know the gradient vector of f at (a,b) then you can find the directional derivative $f_{\vec{u}}(a,b)$ for any unit vector $\vec{u}$.

102. If you know the directional derivative $f_{\vec{u}}(a,b)$ for all unit vectors $\vec{u}$ then you can find the gradient vector of f at (a,b).

103. The directional derivative $f_{\vec{u}}(a,b)$ is parallel to $\vec{u}$.

104. The gradient $\operatorname{grad} f(3,4)$ is perpendicular to the vector $3\vec{i} + 4\vec{j}$.

105. If $\operatorname{grad} f(1,2) = \vec{i}$, then f decreases in the $-\vec{i}$ direction at $(1,2)$.

106. If $\operatorname{grad} f(1,2) = \vec{i}$, then $f(10,2) > f(1,2)$.

107. At the point $(3,0)$, the function $g(x,y) = x^2 + y^2$ has the same maximal rate of increase as that of the function $h(x,y) = 2xy$.

108. If $f(x,y) = e^{x+y}$, then the directional derivative in any direction $\vec{u}$ (with $\|\vec{u}\| = 1$) at the point $(0,0)$ is always less than or equal to $\sqrt{2}$.

Assume that $f(x,y)$ is a differentiable function. Are the statements in Problems 109–113 true or false? Explain your answer.

109. $f_{\vec{u}}(x_0, y_0)$ is a scalar.

110. $f_{\vec{u}}(a,b) = \|\nabla f(a,b)\|$

111. If $\vec{u}$ is tangent to the level curve of f at some point, then $\operatorname{grad} f \cdot \vec{u} = 0$ there.

112. There is always a direction in which the rate of change of f at (a,b) is 0.

113. There is a function with a point in its domain where $\|\operatorname{grad} f\| = 0$ and where there is a nonzero directional derivative.

14.5 GRADIENTS AND DIRECTIONAL DERIVATIVES IN SPACE

Directional Derivatives of Functions of Three Variables

We calculate directional derivatives of a function of three variables in the same way as for a function of two variables. If the function f is differentiable at the point (a,b,c), then the rate of change of $f(x,y,z)$ at the point (a,b,c) in the direction of a unit vector $\vec{u} = u_1\vec{i} + u_2\vec{j} + u_3\vec{k}$ is

$$f_{\vec{u}}(a,b,c) = f_x(a,b,c)u_1 + f_y(a,b,c)u_2 + f_z(a,b,c)u_3.$$

This can be justified using local linearity in the same way as for functions of two variables.

Example 1   Find the directional derivative of $f(x,y,z) = xy + z$ at the point $(-1,0,1)$ in the direction of the vector $\vec{v} = 2\vec{i} + \vec{k}$.

Solution   The magnitude of $\vec{v}$ is $\|\vec{v}\| = \sqrt{2^2 + 1} = \sqrt{5}$, so a unit vector in the same direction as $\vec{v}$ is

$$\vec{u} = \frac{\vec{v}}{\|\vec{v}\|} = \frac{2}{\sqrt{5}}\vec{i} + 0\vec{j} + \frac{1}{\sqrt{5}}\vec{k}.$$

The partial derivatives of f are $f_x(x, y, z) = y$ and $f_y(x, y, z) = x$ and $f_z(x, y, z) = 1$. Thus,

$$f_{\vec{u}}(-1, 0, 1) = f_x(-1, 0, 1)u_1 + f_y(-1, 0, 1)u_2 + f_z(-1, 0, 1)u_3$$
$$= (0)\left(\frac{2}{\sqrt{5}}\right) + (-1)(0) + (1)\left(\frac{1}{\sqrt{5}}\right) = \frac{1}{\sqrt{5}}.$$

The Gradient Vector of a Function of Three Variables

The gradient of a function of three variables is defined in the same way as for two variables:

$$\operatorname{grad} f(a, b, c) = f_x(a, b, c)\vec{i} + f_y(a, b, c)\vec{j} + f_z(a, b, c)\vec{k}.$$

Directional derivatives are related to gradients in the same way as for functions of two variables:

$$f_{\vec{u}}(a, b, c) = f_x(a, b, c)u_1 + f_y(a, b, c)u_2 + f_z(a, b, c)u_3 = \operatorname{grad} f(a, b, c) \cdot \vec{u}.$$

Since $\operatorname{grad} f(a, b, c) \cdot \vec{u} = \|\operatorname{grad} f(a, b, c)\| \cos\theta$, where θ is the angle between $\operatorname{grad} f(a, b, c)$ and $\vec{u}$, the value of $f_{\vec{u}}(a, b, c)$ is largest when $\theta = 0$, that is, when $\vec{u}$ is in the same direction as $\operatorname{grad} f(a, b, c)$. In addition, $f_{\vec{u}}(a, b, c) = 0$ when $\theta = \pi/2$, so $\operatorname{grad} f(a, b, c)$ is perpendicular to the level surface of f. The properties of gradients in space are similar to those in the plane:

Properties of the Gradient Vector in Space

If f is differentiable at (a, b, c) and $\vec{u}$ is a unit vector, then

$$f_{\vec{u}}(a, b, c) = \operatorname{grad} f(a, b, c) \cdot \vec{u}.$$

If, in addition, $\operatorname{grad} f(a, b, c) \neq \vec{0}$, then
- $\operatorname{grad} f(a, b, c)$ is in the direction of the greatest rate of increase of f
- $\operatorname{grad} f(a, b, c)$ is perpendicular to the level surface of f at (a, b, c)
- $\|\operatorname{grad} f(a, b, c)\|$ is the maximum rate of change of f at (a, b, c).

Example 2   Let $f(x, y, z) = x^2 + y^2$ and $g(x, y, z) = -x^2 - y^2 - z^2$. What can we say about the direction of the following vectors?
(a) $\operatorname{grad} f(0, 1, 1)$ (b) $\operatorname{grad} f(1, 0, 1)$ (c) $\operatorname{grad} g(0, 1, 1)$ (d) $\operatorname{grad} g(1, 0, 1)$.

Solution   The cylinder $x^2 + y^2 = 1$ in Figure 14.43 is a level surface of f and contains both the points $(0, 1, 1)$ and $(1, 0, 1)$. Since the value of f does not change at all in the z-direction, all the gradient vectors are horizontal. They are perpendicular to the cylinder and point outward because the value of f increases as we move out.

Similarly, the points $(0, 1, 1)$ and $(1, 0, 1)$ also lie on the same level surface of g, namely $g(x, y, z) = -x^2 - y^2 - z^2 = -2$, which is the sphere $x^2 + y^2 + z^2 = 2$. Part of this level surface is shown in Figure 14.44. This time the gradient vectors point inward, since the negative signs mean that the function increases (from large negative values to small negative values) as we move inward.

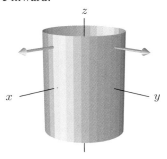

Figure 14.43: The level surface $f(x, y, z) = x^2 + y^2 = 1$ with two gradient vectors

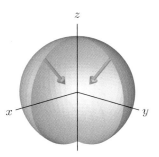

Figure 14.44: The level surface $g(x, y, z) = -x^2 - y^2 - z^2 = -2$ with two gradient vectors

Example 3

Consider the functions $f(x, y) = 4 - x^2 - 2y^2$ and $g(x, y) = 4 - x^2$. Calculate a vector perpendicular to each of the following:

(a) The level curve of f at the point $(1, 1)$
(b) The surface $z = f(x, y)$ at the point $(1, 1, 1)$
(c) The level curve of g at the point $(1, 1)$
(d) The surface $z = g(x, y)$ at the point $(1, 1, 3)$

Solution

(a) The vector we want is a 2-vector in the plane. Since $\operatorname{grad} f = -2x\vec{i} - 4y\vec{j}$, we have

$$\operatorname{grad} f(1, 1) = -2\vec{i} - 4\vec{j}.$$

Any nonzero multiple of this vector is perpendicular to the level curve at the point $(1, 1)$.

(b) In this case we want a 3-vector in space. To find it we rewrite $z = 4 - x^2 - 2y^2$ as the level surface of the function F, where

$$F(x, y, z) = 4 - x^2 - 2y^2 - z = 0.$$

Then

$$\operatorname{grad} F = -2x\vec{i} - 4y\vec{j} - \vec{k},$$

so

$$\operatorname{grad} F(1, 1, 1) = -2\vec{i} - 4\vec{j} - \vec{k},$$

and $\operatorname{grad} F(1, 1, 1)$ is perpendicular to the surface $z = 4 - x^2 - 2y^2$ at the point $(1, 1, 1)$. Notice that $-2\vec{i} - 4\vec{j} - \vec{k}$ is not the only possible answer: any multiple of this vector will do.

(c) We are looking for a 2-vector. Since $\operatorname{grad} g = -2x\vec{i} + 0\vec{j}$, we have

$$\operatorname{grad} g(1, 1) = -2\vec{i}.$$

Any multiple of this vector is perpendicular to the level curve also.

(d) We are looking for a 3-vector. We rewrite $z = 4 - x^2$ as the level surface of the function G, where

$$G(x, y, z) = 4 - x^2 - z = 0.$$

Then

$$\operatorname{grad} G = -2x\vec{i} - \vec{k}$$

So

$$\operatorname{grad} G(1, 1, 3) = -2\vec{i} - \vec{k},$$

and any multiple of $\operatorname{grad} G(1, 1, 3)$ is perpendicular to the surface $z = 4 - x^2$ at this point.

Example 4   (a) A hiker on the surface $f(x, y) = 4 - x^2 - 2y^2$ at the point $(1, -1, 1)$ starts to climb along the path of steepest ascent. What is the relation between the vector $\text{grad } f(1, -1)$ and a vector tangent to the path at the point $(1, -1, 1)$ and pointing uphill?

(b) At the point $(1, -1, 1)$ on the surface $f(x, y) = 4 - x^2 - 2y^2$, calculate a vector, $\vec{n}$, perpendicular to the surface and a vector, $\vec{T}$, tangent to the curve of steepest ascent.

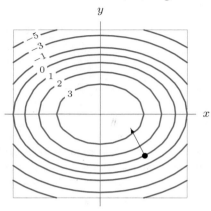

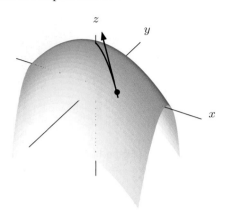

Figure 14.45: Contour diagram for $z = f(x, y) = 4 - x^2 - 2y^2$ showing direction of $\text{grad } f(1, -1)$

Figure 14.46: Graph of $f(x, y) = 4 - x^2 - 2y^2$ showing path of steepest ascent from the point $(1, -1, 1)$

Solution   (a) The hiker at the point $(1, -1, 1)$ lies directly above the point $(1, -1)$ in the xy-plane. The vector $\text{grad } f(1, -1)$ lies in 2-space, pointing like a compass in the direction in which f increases most rapidly. Therefore, $\text{grad } f(1, -1)$ lies directly under a vector tangent to the hiker's path at $(1, -1, 1)$ and pointing uphill. (See Figures 14.45 and 14.46.)

(b) The surface is represented by $F(x, y, z) = 4 - x^2 - 2y^2 - z = 0$. Since $\text{grad } F = -2x\vec{i} - 4y\vec{j} - \vec{k}$, the normal, $\vec{n}$, to the surface is given by

$$\vec{n} = \text{grad } F(1, -1, 1) = -2(1)\vec{i} - 4(-1)\vec{j} - \vec{k} = -2\vec{i} + 4\vec{j} - \vec{k}.$$

We take the $\vec{i}$ and $\vec{j}$ components of $\vec{T}$ to be the vector $\text{grad } f(1, -1) = -2\vec{i} + 4\vec{j}$. Thus, we have that, for some $a > 0$,

$$\vec{T} = -2\vec{i} + 4\vec{j} + a\vec{k}$$

We want $\vec{n} \cdot \vec{T} = 0$, so

$$\vec{n} \cdot \vec{T} = (-2\vec{i} + 4\vec{j} - \vec{k}) \cdot (-2\vec{i} + 4\vec{j} + a\vec{k}) = 4 + 16 - a = 0$$

So $a = 20$ and hence

$$\vec{T} = -2\vec{i} + 4\vec{j} + 20\vec{k}.$$

Example 5   Find the equation of the tangent plane to the sphere $x^2 + y^2 + z^2 = 14$ at the point $(1, 2, 3)$.

Solution   We write the sphere as a level surface as follows:

$$f(x, y, z) = x^2 + y^2 + z^2 = 14.$$

We have

$$\text{grad } f = 2x\vec{i} + 2y\vec{j} + 2z\vec{k},$$

so the vector

$$\text{grad } f(1, 2, 3) = 2\vec{i} + 4\vec{j} + 6\vec{k}$$

is perpendicular to the sphere at the point $(1, 2, 3)$. Since the vector $\text{grad } f(1, 2, 3)$ is normal to the tangent plane, the equation of the plane is

$$2x + 4y + 6z = 2 \cdot 1 + 4 \cdot 2 + 6 \cdot 3 = 28 \quad \text{or} \quad x + 2y + 3z = 14.$$

We could also try to find the tangent plane to the level surface $f(x, y, z) = k$ by solving algebraically for z and using the method of Section 14.3, page 791. (See Problem 47.) Solving for z can be difficult or impossible, however, so the method of Example 5 is preferable.

Tangent Plane to a Level Surface

If $f(x, y, z)$ is differentiable at (a, b, c), then an equation for the tangent plane to the level surface of f at the point (a, b, c) is

$$f_x(a, b, c)(x - a) + f_y(a, b, c)(y - b) + f_z(a, b, c)(z - c) = 0.$$

Caution: Scale on the Axis and the Geometric Interpretation of the Gradient

When we interpreted the gradient of a function geometrically (page 801), we tacitly assumed that the units and scales along the x and y axes were the same. If the scales are not the same, the gradient vector may not look perpendicular to the contours. Consider the function $f(x, y) = x^2 + y$ with gradient vector grad $f = 2x\vec{i} + \vec{j}$. Figure 14.47 shows the gradient vector at $(1, 1)$ using the same scales in the x and y directions. As expected, the gradient vector is perpendicular to the contour line. Figure 14.48 shows contours of the same function with unequal scales on the two axes. Notice that the gradient vector no longer appears perpendicular to the contour lines. Thus, we see that the geometric interpretation of the gradient vector requires that the same scale be used on both axes.

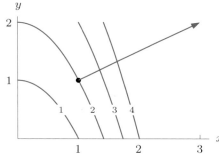

Figure 14.47: The gradient vector with x and y scales equal

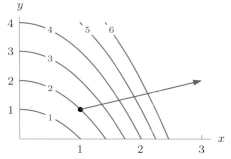

Figure 14.48: The gradient vector with x and y scales unequal

Exercises and Problems for Section 14.5

Exercises

In Exercises 1–13, find the gradient of the function.

1. $f(x, y, z) = x^2$

2. $f(x, y, z) = z^3$

3. $f(x, y, z) = x^2 + y^3 - z^4$

4. $f(x, y, z) = e^{x+y+z}$

5. $f(x, y, z) = \cos(x + y) + \sin(y + z)$

6. $f(x, y, z) = yz^2/(1 + x^2)$

7. $f(x, y, z) = 1/(x^2 + y^2 + z^2)$

8. $f(x, y, z) = \sqrt{x^2 + y^2 + z^2}$

9. $f(x, y, z) = xe^y \sin z$

10. $f(x, y, z) = xy + \sin(e^z)$

11. $f(x_1, x_2, x_3) = x_1^2 x_2^3 x_3^4$

12. $f(p, q, r) = e^p + \ln q + e^{r^2}$

13. $f(x, y, z) = e^{z^2} + y \ln(x^2 + 5)$

In Exercises 14–19, find the gradient at the point.

14. $f(x, y, z) = zy^2$, at $(1, 0, 1)$

15. $f(x, y, z) = 2x + 3y + 4z$, at $(1, 1, 1)$

16. $f(x, y, z) = x^2 + y^2 - z^4$, at $(3, 2, 1)$

17. $f(x, y, z) = xyz$, at $(1, 2, 3)$

18. $f(x, y, z) = \sin(xy) + \sin(yz)$, at $(1, \pi, -1)$

19. $f(x, y, z) = x \ln(yz)$, at $(2, 1, e)$

In Exercises 20–25, find the directional derivative using $f(x, y, z) = xy + z^2$.

20. At $(1, 2, 3)$ in the direction of $\vec{i} + \vec{j} + \vec{k}$.

21. At $(1, 1, 1)$ in the direction of $\vec{i} + 2\vec{j} + 3\vec{k}$.

22. As you leave the point $(1, 1, 0)$ heading in the direction of the point $(0, 1, 1)$.

23. As you arrive at $(0, 1, 1)$ from the direction of $(1, 1, 0)$.

24. At the point $(2, 3, 4)$ in the direction of a vector making an angle of $3\pi/4$ with grad $f(2, 3, 4)$.

25. At the point $(2, 3, 4)$ in the direction of the maximum rate of change of f.

In Exercises 26–31, check that the point $(-1, 1, 2)$ lies on the given surface. Then, viewing the surface as a level surface for a function $f(x, y, z)$, find a vector normal to the surface and an equation for the tangent plane to the surface at $(-1, 1, 2)$.

26. $x^2 - y^2 + z^2 = 4$ **27.** $z = x^2 + y^2$

28. $y^2 = z^2 - 3$ **29.** $x^2 - xyz = 3$

30. $\cos(x + y) = e^{xz+2}$ **31.** $y = 4/(2x + 3z)$

32. For $f(x, y, z) = 3x^2y^2 + 2yz$, find the directional derivative at the point $(-1, 0, 4)$ in the direction of
(a) $\vec{i} - \vec{k}$ (b) $-\vec{i} + 3\vec{j} + 3\vec{k}$

33. If $f(x, y, z) = x^2 + 3xy + 2z$, find the directional derivative at the point $(2, 0, -1)$ in the direction of $2\vec{i} + \vec{j} - 2\vec{k}$.

34. (a) Let $f(x, y, z) = x^2 + y^2 - xyz$. Find grad f.
(b) Find the equation for the tangent plane to the surface $f(x, y, z) = 7$ at the point $(2, 3, 1)$.

35. Find the equation of the tangent plane at the point $(3, 2, 2)$ to $z = \sqrt{17 - x^2 - y^2}$.

36. Find the equation of the tangent plane to $z = 8/(xy)$ at the point $(1, 2, 4)$.

37. Find an equation of the tangent plane and of a normal vector to the surface $x = y^3z^7$ at the point $(1, -1, -1)$.

In Exercises 38–39, the gradient of f and a point P on the level surface $f(x, y, z) = 0$ are given. Find an equation for the tangent plane to the surface at the point P.

38. grad $f = yz\vec{i} + xz\vec{j} + xy\vec{k}$, $P = (1, 2, 3)$

39. grad $f = 2x\vec{i} + z^2\vec{j} + 2yz\vec{k}$, $P = (10, -10, 30)$

In Exercises 40–44, find an equation of the tangent plane to the surface at the given point.

40. $x^2 + y^2 + z^2 = 17$ at the point $(2, 3, 2)$
41. $x^2 + y^2 = 1$ at the point $(1, 0, 0)$
42. $z = 9/(x + 4y)$ at the point where $x = 1$ and $y = 2$
43. $z = 2x + y + 3$ at the point $(0, 0, 3)$
44. $3x^2 - 4xy + z^2 = 0$ at the point (a, a, a), where $a \neq 0$

Problems

45. Consider the surface $g(x, y) = 4 - x^2$. What is the relation between grad $g(-1, -1)$ and a vector tangent to the path of steepest ascent at $(-1, -1, 3)$? Illustrate your answer with a sketch.

46. Match the functions $f(x, y, z)$ in (a)–(d) with the descriptions of their gradients in (I)–(IV). No reasons needed.

(a) $x^2 + y^2 + z^2$ (b) $x^2 + y^2$

(c) $\dfrac{1}{x^2 + y^2 + z^2}$ (d) $\dfrac{1}{x^2 + y^2}$

  I Points radially outward from the z-axis.
  II Points radially inward toward the z-axis.
  III Points radially outward from the origin.
  IV Points radially inward toward the origin.

47. Find the equation of the tangent plane at $(2, 3, 1)$ to the surface $x^2 + y^2 - xyz = 7$. Do this in two ways:

(a) Viewing the surface as the level set of a function of three variables, $F(x, y, z)$.
(b) Viewing the surface as the graph of a function of two variables $z = f(x, y)$.

48. Let $f(x, y, z) = x^2 + y^2 + z^2$. At the point $(1, 2, 1)$, find the rate of change of f in the direction perpendicular to

the plane $x + 2y + 3z = 8$ and moving away from the origin.

49. Let $f(x, y) = \cos x \sin y$ and let S be the surface $z = f(x, y)$.

(a) Find a normal vector to the surface S at the point $(0, \pi/2, 1)$.
(b) What is the equation of the tangent plane to the surface S at the point $(0, \pi/2, 1)$?

50. Let $f(x, y, z) = \sin(x^2 + y^2 + z^2)$.

(a) Describe in words the shape of the level surfaces of f.
(b) Find grad f.
(c) Consider the two vectors $\vec{r} = x\vec{i} + y\vec{j} + z\vec{k}$ and grad f at a point (x, y, z) where $\sin(x^2 + y^2 + z^2) \neq 0$. What is (are) the possible values(s) of the angle between these vectors?

51. Each diagram (I) – (IV) in Figure 14.49 represents the level curves of a function $f(x, y)$. For each function f, consider the point above P on the surface $z = f(x, y)$ and choose from the lists which follow:

(a) A vector which could be the normal to the surface at that point;
(b) An equation which could be the equation of the tangent plane to the surface at that point.

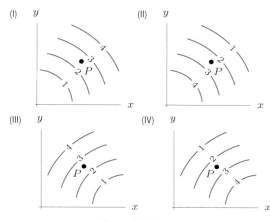

Figure 14.49

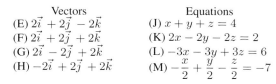

Vectors	Equations
(E) $2\vec{i} + 2\vec{j} - 2\vec{k}$	(J) $x + y + z = 4$
(F) $2\vec{i} + 2\vec{j} + 2\vec{k}$	(K) $2x - 2y - 2z = 2$
(G) $2\vec{i} - 2\vec{j} + 2\vec{k}$	(L) $-3x - 3y + 3z = 6$
(H) $-2\vec{i} + 2\vec{j} + 2\vec{k}$	(M) $-\dfrac{x}{2} + \dfrac{y}{2} - \dfrac{z}{2} = -7$

52. The surface S is represented by the equation $F = 0$ where $F(x, y, z) = x^2 - (y/z^2)$.

(a) Find the unit vectors $\vec{u}_1$ and $\vec{u}_2$ pointing in the direction of maximum increase of F at the points $(0, 0, 1)$ and $(1, 1, 1)$ respectively.

(b) Find the tangent plane to S at the points $(0, 0, 1)$ and $(1, 1, 1)$.

(c) Find all points on S where a normal vector is parallel to the xy-plane.

53. Consider the function $f(x, y) = (e^x - x) \cos y$. Suppose S is the surface $z = f(x, y)$.

(a) Find a vector which is perpendicular to the level curve of f through the point $(2, 3)$ in the direction in which f decreases most rapidly.

(b) Suppose $\vec{v} = 5\vec{i} + 4\vec{j} + a\vec{k}$ is a vector in 3-space which is tangent to the surface S at the point P lying on the surface above $(2, 3)$. What is a?

54. (a) Find the tangent plane to the surface $x^2 + y^2 + 3z^2 = 4$ at the point $(0.6, 0.8, 1)$.

(b) Is there a point on the surface $x^2 + y^2 + 3z^2 = 4$ at which the tangent plane is parallel to the plane $8x + 6y + 30z = 1$? If so, find it. If not, explain why not.

55. Your house lies on the surface $z = f(x, y) = 2x^2 - y^2$ directly above the point $(4, 3)$ in the xy-plane.

(a) How high above the xy-plane do you live?

(b) What is the slope of your lawn as you look from your house directly toward the z-axis (that is, along the vector $-4\vec{i} - 3\vec{j}$)?

(c) When you wash your car in the driveway, on this surface above the point $(4, 3)$, which way does the water run off? (Give your answer as a two-dimensional vector.)

(d) What is the equation of the tangent plane to this surface at your house?

56. (a) Sketch the contours of $z = y - \sin x$ for $z = -1, 0, 1, 2$.

(b) A bug starts on the surface at the point $(\pi/2, 1, 0)$ and walks on the surface $z = y - \sin x$ in the direction parallel to the y-axis, in the direction of increasing y. Is the bug walking in a valley or on top of a ridge? Explain.

(c) On the contour $z = 0$ in your sketch for part (a), draw the gradients of z at $x = 0$, $x = \pi/2$, and $x = \pi$.

57. At what point on the surface $z = 1 + x^2 + y^2$ is its tangent plane parallel to the following planes?

(a) $z = 5$ **(b)** $z = 5 + 6x - 10y$.

58. The concentration of salt in a fluid at (x, y, z) is given by $F(x, y, z) = x^2 + y^4 + x^2 z^2$ mg/cm^3. You are at the point $(-1, 1, 1)$.

(a) In which direction should you move if you want the concentration to increase the fastest?

(b) You start to move in the direction you found in part (a) at a speed of 4 cm/sec. How fast is the concentration changing?

59. Let $g_x(2, 1, 7) = 3$, $g_y(2, 1, 7) = 10$, $g_z(2, 1, 7) = -5$. Find the equation of the tangent plane to $g(x, y, z) = 0$ at the point $(2, 1, 7)$.

60. The vector ∇f at point P and four unit vectors $\vec{u}_1, \vec{u}_2, \vec{u}_3, \vec{u}_4$ are shown in Figure 14.50. Arrange the following quantities in ascending order

$$f_{\vec{u}_1}, \quad f_{\vec{u}_2}, \quad f_{\vec{u}_3}, \quad f_{\vec{u}_4}, \quad \text{the number } 0.$$

The directional derivatives are all evaluated at the point P and the function $f(x, y)$ is differentiable at P.

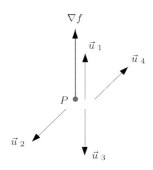

Figure 14.50

61. The temperature of a gas at the point (x, y, z) is given by $G(x, y, z) = x^2 - 5xy + y^2 z$.

    (a) What is the rate of change in the temperature at the point $(1, 2, 3)$ in the direction $\vec{v} = 2\vec{i} + \vec{j} - 4\vec{k}$?

    (b) What is the direction of maximum rate of change of temperature at the point $(1, 2, 3)$?

    (c) What is the maximum rate of change at the point $(1, 2, 3)$?

62. The temperature at the point (x, y, z) in 3-space is given, in degrees Celsius, by $T(x, y, z) = e^{-(x^2 + y^2 + z^2)}$.

    (a) Describe in words the shape of surfaces on which the temperature is constant.

    (b) Find grad T.

    (c) You travel from the point $(1, 0, 0)$ to the point $(2, 1, 0)$ at a speed of 3 units per second. Find the instantaneous rate of change of the temperature as you leave the point $(1, 0, 0)$. Give units.

63. Let $\vec{r} = x\vec{i} + y\vec{j} + z\vec{k}$ and $\vec{a}$ be a constant vector. For each of the quantities in (a)–(c), choose the statement in (I)–(V) that describes it. No reasons needed.

    (a) grad$(\vec{r} + \vec{a})$ **(b)** grad$(\vec{r} \cdot \vec{a})$ **(c)** grad$(\vec{r} \times \vec{a})$

    I Scalar, independent of $\vec{a}$.
    II Scalar, depends on $\vec{a}$.
    III Vector, independent of $\vec{a}$.
    IV Vector, depends on $\vec{a}$.
    V Not defined.

64. The earth has mass M and is located at the origin in 3-space, while the moon has mass m. Newton's Law of Gravitation states that if the moon is located at the point (x, y, z) then the attractive force exerted by the earth on the moon is given by the vector

$$\vec{F} = -GMm \frac{\vec{r}}{\|\vec{r}\|^3},$$

where $\vec{r} = x\vec{i} + y\vec{j} + z\vec{k}$. Show that $\vec{F} = \text{grad } \varphi$, where φ is the function given by

$$\varphi(x, y, z) = \frac{GMm}{\|\vec{r}\|}.$$

65. Two surfaces are said to be *tangential* at a point P if they have the same tangent plane at P. Show that the surfaces $z = \sqrt{2x^2 + 2y^2 - 25}$ and $z = \frac{1}{5}(x^2 + y^2)$ are tangential at the point $(4, 3, 5)$.

66. Two surfaces are said to be *orthogonal* to each other at a point P if the normals to their tangent planes are perpendicular at P. Show that the surfaces $z = \frac{1}{2}(x^2 + y^2 - 1)$ and $z = \frac{1}{2}(1 - x^2 - y^2)$ are orthogonal at all points of intersection.

67. Let $\vec{r}$ be the position vector of the point (x, y, z). If $\vec{\mu} = \mu_1 \vec{i} + \mu_2 \vec{j} + \mu_3 \vec{k}$ is a constant vector, show that

$$\text{grad}(\vec{\mu} \cdot \vec{r}) = \vec{\mu}.$$

68. Let $\vec{r}$ be the position vector of the point (x, y, z). Show that, if a is a constant,

$$\text{grad}(\|\vec{r}\|^a) = a\|\vec{r}\|^{a-2}\vec{r}, \qquad \vec{r} \neq \vec{0}.$$

69. Let f and g be functions on 3-space. Suppose f is differentiable and that

$$\text{grad } f(x, y, z) = (x\vec{i} + y\vec{j} + z\vec{k})g(x, y, z).$$

Explain why f must be constant on any sphere centered at the origin.

Strengthen Your Understanding

In Problems 70–71, explain what is wrong with the statement.

70. The gradient vector grad $f(x, y)$ points in the direction perpendicular to the surface $z = f(x, y)$.

71. The tangent plane at the origin to a surface $f(x, y, z) = 1$ that contains the point $(0, 0, 0)$ has equation

$$f_x(0, 0, 0)x + f_y(0, 0, 0)y + f_z(0, 0, 0)z + 1 = 0.$$

In Problems 72–74, give an example of:

72. A surface $z = f(x, y)$ such that the vector $\vec{i} - 2\vec{j} - \vec{k}$ is normal to the tangent plane at the point where $(x, y) = (0, 0)$.

73. A function $f(x, y, z)$ such that grad $f = 2\vec{i} + 3\vec{j} + 4\vec{k}$.

74. Two nonparallel unit vectors $\vec{u}$ and $\vec{v}$ such that $f_{\vec{u}}(0, 0, 0) = f_{\vec{v}}(0, 0, 0) = 0$, where $f(x, y, z) = 2x - 3y$.

Are the statements in Problems 75–78 true or false? Give reasons for your answer.

75. An equation for the tangent plane to the surface $z = x^2 + y^3$ at $(1, 1)$ is $z = 2 + 2x(x - 1) + 3y^2(y - 1)$.

76. There is a function $f(x, y)$ which has a tangent plane with equation $z = 0$ at a point (a, b).

77. There is a function with $\|\text{grad } f\| = 4$ and $f_{\vec{k}} = 5$ at some point.

78. There is a function with $\|\text{grad } f\| = 5$ and $f_{\vec{k}} = -3$ at some point.

79. Let $f(x, y, z)$ represent the temperature in °C at the point (x, y, z) with x, y, z in meters. Let $\vec{v}$ be your velocity in meters per second. Give units and an interpretation of each of the following quantities.

    (a) $\|\text{grad } f\|$ **(b)** grad $f \cdot \vec{v}$ **(c)** $\|\text{grad } f\| \cdot \|\vec{v}\|$

14.6 THE CHAIN RULE

Composition of Functions of Many Variables and Rates of Change

The chain rule enables us to differentiate *composite functions*. If we have a function of two variables $z = f(x, y)$ and we substitute $x = g(t), y = h(t)$ into $z = f(x, y)$, then we have a composite function in which z is a function of t:

$$z = f(g(t), h(t)).$$

If, on the other hand, we substitute $x = g(u, v), y = h(u, v)$, then we have a different composite function in which z is a function of u and v:

$$z = f(g(u, v), h(u, v)).$$

The next example shows how to calculate the rate of change of a composite function.

Example 1   Corn production, C, depends on annual rainfall, R, and average temperature, T, so $C = f(R, T)$. Global warming predicts that both rainfall and temperature depend on time. Suppose that according to a particular model of global warming, rainfall is decreasing at 0.2 cm per year and temperature is increasing at $0.1°C$ per year. Use the fact that at current levels of production, $f_R = 3.3$ and $f_T = -5$ to estimate the current rate of change, dC/dt.

Solution   By local linearity, we know that changes ΔR and ΔT generate a change, ΔC, in C given approximately by

$$\Delta C \approx f_R \Delta R + f_T \Delta T = 3.3 \Delta R - 5 \Delta T.$$

We want to know how ΔC depends on the time increment, Δt. A change Δt causes changes ΔR and ΔT, which in turn cause a change ΔC. The model of global warming tells us that

$$\frac{dR}{dt} = -0.2 \quad \text{and} \quad \frac{dT}{dt} = 0.1.$$

Thus, a time increment, Δt, generates changes of ΔR and ΔT given by

$$\Delta R \approx -0.2 \Delta t \quad \text{and} \quad \Delta T \approx 0.1 \Delta t.$$

Substituting for ΔR and ΔT in the expression for ΔC gives us

$$\Delta C \approx 3.3(-0.2 \Delta t) - 5(0.1 \Delta t) = -1.16 \Delta t.$$

Thus,

$$\frac{\Delta C}{\Delta t} \approx -1.16 \quad \text{and, therefore,} \quad \frac{dC}{dt} \approx -1.16.$$

The relationship between ΔC and Δt, which gives the value of dC/dt, is an example of the *chain rule*. The argument in Example 1 leads to more general versions of the chain rule.

The Chain Rule for $z = f(x, y)$, $x = g(t)$, $y = h(t)$

Since $z = f(g(t), h(t))$ is a function of t, we can consider the derivative dz/dt. The chain rule gives dz/dt in terms of the derivatives of $f, g,$ and h. Since dz/dt represents the rate of change of z with t, we look at the change Δz generated by a small change, Δt.

We substitute the local linearisations

$$\Delta x \approx \frac{dx}{dt} \Delta t \quad \text{and} \quad \Delta y \approx \frac{dy}{dt} \Delta t$$

into the local linearisation

$$\Delta z \approx \frac{\partial z}{\partial x} \Delta x + \frac{\partial z}{\partial y} \Delta y,$$

yielding

$$\Delta z \approx \frac{\partial z}{\partial x} \frac{dx}{dt} \Delta t + \frac{\partial z}{\partial y} \frac{dy}{dt} \Delta t$$

$$= \left(\frac{\partial z}{\partial x} \frac{dx}{dt} + \frac{\partial z}{\partial y} \frac{dy}{dt} \right) \Delta t.$$

Thus,

$$\frac{\Delta z}{\Delta t} \approx \frac{\partial z}{\partial x} \frac{dx}{dt} + \frac{\partial z}{\partial y} \frac{dy}{dt}.$$

Taking the limit as $\Delta t \to 0$, we get the following result.

If f, g, and h are differentiable and if $z = f(x, y)$, and $x = g(t)$, and $y = h(t)$, then

$$\frac{dz}{dt} = \frac{\partial z}{\partial x} \frac{dx}{dt} + \frac{\partial z}{\partial y} \frac{dy}{dt}.$$

Visualising the Chain Rule with a Diagram

The diagram in Figure 14.51 provides a way of remembering the chain rule. It shows the chain of dependence: z depends on x and y, which in turn depend on t. Each line in the diagram is labelled with a derivative relating the variables at its ends.

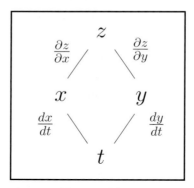

Figure 14.51: Diagram for $z = f(x, y)$, $x = g(t)$, $y = h(t)$. Lines represent dependence of z on x and y, and of x and y on t

The diagram keeps track of how a change in t propagates through the chain of composed functions. There are two paths from t to z, one through x and one through y. For each path, we multiply together the derivatives along the path. Then, to calculate dz/dt, we add the contributions from the two paths.

Example 2   Suppose that $z = f(x, y) = x \sin y$, where $x = t^2$ and $y = 2t + 1$. Let $z = g(t)$. Compute $g'(t)$ directly and using the chain rule.

Solution   Since $z = g(t) = f(t^2, 2t + 1) = t^2 \sin(2t + 1)$, it is possible to compute $g'(t)$ directly by one-variable methods:

$$g'(t) = t^2 \frac{d}{dt}(\sin(2t + 1)) + \left(\frac{d}{dt}(t^2) \right) \sin(2t + 1) = 2t^2 \cos(2t + 1) + 2t \sin(2t + 1).$$

The chain rule provides an alternative route to the same answer. We have

$$\frac{dz}{dt} = \frac{\partial z}{\partial x}\frac{dx}{dt} + \frac{\partial z}{\partial y}\frac{dy}{dt} = (\sin y)(2t) + (x\cos y)(2) = 2t\sin(2t+1) + 2t^2\cos(2t+1).$$

Example 3   The capacity, C, of a communication channel, such as a telephone line, to carry information depends on the ratio of the signal strength, S, to the noise, N. For some positive constant k,

$$C = k\ln\left(1 + \frac{S}{N}\right).$$

Suppose that the signal and noise are given as a function of time, t in seconds, by

$$S(t) = 4 + \cos(4\pi t) \qquad N(t) = 2 + \sin(2\pi t).$$

What is dC/dt one second after transmission started? Is the capacity increasing or decreasing at that instant?

Solution   By the chain rule

$$\frac{dC}{dt} = \frac{\partial C}{\partial S}\frac{dS}{dt} + \frac{\partial C}{\partial N}\frac{dN}{dt}$$

$$= \frac{k}{1 + S/N} \cdot \frac{1}{N}(-4\pi\sin 4\pi t) + \frac{k}{1 + S/N}\left(-\frac{S}{N^2}\right)(2\pi\cos 2\pi t).$$

When $t = 1$, the first term is zero, $S(1) = 5$, and $N(1) = 2$, so

$$\frac{dC}{dt} = \frac{k}{1 + S(1)/N(1)}\left(-\frac{S(1)}{(N(1))^2}\right)\cdot 2\pi = \frac{k}{1 + \frac{5}{2}}\left(-\frac{5}{4}\right)\cdot 2\pi.$$

Since dC/dt is negative, the capacity is decreasing at time $t = 1$ second.

How to Formulate a General Chain Rule

A diagram can be used to write the chain rule for general compositions.

> To find the rate of change of one variable with respect to another in a chain of composed differentiable functions:
> - Draw a diagram expressing the relationship between the variables, and label each link in the diagram with the derivative relating the variables at its ends.
> - For each path between the two variables, multiply together the derivatives from each step along the path.
> - Add the contributions from each path.

The diagram keeps track of all the ways in which a change in one variable can cause a change in another; the diagram generates all the terms we would get from the appropriate substitutions into the local linearisations.

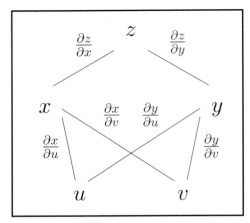

Figure 14.52: Diagram for $z = f(x, y)$, $x = g(u, v)$, $y = h(u, v)$. Lines represent dependence of z on x and y, and of x and y on u and v

For example, we can use Figure 14.52 to find formulas for $\partial z/\partial u$ and $\partial z/\partial v$. Adding the contributions for the two paths from z to u, we get the following results:

If f, g, h are differentiable and if $z = f(x, y)$, with $x = g(u, v)$ and $y = h(u, v)$, then

$$\frac{\partial z}{\partial u} = \frac{\partial z}{\partial x}\frac{\partial x}{\partial u} + \frac{\partial z}{\partial y}\frac{\partial y}{\partial u},$$

$$\frac{\partial z}{\partial v} = \frac{\partial z}{\partial x}\frac{\partial x}{\partial v} + \frac{\partial z}{\partial y}\frac{\partial y}{\partial v}.$$

Example 4　Let $w = x^2 e^y$, $x = 4u$, and $y = 3u^2 - 2v$. Compute $\partial w/\partial u$ and $\partial w/\partial v$ using the chain rule.

Solution　Using the previous result, we have

$$\frac{\partial w}{\partial u} = \frac{\partial w}{\partial x}\frac{\partial x}{\partial u} + \frac{\partial w}{\partial y}\frac{\partial y}{\partial u} = 2xe^y(4) + x^2 e^y(6u) = (8x + 6x^2 u)e^y$$

$$= (32u + 96u^3)e^{3u^2 - 2v}.$$

Similarly,

$$\frac{\partial w}{\partial v} = \frac{\partial w}{\partial x}\frac{\partial x}{\partial v} + \frac{\partial w}{\partial y}\frac{\partial y}{\partial v} = 2xe^y(0) + x^2 e^y(-2) = -2x^2 e^y$$

$$= -32u^2 e^{3u^2 - 2v}.$$

Example 5　A quantity z can be expressed either as a function of x and y, so that $z = f(x, y)$, or as a function of u and v, so that $z = g(u, v)$. The two coordinate systems are related by

$$x = u + v, \quad y = u - v.$$

(a) Use the chain rule to express $\partial z/\partial u$ and $\partial z/\partial v$ in terms of $\partial z/\partial x$ and $\partial z/\partial y$.
(b) Solve the equations in part (a) for $\partial z/\partial x$ and $\partial z/\partial y$.
(c) Show that the expressions we get in part (b) are the same as we get by expressing u and v in terms of x and y and using the chain rule.

Solution　(a) We have $\partial x/\partial u = 1$ and $\partial x/\partial v = 1$, and also $\partial y/\partial u = 1$ and $\partial y/\partial v = -1$. Thus,

$$\frac{\partial z}{\partial u} = \frac{\partial z}{\partial x}(1) + \frac{\partial z}{\partial y}(1) = \frac{\partial z}{\partial x} + \frac{\partial z}{\partial y}$$

and

$$\frac{\partial z}{\partial v} = \frac{\partial z}{\partial x}(1) + \frac{\partial z}{\partial y}(-1) = \frac{\partial z}{\partial x} - \frac{\partial z}{\partial y}.$$

(b) Adding together the equations for $\partial z/\partial u$ and $\partial z/\partial v$, we get

$$\frac{\partial z}{\partial u} + \frac{\partial z}{\partial v} = 2\frac{\partial z}{\partial x}, \quad \text{so} \quad \frac{\partial z}{\partial x} = \frac{1}{2}\frac{\partial z}{\partial u} + \frac{1}{2}\frac{\partial z}{\partial v}.$$

Similarly, subtracting the equations for $\partial z/\partial u$ and $\partial z/\partial v$ yields

$$\frac{\partial z}{\partial y} = \frac{1}{2}\frac{\partial z}{\partial u} - \frac{1}{2}\frac{\partial z}{\partial v}.$$

(c) Alternatively, we can solve the equations

$$x = u + v, \quad y = u - v$$

for u and v, which yields

$$u = \frac{1}{2}x + \frac{1}{2}y, \quad v = \frac{1}{2}x - \frac{1}{2}y.$$

Now we can think of z as a function of u and v, and u and v as functions of x and y, and apply the chain rule again. This gives us

$$\frac{\partial z}{\partial x} = \frac{\partial z}{\partial u}\frac{\partial u}{\partial x} + \frac{\partial z}{\partial v}\frac{\partial v}{\partial x} = \frac{1}{2}\frac{\partial z}{\partial u} + \frac{1}{2}\frac{\partial z}{\partial v}$$

and

$$\frac{\partial z}{\partial y} = \frac{\partial z}{\partial u}\frac{\partial u}{\partial y} + \frac{\partial z}{\partial v}\frac{\partial v}{\partial y} = \frac{1}{2}\frac{\partial z}{\partial u} - \frac{1}{2}\frac{\partial z}{\partial v}.$$

These are the same expressions we got in part (b).

An Application to Physical Chemistry

A chemist investigating the properties of a gas such as carbon dioxide may want to know how the internal energy U of a given quantity of the gas depends on its temperature, T, pressure, P, and volume, V. The three quantities T, P, and V are not independent, however. For instance, according to the ideal gas law, they satisfy the equation

$$PV = kT$$

where k is a constant which depends only upon the quantity of the gas. The internal energy can then be thought of as a function of any two of the three quantities T, P, and V:

$$U = U_1(T, P) = U_2(T, V) = U_3(P, V).$$

The chemist writes, for example, $\left(\frac{\partial U}{\partial T}\right)_P$ to indicate the partial derivative of U with respect to T *holding P constant*, signifying that for this computation U is viewed as a function of T and P. Thus, we interpret $\left(\frac{\partial U}{\partial T}\right)_P$ as

$$\left(\frac{\partial U}{\partial T}\right)_P = \frac{\partial U_1(T, P)}{\partial T}.$$

If U is to be viewed as a function of T and V, the chemist writes $\left(\frac{\partial U}{\partial T}\right)_V$ for the partial derivative of U with respect to T holding V constant: thus, $\left(\frac{\partial U}{\partial T}\right)_V = \frac{\partial U_2(T, V)}{\partial T}$.

Each of the functions U_1, U_2, U_3 gives rise to one of the following formulas for the differential dU:

$$dU = \left(\frac{\partial U}{\partial T}\right)_P dT + \left(\frac{\partial U}{\partial P}\right)_T dP \qquad \text{corresponds to } U_1,$$

$$dU = \left(\frac{\partial U}{\partial T}\right)_V dT + \left(\frac{\partial U}{\partial V}\right)_T dV \qquad \text{corresponds to } U_2,$$

$$dU = \left(\frac{\partial U}{\partial P}\right)_V dP + \left(\frac{\partial U}{\partial V}\right)_P dV \qquad \text{corresponds to } U_3.$$

All the six partial derivatives appearing in formulas for dU have physical meaning, but they are not all equally easy to measure experimentally. A relationship among the partial derivatives, usually derived from the chain rule, may make it possible to evaluate one of the partials in terms of others that are more easily measured.

Example 6 Suppose a gas satisfies the equation $PV = 2T$ and $P = 3$ when $V = 4$. If $\left(\frac{\partial U}{\partial P}\right)_V = 7$ and $\left(\frac{\partial U}{\partial V}\right)_P = 8$, find the values of $\left(\frac{\partial U}{\partial P}\right)_T$ and $\left(\frac{\partial U}{\partial T}\right)_P$.

Solution Since we know the values of $\left(\frac{\partial U}{\partial P}\right)_V$ and $\left(\frac{\partial U}{\partial V}\right)_P$, we think of U as a function of P and V and use the function U_3 to write

$$dU = \left(\frac{\partial U}{\partial P}\right)_V dP + \left(\frac{\partial U}{\partial V}\right)_P dV$$
$$dU = 7dP + 8dV.$$

To calculate $\left(\frac{\partial U}{\partial P}\right)_T$ and $\left(\frac{\partial U}{\partial T}\right)_P$, we think of U as a function of T and P. Thus, we want to substitute for dV in terms of dT and dP. Since $PV = 2T$, we have

$$P\,dV + V\,dP = 2dT,$$
$$3dV + 4dP = 2dT.$$

Solving gives $dV = (2dT - 4dP)/3$, so

$$dU = 7dP + 8\left(\frac{2dT - 4dP}{3}\right)$$
$$dU = -\frac{11}{3}dP + \frac{16}{3}dT.$$

Comparing with the formula for dU obtained from U_1,

$$dU = \left(\frac{\partial U}{\partial T}\right)_P dT + \left(\frac{\partial U}{\partial P}\right)_T dP,$$

we have

$$\left(\frac{\partial U}{\partial T}\right)_P = \frac{16}{3} \qquad \text{and} \qquad \left(\frac{\partial U}{\partial P}\right)_T = -\frac{11}{3}.$$

In Example 6, we could have substituted for dP instead of dV, leading to values of $\left(\frac{\partial U}{\partial T}\right)_V$ and $\left(\frac{\partial U}{\partial V}\right)_T$. See Problem 42.

In general, if for some particular P, V, and T, we can measure two of the six quantities $\left(\frac{\partial U}{\partial P}\right)_V$, $\left(\frac{\partial U}{\partial V}\right)_P$, $\left(\frac{\partial U}{\partial P}\right)_T$, $\left(\frac{\partial U}{\partial T}\right)_P$, $\left(\frac{\partial U}{\partial V}\right)_T$, $\left(\frac{\partial U}{\partial T}\right)_V$, then we can compute the other four using the relationship between dP, dV, and dT given by the gas law. General formulas for each partial derivative in terms of others can be obtained in the same way. See the following example and Problem 42.

Example 7   Express $\left(\dfrac{\partial U}{\partial T}\right)_P$ in terms of $\left(\dfrac{\partial U}{\partial T}\right)_V$ and $\left(\dfrac{\partial U}{\partial V}\right)_T$ and $\left(\dfrac{\partial V}{\partial T}\right)_P$.

Solution   Since we are interested in the derivatives $\left(\dfrac{\partial U}{\partial T}\right)_V$ and $\left(\dfrac{\partial U}{\partial V}\right)_T$, we think of U as a function of T and V and use the formula

$$dU = \left(\dfrac{\partial U}{\partial T}\right)_V dT + \left(\dfrac{\partial U}{\partial V}\right)_T dV \qquad \text{corresponding to } U_2.$$

We want to find a formula for $\left(\dfrac{\partial U}{\partial T}\right)_P$, which means thinking of U as a function of T and P. Thus, we want to substitute for dV. Since V is a function of T and P, we have

$$dV = \left(\dfrac{\partial V}{\partial T}\right)_P dT + \left(\dfrac{\partial V}{\partial P}\right)_T dP.$$

Substituting for dV into the formula for dU corresponding to U_2 gives

$$dU = \left(\dfrac{\partial U}{\partial T}\right)_V dT + \left(\dfrac{\partial U}{\partial V}\right)_T \left(\left(\dfrac{\partial V}{\partial T}\right)_P dT + \left(\dfrac{\partial V}{\partial P}\right)_T dP\right).$$

Collecting the terms containing dT and the terms containing dP gives

$$dU = \left(\left(\dfrac{\partial U}{\partial T}\right)_V + \left(\dfrac{\partial U}{\partial V}\right)_T \left(\dfrac{\partial V}{\partial T}\right)_P\right) dT + \left(\dfrac{\partial U}{\partial V}\right)_T \left(\dfrac{\partial V}{\partial P}\right)_T dP.$$

But we also have the formula

$$dU = \left(\dfrac{\partial U}{\partial T}\right)_P dT + \left(\dfrac{\partial U}{\partial P}\right)_T dP \qquad \text{corresponding to } U_1.$$

We now have two formulas for dU in terms of dT and dP. The coefficients of dT must be identical, so we conclude

$$\left(\dfrac{\partial U}{\partial T}\right)_P = \left(\dfrac{\partial U}{\partial T}\right)_V + \left(\dfrac{\partial U}{\partial V}\right)_T \left(\dfrac{\partial V}{\partial T}\right)_P.$$

Example 7 expresses $\left(\dfrac{\partial U}{\partial T}\right)_P$ in terms of three other partial derivatives. Two of them, namely $\left(\dfrac{\partial U}{\partial T}\right)_V$, the constant-volume heat capacity, and $\left(\dfrac{\partial V}{\partial T}\right)_P$, the expansion coefficient, can be easily measured experimentally. The third, the internal pressure, $\left(\dfrac{\partial U}{\partial V}\right)_T$, cannot be measured directly but can be related to $\left(\dfrac{\partial P}{\partial T}\right)_V$, which is measurable. Thus, $\left(\dfrac{\partial U}{\partial T}\right)_P$ can be determined indirectly using this identity.

Exercises and Problems for Section 14.6

Exercises

For Exercises 1–6, find dz/dt using the chain rule. Assume the variables are restricted to domains on which the functions are defined.

1. $z = xy^2$, $x = e^{-t}$, $y = \sin t$

2. $z = \ln(x^2 + y^2)$, $x = 1/t$, $y = \sqrt{t}$

3. $z = xe^y$, $x = 2t$, $y = 1 - t^2$

4. $z = (x + y)e^y$, $x = 2t$, $y = 1 - t^2$

5. $z = x \sin y + y \sin x$, $x = t^2$, $y = \ln t$

6. $z = \sin(x/y)$, $x = 2t$, $y = 1 - t^2$

For Exercises 7–15, find $\partial z/\partial u$ and $\partial z/\partial v$. The variables are restricted to domains on which the functions are defined.

7. $z = xe^y$, $x = \ln u$, $y = v$

8. $z = (x + y)e^y$, $x = \ln u$, $y = v$

9. $z = xe^y$, $x = u^2 + v^2$, $y = u^2 - v^2$

10. $z = (x + y)e^y$, $x = u^2 + v^2$, $y = u^2 - v^2$

11. $z = \sin(x/y)$, $x = \ln u$, $y = v$

12. $z = \ln(xy)$, $x = (u^2 + v^2)^2$, $y = (u^3 + v^3)^2$

13. $z = \tan^{-1}(x/y)$, $x = u^2 + v^2$, $y = u^2 - v^2$

14. $z = xe^{-y} + ye^{-x}$, $x = u \sin v$, $y = v \cos u$

15. $z = \cos(x^2 + y^2)$, $x = u \cos v$, $y = u \sin v$

Problems

16. Use the chain rule to find dz/dt, and check the result by expressing z as a function of t and differentiating directly.
$$z = x^3 y^2, \quad x = t^3, \quad y = t^2$$

17. Use the chain rule to find $\partial w/\partial \rho$ and $\partial w/\partial \theta$, given that
$$w = x^2 + y^2 - z^2,$$
and
$$x = \rho \sin \phi \cos \theta, \quad y = \rho \sin \phi \sin \theta, \quad z = \rho \cos \phi.$$

18. A bison is charging across the plain one morning. His path takes him to location (x, y) at time t where x and y are functions of t and north is in the direction of increasing y. The temperature is always colder farther north. As time passes, the sun rises in the sky, sending out more heat, and a cold front blows in from the east. At time t the air temperature H near the bison is given by $H = f(x, y, t)$. The chain rule expresses the derivative dH/dt as a sum of three terms:
$$\frac{dH}{dt} = \frac{\partial f}{\partial x}\frac{dx}{dt} + \frac{\partial f}{\partial y}\frac{dy}{dt} + \frac{\partial f}{\partial t}.$$

Identify the term that gives the contribution to the change in temperature experienced by the bison that is due to

(a) The rising sun.

(b) The coming cold front.

(c) The bison's change in latitude.

19. Let $z = f(x, y)$ where $x = g(t)$, $y = h(t)$ and f, g, h are all differentiable functions. Given the information in the table, find $\left.\dfrac{\partial z}{\partial t}\right|_{t=1}$.

$f(3, 10) = 7$	$f(4, 11) = -20$
$f_x(3, 10) = 100$	$f_y(3, 10) = 0.1$
$f_x(4, 11) = 200$	$f_y(4, 11) = 0.2$
$f(3, 4) = -10$	$f(10, 11) = -1$
$g(1) = 3$	$h(1) = 10$
$g'(1) = 4$	$h'(1) = 11$

20. The voltage, V, (in volts) across a circuit is given by Ohm's law: $V = IR$, where I is the current (in amps) flowing through the circuit and R is the resistance (in ohms). If we place two circuits, with resistance R_1 and R_2, in parallel, then their combined resistance, R, is given by
$$\frac{1}{R} = \frac{1}{R_1} + \frac{1}{R_2}.$$

Suppose the current is 2 amps and increasing at 10^{-2} amp/sec and R_1 is 3 ohms and increasing at 0.5 ohm/sec, while R_2 is 5 ohms and decreasing at 0.1 ohm/sec. Calculate the rate at which the voltage is changing.

21. The air pressure is decreasing at a rate of 2 pascals per kilometer in the eastward direction. In addition, the air pressure is dropping at a constant rate with respect to time everywhere. A ship sailing eastward at 10 km/hour past an island takes barometer readings and records a pressure drop of 50 pascals in 2 hours. Estimate the time rate of change of air pressure on the island. (A pascal is a unit of air pressure.)

22. A steel bar with square cross sections 5 cm by 5 cm and length 3 meters is being heated. For each dimension, the bar expands $13 \cdot 10^{-6}$ meters for each $1°$C rise in temperature.[2] What is the rate of change in the volume of the steel bar?

23. Suppose that f is any differentiable function of one variable. Define V, a function of two variables, by
$$V(x, t) = f(x + ct).$$

Show that V satisfies the equation
$$\frac{\partial V}{\partial t} = c\frac{\partial V}{\partial x}.$$

24. Corn production, C, is a function of rainfall, R, and temperature, T. (See Example 1 on page 815.) Figures 14.53 and 14.54 show how rainfall and temperature are predicted to vary with time because of global warming. Suppose we know that $\Delta C \approx 1.1\Delta R - 5\Delta T$. Use this to estimate the change in corn production between the year 2020 and the year 2021. Hence, estimate dC/dt when $t = 2020$.

---
[2]http://www.engineeringtoolbox.com/

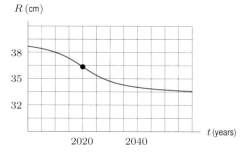

Figure 14.53: Rainfall as a function of time

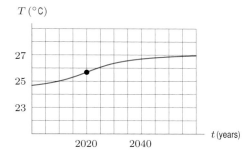

Figure 14.54: Temperature as a function of time

25. The function $g(\rho)$ is graphed in Figure 14.55. Let $\rho = \sqrt{x^2 + y^2 + z^2}$. Define f, a function of x, y, z by $f(x, y, z) = g\left(\sqrt{x^2 + y^2 + z^2}\right)$. Let $\vec{F} = \text{grad } f$.

(a) Describe precisely in words the level surfaces of f.
(b) Give a unit vector in the direction of $\vec{F}$ at the point $(1, 2, 2)$.
(c) Estimate $\|\vec{F}\|$ at the point $(1, 2, 2)$.
(d) Estimate $\vec{F}$ at the point $(1, 2, 2)$.
(e) The points $(1, 2, 2)$ and $(3, 0, 0)$ are both on the sphere $x^2 + y^2 + z^2 = 9$. Estimate $\vec{F}$ at $(3, 0, 0)$.
(f) If P and Q are any two points on the sphere $x^2 + y^2 + z^2 = k^2$:

   (i) Compare the magnitudes of $\vec{F}$ at P and at Q.
   (ii) Describe the directions of $\vec{F}$ at P and at Q.

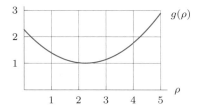

Figure 14.55

26. Let $z = g(u, v, w)$ and $u = u(s, t), v = v(s, t), w = w(s, t)$. How many terms are there in the expression for $\partial z/\partial t$?

27. Suppose $w = f(x, y, z)$ and that x, y, z are functions of u and v. Use a tree diagram to write down the chain rule formula for $\partial w/\partial u$ and $\partial w/\partial v$.

28. Suppose $w = f(x, y, z)$ and that x, y, z are all functions of t. Use a tree diagram to write down the chain rule for dw/dt.

29. Let $F(u, v)$ be a function of two variables. Find $f'(x)$ if

(a) $f(x) = F(x, 3)$ **(b)** $f(x) = F(3, x)$
(c) $f(x) = F(x, x)$ **(d)** $f(x) = F(5x, x^2)$

In Problems 30–31, let $z = f(x, y)$, $x = x(u, v)$, $y = y(u, v)$ and $x(1, 2) = 5$, $y(1, 2) = 3$, calculate the partial derivative in terms of some of the numbers a, b, c, d, e, k, p, q:

$$f_x(1, 2) = a \quad f_y(1, 2) = c \quad x_u(1, 2) = e \quad y_u(1, 2) = p$$
$$f_x(5, 3) = b \quad f_y(5, 3) = d \quad x_v(1, 2) = k \quad y_v(1, 2) = q$$

30. $z_u(1, 2)$ **31.** $z_v(1, 2)$

32. The equation $f(x, y) = f(a, b)$ defines a level curve through a point (a, b) where grad $f(a, b) \neq \vec{0}$. Use implicit differentiation and the chain rule to show that the slope of the line tangent to this curve at the point (a, b) is $-f_x(a, b)/f_y(a, b)$ if $f_y(a, b) \neq 0$.

33. Let $z = f(t)g(t)$. Use the chain rule applied to $h(x, y) = f(x)g(y)$ to show that $dz/dt = f'(t)g(t) + f(t)g'(t)$. The one-variable product rule for differentiation is a special case of the two-variable chain rule.

34. A function $f(x, y)$ is *homogeneous of degree* p if $f(tx, ty) = t^p f(x, y)$ for all t. Show that any differentiable, homogeneous function of degree p satisfies Euler's Theorem:

$$x f_x(x, y) + y f_y(x, y) = p f(x, y).$$

[Hint: Define $g(t) = f(tx, ty)$ and compute $g'(1)$.]

35. Let $F(x, y, z)$ be a function and define a function $z = f(x, y)$ implicitly by letting $F(x, y, f(x, y)) = 0$. Use the chain rule to show that

$$\frac{\partial z}{\partial x} = -\frac{\partial F/\partial x}{\partial F/\partial z} \quad \text{and} \quad \frac{\partial z}{\partial y} = -\frac{\partial F/\partial y}{\partial F/\partial z}.$$

In Problems 36–37, let $z = f(x, y)$, $x = x(u, v)$, $y = y(u, v)$ and $x(4, 5) = 2$, $y(4, 5) = 3$. Calculate the partial derivative in terms of $a, b, c, d, e, k, p, q, r, s, t, w$:

$$f_x(4, 5) = a \quad f_y(4, 5) = c \quad x_u(4, 5) = e \quad y_u(4, 5) = p$$
$$f_x(2, 3) = b \quad f_y(2, 3) = d \quad x_v(4, 5) = k \quad y_v(4, 5) = q$$
$$x_u(2, 3) = r \quad y_u(2, 3) = s \quad x_v(2, 3) = t \quad y_v(2, 3) = w$$

36. $z_u(4, 5)$ **37.** $z_v(4, 5)$

For Problems 38–39, suppose that $x > 0, y > 0$ and that z can be expressed either as a function of Cartesian coordinates (x, y) or as a function of polar coordinates (r, θ), so that $z = f(x, y) = g(r, \theta)$. [Recall that $x = r\cos\theta, y = r\sin\theta, r = \sqrt{x^2 + y^2}$, and, for $x > 0, y > 0, \theta = \arctan(y/x)$]

38. **(a)** Use the chain rule to find $\partial z/\partial r$ and $\partial z/\partial\theta$ in terms of $\partial z/\partial x$ and $\partial z/\partial y$.

   (b) Solve the equations you have just written down for $\partial z/\partial x$ and $\partial z/\partial y$ in terms of $\partial z/\partial r$ and $\partial z/\partial\theta$.

   (c) Show that the expressions you get in part (b) are the same as you would get by using the chain rule to find $\partial z/\partial x$ and $\partial z/\partial y$ in terms of $\partial z/\partial r$ and $\partial z/\partial\theta$.

39. Show that

$$\left(\frac{\partial z}{\partial x}\right)^2 + \left(\frac{\partial z}{\partial y}\right)^2 = \left(\frac{\partial z}{\partial r}\right)^2 + \frac{1}{r^2}\left(\frac{\partial z}{\partial\theta}\right)^2.$$

Problems 40–45 are continuations of the physical chemistry example on page 821.

40. Write $\left(\frac{\partial U}{\partial P}\right)_V$ as a partial derivative of one of the functions $U_1, U_2,$ or U_3.

41. Write $\left(\frac{\partial U}{\partial P}\right)_T$ as a partial derivative of one of the functions U_1, U_2, U_3.

42. For the gas in Example 6, find $\left(\frac{\partial U}{\partial T}\right)_V$ and $\left(\frac{\partial U}{\partial V}\right)_T$. [Hint: Use the same method as the example, but substitute for dP instead of dV.]

43. Show that $\left(\frac{\partial T}{\partial V}\right)_P = 1 \Big/ \left(\frac{\partial V}{\partial T}\right)_P$.

44. Use Example 7 and Problem 43 to show that

$$\left(\frac{\partial U}{\partial V}\right)_P = \left(\frac{\partial U}{\partial V}\right)_T + \frac{\left(\frac{\partial U}{\partial T}\right)_V}{\left(\frac{\partial V}{\partial T}\right)_P}.$$

45. In Example 6, we calculated values of $(\partial U/\partial T)_P$ and $(\partial U/\partial P)_T$ using the relationship $PV = 2T$ for a specific gas. In this problem, you will derive general relationships for these two partial derivatives.

   (a) Think of V as a function of P and T and write an expression for dV.

   (b) Substitute for dV into the following formula for dU (thinking of U as a function of P and V):

$$dU = \left(\frac{\partial U}{\partial P}\right)_V dP + \left(\frac{\partial U}{\partial V}\right)_P dV.$$

   (c) Thinking of U as a function of P and T, write an expression for dU.

   (d) By comparing coefficients of dP and dT in your answers to parts (b) and (c), show that

$$\left(\frac{\partial U}{\partial T}\right)_P = \left(\frac{\partial U}{\partial V}\right)_P \cdot \left(\frac{\partial V}{\partial T}\right)_P$$

$$\left(\frac{\partial U}{\partial P}\right)_T = \left(\frac{\partial U}{\partial P}\right)_V + \left(\frac{\partial U}{\partial V}\right)_P \cdot \left(\frac{\partial V}{\partial P}\right)_T.$$

Problems 46–48 concern differentiating an integral in one variable, y, which also involves another variable x, either in the integrand, or in the limits, or both:

$$\int_0^5 (x^2 y + 4)\, dy \quad \text{or} \quad \int_0^x (y+4)\, dy \quad \text{or} \quad \int_0^x (xy+4)\, dy.$$

To differentiate the first integral with respect to x, it can be shown that in most cases we can differentiate with respect to x inside the integral:

$$\frac{d}{dx}\left(\int_0^5 (x^2 y + 4)\, dy\right) = \int_0^5 2xy\, dy.$$

Differentiating the second integral with respect to x uses the Fundamental Theorem of Calculus:

$$\frac{d}{dx}\int_0^x (y+4)\, dy = x + 4.$$

Differentiating the third integral involves the chain rule, as shown in Problem 48. Assume that the function F is continuously differentiable and b is constant throughout.

46. Let $f(x) = \int_0^b F(x, y)\, dy$. Find $f'(x)$.

47. Let $f(x) = \int_0^x F(b, y)\, dy$. Find $f'(x)$.

48. Let $f(x) = \int_0^x F(x, y)\, dy$. Use Problem 46 and Problem 47 to find $f'(x)$ by the following steps:

   (a) Let $G(u, w) = \int_0^w F(u, y)\, dy$. Find $G_u(u, w)$ and $G_w(u, w)$.

   (b) Use part (a) and the chain rule applied to $G(x, x) = f(x)$ to show:

$$f'(x) = \int_0^x F_x(x, y)\, dy + F(x, x).$$

Strengthen Your Understanding

In Problems 49–51, explain what is wrong with the statement.

49. If $z = f(g(t), h(t))$, then $dz/dt = f(g'(t), h(t)) + f(g(t), h'(t))$.

50. If $C = C(R, T), R = R(x, y), T = T(x, y)$ and $R(0, 2) = 5, T(0, 2) = 1$, then $C_x(0, 2) = C_R(0, 2)R_x(0, 2) + C_T(0, 2)T_x(0, 2)$.

51. If $z = f(x, y)$ and $x = g(t), y = h(t)$ with $g(0) = 2$ and $h(0) = 3$, then

$$\left.\frac{dz}{dt}\right|_{t=0} = f_x(0, 0)g'(0) + f_y(0, 0)h'(0).$$

In Problems 52–56, give an example of:

52. Functions $x = g(t)$ and $y = h(t)$ such that $(dz/dt)|_{t=0} = 9$, given that $z = x^2 y$.

53. A function $z = f(x, y)$ such that $dz/dt|_{t=0} = 10$, given that $x = e^{2t}$ and $y = \sin t$.

54. Functions z, x and y where you need to follow the diagram in order to answer questions about the derivative of z with respect to the other variables.

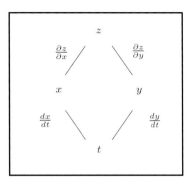

55. Functions w, u and v where you need to follow the diagram in order to answer questions about the derivative of w with respect to the other variables.

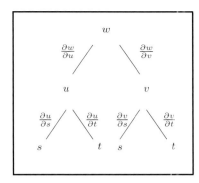

56. Function $z = f(x, y)$ where x and y are functions of one variable, t, for which $\dfrac{\partial z}{\partial t} = 2$.

57. Let $z = g(u, v)$ and $u = u(x, y, t)$, $v = v(x, y, t)$ and $x = x(t)$, $y = y(t)$. Then the expression for dz/dt has

(a) Three terms   (b) Four terms

(c) Six terms   (d) Seven terms

(e) Nine terms   (f) None of the above

14.7   SECOND-ORDER PARTIAL DERIVATIVES

Since the partial derivatives of a function are themselves functions, we can differentiate them, giving *second-order partial derivatives*. A function $z = f(x, y)$ has two first-order partial derivatives, f_x and f_y, and four second-order partial derivatives.

The Second-Order Partial Derivatives of $z = f(x, y)$

$$\frac{\partial^2 z}{\partial x^2} = f_{xx} = (f_x)_x, \qquad \frac{\partial^2 z}{\partial x \partial y} = f_{yx} = (f_y)_x,$$

$$\frac{\partial^2 z}{\partial y \partial x} = f_{xy} = (f_x)_y, \qquad \frac{\partial^2 z}{\partial y^2} = f_{yy} = (f_y)_y.$$

It is usual to omit the parentheses, writing f_{xy} instead of $(f_x)_y$ and $\dfrac{\partial^2 z}{\partial y \, \partial x}$ instead of $\dfrac{\partial}{\partial y}\left(\dfrac{\partial z}{\partial x}\right)$.

Example 1   Compute the four second-order partial derivatives of $f(x, y) = xy^2 + 3x^2 e^y$.

Solution   From $f_x(x, y) = y^2 + 6xe^y$ we get

$$f_{xx}(x, y) = \frac{\partial}{\partial x}(y^2 + 6xe^y) = 6e^y \quad \text{and} \quad f_{xy}(x, y) = \frac{\partial}{\partial y}(y^2 + 6xe^y) = 2y + 6xe^y.$$

From $f_y(x, y) = 2xy + 3x^2 e^y$ we get

$$f_{yx}(x, y) = \frac{\partial}{\partial x}(2xy + 3x^2 e^y) = 2y + 6xe^y \quad \text{and} \quad f_{yy}(x, y) = \frac{\partial}{\partial y}(2xy + 3x^2 e^y) = 2x + 3x^2 e^y.$$

Observe that $f_{xy} = f_{yx}$ in this example.

Example 2   Use the values of the function $f(x, y)$ in Table 14.6 to estimate $f_{xy}(1, 2)$ and $f_{yx}(1, 2)$.

Table 14.6   *Values of $f(x, y)$*

$y \backslash x$	0.9	1.0	1.1
1.8	4.72	5.83	7.06
2.0	6.48	8.00	9.60
2.2	8.62	10.65	12.88

Solution   Since $f_{xy} = (f_x)_y$, we first estimate f_x

$$f_x(1, 2) \approx \frac{f(1.1, 2) - f(1, 2)}{0.1} = \frac{9.60 - 8.00}{0.1} = 16.0,$$

$$f_x(1, 2.2) \approx \frac{f(1.1, 2.2) - f(1, 2.2)}{0.1} = \frac{12.88 - 10.65}{0.1} = 22.3.$$

Thus,

$$f_{xy}(1, 2) \approx \frac{f_x(1, 2.2) - f_x(1, 2)}{0.2} = \frac{22.3 - 16.0}{0.2} = 31.5.$$

Similarly,

$$f_{yx}(1, 2) \approx \frac{f_y(1.1, 2) - f_y(1, 2)}{0.1} \approx \frac{1}{0.1}\left(\frac{f(1.1, 2.2) - f(1.1, 2)}{0.2} - \frac{f(1, 2.2) - f(1, 2)}{0.2} \right)$$

$$= \frac{1}{0.1}\left(\frac{12.88 - 9.60}{0.2} - \frac{10.65 - 8.00}{0.2} \right) = 31.5.$$

Observe that in this example also, $f_{xy} = f_{yx}$.

The Mixed Partial Derivatives Are Equal

It is not an accident that the estimates for $f_{xy}(1, 2)$ and $f_{yx}(1, 2)$ are equal in Example 2, because the same values of the function are used to calculate each one. The fact that $f_{xy} = f_{yx}$ in Examples 1 and 2 corroborates the following general result; Problem 44 suggests why you might expect it to be true.[3]

Theorem 14.1: Equality of Mixed Partial Derivatives

If f_{xy} and f_{yx} are continuous at (a, b), an interior point of their domain, then

$$f_{xy}(a, b) = f_{yx}(a, b).$$

For most functions f we encounter and most points (a, b) in their domains, not only are f_{xy} and f_{yx} continuous at (a, b), but all their higher-order partial derivatives (such as f_{xxy} or f_{xyyy}) exist and are continuous at (a, b). In that case we say f is *smooth* at (a, b). We say f is smooth on a region R if it is smooth at every point of R.

What Do the Second-Order Partial Derivatives Tell Us?

Example 3   Let us return to the guitar string of Example 4, page 785. The string is 1 metre long and at time t seconds, the point x metres from one end is displaced $f(x, t)$ metres from its rest position, where

$$f(x, t) = 0.003 \sin(\pi x) \sin(2765t).$$

Compute the four second-order partial derivatives of f at the point $(x, t) = (0.3, 1)$ and describe the meaning of their signs in practical terms.

---
[3] For a proof, see M. Spivak, *Calculus on Manifolds*, p. 26 (New York: Benjamin, 1965).

Solution   First we compute $f_x(x,t) = 0.003\pi \cos(\pi x) \sin(2765t)$, from which we get

$$f_{xx}(x,t) = \frac{\partial}{\partial x}(f_x(x,t)) = -0.003\pi^2 \sin(\pi x)\sin(2765t), \qquad \text{so} \qquad f_{xx}(0.3,1) \approx -0.01;$$

and

$$f_{xt}(x,t) = \frac{\partial}{\partial t}(f_x(x,t)) = (0.003)(2765)\pi \cos(\pi x)\cos(2765t), \qquad \text{so} \qquad f_{xt}(0.3,1) \approx 14.$$

On page 785 we saw that $f_x(x,t)$ gives the slope of the string at any point and time. Therefore, $f_{xx}(x,t)$ measures the concavity of the string. The fact that $f_{xx}(0.3,1) < 0$ means the string is concave down at the point $x = 0.3$ when $t = 1$. (See Figure 14.56.)

On the other hand, $f_{xt}(x,t)$ is the rate of change of the slope of the string with respect to time. Thus, $f_{xt}(0.3,1) > 0$ means that at time $t = 1$ the slope at the point $x = 0.3$ is increasing. (See Figure 14.57.)

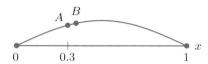

Figure 14.56: Interpretation of $f_{xx}(0.3,1) < 0$: The concavity of the string at $t = 1$

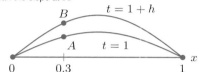

Figure 14.57: Interpretation of $f_{xt}(0.3,1) > 0$: The slope of one point on the string at two different times

Now we compute $f_t(x,t) = (0.003)(2765)\sin(\pi x)\cos(2765t)$, from which we get

$$f_{tx}(x,t) = \frac{\partial}{\partial x}(f_t(x,t)) = (0.003)(2765)\pi \cos(\pi x)\cos(2765t), \qquad \text{so} \quad f_{tx}(0.3,1) \approx 14$$

and

$$f_{tt}(x,t) = \frac{\partial}{\partial t}(f_t(x,t)) = -(0.003)(2765)^2 \sin(\pi x)\sin(2765t), \qquad \text{so} \quad f_{tt}(0.3,1) \approx -7200.$$

On page 785 we saw that $f_t(x,t)$ gives the velocity of the string at any point and time. Therefore, $f_{tx}(x,t)$ and $f_{tt}(x,t)$ will both be rates of change of velocity. That $f_{tx}(0.3,1) > 0$ means that at time $t = 1$ the velocities of points just to the right of $x = 0.3$ are greater than the velocity at $x = 0.3$. (See Figure 14.58.) That $f_{tt}(0.3,1) < 0$ means that the velocity of the point $x = 0.3$ is decreasing at time $t = 1$. Thus, $f_{tt}(0.3,1) = -7200$ m/sec^2 is the acceleration of this point. (See Figure 14.59.)

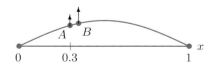

Figure 14.58: Interpretation of $f_{tx}(0.3,1) > 0$: The velocity of different points on the string at $t = 1$

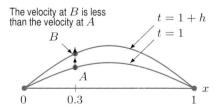

Figure 14.59: Interpretation of $f_{tt}(0.3,1) < 0$: Negative acceleration. The velocity of one point on the string at two different times

Taylor Approximations

We use second derivatives to construct quadratic Taylor approximations. In Section 14.3, we saw how to approximate $f(x, y)$ by a linear function (its local linearisation). We now see how to improve this approximation of $f(x, y)$ using a quadratic function.

Linear and Quadratic Approximations Near (0,0)

For a function of one variable, local linearity tells us that the best *linear* approximation is the degree-1 Taylor polynomial

$$f(x) \approx f(a) + f'(a)(x - a) \quad \text{for } x \text{ near } a.$$

A better approximation to $f(x)$ is given by the degree-2 Taylor polynomial:

$$f(x) \approx f(a) + f'(a)(x - a) + \frac{f''(a)}{2}(x - a)^2 \quad \text{for } x \text{ near } a.$$

For a function of two variables the local linearisation for (x, y) near (a, b) is

$$f(x, y) \approx L(x, y) = f(a, b) + f_x(a, b)(x - a) + f_y(a, b)(y - b).$$

In the case $(a, b) = (0, 0)$, we have:

Taylor Polynomial of Degree 1 Approximating $f(x, y)$ for (x, y) near (0,0)

If f has continuous first-order partial derivatives, then

$$f(x, y) \approx L(x, y) = f(0, 0) + f_x(0, 0)x + f_y(0, 0)y.$$

We get a better approximation to f by using a quadratic polynomial. We choose a quadratic polynomial $Q(x, y)$, with the same partial derivatives as the original function f. You can check that the following Taylor polynomial of degree 2 has this property.

Taylor Polynomial of Degree 2 Approximating $f(x, y)$ for (x, y) near (0,0)

If f has continuous second-order partial derivatives, then

$$f(x, y) \approx Q(x, y)$$
$$= f(0, 0) + f_x(0, 0)x + f_y(0, 0)y + \frac{f_{xx}(0, 0)}{2}x^2 + f_{xy}(0, 0)xy + \frac{f_{yy}(0, 0)}{2}y^2.$$

Example 4 Let $f(x, y) = \cos(2x + y) + 3\sin(x + y)$

(a) Compute the linear and quadratic Taylor polynomials, L and Q, approximating f near $(0, 0)$.

(b) Explain why the contour plots of L and Q for $-1 \le x \le 1$, $-1 \le y \le 1$ look the way they do.

Solution

(a) We have $f(0, 0) = 1$. The derivatives we need are as follows:

$$
\begin{aligned}
f_x(x, y) &= -2\sin(2x + y) + 3\cos(x + y) &\quad \text{so} \quad f_x(0, 0) &= 3, \\
f_y(x, y) &= -\sin(2x + y) + 3\cos(x + y) &\quad \text{so} \quad f_y(0, 0) &= 3, \\
f_{xx}(x, y) &= -4\cos(2x + y) - 3\sin(x + y) &\quad \text{so} \quad f_{xx}(0, 0) &= -4, \\
f_{xy}(x, y) &= -2\cos(2x + y) - 3\sin(x + y) &\quad \text{so} \quad f_{xy}(0, 0) &= -2, \\
f_{yy}(x, y) &= -\cos(2x + y) - 3\sin(x + y) &\quad \text{so} \quad f_{yy}(0, 0) &= -1.
\end{aligned}
$$

Thus, the linear approximation, $L(x, y)$, to $f(x, y)$ at $(0, 0)$ is given by

$$f(x, y) \approx L(x, y) = f(0, 0) + f_x(0, 0)x + f_y(0, 0)y = 1 + 3x + 3y.$$

The quadratic approximation, $Q(x, y)$, to $f(x, y)$ near $(0, 0)$ is given by

$$f(x, y) \approx Q(x, y)$$
$$= f(0, 0) + f_x(0, 0)x + f_y(0, 0)y + \frac{f_{xx}(0, 0)}{2}x^2 + f_{xy}(0, 0)xy + \frac{f_{yy}(0, 0)}{2}y^2$$
$$= 1 + 3x + 3y - 2x^2 - 2xy - \frac{1}{2}y^2.$$

Notice that the linear terms in $Q(x, y)$ are the same as the linear terms in $L(x, y)$. The quadratic terms in $Q(x, y)$ can be thought of as "correction terms" to the linear approximation.
(b) The contour plots of $f(x, y)$, $L(x, y)$, and $Q(x, y)$ are in Figures 14.60–14.62.

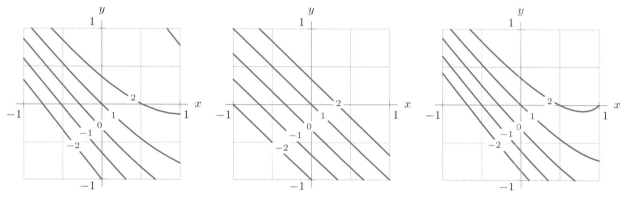

Figure 14.60: Original function, $f(x, y)$

Figure 14.61: Linear approximation, $L(x, y)$

Figure 14.62: Quadratic approximation, $Q(x, y)$

Notice that the contour plot of Q is more similar to the contour plot of f than is the contour plot of L. Since L is linear, the contour plot of L consists of parallel, equally spaced lines.

An alternative, and much quicker, way to find the Taylor polynomial in the previous example is to use the single-variable approximations. For example, since

$$\cos u = 1 - \frac{u^2}{2!} + \frac{u^4}{4!} + \cdots \quad \text{and} \quad \sin v = v - \frac{v^3}{3!} + \cdots,$$

we can substitute $u = 2x + y$ and $v = x + y$ and expand. We discard terms beyond the second (since we want the quadratic polynomial), getting

$$\cos(2x + y) = 1 - \frac{(2x + y)^2}{2!} + \frac{(2x + y)^4}{4!} + \cdots \approx 1 - \frac{1}{2}(4x^2 + 4xy + y^2) = 1 - 2x^2 - 2xy - \frac{1}{2}y^2$$

and

$$\sin(x + y) = (x + y) - \frac{(x + y)^3}{3!} + \cdots \approx x + y.$$

Combining these results, we get

$$\cos(2x + y) + 3\sin(x + y) \approx 1 - 2x^2 - 2xy - \frac{1}{2}y^2 + 3(x + y) = 1 + 3x + 3y - 2x^2 - 2xy - \frac{1}{2}y^2.$$

Linear and Quadratic Approximations near (a, b)

The local linearisation for a function $f(x, y)$ at a point (a, b) is

> **Taylor Polynomial of Degree 1 Approximating $f(x, y)$ for (x, y) near (a, b)**
> If f has continuous first-order partial derivatives, then
>
> $$f(x, y) \approx L(x, y) = f(a, b) + f_x(a, b)(x - a) + f_y(a, b)(y - b).$$

This suggests that a quadratic polynomial approximation $Q(x, y)$ for $f(x, y)$ near a point (a, b) should be written in terms of $(x - a)$ and $(y - b)$ instead of x and y. If we require that $Q(a, b) = f(a, b)$ and that the first- and second-order partial derivatives of Q and f at (a, b) be equal, then we get the following polynomial:

> **Taylor Polynomial of Degree 2 Approximating $f(x, y)$ for (x, y) near (a, b)**
> If f has continuous second-order partial derivatives, then
>
> $$f(x, y) \approx Q(x, y)$$
> $$= f(a, b) + f_x(a, b)(x - a) + f_y(a, b)(y - b)$$
> $$+ \frac{f_{xx}(a, b)}{2}(x - a)^2 + f_{xy}(a, b)(x - a)(y - b) + \frac{f_{yy}(a, b)}{2}(y - b)^2.$$

These coefficients are derived in exactly the same way as for $(a, b) = (0, 0)$.

Example 5   Find the Taylor polynomial of degree 2 at the point $(1, 2)$ for the function $f(x, y) = \dfrac{1}{xy}$.

Solution   Table 14.7 contains the partial derivatives and their values at the point $(1, 2)$.

Table 14.7   *Partial derivatives of $f(x, y) = 1/(xy)$*

Derivative	Formula	Value at $(1, 2)$	Derivative	Formula	Value at $(1, 2)$
$f(x, y)$	$1/(xy)$	$1/2$	$f_{xx}(x, y)$	$2/(x^3 y)$	1
$f_x(x, y)$	$-1/(x^2 y)$	$-1/2$	$f_{xy}(x, y)$	$1/(x^2 y^2)$	$1/4$
$f_y(x, y)$	$-1/(xy^2)$	$-1/4$	$f_{yy}(x, y)$	$2/(xy^3)$	$1/4$

So, the quadratic Taylor polynomial for f near $(1, 2)$ is

$$\frac{1}{xy} \approx Q(x, y)$$

$$= \frac{1}{2} - \frac{1}{2}(x - 1) - \frac{1}{4}(y - 2) + \frac{1}{2}(1)(x - 1)^2 + \frac{1}{4}(x - 1)(y - 2) + \left(\frac{1}{2}\right)\left(\frac{1}{4}\right)(y - 2)^2$$

$$= \frac{1}{2} - \frac{x - 1}{2} - \frac{y - 2}{4} + \frac{(x - 1)^2}{2} + \frac{(x - 1)(y - 2)}{4} + \frac{(y - 2)^2}{8}.$$

Exercises and Problems for Section 14.7

Exercises

In Exercises 1–11, calculate all four second-order partial derivatives and check that $f_{xy} = f_{yx}$. Assume the variables are restricted to a domain on which the function is defined.

1. $f(x, y) = 3x^2 y + 5xy^3$

2. $f(x, y) = (x + y)^2$

3. $f(x, y) = (x + y)^3$

4. $f(x, y) = xe^y$

5. $f(x, y) = e^{2xy}$

6. $f(x, y) = (x + y)e^y$

7. $f(x, y) = \sqrt{x^2 + y^2}$

8. $f(x, y) = \sin(x/y)$

9. $f(x, y) = 5x^3y^2 - 7xy^3 + 9x^2 + 11$

10. $f(x, y) = \sin(x^2 + y^2)$

11. $f(x, y) = 3\sin 2x \cos 5y$

In Exercises 12–21, use the level curves of the function $z = f(x, y)$ to decide the sign (positive, negative, or zero) of each of the following partial derivatives at the point P. Assume the x- and y-axes are in the usual positions.

(a) $f_x(P)$ **(b)** $f_y(P)$ **(c)** $f_{xx}(P)$

(d) $f_{yy}(P)$ **(e)** $f_{xy}(P)$

12.

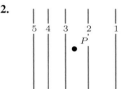

13.

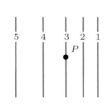

14.

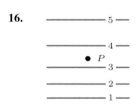

15.

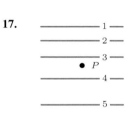

16.

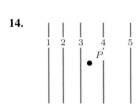

17.

18.

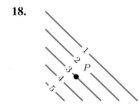

19.

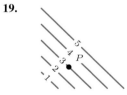

20.

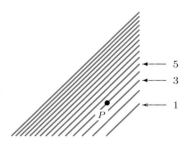

21.

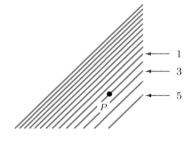

In Exercises 22–29, find the quadratic Taylor polynomials about $(0, 0)$ for the function.

22. $(y - 1)(x + 1)^2$ **23.** $(x - y + 1)^2$

24. $e^{-2x^2 - y^2}$ **25.** $e^x \cos y$

26. $1/(1 + 2x - y)$ **27.** $\cos(x + 3y)$

28. $\sin 2x + \cos y$ **29.** $\ln(1 + x^2 - y)$

In Exercises 30–31, find the best quadratic approximation for $f(x, y)$ for (x, y) near $(0, 0)$.

30. $f(x, y) = \ln(1 + x - 2y)$

31. $f(x, y) = \sqrt{1 + 2x - y}$

Problems

In Problems 32–35, find the linear, $L(x, y)$, and quadratic, $Q(x, y)$, Taylor polynomials valid near $(1, 0)$. Compare the values of the approximations $L(0.9, 0.2)$ and $Q(0.9, 0.2)$ with the exact value of the function $f(0.9, 0.2)$.

32. $f(x, y) = x^2y$ **33.** $f(x, y) = \sqrt{x + 2y}$

34. $f(x, y) = xe^{-y}$

35. $f(x, y) = \sin(x - 1)\cos y$

In Problems 36–38, show that the function satisfies Laplace's equation, $F_{xx} + F_{yy} = 0$.

36. $F(x, y) = e^{-x} \sin y$

37. $F(x, y) = \arctan(y/x)$

38. $F(x, y) = e^x \sin y + e^y \sin x$

39. If $u(x, t) = e^{at} \sin(bx)$ satisfies the heat equation $u_t = u_{xx}$, find the relationship between a and b.

40. (a) Check that $u(x, t)$ satisfies the heat equation $u_t = u_{xx}$ for $t > 0$ and all x, where

$$u(x, t) = \frac{1}{2\sqrt{\pi t}} e^{-x^2/(4t)}$$

(b) Graph $u(x, t)$ against x for $t = 0.01, 0.1, 1, 10$. These graphs represent the temperature in an infinitely long insulated rod that at $t = 0$ is $0°C$ everywhere except at the origin $x = 0$, and that is infinitely hot at $t = 0$ at the origin.

41. Figure 14.63 shows a graph of $z = f(x, y)$. Is $f_{xx}(0, 0)$ positive or negative? Is $f_{yy}(0, 0)$ positive or negative? Give reasons for your answers.

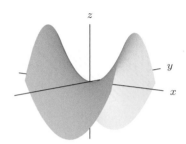

Figure 14.63

42. If $z = f(x) + yg(x)$, what can you say about z_{yy}? Explain your answer.

43. If $z_{xy} = 4y$, what can you say about the value of
(a) z_{yx}?   **(b)** z_{xyx}?   **(c)** z_{xyy}?

44. Give an explanation of why you might expect $f_{xy}(a, b) = f_{yx}(a, b)$ using the following steps.

(a) Write the definition of $f_x(a, b)$.
(b) Write a definition of $f_{xy}(a, b)$ as $(f_x)_y$.
(c) Substitute for f_x in the definition of f_{xy}.
(d) Write an expression for f_{yx} similar to the one for f_{xy} you obtained in part (c).
(e) Compare your answers to parts (c) and (d). What do you have to assume to conclude that f_{xy} and f_{yx} are equal?

45. A contour diagram for the smooth function $z = f(x, y)$ is in Figure 14.64.

(a) Is z an increasing or decreasing function of x? Of y?
(b) Is f_x positive or negative? How about f_y?
(c) Is f_{xx} positive or negative? How about f_{yy}?
(d) Sketch the direction of grad f at points P and Q.
(e) Is grad f longer at P or at Q? How do you know?

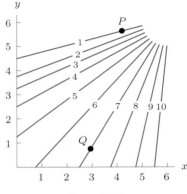

Figure 14.64

Problems 46–49 give tables of values of quadratic polynomials $P(x, y) = a + bx + cy + dx^2 + exy + fy^2$. Determine whether each of the coefficients d, e and f of the quadratic terms is positive, negative, or zero.

46.

		x		
		10	12	14
	10	35	37	39
y	15	45	47	49
	20	55	57	59

47.

		x		
		10	12	14
	10	26	36	54
y	15	31	41	59
	20	36	46	64

48.

		x		
		10	12	14
	10	90	82	74
y	15	75	87	99
	20	10	42	74

49.

		x		
		10	12	14
	10	13	33	61
y	15	28	28	36
	20	93	73	61

50. You are hiking on a level trail going due east and planning to strike off cross country up the mountain to your left. The slope up to the left is too steep now and seems to be gentler the further you go along the trail, so you decide to wait before turning off.

(a) Sketch a topographical contour map that illustrates this story.
(b) What information does the story give about partial derivatives? Define all variables and functions that you use.
(c) What partial derivative influenced your decision to wait before turning?

51. The weekly production, Y, in factories that manufacture a certain item is modeled as a function of the quantity of capital, K, and quantity of labor, L, at the factory. Data shows that hiring a few extra workers increases production. Moreover, for two factories with the same number of workers, hiring a few extra workers increases production more for the factory with more capital. (With more equipment, additional labor can be used more effectively.) What does this tell you about the sign of

(a) $\partial Y/\partial L$?
(b) $\partial^2 Y/(\partial K \partial L)$?

52. Data suggests that human surface area, S, can reasonably be modeled as a function of height, h, and weight, w. In the Dubois model, we have $\partial^2 S/\partial w^2 < 0$ and $\partial^2 S/(\partial h \partial w) > 0$. Two people A and B each gain 1 pound. Which experiences the greater increase in surface area if

(a) They have the same weight but A is taller?
(b) They have the same height, but A is heavier?

53. Figure 14.65 shows the level curves of a function $f(x, y)$ around a maximum or minimum, M. One of the points P and Q has coordinates (x_1, y_1) and the other has coordinates (x_2, y_2). Suppose $b > 0$ and $c > 0$. Consider the two linear approximations to f given by

$$f(x, y) \approx a + b(x - x_1) + c(y - y_1)$$
$$f(x, y) \approx k + m(x - x_2) + n(y - y_2).$$

(a) What is the relationship between the values of a and k?
(b) What are the coordinates of P?
(c) Is M a maximum or a minimum?
(d) What can you say about the sign of the constants m and n?

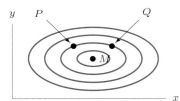

Figure 14.65

54. Consider the function $f(x, y) = (\sin x)(\sin y)$.

(a) Find the Taylor polynomials of degree 2 for f about the points $(0, 0)$ and $(\pi/2, \pi/2)$.
(b) Use the Taylor polynomials to sketch the contours of f close to each of the points $(0, 0)$ and $(\pi/2, \pi/2)$.

55. You plan to buy a used car. You are debating between a 5-year old car and a 10-year old car and thinking about the price. Experts report that the original price matters more when buying a 5-year old car than a 10-year old car. This suggests that we model the average market price, P, in dollars as a function of two variables: the original price, C, in dollars, and the age of the car, A, in years.

(a) Give units for the following partial derivatives and say whether you think they are positive or negative. Explain your reasoning.

(a) $\partial P/\partial A$
(b) $\partial P/\partial C$

(b) Express the experts' report in terms of partial derivatives.
(c) Using a quadratic polynomial to model P, we have

$$P = a + bC + cA + dC^2 + eCA + fA^2.$$

Which term in this polynomial is most relevant to the experts' report?

56. The tastiness, T, of a soup depends on the volume, V, of the soup in the pot and the quantity, S, of salt in the soup. If you have more soup, you need more salt to make it taste good. Match the three stories (a)–(c) to the three statements (I)–(III) about partial derivatives.

(a) I started adding salt to the soup in the pot. At first the taste improved, but eventually the soup became too salty and continuing to add more salt made it worse.
(b) The soup was too salty, so I started adding unsalted soup. This improved the taste at first, but eventually there was too much soup for the salt, and continuing to add unsalted soup just made it worse.
(c) The soup was too salty, so adding more salt would have made it taste worse. I added a quart of unsalted soup instead. Now it is not salty enough, but I can improve the taste by adding salt.

(I) $\partial^2 T/\partial V^2 < 0$
(II) $\partial^2 T/\partial S^2 < 0$
(III) $\partial^2 T/\partial V \partial S > 0$

57. Let $f(x, y) = \sqrt{x + 2y + 1}$.

(a) Compute the local linearization of f at $(0, 0)$.
(b) Compute the quadratic Taylor polynomial for f at $(0, 0)$.
(c) Compare the values of the linear and quadratic approximations in part (a) and part (b) with the true values for $f(x, y)$ at the points $(0.1, 0.1)$, $(-0.1, 0.1)$, $(0.1, -0.1)$, $(-0.1, -0.1)$. Which approximation gives the closest values?

58. Using a computer and your answer to Problem 57, draw the six contour diagrams of $f(x, y) = \sqrt{x + 2y + 1}$ and its linear and quadratic approximations, $L(x, y)$ and $Q(x, y)$, in the two windows $[-0.6, 0.6] \times [-0.6, 0.6]$ and $[-2, 2] \times [-2, 2]$. Explain the shape of the contours, their spacing, and the relationship between the contours of f, L, and Q.

59. Suppose that $f(x, y)$ has continuous partial derivatives f_x and f_y. Using the Fundamental Theorem of Calculus to evaluate the integrals, show that

$$f(a, b) = f(0, 0) + \int_{t=0}^{a} f_x(t, 0)dt + \int_{t=0}^{b} f_y(a, t)dt.$$

60. Suppose that $f(x, y)$ has continuous partial derivatives and that $f(0, 0) = 0$ and $|f_x(x, y)| \leq A$ and $|f_y(x, y)| \leq B$ for all points (x, y) in the plane. Use Problem 59 to show that $|f(x, y)| \leq A|x| + B|y|$.

This inequality shows how bounds on the partial derivatives of f limit the growth of f.

Strengthen Your Understanding

In Problems 61–62, explain what is wrong with the statement.

61. If $f(x, y) \neq 0$, then the Taylor polynomial of degree 2 approximating $f(x, y)$ near $(0, 0)$ is also nonzero.

62. There is a function $f(x, y)$ with partial derivatives $f_x = xy$ and $f_y = y^2$.

In Problems 63–65, give an example of:

63. A function $f(x, y)$ such that $f_{xx} \neq 0$, $f_{yy} \neq 0$, and $f_{xy} = 0$.

64. Formulas for two different functions $f(x, y)$ and $g(x, y)$ with the same quadratic approximation near $(0, 0)$.

65. Contour diagrams for two different functions $f(x, y)$ and $g(x, y)$ that have the same quadratic approximations near $(0, 0)$.

14.8 DIFFERENTIABILITY

In Section 14.3 we gave an informal introduction to the concept of differentiability. We called a function $f(x, y)$ *differentiable* at a point (a, b) if it is well approximated by a linear function near (a, b). This section focuses on the precise meaning of the phrase "well approximated." By looking at examples, we shall see that local linearity requires the existence of partial derivatives, but they do not tell the whole story. In particular, existence of partial derivatives at a point is not sufficient to guarantee local linearity at that point.

We begin by discussing the relation between continuity and differentiability. As an illustration, take a sheet of paper, crumple it into a ball and smooth it out again. Wherever there is a crease it would be difficult to approximate the surface by a plane—these are points of nondifferentiability of the function giving the height of the paper above the floor. Yet the sheet of paper models a graph which is continuous—there are no breaks. As in the case of one-variable calculus, continuity does not imply differentiability. But differentiability does *require* continuity: there cannot be linear approximations to a surface at points where there are abrupt changes in height.

Differentiability for Functions of Two Variables

For a function of two variables, as for a function of one variable, we define differentiability at a point in terms of the error and the distance from the point. If the point is (a, b) and a nearby point is $(a + h, b + k)$, the distance between them is $\sqrt{h^2 + k^2}$. (See Figure 14.66.)

A function $f(x, y)$ is **differentiable at the point** (a, b) if there is a linear function $L(x, y) = f(a, b) + m(x - a) + n(y - b)$ such that if the *error* $E(x, y)$ is defined by

$$f(x, y) = L(x, y) + E(x, y),$$

and if $h = x - a, k = y - b$, then the *relative error* $E(a + h, b + k)/\sqrt{h^2 + k^2}$ satisfies

$$\lim_{\substack{h \to 0 \\ k \to 0}} \frac{E(a + h, b + k)}{\sqrt{h^2 + k^2}} = 0.$$

The function f is **differentiable on a region** R if it is differentiable at each point of R. The function $L(x, y)$ is called the *local linearisation* of $f(x, y)$ near (a, b).

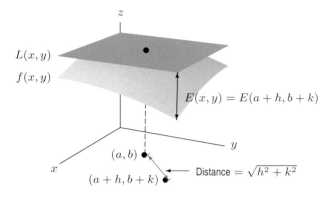

Figure 14.66: Graph of function $z = f(x, y)$ and its local linearisation $z = L(x, y)$ near the point (a, b)

Partial Derivatives and Differentiability

In the next example, we show that this definition of differentiability is consistent with our previous notion — that is, that $m = f_x$ and $n = f_y$ and that the graph of $L(x, y)$ is the tangent plane.

Example 1   Show that if f is a differentiable function with local linearisation $L(x, y) = f(a, b) + m(x - a) + n(y - b)$, then $m = f_x(a, b)$ and $n = f_y(a, b)$.

Solution   Since f is differentiable, we know that the relative error in $L(x, y)$ tends to 0 as we get close to (a, b). Suppose $h > 0$ and $k = 0$. Then we know that

$$0 = \lim_{h \to 0} \frac{E(a + h, b + k)}{\sqrt{h^2 + k^2}} = \lim_{h \to 0} \frac{E(a + h, b)}{h} = \lim_{h \to 0} \frac{f(a + h, b) - L(a + h, b)}{h}$$

$$= \lim_{h \to 0} \frac{f(a + h, b) - f(a, b) - mh}{h}$$

$$= \lim_{h \to 0} \left(\frac{f(a + h, b) - f(a, b)}{h} \right) - m = f_x(a, b) - m.$$

A similar result holds if $h < 0$, so we have $m = f_x(a, b)$. The result $n = f_y(a, b)$ is found in a similar manner.

The previous example shows that if a function is differentiable at a point, it has partial derivatives there. Therefore, if any of the partial derivatives fail to exist, then the function cannot be differentiable. This is what happens in the following example of a cone.

Example 2   Consider the function $f(x, y) = \sqrt{x^2 + y^2}$. Is f differentiable at the origin?

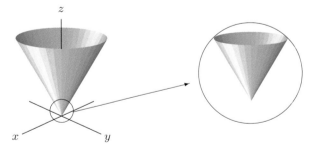

Figure 14.67: The function $f(x, y) = \sqrt{x^2 + y^2}$ is not locally linear at $(0, 0)$: Zooming in around $(0, 0)$ does not make the graph look like a plane

Solution   If we zoom in on the graph of the function $f(x, y) = \sqrt{x^2 + y^2}$ at the origin, as shown in Figure 14.67, the sharp point remains; the graph never flattens out to look like a plane. Near its vertex, the graph does not look as if is well approximated (in any reasonable sense) by any plane.

Judging from the graph of f, we would not expect f to be differentiable at $(0, 0)$. Let us check this by trying to compute the partial derivatives of f at $(0, 0)$:

$$f_x(0, 0) = \lim_{h \to 0} \frac{f(h, 0) - f(0, 0)}{h} = \lim_{h \to 0} \frac{\sqrt{h^2 + 0} - 0}{h} = \lim_{h \to 0} \frac{|h|}{h}.$$

Since $|h|/h = \pm 1$, depending on whether h approaches 0 from the left or right, this limit does not exist and so neither does the partial derivative $f_x(0, 0)$. Thus, f cannot be differentiable at the origin. If it were, both of the partial derivatives, $f_x(0, 0)$ and $f_y(0, 0)$, would exist.

Alternatively, we could show directly that there is no linear approximation near $(0, 0)$ that satisfies the small relative error criterion for differentiability. Any plane passing through the point $(0, 0, 0)$ has the form $L(x, y) = mx + ny$ for some constants m and n. If $E(x, y) = f(x, y) - L(x, y)$, then

$$E(x, y) = \sqrt{x^2 + y^2} - mx - ny.$$

Then for f to be differentiable at the origin, we would need to show that

$$\lim_{\substack{h \to 0 \\ k \to 0}} \frac{\sqrt{h^2 + k^2} - mh - nk}{\sqrt{h^2 + k^2}} = 0.$$

Taking $k = 0$ gives

$$\lim_{h \to 0} \frac{|h| - mh}{|h|} = 1 - m \lim_{h \to 0} \frac{h}{|h|}.$$

This limit exists only if $m = 0$ for the same reason as before. But then the value of the limit is 1 and not 0 as required. Thus, we again conclude f is not differentiable.

___

In Example 2 the partial derivatives f_x and f_y did not exist at the origin and this was sufficient to establish nondifferentiability there. We might expect that if both partial derivatives do exist, then f *is* differentiable. But the next example shows that this not necessarily true: the existence of both partial derivatives at a point is *not* sufficient to guarantee differentiability.

___

Example 3   Consider the function $f(x, y) = x^{1/3}y^{1/3}$. Show that the partial derivatives $f_x(0, 0)$ and $f_y(0, 0)$ exist, but that f is not differentiable at $(0, 0)$.

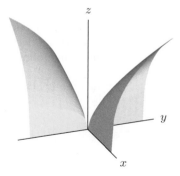

Figure 14.68: Graph of $z = x^{1/3}y^{1/3}$ for $z \geq 0$

Solution   See Figure 14.68 for the part of the graph of $z = x^{1/3}y^{1/3}$ when $z \geq 0$. We have $f(0, 0) = 0$ and we compute the partial derivatives using the definition:

$$f_x(0, 0) = \lim_{h \to 0} \frac{f(h, 0) - f(0, 0)}{h} = \lim_{h \to 0} \frac{0 - 0}{h} = 0,$$

and similarly

$$f_y(0, 0) = 0.$$

So, if there did exist a linear approximation near the origin, it would have to be $L(x, y) = 0$. But we can show that this choice of $L(x, y)$ does not result in the small relative error that is required for differentiability. In fact, since $E(x, y) = f(x, y) - L(x, y) = f(x, y)$, we need to look at the limit

$$\lim_{\substack{h \to 0 \\ k \to 0}} \frac{h^{1/3}k^{1/3}}{\sqrt{h^2 + k^2}}.$$

If this limit exists, we get the same value no matter how h and k approach 0. Suppose we take $k = h > 0$. Then the limit becomes

$$\lim_{h \to 0} \frac{h^{1/3}h^{1/3}}{\sqrt{h^2 + h^2}} = \lim_{h \to 0} \frac{h^{2/3}}{h\sqrt{2}} = \lim_{h \to 0} \frac{1}{h^{1/3}\sqrt{2}}.$$

But this limit does not exist, since small values for h will make the fraction arbitrarily large. So the only possible candidate for a linear approximation at the origin does not have a sufficiently small relative error. Thus, this function is *not* differentiable at the origin, even though the partial derivatives $f_x(0, 0)$ and $f_y(0, 0)$ exist. Figure 14.68 confirms that near the origin the graph of $z = f(x, y)$ is not well approximated by any plane.

In summary,

> - If a function is differentiable at a point, then both partial derivatives exist there.
> - Having both partial derivatives at a point does not guarantee that a function is differentiable there.

Continuity and Differentiability

We know that differentiable functions of one variable are continuous. Similarly, it can be shown that if a function of two variables is differentiable at a point, then the function is continuous there.

In Example 3 the function f was continuous at the point where it was not differentiable. Example 4 shows that even if the partial derivatives of a function exist at a point, the function is not necessarily continuous at that point if it is not differentiable there.

Example 4   Suppose that f is the function of two variables defined by

$$f(x, y) = \begin{cases} \dfrac{xy}{x^2 + y^2}, & (x, y) \neq (0, 0), \\ 0, & (x, y) = (0, 0). \end{cases}$$

Problem 23 on page 727 showed that $f(x, y)$ is not continuous at the origin. Show that the partial derivatives $f_x(0, 0)$ and $f_y(0, 0)$ exist. Could f be differentiable at $(0, 0)$?

Solution   From the definition of the partial derivative we see that

$$f_x(0, 0) = \lim_{h \to 0} \frac{f(h, 0) - f(0, 0)}{h} = \lim_{h \to 0} \left(\frac{1}{h} \cdot \frac{0}{h^2 + 0^2} \right) = \lim_{h \to 0} \frac{0}{h} = 0,$$

and similarly

$$f_y(0, 0) = 0.$$

So, the partial derivatives $f_x(0, 0)$ and $f_y(0, 0)$ exist. However, f cannot be differentiable at the origin since it is not continuous there.

In summary,

- If a function is differentiable at a point, then it is continuous there.
- Having both partial derivatives at a point does not guarantee that a function is continuous there.

How Do We Know If a Function Is Differentiable?

Can we use partial derivatives to tell us if a function is differentiable? As we see from Examples 3 and 4, it is not enough that the partial derivatives exist. However, the following theorem gives conditions that *do* guarantee differentiability[4]:

Theorem 14.2: Continuity of Partial Derivatives Implies Differentiability

If the partial derivatives, f_x and f_y, of a function f exist and are continuous on a small disk centred at the point (a, b), then f is differentiable at (a, b).

We will not prove this theorem, although it provides a criterion for differentiability which is often simpler to use than the definition. It turns out that the requirement of continuous partial derivatives is more stringent than that of differentiability, so there exist differentiable functions which do not have continuous partial derivatives. However, most functions we encounter will have continuous partial derivatives. The class of functions with continuous partial derivatives is given the name C^1.

Example 5   Show that the function $f(x, y) = \ln(x^2 + y^2)$ is differentiable everywhere in its domain.

Solution   The domain of f is all of 2-space except for the origin. We shall show that f has continuous partial derivatives everywhere in its domain (that is, the function f is in C^1). The partial derivatives are

$$f_x = \frac{2x}{x^2 + y^2} \quad \text{and} \quad f_y = \frac{2y}{x^2 + y^2}.$$

Since each of f_x and f_y is the quotient of continuous functions, the partial derivatives are continuous everywhere except the origin (where the denominators are zero). Thus, f is differentiable everywhere in its domain.

Most functions built up from elementary functions have continuous partial derivatives, except perhaps at a few obvious points. Thus, in practice, we can often identify functions as being C^1 without explicitly computing the partial derivatives.

Exercises and Problems for Section 14.8

Exercises

In Exercises 1–10, list the points in the xy-plane, if any, at which the function $z = f(x, y)$ is not differentiable.

1. $z = -\sqrt{x^2 + y^2}$

2. $z = \sqrt{(x + 1)^2 + y^2}$

3. $z = e^{-(x^2 + y^2)}$

4. $z = |x| + |y|$

5. $z = x^{1/3} + y^2$

6. $z = |x + 2| - |y - 3|$

7. $z = (\sin x)(\cos |y|)$

8. $z = |x - 3|^2 + y^3$

9. $z = 4 + \sqrt{(x - 1)^2 + (y - 2)^2}$

10. $z = 1 + \left((x - 1)^2 + (y - 2)^2\right)^2$

---

[4]For a proof, see M. Spivak, *Calculus on Manifolds*, p. 31 (New York: Benjamin, 1965).

Problems

In Problems 11–14, a functions f is given.

(a) Use a computer to draw a contour diagram for f.

(b) Is f differentiable at all points $(x, y) \neq (0, 0)$?

(c) Do the partial derivatives f_x and f_y exist and are they continuous at all points $(x, y) \neq (0, 0)$?

(d) Is f differentiable at $(0, 0)$?

(e) Do the partial derivatives f_x and f_y exist and are they continuous at $(0, 0)$?

11. $f(x, y) = \begin{cases} \dfrac{x}{y} + \dfrac{y}{x}, & x \neq 0 \text{ and } y \neq 0, \\ 0, & x = 0 \text{ or } y = 0. \end{cases}$

12. $f(x, y) = \begin{cases} \dfrac{2xy}{(x^2 + y^2)^2}, & (x, y) \neq (0, 0), \\ 0, & (x, y) = (0, 0). \end{cases}$

13. $f(x, y) = \begin{cases} \dfrac{xy}{\sqrt{x^2 + y^2}}, & (x, y) \neq (0, 0), \\ 0, & (x, y) = (0, 0). \end{cases}$

14. $f(x, y) = \begin{cases} \dfrac{x^2 y}{x^4 + y^2}, & (x, y) \neq (0, 0), \\ 0, & (x, y) = (0, 0). \end{cases}$

15. Consider the function

$$f(x, y) = \begin{cases} \dfrac{xy^2}{x^2 + y^2}, & (x, y) \neq (0, 0), \\ 0, & (x, y) = (0, 0). \end{cases}$$

(a) Use a computer to draw the contour diagram for f.

(b) Is f differentiable for $(x, y) \neq (0, 0)$?

(c) Show that $f_x(0, 0)$ and $f_y(0, 0)$ exist.

(d) Is f differentiable at $(0, 0)$?

(e) Suppose $x(t) = at$ and $y(t) = bt$, where a and b are constants, not both zero. If $g(t) = f(x(t), y(t))$, show that

$$g'(0) = \frac{ab^2}{a^2 + b^2}.$$

(f) Show that

$$f_x(0, 0)x'(0) + f_y(0, 0)y'(0) = 0.$$

Does the chain rule hold for the composite function $g(t)$ at $t = 0$? Explain.

(g) Show that the directional derivative $f_{\vec{u}}(0, 0)$ exists for each unit vector $\vec{u}$. Does this imply that f is differentiable at $(0, 0)$?

16. Consider the function

$$f(x, y) = \begin{cases} \dfrac{xy^2}{x^2 + y^4}, & (x, y) \neq (0, 0), \\ 0, & (x, y) = (0, 0). \end{cases}$$

(a) Use a computer to draw the contour diagram for f.

(b) Show that the directional derivative $f_{\vec{u}}(0, 0)$ exists for each unit vector $\vec{u}$.

(c) Is f continuous at $(0, 0)$? Is f differentiable at $(0, 0)$? Explain.

17. Consider the function $f(x, y) = \sqrt{|xy|}$.

(a) Use a computer to draw the contour diagram for f. Does the contour diagram look like that of a plane when we zoom in on the origin?

(b) Use a computer to draw the graph of f. Does the graph look like a plane when we zoom in on the origin?

(c) Is f differentiable for $(x, y) \neq (0, 0)$?

(d) Show that $f_x(0, 0)$ and $f_y(0, 0)$ exist.

(e) Is f differentiable at $(0, 0)$? [Hint: Consider the directional derivative $f_{\vec{u}}(0, 0)$ for $\vec{u} = (\vec{i} + \vec{j})/\sqrt{2}$.]

18. Suppose a function f is differentiable at the point (a, b). Show that f is continuous at (a, b).

19. Suppose $f(x, y)$ is a function such that $f_x(0, 0) = 0$ and $f_y(0, 0) = 0$, and $f_{\vec{u}}(0, 0) = 3$ for $\vec{u} = (\vec{i} + \vec{j})/\sqrt{2}$.

(a) Is f differentiable at $(0, 0)$? Explain.

(b) Give an example of a function f defined on 2-space which satisfies these conditions. [Hint: The function f does not have to be defined by a single formula valid over all of 2-space.]

20. Consider the following function:

$$f(x, y) = \begin{cases} \dfrac{xy(x^2 - y^2)}{x^2 + y^2}, & (x, y) \neq (0, 0), \\ 0, & (x, y) = (0, 0). \end{cases}$$

The graph of f is shown in Figure 14.69, and the contour diagram of f is shown in Figure 14.70.

(a) Find $f_x(x, y)$ and $f_y(x, y)$ for $(x, y) \neq (0, 0)$.

(b) Show that $f_x(0, 0) = 0$ and $f_y(0, 0) = 0$.

(c) Are the functions f_x and f_y continuous at $(0, 0)$?

(d) Is f differentiable at $(0, 0)$?

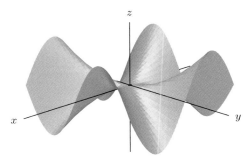

Figure 14.69: Graph of $\dfrac{xy(x^2 - y^2)}{x^2 + y^2}$

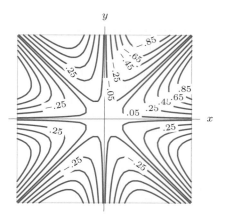

Figure 14.70: Contour diagram of
$$\frac{xy(x^2 - y^2)}{x^2 + y^2}$$

Strengthen Your Understanding

In Problems 21–22, explain what is wrong with the statement.

21. If $f(x, y)$ is continuous at the origin, then it is differentiable at the origin.

22. If the partial derivatives $f_x(0, 0)$ and $f_y(0, 0)$ both exist, then $f(x, y)$ is differentiable at the origin.

In Problems 23–24, give an example of:

23. A continuous function $f(x, y)$ that is not differentiable at the origin.

24. A continuous function $f(x, y)$ that is not differentiable on the line $x = 1$.

25. Which of the following functions $f(x, y)$ is differentiable at the given point?

(a) $\sqrt{1 - x^2 - y^2}$ at $(0, 0)$ (b) $\sqrt{4 - x^2 - y^2}$ at $(2, 0)$

(c) $-\sqrt{x^2 + 2y^2}$ at $(0, 0)$ (d) $-\sqrt{x^2 + 2y^2}$ at $(2, 0)$

CHAPTER SUMMARY (see also Ready Reference at the end of the book)

- **Partial Derivatives**
  Definition as a difference quotient, visualizing on a graph, estimating from a contour diagram, computing from a formula, interpreting units, alternative notation.

- **Local Linearity**
  Zooming on a surface, contour diagram, or table to see local linearity, the idea of tangent plane, formula for a tangent plane in terms of partials, the differential.

- **Directional Derivatives**
  Definition as a difference quotient, interpretation as a rate of change, computation using partial derivatives.

- **Gradient Vector**

  Definition in terms of partial derivatives, geometric properties of gradient's length and direction, relation to directional derivative, relation to contours and level surfaces.

- **Chain Rule**
  Local linearity and differentials for composition of functions, tree diagrams, chain rule in general, application to physical chemistry.

- **Second- and Higher-Order Partial Derivatives**
  Interpretations, mixed partials are equal.

- **Taylor Approximations**
  Linear and quadratic polynomial approximations to functions near a point.

REVIEW EXERCISES AND PROBLEMS FOR CHAPTER FOURTEEN

Exercises

Are the quantities in Exercises 1–4 vectors or scalars? Calculate them.

1. $\operatorname{grad}(x^3 e^{-y/2})$ at $(1, 2)$

2. The directional derivative of $f(x, y) = x^2 y^3$ at the point $(1, 1)$ in the direction of $(\vec{i} + \vec{j})/\sqrt{2}$.

3. $\operatorname{grad}((\cos x)e^y + z)$

4. $\dfrac{\partial^2 f}{\partial x^2}$ when $f(x, y) = y e^{x^2 y}$

For Exercises 5–27, find the partial derivatives. Assume the variables are restricted to a domain on which the function is defined.

5. f_x and f_y if $f(x, y) = 5x^2y^3 + 8xy^2 - 3x^2$

6. $\dfrac{\partial w}{\partial h}$ if $w = 320\pi gh^2(20 - h)$

7. $\dfrac{\partial T}{\partial l}$ if $T = 2\pi\sqrt{\dfrac{l}{g}}$

8. $\dfrac{\partial B}{\partial t}$ and $\dfrac{\partial B}{\partial r}$ when $B = P(1 + r)^t$.

9. f_x and f_y if $f(x, y) = \dfrac{x^2y}{x^2 + y^2}$

10. $\dfrac{\partial F}{\partial r}$ and $\dfrac{\partial F}{\partial r}$ if $F = \dfrac{G\mu my}{(r^2 + y^2)^{3/2}}$

11. f_p and f_q if $f(p, q) = e^{p/q}$

12. $z_x(2, 3)$ if $z = (\cos x) + y$

13. f_N if $f(N, V) = cN^\alpha V^\beta$

14. f_x and f_y if $f(x, y) = \sqrt{(x - a)^2 + (y - b)^2}$

15. $\dfrac{\partial}{\partial\omega}\left(\tan\sqrt{\omega x}\right)$

16. $\dfrac{\partial y}{\partial t}$ if $y = \sin(ct - 5x)$

17. $\dfrac{\partial \alpha}{\partial \beta}$ if $\alpha = \dfrac{e^{x\beta - 3}}{2y\beta + 5}$

18. $\dfrac{\partial}{\partial w}\left(\sqrt{2\pi xyw - 13x^7y^3v}\right)$

19. $\dfrac{\partial}{\partial \lambda}\left(\dfrac{x^2y\lambda - 3\lambda^5}{\sqrt{\lambda^2 - 3\lambda + 5}}\right)$

20. $\dfrac{\partial}{\partial w}\left(\dfrac{x^2yw - xy^3w^7}{w - 1}\right)^{-7/2}$

21. $\dfrac{\partial}{\partial x}(e^x\cos(xy) + ay^2)$, $\dfrac{\partial}{\partial y}(e^x\cos(xy) + ay^2)$, $\dfrac{\partial}{\partial a}(e^x\cos(xy) + ay^2)$

22. $\dfrac{\partial f_0}{\partial L}$ if $f_0 = \dfrac{1}{2\pi\sqrt{LC}}$

23. f_{xx} and f_{xy} if $f(x, y) = 1/\sqrt{x^2 + y^2}$

24. u_{xx} and u_{yy} if $u = e^x\sin y$

25. V_{rr} and V_{rh} if $V = \pi r^2 h$

26. f_{xxy} and f_{yxx} if $f(x, y) = \sin(x - 2y)$

27. $\dfrac{\partial^2}{\partial x^2}(e^{ax - bt}) + \dfrac{\partial^2}{\partial t^2}(e^{ax - bt})$

In Exercises 28–38, find the gradient of the function.

28. $f(x, y, z) = x^2 + y^2 + y^3$

29. $f(x, y, z) = x^3 + z^3 - xyz$

30. $f(x, y, z) = 1/(xyz)$

31. $f(x, y) = \sin(y^2 - xy)$

32. $f(x, y) = \ln(x^2 + y^2)$

33. $f(x, y, z) = xe^y + \ln(xz)$

34. $f(x, y, z) = \sin(x^2 + y^2 + z^2)$

35. $f(\rho, \phi, \theta) = \rho\sin\phi\cos\theta$

36. $f(s, t) = \dfrac{1}{\sqrt{s}}(t^2 - 2t + 4)$

37. $f(\alpha, \beta) = \sqrt{5\alpha^2 + \beta}$

38. $f(x, y) = \sin(xy) + \cos(xy)$

In Exercises 39–40, find the gradient of f at the point.

39. $f(x, y, z) = x^2$ at $(0, 0, 0)$

40. $f(x, y, z) = x^2z$, at $(1, 1, 1)$

In Exercises 41–46, find the directional derivative of the function.

41. $f(x, y) = x^3 - y^3$ at $(2, -1)$ in the direction of $\vec{i} - \vec{j}$

42. $f(x, y) = xe^y$ at $(3, 0)$ in the direction of $4\vec{i} - 3\vec{j}$

43. $f(x, y, z) = x^2 + y^2 - z^2$ at $(2, 3, 4)$ in the direction of $2\vec{i} - 2\vec{j} + \vec{k}$

44. $f(x, y, z) = 3x^2y^2 + 2yz$ at $(-1, 0, 4)$ in the direction of $\vec{i} - \vec{k}$

45. $f(x, y, z) = 3x^2y^2 + 2yz$ at $(-1, 0, 4)$ in the direction of $-\vec{i} + 3\vec{j} + 3\vec{k}$

46. $f(x, y, z) = e^{x+z}\cos y$ at $(1, 0, -1)$ in the direction of $\vec{i} + \vec{j} + \vec{k}$

In Exercises 47–49, find a vector normal to the curve or surface at the point.

47. $x^2 - y^2 = 3$ at $(2, 1)$

48. $xy + xz + yz = 11$ at $(1, 2, 3)$

49. $z^2 - 2xyz = x^2 + y^2$ at $(1, 2, -1)$

50. Find an equation of the tangent plane to the surface $z^2 - 4x^2 - 3y^2 = 9$ at the point $(1, 1, 4)$.

51. Find an equation of the tangent plane to the surface $x^3 = 2y^2 - z$ at the point $(1, 0, -1)$.

52. Find an equation of the tangent plane to the surface $z - 1/(xy) = 0$ at the point $(1, 1, 1)$.

53. Compute all four second-order partial derivatives of $f(x, y) = x^2y^2 - 5xy^3$.

In Exercises 54–59, find dz/dt using the chain rule. Assume the variables are restricted to domains on which the functions are defined.

54. $z = x\sin y$, $x = \sin t$, $y = \cos t$

55. $z = \sin(x^2 + y^2)$, $x = 2t$, $y = t^2$

56. $z = (x^2 + y)^2$, $x = 2t$, $y = t^2$

57. $z = (x + y)e^x$, $x = t^2$, $y = 1 - t^2$

58. $z = \ln y + \ln x$, $x = t^3$, $y = t^2 + 1$

59. $z = \sin(pq)$, $p = \sin t$, $q = \cos t^2$

In Exercises 60–62, find the quadratic Taylor polynomial for the function.

60. $f(x, y) = (x + 1)^3(y + 2)$ about $(0, 0)$

61. $f(x, y) = \cos x\cos 3y$ about $(0, 0)$

62. $f(x, y) = \sqrt{2x - y}$ about $(3, 5)$

Problems

63. Match each function $f(x, y, z)$ in (a)–(d) with the description of its gradient in (I)–(VI).

(a) $x^2 + y^2 + z^2$ (b) $\sqrt{x^2 + y^2 + z^2}$

(c) $3x + 4y$ (d) $3x + 4z$

   I Constant, parallel to xy-plane.
  II Constant, parallel to xz-plane.
 III Constant, parallel to yz-plane.
 IV Radial, increasing in magnitude away from the origin.
  V Radial, constant magnitude.
 VI Radial, decreasing in magnitude away from the origin.

64. (a) Find an equation of the tangent plane to the surface $2x^2 - 2xy^2 + az = a$ at the point $(1, 1, 1)$.
(b) For which value of a does the tangent plane pass through the origin?

65. The monthly mortgage payment in dollars, P, for a house is a function of three variables:

$$P = f(A, r, N),$$

where A is the amount borrowed in dollars, r is the interest rate, and N is the number of years before the mortgage is paid off.

(a) $f(92000, 14, 30) = 1090.08$. What does this tell you, in financial terms?

(b) $\left.\dfrac{\partial P}{\partial r}\right|_{(92000,14,30)} = 72.82$. What is the financial significance of the number 72.82?

(c) Would you expect $\partial P / \partial A$ to be positive or negative? Why?

(d) Would you expect $\partial P / \partial N$ to be positive or negative? Why?

66. Figure 14.71 is a contour diagram of $f(x, y)$. In each of the following cases, list the marked points in the diagram (there may be none or more than one) at which

(a) $f_x < 0$ (b) $f_y > 0$
(c) $f_{xx} > 0$ (d) $f_{yy} < 0$

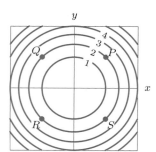

Figure 14.71

67. Figure 14.72 gives a contour diagram for the number n of foxes per square kilometer in southwestern England. Estimate $\partial n / \partial x$ and $\partial n / \partial y$ at the points A, B, and C, where x is kilometers east and y is kilometers north.

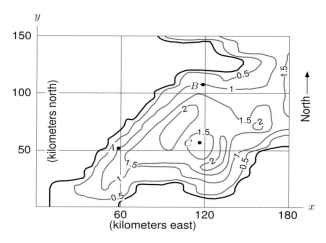

Figure 14.72

68. The cost of producing one unit of a product is given by

$$c = a + bx + ky,$$

where x is the amount of labor used (in man hours) and y is the amount of raw material used (by weight) and a and b and k are constants. What does $\partial c / \partial x = b$ mean? What is the practical interpretation of b?

69. (a) Let $f(u, v) = u(u^2 + v^2)^{3/2}$. Use a difference quotient to approximate $f_u(1, 3)$ with $h = 0.001$.
(b) Now evaluate $f_u(1, 3)$ exactly. Was the approximation in part (a) reasonable?

70. The gravitational force, F newtons, exerted on a mass of m kg at a distance of r meters from the center of the earth is given by

$$F = \frac{GMm}{r^2}$$

where the mass of the earth $M = 6 \cdot 10^{24}$ kilograms, and $G = 6.67 \cdot 10^{-11}$. Find the gravitational force on a person with mass 70 kg at the surface of the earth ($r = 6.4 \cdot 10^6$). Calculate $\partial F / \partial m$ and $\partial F / \partial r$ for these values of m and r. Interpret these partial derivatives in terms of gravitational force.

71. (a) Write a formula for the number π using only the perimeter L and the area A of a circle.
(b) Suppose that L and A are determined experimentally. Show that if the relative, or percent, errors in the measured values of L and A are λ and μ, respectively, then the resulting relative, or percent, error in π is $2\lambda - \mu$.

72. A company uses x hours of unskilled labor and y hours of skilled labor to produce $F(x, y) = 60x^{2/3}y^{1/3}$ units of output. It currently employs 400 hours of unskilled labor and 50 hours of skilled labor. The company is planning to hire an additional 5 hours of skilled labor.

   (a) Use a linear approximation to decide by about how much the company can reduce its use of unskilled labor and keep its output at current level.
   (b) Calculate the exact value of the reduction.

73. One mole of ammonia gas is contained in a vessel which is capable of changing its volume (a compartment sealed by a piston, for example). The total energy U (in joules) of the ammonia is a function of the volume V (in m^3) of the container, and the temperature T (in K) of the gas. The differential dU is given by

$$dU = 840 \, dV + 27.32 \, dT.$$

   (a) How does the energy change if the volume is held constant and the temperature is increased slightly?
   (b) How does the energy change if the temperature is held constant and the volume is increased slightly?
   (c) Find the approximate change in energy if the gas is compressed by 100 cm^3 and heated by 2 K.

74. Figure 14.73 shows grad $f(x, y)$. In each of the following cases, list the marked points (if any) at which

   (a) $f_x > 0$ (b) $f_y < 0$
   (c) $f_{xx} > 0$ (d) $f_{yy} < 0$

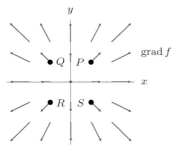

Figure 14.73

In Problems 75–80, use the contour diagram for $f(x, y)$ in Figure 14.74 to estimate the directional derivative of $f(x, y)$ in the direction $\vec{v}$ at the point given.

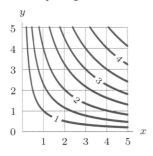

Figure 14.74

75. $\vec{v} = \vec{i}$ at $(1, 1)$ 76. $\vec{v} = \vec{j}$ at $(1, 1)$

77. $\vec{v} = \vec{i} + \vec{j}$ at $(1, 1)$ 78. $\vec{v} = \vec{i} + \vec{j}$ at $(4, 1)$

79. $\vec{v} = -2\vec{i} + \vec{j}$ at $(3, 3)$ 80. $\vec{v} = -2\vec{i} + \vec{j}$ at $(4, 1)$

81. Figure 14.74 shows the level curves of $f(x, y)$. At the points $(1, 1)$ and $(1, 4)$, draw a vector representing grad f. Explain how you know the direction and length of each vector.

82. Find the directional derivative of $z = x^2y$ at $(1, 2)$ in the direction making an angle of $5\pi/4$ with the x-axis. In which direction is the directional derivative the largest?

83. An ant is crawling across a heated plate with velocity $\vec{v}$ cm/sec, and the temperature of the plate at position (x, y) is $H(x, y)$ degrees, where x and y are in centimeters. Which of the following (if any) is correct? The rate of change in deg/sec of the temperature felt by the ant is:

   (a) $\| \text{grad } H \| \| \vec{v} \|$, because it is the product of the ant's speed and the rate of change of H with respect to distance.
   (b) $\text{grad } H \cdot \vec{v}$, because it is the product of the ant's speed and the directional derivative of H in the direction of $\vec{v}$.
   (c) $H_{\vec{u}}$, where $\vec{u} = \vec{v}/\|\vec{v}\|$, because it is the rate of change of H in the direction of $\vec{v}$.

84. The depth, in meters, of a lake at a point x kilometers east and y kilometers north of a buoy is given by

$$h(x, y) = 150 - 30x^2 - 20y^2.$$

   (a) A rowboat is 1 kilometer east and 2 kilometers south of the buoy. At what rate is the depth changing with respect to distance in the direction of the buoy?
   (b) The boat starts moving toward the buoy at a rate of 3 km/h. At what rate is the depth of the lake beneath the boat changing with respect to time?

85. A differentiable function $f(x, y)$ has the property that $f(1, 3) = 7$ and grad $f(1, 3) = 2\vec{i} - 5\vec{j}$.

   (a) Find the equation of the tangent line to the level curve of f through the point $(1, 3)$.
   (b) Find the equation of the tangent plane to the surface $z = f(x, y)$ at the point $(1, 3, 7)$.

86. Let x, y, z be in meters. At the point (x, y, z) in space, the temperature, H, in $^\circ$C, is given by

$$H = e^{-(x^2 + 2y^2 + 3z^2)}.$$

   (a) A particle at the point $(2, 1, 5)$ starts to move in the direction of increasing x. How fast is the temperature changing with respect to distance? Give units.
   (b) If the particle in part (a) moves at 10 meters/sec, how fast is the temperature changing with respect to time? Give units.
   (c) What is the maximum rate of change of temperature with respect to distance at the point $(2, 1, 5)$?

87. A differentiable function $f(x, y)$ has the property that $f(4, 1) = 3$ and $f_x(4, 1) = 2$ and $f_y(4, 1) = -1$. Find the equation of the tangent plane at the point on the surface $z = f(x, y)$ where $x = 4$, $y = 1$.

88. The temperature at (x, y) is $T(x, y) = 100 - x^2 - y^2$. In which direction should a heat-seeking bug move from the point (x, y) to increase its temperature fastest?

89. Do the level curves of $f(x, y) = \sqrt{x^2 + y^2} + x$ and $g(x, y) = \sqrt{x^2 + y^2} - x$ cross at right angles?

90. At any point (x, y, z) outside a spherically symmetric mass m located at the point (x_0, y_0, z_0), the gravitational potential, V, is defined by $V = -Gm/r$, where r is the distance from (x, y, z) to (x_0, y_0, z_0) and G is a constant. Show that, for all points outside the mass, V satisfies Laplace's equation:

$$\frac{\partial^2 V}{\partial x^2} + \frac{\partial^2 V}{\partial y^2} + \frac{\partial^2 V}{\partial z^2} = 0.$$

91. Given $z = u^2 - ue^v$, $u = x + 2y$, $v = 2x - y$, use the chain rule to find:

(a) $\partial z / \partial x |_{(x,y)=(1,2)}$ (b) $\partial z / \partial y |_{(x,y)=(1,2)}$

92. Let $F(u, v, w)$ be a function of three variables. Find $G_x(x, y)$ if

(a) $G(x, y) = F(x, y, 3)$ (b) $G(x, y) = F(3, y, x)$

(c) $G(x, y) = F(x, y, x)$ (d) $G(x, y) = F(x, y, xy)$

93. In analyzing a factory and deciding whether or not to hire more workers, it is useful to know under what circumstances productivity increases. Suppose $P = f(x_1, x_2, x_3)$ is the total quantity produced as a function of x_1, the number of workers, and any other variables x_2, x_3. We define the average productivity of a worker as P/x_1. Show that the average productivity increases as x_1 increases when marginal production, $\partial P/\partial x_1$, is greater than the average productivity, P/x_1.

94. For the Cobb-Douglas function $P = 40L^{0.25}K^{0.75}$, find the differential dP when $L = 2$ and $K = 16$.

95. The period, T, of a pendulum is $T = 2\pi\sqrt{l/g}$. If the approximate length is $l = 2$ meters, find the approximate error in T if the true length is $l = 1.99$ and we take $g = 9.8$ as an approximation for $g = 9.81$ m/s^2.

96. Figure 14.75 shows the monthly payment, m, on a 5-year car loan if you borrow P dollars at r percent interest. Find a formula for a linear function which approximates m. What is the practical significance of the constants in your formula?

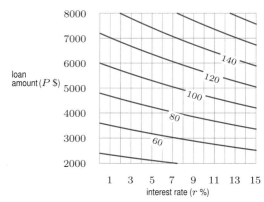

Figure 14.75

In Problems 97–104, the function f is differentiable and $f_x(2, 1) = -3$, $f_y(2, 1) = 4$, and $f(2, 1) = 7$.

97. (a) Give an equation for the tangent plane to the graph of f at $x = 2$, $y = 1$.
(b) Give an equation for the tangent line to the contour for f at $x = 2$, $y = 1$.

98. (a) Find a vector perpendicular to the tangent plane to the graph of f at $x = 2$, $y = 1$.
(b) Find a vector perpendicular to the tangent line to the contour for f at $x = 2$, $y = 1$.

99. Near $x = 2$ and $y = 1$, how far apart are the contours $f(x, y) = 7$ and $f(x, y) = 7.3$?

100. Give an approximate table of values of f for $x = 1.8, 2.0, 2.2$ and $y = 0.9, 1.0, 1, 1$.

101. Give an approximate contour diagram for f for $1 \le x \le 3$, $0 \le y \le 2$, using contour values $\ldots 5, 6, 7, 8, 9 \ldots$.

102. The function f gives temperature in $^\circ$C and x and y are in centimeters. A worm leaves $(2, 1)$ at 3 cm/min so that it cools off as fast as possible. In which direction does the worm head? At what rate does it cool off, in $^\circ$C/min?

103. Find $f_r(2, 1)$ and $f_\theta(2, 1)$, where r and θ are polar coordinates, $x = r\cos\theta$ and $y = r\sin\theta$. If $\vec{u}$ is the unit vector in the direction $2\vec{i} + \vec{j}$, show that $f_{\vec{u}}(2, 1) = f_r(2, 1)$ and explain why this should be the case.

104. Find approximately the largest value of f on or inside the circle of radius 0.1 about the point $(2, 1)$. At what point does f achieve this value?

105. Values of the function $f(x, y)$ near the point $x = 2$, $y = 3$ are given in Table 14.8. Estimate the following.

(a) $\dfrac{\partial f}{\partial x}\bigg|_{(2,3)}$ and $\dfrac{\partial f}{\partial y}\bigg|_{(2,3)}$.

(b) The rate of change of f at $(2, 3)$ in the direction of the vector $\vec{i} + 3\vec{j}$.

(c) The maximum possible rate of change of f as you move away from the point $(2, 3)$. In which direction should you move to obtain this rate of change?

(d) Write an equation for the level curve through the point $(2, 3)$.

(e) Find a vector tangent to the level curve of f through the point $(2, 3)$.

(f) Find the differential of f at the point $(2, 3)$. If $dx = 0.03$ and $dy = 0.04$, find df. What does df represent in this case?

Table 14.8

	x	
	2.00	**2.01**
3.00	7.56	7.42
3.02	7.61	7.47

(with y labeling the rows 3.00, 3.02)

106. Find the quadratic Taylor polynomial about $(0, 0)$ for $f(x, y) = \cos(x + 2y) \sin(x - y)$.

107. Suppose $f(x, y) = e^{(x-1)^2 + (y-3)^2}$.

(a) Find the first-order Taylor polynomial about $(0, 0)$.

(b) Find the second-order (quadratic) Taylor polynomial about the point $(1, 3)$.

(c) Find a 2-vector perpendicular to the level curve through $(0, 0)$.

(d) Find a 3-vector perpendicular to the surface $z = f(x, y)$ at the point $(0, 0)$.

108. Figure 14.76 shows a contour diagram for a vibrating string function, $f(x, t)$.

(a) Is $f_t(\pi/2, \pi/2)$ positive or negative? How about $f_t(\pi/2, \pi)$? What does the sign of $f_t(\pi/2, b)$ tell you about the motion of the point on the string at $x = \pi/2$ when $t = b$?

(b) Find all t for which f_t is positive, for $0 \le t \le 5\pi/2$.

(c) Find all x and t such that f_x is positive.

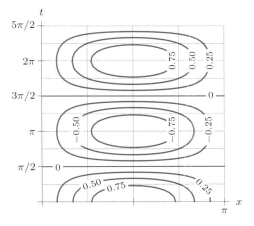

Figure 14.76

CAS Challenge Problems

109. (a) Find the quadratic Taylor polynomial about $(0, 0)$ of
$$f(x, y) = \frac{e^x(1 + \sin(3y))^2}{5 + e^{2x}}.$$

(b) Find the quadratic Taylor polynomial about 0 of the one-variable functions $g(x) = e^x/(5 + e^{2x})$ and $h(y) = (1 + \sin(3y))^2$. Multiply these polynomials together and compare with your answer to part (a).

(c) Show that if $f(x, y) = g(x)h(y)$, then the quadratic Taylor polynomial of f about $(0, 0)$ is the product of the quadratic Taylor polynomials of $g(x)$ and $h(y)$ about 0. [If you use a computer algebra system, make sure that f, g and h do not have any previously assigned formula.]

110. Let
$$f(x, y) = A_0 + A_1 x + A_2 y + A_3 x^2 + A_4 xy + A_5 y^2,$$
$$g(t) = 1 + B_1 t + B_2 t^2,$$
$$h(t) = 2 + C_1 t + C_2 t^2.$$

(a) Find $L(x, y)$, the linear approximation to $f(x, y)$ at the point $(1, 2)$. Also find $m(t)$ and $n(t)$, the linear approximations to $g(t)$ and $h(t)$ at $t = 0$.

(b) Calculate $(d/dt)f(g(t), h(t))|_{t=0}$ and $(d/dt)L(m(t), n(t))|_{t=0}$. Describe what you notice and explain it in terms of the chain rule.

111. Let $f(x, y) = A_0 + A_1 x + A_2 y + A_3 x^2 + A_4 xy + A_5 y^2$.

(a) Find the quadratic Taylor approximation for $f(x, y)$ near the point $(1, 2)$, and expand the result in powers of x and y.

(b) Explain what you notice in part (a) and formulate a generalization to points other than $(1, 2)$.

(c) Repeat part (a) for the linear approximation. How does it differ from the quadratic?

112. Suppose that $w = f(x, y, z)$, that x and y are functions of u and v, and that z, u, and v are functions of t. Use a computer algebra system to find the derivative
$$\frac{d}{dt}f(x(u(t), v(t)), y(u(t), v(t)), z(t))$$
and explain the answer using a tree diagram.

PROJECTS FOR CHAPTER FOURTEEN

1. Heat Equation

The function $T(x, y, z, t)$ is a solution to the *heat equation*

$$T_t = K(T_{xx} + T_{yy} + T_{zz})$$

and gives the temperature at the point (x, y, z) in 3-space and time t. The constant K is the *thermal conductivity* of the medium through which the heat is flowing.

(a) Show that the function

$$T(x, y, z, t) = \frac{1}{(4\pi Kt)^{3/2}} e^{-(x^2+y^2+z^2)/4Kt}$$

is a solution to the heat equation for all (x, y, z) in 3-space and $t > 0$.

(b) For each fixed time t, what are the level surfaces of the function $T(x, y, z, t)$ in 3-space?

(c) Regard t as fixed and compute $\operatorname{grad} T(x, y, z, t)$. What does $\operatorname{grad} T(x, y, z, t)$ tell us about the direction and magnitude of the heat flow?

2. Matching Birthdays

Consider a class of m students and a year with n days. Let $q(m, n)$ denote the probability, expressed as a number between 0 and 1, that at least two students have the same birthday. Surprisingly, $q(23, 365) \approx 0.5073$. (This means that there is slightly better than an even chance that at least two students in a class of 23 have the same birthday.) A general formula for q is complicated, but it can be shown that

$$\frac{\partial q}{\partial m} \approx +\frac{m}{n}(1 - q) \qquad \text{and} \qquad \frac{\partial q}{\partial n} \approx -\frac{m^2}{2n^2}(1 - q).$$

(These approximations hold when n is a good deal larger than m, and m is a good deal larger than 1.)

(a) Explain why the $+$ and $-$ signs in the approximations for $\partial q/\partial m$ and $\partial q/\partial n$ are to be expected.

(b) Suppose there are 21 students in a class. What is the approximate probability that at least two students in the class have the same birthday? (Assume that a year always has 365 days.)

(c) Suppose there is a class of 24 students and you know that no one was born in the first week of the year. (This has the effect of making $n = 358$.) What is the approximate value of q for this class?

(d) If you want to bet that a certain class of 23 students has at least two matching birthdays, would you prefer to have two more students added to the class or to be told that no one in the class was born in December?

(e) (Optional) Find the actual formula for q. [Hint: It's easier to find $1 - q$. There are $n \cdot n \cdot n \cdots n = n^m$ different choices for the students' birthdays. How many such choices have no matching birthdays?]

Chapter Fifteen

OPTIMIZATION: LOCAL AND GLOBAL EXTREMA

Contents

15.1 CRITICAL POINTS: LOCAL EXTREMA AND SADDLE POINTS

Functions of several variables, like functions of one variable, can have *local* and *global* extrema. (That is, local and global maxima and minima.) A function has a local extremum at a point where it takes on the largest or smallest value in a small region around the point. Global extrema are the largest or smallest values anywhere on the domain under consideration. (See Figures 15.1 and 15.2.)

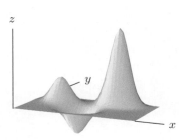

Figure 15.1: Local and global extrema for a function of two variables on $0 \leq x \leq a$, $0 \leq y \leq b$

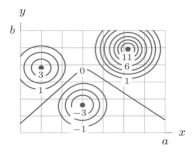

Figure 15.2: Contour map of the function in Figure 15.1

More precisely, considering only points at which f is defined, we say:

> - f has a **local maximum** at the point P_0 if $f(P_0) \geq f(P)$ for all points P near P_0.
> - f has a **local minimum** at the point P_0 if $f(P_0) \leq f(P)$ for all points P near P_0.

How Do We Detect a Local Maximum or Minimum?

Recall that if the gradient vector of a function is defined and nonzero, then it points in a direction in which the function increases. Suppose that a function f has a local maximum at a point P_0 which is not on the boundary of the domain. If the vector $\operatorname{grad} f(P_0)$ were defined and nonzero, then we could increase f by moving in the direction of $\operatorname{grad} f(P_0)$. Since f has a local maximum at P_0, there is no direction in which f is increasing. Thus, if $\operatorname{grad} f(P_0)$ is defined, we must have

$$\operatorname{grad} f(P_0) = \vec{0}.$$

Similarly, suppose f has a local minimum at the point P_0. If $\operatorname{grad} f(P_0)$ were defined and nonzero, then we could decrease f by moving in the direction opposite to $\operatorname{grad} f(P_0)$, and so we must again have $\operatorname{grad} f(P_0) = \vec{0}$. Therefore, we make the following definition:

> Points where the gradient is either $\vec{0}$ or undefined are called **critical points** of the function.

If a function has a local maximum or minimum at a point P_0, not on the boundary of its domain, then P_0 is a critical point. For a function of two variables, we can also see that the gradient vector must be zero or undefined at a local maximum by looking at its contour diagram and a plot of its gradient vectors. (See Figures 15.3 and 15.4.) Around the maximum the vectors are all pointing inward, perpendicularly to the contours. At the maximum the gradient vector must be zero or undefined. A similar argument shows that the gradient must be zero or undefined at a local minimum.

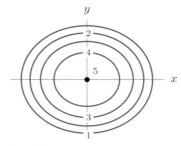

Figure 15.3: Contour diagram around a local maximum of a function

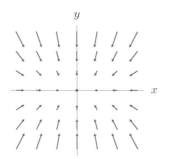

Figure 15.4: Gradients pointing toward the local maximum of the function in Figure 15.3

Finding and Analyzing Critical Points

To find critical points of f we set $\operatorname{grad} f = f_x \vec{i} + f_y \vec{j} + f_z \vec{k} = \vec{0}$, which means setting all the partial derivatives of f equal to zero. We must also look for the points where one or more of the partial derivatives is undefined.

Example 1 Find and analyse the critical points of $f(x, y) = x^2 - 2x + y^2 - 4y + 5$.

Solution To find the critical points, we set both partial derivatives equal to zero:

$$f_x(x, y) = 2x - 2 = 0$$
$$f_y(x, y) = 2y - 4 = 0.$$

Solving these equations gives $x = 1$, $y = 2$. Hence, f has only one critical point, namely $(1, 2)$. To see the behaviour of f near $(1, 2)$, look at the values of the function in Table 15.1.

Table 15.1 *Values of $f(x, y)$ near the point $(1, 2)$*

		\multicolumn{5}{c}{x}			
		0.8	0.9	1.0	1.1	1.2
	1.8	0.08	0.05	0.04	0.05	0.08
	1.9	0.05	0.02	0.01	0.02	0.05
y	2.0	0.04	0.01	0.00	0.01	0.04
	2.1	0.05	0.02	0.01	0.02	0.05
	2.2	0.08	0.05	0.04	0.05	0.08

The table suggests that the function has a local minimum value of 0 at $(1, 2)$. We can confirm this by completing the square:

$$f(x, y) = x^2 - 2x + y^2 - 4y + 5 = (x - 1)^2 + (y - 2)^2.$$

Figure 15.5 shows that the graph of f is a paraboloid with vertex at the point $(1, 2, 0)$. It is the same shape as the graph of $z = x^2 + y^2$ (see Figure 12.12 on page 693), except that the vertex has been shifted to $(1, 2)$. So the point $(1, 2)$ is a local minimum of f (as well as a global minimum).

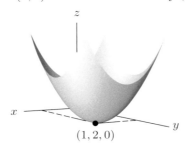

Figure 15.5: The graph of $f(x, y) = x^2 - 2x + y^2 - 4y + 5$ with a local minimum at the point $(1, 2)$

Example 2   Find and analyse any critical points of $f(x, y) = -\sqrt{x^2 + y^2}$.

Solution   We look for points where $\text{grad } f = \vec{0}$ or is undefined. The partial derivatives are given by

$$f_x(x, y) = -\frac{x}{\sqrt{x^2 + y^2}},$$

$$f_y(x, y) = -\frac{y}{\sqrt{x^2 + y^2}}.$$

These partial derivatives are never simultaneously zero, but they are undefined at $x = 0$, $y = 0$. Thus, $(0, 0)$ is a critical point and a possible extreme point. The graph of f (see Figure 15.6) is a cone, with vertex at $(0, 0)$. So f has a local and global maximum at $(0, 0)$.

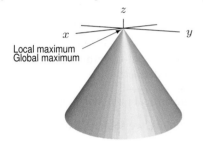

Figure 15.6: Graph of $f(x, y) = -\sqrt{x^2 + y^2}$

Example 3   Find and analyze any critical points of $g(x, y) = x^2 - y^2$.

Solution   To find the critical points, we look for points where both partial derivatives are zero:

$$g_x(x, y) = 2x = 0$$

$$g_y(x, y) = -2y = 0.$$

Solving gives $x = 0$, $y = 0$, so the origin is the only critical point.

Figure 15.7 shows that near the origin g takes on both positive and negative values. Since $g(0, 0) = 0$, the origin is a critical point which is neither a local maximum nor a local minimum. The graph of g looks like a saddle.

The previous examples show that critical points can occur at local maxima or minima, or at points which are neither: The functions g and h in Figures 15.7 and 15.8 both have critical points at the origin. Figure 15.9 shows level curves of g. They are hyperbolas showing both positive and negative values of g near $(0, 0)$. Contrast this with the level curves of h near the local minimum in Figure 15.10.

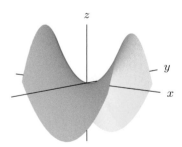

Figure 15.7: Graph of $g(x, y) = x^2 - y^2$, showing saddle shape at the origin

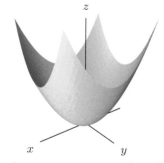

Figure 15.8: Graph of $h(x, y) = x^2 + y^2$, showing minimum at the origin

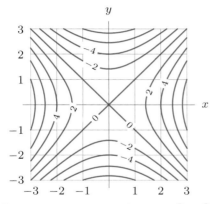

Figure 15.9: Contours of $g(x, y) = x^2 - y^2$, showing a saddle shape at the origin

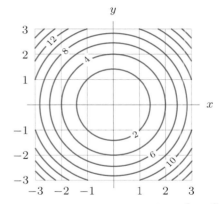

Figure 15.10: Contours of $h(x, y) = x^2 + y^2$, showing a local minimum at the origin

Example 4   Find the local extrema of the function $f(x, y) = 8y^3 + 12x^2 - 24xy$.

Solution   We begin by looking for critical points:

$$f_x(x, y) = 24x - 24y,$$
$$f_y(x, y) = 24y^2 - 24x.$$

Setting these expressions equal to zero gives the system of equations

$$x = y, \qquad x = y^2,$$

which has two solutions, $(0, 0)$ and $(1, 1)$. Are these local maxima, local minima or neither? Figure 15.11 shows contours of f near the points. Notice that $f(1, 1) = -4$ and that there is no other -4 contour. The contours near $(1, 1)$ suggest that f has a local minimum at the point $(1, 1)$.

We have $f(0, 0) = 0$ and the contours near $(0, 0)$ show that f takes both positive and negative values nearby. This suggests that $(0, 0)$ is a critical point which is neither a local maximum nor a local minimum.

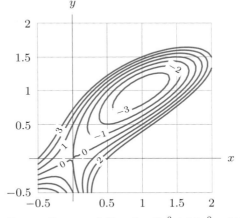

Figure 15.11: Contour diagram of $f(x, y) = 8y^3 + 12x^2 - 24xy$ showing critical points at $(0, 0)$ and $(1, 1)$

Classifying Critical Points

We can see whether a critical point of a function, f, is a local maximum, local minimum, or neither by looking at the contour diagram. There is also an analytic method for making this distinction.

Quadratic Functions of the Form $f(x, y) = ax^2 + bxy + cy^2$

Near most critical points, a function has the same behaviour as its quadratic Taylor approximation about that point. Thus, we start by investigating critical points of quadratic functions of the form $f(x, y) = ax^2 + bxy + cy^2$, where a, b and c are constants.

Example 5   Find and analyse the local extrema of the function $f(x, y) = x^2 + xy + y^2$.

Solution   To find critical points, we set

$$f_x(x, y) = 2x + y = 0,$$
$$f_y(x, y) = x + 2y = 0.$$

The only critical point is $(0, 0)$, and the value of the function there is $f(0, 0) = 0$. If f is always positive or zero near $(0, 0)$, then $(0, 0)$ is a local minimum; if f is always negative or zero near $(0, 0)$, it is a local maximum; if f takes both positive and negative values, it is neither. The graph in Figure 15.12 suggests that $(0, 0)$ is a local minimum.

How can we be sure that $(0, 0)$ is a local minimum? We complete the square. Writing

$$f(x, y) = x^2 + xy + y^2 = \left(x + \frac{1}{2}y\right)^2 + \frac{3}{4}y^2,$$

shows that $f(x, y)$ is a sum of two nonnegative terms, so it is always greater than or equal to zero. Thus, the critical point is both a local and a global minimum.

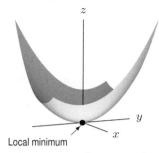
Local minimum

Figure 15.12: Graph of $f(x, y) = x^2 + xy + y^2 = (x + \frac{1}{2}y)^2 + \frac{3}{4}y^2$
showing local minimum at the origin

The Shape of the Graph of $f(x, y) = ax^2 + bxy + cy^2$

In general, a function of the form $f(x, y) = ax^2 + bxy + cy^2$ has one critical point at $(0, 0)$. Assuming $a \neq 0$, we complete the square and write

$$ax^2 + bxy + cy^2 = a\left[x^2 + \frac{b}{a}xy + \frac{c}{a}y^2\right] = a\left[\left(x + \frac{b}{2a}y\right)^2 + \left(\frac{c}{a} - \frac{b^2}{4a^2}\right)y^2\right]$$

$$= a\left[\left(x + \frac{b}{2a}y\right)^2 + \left(\frac{4ac - b^2}{4a^2}\right)y^2\right].$$

The shape of the graph of f depends on whether the coefficient of y^2 is positive, negative, or zero. The sign of the *discriminant*, $D = 4ac - b^2$, determines the sign of the coefficient of y^2.

- If $D > 0$, then the expression inside the square brackets is positive or zero, so the function has a local maximum or a local minimum.

  - If $a > 0$, the function has a local minimum, since the graph is a right-side-up paraboloid, like $z = x^2 + y^2$. (See Figure 15.13.)
  - If $a < 0$, the function has a local maximum, since the graph is an upside-down paraboloid, like $z = -x^2 - y^2$. (See Figure 15.14.)

- If $D < 0$, then the function goes up in some directions and goes down in others, like $z = x^2 - y^2$. We say the function has a *saddle point*. (See Figure 15.15.)

- If $D = 0$, then the quadratic function is $a(x + by/2a)^2$, whose graph is a parabolic cylinder. (See Figure 15.16.)

Figure 15.13: Local minimum: $D > 0$ and $a > 0$

Figure 15.14: Local maximum: $D > 0$ and $a < 0$

Figure 15.15: Saddle point: $D < 0$

Figure 15.16: Parabolic cylinder: $D = 0$

More generally, the graph of $g(x, y) = a(x - x_0)^2 + b(x - x_0)(y - y_0) + c(y - y_0)^2$ has the same shape as the graph of $f(x, y) = ax^2 + bxy + cy^2$, except that the critical point is at (x_0, y_0) rather than $(0, 0)$.[1]

Classifying the Critical Points of a Function

Suppose that f is any function with $\operatorname{grad} f(0, 0) = \vec{0}$. Its quadratic Taylor polynomial near $(0, 0)$,

$$f(x, y) \approx f(0, 0) + f_x(0, 0)x + f_y(0, 0)y$$
$$+ \frac{1}{2}f_{xx}(0, 0)x^2 + f_{xy}(0, 0)xy + \frac{1}{2}f_{yy}(0, 0)y^2,$$

can be simplified using $f_x(0, 0) = f_y(0, 0) = 0$, which gives

$$f(x, y) - f(0, 0) \approx \frac{1}{2}f_{xx}(0, 0)x^2 + f_{xy}(0, 0)xy + \frac{1}{2}f_{yy}(0, 0)y^2.$$

The discriminant of this quadratic polynomial is

$$D = 4ac - b^2 = 4\left(\frac{1}{2}f_{xx}(0, 0)\right)\left(\frac{1}{2}f_{yy}(0, 0)\right) - (f_{xy}(0, 0))^2,$$

which simplifies to

$$D = f_{xx}(0, 0)f_{yy}(0, 0) - (f_{xy}(0, 0))^2.$$

There is a similar formula for D if the critical point is at (x_0, y_0). An analogy with quadratic functions suggests the following test for classifying a critical point of a function of two variables:

Second-Derivative Test for Functions of Two Variables

Suppose (x_0, y_0) is a point where $\operatorname{grad} f(x_0, y_0) = \vec{0}$. Let

$$D = f_{xx}(x_0, y_0)f_{yy}(x_0, y_0) - (f_{xy}(x_0, y_0))^2.$$

- If $D > 0$ and $f_{xx}(x_0, y_0) > 0$, then f has a local minimum at (x_0, y_0).
- If $D > 0$ and $f_{xx}(x_0, y_0) < 0$, then f has a local maximum at (x_0, y_0).
- If $D < 0$, then f has a saddle point at (x_0, y_0).
- If $D = 0$, anything can happen: f can have a local maximum, or a local minimum, or a saddle point, or none of these, at (x_0, y_0).

Example 6   Find the local maxima, minima, and saddle points of $f(x, y) = \frac{1}{2}x^2 + 3y^3 + 9y^2 - 3xy + 9y - 9x$.

Solution   Setting the partial derivatives of f to zero gives

$$f_x(x, y) = x - 3y - 9 = 0,$$
$$f_y(x, y) = 9y^2 + 18y + 9 - 3x = 0.$$

---

[1] We assumed that $a \neq 0$. If $a = 0$ and $c \neq 0$, the same argument works. If both $a = 0$ and $c = 0$, then $f(x, y) = bxy$, which has a saddle point.

Eliminating x gives $9y^2 + 9y - 18 = 0$, with solutions $y = -2$ and $y = 1$. The corresponding values of x are $x = 3$ and $x = 12$, so the critical points of f are $(3, -2)$ and $(12, 1)$. The discriminant is

$$D(x, y) = f_{xx}f_{yy} - f_{xy}^2 = (1)(18y + 18) - (-3)^2 = 18y + 9.$$

Since $D(3, -2) = -36 + 9 < 0$, we know that $(3, -2)$ is a saddle point of f. Since $D(12, 1) = 18 + 9 > 0$ and $f_{xx}(12, 1) = 1 > 0$, we know that $(12, 1)$ is a local minimum of f.

The second-derivative test does not give any information if $D = 0$. However, as the following example illustrates, we may still be able to classify the critical points.

Example 7   Classify the critical points of $f(x, y) = x^4 + y^4$, and $g(x, y) = -x^4 - y^4$, and $h(x, y) = x^4 - y^4$.

Solution   Each of these functions has a critical point at $(0, 0)$. Since all the second partial derivatives are 0 there, each function has $D = 0$. Near the origin, the graphs of f, g and h look like the surfaces in Figures 15.13–15.15, respectively, so f has a local minimum at $(0, 0)$, and g has a local maximum at $(0, 0)$, and h is saddle-shaped at $(0, 0)$.

We can get the same results algebraically. Since $f(0, 0) = 0$ and $f(x, y) > 0$ elsewhere, f has a local minimum at the origin. Since $g(0, 0) = 0$ and $g(x, y) < 0$ elsewhere, g has a local maximum at the origin. Lastly, h is saddle-shaped at the origin since $h(0, 0) = 0$ and, away from the origin, $h(x, y) > 0$ on the x-axis and $h(x, y) < 0$ on the y-axis.

Exercises and Problems for Section 15.1

Exercises

1. Which of the points A, B, C in Figure 15.17 appear to be critical points? Classify those that are critical points.

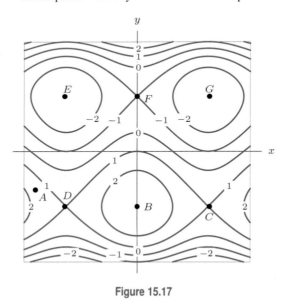

Figure 15.17

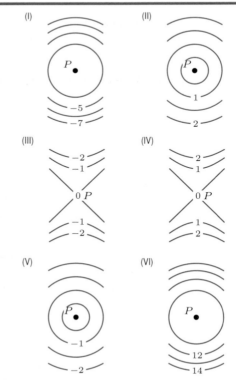

2. Figures (I)–(VI) show level curves of six functions around a critical point P. Does each function have a local maximum, a local minimum, or a saddle point at P?

3. Which of the points D–G in Figure 15.17 appear to be

    (a) Local maxima?
    (b) Local minima?
    (c) Saddle points?

Each function in Exercises 4–8 has a critical point at $(0,0)$. What sort of critical point is it?

4. $m(x,y) = x^5 + y^8$ **5.** $g(x,y) = x^4 + y^3$

6. $f(x,y) = x^6 + y^6$ **7.** $k(x,y) = \sin x \sin y$

8. $h(x,y) = \cos x \cos y$

For Exercises 9–21, find the critical points and classify them as local maxima, local minima, saddle points, or none of these.

9. $f(x,y) = x^2 - 2xy + 3y^2 - 8y$

10. $f(x,y) = 5 + 6x - x^2 + xy - y^2$

11. $f(x,y) = x^2 - y^2 + 4x + 2y$

12. $f(x,y) = 400 - 3x^2 - 4x + 2xy - 5y^2 + 48y$

13. $f(x,y) = 15 - x^2 + 2y^2 + 6x - 8y$

14. $f(x,y) = x^2 y + 2y^2 - 2xy + 6$

15. $f(x,y) = x^3 + y^2 - 3x^2 + 10y + 6$

16. $f(x,y) = x^3 - 3x + y^3 - 3y$

17. $f(x,y) = x^3 + y^3 - 3x^2 - 3y + 10$

18. $f(x,y) = x^3 + y^3 - 6y^2 - 3x + 9$

19. $f(x,y) = (x+y)(xy+1)$

20. $f(x,y) = 8xy - \frac{1}{4}(x+y)^4$

21. $f(x,y) = e^{2x^2+y^2}$

Problems

22. Find A and B so that $f(x,y) = x^2 + Ax + y^2 + B$ has a local minimum value of 20 at $(1,0)$.

23. For $f(x,y) = x^2 + xy + y^2 + ax + by + c$, find values of a, b, and c giving a local minimum at $(2,5)$ and so that $f(2,5) = 11$.

24. **(a)** Find critical points for $f(x,y) = e^{-(x-a)^2-(y-b)^2}$.
    (b) Find a and b such that the critical point is at $(-1,5)$.
    (c) For the values of a and b in part (b), is $(-1,5)$ a local maximum, local minimum, or a saddle point?

25. Let $f(x,y) = (ax+by)e^{-(x^2+y^2)}$, with $(a,b) \neq (0,0)$.

    (a) Find the partial derivatives f_x and f_y of f.
    (b) What are the critical points of f?
    (c) What is the nature of the critical points?

26. Let $f(x,y) = kx^2 + y^2 - 4xy$. Determine the values of k (if any) for which the critical point at $(0,0)$ is:

    (a) A saddle point
    (b) A local maximum
    (c) A local minimum

For Problems 27–30, find critical points and classify them as local maxima, local minima, saddle points, or none of these.

27. $f(x,y) = x^3 + e^{-y^2}$

28. $f(x,y) = \sin x \sin y$

29. $f(x,y) = 1 - \cos x + y^2/2$

30. $f(x,y) = e^x(1 - \cos y)$

31. At the point $(1,3)$, suppose that $f_x = f_y = 0$ and $f_{xx} > 0$, $f_{yy} > 0$, $f_{xy} = 0$.

    (a) What can you conclude about the behavior of the function near the point $(1,3)$?
    (b) Sketch a possible contour diagram.

32. At the point (a,b), suppose that $f_x = f_y = 0$, $f_{xx} > 0$, $f_{yy} = 0$, $f_{xy} > 0$.

    (a) What can you conclude about the shape of the graph of f near the point (a,b)?
    (b) Sketch a possible contour diagram.

For Problems 33–35, use the contours of f in Figure 15.18.

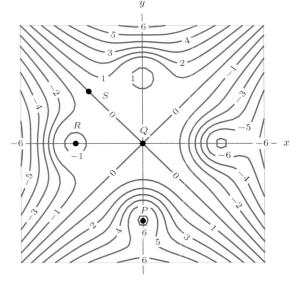

Figure 15.18

33. Decide whether you think each point is a local maximum, local minimum, saddle point, or none of these.

    (a) P **(b)** Q **(c)** R **(d)** S

34. Sketch the direction of ∇f at several points around each of P, Q, and R.

35. At the points where $\|\nabla f\|$ is largest, put arrows showing the direction of ∇f.

36. Draw a possible contour diagram of f such that $f_x(-1, 0) = 0$, $f_y(-1, 0) < 0$, $f_x(3, 3) > 0$, $f_y(3, 3) > 0$, and f has a local maximum at $(3, -3)$.

37. Draw a possible contour diagram of a function with a saddle point at $(2, 1)$, a local minimum at $(2, 4)$, and no other critical points. Label the contours.

38. For constants a and b with $ab \neq 0$ and $ab \neq 1$, let

$$f(x, y) = ax^2 + by^2 - 2xy - 4x - 6y.$$

   (a) Find the x- and y-coordinates of the critical point. Your answer will be in terms of a and b.
   (b) If $a = b = 2$, is the critical point a local maximum, a local minimum, or neither? Give a reason for your answer.
   (c) Classify the critical point for all values of a and b with $ab \neq 0$ and $ab \neq 1$.

39. **(a)** Find the critical point of $f(x, y) = (x^2 - y)(x^2 + y)$.
   (b) Show that at the critical point, the discriminant $D = 0$, so the second-derivative test gives no information about the nature of the critical point.
   (c) Sketch contours near the critical point to determine whether it is a local maximum, a local minimum, a saddle point, or none of these.

40. On a computer, draw contour diagrams for functions

$$f(x, y) = k(x^2 + y^2) - 2xy$$

for $k = -2, -1, 0, 1, 2$. Use these figures to classify the critical point at $(0, 0)$ for each value of k. Explain your observations using the discriminant, D.

41. The behavior of a function can be complicated near a critical point where $D = 0$. Suppose that

$$f(x, y) = x^3 - 3xy^2.$$

Show that there is one critical point at $(0, 0)$ and that $D = 0$ there. Show that the contour for $f(x, y) = 0$ consists of three lines intersecting at the origin and that these lines divide the plane into six regions around the origin

where f alternates from positive to negative. Sketch a contour diagram for f near $(0, 0)$. The graph of this function is called a *monkey saddle*.

42. The contour diagrams for four functions $z = f(x, y)$ are in (a)–(d). Each function has a critical point with $z = 0$ at the origin. Graphs (I)–(IV) show the value of z for these four functions on a small circle around the origin, expressed as function of θ, the angle between the positive x-axis and a line through the origin. Match the contour diagrams (a)–(d) with the graphs (I)–(IV). Classify the critical points as local maxima, local minima or saddle points.

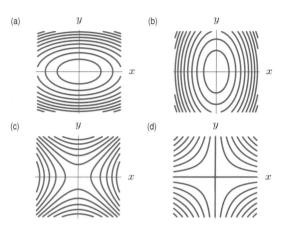

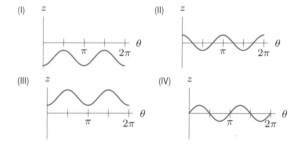

Strengthen Your Understanding

In Problems 43–45, explain what is wrong with the statement.

43. If $f_x = f_y = 0$ at $(1, 3)$, then f has a local maximum or local minimum at $(1, 3)$.

44. For $f(x, y)$, if $D = f_{xx}f_{yy} - (f_{xy})^2 = 0$ at (a, b), then (a, b) is a saddle point.

45. A critical point (a, b) for the function f must be a local minimum if both cross-sections for $x = a$ and $y = b$ are concave up.

In Problems 46–47, give an example of:

46. A nonlinear function having no critical points

47. A function $f(x, y)$ with a local maximum at $(2, -3, 4)$.

Are the statements in Problems 48–58 true or false? Give reasons for your answer.

48. If $f_x(P_0) = f_y(P_0) = 0$, then P_0 is a critical point of f.

49. If $f_x(P_0) = f_y(P_0) = 0$, then P_0 is a local maximum or local minimum of f.

50. If P_0 is a critical point of f, then P_0 is either a local maximum or local minimum of f.

51. If P_0 is a local maximum or local minimum of f, and not on the boundary of the domain of f, then P_0 is a critical point of f.

52. The function $f(x,y) = \sqrt{x^2 + y^2}$ has a local minimum at the origin.

53. The function $f(x,y) = x^2 - y^2$ has a local minimum at the origin.

54. If f has a local minimum at P_0 then so does the function

$$g(x,y) = f(x,y) + 5.$$

55. If f has a local minimum at P_0 then the function $g(x,y) = -f(x,y)$ has a local maximum at P_0.

56. Every function has at least one local maximum.

57. If P_0 is a local maximum of f, then $f(a,b) \le f(P_0)$ for all points (a,b) in 2-space.

58. If P_0 is a local maximum of f, then P_0 is also a global maximum of f.

15.2 OPTIMIZATION

Suppose we want to find the highest and the lowest points in Colorado. A contour map is shown in Figure 15.19. The highest point is the top of a mountain peak (point A on the map, Mt. Elbert, 4,373 metres high). What about the lowest point? Colorado does not have large pits without drainage, like Death Valley in California. A drop of rain falling at any point in Colorado will eventually flow out of the state. If there is no local minimum inside the state, where is the lowest point? It must be on the state boundary at a point where a river is flowing out of the state (point B where the Arikaree River leaves the state, 1,030 metres high). The highest point in Colorado is a global maximum for the elevation function in Colorado and the lowest point is the global minimum.

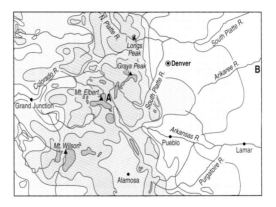

Figure 15.19: The highest and lowest points in the state of Colorado

In general, if we are given a function f defined on a region R, we say:

- f has a **global maximum on** R at the point P_0 if $f(P_0) \ge f(P)$ for all points P in R.
- f has a **global minimum on** R at the point P_0 if $f(P_0) \le f(P)$ for all points P in R.

The process of finding a global maximum or minimum for a function f on a region R is called *optimisation*. If the region R is not stated explicitly, we take it to be the whole xy-plane unless the context of the problem suggests otherwise.

How Do We Find Global Maxima and Minima?

As the Colorado example illustrates, a global extremum can occur either at a critical point inside the region or at a point on the boundary of the region. This is analogous to single-variable calculus, where a function achieves its global extrema on an interval either at a critical point inside the interval or at an endpoint of the interval.

> To locate **global maxima and minima**:
>
> - Find the critical points.
> - Investigate whether the critical points give global maxima or minima.

Not all functions have a global maximum or minimum: it depends on the function and the region. First, we consider applications in which global extrema are expected from practical considerations. At the end of this section, we examine the conditions that lead to global extrema. In general, the fact that a function has a single local maximum or minimum does not guarantee that the point is the global maximum or minimum. (See Problem 27.) An exception is if the function is quadratic, in which case the local maximum or minimum is the global maximum or minimum. (See Example 1 on page 849 and Example 5 on page 852.)

Maximising Profit and Minimising Cost

In planning production of an item, a company often chooses the combination of price and quantity that maximises its profit. We use

$$\text{Profit} = \text{Revenue} - \text{Cost},$$

and, provided the price is constant,

$$\text{Revenue} = \text{Price} \cdot \text{Quantity} = pq.$$

In addition, we need to know how the cost and price depend on quantity.

Example 1 A company manufactures two items which are sold in two separate markets where it has a monopoly. The quantities, q_1 and q_2, demanded by consumers, and the prices, p_1 and p_2 (in dollars), of each item are related by

$$p_1 = 600 - 0.3q_1 \quad \text{and} \quad p_2 = 500 - 0.2q_2.$$

Thus, if the price for either item increases, the demand for it decreases. The company's total production cost is given by

$$C = 16 + 1.2q_1 + 1.5q_2 + 0.2q_1q_2.$$

To maximise its total profit, how much of each product should be produced? What is the maximum profit? [2]

Solution The total revenue, R, is the sum of the revenues, p_1q_1 and p_2q_2, from each market. Substituting for p_1 and p_2, we get

$$R = p_1q_1 + p_2q_2 = (600 - 0.3q_1)q_1 + (500 - 0.2q_2)q_2$$
$$= 600q_1 - 0.3q_1^2 + 500q_2 - 0.2q_2^2.$$

Thus, the total profit P is given by

$$P = R - C = 600q_1 - 0.3q_1^2 + 500q_2 - 0.2q_2^2 - (16 + 1.2q_1 + 1.5q_2 + 0.2q_1q_2)$$
$$= -16 + 598.8q_1 - 0.3q_1^2 + 498.5q_2 - 0.2q_2^2 - 0.2q_1q_2.$$

Since q_1 and q_2 cannot be negative,[3] the region we consider is the first quadrant with boundary $q_1 = 0$ and $q_2 = 0$.

[2] Adapted from M. Rosser, *Basic Mathematics for Economists*, p. 316 (New York: Routledge, 1993).
[3] Restricting prices to be nonnegative further restricts the region but does not alter the solution.

To maximise P, we look for critical points by setting the partial derivatives equal to 0:

$$\frac{\partial P}{\partial q_1} = 598.8 - 0.6q_1 - 0.2q_2 = 0,$$

$$\frac{\partial P}{\partial q_2} = 498.5 - 0.4q_2 - 0.2q_1 = 0.$$

Since $\operatorname{grad} P$ is defined everywhere, the only critical points of P are those where $\operatorname{grad} P = \vec{0}$. Thus, solving for q_1, and q_2, we find that

$$q_1 = 699.1 \quad \text{and} \quad q_2 = 896.7.$$

The corresponding prices are

$$p_1 = 390.27 \quad \text{and} \quad p_2 = 320.66.$$

To see whether or not we have found a local maximum, we compute second partial derivatives:

$$\frac{\partial^2 P}{\partial q_1^2} = -0.6, \quad \frac{\partial^2 P}{\partial q_2^2} = -0.4, \quad \frac{\partial^2 P}{\partial q_1 \partial q_2} = -0.2,$$

so,

$$D = \frac{\partial^2 P}{\partial q_1^2}\frac{\partial^2 P}{\partial q_2^2} - \left(\frac{\partial^2 P}{\partial q_1 \partial q_2}\right)^2 = (-0.6)(-0.4) - (-0.2)^2 = 0.2.$$

Therefore we have found a local maximum. The graph of P is an upside-down paraboloid, so $(699.1, 896.7)$ is a global maximum. This point is within the region, so points on the boundary give smaller values of P.

The company should produce 699.1 units of the first item priced at \$390.27 per unit, and 896.7 units of the second item priced at \$320.66 per unit. The maximum profit $P(699.1, 896.7) \approx$ \$433,000.

Example 2   A delivery of 480 cubic metres of gravel is to be made to a landfill. The trucker plans to purchase an open-top box in which to transport the gravel in numerous trips. The total cost to the trucker is the cost of the box plus \$80 per trip. The box must have height 2 metres, but the trucker can choose the length and width. The cost of the box is \$100/m² for the ends, \$50/m² for the sides and \$200/m² for the bottom. Notice the tradeoff: A smaller box is cheaper to buy but requires more trips. What size box should the trucker buy to minimise the total cost? [4]

Solution   We first get an algebraic expression for the trucker's cost. Let the length of the box be x metres and the width be y metres; the height is 2 metres. (See Figure 15.20.)

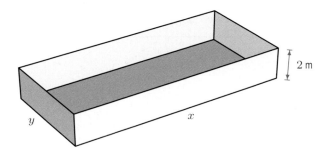

Table 15.2   *Trucker's itemised cost*

Expense	Cost in dollars
Travel: $480/(2xy)$ at \$80/trip	$(240 \cdot 80)/(xy)$
Ends: 2 at \$100/m² $\cdot 2y$ m²	$400y$
Sides: 2 at \$50/m² $\cdot 2x$ m²	$200x$
Bottom: 1 at \$200/m² $\cdot xy$ m²	$200xy$

Figure 15.20: The box for transporting gravel

---

[4]Adapted from Claude McMillan, Jr., *Mathematical Programming*, 2nd ed., p. 156-157 (New York: Wiley, 1978).

The volume of the box is $2xy$ m^3, so delivery of 480 m^3 of gravel requires $480/(2xy)$ trips. The number of trips is a whole number; however, we treat it as continuous so that we can optimise using derivatives. The trucker's cost is itemised in Table 15.2. The problem is to minimise

$$\text{Total cost} = \frac{240 \cdot 80}{xy} + 400y + 200x + 200xy = 200\left(\frac{96}{xy} + 2y + x + xy\right).$$

The length and width of the box must be positive. Thus, the region is the first quadrant but it does not contain the boundary, $x = 0$ and $y = 0$.

Our problem is to minimize

$$f(x, y) = \frac{96}{xy} + 2y + x + xy.$$

The critical points of this function occur where

$$f_x(x, y) = -\frac{96}{x^2 y} + 1 + y = 0$$

$$f_y(x, y) = -\frac{96}{xy^2} + 2 + x = 0.$$

We put the $96/(x^2y)$ and $96/(xy^2)$ terms on the other side of the the equation, divide, and simplify:

$$\frac{96/(x^2 y)}{96/(xy^2)} = \frac{1+y}{2+x} \quad \text{so} \quad \frac{y}{x} = \frac{1+y}{2+x} \quad \text{giving} \quad 2y = x.$$

Substituting $x = 2y$ in the equation $f_y(x, y) = 0$ gives

$$-\frac{96}{2y \cdot y^2} + 2 + 2y = 0$$

$$y^4 + y^3 - 24 = 0.$$

The only positive solution to this equation is $y = 2$, so the only critical point in the region is $(4, 2)$. To check that the critical point is a local minimum, we use the second-derivative test. Since

$$D(4, 2) = f_{xx}f_{yy} - (f_{xy})^2 = \frac{192}{4^3 \cdot 2} \cdot \frac{192}{4 \cdot 2^3} - \left(\frac{96}{4^2 \cdot 2^2} + 1\right)^2 = 9 - \frac{25}{4} > 0$$

and $f_{xx}(4, 2) > 0$, the point $(4, 2)$ is a local minimum. Since the value of f increases without bound as x or y increases without bound and as $x \to 0^+$ and $y \to 0^+$, it can be shown that $(4, 2)$ is a global minimum. (See Problem 30.) Thus, the optimal box is 4 metres long and 2 metres wide.

Fitting a Line to Data: Least Squares

Suppose we want to fit the "best" line to some data in the plane. We measure the distance from a line to the data points by adding the squares of the vertical distances from each point to the line. The smaller this sum of squares is, the better the line fits the data. The line with the minimum sum of square distances is called the *least squares line*, or the *regression line*. If the data is nearly linear, the least squares line is a good fit; otherwise it may not be. (See Figure 15.21.)

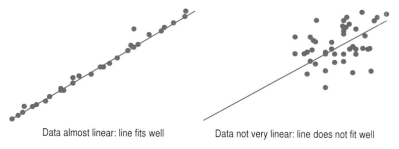

Data almost linear: line fits well   Data not very linear: line does not fit well

Figure 15.21: Fitting lines to data points

Example 3   Find a least squares line for the following data points: $(1, 1)$, $(2, 1)$, and $(3, 3)$.

Solution   Suppose the line has equation $y = b + mx$. If we find b and m then we have found the line. So, for this problem, b and m are the two variables. Any values of m and b are possible, so this is an unconstrained problem. We want to minimise the function $f(b, m)$ that gives the sum of the three squared vertical distances from the points to the line in Figure 15.22.

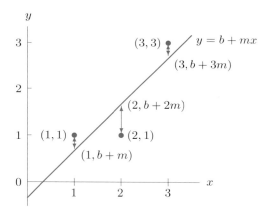

Figure 15.22: The least squares line minimises the sum of the squares of these vertical distances

The vertical distance from the point $(1, 1)$ to the line is the difference in the y-coordinates $1 - (b + m)$; similarly for the other points. Thus, the sum of squares is

$$f(b, m) = (1 - (b + m))^2 + (1 - (b + 2m))^2 + (3 - (b + 3m))^2.$$

To minimise f we look for critical points. First we differentiate f with respect to b:

$$\frac{\partial f}{\partial b} = -2(1 - (b + m)) - 2(1 - (b + 2m)) - 2(3 - (b + 3m))$$
$$= -2 + 2b + 2m - 2 + 2b + 4m - 6 + 2b + 6m$$
$$= -10 + 6b + 12m.$$

Now we differentiate with respect to m:

$$\frac{\partial f}{\partial m} = 2(1 - (b + m))(-1) + 2(1 - (b + 2m))(-2) + 2(3 - (b + 3m))(-3)$$
$$= -2 + 2b + 2m - 4 + 4b + 8m - 18 + 6b + 18m$$
$$= -24 + 12b + 28m.$$

The equations $\frac{\partial f}{\partial b} = 0$ and $\frac{\partial f}{\partial m} = 0$ give a system of two linear equations in two unknowns:

$$-10 + 6b + 12m = 0,$$
$$-24 + 12b + 28m = 0.$$

The solution to this pair of equations is the critical point $b = -1/3$ and $m = 1$. Since

$$D = f_{bb}f_{mm} - (f_{mb})^2 = (6)(28) - 12^2 = 24 \quad \text{and} \quad f_{bb} = 6 > 0,$$

we have found a local minimum. The graph of $f(b, m)$ is a parabolic bowl, so the local minimum is the global minimum of f. Thus, the least squares line is

$$y = x - \frac{1}{3}.$$

As a check, notice that the line $y = x$ passes through the points $(1, 1)$ and $(3, 3)$. It is reasonable that introducing the point $(2, 1)$ moves the y-intercept down from 0 to $-1/3$.

The general formulas for the slope and y-intercept of a least squares line are in Project 2 on page 883. Many calculators have built-in formulas for b and m, as well as for the *correlation coefficient*, which measures how well the data points fit the least squares line.

How Do We Know Whether a Function Has a Global Maximum or Minimum?

Under what circumstances does a function of two variables have a global maximum or minimum? The next example shows that a function may have both a global maximum and a global minimum on a region, or just one, or neither.

Example 4   Investigate the global maxima and minima of the following functions:

(a) $h(x, y) = 1 + x^2 + y^2$ on the disk $x^2 + y^2 \leq 1$.
(b) $f(x, y) = x^2 - 2x + y^2 - 4y + 5$ on the xy-plane.
(c) $g(x, y) = x^2 - y^2$ on the xy-plane.

Solution   (a) The graph of $h(x, y) = 1 + x^2 + y^2$ is a bowl-shaped paraboloid with a global minimum of 1 at $(0, 0)$, and a global maximum of 2 on the edge of the region, $x^2 + y^2 = 1$.

(b) The graph of f in Figure 15.5 on page 849 shows that f has a global minimum at the point $(1, 2)$ and no global maximum (because the value of f increases without bound as $x \to \infty, y \to \infty$).

(c) The graph of g in Figure 15.7 on page 850 shows that g has no global maximum because $g(x, y) \to \infty$ as $x \to \infty$ if y is constant. Similarly, g has no global minimum because $g(x, y) \to -\infty$ as $y \to \infty$ if x is constant.

Sometimes a function is guaranteed to have a global maximum and minimum. For example, a continuous function, $h(x)$, of one variable has a global maximum and minimum on every closed interval $a \leq x \leq b$. On a non-closed interval, such as $a \leq x < b$ or $a < x < b$, or on an unbounded interval, such as $a < x < \infty$, then h may not have a maximum or minimum value.

What is the situation for functions of two variables? As it turns out, a similar result is true for continuous functions defined on regions which are closed and bounded, analogous to the closed and bounded interval $a \leq x \leq b$. In everyday language we say

- A **closed** region is one which contains its boundary;
- A **bounded** region is one which does not stretch to infinity in any direction.

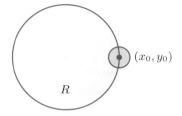

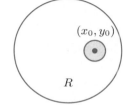

Figure 15.23: Boundary point (x_0, y_0) of R **Figure 15.24:** Interior point (x_0, y_0) of R

More precise definitions follow. Suppose R is a region in 2-space. A point (x_0, y_0) is a *bound-ary point* of R if, for every $r > 0$, the disk $(x - x_0)^2 + (y - y_0)^2 < r^2$ with centre (x_0, y_0) and radius r contains both points which are in R and points which are not in R. See Figure 15.23. A point (x_0, y_0) can be a boundary point of the region R without belonging to R. A point (x_0, y_0) in R is an *interior point* if it is not a boundary point; thus, for small enough $r > 0$, the disk of radius r centred at (x_0, y_0) lies entirely in the region R. See Figure 15.24. The collection of all the boundary points is the *boundary* of R and the collection of all the interior points is the *interior* of R. The region R is *closed* if it contains its boundary; it is *open* if every point in R is an interior point.

A region R in 2-space is *bounded* if the distance between every point (x, y) in R and the origin is less than some constant K. Closed and bounded regions in 3-space are defined in the same way.

Example 5   (a) The square $-1 \leq x \leq 1$, $-1 \leq y \leq 1$ is closed and bounded.
(b) The first quadrant $x \geq 0$, $y \geq 0$ is closed but is not bounded.
(c) The disk $x^2 + y^2 < 1$ is open and bounded, but is not closed.
(d) The half-plane $y > 0$ is open, but is neither closed nor bounded.

The reason that closed and bounded regions are useful is the following theorem, which is also true for functions of three or more variables:[5]

Theorem 15.1: Extreme Value Theorem for Multivariable Functions

If f is a continuous function on a closed and bounded region R, then f has a global maximum at some point (x_0, y_0) in R and a global minimum at some point (x_1, y_1) in R.

If f is not continuous or the region R is not closed and bounded, there is no guarantee that f achieves a global maximum or global minimum on R. In Example 4, the function g is continuous but does not achieve a global maximum or minimum in 2-space, a region which is closed but not bounded. Example 6 illustrates what can go wrong when the region is bounded but not closed.

Example 6   Does the function f have a global maximum or minimum on the region R given by $0 < x^2 + y^2 \leq 1$?

$$f(x, y) = \frac{1}{x^2 + y^2}$$

Solution   The region R is bounded, but it is not closed since it does not contain the boundary point $(0, 0)$. We see from the graph of $z = f(x, y)$ in Figure 15.25 that f has a global minimum on the circle $x^2 + y^2 = 1$. However, $f(x, y) \to \infty$ as $(x, y) \to (0, 0)$, so f has no global maximum.

[5]For a proof, see W. Rudin, *Principles of Mathematical Analysis*, 2nd ed., p. 89 (New York: McGraw-Hill, 1976).

Figure 15.25: Graph showing $f(x,y) = \frac{1}{x^2+y^2}$ has no global maximum on $0 < x^2 + y^2 \le 1$

Exercises and Problems for Section 15.2

Exercises

1. By looking at the weather map in Figure 12.1 on page 684, find the maximum and minimum daily high temperatures in the states of Mississippi, Alabama, Pennsylvania, New York, California, Arizona, and Massachusetts.

In Exercises 2–4, estimate the position and approximate value of the global maxima and minima on the region shown.

2.

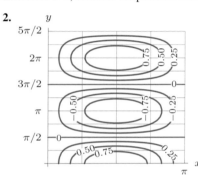

3.

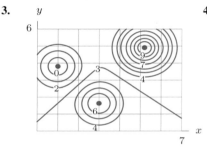

4.

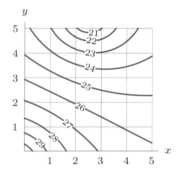

In Exercises 5–10, do the functions have global maxima and minima in the xy-plane?

5. $f(x,y) = x^2 - 2y^2$

6. $g(x,y) = x^2 y^2$

7. $k(x,y) = x^2 y^5$

8. $h(x,y) = x^3 + y^3$

9. $f(x,y) = -2x^2 - 7y^2$

10. $f(x,y) = x^2/2 + 3y^3 + 9y^2 - 3x$

In Exercises 11–13, find the global maximum and minimum of the function on $-1 \le x \le 1$, $-1 \le y \le 1$, and say whether it occurs on the boundary of the square. [Hint: Use graphs.]

11. $z = x^2 + y^2$ 12. $z = -x^2 - y^2$ 13. $z = x^2 - y^2$

Problems

14. (a) Compute and classify the critical points of $f(x,y) = 2x^2 - 3xy + 8y^2 + x - y$.
    (b) By completing the square, plot the contour diagram of f and show that the local extremum found in part (a) is a global one.

15. A missile has a guidance device which is sensitive to both temperature, $t°C$, and humidity, h. The range in km over which the missile can be controlled is given by

$$\text{Range} = 27{,}800 - 5t^2 - 6ht - 3h^2 + 400t + 300h.$$

What are the optimal atmospheric conditions for controlling the missile?

16. A closed rectangular box has volume 32 cm^3. What are the lengths of the edges giving the minimum surface area?

17. A closed rectangular box with faces parallel to the coordinate planes has one bottom corner at the origin and the opposite top corner in the first octant on the plane $3x + 2y + z = 1$. What is the maximum volume of such a box?

18. An international airline has a regulation that each passenger can carry a suitcase having the sum of its width, length and height less than or equal to 135 cm. Find the dimensions of the suitcase of maximum volume that a passenger may carry under this regulation.

19. Design a rectangular milk carton box of width w, length l, and height h which holds 512 cm^3 of milk. The sides of the box cost 1 cent/cm^2 and the top and bottom cost 2 cent/cm^2. Find the dimensions of the box that minimize the total cost of materials used.

20. Find the point on the plane $3x + 2y + z = 1$ that is closest to the origin by minimizing the square of the distance.

21. What is the shortest distance from the surface $xy + 3x + z^2 = 9$ to the origin?

22. For constants a, b, and c, let $f(x, y) = ax + by + c$ be a linear function, and let R be a region in the xy-plane.

   (a) If R is any disk, show that the maximum and minimum values of f on R occur on the boundary of the disk.

   (b) If R is any rectangle, show that the maximum and minimum values of f on R occur at the corners of the rectangle. They may occur at other points of the rectangle as well.

   (c) Use a graph of the plane $z = f(x, y)$ to explain your answers in parts (a) and (b).

23. Two products are manufactured in quantities q_1 and q_2 and sold at prices of p_1 and p_2, respectively. The cost of producing them is given by

$$C = 2q_1^2 + 2q_2^2 + 10.$$

   (a) Find the maximum profit that can be made, assuming the prices are fixed.

   (b) Find the rate of change of that maximum profit as p_1 increases.

24. A company operates two plants which manufacture the same item and whose total cost functions are

$$C_1 = 8.5 + 0.03q_1^2 \quad \text{and} \quad C_2 = 5.2 + 0.04q_2^2,$$

where q_1 and q_2 are the quantities produced by each plant. The company is a monopoly. The total quantity demanded, $q = q_1 + q_2$, is related to the price, p, by

$$p = 60 - 0.04q.$$

How much should each plant produce in order to maximize the company's profit?[6]

25. The quantity of a product demanded by consumers is a function of its price. The quantity of one product demanded may also depend on the price of other products. For example, if the only chocolate shop in town (a monopoly) sells milk and dark chocolates, the price it sets for each affects the demand of the other. The quantities demanded, q_1 and q_2, of two products depend on their prices, p_1 and p_2, as follows:

$$q_1 = 150 - 2p_1 - p_2$$
$$q_2 = 200 - p_1 - 3p_2.$$

   (a) What does the fact that the coefficients of p_1 and p_2 are negative tell you? Give an example of two products that might be related this way.

   (b) If one manufacturer sells both products, how should the prices be set to generate the maximum possible revenue? What is that maximum possible revenue?

26. A company manufactures a product which requires capital and labor to produce. The quantity, Q, of the product manufactured is given by the Cobb-Douglas function

$$Q = AK^a L^b,$$

where K is the quantity of capital; L is the quantity of labor used; and A, a, and b are positive constants with $0 < a < 1$ and $0 < b < 1$. One unit of capital costs $\$k$ and one unit of labor costs $\$\ell$. The price of the product is fixed at $\$p$ per unit.

   (a) If $a + b < 1$, how much capital and labor should the company use to maximize its profit?

   (b) Is there a maximum profit in the case $a + b = 1$? What about $a + b \geq 1$? Explain.

27. Let $f(x, y) = x^2(y + 1)^3 + y^2$. Show that f has only one critical point, namely $(0, 0)$, and that point is a local minimum but not a global minimum. Contrast this with the case of a function with a single local minimum in one-variable calculus.

28. Find the parabola of the form $y = ax^2 + b$ which best fits the points $(1, 0)$, $(2, 2)$, $(3, 4)$ by minimizing the sum of squares, S, given by

$$S = (a + b)^2 + (4a + b - 2)^2 + (9a + b - 4)^2.$$

29. Find the least squares line for the data points $(0, 4)$, $(1, 3)$, $(2, 1)$.

30. Let $f(x, y) = 80/(xy) + 20y + 10x + 10xy$ in the region R where $x, y > 0$.

   (a) Explain why $f(x, y) > f(2, 1)$ at every point in R where

   (i) $x > 20$ (ii) $y > 20$

   (iii) $x < 0.01$ and $y \leq 20$

   (iv) $y < 0.01$ and $x \leq 20$

   (b) Explain why f must have a global minimum at a critical point in R.

   (c) Explain why f must have a global minimum in R at the point $(2, 1)$.

31. Let $f(x, y) = 2/x + 3/y + 4x + 5y$ in the region R where $x, y > 0$.

   (a) Explain why f must have a global minimum at some point in R.

   (b) Find the global minimum.

---

[6]Adapted from M. Rosser, *Basic Mathematics for Economists*, p. 318 (New York: Routledge, 1993).

32. (a) The energy, E, required to compress a gas from a fixed initial pressure P_0 to a fixed final pressure P_F through an intermediate pressure p is[7]

$$E = \left(\frac{p}{P_0}\right)^2 + \left(\frac{P_F}{p}\right)^2 - 1.$$

How should p be chosen to minimize the energy?

(b) Now suppose the compression takes place in two stages with two intermediate pressures, p_1 and p_2. What choices of p_1 and p_2 minimize the energy if

$$E = \left(\frac{p_1}{P_0}\right)^2 + \left(\frac{p_2}{p_1}\right)^2 + \left(\frac{P_F}{p_2}\right)^2 - 2?$$

33. The Dorfman-Steiner rule shows how a company which has a monopoly should set the price, p, of its product and how much advertising, a, it should buy. The price of advertising is p_a per unit. The quantity, q, of the product sold is given by $q = Kp^{-E}a^{\theta}$, where $K > 0$, $E > 1$, and $0 < \theta < 1$ are constants. The cost to the company to make each item is c.

(a) How does the quantity sold, q, change if the price,

p, increases? If the quantity of advertising, a, increases?

(b) Show that the partial derivatives can be written in the form $\partial q/\partial p = -Eq/p$ and $\partial q/\partial a = \theta q/a$.

(c) Explain why profit, π, is given by $\pi = pq - cq - p_a a$.

(d) If the company wants to maximize profit, what must be true of the partial derivatives, $\partial \pi/\partial p$ and $\partial \pi/\partial a$?

(e) Find $\partial \pi/\partial p$ and $\partial \pi/\partial a$.

(f) Use your answers to parts (d) and (e) to show that at maximum profit,

$$\frac{p-c}{p} = \frac{1}{E} \quad \text{and} \quad \frac{p-c}{p_a} = \frac{a}{\theta q}.$$

(g) By dividing your answers in part (f), show that at maximum profit,

$$\frac{p_a a}{pq} = \frac{\theta}{E}.$$

This is the Dorfman-Steiner rule, that the ratio of the advertising budget to revenue does not depend on the price of advertising.

Strengthen Your Understanding

In Problems 34–36, explain what is wrong with the statement.

34. A function having no critical points in a region R cannot have a global maximum in the region.

35. No continuous function has a global minimum on an unbounded region R.

36. If $f(x, y)$ has a local maximum value of 1 at the origin, then the global maximum is 1.

In Problems 37–38, give an example of:

37. A continuous function $f(x, y)$ that has no global maximum and no global minimum on the xy-plane.

38. A function $f(x, y)$ and a region R such that the maximum value of f on R is on the boundary of R.

Are the statements in Problems 39–47 true or false? Give reasons for your answer.

39. If P_0 is a global maximum of f, where f is defined on all of 2-space, then P_0 is also a local maximum of f.

40. Every function has a global maximum.

41. The region consisting of all points (x, y) satisfying $x^2 + y^2 < 1$ is bounded.

42. The region consisting of all points (x, y) satisfying $x^2 + y^2 < 1$ is closed.

43. The function $f(x, y) = x^2 + y^2$ has a global minimum on the region $x^2 + y^2 < 1$.

44. The function $f(x, y) = x^2 + y^2$ has a global maximum on the region $x^2 + y^2 < 1$.

45. If P and Q are two distinct points in 2-space, and f has a global maximum at P, then f cannot have a global maximum at Q.

46. The function $f(x, y) = \sin(1 + e^{xy})$ must have a global minimum in the square region $0 \le x \le 1, 0 \le y \le 1$.

47. If P_0 is a global minimum of f on a closed and bounded region, then P_0 need not be a critical point of f.

15.3 CONSTRAINED OPTIMIZATION: LAGRANGE MULTIPLIERS

Many, perhaps most, real optimisation problems are constrained by external circumstances. For example, a city wanting to build a public transportation system has only a limited number of tax dollars it can spend on the project. In this section, we see how to find an optimum value under such constraints.

In Section 15.2, we saw how to optimize a function $f(x, y)$ on a region R. If the region R is the entire xy-plane, we have *unconstrained optimisation*; if the region R is not the entire xy-plane, that is, if x or y is restricted in some way, then we have *constrained optimisation*.

---

[7] Adapted from Aris Rutherford, *Discrete Dynamic Programming*, p. 35 (New York: Blaisdell, 1964).

Graphical Approach: Maximising Production Subject to a Budget Constraint

Suppose we want to maximise the production under a budget constraint. Suppose production, f, is a function of two variables, x and y, which are quantities of two raw materials, and that

$$f(x, y) = x^{2/3}y^{1/3}.$$

If x and y are purchased at prices of p_1 and p_2 thousands of dollars per unit, what is the maximum production f that can be obtained with a budget of c thousand dollars?

To maximise f without regard to the budget, we simply increase x and y. However, the budget constraint prevents us from increasing x and y beyond a certain point. Exactly how does the budget constrain us? With prices of p_1 and p_2, the amount spent on x is p_1x and the amount spent on y is p_2y, so we must have

$$g(x, y) = p_1x + p_2y \leq c,$$

where $g(x, y)$ is the total cost of the raw materials and c is the budget in thousands of dollars.

Let's look at the case when $p_1 = p_2 = 1$ and $c = 3.78$. Then

$$x + y \leq 3.78.$$

Figure 15.26 shows some contours of f and the budget constraint represented by the line $x+y = 3.78$. Any point on or below the line represents a pair of values of x and y that we can afford. A point on the line completely exhausts the budget, while a point below the line represents values of x and y which can be bought without using up the budget. Any point above the line represents a pair of values that we cannot afford.

To maximise f, we find the point which lies on the level curve with the largest possible value of f *and* which lies within the budget. The point must lie on the budget constraint because we should spend all the available money. Unless we are at the point where the budget constraint is tangent to the contour $f = 2$, we can increase f by moving along the line representing the budget constraint in Figure 15.26. For example, if we are on the line to the left of the point of tangency, moving right on the constraint will increase f; if we are on the line to the right of the point of tangency, moving left will increase f. Thus, the maximum value of f on the budget constraint occurs at the point where the budget constraint is tangent to the contour $f = 2$.

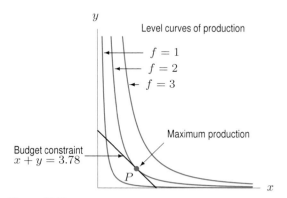

Figure 15.26: Optimal point, P, where budget constraint is tangent to a level of production function

Analytical Solution: Lagrange Multipliers

Figure 15.26 suggests that maximum production is achieved at the point where the budget constraint is tangent to a level curve of the production function. The method of Lagrange multipliers uses this fact in algebraic form. Figure 15.27 shows that at the optimum point, P, the gradient of f and the normal to the budget line $g(x, y) = 3.78$ are parallel. Thus, at P, grad f and grad g are parallel, so for some scalar λ, called the *Lagrange multiplier,*

$$\operatorname{grad} f = \lambda \operatorname{grad} g.$$

Since grad $f = \left(\frac{2}{3}x^{-1/3}y^{1/3}\right)\vec{i} + \left(\frac{1}{3}x^{2/3}y^{-2/3}\right)\vec{j}$ and grad $g = \vec{i} + \vec{j}$, equating components gives

$$\frac{2}{3}x^{-1/3}y^{1/3} = \lambda \quad \text{and} \quad \frac{1}{3}x^{2/3}y^{-2/3} = \lambda.$$

Eliminating λ gives

$$\frac{2}{3}x^{-1/3}y^{1/3} = \frac{1}{3}x^{2/3}y^{-2/3}, \quad \text{which leads to} \quad 2y = x.$$

Since the constraint $x + y = 3.78$ must be satisfied, we have $x = 2.52$ and $y = 1.26$. Then

$$f(2.52, 1.26) = (2.52)^{2/3}(1.26)^{1/3} \approx 2.$$

As before, we see that the maximum value of f is approximately 2. Thus, to maximize production on a budget of \$3780, we should use 2.52 units of one raw material and 1.26 units of the other.

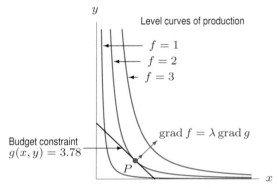

Figure 15.27: At the point, P, of maximum production, the vectors grad f and grad g are parallel

Lagrange Multipliers in General

Suppose we want to optimise an *objective function* $f(x, y)$ subject to a *constraint* $g(x, y) = c$. We look for extrema among the points which satisfy the constraint. We make the following definition.

> Suppose P_0 is a point satisfying the constraint $g(x, y) = c$.
> - f has a **local maximum** at P_0 **subject to the constraint** if $f(P_0) \geq f(P)$ for all points P near P_0 satisfying the constraint.
> - f has a **global maximum** at P_0 **subject to the constraint** if $f(P_0) \geq f(P)$ for all points P satisfying the constraint.
>
> Local and global minima are defined similarly.

As we saw in the production example, constrained extrema occur at points of tangency of contours of f and g; they can also occur at endpoints of constraints. At a point of tangency, grad f is perpendicular to the constraint and so parallel to grad g. At interior points on the constraint where grad f is not perpendicular to the constraint, the value of f can be increased or decreased by moving along the constraint. Therefore constrained extrema occur only at points where grad f and grad g are parallel or at endpoints of the constraint. (See Figure 15.28.) At points where the gradients are parallel, provided grad $g \neq \vec{0}$, there is a constant λ such that grad $f = \lambda \operatorname{grad} g$.

Optimising f Subject to the Constraint $g = c$:
If a smooth function, f, has a maximum or minimum subject to a smooth constraint $g = c$ at a point P_0, then either P_0 satisfies the equations

$$\text{grad } f = \lambda \text{ grad } g \quad \text{and} \quad g = c,$$

or P_0 is an endpoint of the constraint, or $\text{grad } g(P_0) = \vec{0}$. To find P_0, compare values of f at the points satisfying these three conditions. The number λ is called the **Lagrange multiplier**.

If the set of points satisfying the constraint is closed and bounded, such as a circle or line segment, then there must be a global maximum and minimum of f subject to the constraint. If the constraint is not closed and bounded, such as a line or hyperbola, then there may or may not be a global maximum and minimum.

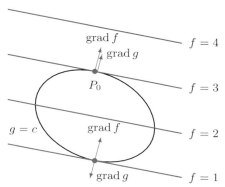

Figure 15.28: Maximum and minimum values of $f(x, y)$ on $g(x, y) = c$ are at points where grad f is parallel to grad g

Example 1   Find the maximum and minimum values of $x + y$ on the circle $x^2 + y^2 = 4$.

Solution   The objective function is

$$f(x, y) = x + y,$$

and the constraint is

$$g(x, y) = x^2 + y^2 = 4.$$

Since $\text{grad } f = f_x \vec{i} + f_y \vec{j} = \vec{i} + \vec{j}$ and $\text{grad } g = g_x \vec{i} + g_y \vec{j} = 2x\vec{i} + 2y\vec{j}$, the condition $\text{grad } f = \lambda \text{ grad } g$ gives

$$1 = 2\lambda x \quad \text{and} \quad 1 = 2\lambda y,$$

so

$$x = y.$$

We also know that

$$x^2 + y^2 = 4,$$

giving $x = y = \sqrt{2}$ or $x = y = -\sqrt{2}$. The constraint has no endpoints (it's a circle) and $\text{grad } g \neq \vec{0}$ on the circle, so we compare values of f at $(\sqrt{2}, \sqrt{2})$ and $(-\sqrt{2}, -\sqrt{2})$. Since $f(x, y) = x + y$, the maximum value of f is $f(\sqrt{2}, \sqrt{2}) = 2\sqrt{2}$; the minimum value is $f(-\sqrt{2}, -\sqrt{2}) = -2\sqrt{2}$. (See Figure 15.29.)

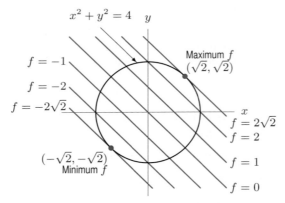

Figure 15.29: Maximum and minimum values of $f(x, y) = x + y$ on the circle
$x^2 + y^2 = 4$ are at points where contours of f are tangent to the circle

How to Distinguish Maxima from Minima

There is a second-derivative test[8] for classifying the critical points of constrained optimisation problems, but it is more complicated than the test in Section 15.1. However, a graph of the constraint and some contours usually shows which points are maxima, which points are minima, and which are neither.

Optimisation with Inequality Constraints

The production problem that we looked at first was to maximise production $f(x, y)$ subject to a budget constraint

$$g(x, y) = p_1 x + p_2 y \leq c.$$

Since the inputs are nonnegative, $x \geq 0$ and $y \geq 0$, we have three inequality constraints, which restrict (x, y) to a region of the plane rather than to a curve in the plane. In principle, we should first check to see whether or not $f(x, y)$ has any critical points in the interior:

$$p_1 x + p_2 y < c, \qquad x > 0 \quad y > 0.$$

However, in the case of a budget constraint, we can see that the maximum of f must occur when the budget is exhausted, so we look for the maximum value of f on the boundary line:

$$p_1 x + p_2 y = c, \qquad x \geq 0 \quad y \geq 0.$$

Strategy for Optimising $f(x, y)$ Subject to the Constraint $g(x, y) \leq c$
- Find all points in the region $g(x, y) < c$ where grad f is zero or undefined.
- Use Lagrange multipliers to find the local extrema of f on the boundary $g(x, y) = c$.
- Evaluate f at the points found in the previous two steps and compare the values.

From Section 15.2 we know that if f is continuous on a closed and bounded region, R, then f is guaranteed to attain its global maximum and minimum values on R.

Example 2   Find the maximum and minimum values of $f(x, y) = (x - 1)^2 + (y - 2)^2$ subject to the constraint $x^2 + y^2 \leq 45$.

Solution   First, we look for all critical points of f in the interior of the region. Setting

$$f_x(x, y) = 2(x - 1) = 0$$
$$f_y(x, y) = 2(y - 2) = 0.$$

---

[8] See J. E. Marsden and A. J. Tromba, *Vector Calculus*, 4th ed., p. 218 (New York: W.H. Freeman, 1996).

we find f has exactly one critical point at $x = 1$, $y = 2$. Since $1^2 + 2^2 < 45$, that critical point is in the interior of the region.

Next, we find the local extrema of f on the boundary curve $x^2 + y^2 = 45$. To do this, we use Lagrange multipliers with constraint $g(x, y) = x^2 + y^2 = 45$. Setting $\operatorname{grad} f = \lambda \operatorname{grad} g$, we get

$$2(x - 1) = \lambda \cdot 2x,$$
$$2(y - 2) = \lambda \cdot 2y.$$

We can't have $x = 0$ since the first equation would become $-2 = 0$. Similarly, $y \neq 0$. So we can solve each equation for λ by dividing by x and y. Setting the expressions for λ equal gives

$$\frac{x - 1}{x} = \frac{y - 2}{y},$$

so

$$y = 2x.$$

Combining this with the constraint $x^2 + y^2 = 45$, we get

$$5x^2 = 45$$

so

$$x = \pm 3.$$

Since $y = 2x$, we have possible local extrema at $x = 3$, $y = 6$ and $x = -3$, $y = -6$.

We conclude that the only candidates for the maximum and minimum values of f in the region occur at $(1, 2)$, $(3, 6)$, and $(-3, -6)$. Evaluating f at these three points, we find

$$f(1, 2) = 0, \qquad f(3, 6) = 20, \qquad f(-3, -6) = 80.$$

Therefore, the minimum value of f is 0 at $(1, 2)$ and the maximum value is 80 at $(-3, -6)$.

The Meaning of λ

In the uses of Lagrange multipliers so far, we never found (or needed) the value of λ. However, λ does have a practical interpretation. In the production example, we wanted to maximise

$$f(x, y) = x^{2/3} y^{1/3}$$

subject to the constraint

$$g(x, y) = x + y = 3.78.$$

We solved the equations

$$\frac{2}{3} x^{-1/3} y^{1/3} = \lambda,$$
$$\frac{1}{3} x^{2/3} y^{-2/3} = \lambda,$$
$$x + y = 3.78,$$

to get $x = 2.52$, $y = 1.26$ and $f(2.52, 1.26) \approx 2$. Continuing to find λ gives us

$$\lambda \approx 0.53.$$

Now we do another, apparently unrelated, calculation. Suppose our budget is increased slightly, from 3.78 to 4.78, giving a new budget constraint of $x + y = 4.78$. Then the corresponding solution is at $x = 3.19$ and $y = 1.59$ and the new maximum value (instead of $f = 2$) is

$$f = (3.19)^{2/3}(1.59)^{1/3} \approx 2.53.$$

Notice that the amount by which f has increased is 0.53, the value of λ. Thus, in this example, the value of λ represents the extra production achieved by increasing the budget by one—in other words, the extra "bang" you get for an extra "buck" of budget. In fact, this is true in general:

- The value of λ is approximately the increase in the optimum value of f when the budget is increased by 1 unit.

More precisely:

- The value of λ represents the rate of change of the optimum value of f as the budget increases.

An Expression for λ

To interpret λ, we look at how the optimum value of the objective function f changes as the value c of the constraint function g is varied. In general, the optimum point (x_0, y_0) depends on the constraint value c. So, provided x_0 and y_0 are differentiable functions of c, we can use the chain rule to differentiate the optimum value $f(x_0(c), y_0(c))$ with respect to c:

$$\frac{df}{dc} = \frac{\partial f}{\partial x}\frac{dx_0}{dc} + \frac{\partial f}{\partial y}\frac{dy_0}{dc}.$$

At the optimum point (x_0, y_0), we have $f_x = \lambda g_x$ and $f_y = \lambda g_y$, and therefore

$$\frac{df}{dc} = \lambda \left(\frac{\partial g}{\partial x}\frac{dx_0}{dc} + \frac{\partial g}{\partial y}\frac{dy_0}{dc} \right) = \lambda \frac{dg}{dc}.$$

But, as $g(x_0(c), y_0(c)) = c$, we see that $dg/dc = 1$, so $df/dc = \lambda$. Thus, we have the following interpretation of the Lagrange multiplier λ:

> The value of λ is the rate of change of the optimum value of f as c increases (where $g(x, y) = c$). If the optimum value of f is written as $f(x_0(c), y_0(c))$, then
> $$\frac{d}{dc}f(x_0(c), y_0(c)) = \lambda.$$

Example 3   The quantity of goods produced according to the function $f(x, y) = x^{2/3}y^{1/3}$ is maximised subject to the budget constraint $x + y \leq 3.78$. The budget is increased to allow for a small increase in production. What is the price of the product if the sale of the additional goods covers the budget increase?

Solution   We know that $\lambda = 0.53$, which tells us that $df/dc = 0.53$. The constraint corresponds to a budget of \$3.78 thousand. Therefore increasing the budget by \$1000 increases production by about 0.53 units. In order to make the increase in budget profitable, the extra goods produced must sell for more than \$1000. Thus, if p is the price of each unit of the good, then $0.53p$ is the revenue from the extra 0.53 units sold. Thus, we need $0.53p \geq 1000$ so $p \geq 1000/0.53 = \$1890$.

The Lagrangian Function

Constrained optimisation problems are frequently solved using a *Lagrangian function*, $\mathcal{L}$. For example, to optimise $f(x, y)$ subject to the constraint $g(x, y) = c$, we use the Lagrangian function

$$\mathcal{L}(x, y, \lambda) = f(x, y) - \lambda(g(x, y) - c).$$

To see how the function $\mathcal{L}$ is used, compute the partial derivatives of $\mathcal{L}$:

$$\frac{\partial \mathcal{L}}{\partial x} = \frac{\partial f}{\partial x} - \lambda \frac{\partial g}{\partial x},$$

$$\frac{\partial \mathcal{L}}{\partial y} = \frac{\partial f}{\partial y} - \lambda \frac{\partial g}{\partial y},$$

$$\frac{\partial \mathcal{L}}{\partial \lambda} = -(g(x, y) - c).$$

Notice that if (x_0, y_0) is an extreme point of $f(x, y)$ subject to the constraint $g(x, y) = c$ and λ_0 is the corresponding Lagrange multiplier, then at the point (x_0, y_0, λ_0) we have

$$\frac{\partial \mathcal{L}}{\partial x} = 0 \quad \text{and} \quad \frac{\partial \mathcal{L}}{\partial y} = 0 \quad \text{and} \quad \frac{\partial \mathcal{L}}{\partial \lambda} = 0.$$

In other words, (x_0, y_0, λ_0) is a critical point for the unconstrained Lagrangian function, $\mathcal{L}(x, y, \lambda)$. Thus, the Lagrangian converts a constrained optimisation problem to an unconstrained problem.

Example 4   A company has a production function with three inputs x, y, and z given by

$$f(x, y, z) = 50x^{2/5}y^{1/5}z^{1/5}.$$

The total budget is \$24,000 and the company can buy x, y, and z at \$80, \$12, and \$10 per unit, respectively. What combination of inputs will maximise production? [9]

Solution   We need to maximise the objective function

$$f(x, y, z) = 50x^{2/5}y^{1/5}z^{1/5},$$

subject to the constraint

$$g(x, y, z) = 80x + 12y + 10z = 24,000.$$

The method for functions of two variables works for functions of three variables, so we construct the Lagrangian function

$$\mathcal{L}(x, y, z, \lambda) = 50x^{2/5}y^{1/5}z^{1/5} - \lambda(80x + 12y + 10z - 24,000),$$

and solve the system of equations we get from $\operatorname{grad} \mathcal{L} = \vec{0}$:

$$\frac{\partial \mathcal{L}}{\partial x} = 20x^{-3/5}y^{1/5}z^{1/5} - 80\lambda = 0,$$

$$\frac{\partial \mathcal{L}}{\partial y} = 10x^{2/5}y^{-4/5}z^{1/5} - 12\lambda = 0,$$

$$\frac{\partial \mathcal{L}}{\partial z} = 10x^{2/5}y^{1/5}z^{-4/5} - 10\lambda = 0,$$

$$\frac{\partial \mathcal{L}}{\partial \lambda} = -(80x + 12y + 10z - 24,000) = 0.$$

---

[9]Adapted from M. Rosser, *Basic Mathematics for Economists*, p. 363 (New York: Routledge, 1993).

Simplifying this system gives

$$\lambda = \frac{1}{4}x^{-3/5}y^{1/5}z^{1/5},$$

$$\lambda = \frac{5}{6}x^{2/5}y^{-4/5}z^{1/5},$$

$$\lambda = x^{2/5}y^{1/5}z^{-4/5},$$

$$80x + 12y + 10z = 24{,}000.$$

Eliminating z from the first two equations gives $x = 0.3y$. Eliminating x from the second and third equations gives $z = 1.2y$. Substituting for x and z into $80x + 12y + 10z = 24{,}000$ gives

$$80(0.3y) + 12y + 10(1.2y) = 24{,}000,$$

so $y = 500$. Then $x = 150$ and $z = 600$, and $f(150, 500, 600) = 4{,}622$ units.

The graph of the constraint, $80x + 12y + 10z = 24{,}000$, is a plane. Since the inputs x, y, z must be nonnegative, the graph is a triangle in the first octant, with edges on the coordinate planes. On the boundary of the triangle, one (or more) of the variables x, y, z is zero, so the function f is zero. Thus production is maximized within the budget using $x = 150$, $y = 500$, and $z = 600$.

Exercises and Problems for Section 15.3

Exercises

1. Decide whether each point appears to be a maximum, minimum, or neither for the function f constrained by the loop in Figure 15.30.

   (a) P (b) Q (c) R (d) S

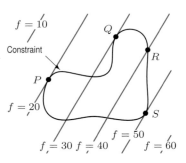

Figure 15.30

In Exercises 2–18, use Lagrange multipliers to find the maximum and minimum values of f subject to the given constraint, if such values exist.

2. $f(x, y) = x + y, \quad x^2 + y^2 = 1$

3. $f(x, y) = x + 3y + 2, \quad x^2 + y^2 = 10$

4. $f(x, y) = (x - 1)^2 + (y + 2)^2, \quad x^2 + y^2 = 5$

5. $f(x, y) = x^3 + y, \quad 3x^2 + y^2 = 4$

6. $f(x, y) = 3x - 2y, \quad x^2 + 2y^2 = 44$

7. $f(x, y) = 2xy, \quad 5x + 4y = 100$

8. $f(x_1, x_2) = x_1{}^2 + x_2{}^2, \quad x_1 + x_2 = 1$

9. $f(x, y) = x^2 + y, \quad x^2 - y^2 = 1$

10. $f(x, y, z) = x + 3y + 5z, \quad x^2 + y^2 + z^2 = 1$

11. $f(x, y, z) = x^2 - y^2 - 2z, \quad x^2 + y^2 = z$

12. $f(x, y, z) = xyz, \quad x^2 + y^2 + 4z^2 = 12$

13. $f(x, y) = x^2 + 2y^2, \quad x^2 + y^2 \le 4$

14. $f(x, y) = x + 3y, \quad x^2 + y^2 \le 2$

15. $f(x, y) = xy, \quad x^2 + 2y^2 \le 1$

16. $f(x, y) = x^3 + y, \quad x + y \ge 1$

17. $f(x, y) = (x + 3)^2 + (y - 3)^2, \quad x^2 + y^2 \le 2$

18. $f(x, y) = x^2 y + 3y^2 - y, \quad x^2 + y^2 \le 10$

Problems

19. Figure 15.31 shows contours of f. Does f have a maximum value subject to the constraint $g(x, y) = c$ for $x \ge 0$, $y \ge 0$? If so, approximately where is it and what is its value? Does f have a minimum value subject to the constraint? If so, approximately where and what?

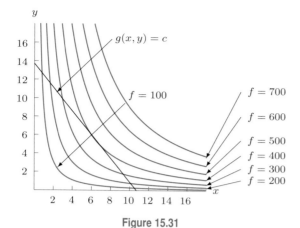

Figure 15.31

20. Each person tries to balance his or her time between leisure and work. The tradeoff is that as you work less your income falls. Therefore each person has *indifference curves* which connect the number of hours of leisure, l, and income, s. If, for example, you are indifferent between 0 hours of leisure and an income of \$1125 a week on the one hand, and 10 hours of leisure and an income of \$750 a week on the other hand, then the points $l = 0$, $s = 1125$, and $l = 10$, $s = 750$ both lie on the same indifference curve. Table 15.3 gives information on three indifference curves, I, II, and III.

Table 15.3

Weekly income			Weekly leisure hours		
I	II	III	I	II	III
1125	1250	1375	0	20	40
750	875	1000	10	30	50
500	625	750	20	40	60
375	500	625	30	50	70
250	375	500	50	70	90

(a) Graph the three indifference curves.
(b) You have 100 hours a week available for work and leisure combined, and you earn \$10/hour. Write an equation in terms of l and s which represents this constraint.
(c) On the same axes, graph this constraint.
(d) Estimate from the graph what combination of leisure hours and income you would choose under these circumstances. Give the corresponding number of hours per week you would work.

21. Figure 15.32 shows ∇f for a function $f(x, y)$ and two curves $g(x, y) = 1$ and $g(x, y) = 2$. Mark the following:

(a) The point(s) A where f has a local maximum.

(b) The point(s) B where f has a saddle point.
(c) The point C where f has a maximum on $g = 1$.
(d) The point D where f has a minimum on $g = 1$.
(e) If you used Lagrange multipliers to find C, what would the sign of λ be? Why?

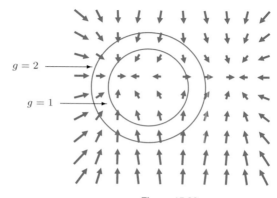

Figure 15.32

22. Find the maximum value of $f(x, y) = x + y - (x - y)^2$ on the triangular region $x \geq 0$, $y \geq 0$, $x + y \leq 1$.

23. (a) Draw contours of $f(x, y) = 2x + y$ for $z = -7, -5, -3, -1, 1, 3, 5, 7$.
(b) On the same axes, graph the constraint $x^2 + y^2 = 5$.
(c) Use the graph to approximate the points at which f has a maximum or a minimum value subject to the constraint $x^2 + y^2 = 5$.
(d) Use Lagrange multipliers to find the maximum and minimum values of $f(x, y) = 2x + y$ subject to $x^2 + y^2 = 5$.

24. Let $f(x, y) = x^\alpha y^{1-\alpha}$ for $0 < \alpha < 1$. Find the value of α such that the maximum value of f on the line $2x + 3y = 6$ occurs at $(1.5, 1)$.

25. If the right side of the constraint in Exercise 6 is changed by the small amount Δc, by approximately how much do the maximum and minimum values change?

26. If the right side of the constraint in Exercise 15 is changed by the small amount Δc, by approximately how much do the maximum and minimum values change?

27. The function $P(x, y)$ gives the number of units produced and $C(x, y)$ gives the cost of production.

(a) A company wishes to maximize production at a fixed cost of \$50,000. What is the objective function f? What is the constraint equation? What is the meaning of λ in this situation?
(b) A company wishes to minimize costs at a fixed production level of 2000 units. What is the objective function f? What is the constraint equation? What is the meaning of λ in this situation?

28. Design a closed cylindrical container which holds 100 cm^3 and has the minimal possible surface area. What should its dimensions be?

29. A company manufactures x units of one item and y units of another. The total cost in dollars, C, of producing these two items is approximated by the function

$$C = 5x^2 + 2xy + 3y^2 + 800.$$

(a) If the production quota for the total number of items (both types combined) is 39, find the minimum production cost.

(b) Estimate the additional production cost or savings if the production quota is raised to 40 or lowered to 38.

30. An international organization must decide how to spend the $2000 they have been allotted for famine relief in a remote area. They expect to divide the money between buying rice at $5/sack and beans at $10/sack. The number, P, of people who would be fed if they buy x sacks of rice and y sacks of beans is given by

$$P = x + 2y + \frac{x^2 y^2}{2 \cdot 10^8}.$$

What is the maximum number of people that can be fed, and how should the organization allocate its money?

31. The quantity, q, of a product manufactured depends on the number of workers, W, and the amount of capital invested, K, and is given by

$$q = 6W^{3/4} K^{1/4}.$$

Labor costs are $10 per worker and capital costs are $20 per unit, and the budget is $3000.

(a) What are the optimum number of workers and the optimum number of units of capital?

(b) Show that at the optimum values of W and K, the ratio of the marginal productivity of labor ($\partial q/\partial W$) to the marginal productivity of capital ($\partial q/\partial K$) is the same as the ratio of the cost of a unit of labor to the cost of a unit of capital.

(c) Recompute the optimum values of W and K when the budget is increased by one dollar. Check that increasing the budget by $1 allows the production of λ extra units of the good, where λ is the Lagrange multiplier.

32. A neighborhood health clinic has a budget of $600,000 per quarter. The director of the clinic wants to allocate the budget to maximize the number of patient visits, V, which is given as a function of the number of doctors, D, and the number of nurses, N, by

$$V = 1000 D^{0.6} N^{0.3}.$$

A doctor gets $40,000 per quarter; nurses get $10,000 per quarter.

(a) Set up the director's constrained optimization problem.

(b) Describe, in words, the conditions which must be satisfied by $\partial V/\partial D$ and $\partial V/\partial N$ for V to have an optimum value.

(c) Solve the problem formulated in part (a).

(d) Find the value of the Lagrange multiplier and interpret its meaning in this problem.

(e) At the optimum point, what is the marginal cost of a patient visit (that is, the cost of an additional visit)? Will that marginal cost rise or fall with the number of visits? Why?

33. (a) In Problem 31, does the value of λ change if the budget changes from $3000 to $4000?

(b) In Problem 32, does the value of λ change if the budget changes from $600,000 to $700,000?

(c) What condition must a Cobb-Douglas production function, $Q = cK^a L^b$, satisfy to ensure that the marginal increase of production (that is, the rate of increase of production with budget) is not affected by the size of the budget?

34. The production function $P(K, L)$ gives the number of pairs of skis produced per week at a factory operating with K units of capital and L units of labor. The contour diagram for P is in Figure 15.33; the parallel lines are budget constraints for budgets, B, in dollars.

(a) On each budget constraint, mark the point that gives the maximum production.

(b) Complete the table, where the budget, B, is in dollars and the maximum production is the number of pairs of skis to be produced each week.

B	2000	4000	6000	8000	10000
M					

(c) Estimate the Lagrange multiplier $\lambda = dM/dB$ at a budget of $6000. Give units for the multiplier.

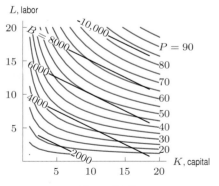

Figure 15.33

35. A doctor wants to schedule visits for two patients who have been operated on for tumors so as to minimize the expected delay in detecting a new tumor. Visits for patients 1 and 2 are scheduled at intervals of x_1 and x_2 weeks, respectively. A total of m visits per week is available for both patients combined.

The recurrence rates for tumors for patients 1 and 2 are judged to be v_1 and v_2 tumors per week, respectively. Thus, $v_1/(v_1 + v_2)$ and $v_2/(v_1 + v_2)$ are the probabilities that patient 1 and patient 2, respectively, will have the next tumor. It is known that the expected delay in detecting a tumor for a patient checked every x weeks is $x/2$. Hence, the expected detection delay for both patients combined is given by[10]

$$f(x_1, x_2) = \frac{v_1}{v_1 + v_2} \cdot \frac{x_1}{2} + \frac{v_2}{v_1 + v_2} \cdot \frac{x_2}{2}.$$

Find the values of x_1 and x_2 in terms of v_1 and v_2 that minimize $f(x_1, x_2)$ subject to the fact that m, the number of visits per week, is fixed.

36. What is the value of the Lagrange multiplier in Problem 35? What are the units of λ? What is its practical significance to the doctor?

37. Figure 15.34 shows two weightless springs with spring constants k_1 and k_2 attached between a ceiling and floor without tension or compression. A mass m is placed between the springs which settle into equilibrium as in Figure 15.35. The magnitudes f_1 and f_2 of the forces of the springs on the mass minimize the complementary energy

$$\frac{f_1^2}{2k_1} + \frac{f_2^2}{2k_2}$$

subject to the force balance constraint $f_1 + f_2 = mg$.

(a) Determine f_1 and f_2 by the method of Lagrange multipliers.
(b) If you are familiar with Hooke's law, find the meaning of λ.

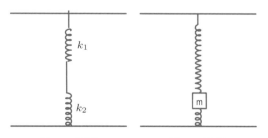

Figure 15.34   Figure 15.35

38. (a) If $\sum_{i=1}^{3} x_i = 1$, find the values of x_1, x_2, x_3 making $\sum_{i=1}^{3} x_i^2$ minimum.
(b) Generalize the result of part (a) to find the minimum value of $\sum_{i=1}^{n} x_i^2$ subject to $\sum_{i=1}^{n} x_i = 1$.

39. Let $f(x, y) = ax^2 + bxy + cy^2$. Show that the maximum value of $f(x, y)$ subject to the constraint $x^2 + y^2 = 1$ is equal to λ, the Lagrange multiplier.

40. Find the minimum distance from the point $(1, 2, 10)$ to the paraboloid given by the equation $z = x^2 + y^2$. Give a geometric justification for your answer.

41. A company produces one product from two inputs (for example, capital and labor). Its production function $g(x, y)$ gives the quantity of the product that can be produced with x units of the first input and y units of the second. The *cost function* (or *expenditure function*) is the three-variable function $C(p, q, u)$ where p and q are the unit prices of the two inputs. For fixed p, q, and u, the value $C(p, q, u)$ is the minimum of $f(x, y) = px + qy$ subject to the constraint $g(x, y) = u$.

(a) What is the practical meaning of $C(p, q, u)$?
(b) Find a formula for $C(p, q, u)$ if $g(x, y) = xy$.

42. A *utility function* $U(x, y)$ for two items gives the utility (benefit) to a consumer of x units of item 1 and y units of item 2. The *indirect utility function* is the three-variable function $V(p, q, I)$ where p and q are the unit prices of the two items. For fixed p, q, and I, the value $V(p, q, I)$ is the maximum of $U(x, y)$ subject to the constraint $px + qy = I$.

(a) What is the practical meaning of $V(p, q, I)$?
(b) The Lagrange multiplier λ that arises in the maximization defining V is called the *marginal utility of money*. What is its practical meaning?
(c) Find formulas for $V(p, q, I)$ and λ if $U(x, y) = xy$.

43. The function $h(x, y) = x^2 + y^2 - \lambda(2x + 4y - 15)$ has a minimum value $m(\lambda)$ for each value of λ.

(a) Find $m(\lambda)$.
(b) For which value of λ is $m(\lambda)$ the largest and what is that maximum value?
(c) Find the minimum value of $f(x, y) = x^2 + y^2$ subject to the constraint $2x + 4y = 15$ using the method of Lagrange multipliers and evaluate λ.
(d) Compare your answers to parts (b) and (c).

44. Let f be differentiable and grad $f(2, 1) = -3\vec{i} + 4\vec{j}$. You want to see if $(2, 1)$ is a candidate for the maximum and minimum values of f subject to a constraint satisfied by the point $(2, 1)$.

(a) Show $(2, 1)$ is not a candidate if the constraint is $x^2 + y^2 = 5$.
(b) Show $(2, 1)$ is a candidate if the constraint is $(x - 5)^2 + (y + 3)^2 = 25$. From a sketch of the contours for f near $(2, 1)$ and the constraint, decide whether $(2, 1)$ is a candidate for a maximum or minimum.
(c) Do the same as part (b), but using the constraint $(x + 1)^2 + (y - 5)^2 = 25$.

[10]Adapted from Daniel Kent, Ross Shachter, *et al.*, Efficient Scheduling of Cystoscopies in Monitoring for Recurrent Bladder Cancer in *Medical Decision Making* (Philadelphia: Hanley and Belfus, 1989).

45. A person's satisfaction from consuming a quantity x_1 of one item and a quantity x_2 of another item is given by

$$S = u(x_1, x_2) = a \ln x_1 + (1 - a) \ln x_2,$$

where a is a constant, $0 < a < 1$. The prices of the two items are p_1 and p_2 respectively, and the budget is b.

(a) Express the maximum satisfaction that can be achieved as a function of p_1, p_2, and b.

(b) Find the amount of money that must be spent to achieve a particular level of satisfaction, c, as a function of p_1, p_2, and c.

46. This problem illustrates the Envelope Theorem, which relates the maxima of $z = f(x, y)$ subject to the constraint $x = c$ to the contour diagram in Figure 15.36 and the cross-sections in Figure 15.37.

(a) For each value c, there is a maximum value of $f(x, y)$ with $x = c$. On Figure 15.36, sketch the curve that goes through the points where the maxima are achieved.

(b) On Figure 15.37, sketch the curve going through the points corresponding to the same maximum values in part (a). This curve is called the envelope of the cross-sections.

(c) Show that the Lagrange multiplier λ for this constrained optimization problem is the slope of the envelope curve in part (b).

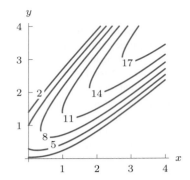

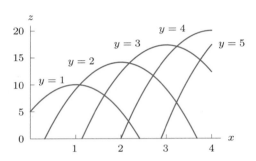

Figure 15.36: Contour diagram of f

Figure 15.37: Cross-sections of f

Strengthen Your Understanding

In Problems 47–48, explain what is wrong with the statement.

47. The function $f(x, y) = xy$ has a maximum of 2 on the constraint $x + y = 2$.

48. If the level curves of $f(x, y)$ and the level curves of $g(x, y)$ are not tangent at any point on the constraint $g(x, y) = c$, $x \geq 0$, $y \geq 0$, then f has no maximum on the constraint.

In Problems 49–53, give an example of:

49. A function $f(x, y)$ whose maximum subject to the constraint $x^2 + y^2 = 5$ is at $(3, 4)$.

50. A function $f(x, y)$ to be optimized with constraint $x^2 + 2y^2 \leq 1$ such that the minimum value does not change when the constraint is changed to $x^2 + y^2 \leq 1 + c$ for $c > 0$.

51. A function $f(x, y)$ with a minimum at $(1, 1)$ on the constraint $x + y = 2$.

52. A function $f(x, y)$ that has a maximum but no minimum on the constraint $x + y = 4$.

53. A contour diagram of a function f whose maximum value on the constraint $x + 2y = 6$, $x \geq 0$, $y \geq 0$ occurs at one of the endpoints.

For Problems 54–55, use Figure 15.38. The grid lines are one unit apart.

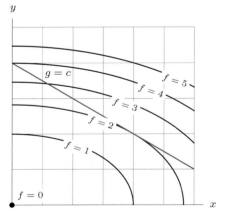

Figure 15.38

54. Find the maximum and minimum values of f on $g = c$. At which points do they occur?

55. Find the maximum and minimum values of f on the triangular region below $g = c$ in the first quadrant.

Are the statements in Problems 56–60 true or false? Give reasons for your answer.

56. If $f(x, y)$ has a local maximum at (a, b) subject to the constraint $g(x, y) = c$, then $g(a, b) = c$.

57. If $f(x, y)$ has a local maximum at (a, b) subject to the constraint $g(x, y) = c$, then $\operatorname{grad} f(a, b) = \vec{0}$.

58. The function $f(x, y) = x + y$ has no global maximum subject to the constraint $x - y = 0$.

59. The point $(2, -1)$ is a local minimum of $f(x, y) = x^2 + y^2$ subject to the constraint $x + 2y = 0$.

60. If $\operatorname{grad} f(a, b)$ and $\operatorname{grad} g(a, b)$ point in opposite directions, then (a, b) is a local minimum of $f(x, y)$ constrained by $g(x, y) = c$.

In Problems 61–68, suppose that M and m are the maximum and minimum values of $f(x, y)$ subject to the constraint $g(x, y) = c$ and that (a, b) satisfies $g(a, b) = c$. Decide whether the statements are true or false. Give an explanation for your answer.

61. If $f(a, b) = M$, then $f_x(a, b) = f_y(a, b) = 0$.

62. If $f(a, b) = M$, then $f(a, b) = \lambda g(a, b)$ for some value of λ.

63. If $\operatorname{grad} f(a, b) = \lambda \operatorname{grad} g(a, b)$, then $f(a, b) = M$ or $f(a, b) = m$.

64. If $f(a, b) = M$ and $f_x(a, b)/f_y(a, b) = 5$, then $g_x(a, b)/g_y(a, b) = 5$.

65. If $f(a, b) = m$ and $g_x(a, b) = 0$, then $f_x(a, b) = 0$.

66. Increasing the value of c increases the value of M.

67. Suppose that $f(a, b) = M$ and that $\operatorname{grad} f(a, b) = 3 \operatorname{grad} g(a, b)$. Then increasing the value of c by 0.02 increases the value of M by about 0.06.

68. Suppose that $f(a, b) = m$ and that $\operatorname{grad} f(a, b) = 3 \operatorname{grad} g(a, b)$. Then increasing the value of c by 0.02 decreases the value of m by about 0.06.

CHAPTER SUMMARY (see also Ready Reference at the end of the book)

- **Critical Points**
  Definitions of local extrema, critical points, saddle points, finding critical points algebraically, behavior of contours near critical points, discriminant, second-derivative test to classify critical points.

- **Unconstrained Optimization**
  Definitions of global extrema, method of least squares,

closed and bounded regions, existence of global extrema.

- **Constrained Optimization**
  Objective function and constraint, definitions of extrema subject to a constraint, geometric interpretation of Lagrange multiplier method, solving Lagrange multiplier problems algebraically, inequality constraints, meaning of the Lagrange multiplier λ, the Lagrangian function.

REVIEW EXERCISES AND PROBLEMS FOR CHAPTER FIFTEEN

Exercises

For Exercises 1–5, find the critical points of the given function and classify them as local maxima, local minima, saddle points, or none of these.

1. $f(x, y) = 2xy^2 - x^2 - 2y^2 + 1$

2. $f(x, y) = 2x^3 - 3x^2 y + 6x^2 - 6y^2$

3. $f(x, y) = x^2 y + 2y^2 - 2xy + 6$

4. $f(x, y) = \dfrac{80}{xy} + 10x + 10xy + 20y$

5. $f(x, y) = \sin x + \sin y + \sin(x + y)$, $0 < x < \pi$, $0 < y < \pi$.

For Exercises 6–9, find the local maxima, local minima, and saddle points of the function. Decide if the local maxima or minima are global maxima or minima. Explain.

6. $f(x, y) = 10 + 12x + 6y - 3x^2 - y^2$

7. $f(x, y) = x^2 + y^3 - 3xy$

8. $f(x, y) = x + y + \dfrac{1}{x} + \dfrac{4}{y}$

9. $f(x, y) = xy + \ln x + y^2 - 10$, $x > 0$

For Exercises 10–21, use Lagrange multipliers to find the maximum and minimum values of f subject to the constraint.

10. $f(x, y) = 3x - 4y$, $x^2 + y^2 = 5$

11. $f(x, y) = x^2 + y^2$, $x^4 + y^4 = 2$

12. $f(x, y) = x^2 + y^2$, $4x - 2y = 15$

13. $f(x, y) = x^2 - xy + y^2$, $x^2 - y^2 = 1$

14. $f(x, y) = x^2 + 2y^2$, $3x + 5y = 200$

15. $f(x, y) = xy$, $4x^2 + y^2 = 8$

16. $f(x, y) = -3x^2 - 2y^2 + 20xy$, $x + y = 100$

17. $f(x, y, z) = x^2 - 2y + 2z^2$, $x^2 + y^2 + z^2 = 1$

18. $f(x, y, z) = 2x + y + 4z$, $x^2 + y + z^2 = 16$

19. $z = 4x^2 - xy + 4y^2$, $x^2 + y^2 \le 2$

20. $f(x, y) = x^2 - y^2$, $x^2 \ge y$

21. $f(x, y) = x^3 - y^2$, $x^2 + y^2 \le 1$

In Exercises 22–25, does $f(x, y) = x^2 - y^2$ have a maximum, a minimum, neither, or both when subject to the constraint?

22. $x = 10$

23. $y = 10$

24. $x^2 + y^2 = 10$

25. $xy = 10$

Problems

26. Maximize $0.3 \ln x + 0.7 \ln y$ on $2x + 3y = 6$.

27. (a) Write an expression for the distance between the points $(3, 4)$ and (x, y).
(b) Minimize this distance if (x, y) lies on the unit circle centered at the origin. At what point does the minimum occur?
(c) What is the maximum distance? At what point does it occur?

28. If the right side of the constraint in Exercise 11 is changed by the small amount Δc, by approximately how much do the maximum and minimum values change?

29. Compute the regression line for the points $(-1, 2)$, $(0, -1)$, $(1, 1)$ using least squares.

30. At the point $(1, 3)$, suppose $f_x = f_y = 0$ and $f_{xx} < 0$, $f_{yy} < 0$, $f_{xy} = 0$. Draw a possible contour diagram.

31. For $f(x, y) = A - (x^2 + Bx + y^2 + Cy)$, what values of A, B, and C give f a local maximum value of 15 at the point $(-2, 1)$?

32. A biological rule of thumb states that as the area A of an island increases tenfold, the number of animal species, N, living on it doubles. The table contains data for islands in the West Indies. Assume that N is a power function of A.

(a) Use the biological rule of thumb to find
    (i) N as a function of A
    (ii) $\ln N$ as a function of $\ln A$
(b) Using the data given, tabulate $\ln N$ against $\ln A$ and find the line of best fit. Does your answer agree with the biological rule of thumb?

Island	Area (sq km)	Number of species
Redonda	3	5
Saba	20	9
Montserrat	192	15
Puerto Rico	8858	75
Jamaica	10854	70
Hispaniola	75571	130
Cuba	113715	125

33. A firm manufactures a commodity at two different factories. The total cost of manufacturing depends on the quantities, q_1 and q_2, supplied by each factory, and is expressed by the *joint cost function*,

$$C = f(q_1, q_2) = 2q_1^2 + q_1 q_2 + q_2^2 + 500.$$

The company's objective is to produce 200 units, while minimizing production costs. How many units should be supplied by each factory?

34. (a) Let $f(x, y) = x^2 + 2y^2$. Find the minimum value $m(c)$ of f on the line $x + y = c$ as a function of c.
(b) Give the value of the Lagrange multiplier λ at this minimum.
(c) What is the relation between your answers in parts (a) and (b)?

35. The maximum value of $f(x, y)$ subject to the constraint $g(x, y) = 240$ is 6300. The method of Lagrange multipliers gives $\lambda = 20$. Find an approximate value for the maximum of $f(x, y)$ subject to the constraint $g(x, y) = 242$.

36. An industry manufactures a product from two raw materials. The quantity produced, Q, can be given by the Cobb-Douglas function:

$$Q = cx^a y^b,$$

where x and y are quantities of each of the two raw materials used and a, b, and c are positive constants. The first raw material costs $\$P_1$ per unit and the second costs $\$P_2$ per unit. Find the maximum production possible if no more than $\$K$ can be spent on raw materials.

37. The quantity, Q, of a product manufactured by a company is given by

$$Q = aK^{0.6} L^{0.4},$$

where a is a positive constant, K is the quantity of capital and L is the quantity of labor used. Capital costs are $\$20$ per unit, labor costs are $\$10$ per unit, and the company wants costs for capital and labor combined to be no higher than $\$150$. Suppose you are asked to consult for the company, and learn that 5 units each of capital and labor are being used.

(a) What do you advise? Should the company use more or less labor? More or less capital? If so, by how much?
(b) Write a one-sentence summary that could be used to sell your advice to the board of directors.

38. Figure 15.39 shows contours labeled with values of $f(x, y)$ and a constraint $g(x, y) = c$. Mark the approximate points at which:

(a) $\operatorname{grad} f = \lambda \operatorname{grad} g$
(b) f has a maximum
(c) f has a maximum on the constraint $g = c$.

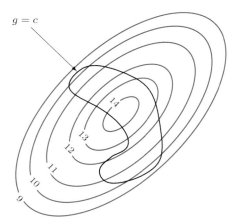

$g = c$

14

13

12

11

10

9

Figure 15.39

39. A mountain climber at the summit of a mountain wants to descend to a lower altitude as fast as possible. The altitude of the mountain is given approximately by

$$h(x, y) = 3000 - \frac{1}{10,000}(5x^2 + 4xy + 2y^2) \text{ meters,}$$

where x, y are horizontal coordinates on the earth (in meters), with the mountain summit located above the origin. In thirty minutes, the climber can reach any point (x, y) on a circle of radius 1000 m. In which direction should she travel in order to descend as far as possible?

40. Let $f(x, y, z) = \sqrt{(x - a)^2 + (y - b)^2 + (z - c)^2}$. Minimize f subject to $Ax + By + Cz + D = 0$. What is the geometric meaning of your solution?

41. A company with monopoly pricing power, like the only grocery store in a sparsely populated area or all-night convenience store, sells two products which are partial substitutes for each other, such as coffee and tea. If the price of one product rises, then the demand for the other product rises. The quantities demanded, q_1 and q_2, are given as a function of the prices, p_1 and p_2, by

$$q_1 = 517 - 3.5p_1 + 0.8p_2, \quad q_2 = 770 - 4.4p_2 + 1.4p_1.$$

What prices should the company charge in order to maximize the total sales revenue? [11]

42. The quantity, Q, of a certain product manufactured depends on the quantity of labor, L, and of capital, K, used according to the function

$$Q = 900L^{1/2}K^{2/3}.$$

Labor costs \$100 per unit and capital costs \$200 per unit. What combination of labor and capital should be used to produce 36,000 units of the goods at minimum cost? What is that minimum cost?

43. A company manufactures a product using inputs x, y, and z according to the production function

$$Q(x, y, z) = 20x^{1/2}y^{1/4}z^{2/5}.$$

The prices per unit are \$20 for x, and \$10 for y, and \$5 for z. What quantity of each input should be used in order to manufacture 1,200 units at minimum cost? [12]

44. The Cobb-Douglas function models the quantity, q, of a commodity produced as a function of the number of workers, W, and the amount of capital invested, K:

$$q = cW^{1-a}K^a,$$

where a and c are positive constants. Labor costs are \$$p_1$ per worker, capital costs are \$$p_2$ per unit, and there is a fixed budget of \$$b$. Show that when W and K are at their optimal levels, the ratio of marginal productivity of labor to marginal productivity of capital equals the ratio of the cost of one unit of labor to one unit of capital.

45. An electrical current I flows through a circuit containing the resistors R_1 and R_2 in Figure 15.40. The currents i_1 and i_2 through the individual resistors minimize energy loss $i_1^2 R_1 + i_2^2 R_2$ subject to the constraint $i_1 + i_2 = I$ given by Kirchoff's current law.

(a) Find the currents i_1 and i_2 by the method of Lagrange multipliers.
(b) If you are familiar with Ohm's law, find the meaning of λ.

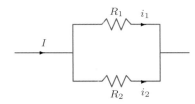

R_1 i_1

I

R_2 i_2

Figure 15.40

46. An open rectangular box has volume 32 cm^3. What are the lengths of the edges giving the minimum surface area?

[11] Adapted from M. Rosser, *Basic Mathematics for Economists*, p. 318 (New York: Routledge, 1993).
[12] Adapted from M. Rosser, *Basic Mathematics for Economists*, p. 363 (New York: Routledge, 1993).

47. The function $f(x, y)$ is defined for all x, y and the origin does not lie on the surface representing $z = f(x, y)$. There is a unique point $P = (a, b, c)$ on the surface which is closest to the origin. Explain why the position vector from the origin to P must be perpendicular to the surface at that point.

48. In a curious game, you and your opponent will choose three real numbers. The rules say that you must first choose a value λ, then your opponent is free to choose any values for x and y. Your goal is to make the value of $\mathcal{L}(x, y, \lambda) = 10 - x^2 - y^2 - 2x - \lambda(2x + 2y)$ as small as you can, and your opponent's goal is to make it as large as possible. What value of λ should you choose (assuming you have a brilliant opponent who never makes mistakes)?

49. An irrigation canal has a trapezoidal cross-section of area

50 m^2, as in Figure 15.41. The average flow rate in the canal is inversely proportional to the wetted perimeter, p, of the canal, that is, to the perimeter of the trapezoid in Figure 15.41, excluding the top. Thus, to maximize the flow rate we must minimize p. Find the depth d, base width w, and angle θ that give the maximum flow rate.[13]

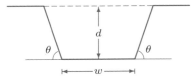

Figure 15.41

CAS Challenge Problems

50. Let $f(x, y) = \dfrac{\sqrt{a + x + y}}{1 + y + \sqrt{a + x}}$, for $x, y > 0$, where a is a positive constant.

    (a) Find the critical points of f and classify them as local maxima, local minima, saddle points, or none of these.

    (b) Describe how the position and type of the critical points changes with respect to a, and explain this in terms of the graph of f.

51. Students are asked to find the global maximum of $f(x, y) = x^2 + y$ subject to the constraint $g(x, y) = x^2 + 2xy + y^2 - 9 = 0$. Student A uses the method of Lagrange multipliers with the help of a computer algebra

system, and says that the global maximum is 11/4. Student B looks at a contour diagram of f and a graph of $g = 0$ and says there is no global maximum. Which student is correct and what mistake is the other one making?

52. Let $f(x, y) = 3x + 2y + 5$, $g(x, y) = 2x^2 - 4xy + 5y^2$.

    (a) Find the maximum of f subject to the constraint $g = 20$.

    (b) Using the value of λ in part (a), estimate the maximum of f subject to each of the constraints $g = 20.5$ and $g = 20.2$.

    (c) Use Lagrange multipliers to find the two maxima in part (b) exactly. Compare them with the estimates.

PROJECTS FOR CHAPTER FIFTEEN

1. Optimization in Manufacturing

A recycling company makes paper from a combination of two raw materials A and B where A is wood pulp from a timber company and B is waste paper from a recycling depot. The production function is $Q = f(x, y)$ where Q is the quantity of paper the company can make using x units of A and y units of B. The cost of acquiring these materials is given by the cost function $C = g(x, y) = px + qy$, where p is the unit price of A and q is the unit price of B.

    (a) If the company decides to reduce the amount of A that it buys, then it can use the money saved to buy additional B. The *economic rate of substitution,* or *ERS,* of B for A tells how much additional B can be bought for the cost of a unit of A. Show that the *ERS* is $g_x/g_y = p/q$.

    (b) If the company decides to reduce slightly the amount of A that it buys, then it must buy additional B in order to maintain a constant level of production. The *technical rate of substitution,* or *TRS,* of B for A tells how much additional B to buy per unit reduction in A. Show that the *TRS* is f_x/f_y. [Hint: The *TRS* is the rate at which y increases with respect to x as a point (x, y) slides in the direction of decreasing x along a fixed contour $f(x, y) = Q$ of the production function.]

[13] Adapted from Robert M. Stark and Robert L. Nichols, *Mathematical Foundations of Design: Civil Engineering Systems* (New York: McGraw-Hill, 1972).

(c) Show that to maximize the quantity of paper produced with a fixed budget the company should use raw materials A and B in quantities such that $ERS = TRS$.

(d) Show that to minimize the cost of producing a fixed quantity of paper the company should use raw materials A and B in quantities such that $ERS = TRS$.

2. Fitting a Line to Data Using Least Squares

In this problem you will derive the general formulas for the slope and y-intercept of a least squares line. Assume that you have n data points $(x_1, y_1), (x_2, y_2), \ldots, (x_n, y_n)$. Let the equation of the least squares line be $y = b + mx$.

(a) For each data point (x_i, y_i), show that the corresponding point directly above or below it on the least squares line has y-coordinate $b + mx_i$.

(b) For each data point (x_i, y_i), show that the square of the vertical distance from it to the point found in part (a) is $(y_i - (b + mx_i))^2$.

(c) Form the function $f(b, m)$ which is the sum of all of the n squared distances found in part (b). That is,

$$f(b, m) = \sum_{i=1}^{n} (y_i - (b + mx_i))^2.$$

Show that the partial derivatives $\dfrac{\partial f}{\partial b}$ and $\dfrac{\partial f}{\partial m}$ are given by

$$\frac{\partial f}{\partial b} = -2 \sum_{i=1}^{n} (y_i - (b + mx_i))$$

and

$$\frac{\partial f}{\partial m} = -2 \sum_{i=1}^{n} (y_i - (b + mx_i)) \cdot x_i.$$

(d) Show that the critical point equations $\dfrac{\partial f}{\partial b} = 0$ and $\dfrac{\partial f}{\partial m} = 0$ lead to a pair of simultaneous linear equations in b and m:

$$nb + \left(\sum x_i \right) m = \sum y_i$$

$$\left(\sum x_i \right) b + \left(\sum x_i^2 \right) m = \sum x_i y_i$$

(e) Solve the equations in part (d) for b and m, getting

$$b = \left(\sum_{i=1}^{n} x_i^2 \sum_{i=1}^{n} y_i - \sum_{i=1}^{n} x_i \sum_{i=1}^{n} x_i y_i \right) \Bigg/ \left(n \sum_{i=1}^{n} x_i^2 - \left(\sum_{i=1}^{n} x_i \right)^2 \right)$$

$$m = \left(n \sum_{i=1}^{n} x_i y_i - \sum_{i=1}^{n} x_i \sum_{i=1}^{n} y_i \right) \Bigg/ \left(n \sum_{i=1}^{n} x_i^2 - \left(\sum_{i=1}^{n} x_i \right)^2 \right)$$

(f) Apply the formulas of part (e) to the data points $(1, 1), (2, 1), (3, 3)$ to check that you get the same result as in Example 3 on page 861.

3. Hockey and Entropy

Thirty teams compete for the Stanley Cup in the National Hockey League (after expansion in 2000). At the beginning of the season an experienced fan estimates that the probability that team i will win is some number p_i, where $0 \leq p_i \leq 1$ and

$$\sum_{i=1}^{30} p_i = 1.$$

Exactly one team will actually win, so the probabilities have to add to 1. If one of the teams, say team i, is certain to win then p_i is equal to 1 and all the other p_j are zero. Another extreme case occurs if all the teams are equally likely to win, so all the p_i are equal to $1/30$, and the outcome of the hockey season is completely unpredictable. Thus, the *uncertainty* in the outcome of the hockey season depends on the probabilities $p_1, \ldots, p_{30}$. In this problem we measure this uncertainty quantitatively using the following function:

$$S(p_1, \ldots, p_{30}) = -\sum_{i=1}^{30} p_i \frac{\ln p_i}{\ln 2}.$$

Note that as $p_i \leq 1$, we have $-\ln p_i \geq 0$ and hence $S \geq 0$.

(a) Show that $\lim_{p \to 0} p \ln p = 0$. (This means that S is a continuous function of the p_i, where $0 \leq p_i \leq 1$ and $1 \leq i \leq 30$, if we set $p \ln p|_{p=0}$ equal to zero. Since S is then a continuous function on a closed and bounded region, it attains a maximum and a minimum value on this region.)

(b) Find the maximum value of $S(p_1, \ldots, p_{30})$ subject to the constraint $p_1 + \cdots + p_{30} = 1$. What are the values of p_i in this case? What does your answer mean in terms of the uncertainty in the outcome of the hockey season?

(c) Find the minimum value of $S(p_1, \ldots, p_{30})$, subject to the constraint $p_1 + \cdots + p_{30} = 1$. What are the values of p_i in this case? What does your answer mean in terms of the uncertainty in the outcome of the hockey season?

[Note: The function S is an example of an *entropy* function; the concept of entropy is used in information theory, statistical mechanics, and thermodynamics when measuring the uncertainty in an experiment (the hockey season in this problem) or physical system.]

Chapter Sixteen

INTEGRATING FUNCTIONS OF SEVERAL VARIABLES

Contents

16.1 THE DEFINITE INTEGRAL OF A FUNCTION OF TWO VARIABLES

The definite integral of a continuous one-variable function, f, is a limit of Riemann sums:

$$\int_a^b f(x)\,dx = \lim_{\Delta x \to 0} \sum_i f(x_i)\,\Delta x,$$

where x_i is a point in the ith subdivision of the interval $[a, b]$. In this section we extend this definition to functions of two variables. We start by considering how to estimate total population from a two-variable population density.

Population Density of Foxes in England

The fox population in parts of England is important to public health officials concerned about the disease rabies, which is spread by animals. Biologists use a contour diagram to display the fox population density, D; see Figure 16.1, where D is in foxes per square kilometre.[1] The bold contour is the coastline, which may be thought of as the $D = 0$ contour; clearly the density is zero outside it. We can think of D as a function of position, $D = f(x, y)$ where x and y are in kilometres from the southwest corner of the map.

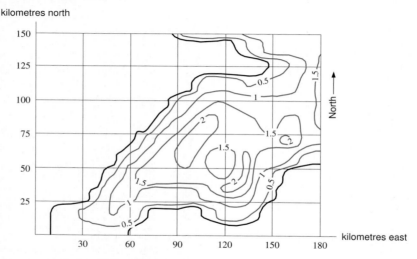

Figure 16.1: Population density of foxes in southwestern England

Example 1   Estimate the total fox population in the region represented by the map in Figure 16.1.

Solution   We subdivide the map into the rectangles shown in Figure 16.1 and estimate the population in each rectangle. For simplicity, we use the population density at the northeast corner of each rectangle. For example, in the bottom left rectangle, the density is 0 at the northeast corner, in the next rectangle to the east (right), the density in the northeast corner is 1. Continuing in this way, we get the estimates in Table 16.1. To estimate the population in a rectangle, we multiply the density by the area of the rectangle, $30 \cdot 25 = 750 \text{ km}^2$. Adding the results, we obtain

$$\text{Estimate of population} = (0.2 + 0.7 + 1.2 + 1.2 + 0.1 + 1.6 + 0.5 + 1.4$$
$$+ 1.1 + 1.6 + 1.5 + 1.8 + 1.5 + 1.3 + 1.1 + 2.0$$
$$+ 1.4 + 1.0 + 1.0 + 0.6 + 1.2)750 = 18{,}000 \text{ foxes.}$$

Taking the upper and lower bounds for the population density on each rectangle enables us to find upper and lower estimates for the population. Using the same rectangles, the upper estimate is approximately 35,000 and the lower estimate is 4,000. There is a wide discrepancy between the upper and lower estimates; we could make them closer by taking finer subdivisions.

[1] Adapted from J. D. Murray, *Mathematical Biology*, Springer-Verlag, 1989.

Table 16.1 *Estimates of population density (northeast corner)*

0.0	0.0	0.2	0.7	1.2	1.2
0.0	0.0	0.0	0.0	0.1	1.6
0.0	0.0	0.5	1.4	1.1	1.6
0.0	0.0	1.5	1.8	1.5	1.3
0.0	1.1	2.0	1.4	1.0	0.0
0.0	1.0	0.6	1.2	0.0	0.0

Definition of the Definite Integral

The sums used to approximate the fox population are Riemann sums. We now define the definite integral for a function f of two variables on a rectangular region. Given a continuous function $f(x, y)$ defined on a region $a \leq x \leq b$ and $c \leq y \leq d$, we subdivide each of the intervals $a \leq x \leq b$ and $c \leq y \leq d$ into n and m equal subintervals respectively, giving nm subrectangles. (See Figure 16.2.)

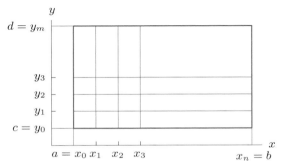

Figure 16.2: Subdivision of a rectangle into nm subrectangles

The area of each subrectangle is $\Delta A = \Delta x\, \Delta y$, where $\Delta x = (b - a)/n$ is the width of each subdivision on the x-axis, and $\Delta y = (d - c)/m$ is the width of each subdivision on the y-axis. To compute the Riemann sum, we multiply the area of each subrectangle by the value of the function at a point in the rectangle and add the resulting numbers. Choosing the point which gives the maximum value, M_{ij}, of the function on each rectangle, we get the *upper sum*, $\sum_{i,j} M_{ij} \Delta x \Delta y$.

The *lower sum*, $\sum_{i,j} L_{ij} \Delta x \Delta y$, is obtained by taking the minimum value on each rectangle. If (u_{ij}, v_{ij}) is any point in the ij-th subrectangle, any other Riemann sum satisfies

$$\sum_{i,j} L_{ij} \Delta x \Delta y \leq \sum_{i,j} f(u_{ij}, v_{ij})\, \Delta x\, \Delta y \leq \sum_{i,j} M_{ij} \Delta x \Delta y.$$

We define the definite integral by taking the limit as the numbers of subdivisions, n and m, tend to infinity. By comparing upper and lower sums, as we did for the fox population, it can be shown that the limit exists when the function, f, is continuous. We get the same limit by letting Δx and Δy tend to 0. Thus, we have the following definition:

Suppose the function f is continuous on R, the rectangle $a \leq x \leq b$, $c \leq y \leq d$. If (u_{ij}, v_{ij}) is any point in the ij-th subrectangle, we define the **definite integral** of f over R

$$\int_R f\, dA = \lim_{\Delta x, \Delta y \to 0} \sum_{i,j} f(u_{ij}, v_{ij}) \Delta x \Delta y.$$

Such an integral is called a **double integral**.

The case when R is not rectangular is considered on page 890. Sometimes we think of dA as being the area of an infinitesimal rectangle of length dx and height dy, so that $dA = dx\,dy$. Then we use the notation[2]

$$\int_R f\,dA = \int_R f(x,y)\,dx\,dy.$$

For this definition, we used a particular type of Riemann sum with equal-sized rectangular subdivisions. In a general Riemann sum, the subdivisions do not all have to be the same size.

Interpretation of the Double Integral as Volume

Just as the definite integral of a positive one-variable function can be interpreted as an area, so the double integral of a positive two-variable function can be interpreted as a volume. In the one-variable case we visualise the Riemann sums as the total area of rectangles above the subdivisions. In the two-variable case we get solid bars instead of rectangles. As the number of subdivisions grows, the tops of the bars approximate the surface better, and the volume of the bars gets closer to the volume under the graph of the function. (See Figure 16.3.)

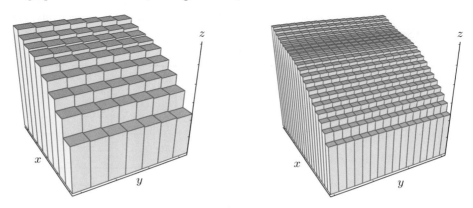

Figure 16.3: Approximating volume under a graph with finer and finer Riemann sums

Thus, we have the following result:

> If x, y, z represent length and f is positive, then
>
> $$\begin{array}{c}\text{Volume under graph}\\\text{of } f \text{ above region } R\end{array} = \int_R f\,dA.$$

Example 2   Let R be the rectangle $0 \le x \le 1$ and $0 \le y \le 1$. Use Riemann sums to make upper and lower estimates of the volume of the region above R and under the graph of $z = e^{-(x^2+y^2)}$.

Solution   If R is the rectangle $0 \le x \le 1, 0 \le y \le 1$, the volume we want is given by

$$\text{Volume} = \int_R e^{-(x^2+y^2)}\,dA.$$

We divide R into 16 subrectangles by dividing each edge into four parts. Figure 16.4 shows that $f(x,y) = e^{-(x^2+y^2)}$ decreases as we move away from the origin. Thus, to get an upper sum we evaluate f on each subrectangle at the corner nearest the origin. For example, in the rectangle $0 \le x \le 0.25, 0 \le y \le 0.25$, we evaluate f at $(0,0)$. Using Table 16.2, we find that

---

[2] Another common notation for the double integral is $\int\int_R f\,dA$.

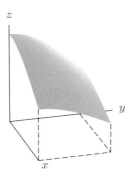

Figure 16.4: Graph of $e^{-(x^2+y^2)}$ above the rectangle R

$$\text{Upper sum} = (\ 1 + 0.9394 + 0.7788 + 0.5698$$
$$+ 0.9394 + 0.8825 + 0.7316 + 0.5353$$
$$+ 0.7788 + 0.7316 + 0.6065 + 0.4437$$
$$+ 0.5698 + 0.5353 + 0.4437 + 0.3247)(0.0625) = 0.68.$$

To get a lower sum, we evaluate f at the opposite corner of each rectangle because the surface slopes down in both the x and y directions. This yields a lower sum of 0.44. Thus,

$$0.44 \leq \int_R e^{-(x^2+y^2)}\, dA \leq 0.68.$$

To get a better approximation, we use more subdivisions. See Table 16.3.

Table 16.2 *Values of $f(x,y) = e^{-(x^2+y^2)}$ on the rectangle R*

		y				
		0.0	0.25	0.50	0.75	1.00
	0.0	1	0.9394	0.7788	0.5698	0.3679
	0.25	0.9394	0.8825	0.7316	0.5353	0.3456
x	0.50	0.7788	0.7316	0.6065	0.4437	0.2865
	0.75	0.5698	0.5353	0.4437	0.3247	0.2096
	1.00	0.3679	0.3456	0.2865	0.2096	0.1353

Table 16.3 *Riemann sum approximations to $\int_R e^{-(x^2+y^2)}\, dA$*

	Number of subdivisions in x and y directions			
	8	16	32	64
Upper	0.6168	0.5873	0.5725	0.5651
Lower	0.4989	0.5283	0.5430	0.5504

The true value of the double integral, $0.5577\ldots$, is trapped between the lower and upper sums. Notice that the lower sum increases and the upper sum decreases as the number of subdivisions increases. However, even with 64 subdivisions, the lower and upper sums agree with the true value of the integral only in the first decimal place.

Interpretation of the Double Integral as Area

In the special case that $f(x,y) = 1$ for all points (x,y) in the region R, each term in the Riemann sum is of the form $1 \cdot \Delta A = \Delta A$ and the double integral gives the area of the region R:

$$\text{Area}(R) = \int_R 1\, dA = \int_R dA$$

Interpretation of the Double Integral as Average Value

As in the one-variable case, the definite integral can be used to compute the average value of a function:

$$\begin{array}{c}\text{Average value of } f \\ \text{on the region } R\end{array} = \frac{1}{\text{Area of } R}\int_R f\, dA$$

We can rewrite this as

$$\text{Average value} \times \text{Area of } R = \int_R f\, dA.$$

If we interpret the integral as the volume under the graph of f, then we can think of the average value of f as the height of the box with the same volume that is on the same base. (See Figure 16.5.) Imagine that the volume under the graph is made out of wax. If the wax melted within the perimeter of R, then it would end up box-shaped with height equal to the average value of f.

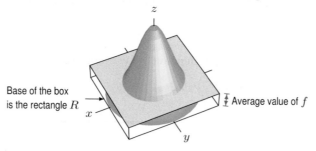

Figure 16.5: Volume and average value

Integral over Regions that Are Not Rectangles

We defined the definite integral $\int_R f(x, y)\, dA$, for a rectangular region R. Now we extend the definition to regions of other shapes, including triangles, circles, and regions bounded by the graphs of piecewise continuous functions.

To approximate the definite integral over a region, R, which is not rectangular, we use a grid of rectangles approximating the region. We obtain this grid by enclosing R in a large rectangle and subdividing that rectangle; we consider just the subrectangles which are inside R.

As before, we pick a point (u_{ij}, v_{ij}) in each subrectangle and form a Riemann sum

$$\sum_{i,j} f(u_{ij}, v_{ij})\Delta x \Delta y.$$

This time, however, the sum is over only those subrectangles within R. For example, in the case of the fox population we can use the rectangles which are entirely on land. As the subdivisions become finer, the grid approximates the region R more closely. For a function, f, which is continuous on R, we define the definite integral as follows:

$$\int_R f\, dA = \lim_{\Delta x, \Delta y \to 0} \sum_{i,j} f(u_{ij}, v_{ij})\Delta x \Delta y$$

where the Riemann sum is taken over the subrectangles inside R.

You may wonder why we can leave out the rectangles which cover the edge of R—if we included them, might we get a different value for the integral? The answer is that for any region that we are likely to meet, the area of the subrectangles covering the edge tends to 0 as the grid becomes finer. Therefore, omitting these rectangles does not affect the limit.

Convergence of Upper and Lower Sums to Same Limit

We have said that if f is continuous on the rectangle R, then the difference between upper and lower sums for f converges to 0 as Δx and Δy approach 0. In the following example, we show this in a particular case. The ideas in this example can be used in a general proof.

Example 3   Let $f(x, y) = x^2 y$ and let R be the rectangle $0 \leq x \leq 1, 0 \leq y \leq 1$. Show that the difference between upper and lower Riemann sums for f on R converges to 0, as Δx and Δy approach 0.

Solution   The difference between the sums is

$$\sum M_{ij} \Delta x \Delta y - \sum L_{ij} \Delta x \Delta y = \sum (M_{ij} - L_{ij}) \Delta x \Delta y,$$

where M_{ij} and L_{ij} are the maximum and minimum of f on the ij-th subrectangle. Since f increases in both the x and y directions, M_{ij} occurs at the corner of the subrectangle farthest from the origin and L_{ij} at the closest. Moreover, since the slopes in the x and y directions don't decrease as x and y increase, the difference $M_{ij} - L_{ij}$ is largest in the subrectangle R_{nm} which is farthest from the origin. Thus,

$$\sum (M_{ij} - L_{ij}) \Delta x \Delta y \leq (M_{nm} - L_{nm}) \sum \Delta x \Delta y = (M_{nm} - L_{nm}) \text{Area}(R).$$

Thus, the difference converges to 0 as long as $(M_{nm} - L_{nm})$ does. The maximum M_{nm} of f on the nm-th subrectangle occurs at $(1, 1)$, the subrectangle's top right corner, and the minimum L_{nm} occurs at the opposite corner, $(1 - 1/n, 1 - 1/m)$. Substituting into $f(x, y) = x^2 y$ gives

$$M_{nm} - L_{nm} = (1)^2(1) - \left(1 - \frac{1}{n}\right)^2 \left(1 - \frac{1}{m}\right) = \frac{2}{n} - \frac{1}{n^2} + \frac{1}{m} - \frac{2}{nm} + \frac{1}{n^2 m}.$$

The right-hand side converges to 0 as $n, m \to \infty$, that is, as $\Delta x, \Delta y \to 0$.

Exercises and Problems for Section 16.1

Exercises

1. Values of $f(x, y)$ are in Table 16.4. Let R be the rectangle $1 \leq x \leq 1.2, 2 \leq y \leq 2.4$. Find Riemann sums which are reasonable over and underestimates for $\int_R f(x, y) \, dA$ with $\Delta x = 0.1$ and $\Delta y = 0.2$.

Table 16.4

		x	
	1.0	1.1	1.2
y 2.0	5	7	10
2.2	4	6	8
2.4	3	5	4

2. Table 16.5 gives values of the function $f(x, y)$, which is increasing in x and decreasing in y on the region $R : 0 \leq x \leq 6, 0 \leq y \leq 1$. Make the best possible upper and lower estimates of $\int_R f(x, y) \, dA$.

Table 16.5

		x		
		0	3	6
y	0	5	7	10
	0.5	4	5	7
	1	3	4	6

3. Figure 16.6 shows contours of $g(x, y)$ on the region R, with $5 \leq x \leq 11$ and $4 \leq y \leq 10$. Using $\Delta x = \Delta y = 2$, find an overestimate and an underestimate for $\int_R g(x, y) dA$.

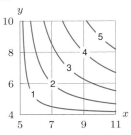

Figure 16.6

4. Figure 16.7 shows contours of $f(x, y)$ on the rectangle R with $0 \leq x \leq 30$ and $0 \leq y \leq 15$. Using $\Delta x = 10$ and $\Delta y = 5$, find an overestimate and an underestimate for $\int_R f(x, y) dA$.

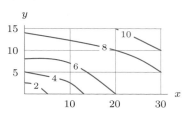

Figure 16.7

5. Let R be the rectangle with vertices $(0, 0)$, $(4, 0)$, $(4, 4)$, and $(0, 4)$ and let $f(x, y) = \sqrt{xy}$.

(a) Find reasonable upper and lower bounds for $\int_R f\, dA$ without subdividing R.

(b) Estimate $\int_R f\, dA$ by partitioning R into four sub-rectangles and evaluating f at its maximum and minimum values on each subrectangle.

6. Figure 16.8 shows a contour plot of population density, people per square kilometer, in a rectangle of land 3 km by 2 km. Estimate the population in the region represented by Figure 16.8.

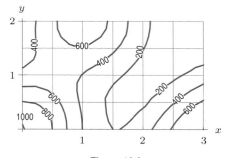

Figure 16.8

Problems

In Problems 7–13, decide (without calculation) whether the integrals are positive, negative, or zero. Let D be the region inside the unit circle centered at the origin, let R be the right half of D and let B be the bottom half of D.

7. $\int_D dA$

8. $\int_R 5x\, dA$

9. $\int_B 5x\, dA$

10. $\int_D (y^3 + y^5)\, dA$

11. $\int_B (y^3 + y^5)\, dA$

12. $\int_D (y - y^3)\, dA$

13. $\int_B (y - y^3)\, dA$

14. Figure 16.9 shows contours of $f(x, y)$. Let R be the square $-0.5 \leq x \leq 1$, $-0.5 \leq y \leq 1$. Is the integral $\int_R f\, dA$ positive or negative? Explain your reasoning.

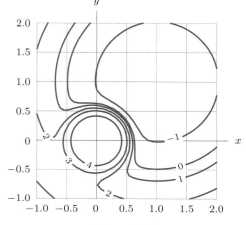

Figure 16.9

15. Table 16.6 gives values of $f(x, y)$, the number of milligrams of mosquito larvae per square meter in a swamp.

If x and y are in meters and R is the rectangle $0 \leq x \leq 8$, $0 \leq y \leq 6$, estimate $\int_R f(x, y) dA$. Give units and interpret your answer.

Table 16.6

		x		
		0	4	8
	0	1	3	6
y	3	2	5	9
	6	4	9	15

16. Figure 16.10 shows the temperature, in °C, in a 5 meter by 5 meter heated room. Using Riemann sums, estimate the average temperature in the room.

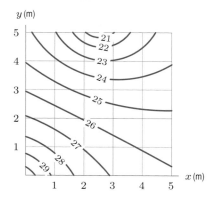

Figure 16.10

17. Use four subrectangles to approximate the volume of the object whose base is the region $0 \leq x \leq 4$ and $0 \leq y \leq 6$, and whose height is given by $f(x, y) = x + y$. Find an overestimate and an underestimate and average the two.

Strengthen Your Understanding

In Problems 18–19, explain what is wrong with the statement.

18. For all f, the integral $\int_R f(x, y)\, dA$ gives the volume of the solid under the graph of f over the region R.

19. If R is a region in the third quadrant where $x < 0, y < 0$, then $\int_R f(x, y)\, dA$ is negative.

In Problems 20–21, give an example of:

20. A function $f(x, y)$ and rectangle R such that the Riemann sums obtained using the lower left-hand corner of each subrectangle are an overestimate.

21. A function $f(x, y)$ whose average value over the square $0 \leq x \leq 1, 0 \leq y \leq 1$ is negative.

Are the statements in Problems 22–31 true or false? Give reasons for your answer.

22. The double integral $\int_R f\, dA$ is always positive.

23. If $f(x, y) = k$ for all points (x, y) in a region R then $\int_R f\, dA = k \cdot \text{Area}(R)$.

24. If R is the rectangle $0 \leq x \leq 1, 0 \leq y \leq 1$ then $\int_R e^{xy}\, dA > 3$.

25. If R is the rectangle $0 \leq x \leq 2, 0 \leq y \leq 3$ and S is the rectangle $-2 \leq x \leq 0, -3 \leq y \leq 0$ then $\int_R f\, dA = -\int_S f\, dA$.

26. Let $\rho(x, y)$ be the population density of a city, in people per km^2. If R is a region in the city, then $\int_R \rho\, dA$ gives the total number of people in the region R.

27. If $\int_R f\, dA = 0$ then $f(x, y) = 0$ at all points of R.

28. If $g(x, y) = kf(x, y)$, where k is constant, then $\int_R g\, dA = k \int_R f\, dA$.

29. If f and g are two functions continuous on a region R, then $\int_R f \cdot g\, dA = \int_R f\, dA \cdot \int_R g\, dA$.

30. If R is the rectangle $0 \leq x \leq 1, 0 \leq y \leq 2$ and S is the square $0 \leq x \leq 1, 0 \leq y \leq 1$, then $\int_R f\, dA = 2 \int_S f\, dA$.

31. If R is the rectangle $2 \leq x \leq 4, 5 \leq y \leq 9$, $f(x, y) = 2x$ and $g(x, y) = x + y$, then the average value of f on R is less than the average value of g on R.

16.2 ITERATED INTEGRALS

In Section 16.1 we approximated double integrals using Riemann sums. In this section we see how to compute double integrals exactly using one-variable integrals.

The Fox Population Again: Expressing a Double Integral as an Iterated Integral

To estimate the fox population, we computed a sum of the form

$$\text{Total population} \approx \sum_{i,j} f(u_{ij}, v_{ij}) \Delta x\, \Delta y,$$

where $1 \leq i \leq n$ and $1 \leq j \leq m$ and the values $f(u_{ij}, v_{ij})$ can be arranged as in Table 16.7.

Table 16.7   *Estimates for fox population densities for $n = m = 6$*

0.0	0.0	0.2	0.7	1.2	1.2
0.0	0.0	0.0	0.0	0.1	1.6
0.0	0.0	0.5	1.4	1.1	1.6
0.0	0.0	1.5	1.8	1.5	1.3
0.0	1.1	2.0	1.4	1.0	0.0
0.0	1.0	0.6	1.2	0.0	0.0

For any values of n and m, we can either add across the rows first or add down the columns first. If we add rows first, we can write the sum in the form

$$\text{Total population} \approx \sum_{j=1}^{m} \left(\sum_{i=1}^{n} f(u_{ij}, v_{ij}) \Delta x \right) \Delta y.$$

The inner sum, $\sum_{i=1}^{n} f(u_{ij}, v_{ij}) \Delta x$, approximates the integral $\int_0^{180} f(x, v_{ij}) \, dx$. Thus, we have

$$\text{Total population} \approx \sum_{j=1}^{m} \left(\int_0^{180} f(x, v_{ij}) \, dx \right) \Delta y.$$

The outer Riemann sum approximates another integral, this time with integrand $\int_0^{180} f(x, y) \, dx$, which is a function of y. Thus, we can write the total population in terms of nested, or *iterated*, one-variable integrals:

$$\text{Total population} = \int_0^{150} \left(\int_0^{180} f(x, y) \, dx \right) dy.$$

Since the total population is represented by $\int_R f \, dA$, this suggests the method of computing double integrals in the following theorem:[3]

Theorem 16.1: Writing a Double Integral as an Iterated Integral

If R is the rectangle $a \leq x \leq b$, $c \leq y \leq d$ and f is a continuous function on R, then the integral of f over R exists and is equal to the **iterated integral**

$$\int_R f \, dA = \int_{y=c}^{y=d} \left(\int_{x=a}^{x=b} f(x, y) \, dx \right) dy.$$

The expression $\int_{y=c}^{y=d} \left(\int_{x=a}^{x=b} f(x, y) \, dx \right) dy$ can be written $\int_c^d \int_a^b f(x, y) \, dx \, dy$.

To evaluate the iterated integral, first perform the inside integral with respect to x, holding y constant; then integrate the result with respect to y.

Example 1   A building is 8 metres wide and 16 metres long. It has a flat roof that is 12 metres high at one corner, and 10 metres high at each of the adjacent corners. What is the volume of the building?

Solution   If we put the high corner on the z-axis, the long side along the y-axis, and the short side along the x-axis, as in Figure 16.11, then the roof is a plane with z-intercept 12, and x slope $(-2)/8 = -1/4$, and y slope $(-2)/16 = -1/8$. Hence, the equation of the roof is

$$z = 12 - \tfrac{1}{4}x - \tfrac{1}{8}y.$$

The volume is given by the double integral

$$\text{Volume} = \int_R \left(12 - \tfrac{1}{4}x - \tfrac{1}{8}y\right) dA,$$

where R is the rectangle $0 \leq x \leq 8$, $0 \leq y \leq 16$. Setting up an iterated integral, we get

$$\text{Volume} = \int_0^{16} \int_0^8 \left(12 - \tfrac{1}{4}x - \tfrac{1}{8}y\right) dx \, dy.$$

The inside integral is

$$\int_0^8 \left(12 - \tfrac{1}{4}x - \tfrac{1}{8}y\right) dx = \left(12x - \tfrac{1}{8}x^2 - \tfrac{1}{8}xy\right) \Big|_{x=0}^{x=8} = 88 - y.$$

Then the outside integral gives

$$\text{Volume} = \int_0^{16} (88 - y) \, dy = \left(88y - \tfrac{1}{2}y^2\right) \Big|_0^{16} = 1280.$$

The volume of the building is 1280 cubic metres.

---

[3] For a proof, see M. Spivak, *Calculus on Manifolds*, pp. 53 and 58 (New York: Benjamin, 1965).

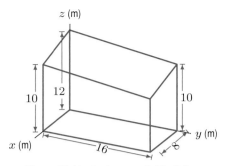

Figure 16.11: A slant-roofed building

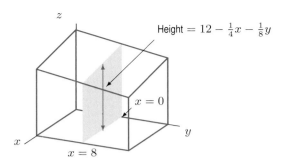

Figure 16.12: Cross-section of a building

Notice that the inner integral $\int_0^8 (12 - \frac{1}{4}x - \frac{1}{8}y)\,dx$ in Example 1 gives the area of the cross section of the building perpendicular to the y-axis in Figure 16.12.

The iterated integral $\int_0^{16} \int_0^8 (12 - \frac{1}{4}x - \frac{1}{8}y)\,dx\,dy$ thus calculates the volume by adding the volumes of thin cross-sectional slabs.

The Order of Integration

In computing the fox population, we could have chosen to add columns (fixed x) first, instead of the rows. This leads to an iterated integral where x is constant in the inner integral instead of y. Thus,

$$\int_R f(x,y)\,dA = \int_a^b \left(\int_c^d f(x,y)\,dy \right) dx$$

where R is the rectangle $a \le x \le b$ and $c \le y \le d$.

For any function we are likely to meet, it does not matter in which order we integrate over a rectangular region R; we get the same value for the double integral either way.

$$\int_R f\,dA = \int_c^d \left(\int_a^b f(x,y)\,dx \right) dy = \int_a^b \left(\int_c^d f(x,y)\,dy \right) dx$$

Example 2   Compute the volume of Example 1 as an iterated integral by integrating with respect to y first.

Solution   Rewriting the integral, we have

$$\text{Volume} = \int_0^8 \left(\int_0^{16} (12 - \frac{1}{4}x - \frac{1}{8}y)\,dy \right) dx = \int_0^8 \left((12y - \frac{1}{4}xy - \frac{1}{16}y^2) \Big|_{y=0}^{y=16} \right) dx$$

$$= \int_0^8 (176 - 4x)\,dx = (176x - 2x^2) \Big|_0^8 = 1280 \text{ meter}^3.$$

Iterated Integrals Over Non-Rectangular Regions

Example 3   The density at the point (x,y) of a triangular metal plate, as shown in Figure 16.13, is $\delta(x,y)$. Express its mass as an iterated integral.

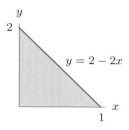

Figure 16.13: A triangular metal plate with density $\delta(x, y)$ at the point (x, y)

Solution

Approximate the triangular region using a grid of small rectangles of sides Δx and Δy. The mass of one rectangle is given by

$$\text{Mass of rectangle} \approx \text{Density} \cdot \text{Area} \approx \delta(x, y)\Delta x \Delta y.$$

Summing over all rectangles gives a Riemann sum which approximates the double integral:

$$\text{Mass} = \int_R \delta(x, y)\, dA,$$

where R is the triangle. We want to compute this integral using an iterated integral.

Think about how the iterated integral over the rectangle $a \leq x \leq b$, $c \leq y \leq d$ works:

$$\int_a^b \int_c^d f(x, y)\, dy\, dx.$$

The inside integral with respect to y is along vertical strips which begin at the horizontal line $y = c$ and end at the line $y = d$. There is one such strip for each x between $x = a$ and $x = b$. (See Figure 16.14.)

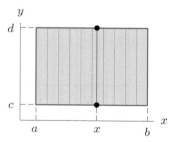

Figure 16.14: Integrating over a rectangle using vertical strips

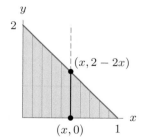

Figure 16.15: Integrating over a triangle using vertical strips

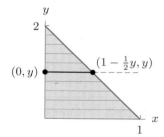

Figure 16.16: Integrating over a triangle using horizontal strips

For the triangular region in Figure 16.13, the idea is the same. The only difference is that the individual vertical strips no longer all go from $y = c$ to $y = d$. The vertical strip that starts at the point $(x, 0)$ ends at the point $(x, 2 - 2x)$, because the top edge of the triangle is the line $y = 2 - 2x$. See Figure 16.15. On this vertical strip, y goes from 0 to $2 - 2x$. Hence, the inside integral is

$$\int_0^{2-2x} \delta(x, y)\, dy.$$

Finally, since there is a vertical strip for each x between 0 and 1, the outside integral goes from $x = 0$ to $x = 1$. Thus, the iterated integral we want is

$$\text{Mass} = \int_0^1 \int_0^{2-2x} \delta(x, y)\, dy\, dx.$$

We could have chosen to integrate in the opposite order, keeping y fixed in the inner integral instead of x. The limits are formed by looking at horizontal strips instead of vertical ones, and expressing the x-values at the end points in terms of y. To find the right endpoint of the strip, we use the equation of the top edge of the triangle in the form $x = 1 - \frac{1}{2}y$. Thus, a horizontal strip goes from $x = 0$ to $x = 1 - \frac{1}{2}y$. Since there is a strip for every y from 0 to 2, the iterated integral is

$$\text{Mass} = \int_0^2 \int_0^{1 - \frac{1}{2}y} \delta(x, y)\, dx\, dy.$$

Limits on Iterated Integrals

- The limits on the outer integral must be constants.

- The limits on the inner integral can involve only the variable in the outer integral. For example, if the inner integral is with respect to x, its limits can be functions of y.

Example 4   Find the mass M of a metal plate R bounded by $y = x$ and $y = x^2$, with density given by $\delta(x, y) = 1 + xy$ kg/metre^2. (See Figure 16.17.)

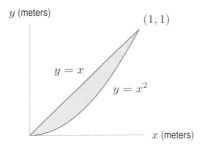

Figure 16.17: A metal plate with density $\delta(x, y)$

Solution   The mass is given by

$$M = \int_R \delta(x, y)\, dA.$$

We integrate along vertical strips first; this means we do the y integral first, which goes from the bottom boundary $y = x^2$ to the top boundary $y = x$. The left edge of the region is at $x = 0$ and the right edge is at the intersection point of $y = x$ and $y = x^2$, which is $(1, 1)$. Thus, the x-coordinate of the vertical strips can vary from $x = 0$ to $x = 1$, and so the mass is given by

$$M = \int_0^1 \int_{x^2}^x \delta(x, y)\, dy\, dx = \int_0^1 \int_{x^2}^x (1 + xy)\, dy\, dx.$$

Calculating the inner integral first gives

$$M = \int_0^1 \int_{x^2}^x (1 + xy)\, dy\, dx = \int_0^1 \left(y + x\frac{y^2}{2} \right) \Bigg|_{y=x^2}^{y=x} dx$$

$$= \int_0^1 \left(x - x^2 + \frac{x^3}{2} - \frac{x^5}{2} \right) dx = \left(\frac{x^2}{2} - \frac{x^3}{3} + \frac{x^4}{8} - \frac{x^6}{12} \right) \Bigg|_0^1 = \frac{5}{24} = 0.208 \text{ kg.}$$

Example 5   A city occupies a semicircular region of radius 3 km bordering on the ocean. Find the average distance from points in the city to the ocean.

Solution   Think of the ocean as everything below the x-axis in the xy-plane and think of the city as the upper half of the circular disk of radius 3 bounded by $x^2 + y^2 = 9$. (See Figure 16.18.)

The distance from any point (x, y) in the city to the ocean is the vertical distance to the x-axis, namely y. Thus, we want to compute

$$\text{Average distance} = \frac{1}{\text{Area}(R)} \int_R y \, dA,$$

where R is the region between the upper half of the circle $x^2 + y^2 = 9$ and the x-axis. The area of R is $\pi 3^2/2 = 9\pi/2$.

To compute the integral, let's take the inner integral with respect to y. A vertical strip goes from the x-axis, namely $y = 0$, to the semicircle. The upper limit must be expressed in terms of x, so we solve $x^2 + y^2 = 9$ to get $y = \sqrt{9 - x^2}$. Since there is a strip for every x from -3 to 3, the integral is:

$$\int_R y \, dA = \int_{-3}^{3} \left(\int_0^{\sqrt{9-x^2}} y \, dy \right) dx = \int_{-3}^{3} \left(\frac{y^2}{2} \Big|_{y=0}^{y=\sqrt{9-x^2}} \right) dx$$

$$= \int_{-3}^{3} \frac{1}{2}(9 - x^2) \, dx = \frac{1}{2} \left(9x - \frac{x^3}{3} \right) \Big|_{-3}^{3} = \frac{1}{2}(18 - (-18)) = 18.$$

Therefore, the average distance is $18/(9\pi/2) = 4/\pi = 1.273$ km.

What if we choose the inner integral with respect to x? Then we get the limits by looking at horizontal strips, not vertical, and we solve $x^2 + y^2 = 9$ for x in terms of y. We get $x = -\sqrt{9 - y^2}$ at the left end of the strip and $x = \sqrt{9 - y^2}$ at the right. There is a strip for every y from 0 to 3, so

$$\int_R y \, dA = \int_0^3 \left(\int_{-\sqrt{9-y^2}}^{\sqrt{9-y^2}} y \, dx \right) dy = \int_0^3 \left(yx \Big|_{x=-\sqrt{9-y^2}}^{x=\sqrt{9-y^2}} \right) dy = \int_0^3 2y\sqrt{9 - y^2} \, dy$$

$$= -\frac{2}{3}(9 - y^2)^{3/2} \Big|_0^3 = -\frac{2}{3}(0 - 27) = 18.$$

We get the same result as before. The average distance to the ocean is $(2/(9\pi))18 = 4/\pi = 1.273$ km.

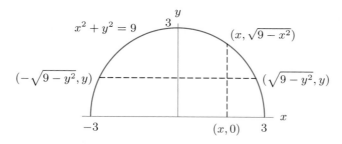

Figure 16.18: The city by the ocean showing a typical vertical strip and a typical horizontal strip

In the examples so far, a region was given and the problem was to determine the limits for an iterated integral. Sometimes the limits are known and we want to determine the region.

Example 6   Sketch the region of integration for the iterated integral $\displaystyle\int_0^6 \int_{x/3}^2 x\sqrt{y^3+1}\, dy\, dx.$

Solution   The inner integral is with respect to y, so we imagine the region built of vertical strips. The bottom of each strip is on the line $y = x/3$, and the top is on the horizontal line $y = 2$. Since the limits of the outer integral are 0 and 6, the whole region is contained between the vertical lines $x = 0$ and $x = 6$. Notice that the lines $y = 2$ and $y = x/3$ meet where $x = 6$. See Figure 16.19.

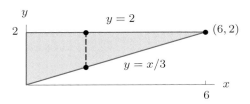

Figure 16.19: The region of integration for Example 6, showing the vertical strip

Reversing the Order of Integration

It is sometimes helpful to reverse the order of integration in an iterated integral. An integral which is difficult or impossible with the integration in one order can be quite straightforward in the other. The next example is such a case.

Example 7   Evaluate $\displaystyle\int_0^6 \int_{x/3}^2 x\sqrt{y^3+1}\, dy\, dx$ using the region sketched in Figure 16.19.

Solution   Since $\sqrt{y^3+1}$ has no elementary antiderivative, we cannot calculate the inner integral symbolically. We try reversing the order of integration. From Figure 16.19, we see that horizontal strips go from $x = 0$ to $x = 3y$ and that there is a strip for every y from 0 to 2. Thus, when we change the order of integration we get

$$\int_0^6 \int_{x/3}^2 x\sqrt{y^3+1}\, dy\, dx = \int_0^2 \int_0^{3y} x\sqrt{y^3+1}\, dx\, dy.$$

Now we can at least do the inner integral because we know the antiderivative of x. What about the outer integral?

$$\int_0^2 \int_0^{3y} x\sqrt{y^3+1}\, dx\, dy = \int_0^2 \left(\frac{x^2}{2}\sqrt{y^3+1}\right)\Bigg|_{x=0}^{x=3y} dy = \int_0^2 \frac{9y^2}{2}(y^3+1)^{1/2}\, dy$$

$$= (y^3+1)^{3/2}\Big|_0^2 = 27 - 1 = 26.$$

Thus, reversing the order of integration made the integral in the previous problem much easier. Notice that to reverse the order it is essential first to sketch the region over which the integration is being performed.

Exercises and Problems for Section 16.2

Exercises

In Exercises 1–4, sketch the region of integration.

1. $\int_0^\pi \int_0^x y \sin x \, dy \, dx$

2. $\int_0^1 \int_{x-2}^{\cos \pi x} y \, dy \, dx$

3. $\int_0^1 \int_{y^2}^y xy \, dx \, dy$

4. $\int_0^2 \int_0^{y^2} y^2 x \, dx \, dy$

For Exercises 5–12, evaluate the integral.

5.
$$\int_0^3 \int_0^4 (4x + 3y) \, dx \, dy$$

6.
$$\int_0^2 \int_0^3 (x^2 + y^2) \, dy \, dx$$

7. $\int_0^3 \int_0^2 6xy \, dy \, dx$

8. $\int_0^1 \int_0^2 x^2 y \, dy \, dx$

9. $\int_0^1 \int_0^1 y e^{xy} \, dx \, dy$

10. $\int_0^2 \int_0^y y \, dx \, dy$

11. $\int_0^3 \int_0^y \sin x \, dx \, dy$

12. $\int_0^{\pi/2} \int_0^{\sin x} x \, dy \, dx$

In Exercises 13–18, write $\int_R f \, dA$ as an iterated integral for the shaded region R.

13.

14.
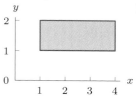

15.
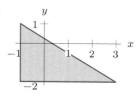

16.
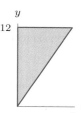

17.

18.
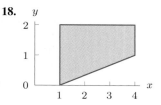

For Exercises 19–25, sketch the region of integration and evaluate the integral.

19. $\int_1^3 \int_0^4 e^{x+y} \, dy \, dx$

20. $\int_0^2 \int_0^x e^{x^2} \, dy \, dx$

21. $\int_1^5 \int_x^{2x} \sin x \, dy \, dx$

22. $\int_1^4 \int_{\sqrt{y}}^y x^2 y^3 \, dx \, dy$

23. $\int_0^3 \int_0^{2x} (x^2 + y^2) \, dy \, dx$

24. $\int_0^\pi \int_0^x \sin x \, dy \, dx$

25. $\int_{-2}^0 \int_{-\sqrt{9-x^2}}^0 2xy \, dy \, dx$

For Exercises 26–30, evaluate the integral.

26. $\int_R \sqrt{x+y} \, dA$, where R is the rectangle $0 \le x \le 1$, $0 \le y \le 2$.

27. Calculate the integral in Exercise 26 using the other order of integration.

28. $\int_R (5x^2 + 1) \sin 3y \, dA$, where R is the rectangle $-1 \le x \le 1$, $0 \le y \le \pi/3$.

29. $\int_R xy \, dA$, where R is the triangle $x + y \le 1$, $x \ge 0$, $y \ge 0$.

30. $\int_R (2x + 3y)^2 \, dA$, where R is the triangle with vertices at $(-1, 0)$, $(0, 1)$, and $(1, 0)$.

Problems

In Problems 31–34, integrate $f(x, y) = xy$ over the region R.

31.
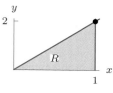

32.
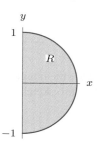

33.
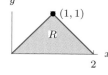

34.
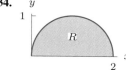

35. (a) Use four subrectangles to approximate the volume of the object whose base is the region $0 \leq x \leq 4$ and $0 \leq y \leq 6$, and whose height is given by $f(x, y) = xy$. Find an overestimate and an underestimate and average the two.

   (b) Integrate to find the exact volume of the three-dimensional object described in part (a).

In Problems 36–40, evaluate the integral by reversing the order of integration.

36. $\displaystyle\int_0^1 \int_y^1 e^{x^2}\, dx\, dy$

37. $\displaystyle\int_0^1 \int_y^1 \sin\left(x^2\right)\, dx\, dy$

38. $\displaystyle\int_0^1 \int_{\sqrt{y}}^1 \sqrt{2 + x^3}\, dx\, dy$

39. $\displaystyle\int_0^3 \int_{y^2}^9 y \sin(x^2)\, dx\, dy$

40. $\displaystyle\int_0^1 \int_{e^y}^e \frac{x}{\ln x}\, dx\, dy$

41. Find the volume under the graph of the function $f(x, y) = 6x^2 y$ over the region shown in Figure 16.20.

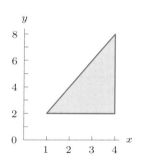

Figure 16.20

42. (a) Find the volume below the surface $z = x^2 + y^2$ and above the xy-plane for $-1 \leq x \leq 1, -1 \leq y \leq 1$.

   (b) Find the volume above the surface $z = x^2 + y^2$ and below the plane $z = 2$ for $-1 \leq x \leq 1$, $-1 \leq y \leq 1$.

43. Compute the integral

$$\int\int_R (2x^2 + y)\, dA,$$

where R is the triangular region with vertices at $(0, 1)$, $(-2, 3)$ and $(2, 3)$.

44. (a) Sketch the region in the xy-plane bounded by the x-axis, $y = x$, and $x + y = 1$.

   (b) Express the integral of $f(x, y)$ over this region in terms of iterated integrals in two ways. (In one, use $dx\, dy$; in the other, use $dy\, dx$.)

   (c) Using one of your answers to part (b), evaluate the integral exactly with $f(x, y) = x$.

45. Let $f(x, y) = x^2 e^{x^2}$ and let R be the triangle bounded by the lines $x = 3$, $x = y/2$, and $y = x$ in the xy-plane.

   (a) Express $\int_R f\, dA$ as a double integral in two different ways.

   (b) Evaluate one of them.

46. Find the average value of $f(x, y) = x^2 + 4y$ on the rectangle $0 \leq x \leq 3$ and $0 \leq y \leq 6$.

47. Find the average value of $f(x, y) = xy^2$ on the rectangle $0 \leq x \leq 4, 0 \leq y \leq 3$.

In Problems 48–50 set up, but do not evaluate, an iterated integral for the volume of the solid.

48. Under the graph of $f(x, y) = 25 - x^2 - y^2$ and above the xy-plane.

49. Below the graph of $f(x, y) = 25 - x^2 - y^2$ and above the plane $z = 16$.

50. The three-sided pyramid whose base is on the xy-plane and whose three sides are the vertical planes $y = 0$ and $y - x = 4$, and the slanted plane $2x + y + z = 4$.

In Problems 51–56, find the volume of the solid region.

51. Under the graph of $f(x, y) = xy$ and above the square $0 \leq x \leq 2, 0 \leq y \leq 2$ in the xy-plane.

52. Under the graph of $f(x, y) = x^2 + y^2$ and above the triangle $0 \leq y \leq x, 0 \leq x \leq 1$.

53. Under the graph of $f(x, y) = x + y$ and above the region $y^2 \leq x, 0 \leq x \leq 9, y \geq 0$.

54. Under the graph of $2x + y + z = 4$ in the first octant.

55. The solid between the planes $z = 3x + 2y + 1$ and $z = x + y$, and above the triangle with vertices $(1, 0, 0)$, $(2, 2, 0)$, and $(0, 1, 0)$ in the xy-plane. See Figure 16.21.

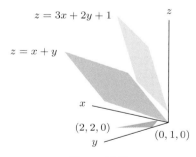

Figure 16.21

56. The solid region R bounded by the coordinate planes and the graph of $ax + by + cz = 1$. Assume a, b, and $c > 0$.

57. If R is the region $x + y \geq a, x^2 + y^2 \leq a^2$, with $a > 0$, evaluate the integral

$$\int_R xy \, dA.$$

58. The region W lies below the surface $f(x, y) = 2e^{-(x-1)^2 - y^2}$ and above the disk $x^2 + y^2 \leq 4$ in the xy-plane.

(a) Describe in words the contours of f, using $f(x, y) = 1$ as an example.

(b) Write an integral giving the area of the cross-section of W in the plane $x = 1$.

(c) Write an iterated double integral giving the volume of W.

59. Find the average distance to the x-axis for points in the region bounded by the x-axis and the graph of $y = x - x^2$.

60. Give the contour diagram of a function f whose average value on the square $0 \leq x \leq 1, 0 \leq y \leq 1$ is

(a) Greater than the average of the values of f at the four corners of the square.

(b) Less than the average of the values of f at the four corners of the square.

61. The function $f(x, y) = ax + by$ has an average value of 20 on the rectangle $0 \leq x \leq 2, 0 \leq y \leq 3$.

(a) What can you say about the constants a and b?

(b) Find two different choices for f that have average value 20 on the rectangle, and give their contour diagrams on the rectangle.

62. The function $f(x, y) = ax^2 + bxy + cy^2$ has an average value of 20 on the square $0 \leq x \leq 2, 0 \leq y \leq 2$.

(a) What can you say about the constants a, b, and c?

(b) Find two different choices for f that have average value 20 on the square, and give their contour diagrams on the square.

63. Show that for a right triangle the average distance from any point in the triangle to one of the legs is one-third the length of the other leg. (The legs of a right triangle are the two sides that are not the hypotenuse.)

64. A rectangular plate of sides a and b is subjected to a normal force (that is, perpendicular to the plate). The pressure, p, at any point on the plate is proportional to the square of the distance of that point from one corner. Find the total force on the plate. [Note that pressure is force per unit area.]

Strengthen Your Understanding

In Problems 65–66, explain what is wrong with the statement.

65. $\int_0^1 \int_0^x f(x, y) \, dy \, dx = \int_0^1 \int_0^y f(x, y) \, dx \, dy$

66. $\int_0^1 \int_0^y xy \, dx \, dy = \int_0^y \int_0^1 xy \, dy \, dx$

In Problems 67–69, give an example of:

67. An iterated double integral, with limits of integration, giving the volume of a cylinder standing vertically with a circular base in the xy-plane.

68. A nonconstant function, f, whose integral is 4 over the triangular region with vertices $(0, 0), (1, 0), (1, 1)$.

69. A double integral representing the volume of a triangular prism of base area 6.

Are the statements in Problems 70–77 true or false? Give reasons for your answer.

70. The iterated integral $\int_0^1 \int_5^{12} f \, dx \, dy$ is computed over the rectangle $0 \leq x \leq 1, 5 \leq y \leq 12$.

71. If R is the region inside the triangle with vertices $(0, 0), (1, 1)$ and $(0, 2)$, then the double integral $\int_R f \, dA$ can be evaluated by an iterated integral of the form $\int_0^2 \int_0^1 f \, dx \, dy$.

72. The region of integration of the iterated integral $\int_1^2 \int_{x^2}^{x^3} f \, dy \, dx$ lies completely in the first quadrant (that is, $x \geq 0, y \geq 0$).

73. If the limits a, b, c and d in the iterated integral $\int_a^b \int_c^d f \, dy \, dx$ are all positive, then the value of $\int_a^b \int_c^d f \, dy \, dx$ is also positive.

74. If $f(x, y)$ is a function of y only, then $\int_a^b \int_0^1 f \, dx \, dy = \int_a^b f \, dy$.

75. If R is the region inside a circle of radius a, centered at the origin, then $\int_R f \, dA = \int_{-a}^a \int_0^{\sqrt{a^2 - x^2}} f \, dy \, dx$.

76. If $f(x, y) = g(x) \cdot h(y)$, where g and h are single-variable functions, then

$$\int_a^b \int_c^d f \, dy \, dx = \left(\int_a^b g(x) \, dx \right) \cdot \left(\int_c^d h(y) \, dy \right).$$

77. If $f(x, y) = g(x) + h(y)$, where g and h are single-variable functions, then

$$\int_a^b \int_c^d f \, dx \, dy = \left(\int_a^b g(x) \, dx \right) + \left(\int_c^d h(y) \, dy \right).$$

16.3 TRIPLE INTEGRALS

A continuous function of three variables can be integrated over a solid region W in 3-space in the same way as a function of two variables is integrated over a flat region in 2-space. Again, we start with a Riemann sum. First we subdivide W into smaller regions, then we multiply the volume of each region by a value of the function in that region, and then we add the results. For example, if W is the box $a \leq x \leq b$, $c \leq y \leq d$, $p \leq z \leq q$, then we subdivide each side into n, m, and l pieces, thereby chopping W into nml smaller boxes, as shown in Figure 16.22.

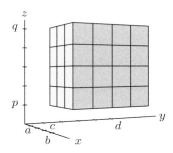

Figure 16.22: Subdividing a three-dimensional box

The volume of each smaller box is

$$\Delta V = \Delta x \Delta y \Delta z,$$

where $\Delta x = (b - a)/n$, and $\Delta y = (d - c)/m$, and $\Delta z = (q - p)/l$. Using this subdivision, we pick a point $(u_{ijk}, v_{ijk}, w_{ijk})$ in the ijk-th small box and construct a Riemann sum

$$\sum_{i,j,k} f(u_{ijk}, v_{ijk}, w_{ijk}) \, \Delta V.$$

If f is continuous, as Δx, Δy, and Δz approach 0, this Riemann sum approaches the definite integral, $\int_W f \, dV$, called a *triple integral*, which is defined as

$$\int_W f \, dV = \lim_{\Delta x, \Delta y, \Delta z \to 0} \sum_{i,j,k} f(u_{ijk}, v_{ijk}, w_{ijk}) \, \Delta x \, \Delta y \, \Delta z.$$

As in the case of a double integral, we can evaluate this integral as an iterated integral:

Triple integral as an iterated integral

$$\int_W f \, dV = \int_p^q \left(\int_c^d \left(\int_a^b f(x, y, z) \, dx \right) dy \right) dz,$$

where y and z are treated as constants in the innermost (dx) integral, and z is treated as a constant in the middle (dy) integral. Other orders of integration are possible.

Example 1   A cube C has sides of length 4 cm and is made of a material of variable density. If one corner is at the origin and the adjacent corners are on the positive x, y, and z axes, then the density at the point (x, y, z) is $\delta(x, y, z) = 1 + xyz$ gm/cm^3. Find the mass of the cube.

Solution   Consider a small piece ΔV of the cube, small enough so that the density remains close to constant over the piece. Then

$$\text{Mass of small piece} = \text{Density} \cdot \text{Volume} \approx \delta(x, y, z)\, \Delta V.$$

To get the total mass, we add the masses of the small pieces and take the limit as $\Delta V \to 0$. Thus, the mass is the triple integral

$$M = \int_C \delta\, dV = \int_0^4 \int_0^4 \int_0^4 (1 + xyz)\, dx\, dy\, dz = \int_0^4 \int_0^4 \left(x + \frac{1}{2}x^2 yz\right)\Bigg|_{x=0}^{x=4} dy\, dz$$

$$= \int_0^4 \int_0^4 (4 + 8yz)\, dy\, dz = \int_0^4 \left(4y + 4y^2 z\right)\Bigg|_{y=0}^{y=4} dz = \int_0^4 (16 + 64z)\, dz = 576\,\text{gm}.$$

Example 2   Express the volume of the building described in Example 1 on page 894 as a triple integral.

Solution   The building is given by $0 \le x \le 8, 0 \le y \le 16$, and $0 \le z \le 12 - x/4 - y/8$. (See Figure 16.23.) To find its volume, divide it into small cubes of volume $\Delta V = \Delta x\, \Delta y\, \Delta z$ and add. First, make a vertical stack of cubes above the point $(x, y, 0)$. This stack goes from $z = 0$ to $z = 12 - x/4 - y/8$, so

$$\text{Volume of vertical stack} \approx \sum_z \Delta V = \sum_z \Delta x\, \Delta y\, \Delta z = \left(\sum_z \Delta z\right)\Delta x\, \Delta y.$$

Next, line up these stacks parallel to the y-axis to form a slice from $y = 0$ to $y = 16$. So

$$\text{Volume of slice} \approx \left(\sum_y \sum_z \Delta z\, \Delta y\right)\Delta x.$$

Finally, line up the slices along the x-axis from $x = 0$ to $x = 8$ and add up their volumes, to get

$$\text{Volume of building} \approx \sum_x \sum_y \sum_z \Delta z\, \Delta y\, \Delta x.$$

Thus, in the limit,

$$\text{Volume of building} = \int_0^8 \int_0^{16} \int_0^{12 - x/4 - y/8} 1\, dz\, dy\, dx.$$

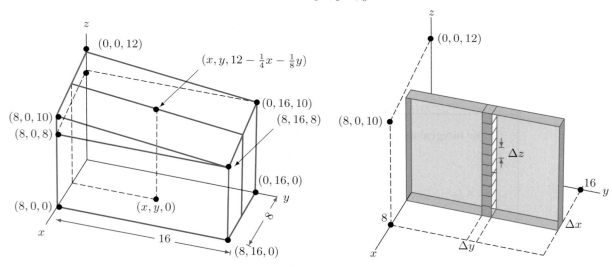

Figure 16.23: Volume of building (shown to left) divided into blocks and slabs for a triple integral

---

Example 3   Set up an iterated integral to compute the mass of the solid cone bounded by $z = \sqrt{x^2 + y^2}$ and $z = 3$, if the density is given by $\delta(x, y, z) = z$.

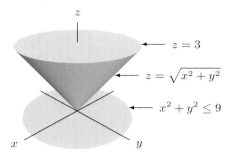

Figure 16.24

Solution   We break the cone in Figure 16.24 into small cubes of volume $\Delta V = \Delta x\, \Delta y\, \Delta z$, on which the density is approximately constant, and approximate the mass of each cube by $\delta(x, y, z)\, \Delta x\, \Delta y\, \Delta z$. Stacking the cubes vertically above the point $(x, y, 0)$, starting on the cone at height $z = \sqrt{x^2 + y^2}$ and going up to $z = 3$, tells us that the inner integral is

$$\int_{\sqrt{x^2+y^2}}^{3} \delta(x, y, z)\, dz = \int_{\sqrt{x^2+y^2}}^{3} z\, dz.$$

There is a stack for every point in the xy-plane in the shadow of the cone. The cone $z = \sqrt{x^2 + y^2}$ intersects the horizontal plane $z = 3$ in the circle $x^2 + y^2 = 9$, so there is a stack for all (x, y) in the region $x^2 + y^2 \leq 9$. Lining up the stacks parallel to the y-axis gives a slice from $y = -\sqrt{9 - x^2}$ to $y = \sqrt{9 - x^2}$, for each fixed value of x. Thus, the limits on the middle integral are

$$\int_{-\sqrt{9-x^2}}^{\sqrt{9-x^2}} \int_{\sqrt{x^2+y^2}}^{3} z\, dz\, dy.$$

Finally, there is a slice for each x between -3 and 3, so the integral we want is

$$\text{Mass} = \int_{-3}^{3} \int_{-\sqrt{9-x^2}}^{\sqrt{9-x^2}} \int_{\sqrt{x^2+y^2}}^{3} z\, dz\, dy\, dx.$$

Notice that setting up the limits on the two outer integrals was just like setting up the limits for a double integral over the region $x^2 + y^2 \leq 9$.

As the previous example illustrates, for a region W contained between two surfaces, the inner-most limits correspond to these surfaces. The middle and outer limits ensure that we integrate over the "shadow" of W in the xy-plane.

Limits on Triple Integrals

- The limits for the outer integral are constants.
- The limits for the middle integral can involve only one variable (that in the outer integral).
- The limits for the inner integral can involve two variables (those on the two outer integrals).

Exercises and Problems for Section 16.3

Exercises

Sketch the region of integration in Exercises 1–9.

1. $\displaystyle\int_{0}^{1} \int_{-1}^{1} \int_{0}^{\sqrt{1-x^2}} f(x, y, z)\, dz\, dx\, dy$

2. $\displaystyle\int_{0}^{1} \int_{-1}^{1} \int_{0}^{\sqrt{1-z^2}} f(x, y, z)\, dy\, dz\, dx$

3. $\displaystyle\int_{0}^{1} \int_{-1}^{1} \int_{-\sqrt{1-x^2}}^{\sqrt{1-x^2}} f(x, y, z)\, dz\, dx\, dy$

4. $\int_{-1}^{1}\int_{0}^{1}\int_{-\sqrt{1-z^2}}^{\sqrt{1-z^2}} f(x,y,z)\,dy\,dz\,dx$

5. $\int_{-1}^{1}\int_{-\sqrt{1-x^2}}^{\sqrt{1-x^2}}\int_{0}^{\sqrt{1-x^2-z^2}} f(x,y,z)\,dy\,dz\,dx$

6. $\int_{0}^{1}\int_{-\sqrt{1-z^2}}^{\sqrt{1-z^2}}\int_{0}^{\sqrt{1-x^2-z^2}} f(x,y,z)\,dy\,dx\,dz$

7. $\int_{0}^{1}\int_{0}^{\sqrt{1-y^2}}\int_{-\sqrt{1-x^2-y^2}}^{\sqrt{1-x^2-y^2}} f(x,y,z)\,dz\,dx\,dy$

8. $\int_{0}^{1}\int_{-\sqrt{1-z^2}}^{\sqrt{1-z^2}}\int_{-\sqrt{1-y^2-z^2}}^{\sqrt{1-y^2-z^2}} f(x,y,z)\,dx\,dy\,dz$

9. $\int_{0}^{1}\int_{0}^{\sqrt{1-z^2}}\int_{-\sqrt{1-x^2-z^2}}^{\sqrt{1-x^2-z^2}} f(x,y,z)\,dy\,dx\,dz$

In Exercises 10–13, find the triple integrals of the function over the region W.

10. $f(x,y,z) = x^2 + 5y^2 - z$, W is the rectangular box $0 \le x \le 2, -1 \le y \le 1, 2 \le z \le 3$.

11. $h(x,y,z) = ax + by + cz$, W is the rectangular box $0 \le x \le 1, 0 \le y \le 1, 0 \le z \le 2$.

12. $f(x,y,z) = \sin x \cos(y+z)$, W is the cube $0 \le x \le \pi$, $0 \le y \le \pi, 0 \le z \le \pi$.

13. $f(x,y,z) = e^{-x-y-z}$, W is the rectangular box with corners at $(0,0,0)$, $(a,0,0)$, $(0,b,0)$, and $(0,0,c)$.

Problems

In Problems 14–19, write a triple integral, including limits of integration, that gives the specified volume.

14. Between $z = x + y$ and $z = 1 + 2x + 2y$ and above $0 \le x \le 1, 0 \le y \le 2$.

15. Between the paraboloid $z = x^2 + y^2$ and the sphere $x^2 + y^2 + z^2 = 4$ and above the disk $x^2 + y^2 \le 1$.

16. Between $2x + 2y + z = 6$ and $3x + 4y + z = 6$ and above $x + y \le 1, x \ge 0, y \ge 0$.

17. Under the sphere $x^2 + y^2 + z^2 = 9$ and above the region between $y = x$ and $y = 2x - 2$ in the xy-plane in the first quadrant.

18. Between the top portion of the sphere $x^2 + y^2 + z^2 = 9$ and the plane $z = 2$.

19. Under the sphere $x^2 + y^2 + z^2 = 4$ and above the region $x^2 + y^2 \le 4, 0 \le x \le 1, 0 \le y \le 2$ in the xy-plane.

In Problems 20–23, write limits of integration for the integral $\int_W f(x,y,z)\,dV$ where W is the quarter or half sphere or cylinder shown.

20.

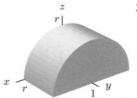

21.

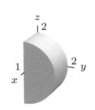

22.

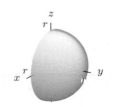

23.

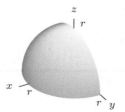

In Problems 24–28, decide whether the integrals are positive, negative, or zero. Let S be the solid sphere $x^2 + y^2 + z^2 \le 1$, and T be the top half of this sphere (with $z \ge 0$), and B be the bottom half (with $z \le 0$), and R be the right half of the sphere (with $x \ge 0$), and L be the left half (with $x \le 0$).

24. $\int_T e^z\,dV$

25. $\int_B e^z\,dV$

26. $\int_S \sin z\,dV$

27. $\int_T \sin z\,dV$

28. $\int_R \sin z\,dV$

Let W be the solid cone bounded by $z = \sqrt{x^2 + y^2}$ and $z = 2$. For Problems 29–37, decide (without calculating its value) whether the integral is positive, negative, or zero.

29. $\int_W y\,dV$

30. $\int_W x\,dV$

31. $\int_W z\,dV$

32. $\int_W xy\,dV$

33. $\int_W xyz\,dV$

34. $\int_W (z-2)\,dV$

35. $\int_W \sqrt{x^2 + y^2}\,dV$

36. $\int_W e^{-xyz}\,dV$

37. $\int_W (z - \sqrt{x^2 + y^2})\,dV$

38. Find the volume of the region bounded by the planes $z = 3y, z = y, y = 1, x = 1$, and $x = 2$.

39. Find the volume of the region bounded by $z = x^2$, $0 \le x \le 5$, and the planes $y = 0, y = 3$, and $z = 0$.

40. Find the volume of the region in the first octant bounded by the coordinate planes and the surface $x + y + z = 2$.

41. A trough with triangular cross-section lies along the x-axis for $0 \leq x \leq 10$. The slanted sides are given by $z = y$ and $z = -y$ for $0 \leq z \leq 1$ and the ends by $x = 0$ and $x = 10$, where x, y, z are in meters. The trough contains a sludge whose density at the point (x, y, z) is $\delta = e^{-3x}$ kg per m³.

    (a) Express the total mass of sludge in the trough in terms of triple integrals.
    (b) Find the mass.

42. Find the volume of the region bounded by $z = x+y$, $z = 10$, and the planes $x = 0$, $y = 0$.

43. Find the volume of the region between the plane $z = x$ and the surface $z = x^2$, and the planes $y = 0$, and $y = 3$.

44. Find the volume of the region bounded by $z = x + y$, $0 \leq x \leq 5$, $0 \leq y \leq 5$, and the planes $x = 0$, $y = 0$, and $z = 0$.

45. Find the volume of the pyramid with base in the plane $z = -6$ and sides formed by the three planes $y = 0$ and $y - x = 4$ and $2x + y + z = 4$.

46. Find the volume between the planes $z = 1 + x + y$ and $x + y + z = 1$ and above the triangle $x + y \leq 1$, $x \geq 0$, $y \geq 0$ in the xy-plane.

47. Find the volume between the plane $x + y + z = 1$ and the xy-plane, for $x + y \leq 2$, $x \geq 0$, $y \geq 0$.

48. A solid shaped like a wedge of cheese has as its base the xy-plane, bounded by the x-axis, the line $y = x$ and the line $x + y = 1$. Its sides are vertical, and its top is the plane $x + y + z = 2$. At any point, the density of the solid is four times the distance from the xy-plane.

    (a) Express the mass of the region in terms of triple integrals.
    (b) Find the mass.

49. Find the mass of a triangular-shaped solid bounded by the planes $z = 1 + x$, $z = 1 - x$, $z = 0$, and with $0 \leq y \leq 3$. The density is $\delta = 10 - z$ gm/(cm)³, and x, y, z are in cm.

50. Find the mass of the solid bounded by the xy-plane, yz-plane, xz-plane, and the plane $(x/3) + (y/2) + (z/6) = 1$, if the density of the solid is given by $\delta(x, y, z) = x+y$.

51. Find the mass of the pyramid with base in the plane $z = -6$ and sides formed by the three planes $y = 0$ and $y - x = 4$ and $2x + y + z = 4$, if the density of the solid is given by $\delta(x, y, z) = y$.

52. Let E be the solid pyramid bounded by the planes $x + z = 6$, $x - z = 0$, $y + z = 6$, $y - z = 0$, and above the plane $z = 0$ (see Figure 16.25). The density at any point in the pyramid is given by $\delta(x, y, z) = z$ grams per cm³, where x, y, and z are measured in cm.

    (a) Explain in practical terms what the triple integral $\int_E z \, dV$ represents.

(b) In evaluating the integral from part (a), how many separate triple integrals would be required if we chose to integrate in the z-direction first?
(c) Evaluate the triple integral from part (a) by integrating in a well-chosen order.

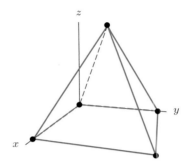

Figure 16.25

53. (a) What is the equation of the plane passing through the points $(1, 0, 0)$, $(0, 1, 0)$, and $(0, 0, 1)$?
    (b) Find the volume of the region bounded by this plane and the planes $x = 0$, $y = 0$, and $z = 0$.

54. Figure 16.26 shows part of a spherical ball of radius 5 cm. Write an iterated triple integral which represents the volume of this region.

2 cm

Figure 16.26

55. A solid region D is a half cylinder of radius 1 lying horizontally with its rectangular base in the xy-plane and its axis along the y-axis from $y = 0$ to $y = 10$. (The region is above the xy-plane.)

    (a) What is the equation of the curved surface of this half cylinder?
    (b) Write the limits of integration of the integral $\int_D f(x, y, z) \, dV$ in Cartesian coordinates.

56. Set up, but do not evaluate, an iterated integral for the volume of the solid formed by the intersections of the cylinders $x^2 + z^2 = 1$ and $y^2 + z^2 = 1$.

Problems 57–59 refer to Figure 16.27, which shows E, the region in the first octant bounded by the parabolic cylinder $z = 6y^2$ and the elliptical cylinder $x^2 + 3y^2 = 12$. For the given order of integration, write an iterated integral equivalent to the triple integral $\int_E f(x, y, z) \, dV$.

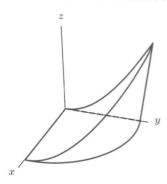

Figure 16.27

57. $dz \, dx \, dy$ **58.** $dx \, dz \, dy$ **59.** $dy \, dz \, dx$

60. Find the average value of the sum of the squares of three numbers x, y, z, where each number is between 0 and 2.

61. Let E be the region in the first octant bounded between the plane $x + 2y + z = 4$, the parabolic cylinder $x = 2y^2$, and the coordinate planes (see Figure 16.28). For each of the following orders of integration, write down an iterated integral equivalent to the triple integral $\int_E f(x, y, z) \, dV$.

(a) $dz \, dy \, dx$
(b) $dy \, dz \, dx$

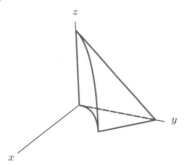

Figure 16.28

Problems 62–63 concern the *center of mass*, the point at which the mass of a solid body in motion can be considered to be concentrated. If the object has density $\rho(x, y, z)$ at the point (x, y, z) and occupies a region W, then the coordinates $(\bar{x}, \bar{y}, \bar{z})$ of the center of mass are given by

$$\bar{x} = \frac{1}{m} \int_W x\rho \, dV \quad \bar{y} = \frac{1}{m} \int_W y\rho \, dV \quad \bar{z} = \frac{1}{m} \int_W z\rho \, dV$$

where $m = \int_W \rho \, dV$ is the total mass of the body.

62. A solid is bounded below by the square $z = 0, 0 \le x \le 1, 0 \le y \le 1$ and above by the surface $z = x + y + 1$. Find the total mass and the coordinates of the center of mass if the density is 1 gm/cm^3 and x, y, z are measured in centimeters.

63. Find the center of mass of the tetrahedron that is bounded by the xy, yz, xz planes and the plane $x + 2y + 3z = 1$. Assume the density is 1 gm/cm^3 and x, y, z are in centimeters.

Problems 64–66 concern a rotating solid body and its *moment of inertia* about an axis; this moment relates angular acceleration to torque (an analogue of force). For a body of constant density and mass m occupying a region W of volume V, the moments of inertia about the coordinate axes are

$$I_x = \frac{m}{V} \int_W (y^2 + z^2) \, dV \qquad I_y = \frac{m}{V} \int_W (x^2 + z^2) \, dV$$

$$I_z = \frac{m}{V} \int_W (x^2 + y^2) \, dV.$$

64. Find the moment of inertia about the z-axis of the rectangular solid of mass m given by $0 \le x \le 1, 0 \le y \le 2, 0 \le z \le 3$.

65. Find the moment of inertia about the x-axis of the rectangular solid $-a \le x \le a, -b \le y \le b$ and $-c \le z \le c$ of mass m.

66. Let a, b, and c denote the moments of inertia of a homogeneous solid object about the x, y and z-axes respectively. Explain why $a + b > c$.

Strengthen Your Understanding

In Problems 67–68, explain what is wrong with the statement.

67. Let S be the solid sphere $x^2 + y^2 + z^2 \le 1$ and let U be the upper half of S where $z \ge 0$. Then $\int_S f(x, y, z) \, dV = 2 \int_U f(x, y, z) \, dV$.

68. $\int_0^1 \int_0^x \int_0^y f(x, y, z) \, dz \, dy \, dx = \int_0^1 \int_y^1 \int_0^x f(x, y, z) \, dz \, dx \, dy$

In Problems 69–70, give an example of:

69. A function f such that $\int_R f \, dV = 7$, where R is the cylinder $x^2 + y^2 \le 4, 0 \le z \le 3$.

70. A nonconstant function $f(x, y, z)$ such that if B is the region enclosed by the sphere of radius 1 centered at the origin, the integral $\int_B f(x, y, z) \, dx \, dy \, dz$ is zero.

Are the statements in Problems 71–80 true or false? Give reasons for your answer.

71. If $\rho(x, y, z)$ is mass density of a material in 3-space, then $\int_W \rho(x, y, z) \, dV$ gives the volume of the solid region W.

72. The region of integration of the triple iterated integral

$\int_0^1 \int_0^1 \int_0^x f \, dz \, dy \, dx$ lies above a square in the xy-plane and below a plane.

73. If W is the entire unit ball $x^2 + y^2 + z^2 \leq 1$ then an iterated integral over W has limits $\int_0^1 \int_0^{\sqrt{1-x^2}} \int_0^{\sqrt{1-x^2-y^2}} f \, dz \, dy \, dx$.

74. The iterated integrals $\int_0^1 \int_0^{1-x} \int_0^{1-x-y} f \, dz \, dy \, dx$ and $\int_0^1 \int_0^{1-z} \int_0^{1-y-z} f \, dx \, dy \, dz$ are equal.

75. The iterated integrals $\int_{-1}^1 \int_0^1 \int_0^{1-x^2} f \, dz \, dy \, dx$ and $\int_0^1 \int_0^1 \int_{-\sqrt{1-z}}^{\sqrt{1-z}} f \, dx \, dy \, dz$ are equal.

76. If W is a rectangular solid in 3-space, then $\int_W f \, dV = \int_a^b \int_c^d \int_e^k f \, dz \, dy \, dx$, where $a, b, c, d, e,$ and k are constants.

77. If W is the unit cube $0 \leq x \leq 1, 0 \leq y \leq 1, 0 \leq z \leq 1$ and $\int_W f \, dV = 0$, then $f = 0$ everywhere in the unit cube.

78. If $f > g$ at all points in the solid region W, then $\int_W f \, dV > \int_W g \, dV$.

79. If W_1 and W_2 are solid regions with volume$(W_1) >$ volume(W_2) then $\int_{W_1} f \, dV > \int_{W_2} f \, dV$.

80. Both double and triple integrals can be used to compute volume.

16.4 DOUBLE INTEGRALS IN POLAR COORDINATES

Integration in Polar Coordinates

We started this chapter by putting a rectangular grid on the fox population density map, to estimate the total population using a Riemann sum. However, sometimes a polar grid is more appropriate.

Example 1   A biologist studying insect populations around a circular lake divides the area into the polar sectors in Figure 16.29. The approximate population density in each sector is shown in millions per square km. Estimate the total insect population around the lake.

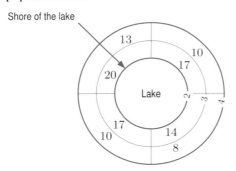

Figure 16.29: An insect-infested lake showing the insect population density by sector

Solution   To get the estimate, we multiply the population density in each sector by the area of that sector. Unlike the rectangles in a rectangular grid, the sectors in this grid do not all have the same area. The inner sectors have area

$$\frac{1}{4}(\pi 3^2 - \pi 2^2) = \frac{5\pi}{4} \approx 3.93 \text{ km}^2,$$

and the outer sectors have area

$$\frac{1}{4}(\pi 4^2 - \pi 3^2) = \frac{7\pi}{4} \approx 5.50 \text{ km}^2,$$

so we estimate

$$\text{Population} \approx (20)(3.93) + (17)(3.93) + (14)(3.93) + (17)(3.93) +$$
$$(13)(5.50) + (10)(5.50) + (8)(5.50) + (10)(5.50)$$
$$= 492.74 \text{ million insects.}$$

What Is dA in Polar Coordinates?

The previous example used a polar grid rather than a rectangular grid. A rectangular grid is constructed from vertical and horizontal lines of the form $x = k$ (a constant) and $y = l$ (another constant). In polar coordinates, $r = k$ gives a circle of radius k centred at the origin and $\theta = l$ gives a ray emanating from the origin (at angle l with the x-axis). A polar grid is built out of these circles and rays. Suppose we want to integrate $f(r, \theta)$ over the region R in Figure 16.30.

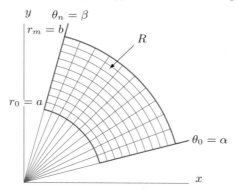

Figure 16.30: Dividing up a region using a polar grid

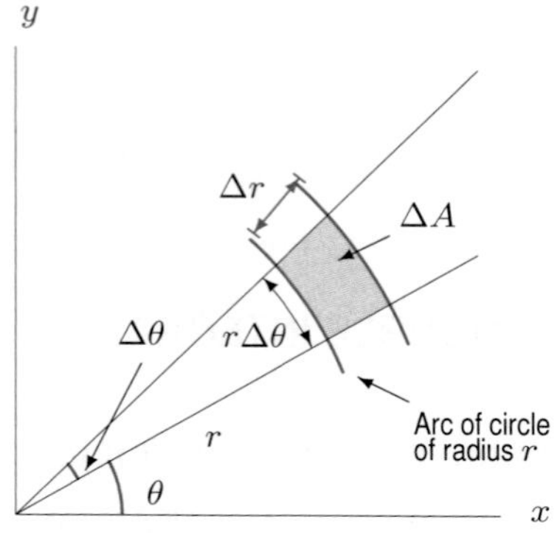

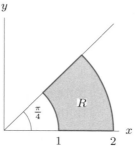

Figure 16.31: Calculating area ΔA in polar coordinates

Choosing (r_{ij}, θ_{ij}) in the ij-th bent rectangle in Figure 16.30 gives a Riemann sum:

$$\sum_{i,j} f(r_{ij}, \theta_{ij}) \, \Delta A.$$

To calculate the area ΔA, look at Figure 16.31. If Δr and $\Delta \theta$ are small, the shaded region is approximately a rectangle with sides $r \, \Delta \theta$ and Δr, so

$$\Delta A \approx r \Delta \theta \Delta r.$$

Thus, the Riemann sum is approximately

$$\sum_{i,j} f(r_{ij}, \theta_{ij}) \, r_{ij} \, \Delta \theta \, \Delta r.$$

If we take the limit as Δr and $\Delta \theta$ approach 0, we obtain

$$\int_R f \, dA = \int_\alpha^\beta \int_a^b f(r, \theta) \, r \, dr \, d\theta.$$

When computing integrals in polar coordinates, use $x = r \cos \theta, y = r \sin \theta, x^2 + y^2 = r^2$.
Put $dA = r \, dr \, d\theta$ or $dA = r \, d\theta \, dr$.

Example 2　　Compute the integral of $f(x, y) = 1/(x^2 + y^2)^{3/2}$ over the region R shown in Figure 16.32.

Figure 16.32: Integrate f over the polar region

Solution   The region R is described by the inequalities $1 \leq r \leq 2,\ 0 \leq \theta \leq \pi/4$. In polar coordinates, $r = \sqrt{x^2 + y^2}$, so we can write f as

$$f(x, y) = \frac{1}{(x^2 + y^2)^{3/2}} = \frac{1}{(r^2)^{3/2}} = \frac{1}{r^3}.$$

Then

$$\int_R f\, dA = \int_0^{\pi/4} \int_1^2 \frac{1}{r^3} r\, dr\, d\theta = \int_0^{\pi/4} \left(\int_1^2 r^{-2}\, dr \right) d\theta$$
$$= \int_0^{\pi/4} \left. -\frac{1}{r} \right|_{r=1}^{r=2} d\theta = \int_0^{\pi/4} \frac{1}{2}\, d\theta = \frac{\pi}{8}.$$

Example 3   For each region in Figure 16.33, decide whether to integrate using polar or Cartesian coordinates. On the basis of its shape, write an iterated integral of an arbitrary function $f(x, y)$ over the region.

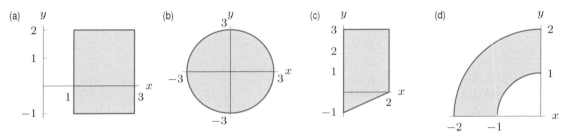

Figure 16.33

Solution   (a) Since this is a rectangular region, Cartesian coordinates are likely to be a better choice. The rectangle is described by the inequalities $1 \leq x \leq 3$ and $-1 \leq y \leq 2$, so the integral is

$$\int_{-1}^2 \int_1^3 f(x, y)\, dx\, dy.$$

(b) A circle is best described in polar coordinates. The radius is 3, so r goes from 0 to 3, and to describe the whole circle, θ goes from 0 to 2π. The integral is

$$\int_0^{2\pi} \int_0^3 f(r\cos\theta, r\sin\theta)\, r\, dr\, d\theta.$$

(c) The bottom boundary of this trapezoid is the line $y = (x/2) - 1$ and the top is the line $y = 3$, so we use Cartesian coordinates. If we integrate with respect to y first, the lower limit of the integral is $(x/2) - 1$ and the upper limit is 3. The x limits are $x = 0$ to $x = 2$. So the integral is

$$\int_0^2 \int_{(x/2)-1}^3 f(x, y)\, dy\, dx.$$

(d) This is another polar region: it is a piece of a ring in which r goes from 1 to 2. Since it is in the second quadrant, θ goes from $\pi/2$ to π. The integral is

$$\int_{\pi/2}^\pi \int_1^2 f(r\cos\theta, r\sin\theta)\, r\, dr\, d\theta.$$

Exercises and Problems for Section 16.4

Exercises

For the regions R in Exercises 1–4, write $\int_R f \, dA$ as an iterated integral in polar coordinates.

1.

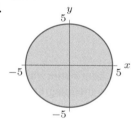

2.

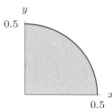

3.

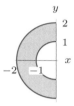

4.

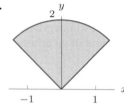

In Exercises 5–8, choose rectangular or polar coordinates to set up an iterated integral of an arbitrary function $f(x, y)$ over the region.

5.

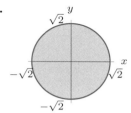

6.

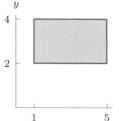

7.

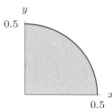

8.   −4 −2 y 2   4

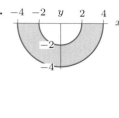

Sketch the region of integration in Exercises 9–15.

9. $\displaystyle\int_{\pi/2}^{\pi} \int_0^1 f(r, \theta)\, r\, dr\, d\theta$

10. $\displaystyle\int_0^4 \int_{-\pi/2}^{\pi/2} f(r, \theta)\, r\, d\theta\, dr$

11. $\displaystyle\int_{\pi/6}^{\pi/3} \int_0^1 f(r, \theta)\, r\, dr\, d\theta$

12. $\displaystyle\int_0^{2\pi} \int_1^2 f(r, \theta)\, r\, dr\, d\theta$

13. $\displaystyle\int_3^4 \int_{3\pi/4}^{3\pi/2} f(r, \theta)\, r\, d\theta\, dr$

14. $\displaystyle\int_0^{\pi/4} \int_0^{1/\cos\theta} f(r, \theta)\, r\, dr\, d\theta$

15. $\displaystyle\int_{\pi/4}^{\pi/2} \int_0^{2/\sin\theta} f(r, \theta)\, r\, dr\, d\theta$

Problems

In Exercises 16–18, evaluate the integral.

16. $\int_R \sin(x^2 + y^2)\, dA$, where R is the disk of radius 2 centered at the origin.

17. $\int_R (x^2 - y^2)\, dA$, where R is the first quadrant region between the circles of radius 1 and radius 2.

18. $\int_R \sqrt{x^2 + y^2}\, dx\, dy$ where R is $4 \le x^2 + y^2 \le 9$.

19. Consider the integral $\int_0^3 \int_{x/3}^1 f(x, y)\, dy\, dx$.

  (a) Sketch the region R over which the integration is being performed.
  (b) Rewrite the integral with the order of integration reversed.
  (c) Rewrite the integral in polar coordinates.

Convert the integrals in Problems 20–22 to polar coordinates and evaluate.

20. $\displaystyle\int_{-1}^0 \int_{-\sqrt{1-x^2}}^{\sqrt{1-x^2}} x\, dy\, dx$

21. $\displaystyle\int_0^{\sqrt6} \int_{-x}^x dy\, dx$

22. $\displaystyle\int_0^{\sqrt2} \int_y^{\sqrt{4-y^2}} xy\, dx\, dy$

23. Find the volume of the region between the graph of $f(x, y) = 25 - x^2 - y^2$ and the xy plane.

24. Find the volume of an ice cream cone bounded by the hemisphere $z = \sqrt{8 - x^2 - y^2}$ and the cone $z = \sqrt{x^2 + y^2}$.

25. A city surrounds a bay as shown in Figure 16.34. The population density of the city (in thousands of people per square km) is $\delta(r, \theta)$, where r and θ are polar coordinates and distances are in km.

  (a) Set up an iterated integral in polar coordinates giving the total population of the city.

(b) The population density decreases the farther you live from the shoreline of the bay; it also decreases the farther you live from the ocean. Which of the following functions best describes this situation?

   (i) $\delta(r, \theta) = (4 - r)(2 + \cos\theta)$
   (ii) $\delta(r, \theta) = (4 - r)(2 + \sin\theta)$
   (iii) $\delta(r, \theta) = (r + 4)(2 + \cos\theta)$

(c) Estimate the population using your answers to parts (a) and (b).

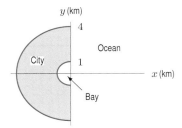

Figure 16.34

26. (a) For $a > 0$, find the volume under the graph of $z = e^{-(x^2+y^2)}$ above the disk $x^2 + y^2 \leq a^2$.

(b) What happens to the volume as $a \to \infty$?

27. A disk of radius 5 cm has density 10 gm/cm^2 at its center and density 0 at its edge, and its density is a linear function of the distance from the center. Find the mass of the disk.

28. (a) Use integration in the following coordinates to find the volume of a solid orange wedge with $x \geq 0$ and cut out by the planes $y = 0$, $y = x/\sqrt{3}$, and a sphere of radius 5 centered at the origin. Which coordinates are the most efficient?

   (i) Spherical coordinates
   (ii) Cylindrical coordinates, in two different ways

(b) Calculate the volume without integration

29. Evaluate the integral by converting it into Cartesian coordinates:
$$\int_0^{\pi/6} \int_0^{2/\cos\theta} r \, dr \, d\theta.$$

30. (a) Sketch the region of integration of
$$\int_0^1 \int_{\sqrt{1-x^2}}^{\sqrt{4-x^2}} x \, dy \, dx + \int_1^2 \int_0^{\sqrt{4-x^2}} x \, dy \, dx$$

(b) Evaluate the quantity in part (a).

31. A circular metal disk of radius 3 lies in the xy-plane with its center at the origin. At a distance r from the origin, the density of the metal per unit area is $\delta = \dfrac{1}{r^2 + 1}$.

(a) Write a double integral giving the total mass of the disk. Include limits of integration.

(b) Evaluate the integral.

32. Electric charge is distributed over the xy-plane, with density inversely proportional to the distance from the origin. Show that the total charge inside a circle of radius R centered at the origin is proportional to R. What is the constant of proportionality?

33. (a) Graph $r = 1/(2\cos\theta)$ for $-\pi/2 \leq \theta \leq \pi/2$ and $r = 1$.

(b) Write an iterated integral representing the area inside the curve $r = 1$ and to the right of $r = 1/(2\cos\theta)$. Evaluate the integral.

34. (a) Sketch the circles $r = 2\cos\theta$ for $-\pi/2 \leq \theta \leq \pi/2$ and $r = 1$.

(b) Write an iterated integral representing the area inside the circle $r = 2\cos\theta$ and outside the circle $r = 1$. Evaluate the integral.

35. Two circular disks, each of radius 1, have centers which are 1 unit apart. Write, but do not evaluate, a double integral, including limits of integration, giving the area of overlap of the disks in

(a) Cartesian coordinates **(b)** Polar coordinates

36. Find the area inside the curve $r = 2 + 3\cos\theta$ and outside the circle $r = 2$.

Strengthen Your Understanding

In Problems 37–38, explain what is wrong with the statement.

37. If R is the region bounded by $x = 1$, $y = 0$, $y = x$, then in polar coordinates $\int_R x \, dA = \int_0^{\pi/4} \int_0^1 r^2 \cos\theta \, dr \, d\theta$.

38. If R is the region $x^2 + y^2 \leq 4$, then $\int_R (x^2 + y^2) \, dA = \int_0^{2\pi} \int_0^2 r^2 \, dr \, d\theta$.

In Problems 39–40, give an example of:

39. A region R of integration in the first quadrant which suggests the use of polar coordinates.

40. An integrand $f(x, y)$ that suggests the use of polar coordinates.

41. Which of the following integrals give the area of the unit circle?

(a) $\int_{-1}^1 \int_{-\sqrt{1-x^2}}^{\sqrt{1-x^2}} dy \, dx$ **(b)** $\int_{-1}^1 \int_{-\sqrt{1-x^2}}^{\sqrt{1-x^2}} x \, dy \, dx$

(c) $\int_0^{2\pi} \int_0^1 r \, dr \, d\theta$ **(d)** $\int_0^{2\pi} \int_0^1 dr \, d\theta$

(e) $\int_0^1 \int_0^{2\pi} r \, d\theta \, dr$ **(f)** $\int_0^1 \int_0^{2\pi} d\theta \, dr$

42. Describe the region of integration for $\int_{\pi/4}^{\pi/2} \int_{1/\sin\theta}^{4/\sin\theta} f(r, \theta) r \, dr \, d\theta$.

16.5   INTEGRALS IN CYLINDRICAL AND SPHERICAL COORDINATES

Some double integrals are easier to evaluate in polar, rather than Cartesian, coordinates. Similarly, some triple integrals are easier in non-Cartesian coordinates.

Cylindrical Coordinates

The cylindrical coordinates of a point (x, y, z) in 3-space are obtained by representing the x and y coordinates in polar coordinates and letting the z-coordinate be the z-coordinate of the Cartesian coordinate system. (See Figure 16.35.)

Relation Between Cartesian and Cylindrical Coordinates

Each point in 3-space is represented using $0 \leq r < \infty, 0 \leq \theta \leq 2\pi, -\infty < z < \infty$.

$$x = r \cos \theta,$$
$$y = r \sin \theta,$$
$$z = z.$$

As with polar coordinates in the plane, note that $x^2 + y^2 = r^2$.

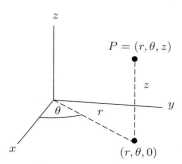

Figure 16.35: Cylindrical
coordinates: (r, θ, z)

A useful way to visualise cylindrical coordinates is to sketch the surfaces obtained by setting one of the coordinates equal to a constant. See Figures 16.36–16.38.

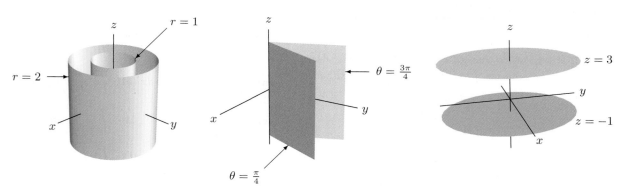

Figure 16.36: The surfaces $r = 1$ and $r = 2$

Figure 16.37: The surfaces $\theta = \pi/4$ and $\theta = 3\pi/4$

Figure 16.38: The surfaces $z = -1$ and $z = 3$

Setting $r = c$ (where c is constant) gives a cylinder around the z-axis whose radius is c. Setting $\theta = c$ gives a half-plane perpendicular to the xy plane, with one edge along the z-axis, making an angle c with the x-axis. Setting $z = c$ gives a horizontal plane $|c|$ units from the xy-plane. We call these *fundamental surfaces*.

The regions that can most easily be described in cylindrical coordinates are those regions whose boundaries are such fundamental surfaces. (For example, vertical cylinders, or wedge-shaped parts of vertical cylinders.)

Example 1   Describe in cylindrical coordinates a wedge of cheese cut from a cylinder 4 cm high and 6 cm in radius; this wedge subtends an angle of $\pi/6$ at the centre. (See Figure 16.39.)

Solution   The wedge is described by the inequalities $0 \leq r \leq 6$, and $0 \leq z \leq 4$, and $0 \leq \theta \leq \pi/6$.

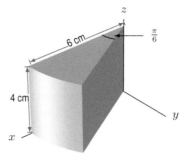

Figure 16.39: A wedge of cheese

Integration in Cylindrical Coordinates

To integrate in polar coordinates, we had to express the area element dA in terms of polar coordinates: $dA = r\, dr\, d\theta$. To evaluate a triple integral $\int_W f\, dV$ in cylindrical coordinates, we need to express the volume element dV in cylindrical coordinates.

In Figure 16.40, consider the volume element ΔV bounded by fundamental surfaces. The area of the base is $\Delta A \approx r\Delta r\Delta\theta$. Since the height is Δz, the volume element is given approximately by $\Delta V \approx r\, \Delta r\, \Delta\theta\, \Delta z$.

> When computing integrals in cylindrical coordinates, put $dV = r\, dr\, d\theta\, dz$. Other orders of integration are also possible.

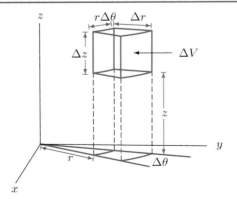

Figure 16.40: Volume element in cylindrical coordinates

Example 2   Find the mass of the wedge of cheese in Example 1, if its density is 1.2 grams/cm^3.

Solution   If the wedge is W, its mass is

$$\int_W 1.2 \, dV.$$

In cylindrical coordinates this integral is

$$\int_0^4 \int_0^{\pi/6} \int_0^6 1.2 \, r \, dr \, d\theta \, dz = \int_0^4 \int_0^{\pi/6} 0.6r^2 \Big|_0^6 d\theta \, dz = 21.6 \int_0^4 \int_0^{\pi/6} d\theta \, dz$$

$$= 21.6 \left(\frac{\pi}{6}\right) 4 = 45.239 \text{ grams}.$$

Example 3   A water tank in the shape of a hemisphere has radius a; its base is its plane face. Find the volume, V, of water in the tank as a function of h, the depth of the water.

Solution   In Cartesian coordinates, a sphere of radius a has the equation $x^2 + y^2 + z^2 = a^2$. (See Figure 16.41.) In cylindrical coordinates, $r^2 = x^2 + y^2$, so this becomes

$$r^2 + z^2 = a^2.$$

Thus, if we want to describe the amount of water in the tank in cylindrical coordinates, we let r go from 0 to $\sqrt{a^2 - z^2}$, we let θ go from 0 to 2π, and we let z go from 0 to h, giving

$$\begin{aligned}
\text{Volume} \atop \text{of water} &= \int_W dV = \int_0^{2\pi} \int_0^h \int_0^{\sqrt{a^2-z^2}} r \, dr \, dz \, d\theta = \int_0^{2\pi} \int_0^h \frac{r^2}{2} \Big|_{r=0}^{r=\sqrt{a^2-z^2}} dz \, d\theta \\
&= \int_0^{2\pi} \int_0^h \frac{1}{2}(a^2 - z^2) \, dz \, d\theta = \int_0^{2\pi} \frac{1}{2}\left(a^2 z - \frac{z^3}{3}\right) \Big|_{z=0}^{z=h} d\theta \\
&= \int_0^{2\pi} \frac{1}{2}\left(a^2 h - \frac{h^3}{3}\right) d\theta = \pi\left(a^2 h - \frac{h^3}{3}\right).
\end{aligned}$$

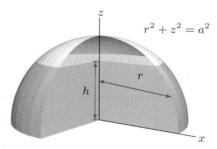

Figure 16.41: Hemispherical water tank with radius a and water of depth h

Spherical Coordinates

In Figure 16.42, the point P has coordinates (x, y, z) in the Cartesian coordinate system. We define spherical coordinates ρ, ϕ, and θ for P as follows: $\rho = \sqrt{x^2 + y^2 + z^2}$ is the distance of P from the origin; ϕ is the angle between the positive z-axis and the line through the origin and the point P; and θ is the same as in cylindrical coordinates.

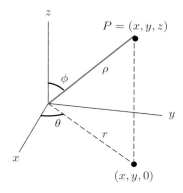

Figure 16.42: Spherical coordinates: (ρ, ϕ, θ)

In cylindrical coordinates,

$$x = r\cos\theta, \quad \text{and} \quad y = r\sin\theta, \quad \text{and} \quad z = z.$$

From Figure 16.42 we have $z = \rho\cos\phi$ and $r = \rho\sin\phi$, giving the following relationship:

Relation Between Cartesian and Spherical Coordinates

Each point in 3-space is represented using $0 \le \rho < \infty$, $0 \le \phi \le \pi$, and $0 \le \theta \le 2\pi$.

$$x = \rho\sin\phi\cos\theta$$
$$y = \rho\sin\phi\sin\theta$$
$$z = \rho\cos\phi.$$

Also, $\rho^2 = x^2 + y^2 + z^2$.

This system of coordinates is useful when there is spherical symmetry with respect to the origin, either in the region of integration or in the integrand. The fundamental surfaces in spherical coordinates are $\rho = k$ (a constant), which is a sphere of radius k centred at the origin, $\theta = k$ (a constant), which is the half-plane with its edge along the z-axis, and $\phi = k$ (a constant), which is a cone if $k \ne \pi/2$ and the xy-plane if $k = \pi/2$. (See Figures 16.43–16.45.)

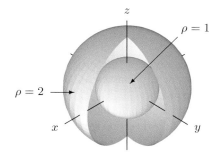

Figure 16.43: The surfaces $\rho = 1$ and $\rho = 2$

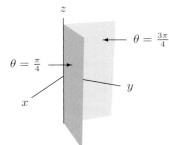

Figure 16.44: The surfaces $\theta = \pi/4$ and $\theta = 3\pi/4$

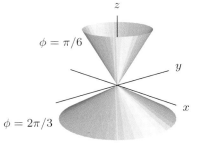

Figure 16.45: The surfaces $\phi = \pi/6$ and $\phi = 2\pi/3$

Integration in Spherical Coordinates

To use spherical coordinates in triple integrals we need to express the volume element, dV, in spherical coordinates. From Figure 16.46, we see that the volume element can be approximated by a box with curved edges. One edge has length $\Delta\rho$. The edge parallel to the xy-plane is an arc of a circle made from rotating the cylindrical radius r ($= \rho\sin\phi$) through an angle $\Delta\theta$, and so has length $\rho\sin\phi\,\Delta\theta$. The remaining edge comes from rotating the radius ρ through an angle $\Delta\phi$, and so has length $\rho\,\Delta\phi$. Therefore, $\Delta V \approx \Delta\rho(\rho\,\Delta\phi)(\rho\sin\phi\,\Delta\theta) = \rho^2\sin\phi\,\Delta\rho\,\Delta\phi\,\Delta\theta$.

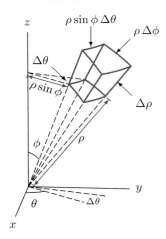

Figure 16.46: Volume element in spherical coordinates

Thus,

> When computing integrals in spherical coordinates, put $dV = \rho^2 \sin \phi \, d\rho \, d\phi \, d\theta$. Other orders of integration are also possible.

Example 4   Use spherical coordinates to derive the formula for the volume of a ball of radius a.

Solution   In spherical coordinates, a ball of radius a is described by the inequalities $0 \leq \rho \leq a$, $0 \leq \theta \leq 2\pi$, and $0 \leq \phi \leq \pi$. Note that θ goes from 0 to 2π, whereas ϕ goes from 0 to π. We find the volume by integrating the constant density function 1 over the ball:

$$\text{Volume} = \int_R 1 \, dV = \int_0^{2\pi} \int_0^{\pi} \int_0^a \rho^2 \sin \phi \, d\rho \, d\phi \, d\theta = \int_0^{2\pi} \int_0^{\pi} \frac{1}{3} a^3 \sin \phi \, d\phi \, d\theta$$

$$= \frac{1}{3} a^3 \int_0^{2\pi} -\cos \phi \Big|_0^{\pi} \, d\theta = \frac{2}{3} a^3 \int_0^{2\pi} d\theta = \frac{4\pi a^3}{3}.$$

Example 5   Find the magnitude of the gravitational force exerted by a solid hemisphere of radius a and constant density δ on a unit mass located at the centre of the base of the hemisphere.

Solution   Assume the base of the hemisphere rests on the xy-plane with centre at the origin. (See Figure 16.47.) Newton's law of gravitation says that the force between two masses m_1 and m_2 at a distance r apart is $F = Gm_1m_2/r^2$, where G is the gravitation constant.

In this example, symmetry shows that the net component of the force on the particle at the origin due to the hemisphere is in the z direction only. Any force in the x or y direction from some part of the hemisphere is cancelled by the force from another part of the hemisphere directly opposite the first.

To compute the net z-component of the gravitational force, we imagine a small piece of the hemisphere with volume ΔV, located at spherical coordinates (ρ, θ, ϕ). This piece has mass $\delta \Delta V$, and exerts a force of magnitude F on the unit mass at the origin. The z-component of this force is given by its projection onto the z-axis, which can be seen from the figure to be $F \cos \phi$. The distance from the mass $\delta \Delta V$ to the unit mass at the origin is the spherical coordinate ρ. Therefore, the z-component of the force due to the small piece ΔV is

$$\begin{array}{c} z\text{-component} \\ \text{of force} \end{array} = \frac{G(\delta \Delta V)(1)}{\rho^2} \cos \phi.$$

Adding the contributions of the small pieces, we get a vertical force with magnitude

Figure 16.47: Gravitational force of hemisphere on mass at origin

$$F = \int_0^{2\pi} \int_0^{\pi/2} \int_0^a \left(\frac{G\delta}{\rho^2}\right)(\cos\phi)\rho^2 \sin\phi \, d\rho \, d\phi \, d\theta = \int_0^{2\pi} \int_0^{\pi/2} G\delta(\cos\phi\sin\phi)\rho \Big|_{\rho=0}^{\rho=a} d\phi \, d\theta$$

$$= \int_0^{2\pi} \int_0^{\pi/2} G\delta a \cos\phi \sin\phi \, d\phi \, d\theta = \int_0^{2\pi} G\delta a \left(-\frac{(\cos\phi)^2}{2}\right)\Big|_{\phi=0}^{\phi=\pi/2} d\theta$$

$$= \int_0^{2\pi} G\delta a \left(\frac{1}{2}\right) d\theta = G\delta a\pi.$$

The integral in this example is improper because the region of integration contains the origin, where the force is undefined. However, it can be shown that the result is nevertheless correct.

Exercises and Problems for Section 16.5

Exercises

1. Match the equations in (a)–(f) with one of the surfaces in (I)–(VII).

  (a) $x = 5$ **(b)** $x^2 + z^2 = 7$ **(c)** $\rho = 5$
  (d) $z = 1$ **(e)** $r = 3$ **(f)** $\theta = 2\pi$

   (I) Cylinder, centered on x-axis.
   (II) Cylinder, centered on y-axis.
   (III) Cylinder, centered on z-axis.
   (IV) Plane, perpendicular to the x-axis.
   (V) Plane, perpendicular to the y-axis.
   (VI) Plane, perpendicular to the z-axis.
   (VII) Sphere.

In Exercises 2–7, find an equation for the surface.

2. The cone $z = \sqrt{x^2 + y^2}$ in cylindrical coordinates.

3. The cone $z = \sqrt{x^2 + y^2}$ in spherical coordinates.

4. The top half of the sphere $x^2 + y^2 + z^2 = 1$ in cylindrical coordinates.

5. The vertical plane $y = x$ in cylindrical coordinates.

6. The plane $z = 4$ in spherical coordinates.

7. The plane $z = 10$ in spherical coordinates.

8. Using Cartesian, cylindrical, or spherical coordinates, write an equation for the following surfaces. Each equation should be of the form "Coordinate = Constant."

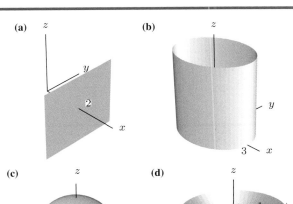

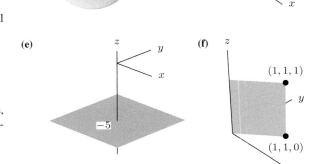

In Exercises 9–10, evaluate the triple integrals in cylindrical coordinates over the region W.

9. $f(x, y, z) = \sin(x^2 + y^2)$, W is the solid cylinder with height 4 and with base of radius 1 centered on the z axis at $z = -1$.

10. $f(x, y, z) = x^2 + y^2 + z^2$, W is the region $0 \le r \le 4$, $\pi/4 \le \theta \le 3\pi/4$, $-1 \le z \le 1$.

In Exercises 11–12, evaluate the triple integrals in spherical coordinates.

11. $f(\rho, \theta, \phi) = \sin\phi$, over the region $0 \le \theta \le 2\pi$, $0 \le \phi \le \pi/4$, $1 \le \rho \le 2$.

12. $f(x, y, z) = 1/(x^2 + y^2 + z^2)^{1/2}$ over the bottom half of the sphere of radius 5 centered at the origin.

For Exercises 13–19, choose coordinates and set up a triple integral, including limits of integration, for a density function f over the region.

13.

14.

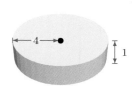

15.

16.

17. A piece of a sphere; angle at the center is $\pi/3$.

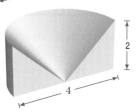

18.

19.

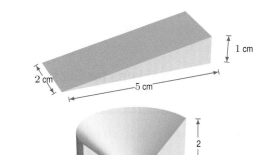

Problems

20. Write a triple integral in cylindrical coordinates giving the volume of a sphere of radius K centered at the origin. Use the order $dz\, dr\, d\theta$.

21. Write a triple integral in spherical coordinates giving the volume of a sphere of radius K centered at the origin. Use the order $d\theta\, d\rho\, d\phi$.

If W is the region in Figure 16.48, what are the limits of integration in Exercises 22–24?

22. $\displaystyle\int_?^? \int_?^? \int_?^? f(r, \theta, z) r\, dz\, dr\, d\theta$

23. $\displaystyle\int_?^? \int_?^? \int_?^? g(\rho, \phi, \theta) \rho^2 \sin\phi\, d\rho\, d\phi\, d\theta$

24. $\displaystyle\int_?^? \int_?^? \int_?^? h(x, y, z)\, dz\, dy\, dx$

For the regions W shown in Problems 25–27, write the limits of integration for $\int_W dV$ in the following coordinates:

(a) Cartesian   **(b)** Cylindrical   **(c)** Spherical

25.

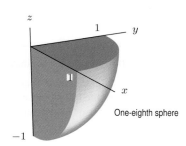

One-eighth sphere

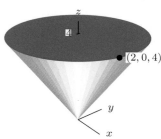

Figure 16.48: Cone with flat top, symmetric about z-axis

26.

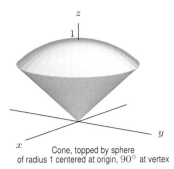

Cone, topped by sphere
of radius 1 centered at origin, $90°$ at vertex

27.

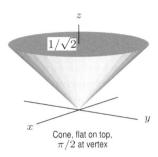

Cone, flat on top,
$\pi/2$ at vertex

28. Write a triple integral representing the volume above the cone $z = \sqrt{x^2 + y^2}$ and below the sphere of radius 2 centered at the origin. Include limits of integration but do not evaluate. Use:

(a) Cylindrical coordinates
(b) Spherical coordinates

29. Write a triple integral representing the volume of the region between spheres of radius 1 and 2, both centered at the origin. Include limits of integration but do not evaluate. Use:

(a) Spherical coordinates.
(b) Cylindrical coordinates. Write your answer as the difference of two integrals.

In Problems 30–35, write a triple integral including limits of integration that gives the specified volume.

30. Under $\rho = 3$ and above $\phi = \pi/3$.

31. Under $\rho = 3$ and above $z = r$.

32. The region between $z = 5$ and $z = 10$, with $2 \leq x^2 + y^2 \leq 3$ and $0 \leq \theta \leq \pi$.

33. Between the cone $z = \sqrt{x^2 + y^2}$ and the first quadrant of the xy-plane, with $x^2 + y^2 \leq 7$.

34. The cap of the solid sphere $x^2 + y^2 + z^2 \leq 10$ cut off by the plane $z = 1$.

35. Below the cone $z = r$, above the xy-plane, and inside the sphere $x^2 + y^2 + z^2 = 8$.

36. (a) Write an integral (including limits of integration) representing the volume of the region inside the cone $z = \sqrt{3(x^2 + y^2)}$ and below the plane $z = 1$.
(b) Evaluate the integral.

37. Find the volume between the cone $z = \sqrt{x^2 + y^2}$ and the plane $z = 10 + x$ above the disk $x^2 + y^2 \leq 1$.

38. Find the volume between the cone $x = \sqrt{y^2 + z^2}$ and the sphere $x^2 + y^2 + z^2 = 4$.

39. The sphere of radius 2 centered at the origin is sliced horizontally at $z = 1$. What is the volume of the cap above the plane $z = 1$?

40. Suppose W is the region outside the cylinder $x^2 + y^2 = 1$ and inside the sphere $x^2 + y^2 + z^2 = 2$. Calculate

$$\int_W (x^2 + y^2)\, dV.$$

41. Write a triple integral representing the volume of a slice of the cylindrical cake of height 2 and radius 5 between the planes $\theta = \pi/6$ and $\theta = \pi/3$. Evaluate this integral.

42. Write a triple integral representing the volume of the cone in Figure 16.49 and evaluate it.

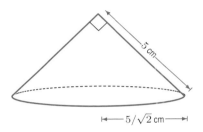

Figure 16.49

Without performing the integration, decide whether each of the integrals in Problems 43–44 is positive, negative, or zero. Give reasons for your decision.

43. W_1 is the unit ball, $x^2 + y^2 + z^2 \leq 1$.

(a) $\int_{W_1} \sin \phi\, dV$ (b) $\int_{W_1} \cos \phi\, dV$

44. W_2 is $0 \leq z \leq \sqrt{1 - x^2 - y^2}$, the top half of the unit ball.

(a) $\int_{W_2} (z^2 - z)\, dV$ (b) $\int_{W_2} (-xz)\, dV$

45. The insulation surrounding a pipe of length l is the region between two cylinders with the same axis. The inner cylinder has radius a, the outer radius of the pipe, and the insulation has thickness h. Write a triple integral, including limits of integration, giving the volume of the insulation. Evaluate the integral.

46. Assume p, q, r are positive constants. Find the volume contained between the coordinate planes and the plane

$$\frac{x}{p} + \frac{y}{q} + \frac{z}{r} = 1.$$

47. A cone stands with its flat base on a table. The cone's circular base has radius a; the vertex (tip) is at a height of h above the center of the base. Write a triple integral, including limits of integration, representing the volume of the cone. Evaluate the integral.

48. A half-melon is approximated by the region between two concentric spheres, one of radius a and the other of radius b, with $0 < a < b$. Write a triple integral, including limits of integration, giving the volume of the half-melon. Evaluate the integral.

49. A bead is made by drilling a cylindrical hole of radius 1 mm through a sphere of radius 5 mm. See Figure 16.50.

(a) Set up a triple integral in cylindrical coordinates representing the volume of the bead.
(b) Evaluate the integral.

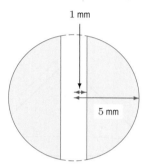

Figure 16.50

50. A pile of hay is approximately in the shape of $0 \leq z \leq 2 - x^2 - y^2$, where x, y, z are in meters. At height z, the density of the hay is $\delta = (16 - 8z)$ kg/m^3.

(a) Write an integral representing the mass of hay in the pile.
(b) Evaluate the integral.

51. Find the mass M of the solid region W given in spherical coordinates by $0 \leq \rho \leq 3, 0 \leq \theta < 2\pi, 0 \leq \phi \leq \pi/4$. The density, $\delta(P)$, at any point P is given by the distance of P from the origin.

52. Write an integral representing the mass of a sphere of radius 3 if the density of the sphere at any point is twice the distance of that point from the center of the sphere.

53. A sphere is made of material whose density at each point is proportional to the square of the distance of the point from the z-axis. The density is 2 gm/cm^3 at a distance of 2 cm from the axis. What is the mass of the sphere if it is centered at the origin and has radius 3 cm?

54. The density of a solid sphere at any point is proportional to the square of the distance of the point to the center of the sphere. What is the ratio of the mass of a sphere of radius 1 to a sphere of radius 2?

55. A spherical shell centered at the origin has an inner radius of 6 cm and an outer radius of 7 cm. The density, δ, of the material increases linearly with the distance from the center. At the inner surface, $\delta = 9$ gm/cm^3; at the outer surface, $\delta = 11$ gm/cm^3.

(a) Using spherical coordinates, write the density, δ, as a function of radius, ρ.
(b) Write an integral giving the mass of the shell.
(c) Find the mass of the shell.

56. (a) Write an iterated integral which represents the mass of a solid ball of radius a. The density at each point in the ball is k times the distance from that point to a fixed plane passing through the center of the ball.
(b) Evaluate the integral.

57. Use appropriate coordinates to find the average distance to the origin for points in the ice cream cone region bounded by the hemisphere $z = \sqrt{8 - x^2 - y^2}$ and the cone $z = \sqrt{x^2 + y^2}$. [Hint: The volume of this region is computed in Problem 24 on page 912.]

For Problems 58–61, use the definition of center of mass given on page 908. Assume x, y, z are in cm.

58. Let C be a solid cone with both height and radius 1 and contained between the surfaces $z = \sqrt{x^2 + y^2}$ and $z = 1$. If C has constant mass density of 1 gm/cm^3, find the z-coordinate of C's center of mass.

59. The density of the cone C in Problem 58 is given by $\delta(z) = z^2$ gm/cm^3. Find

(a) The mass of C.
(b) The z-coordinate of C's center of mass.

60. For $a > 0$, consider the family of solids bounded below by the paraboloid $z = a(x^2 + y^2)$ and above by the plane $z = 1$. If the solids all have constant mass density 1 gm/cm^3, show that the z-coordinate of the center of mass is $2/3$ and so independent of the parameter a.

61. Find the location of the center of mass of a hemisphere of radius a and density b gm/cm^3.

For Problems 62–63, use the definition of moment of inertia given on page 908.

62. The moment of inertia of a solid homogeneous ball B of mass 1 and radius a centered at the origin is the same about any of the coordinate axes (due to the symmetry of the ball). It is easier to evaluate the sum of the three integrals involved in computing the moment of inertia about each of the axes than to compute them individually. Find the sum of the moments of inertia about the x, y and z-axes and thus find the individual moments of inertia.

63. Find the moment of inertia about the z-axis of the solid "fat ice cream cone" given in spherical coordinates by $0 \leq \rho \leq a, 0 \leq \phi \leq \frac{\pi}{3}$ and $0 \leq \theta \leq 2\pi$. Assume that the solid is homogeneous with mass m.

Problems 64–65 deal with the energy stored in an electric field. If a region of space W contains an electric field whose magnitude at a point (x, y, z) is $E(x, y, z)$, then

$$\text{Energy stored by field} = \frac{1}{2} \int_W \epsilon E^2 \, dV,$$

where ϵ is a property of the material called the *permittivity*.

64. The region between two concentric spheres, with radii $a < b$, contains an electric field with magnitude $E = q/(4\pi\epsilon\rho^2)$, where ρ is the distance from the center of the spheres and q is the charge on the inner sphere. Assuming the permittivity, ϵ, is constant, find the total energy stored in the region between the two spheres.

65. Figure 16.51 shows a coaxial cable consisting of two cylindrical conductors centered on the same axis, of radii $a < b$. The electric field between the conductors has magnitude $E = q/(2\pi\epsilon r)$, where r is the distance from the axis and q is the charge per unit length on the cable. The permittivity of the material between the conductors is constant.[4] Show that the stored energy per unit length is proportional to $\ln(b/a)$.

Figure 16.51

66. The density, δ, of a gas in the region under $z = 4 - x^2 - y^2$ and above the xy-plane is $\delta = e^{-x-y} \text{gm/cm}^3$, where x, y, z are in cm. Write an integral, with limits of integration, representing the mass of gas.

67. The density, δ, of the cylinder $x^2 + y^2 \le 4, 0 \le z \le 3$ varies with the distance, r, from the z-axis:

$$\delta = 1 + r \text{ gm/cm}^3.$$

Find the mass of the cylinder, assuming x, y, z are in cm.

68. The density of material at a point in a solid cylinder is proportional to the distance of the point from the z-axis. What is the ratio of the mass of the cylinder $x^2 + y^2 \le 1$, $0 \le z \le 2$ to the mass of the cylinder $x^2 + y^2 \le 9$, $0 \le z \le 2$?

69. A region W consists of the points above the xy-plane and outside the sphere of radius 1 centered at the origin and within the sphere of radius 3 centered at $(0, 0, -1)$. Write an expression for the volume of W. Use cylindrical coordinates and include limits of integration.

70. Compute the force of gravity exerted by a solid cylinder of radius R, height H, and constant density δ on a unit mass at the center of the base of the cylinder.

71. Electric charge is distributed throughout 3-space, with density proportional to the distance from the xy-plane. Show that the total charge inside a cylinder of radius R and height h, sitting on the xy-plane and centered along the z-axis, is proportional to $R^2 h^2$.

72. Electric charge is distributed throughout 3-space with density inversely proportional to the distance from the origin. Show that the total charge inside a sphere of radius R is proportional to R^2.

73. Figure 16.52 shows an alternative notation for spherical coordinates, used often in electrical engineering. Write the volume element dV in this coordinate system.

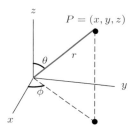

Figure 16.52

Strengthen Your Understanding

74. Which of the following integrals give the volume of the unit sphere?

(a) $\displaystyle\int_0^{2\pi} \int_0^{2\pi} \int_0^1 d\rho \, d\theta \, d\phi$

(b) $\displaystyle\int_0^{\pi} \int_0^{2\pi} \int_0^1 d\rho \, d\theta \, d\phi$

(c) $\displaystyle\int_0^{\pi} \int_0^{2\pi} \int_0^1 \rho^2 \sin\phi \, d\rho \, d\theta \, d\phi$

(d) $\displaystyle\int_0^{\pi} \int_0^{2\pi} \int_0^1 \rho^2 \sin\phi \, d\rho \, d\phi \, d\theta$

(e) $\displaystyle\int_0^{\pi} \int_0^{2\pi} \int_0^1 \rho \, d\rho \, d\phi \, d\theta$

In Problems 75–76, explain what is wrong with the statement.

75. The integral $\displaystyle\int_0^{2\pi} \int_0^{\pi} \int_0^1 1 \, d\rho \, d\phi \, d\theta$ gives the volume inside the sphere of radius 1.

[4] See C. R. Paul and S. A. Nasar, *Introduction to Electromagnetic Fields*, 2nd ed. (New York: McGraw-Hill, 1987).

76. Changing the order of integration gives

$$\int_0^{2\pi} \int_0^{\pi/4} \int_0^{2/\cos\phi} \rho^2 \sin\phi \, d\rho \, d\phi \, d\theta$$

$$= \int_0^{2/\cos\phi} \int_0^{\pi/4} \int_0^{2\pi} \rho^2 \sin\phi \, d\theta \, d\phi \, d\rho.$$

In Problems 77–78, give an example of:

77. An integral in spherical coordinates that gives the volume of a hemisphere.

78. An integral for which it is more convenient to use spherical coordinates than to use Cartesian coordinates.

16.6 APPLICATIONS OF INTEGRATION TO PROBABILITY

To represent how a quantity such as height or weight is distributed throughout a population, we use a density function. To study two or more quantities at the same time and see how they are related, we use a multivariable density function.

Density Functions

Distribution of Weight and Height in Expectant Mothers

Table 16.8 shows the distribution of weight and height in a survey of expectant mothers. The histogram in Figure 16.53 is constructed so that the volume of each bar represents the percentage in the corresponding weight and height range. For example, the bar representing the mothers who weighed 60–70 kg and were 160–165 cm tall has base of area $10 \text{ kg} \cdot 5 \text{ cm} = 50 \text{ kg cm}$. The volume of this bar is 12%, so its height is $12\%/50 \text{ kg cm} = 0.24\%/\text{kg cm}$. Notice that the units on the vertical axis are % per kg cm, so the volume of a bar is a %. The total volume is $100\% = 1$.

Table 16.8 *Distribution of weight and height in a survey of expectant mothers, in %*

	45-50 kg	50-60 kg	60-70 kg	70-80 kg	80-105 kg	Totals by height
150-155 cm	2	4	4	2	1	13
155-160 cm	0	12	8	2	1	23
160-165 cm	1	7	12	4	3	27
165-170 cm	0	8	12	6	2	28
170-180 cm	0	1	3	4	1	9
Totals by weight	3	32	39	18	8	100

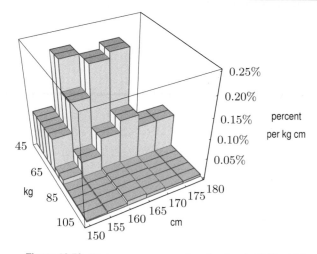

Figure 16.53: Histogram representing the data in Table 16.8

Example 1   Find the percentage of mothers in the survey with height between 170 and 180 cm.

Solution   We add the percentages across the row corresponding to the 170–180 cm height range; this is equivalent to adding the volumes of the corresponding rectangular solids in the histogram.

$$\text{Percentage of mothers} = 0 + 1 + 3 + 4 + 1 = 9\%.$$

Smoothing the Histogram

If we group the data using narrower weight and height groups (and a larger sample), we can draw a smoother histogram and get finer estimates. In the limit, we replace the histogram with a smooth surface, in such a way that the volume under the surface above a rectangle is the percentage of mothers in that rectangle. We define a *density function*, $p(w, h)$, to be the function whose graph is the smooth surface. It has the property that

$$\begin{array}{c} \text{Fraction of sample with} \\ \text{weight between } a \text{ and } b \text{ and} \\ \text{height between } c \text{ and } d \end{array} = \begin{array}{c} \text{Volume under graph of } p \\ \text{over the rectangle} \\ a \le w \le b, c \le h \le d \end{array} = \int_a^b \int_c^d p(w, h) \, dh \, dw.$$

This density also gives the probability that a mother is in these height and weight groups.

Joint Probability Density Functions

We generalise this idea to represent any two characteristics, x and y, distributed throughout a population.

A function $p(x, y)$ is called a **joint probability density function**, or **pdf**, for x and y if

$$\begin{array}{c} \text{Probability that member of} \\ \text{population has } x \text{ between } a \text{ and } b \\ \text{and } y \text{ between } c \text{ and } d \end{array} = \begin{array}{c} \text{Volume under graph of } p \\ \text{above the rectangle} \\ a \le x \le b, c \le y \le d \end{array} = \int_a^b \int_c^d p(x, y) \, dy \, dx$$

where

$$\int_{-\infty}^{\infty} \int_{-\infty}^{\infty} p(x, y) \, dy \, dx = 1 \quad \text{and} \quad p(x, y) \ge 0 \text{ for all } x \text{ and } y.$$

The probability that x falls in an interval of width Δx around x_0 and y falls in an interval of width Δy around y_0 is approximately $p(x_0, y_0) \Delta x \Delta y$.

A joint density function need not be continuous, as in Example 2. In addition, as in Example 4, the integrals involved may be improper and must be computed by methods similar to those used for improper one-variable integrals.

Example 2   Let $p(x, y)$ be defined on the square $0 \le x \le 1, 0 \le y \le 1$ by $p(x, y) = x + y$; let $p(x, y) = 0$ if (x, y) is outside this square. Check that p is a joint density function. In terms of the distribution of x and y in the population, what does it mean that $p(x, y) = 0$ outside the square?

Solution   First, we have $p(x, y) \ge 0$ for all x and y. To check that p is a joint density function, we show that the total volume under the graph is 1:

$$\int_{-\infty}^{\infty} \int_{-\infty}^{\infty} p(x, y) \, dy \, dx = \int_0^1 \int_0^1 (x + y) \, dy \, dx$$

$$= \int_0^1 \left(xy + \frac{y^2}{2} \right) \Big|_0^1 dx = \int_0^1 \left(x + \frac{1}{2} \right) dx = \left(\frac{x^2}{2} + \frac{x}{2} \right) \Big|_0^1 = 1.$$

The fact that $p(x, y) = 0$ outside the square means that the variables x and y never take values outside the interval $[0, 1]$; that is, the value of x and y for any individual in the population is always between 0 and 1.

Example 3   Two variables x and y are distributed in a population according to the density function of Example 2. Find the fraction of the population with $x \leq 1/2$, the fraction with $y \leq 1/2$, and the fraction with both $x \leq 1/2$ and $y \leq 1/2$.

Solution   The fraction with $x \leq 1/2$ is the volume under the graph to the left of the line $x = 1/2$:

$$\int_0^{1/2} \int_0^1 (x + y)\, dy\, dx = \int_0^{1/2} \left(xy + \frac{y^2}{2} \right)\Big|_0^1 dx = \int_0^{1/2} \left(x + \frac{1}{2} \right) dx$$

$$= \left(\frac{x^2}{2} + \frac{x}{2} \right)\Big|_0^{1/2} = \frac{1}{8} + \frac{1}{4} = \frac{3}{8}.$$

Since the function and the regions of integration are symmetric in x and y, the fraction with $y \leq 1/2$ is also $3/8$. Finally, the fraction with both $x \leq 1/2$ and $y \leq 1/2$ is

$$\int_0^{1/2} \int_0^{1/2} (x + y)\, dy\, dx = \int_0^{1/2} \left(xy + \frac{y^2}{2} \right)\Big|_0^{1/2} dx = \int_0^{1/2} \left(\frac{1}{2}x + \frac{1}{8} \right) dx$$

$$= \left(\frac{1}{4}x^2 + \frac{1}{8}x \right)\Big|_0^{1/2} = \frac{1}{16} + \frac{1}{16} = \frac{1}{8}.$$

Recall that a one-variable density function $p(x)$ is a function such that $p(x) \geq 0$ for all x, and $\int_{-\infty}^{\infty} p(x)\, dx = 1$.

Example 4   Let p_1 and p_2 be one-variable density functions for x and y, respectively. Check that $p(x, y) = p_1(x)p_2(y)$ is a joint density function.

Solution   Since both p_1 and p_2 are density functions, they are nonnegative everywhere. Thus, their product $p_1(x)p_2(x) = p(x, y)$ is nonnegative everywhere. Now we must check that the volume under the graph of p is 1. Since $\int_{-\infty}^{\infty} p_2(y)\, dy = 1$ and $\int_{-\infty}^{\infty} p_1(x)\, dx = 1$, we have

$$\int_{-\infty}^{\infty} \int_{-\infty}^{\infty} p(x, y)\, dy\, dx = \int_{-\infty}^{\infty} \int_{-\infty}^{\infty} p_1(x)p_2(y)\, dy\, dx = \int_{-\infty}^{\infty} p_1(x) \left(\int_{-\infty}^{\infty} p_2(y)\, dy \right) dx$$

$$= \int_{-\infty}^{\infty} p_1(x)(1)\, dx = \int_{-\infty}^{\infty} p_1(x)\, dx = 1.$$

Example 5   A machine in a factory is set to produce components 10 cm long and 5 cm in diameter. In fact, there is a slight variation from one component to the next. A component is usable if its length and diameter deviate from the correct values by less than 0.1 cm. With the length, x, in cm and the diameter, y, in cm, the probability density function is

$$p(x, y) = \frac{50\sqrt{2}}{\pi} e^{-100(x-10)^2} e^{-50(y-5)^2}.$$

What is the probability that a component is usable? (See Figure 16.54.)

Figure 16.54: The density function $p(x, y) = \frac{50\sqrt{2}}{\pi} e^{-100(x-10)^2} e^{-50(y-5)^2}$

Solution   We know that

$$\begin{array}{l} \text{Probability that } x \text{ and } y \text{ satisfy} \\ x_0 - \Delta x \leq x \leq x_0 + \Delta x \\ y_0 - \Delta y \leq y \leq y_0 + \Delta y \end{array} = \frac{50\sqrt{2}}{\pi} \int_{y_0 - \Delta y}^{y_0 + \Delta y} \int_{x_0 - \Delta x}^{x_0 + \Delta x} e^{-100(x-10)^2} e^{-50(y-5)^2} \, dx \, dy.$$

Thus,

$$\begin{array}{l} \text{Probability that} \\ \text{component is usable} \end{array} = \frac{50\sqrt{2}}{\pi} \int_{4.9}^{5.1} \int_{9.9}^{10.1} e^{-100(x-10)^2} e^{-50(y-5)^2} \, dx \, dy.$$

The double integral must be evaluated numerically. This yields

$$\begin{array}{l} \text{Probability that} \\ \text{component is usable} \end{array} = \frac{50\sqrt{2}}{\pi} (0.02556) = 0.57530.$$

Thus, there is a 57.530% chance that the component is usable.

Exercises and Problems for Section 16.6

Exercises

In Exercises 1–8, let p be the joint density function such that $p(x, y) = xy$ in R, the rectangle $0 \leq x \leq 2, 0 \leq y \leq 1$, and $p(x, y) = 0$ outside R. Find the fraction of the population satisfying the given constraints.

1. $x \geq 3$

2. $x = 1$

3. $x + y \leq 3$

4. $-1 \leq x \leq 1$

5. $x \geq y$

6. $x + y \leq 1$

7. $0 \leq x \leq 1, 0 \leq y \leq 1/2$

8. Within a distance 1 from the origin

In Exercises 9–14, check whether p is a joint density function. Assume $p(x, y) = 0$ outside the region R.

9. $p(x, y) = 1$, where R is $0 \leq x \leq 1, 0 \leq y \leq 2$

10. $p(x, y) = 1/2$, where R is $4 \leq x \leq 5, -2 \leq y \leq 0$

11. $p(x, y) = x + y$, where R is $-1 \leq x \leq 1, 0 \leq y \leq 1$

12. $p(x, y) = 6(y - x)$, where R is $0 \leq x \leq y \leq 2$

13. $p(x, y) = (2/\pi)(1 - x^2 - y^2)$, where R is $x^2 + y^2 \leq 1$

14. $p(x, y) = xye^{-x-y}$, where R is $x \geq 0, y \geq 0$

Problems

15. Let x and y have joint density function

$$p(x, y) = \begin{cases} \frac{2}{3}(x + 2y) & \text{for } 0 \leq x \leq 1, 0 \leq y \leq 1, \\ 0 & \text{otherwise.} \end{cases}$$

Find the probability that

(a) $x > 1/3$.

(b) $x < (1/3) + y$.

16. The joint density function for x, y is given by

$$f(x, y) = \begin{cases} kxy & \text{for } 0 \leq x \leq y \leq 1, \\ 0 & \text{otherwise.} \end{cases}$$

(a) Determine the value of k.
(b) Find the probability that (x, y) lies in the shaded region in Figure 16.55.

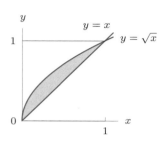

Figure 16.55

17. A joint density function is given by

$$f(x, y) = \begin{cases} kx^2 & \text{for } 0 \leq x \leq 2 \text{ and } 0 \leq y \leq 1, \\ 0 & \text{otherwise.} \end{cases}$$

(a) Find the value of the constant k.
(b) Find the probability that (x, y) satisfies $x + y \leq 2$.
(c) Find the probability that (x, y) satisfies $x \leq 1$ and $y \leq 1/2$.

18. The probability that a radioactive substance will decay at time t is modeled by the density function

$$p(t) = \lambda e^{-\lambda t}$$

for $t \geq 0$, and $p(t) = 0$ for $t < 0$. The positive constant λ depends on the material, and is called the decay rate.

(a) Check that p is a density function.
(b) Two materials with decay rates λ and μ decay independently of each other; their joint density function is the product of the individual density functions. Write the joint density function for the probability that the first material decays at time t and the second at time s.
(c) Find the probability that the first substance decays before the second.

19. A health insurance company wants to know what proportion of its policies are going to cost the company a lot of money because the insured people are over 65 and sick. In order to compute this proportion, the company defines a *disability index*, x, with $0 \leq x \leq 1$, where $x = 0$ represents perfect health and $x = 1$ represents total disability. In addition, the company uses a density function, $f(x, y)$, defined in such a way that the quantity

$$f(x, y) \, \Delta x \, \Delta y$$

approximates the fraction of the population with disability index between x and $x + \Delta x$, and aged between y and $y + \Delta y$. The company knows from experience that a policy no longer covers its costs if the insured person is over 65 and has a disability index exceeding 0.8. Write an expression for the fraction of the company's policies held by people meeting these criteria.

20. A point is chosen at random from the region S in the xy-plane containing all points (x, y) such that $-1 \leq x \leq 1, -2 \leq y \leq 2$ and $x - y \geq 0$ ("at random" means that the density function is constant on S).

(a) Determine the joint density function for x and y.
(b) If T is a subset of S with area α, then find the probability that a point (x, y) is in T.

21. Two independent random numbers x and y between 0 and 1 have joint density function

$$p(x, y) = \begin{cases} 1 & \text{if } 0 \leq x, y \leq 1 \\ 0 & \text{otherwise.} \end{cases}$$

This problem concerns the average $z = (x+y)/2$, which has a one-variable probability density function of its own.

(a) Find $F(t)$, the probability that $z \leq t$. Treat separately the cases $t \leq 0, 0 < t \leq 1/2, 1/2 < t \leq 1$, $1 < t$. Note that $F(t)$ is the cumulative distribution function of z.
(b) Find and graph the probability density function of z.
(c) Are x and y more likely to be near 0, 1/2, or 1? What about z?

22. Figure 16.56 represents a baseball field, with the bases at $(1, 0), (1, 1), (0, 1)$, and home plate at $(0, 0)$. The outer bound of the outfield is a piece of a circle about the origin with radius 4. When a ball is hit by a batter we record the spot on the field where the ball is caught. Let $p(r, \theta)$ be a function in the plane that gives the density of the distribution of such spots. Write an expression that represents the probability that a hit is caught in

(a) The right field (region R).
(b) The center field (region C).

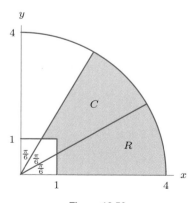

Figure 16.56

Strengthen Your Understanding

In Problems 23–24, explain what is wrong with the statement.

23. If $p_1(x, y)$ and $p_2(x, y)$ are joint density functions, then $p_1(x, y) + p_2(x, y)$ is a joint density function.

24. If $p(w, h)$ is the probability density function of the weight and height of mothers discussed in Section 16.6, then the probability that a mother weighs 60 kg and has a height of 170 cm is $p(60, 170)$.

In Problems 25–26, give an example of:

25. Values for a, b, c and d such that f is a joint density function:

$$f(x, y) = \begin{cases} 1 & \text{for } a \leq x \leq b \text{ and } c \leq y \leq d, \\ 0 & \text{otherwise} \end{cases}$$

26. A one-variable function $g(y)$ such that f is a joint density function:

$$f(x, y) = \begin{cases} g(y) & \text{for } 0 \leq x \leq 2 \text{ and } 0 \leq y \leq 1, \\ 0 & \text{otherwise} \end{cases}$$

For Problems 27–30, let $p(x, y)$ be a joint density function for x and y. Are the following statements true or false?

27. $\int_a^b \int_{-\infty}^{\infty} p(x, y)\, dy\, dx$ is the probability that $a \leq x \leq b$.

28. $0 \leq p(x, y) \leq 1$ for all x.

29. $\int_a^b p(x, y)\, dx$ is the probability that $a \leq x \leq b$.

30. $\int_{-\infty}^{\infty} \int_{-\infty}^{\infty} p(x, y)\, dy\, dx = 1$.

CHAPTER SUMMARY (see also Ready Reference at the end of the book)

- **Double Integral**
  Definition as a limit of Riemann sum; interpretation as volume under graph, as area, as average value, or as total mass from density; estimating from contour diagrams or tables; evaluating using iterated integrals; setting up in polar coordinates.

- **Triple Integral**
  Definition as a limit of Riemann sum; interpretation as volume of solid, as total mass, or as average value; evaluating using iterated integrals; setting up in cylindrical or spherical coordinates.

- **Probability**
  Joint density functions, using integrals to calculate probability.

REVIEW EXERCISES AND PROBLEMS FOR CHAPTER SIXTEEN

Exercises

In Exercises 1–6, sketch the region of integration.

1. $\int_{-1}^{1} \int_{-\sqrt{1-x^2}}^{\sqrt{1-x^2}} f(x,y)\, dy\, dx$

2. $\int_{0}^{2} \int_{-\sqrt{4-y^2}}^{0} f(x,y)\, dx\, dy$

3. $\int_{1}^{4} \int_{-\sqrt{y}}^{\sqrt{y}} f(x,y)\, dx\, dy$

4. $\int_{0}^{1} \int_{0}^{\sin^{-1} y} f(x,y)\, dx\, dy$

5. $\int_{-1}^{1} \int_{-1}^{1} \int_{0}^{\sqrt{1-z^2}} f(x,y,z)\, dy\, dz\, dx$

6. $\int_{0}^{1} \int_{0}^{y} \int_{0}^{x} f(x,y,z)\, dz\, dx\, dy$

In Exercises 7–10, choose coordinates and write a triple integral for a function over the region. Include limits of integration.

7.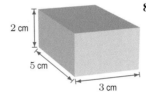

8.

2 cm

3 cm

3 cm

9.

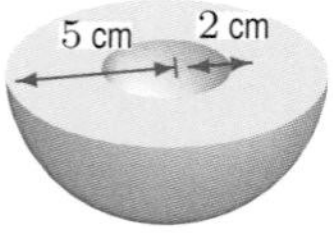

5 cm   2 cm

10.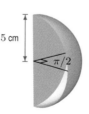

5 cm

$\leq \pi/2$

11. Write $\int_{R} f(x,y)\, dA$ as an iterated integral if R is the region in Figure 16.57.

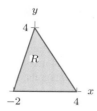

y

4

R

-2 4 x

Figure 16.57

12. Consider the integral $\int_{0}^{4} \int_{0}^{-(y-4)/2} g(x,y)\, dx\, dy$.

   (a) Sketch the region over which the integration is being performed.

   (b) Write the integral with the order of the integration reversed.

13. Evaluate $\int_{R} \sqrt{x^2 + y^2}\, dA$ where R is the region in Figure 16.58.

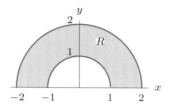

y

2

R

1

-2 -1 1   2 x

Figure 16.58

In Exercises 14–20, calculate the integral exactly.

14. $\int_{0}^{10} \int_{0}^{0.1} x e^{xy}\, dy\, dx$

15. $\int_{0}^{1} \int_{3}^{4} (\sin(2-y)) \cos(3x-7)\, dx\, dy$

16. $\int_{0}^{1} \int_{0}^{y} (\sin^3 x)(\cos x)(\cos y)\, dx\, dy$

17. $\int_{3}^{4} \int_{0}^{1} x^2 y \cos(xy)\, dy\, dx$

18. $\int_{0}^{1} \int_{-\sqrt{1-x^2}}^{\sqrt{1-x^2}} e^{-(x^2+y^2)}\, dy\, dx$

19. $\int_{0}^{1} \int_{0}^{z} \int_{0}^{2} (y+z)^7\, dx\, dy\, dz$

20. $\int_{0}^{1} \int_{0}^{z} \int_{0}^{y} xyz\, dx\, dy\, dz$

If W is the region in Figure 16.59, what are the limits of integration in Problems 21–23?

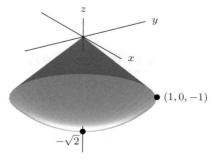

z

y

x

$(1, 0, -1)$

$-\sqrt{2}$

Figure 16.59: Cone with spherical cap

21. $\displaystyle\int_{?}^{?}\int_{?}^{?}\int_{?}^{?} f(\rho,\phi,\theta)\rho^2 \sin\phi \, d\rho \, d\phi \, d\theta$

23. $\displaystyle\int_{?}^{?}\int_{?}^{?}\int_{?}^{?} h(x,y,z) \, dz \, dy \, dx$

24. Set up $\int_R f \, dV$ as an iterated integral in all six possible orders of integration, where R is the hemisphere bounded by the upper half of $x^2 + y^2 + z^2 = 1$ and the xy-plane.

22. $\displaystyle\int_{?}^{?}\int_{?}^{?}\int_{?}^{?} g(r,\theta,z) r \, dz \, dr \, d\theta$

Problems

In Problems 25–33, decide (without calculating its value) whether the integral is positive, negative, or zero. Let W be the solid half-cone bounded by $z = \sqrt{x^2 + y^2}$, $z = 2$ and the yz-plane with $x \geq 0$.

25. $\int_W x \, dV$

26. $\int_W z \, dV$

27. $\int_W \left(z - \sqrt{x^2 + y^2}\right) dV$

28. $\int_W \sqrt{x^2 + y^2} \, dV$

29. $\int_W (z - 2) \, dV$

30. $\int_W y \, dV$

31. $\int_W xy \, dV$

32. $\int_W xyz \, dV$

33. $\int_W e^{-xyz} \, dV$

Problems 34–36 refer to Figure 16.60, which shows triangular portions of the planes $2x + 4y + z = 4$, $3x - 2y = 0$, $z = 2$, and the three coordinate planes $x = 0$, $y = 0$, and $z = 0$. For each solid region E, write down an iterated integral for the triple integral $\int_E f(x, y, z) \, dV$.

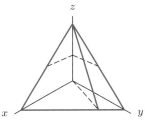

Figure 16.60

34. E is the region bounded by $y = 0$, $z = 0$, $3x - 2y = 0$, and $2x + 4y + z = 4$.

35. E is the region bounded by $x = 0$, $y = 0$, $z = 0$, $z = 2$, and $2x + 4y + z = 4$.

36. E is the region bounded by $x = 0$, $z = 0$, $3x - 2y = 0$, and $2x + 4y + z = 4$.

37. **(a)** Set up a triple integral giving the volume of the tetrahedron bounded by the three coordinate planes and the plane $z - x + y = 2$.
(b) Evaluate the integral.

38. Let B be the solid sphere of radius 1 centered at the origin; let T be the top half of the sphere $(z \geq 0)$; let R be the right half of the sphere $(x \geq 0)$.

(a) Without calculation, decide which of the following integrals are zero. What are the signs of the others?
(i) $\int_B dV$ (ii) $\int_T z dV$ (iii) $\int_R z dV$

(b) Evaluate, numerically where necessary, any of the three integrals that is not zero.

39. Sketch the region R over which the integration is being performed:

$$\int_0^{\pi/2}\int_{\pi/2}^{\pi}\int_0^1 f(\rho,\phi,\theta)\rho^2 \sin\phi \, d\rho \, d\phi \, d\theta.$$

40. **(a)** Convert the following triple integral to spherical coordinates:

$$\int_0^{2\pi}\int_0^3\int_0^r r \, dz dr d\theta.$$

(b) Evaluate either the original integral or your answer to part (a).

In Problems 41–44, sketch the region of integration and write a triple integral, including limits, over the region.

41. Region: $0 \leq z \leq 1 + x$, $0 \leq x \leq 2$, $0 \leq y \leq 1$.

42. Region: $2 \leq z \leq 3$, $5 \leq x^2 + y^2 \leq 6$.

43. Region: $3 \leq x^2 + y^2 + z^2 \leq 4$, $0 \leq \theta \leq \pi$.

44. Region: $x^2 + y^2 + z^2 \leq 9$, $x^2 + y^2 \leq 1$, $z \geq 0$.

In Problems 45–49, is the double integral positive or negative, or is it impossible to tell? The finite regions T, B, R, L are in the xy-plane.

T lies in the region where $y > 0$,
R lies in the region where $x > 0$,
B lies in the region where $y < 0$,
L lies in the region where $x < 0$.

45. $\int_T e^{-x} \, dA$

46. $\int_B y^3 \, dA$

47. $\int_R (x + y^2) \, dA$

48. $\int_L y^3 \, dA$

49. $\int_L (x + y^2) \, dA$

In Problems 50–57, decide (without calculating its value) whether the integral is positive, negative, or zero. Let W be the solid sphere bounded by $x^2 + y^2 + z^2 = 1$.

50. $\int_W z \, dV$

51. $\int_W x \, dV$

52. $\int_W xy \, dV$

53. $\int_W \sin(\frac{\pi}{2}xy) \, dV$

54. $\int_W xyz \, dV$

55. $\int_W e^{-xyz} \, dV$

56. $\int_W (z^2 - 1) \, dV$

57.
$$\int_W \sqrt{x^2 + y^2 + z^2} \, dV$$

In Problems 58–61, evaluate the integral by changing it to cylindrical or spherical coordinates.

58. $\displaystyle\int_{-\sqrt{3}}^{\sqrt{3}} \int_{-\sqrt{3-x^2}}^{\sqrt{3-x^2}} \int_{1}^{4-x^2-y^2} \frac{1}{z^2} \, dz \, dy \, dx$

59. $\displaystyle\int_{0}^{1} \int_{0}^{\sqrt{1-x^2}} \int_{0}^{\sqrt{x^2+y^2}} (z + \sqrt{x^2 + y^2}) \, dz \, dy \, dx$

60. $\displaystyle\int_{0}^{3} \int_{-\sqrt{9-z^2}}^{\sqrt{9-z^2}} \int_{-\sqrt{9-y^2-z^2}}^{\sqrt{9-y^2-z^2}} x^2 \, dx \, dy \, dz$

61. $\displaystyle\int_W \frac{z}{(x^2+y^2)^{3/2}} \, dV$, if W is $1 \le x^2 + y^2 \le 4$,
$$0 \le z \le 4$$

62. (a) Sketch the region of integration of
$$\int_{2}^{\sqrt{8}} \int_{0}^{\sqrt{8-y^2}} e^{-x^2-y^2} \, dx \, dy + \int_{0}^{2} \int_{0}^{y} e^{-x^2-y^2} \, dx \, dy$$
   (b) Evaluate the quantity in part (a).

63. A circular lake 10 km in diameter has a circular island 2 km in diameter at its center. At t kilometers from the island the depth of the lake is $100t(4 - t)$ meters, where $0 \le t \le 4$. What is the volume of water in the lake?

64. A solid region D is a half cylinder with radius 1 lying horizontally with its rectangular base in the xy-plane and its axis along the line $y = 1$ from $x = 0$ to $x = 10$. (The region is above the xy-plane.)
   (a) What is the equation of the curved surface of this half cylinder?
   (b) Write the limits of integration of the integral $\int_D f(x, y, z) \, dV$ in Cartesian coordinates.

65. Find the volume of the region bounded by $z = x + y$, $0 \le x \le 5$, $0 \le y \le 5$, and the planes $x = 0$, $y = 0$, and $z = 0$.

66. (a) Sketch the region of integration, or describe it precisely in words, for the following integral:
$$\int_{-1}^{1} \int_{-1}^{1} \int_{0}^{\sqrt{1-z^2}} f(x, y, z) \, dy \, dz \, dx.$$
   (b) Evaluate the integral with $f(x, y, z) = (y^2 + z^2)^{3/2}$.

67. A thin circular disk of radius 12 cm has density which increases linearly from 1 gm/cm^2 at the center to 25 gm/cm^2 at the rim.
   (a) Write an iterated integral representing the mass of the disk.
   (b) Evaluate the integral.

68. Figure 16.61 shows part of a spherical ball of radius 5 cm. Write an integral in cylindrical coordinates representing the volume of this region and evaluate it.

2 cm

Figure 16.61

69. Figure 16.61 shows part of a spherical ball of radius 5 cm. Write an integral in spherical coordinates representing the volume of this region and evaluate it.

70. Find the mass of the solid bounded by the xy-plane, yz-plane, xz-plane, and the plane $4x + 3y + z = 12$, if the density of the solid is given by $\delta(x, y, z) = x^2$.

71. A forest next to a road has the shape in Figure 16.62. The population density of rabbits is proportional to the distance from the road. It is 0 at the road, and 10 rabbits per square kilometre at the opposite edge of the forest. Find the total rabbit population in the forest.

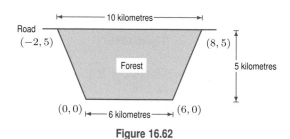

Road
$(-2, 5)$ |←— 10 kilometres —→| $(8, 5)$
Forest   5 kilometres
$(0, 0)$ |←— 6 kilometres —→| $(6, 0)$

Figure 16.62

72. A solid hemisphere of radius 2 cm has density, in gm/cm^3, at each point equal to the distance in centimeters from the point to the center of the base. Write a triple integral representing the total mass of the hemisphere. Evaluate the integral.

73. Find the volume that remains after a cylindrical hole of radius R is bored through a sphere of radius a, where $0 < R < a$, passing through the center of the sphere along the pole.

74. Two spheres, one of radius 1, one of radius $\sqrt{2}$, have centers that are 1 unit apart. Write a triple integral, including limits of integration, giving the volume of the smaller region that is outside one sphere and inside the other. Evaluate the integral.

For Problems 75–76, use the definition of moment of inertia on page 908.

75. Consider a rectangular brick with length 5, width 3, and height 1, and of uniform density 1. Compute the moment of inertia about each of the three axes passing through the center of the brick, perpendicular to one of the sides.

76. Compute the moment of inertia of a ball of radius R about an axis passing through its center. Assume that the ball has a constant density of 1.

77. A particle of mass m is placed at the center of one base of a circular cylindrical shell of inner radius r_1, outer radius r_2, height h, and constant density δ. Find the force of gravity exerted by the cylinder on the particle.

78. Find the area of the crescent-moon shape with circular arcs as edges and the dimensions shown in Figure 16.63.

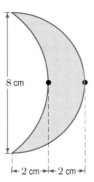

8 cm

2 cm   2 cm

Figure 16.63

79. **(a)** Find the constant k such that $f(x, y) = k(x + y)$ is a probability density function the quarter disk $x^2 + y^2 \leq 100$, $x \geq 0$, $y \geq 0$. [Hint: Use polar coordinates.]
   (b) Find the probability that a point chosen in the quarter disk according to the probability density in part (a) is less than 7 units from the origin.

CAS Challenge Problems

80. Let D be the region inside the triangle with vertices $(0, 0)$, $(1, 1)$ and $(0, 1)$. Express the double integral $\int_D e^{y^2}\, dA$ as an iterated integral in two different ways. Calculate whichever of the two you can do by hand, and calculate the other with a computer algebra system if possible. Compare the answers.

81. Let D be the region inside the circle $x^2 + y^2 = 1$. Express the integral $\int_D \sqrt[3]{x^2 + y^2}\, dA$ as an iterated integral in both Cartesian and polar coordinates. Calculate whichever of the two you can do by hand, and calculate the other with a computer algebra system if possible. Compare the answers.

82. Compute the iterated integrals $\int_0^1 \int_{-1}^0 \frac{x + y}{(x - y)^3}\, dy\, dx$

and $\int_{-1}^0 \int_0^1 \frac{x + y}{(x - y)^3}\, dx\, dy$. Explain why your answers do not contradict Theorem 16.1 on page 894.

83. For each of the following functions, find its average value over the square $-h \leq x \leq h$, $-h \leq y \leq h$, calculate the limit of your answer as $h \to 0$, and compare with the value of the function at $(0, 0)$. Assume a, b, c, d, e, and k are constants.

$$F(x, y) = a + bx^4 + cy^4 + dx^2y^2 + ex^3y^3$$
$$G(x, y) = a\sin(kx) + b\cos(ky) + c$$
$$H(x, y) = ax^2 e^{x+y} + by^2 e^{x-y}$$

Formulate a conjecture from your results and explain why it makes sense.

PROJECTS FOR CHAPTER SIXTEEN

1. **A Connection Between e and π**

   In this problem you will derive one of the remarkable formulas of mathematics, namely that

   $$\int_{-\infty}^{\infty} e^{-x^2}\, dx = \sqrt{\pi}.$$

   (a) Change the following double integral into polar coordinates and evaluate it:

   $$\int_{-\infty}^{\infty} \int_{-\infty}^{\infty} e^{-(x^2+y^2)}\, dx\, dy.$$

   (b) Explain why

   $$\int_{-\infty}^{\infty} \int_{-\infty}^{\infty} e^{-(x^2+y^2)}\, dx\, dy = \left(\int_{-\infty}^{\infty} e^{-x^2}\, dx \right)^2.$$

   (c) Explain why the answers to parts (a) and (b) give the formula we want.

2. **Average Distance Walked to an Airport Gate**

   At airports, departure gates are often lined up in a terminal like points along a line. If you arrive at one gate and proceed to another gate for a connecting flight, what proportion of the length of the terminal will you have to walk, on average?

   (a) One way to model this situation is to randomly choose two numbers, $0 \leq x \leq 1$ and $0 \leq y \leq 1$, and calculate the average value of $|x - y|$. Use a double integral to show that, on average, you have to walk $1/3$ the length of the terminal.

   (b) The terminal gates are not actually located continuously from 0 to 1, as we assumed in part (a). There are only a finite number of gates and they are likely to be equally spaced. Suppose there are $n + 1$ gates located $1/n$ units apart from one end of the terminal ($x_0 = 0$) to the other ($x_n = 1$). Assuming that all pairs (i, j) of arrival and departure gates are equally likely, show that

   $$\text{Average distance between gates} = \frac{1}{(n+1)^2} \cdot \sum_{i=0}^{n} \sum_{j=0}^{n} \left| \frac{i}{n} - \frac{j}{n} \right|.$$

   Identify this sum as approximately (but not exactly) a Riemann sum with n subdivisions for the integrand used in part (a). Compute this sum for $n = 5$ and $n = 10$ and compare to the answer of $1/3$ obtained in part (a).

Chapter Seventeen

PARAMETERIZATION AND VECTOR FIELDS

Contents

17.1 PARAMETERIZED CURVES

A curve in the plane may be parameterized by a pair of equations of the form $x = f(t)$, $y = g(t)$. As the parameter t changes, the point (x, y) traces out the curve. In this section we find parametric equations for curves in three dimensions, and we see how to write parametric equations using position vectors.

Parametric Equations in Three Dimensions

We describe motion in the plane by giving parametric equations for x and y in terms of t. To describe a motion in 3-space parametrically, we need a third equation giving z in terms of t.

Example 1   Find parametric equations for the curve $y = x^2$ in the xy-plane.

Solution   A possible parameterization in two dimensions is $x = t$, $y = t^2$. Since the curve is in the xy-plane, the z-coordinate is zero, so a parameterization in three dimensions is

$$x = t, \quad y = t^2, \quad z = 0.$$

Example 2   Find parametric equations for a particle that starts at $(0, 3, 0)$ and moves around a circle as shown in Figure 17.1.

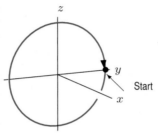

Figure 17.1: Circle of radius 3 in the yz-plane, centered at origin

Solution   Since the motion is in the yz-plane, we have $x = 0$ at all times t. Looking at the yz-plane from the positive x-direction we see motion around a circle of radius 3 in the clockwise direction. Thus,

$$x = 0, \quad y = 3\cos t, \quad z = -3\sin t.$$

Example 3   Describe in words the motion given parametrically by

$$x = \cos t, \quad y = \sin t, \quad z = t.$$

Solution   The particle's x- and y-coordinates give circular motion in the xy-plane, while the z-coordinate increases steadily. Thus, the particle traces out a rising spiral, like a coiled spring. (See Figure 17.2.) This curve is called a *helix*.

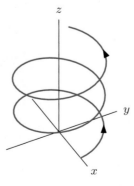

Figure 17.2: The helix $x = \cos t, y = \sin t, z = t$

Example 4   Find parametric equations for the line parallel to the vector $2\vec{i} + 3\vec{j} + 4\vec{k}$ and through the point $(1, 5, 7)$.

Solution   Let's imagine a particle at the point $(1, 5, 7)$ at time $t = 0$ and moving through a displacement of $2\vec{i} + 3\vec{j} + 4\vec{k}$ for each unit of time, t. When $t = 0$, $x = 1$ and x increases by 2 units for every unit of time. Thus, at time t, the x-coordinate of the particle is given by

$$x = 1 + 2t.$$

Similarly, the y-coordinate starts at $y = 5$ and increases at a rate of 3 units for every unit of time. The z-coordinate starts at $y = 7$ and increases by 4 units for every unit of time. Thus, the parametric equations of the line are

$$x = 1 + 2t, \quad y = 5 + 3t, \quad z = 7 + 4t.$$

We can generalise the previous example as follows:

Parametric Equations of a Line through the point (x_0, y_0, z_0) and parallel to the vector $a\vec{i} + b\vec{j} + c\vec{k}$ are

$$x = x_0 + at, \quad y = y_0 + bt, \quad z = z_0 + ct.$$

Notice that the coordinates x, y, and z are linear functions of the parameter t.

Example 5   (a) Describe in words the curve given by the parametric equations $x = 3 + t$, $y = 2t$, $z = 1 - t$.
(b) Find parametric equations for the line through the points $(1, 2, -1)$ and $(3, 3, 4)$.

Solution   (a) The curve is a line through the point $(3, 0, 1)$ and parallel to the vector $\vec{i} + 2\vec{j} - \vec{k}$.
(b) The line is parallel to the vector between the points $P = (1, 2, -1)$ and $Q = (3, 3, 4)$.

$$\overrightarrow{PQ} = (3 - 1)\vec{i} + (3 - 2)\vec{j} + (4 - (-1))\vec{k} = 2\vec{i} + \vec{j} + 5\vec{k}.$$

Thus, using the point P, the parametric equations are

$$x = 1 + 2t, \quad y = 2 + t, \quad z = -1 + 5t.$$

Using the point Q gives the equations $x = 3 + 2t$, $y = 3 + t$, $z = 4 + 5t$, which represent the same line.

Using Position Vectors to Write Parameterised Curves as Vector-Valued Functions

A point in the plane with coordinates (x, y) can be represented by the position vector $\vec{r} = x\vec{i} + y\vec{j}$ in Figure 17.3. Similarly, in 3-space we write $\vec{r} = x\vec{i} + y\vec{j} + z\vec{k}$. (See Figure 17.4.)

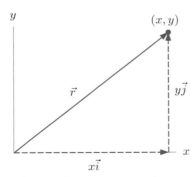

Figure 17.3: Position vector $\vec{r}$ for the point (x, y)

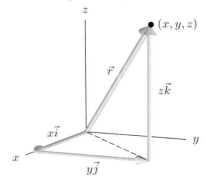

Figure 17.4: Position vector $\vec{r}$ for the point (x, y, z)

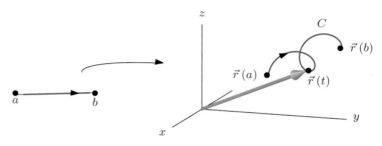

Figure 17.5: The parameterisation sends the interval, $a \leq t \leq b$, to the curve, C, in 3-space

We can write the parametric equations $x = f(t)$, $y = g(t)$, $z = h(t)$ as a single vector equation

$$\vec{r}(t) = f(t)\vec{i} + g(t)\vec{j} + h(t)\vec{k}$$

called a *parameterisation*. As the parameter t varies, the point with position vector $\vec{r}(t)$ traces out a curve in 3-space. For example, the circular motion in the plane

$$x = \cos t, y = \sin t \quad \text{can be written as} \quad \vec{r} = (\cos t)\vec{i} + (\sin t)\vec{j}$$

and the helix in 3-space

$$x = \cos t, y = \sin t, z = t \quad \text{can be written as} \quad \vec{r} = (\cos t)\vec{i} + (\sin t)\vec{j} + t\vec{k}.$$

See Figure 17.5.

Example 6   Use vectors to give a parameterisation for the circle of radius $\frac{1}{2}$ centred at the point $(-1, 2)$.

Solution   The circle of radius 1 centred at the origin is parameterised by the vector-valued function

$$\vec{r}_1(t) = \cos t\vec{i} + \sin t\vec{j}, \quad 0 \leq t \leq 2\pi.$$

The point $(-1, 2)$ has position vector $\vec{r}_0 = -\vec{i} + 2\vec{j}$. The position vector, $\vec{r}(t)$, of a point on the circle of radius $\frac{1}{2}$ centred at $(-1, 2)$ is found by adding $\frac{1}{2}\vec{r}_1$ to $\vec{r}_0$. (See Figures 17.6 and 17.7.) Thus,

$$\vec{r}(t) = \vec{r}_0 + \tfrac{1}{2}\vec{r}_1(t) = -\vec{i} + 2\vec{j} + \tfrac{1}{2}(\cos t\vec{i} + \sin t\vec{j}) = (-1 + \tfrac{1}{2}\cos t)\vec{i} + (2 + \tfrac{1}{2}\sin t)\vec{j},$$

or, equivalently,

$$x = -1 + \tfrac{1}{2}\cos t, \quad y = 2 + \tfrac{1}{2}\sin t, \quad 0 \leq t \leq 2\pi.$$

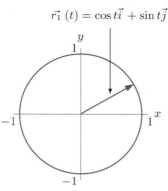

Figure 17.6: The circle $x^2 + y^2 = 1$ parameterised by $\vec{r}_1(t) = \cos t\vec{i} + \sin t\vec{j}$

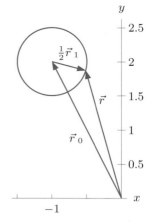

Figure 17.7: The circle of radius $\frac{1}{2}$ and centre $(-1, 2)$ parameterised by $\vec{r}(t) = \vec{r}_0 + \frac{1}{2}\vec{r}_1(t)$

Parametric Equation of a Line

Consider a straight line in the direction of a vector $\vec{v}$ passing through the point (x_0, y_0, z_0) with position vector $\vec{r}_0$. We start at $\vec{r}_0$ and move up and down the line, adding different multiples of $\vec{v}$ to $\vec{r}_0$. (See Figure 17.8.)

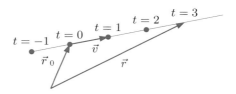

Figure 17.8: The line $\vec{r}(t) = \vec{r}_0 + t\vec{v}$

In this way, every point on the line can be written as $\vec{r}_0 + t\vec{v}$, which yields the following:

Parametric Equation of a Line in Vector Form

The line through the point with position vector $\vec{r}_0 = x_0\vec{i} + y_0\vec{j} + z_0\vec{k}$ in the direction of the vector $\vec{v} = a\vec{i} + b\vec{j} + c\vec{k}$ has parametric equation

$$\vec{r}(t) = \vec{r}_0 + t\vec{v}.$$

Example 7
(a) Find parametric equations for the line passing through the points $(2, -1, 3)$ and $(-1, 5, 4)$.
(b) Represent the line segment from $(2, -1, 3)$ to $(-1, 5, 4)$ parametrically.

Solution
(a) The line passes through $(2, -1, 3)$ and is parallel to the displacement vector $\vec{v} = -3\vec{i} + 6\vec{j} + \vec{k}$ from $(2, -1, 3)$ to $(-1, 5, 4)$. Thus, the parametric equation is

$$\vec{r}(t) = 2\vec{i} - \vec{j} + 3\vec{k} + t(-3\vec{i} + 6\vec{j} + \vec{k}).$$

(b) In the parameterisation in part (a), $t = 0$ corresponds to the point $(2, -1, 3)$ and $t = 1$ corresponds to the point $(-1, 5, 4)$. So the parameterisation of the segment is

$$\vec{r}(t) = 2\vec{i} - \vec{j} + 3\vec{k} + t(-3\vec{i} + 6\vec{j} + \vec{k}), \qquad 0 \le t \le 1.$$

Intersection of Curves and Surfaces

Parametric equations for a curve enable us to find where a curve intersects a surface.

Example 8
Find the points at which the line $x = t$, $y = 2t$, $z = 1 + t$ pierces the sphere of radius 10 centred at the origin.

Solution
The equation for the sphere of radius 10 and centred at the origin is

$$x^2 + y^2 + z^2 = 100.$$

To find the intersection points of the line and the sphere, substitute the parametric equations of the line into the equation of the sphere, giving

$$t^2 + 4t^2 + (1 + t)^2 = 100,$$

so

$$6t^2 + 2t - 99 = 0,$$

which has the two solutions at approximately $t = -4.23$ and $t = 3.90$. Using the parametric equation for the line, $(x, y, z) = (t, 2t, 1 + t)$, we see that the line cuts the sphere at the two points

$$(x, y, z) = (-4.23, 2(-4.23), 1 + (-4.23)) = (-4.23, -8.46, -3.23),$$

and

$$(x, y, z) = (3.90, 2(3.90), 1 + 3.90) = (3.90, 7.80, 4.90).$$

We can also use parametric equations to find the intersection of two curves.

Example 9   Two particles move through space, with equations $\vec{r}_1(t) = t\vec{i} + (1 + 2t)\vec{j} + (3 - 2t)\vec{k}$ and $\vec{r}_2(t) = (-2 - 2t)\vec{i} + (1 - 2t)\vec{j} + (1 + t)\vec{k}$. Do the particles ever collide? Do their paths cross?

Solution   To see if the particles collide, we must find out if they pass through the same point at the same time t. So we must find a solution to the vector equation $\vec{r}_1(t) = \vec{r}_2(t)$, which is the same as finding a common solution to the three scalar equations

$$t = -2 - 2t, \qquad 1 + 2t = 1 - 2t, \qquad 3 - 2t = 1 + t.$$

Separately, the solutions are $t = -2/3, t = 0$, and $t = 2/3$, so there is no common solution, and the particles don't collide. To see if their paths cross, we find out if they pass through the same point at two possibly different times, t_1 and t_2. So we solve the equations

$$t_1 = -2 - 2t_2, \qquad 1 + 2t_1 = 1 - 2t_2, \qquad 3 - 2t_1 = 1 + t_2.$$

We solve the first two equations simultaneously and get $t_1 = 2, t_2 = -2$. Since these values also satisfy the third equation, the paths cross. The position of the first particle at time $t = 2$ is the same as the position of the second particle at time $t = -2$, namely the point $(2, 5, -1)$.

Example 10   Are the lines $x = -1 + t, y = 1 + 2t, z = 5 - t$ and $x = 2 + 2t, y = 4 + t, z = 3 + t$ parallel? Do they intersect?

Solution   In vector form the lines are parameterised by

$$\vec{r} = -\vec{i} + \vec{j} + 5\vec{k} + t(\vec{i} + 2\vec{j} - \vec{k})$$
$$\vec{r} = 2\vec{i} + 4\vec{j} + 3\vec{k} + t(2\vec{i} + \vec{j} + \vec{k})$$

Their direction vectors $\vec{i} + 2\vec{j} - \vec{k}$ and $2\vec{i} + \vec{j} + \vec{k}$ are not multiples of each other, so the lines are not parallel. To find out if they intersect, we see if they pass through the same point at two possibly different times, t_1 and t_2:

$$-1 + t_1 = 2 + 2t_2, \qquad 1 + 2t_1 = 4 + t_2, \qquad 5 - t_1 = 3 + t_2.$$

The first two equations give $t_1 = 1, t_2 = -1$. Since these values do not satisfy the third equation, the paths do not cross, and so the lines do not intersect.

The next example shows how to tell if two different parameterisations give the same line.

Example 11   Show that the following two lines are the same:

$$\vec{r} = -\vec{i} - \vec{j} + \vec{k} + t(3\vec{i} + 6\vec{j} - 3\vec{k})$$
$$\vec{r} = \vec{i} + 3\vec{j} - \vec{k} + t(-\vec{i} - 2\vec{j} + \vec{k})$$

Solution   The direction vectors of the two lines, $3\vec{i} + 6\vec{j} - 3\vec{k}$ and $-\vec{i} - 2\vec{j} + \vec{k}$, are multiples of each other, so the lines are parallel. To see if they are the same, we pick a point on the first line and see if it is on the second line. For example, the point on the first line with $t = 0$ has position vector $-\vec{i} - \vec{j} + \vec{k}$. Solving

$$\vec{i} + 3\vec{j} - \vec{k} + t(-\vec{i} - 2\vec{j} + \vec{k}) = -\vec{i} - \vec{j} + \vec{k}$$

we get $t = 2$, so the two lines have a point in common. Thus, they are the same line, parameterized in two different ways.

Exercises and Problems for Section 17.1

Exercises

In Exercises 1–6, find a parameterization for the curve shown.

1.

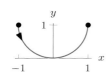

2.

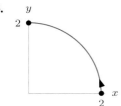

3.

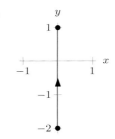

4.

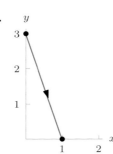

5.

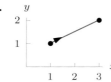

6.

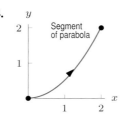

In Exercises 7–17, find parametric equations for the line.

7. The line in the direction of the vector $\vec{i} + 2\vec{j} - \vec{k}$ and through the point $(3, 0, -4)$.

8. The line in the direction of the vector $3\vec{i} - 3\vec{j} + \vec{k}$ and through the point $(1, 2, 3)$.

9. The line in the direction of the vector $2\vec{i} + 2\vec{j} - 3\vec{k}$ and through the point $(-3, 4, -2)$.

10. The line in the direction of the vector $5\vec{j} + 2\vec{k}$ and through the point $(5, -1, 1)$.

11. The line in the direction of the vector $\vec{i} - \vec{k}$ and through the point $(0, 1, 0)$.

12. The line parallel to the z-axis passing through the point $(1, 0, 0)$.

13. The line through the points $(1, 5, 2)$ and $(5, 0, -1)$.

14. The line through the points $(2, 3, -1)$ and $(5, 2, 0)$.

15. The line through $(-3, -2, 1)$ and $(-1, -3, -1)$.

16. The line through $(3, -2, 2)$ and intersecting the y-axis at $y = 2$.

17. The line intersecting the x-axis at $x = 3$ and the z-axis at $z = -5$.

In Exercises 18–34, find a parameterization for the curve.

18. A line segment between $(2, 1, 3)$ and $(4, 3, 2)$.

19. A circle of radius 3 centered on the z-axis and lying in the plane $z = 5$.

20. A line perpendicular to the plane $z = 2x - 3y + 7$ and through the point $(1, 1, 6)$.

21. The circle of radius 2 in the xy-plane, centered at the origin, clockwise.

22. The circle of radius 2 parallel to the xy-plane, centered at the point $(0, 0, 1)$, and traversed counterclockwise when viewed from below.

23. The circle of radius 2 in the xz-plane, centered at the origin.

24. The circle of radius 3 parallel to the xy-plane, centered at the point $(0, 0, 2)$.

25. The circle of radius 3 in the yz-plane, centered at the point $(0, 0, 2)$.

26. The circle of radius 5 parallel to the yz-plane, centered at the point $(-1, 0, -2)$.

27. The curve $y = x^3$ in the xy-plane.

28. The curve $x = y^2$ in the xy-plane.

29. The curve $x = -3z^2$ in the xz-plane.

30. The curve in which the plane $z = 2$ cuts the surface $z = \sqrt{x^2 + y^2}$.

31. The curve $y = 4 - 5x^4$ through the point $(0, 4, 4)$, parallel to the xy-plane.

32. The ellipse of major diameter 6 along the x-axis and minor diameter 4 along the y-axis, centered at the origin.

33. The ellipse of major diameter 5 parallel to the y-axis and minor diameter 2 parallel to the z-axis, centered at $(0, 1, -2)$.

34. The ellipse of major diameter 3 parallel to the x-axis and minor diameter 2 parallel to the z-axis, centered at $(0, 1, -2)$.

In Exercises 35–42, find a parametric equation for the curve segment. (There are many possible answers.)

35. Line from $P_0 = (-1, -3)$ to $P_1 = (5, 2)$.

36. Line from $P_0 = (1, -3, 2)$ to $P_1 = (4, 1, -3)$.

37. Line from $(-1, 2, -3)$ to $(2, 2, 2)$.

38. Semicircle from $(1, 0, 0)$ to $(-1, 0, 0)$ in the xy-plane with $y \geq 0$.

39. Semicircle from $(0, 0, 5)$ to $(0, 0, -5)$ in the yz-plane with $y \geq 0$.

40. Graph of $y = \sqrt{x}$ from $(1, 1)$ to $(16, 4)$.

41. Arc of a circle of radius 5 from $P = (0, 0)$ to $Q = (10, 0)$.

42. Quarter-ellipse from $(4, 0, 3)$ to $(0, -3, 3)$ in the plane $z = 3$.

Problems

In Problems 43–45 two parameterized lines are given. Are they the same line?

43. $\vec{r}_1(t) = (5 - 3t)\vec{i} + 2t\vec{j} + (7 + t)\vec{k}$
$\vec{r}_2(t) = (5 - 6t)\vec{i} + 4t\vec{j} + (7 + 3t)\vec{k}$

44. $\vec{r}_1(t) = (5 - 3t)\vec{i} + (1 + t)\vec{j} + 2t\vec{k}$
$\vec{r}_2(t) = (2 + 6t)\vec{i} + (2 - 2t)\vec{j} + (2 - 4t)\vec{k}$

45. $\vec{r}_1(t) = (5 - 3t)\vec{i} + (1 + t)\vec{j} + 2t\vec{k}$
$\vec{r}_2(t) = (2 + 6t)\vec{i} + (2 - 2t)\vec{j} + (3 - 4t)\vec{k}$

In Problems 46–51, parameterize the line through $P = (2, 5)$ and $Q = (12, 9)$ so that the points P and Q correspond to the given parameter values.

46. $t = 0$ and 1

47. $t = 0$ and 5

48. $t = 20$ and 30

49. $t = 10$ and 11

50. $t = 0$ and -1

51. $t = 10$ and -11

52. At the point where $t = -1$, find an equation for the plane perpendicular to the line

$$x = 5 - 3t, \quad y = 5t - 7, \quad \frac{z}{t} = 6.$$

53. Determine whether the following line is parallel to the plane $2x - 3y + 5z = 5$:

$$x = 5 + 7t, \quad y = 4 + 3t, \quad z = -3 - 2t.$$

54. Show that the equations $x = 3 + t$, $y = 2t$, $z = 1 - t$ satisfy the equations $x + y + 3z = 6$ and $x - y - z = 2$. What does this tell you about the curve parameterized by these equations?

55. **(a)** Explain why the line of intersection of two planes must be parallel to the cross product of a normal vector to the first plane and a normal vector to the second.
(b) Find a vector parallel to the line of intersection of the two planes $x + 2y - 3z = 7$ and $3x - y + z = 0$.
(c) Find parametric equations for the line in part (b).

56. Find an equation for the plane containing the point $(2, 3, 4)$ and the line $x = 1 + 2t, y = 3 - t, z = 4 + t$.

57. **(a)** Find an equation for the line perpendicular to the plane $2x - 3y = z$ and through the point $(1, 3, 7)$.
(b) Where does the line cut the plane?
(c) What is the distance between the point $(1, 3, 7)$ and the plane?

58. Consider two points P_0 and P_1 in 3-space.
(a) Show that the line segment from P_0 to P_1 can be parametrized by

$$\vec{r}(t) = (1 - t)\overrightarrow{OP_0} + t\overrightarrow{OP_1}, \quad 0 \leq t \leq 1.$$

(b) What is represented by the parametric equation

$$\vec{r}(t) = t\overrightarrow{OP_0} + (1 - t)\overrightarrow{OP_1}, \quad 0 \leq t \leq 1?$$

59. **(a)** Find a vector parallel to the line of intersection of the planes $2x - y - 3z = 0$ and $x + y + z = 1$.
(b) Show that the point $(1, -1, 1)$ lies on both planes.
(c) Find parametric equations for the line of intersection.

60. Find the intersection of the line $x = 5 + 7t$, $y = 4 + 3t$, $z = -3 - 2t$ and the plane $2x - 3y + 5z = -7$.

61. If it exists, find the value of c for which the lines $l(t) = (c + t, 1 + t, 5 + t)$ and $m(s) = (s, 1 - s, 3 + s)$ intersect.

62. **(a)** Where does the line $\vec{r} = 2\vec{i} + 5\vec{j} + t(3\vec{i} + \vec{j} + 2\vec{k})$ cut the plane $x + y + z = 1$?
(b) Find a vector perpendicular to the line and lying in the plane.
(c) Find an equation for the line that passes through the point of intersection of the line and plane, is perpendicular to the line, and lies in the plane.

In Problems 63–66, find parametric equations for the line.

63. The line of intersection of the planes $x - y + z = 3$ and $2x + y - z = 5$.

64. The line of intersection of the planes $x + y + z = 3$ and $x - y + 2z = 2$.

65. The line perpendicular to the surface $z = x^2 + y^2$ at the point $(1, 2, 5)$.

66. The line through the point $(-4, 2, 3)$ and parallel to a line in the yz-plane which makes a $45°$ angle with the positive y-axis and the positive z-axis.

67. Is the point $(-3, -4, 2)$ visible from the point $(4, 5, 0)$ if there is an opaque ball of radius 1 centered at the origin?

68. Two particles are traveling through space. At time t the first particle is at the point $(-1 + t, 4 - t, -1 + 2t)$ and the second particle is at $(-7 + 2t, -6 + 2t, -1 + t)$.
(a) Describe the two paths in words.
(b) Do the two particles collide? If so, when and where?
(c) Do the paths of the two particles cross? If so, where?

69. For $t > 0$, a particle moves along the curve $x = a + b \sin kt$, $y = a + b \cos kt$, where a, b, k are positive constants.

(a) Describe the motion in words.
(b) What is the effect on the curve of the following changes?

   (i) Increasing b
   (ii) Increasing a
   (iii) Increasing k
   (iv) Setting a and b equal

70. In the Atlantic Ocean off the coast of Newfoundland, Canada, the temperature and salinity (saltiness) vary throughout the year. Figure 17.9 shows a parametric curve giving the average temperature, T (in °C) and salinity (in grams of salt per kg of water) for t in months, with $t = 1$ corresponding to mid-January.[1]

(a) Why does the parameterized curve form a loop?
(b) When is the water temperature highest?
(c) When is the water saltiest?
(d) Estimate dT/dt at $t = 6$, and give the units. What is the meaning of your answer for seawater?

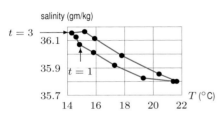

Figure 17.9

71. The concentration (in micrograms/m³) of a pollutant at the point (x, y, z) is given, for x, y, z in meters, by

$$c(x, y, z) = e^{-\left(x^2 + y^2 + z^2\right)}.$$

(a) A particle at point $(0, 1, 0)$ moves in the direction of the point $(-2, 2, -2)$. What is the rate at which the concentration is changing with respect to distance at the moment the particle leaves the point $(0, 1, 0)$?
(b) Now suppose the particle is moving along the path, with time, t, in seconds, given by

$$x = 1 - t^2, \quad y = t, \quad z = 1 - t^2 \quad \text{for } -\infty < t < \infty.$$

When is the concentration a maximum?

72. A light shines on the helix of Example 3 on page 936 from far down each axis. Sketch the shadow the helix casts on each of the coordinate planes: xy, xz, and yz.

73. For a positive constant a and $t \geq 0$, the parametric equations I-V represent the curves described in (a)-(e). Match each description (a)-(e) with its parametric equations and write an equation involving only x and y for the curve.

(a) Line through the origin.
(b) Line not through the origin.
(c) Hyperbola opening along x-axis.
(d) Circle traversed clockwise.
(e) Circle traversed counterclockwise.

I. $x = a \sin t, y = a \cos t$ II. $x = a \sin t, y = a \sin t$
III. $x = a \cos t, y = a \sin t$ IV. $x = a \cos^2 t, y = a \sin^2 t$
V. $x = a / \cos t, y = a \tan t$

74. (a) Find a parametric equation for the line through the point $(2, 1, 3)$ and in the direction of $a\vec{i} + b\vec{j} + c\vec{k}$.
(b) Find conditions on a, b, c so that the line you found in part (a) goes through the origin. Give a reason for your answer.

75. Consider the line $x = 5 - 2t, y = 3 + 7t, z = 4t$ and the plane $ax + by + cz = d$. All the following questions have many possible answers. Find values of a, b, c, d such that:

(a) The plane is perpendicular to the line.
(b) The plane is perpendicular to the line and through the point $(5, 3, 0)$.
(c) The line lies in the plane.

76. Explain the significance of the constants $\alpha > 0$ and $\beta > 0$ in the family of helices given by $\vec{r} = \alpha \cos t \vec{i} + \alpha \sin t \vec{j} + \beta t \vec{k}$.

77. Find parametric equations of the line passing through the points $(1, 2, 3)$, $(3, 5, 7)$ and calculate the shortest distance from the line to the origin.

78. A line has equation $\vec{r} = \vec{a} + t\vec{b}$ where $\vec{r} = x\vec{i} + y\vec{j} + z\vec{k}$ and $\vec{a}$ and $\vec{b}$ are constant vectors such that $\vec{a} \neq \vec{0}, \vec{b} \neq \vec{0}, \vec{b}$ not parallel or perpendicular to $\vec{a}$. For each of the planes (a)–(c), pick the equation (i)–(ix) which represents it. Explain your choice.

(a) A plane perpendicular to the line and through the origin.
(b) A plane perpendicular to the line and not through the origin.
(c) A plane containing the line.

(i) $\vec{a} \cdot \vec{r} = \|\vec{b}\|$ (ii) $\vec{b} \cdot \vec{r} = \|\vec{a}\|$
(iii) $\vec{a} \cdot \vec{r} = \vec{b} \cdot \vec{r}$ (iv) $(\vec{a} \times \vec{b}) \cdot (\vec{r} - \vec{a}) = 0$
(v) $\vec{r} - \vec{a} = \vec{b}$ (vi) $\vec{a} \cdot \vec{r} = 0$
(vii) $\vec{b} \cdot \vec{r} = 0$ (viii) $\vec{a} + \vec{r} = \vec{b}$
(ix) $(\vec{a} \times \vec{b}) \cdot (\vec{r} - \vec{b}) = \|\vec{a}\|$

79. (a) Find a parametric equation for the line through the point $(1, 5, 2)$ and in the direction of the vector $2\vec{i} + 3\vec{j} - \vec{k}$.
(b) By minimizing the square of the distance from a point on the line to the origin, find the exact point on the line which is closest to the origin.

---

[1]Based on http://www.vub.ac.be. Accessed Nov 2011.

80. A plane from Denver, Colorado, (altitude 1650 meters) flies to Bismark, North Dakota (altitude 550 meters). It travels at 650 km/hour along a horizontal line at 8,000 meters above the line joining Denver and Bismark. Bismark is about 850 km in the direction 60° north of east from Denver. Find parametric equations describing the plane's motion. Assume the origin is at sea level beneath Denver, that the x-axis points east and the y-axis points north, and that the earth is flat. Measure distances in kilometers and time in hours.

81. The plane $x + 3y - 2z = 6$ is colored blue and the plane $2x + y + z = 3$ is colored yellow. The planes intersect in a line, which is colored green. You are at the point $P = (1, -2, -1)$.

  (a) You look in the direction $\vec{v} = \vec{i} + 2\vec{j} + \vec{k}$. Do you see the blue plane or the yellow plane?

  (b) In what direction(s) are you looking directly at the green line?

  (c) In what direction(s) should you look to see the yellow plane? The blue plane?

82. The vector $\vec{n}$ is perpendicular to the plane P_1. The vector $\vec{v}$ is parallel to the line L.

  (a) If $\vec{n} \cdot \vec{v} = 0$, what does this tell you about the directions of P_1 and L? (Are they parallel? Perpendicular? Or is it impossible to tell?)

  (b) Suppose $\vec{n} \times \vec{v} \neq \vec{0}$. The plane P_2 has normal $\vec{n} \times \vec{v}$. What can you say about the directions of

    (i) P_1 and P_2?   **(ii)** L and P_2?

83. Figure 17.10 shows the parametric curve $x = x(t), y = y(t)$ for $a \leq t \leq b$.

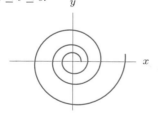

Figure 17.10

(a) Match a graph to each of the parametric curves given, for the same t values, by

  (i) $(-x(t), -y(t))$ **(ii)** $(-x(t), y(t))$
  (iii) $(x(t) + 1, y(t))$ **(iv)** $(x(t) + 1, y(t) + 1)$

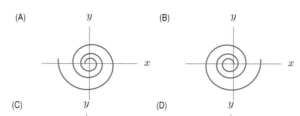
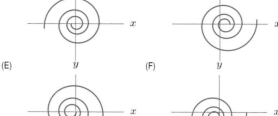
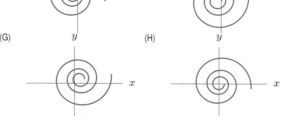

(b) Which of the following could be the formulas for the functions $x(t), y(t)$?

  (i) $x = 10 \cos t \quad y = 10 \sin t$

  (ii) $x = (10 + 8t) \cos t \quad y = (10 + 8t) \sin t$

  (iii) $x = e^{t^2/200} \cos t \quad y = e^{t^2/200} \sin t$

  (iv) $x = (10 - 8t) \cos t \quad y = (10 - 8t) \sin t$

  (v) $x = 10 \cos(t^2 + t) \quad y = 10 \sin(t^2 + t)$

Strengthen Your Understanding

In Problems 84–85, explain what is wrong with the statement.

84. The curve parameterized by $\vec{r}_1(t) = \vec{r}(t - 2)$, defined for all t, is a shift in the $\vec{i}$-direction of the curve parameterized by $\vec{r}(t)$.

85. All points of the curve $\vec{r}(t) = R \cos t\vec{i} + R \sin t\vec{j} + t\vec{k}$ are the same distance, R, from the origin.

In Problems 86–88, give an example of:

86. Parameterizations of two different circles that have the same center and equal radii.

87. Parameterizations of two different lines that intersect at the point $(1, 2, 3)$.

88. A parameterization of the line $x = t, y = 2t, z = 3 + 4t$ that is not given by linear functions.

Are the statements in Problems 89–100 true or false? Give reasons for your answer.

89. The parametric curve $x = 3t + 2, y = -2t$ for $0 \leq t \leq 5$ passes through the origin.

90. The parametric curve $x = t^2, y = t^4$ for $0 \leq t \leq 1$ is a parabola.

91. A parametric curve $x = g(t), y = h(t)$ for $a \leq t \leq b$ is always the graph of a function $y = f(x)$.

92. The parametric curve $x = (3t + 2)^2, y = (3t + 2)^2 - 1$ for $0 \leq t \leq 3$ is a line.

93. The parametric curve $x = -\sin t$, $y = -\cos t$ for $0 \leq t \leq 2\pi$ traces out a unit circle counterclockwise as t increases.

94. A parameterization of the graph of $y = \ln x$ for $x > 0$ is given by $x = e^t$, $y = t$ for $-\infty < t < \infty$.

95. Both $x = -t + 1, y = 2t$ and $x = 2s, y = -4s + 2$ describe the same line.

96. The line of intersection of the two planes $z = x + y$ and $z = 1 - x - y$ can be parameterized by $x = t, y = \frac{1}{2} - t, z = \frac{1}{2}$.

97. The two lines given by $x = t, y = 2 + t, z = 3 + t$ and $x = 2s, y = 1 - s, z = s$ do not intersect.

98. The line parameterized by $x = 1, y = 2t, z = 3 + t$ is parallel to the x-axis.

99. The equation $\vec{r}(t) = 3t\vec{i} + (6t + 1)\vec{j}$ parameterizes a line.

100. The lines parameterized by $\vec{r}_1(t) = t\vec{i} + (-2t + 1)\vec{j}$ and $\vec{r}_2(t) = (2t + 5)\vec{i} + (-t)\vec{j}$ are parallel.

17.2 MOTION, VELOCITY, AND ACCELERATION

In this section we see how to find the vector quantities of velocity and acceleration from a parametric equation for the motion of an object.

The Velocity Vector

The velocity of a moving particle can be represented by a vector with the following properties:

> The **velocity vector** of a moving object is a vector $\vec{v}$ such that:
> - The magnitude of $\vec{v}$ is the speed of the object.
> - The direction of $\vec{v}$ is the direction of motion.
>
> Thus, the speed of the object is $\|\vec{v}\|$ and the velocity vector is tangent to the object's path.

Example 1   A child is sitting on a Ferris wheel of diameter 10 metres, making one revolution every 2 minutes. Find the speed of the child and draw velocity vectors at two different times.

Solution   The child moves at a constant speed around a circle of radius 5 metres, completing one revolution every 2 minutes. One revolution around a circle of radius 5 is a distance of 10π, so the child's speed is $10\pi/2 = 5\pi \approx 15.7$ m/min. Hence, the magnitude of the velocity vector is 15.7 m/min. The direction of motion is tangent to the circle, and hence perpendicular to the radius at that point. Figure 17.11 shows the direction of the vector at two different times.

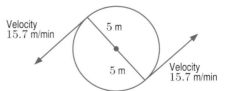

Figure 17.11: Velocity vectors of a child on a Ferris wheel (Note that vectors would be in opposite direction if viewed from the other side)

Computing the Velocity

We find the velocity, as in one-variable calculus, by taking a limit. If the position vector of the particle is $\vec{r}(t)$ at time t, then the displacement vector between its positions at times t and $t + \Delta t$ is $\Delta \vec{r} = \vec{r}(t + \Delta t) - \vec{r}(t)$. (See Figure 17.12.) Over this interval,

$$\text{Average velocity} = \frac{\Delta \vec{r}}{\Delta t}.$$

In the limit as Δt goes to zero we have the instantaneous velocity at time t:

The **velocity vector**, $\vec{v}\,(t)$, of a moving object with position vector $\vec{r}\,(t)$ at time t is

$$\vec{v}\,(t) = \lim_{\Delta t \to 0} \frac{\Delta \vec{r}}{\Delta t} = \lim_{\Delta t \to 0} \frac{\vec{r}\,(t + \Delta t) - \vec{r}\,(t)}{\Delta t},$$

whenever the limit exists. We use the notation $\vec{v} = \dfrac{d\vec{r}}{dt} = \vec{r}\,'(t)$.

Notice that the direction of the velocity vector $\vec{r}\,'(t)$ in Figure 17.12 is approximated by the direction of the vector $\Delta \vec{r}$ and that the approximation gets better as $\Delta t \to 0$.

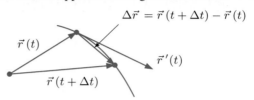

Figure 17.12: The change, $\Delta \vec{r}$, in the position vector for a particle moving on a curve and the velocity vector $\vec{v} = \vec{r}\,'(t)$

The Components of the Velocity Vector

If we represent a curve parametrically by $x = f(t), y = g(t), z = h(t)$, then we can write its position vector as: $\vec{r}\,(t) = f(t)\vec{i} + g(t)\vec{j} + h(t)\vec{k}$. Now we can compute the velocity vector:

$$\begin{aligned} \vec{v}\,(t) &= \lim_{\Delta t \to 0} \frac{\vec{r}\,(t + \Delta t) - \vec{r}\,(t)}{\Delta t} \\ &= \lim_{\Delta t \to 0} \frac{(f(t + \Delta t)\vec{i} + g(t + \Delta t)\vec{j} + h(t + \Delta t)\vec{k}) - (f(t)\vec{i} + g(t)\vec{j} + h(t)\vec{k})}{\Delta t} \\ &= \lim_{\Delta t \to 0} \left(\frac{f(t + \Delta t) - f(t)}{\Delta t}\vec{i} + \frac{g(t + \Delta t) - g(t)}{\Delta t}\vec{j} + \frac{h(t + \Delta t) - h(t)}{\Delta t}\vec{k} \right) \\ &= f'(t)\vec{i} + g'(t)\vec{j} + h'(t)\vec{k} \\ &= \frac{dx}{dt}\vec{i} + \frac{dy}{dt}\vec{j} + \frac{dz}{dt}\vec{k}. \end{aligned}$$

Thus, we have the following result:

The **components of the velocity vector** of a particle moving in space with position vector $\vec{r}\,(t) = f(t)\vec{i} + g(t)\vec{j} + h(t)\vec{k}$ at time t are given by

$$\vec{v}\,(t) = f'(t)\vec{i} + g'(t)\vec{j} + h'(t)\vec{k} = \frac{dx}{dt}\vec{i} + \frac{dy}{dt}\vec{j} + \frac{dz}{dt}\vec{k}.$$

Example 2   Find the components of the velocity vector for the child on the Ferris wheel in Example 1 using a coordinate system which has its origin at the centre of the Ferris wheel and which makes the rotation counterclockwise.

Solution   The Ferris wheel has radius 5 metres and completes 1 revolution counterclockwise every 2 minutes. The motion is parameterised by an equation of the form

$$\vec{r}\,(t) = 5\cos(\omega t)\vec{i} + 5\sin(\omega t)\vec{j},$$

where ω is chosen to make the period 2 minutes. Since the period of $\cos(\omega t)$ and $\sin(\omega t)$ is $2\pi/\omega$, we must have

$$\frac{2\pi}{\omega} = 2, \quad \text{so} \quad \omega = \pi.$$

Thus, the motion is described by the equation

$$\vec{r}(t) = 5\cos(\pi t)\vec{i} + 5\sin(\pi t)\vec{j},$$

where t is in minutes. The velocity is given by

$$\vec{v} = \frac{dx}{dt}\vec{i} + \frac{dy}{dt}\vec{j} = -5\pi\sin(\pi t)\vec{i} + 5\pi\cos(\pi t)\vec{j}.$$

To check, we calculate the magnitude of $\vec{v}$,

$$\|\vec{v}\| = \sqrt{(-5\pi)^2\sin^2(\pi t) + (5\pi)^2\cos^2(\pi t)} = 5\pi\sqrt{\sin^2(\pi t) + \cos^2(\pi t)} = 5\pi \approx 15.7,$$

which agrees with the speed we calculated in Example 1. To see that the direction is correct, we must show that the vector $\vec{v}$ at any time t is perpendicular to the position vector of the particle at time t. To do this, we compute the dot product of $\vec{v}$ and $\vec{r}$:

$$\vec{v} \cdot \vec{r} = (-5\pi\sin(\pi t)\vec{i} + 5\pi\cos(\pi t)\vec{j}) \cdot (5\cos(\pi t)\vec{i} + 5\sin(\pi t)\vec{j})$$
$$= -25\pi\,\sin(\pi t)\,\cos(\pi t) + 25\pi\cos(\pi t)\,\sin(\pi t) = 0.$$

So the velocity vector, $\vec{v}$, is perpendicular to $\vec{r}$ and hence tangent to the circle. (See Figure 17.13.)

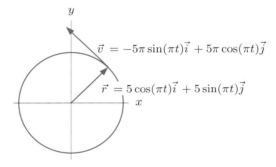

Figure 17.13: Velocity and radius vector of motion around a circle

Velocity Vectors and Tangent Lines

Since the velocity vector is tangent to the path of motion, it can be used to find parametric equations for the tangent line, if there is one.

Example 3   Find the tangent line at the point $(1, 1, 2)$ to the curve defined by the parametric equation

$$\vec{r}(t) = t^2\vec{i} + t^3\vec{j} + 2t\vec{k}.$$

Solution   At time $t = 1$ the particle is at the point $(1, 1, 2)$ with position vector $\vec{r}_0 = \vec{i} + \vec{j} + 2\vec{k}$. The velocity vector at time t is $\vec{r}'(t) = 2t\vec{i} + 3t^2\vec{j} + 2\vec{k}$, so at time $t = 1$ the velocity is $\vec{v} = \vec{r}'(1) = 2\vec{i} + 3\vec{j} + 2\vec{k}$. The tangent line passes through $(1, 1, 2)$ in the direction of $\vec{v}$, so it has the parametric equation

$$\vec{r}(t) = \vec{r}_0 + t\vec{v} = (\vec{i} + \vec{j} + 2\vec{k}) + t(2\vec{i} + 3\vec{j} + 2\vec{k}).$$

The Acceleration Vector

Just as the velocity of a particle moving in 2-space or 3-space is a vector quantity, so is the rate of change of the velocity of the particle, namely its acceleration. Figure 17.14 shows a particle at time t with velocity vector $\vec{v}(t)$ and then a little later at time $t + \Delta t$. The vector $\Delta \vec{v} = \vec{v}(t + \Delta t) - \vec{v}(t)$ is the change in velocity and points approximately in the direction of the acceleration. So,

$$\text{Average acceleration} = \frac{\Delta \vec{v}}{\Delta t}.$$

In the limit as $\Delta t \to 0$, we have the instantaneous acceleration at time t:

> The **acceleration vector** of an object moving with velocity $\vec{v}(t)$ at time t is
>
> $$\vec{a}(t) = \lim_{\Delta t \to 0} \frac{\Delta \vec{v}}{\Delta t} = \lim_{\Delta t \to 0} \frac{\vec{v}(t + \Delta t) - \vec{v}(t)}{\Delta t},$$
>
> if the limit exists. We use the notation $\vec{a} = \dfrac{d\vec{v}}{dt} = \dfrac{d^2 \vec{r}}{dt^2} = \vec{r}''(t)$.

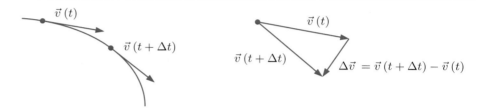

Figure 17.14: Computing the difference between two velocity vectors

Components of the Acceleration Vector

If we represent a curve in space parametrically by $x = f(t)$, $y = g(t)$, $z = h(t)$, we can express the acceleration in components. The velocity vector $\vec{v}(t)$ is given by

$$\vec{v}(t) = f'(t)\vec{i} + g'(t)\vec{j} + h'(t)\vec{k}.$$

From the definition of the acceleration vector, we have

$$\vec{a}(t) = \lim_{\Delta t \to 0} \frac{\vec{v}(t + \Delta t) - \vec{v}(t)}{\Delta t} = \frac{d\vec{v}}{dt}.$$

Using the same method to compute $d\vec{v}/dt$ as we used to compute $d\vec{r}/dt$ on page 946, we obtain

> The **components of the acceleration vector**, $\vec{a}(t)$, at time t of a particle moving in space with position vector $\vec{r}(t) = f(t)\vec{i} + g(t)\vec{j} + h(t)\vec{k}$ at time t are given by
>
> $$\vec{a}(t) = f''(t)\vec{i} + g''(t)\vec{j} + h''(t)\vec{k} = \frac{d^2 x}{dt^2}\vec{i} + \frac{d^2 y}{dt^2}\vec{j} + \frac{d^2 z}{dt^2}\vec{k}.$$

Motion in a Circle and Along a Line

We now consider the velocity and acceleration vectors for two basic motions: uniform motion around a circle, and motion along a straight line.

Example 4   Find the acceleration vector for the child on the Ferris wheel in Examples 1 and 2.

Solution   The child's position vector is given by $\vec{r}(t) = 5\cos(\pi t)\vec{i} + 5\sin(\pi t)\vec{j}$. In Example 2 we saw that the velocity vector is

$$\vec{v}(t) = \frac{dx}{dt}\vec{i} + \frac{dy}{dt}\vec{j} = -5\pi \sin(\pi t)\vec{i} + 5\pi \cos(\pi t)\vec{j}.$$

Thus, the acceleration vector is

$$\vec{a}\,(t) = \frac{d^2x}{dt^2}\vec{i} + \frac{d^2y}{dt^2}\vec{j} = -(5\pi)\cdot\pi\cos(\pi t)\vec{i} - (5\pi)\cdot\pi\sin(\pi t)\vec{j}$$
$$= -5\pi^2\cos(\pi t)\vec{i} - 5\pi^2\sin(\pi t)\vec{j}\,.$$

Notice that $\vec{a}\,(t) = -\pi^2\vec{r}\,(t)$. Thus, the acceleration vector is a multiple of $\vec{r}\,(t)$ and points toward the origin.

The motion of the child on the Ferris wheel is an example of uniform circular motion, whose properties follow. (See Problem 45.)

Uniform Circular Motion: For a particle whose motion is described by

$$\vec{r}\,(t) = R\cos(\omega t)\vec{i} + R\sin(\omega t)\vec{j}$$

- Motion is in a circle of radius R with period $2\pi/|\omega|$.
- Velocity, $\vec{v}$, is tangent to the circle and speed is constant $\|\vec{v}\| = |\omega|R$.
- Acceleration, $\vec{a}$, points toward the centre of the circle with $\|\vec{a}\| = \|\vec{v}\|^2/R$.

In uniform circular motion, the acceleration vector is perpendicular to the velocity vector, $\vec{v}$, because $\vec{v}$ does not change in magnitude, only in direction. There is no acceleration in the direction of $\vec{v}$.

We now look at straight-line motion in which the velocity vector always has the same direction but its magnitude changes. In straight-line motion, the acceleration vector points in the same direction as the velocity vector if the speed is increasing and in the opposite direction to the velocity vector if the speed is decreasing.

Example 5   Consider the motion given by the vector equation

$$\vec{r}\,(t) = 2\vec{i} + 6\vec{j} + (t^3 + t)(4\vec{i} + 3\vec{j} + \vec{k})\,.$$

Show that this is straight-line motion in the direction of the vector $4\vec{i} + 3\vec{j} + \vec{k}$ and relate the acceleration vector to the velocity vector.

Solution   The velocity vector is
$$\vec{v} = (3t^2 + 1)(4\vec{i} + 3\vec{j} + \vec{k})\,.$$

Since $(3t^2 + 1)$ is a positive scalar, the velocity vector $\vec{v}$ always points in the direction of the vector $4\vec{i} + 3\vec{j} + \vec{k}$. In addition,

$$\text{Speed} = \|\vec{v}\| = (3t^2 + 1)\sqrt{4^2 + 3^2 + 1^2} = \sqrt{26}(3t^2 + 1)\,.$$

Notice that the speed is decreasing until $t = 0$, then starts increasing. The acceleration vector is

$$\vec{a} = 6t(4\vec{i} + 3\vec{j} + \vec{k})\,.$$

For $t > 0$, the acceleration vector points in the same direction as $4\vec{i} + 3\vec{j} + \vec{k}$, which is the same direction as $\vec{v}$. This makes sense because the object is speeding up. For $t < 0$, the acceleration vector $6t(4\vec{i} + 3\vec{j} + \vec{k})$ points in the opposite direction to $\vec{v}$ because the object is slowing down.

Motion in a Straight Line: For a particle whose motion is described by

$$\vec{r}(t) = \vec{r}_0 + f(t)\vec{v}$$

- Motion is along a straight line through the point with position vector $\vec{r}_0$ parallel to $\vec{v}$.
- Velocity, $\vec{v}$, and acceleration, $\vec{a}$, are parallel to the line.

If $f(t) = t$, then we have $\vec{r}(t) = \vec{r}_0 + t\vec{v}$, the equation of a line obtained on page 939.

The Length of a Curve

The speed of a particle is the magnitude of its velocity vector:

$$\text{Speed} = \|\vec{v}\| = \sqrt{\left(\frac{dx}{dt}\right)^2 + \left(\frac{dy}{dt}\right)^2 + \left(\frac{dz}{dt}\right)^2}.$$

As in one dimension, we can find the distance travelled by a particle along a curve by integrating its speed. Thus,

$$\text{Distance travelled} = \int_a^b \|\vec{v}(t)\|\, dt.$$

If the particle never stops or reverses its direction as it moves along the curve, the distance it travels will be the same as the length of the curve. This suggests the following formula, which is justified in Problem 54:

If the curve C is given parametrically for $a \le t \le b$ by smooth functions and if the velocity vector $\vec{v}$ is not $\vec{0}$ for $a < t < b$, then

$$\text{Length of } C = \int_a^b \|\vec{v}\|\, dt.$$

Example 6   Find the circumference of the ellipse given by the parametric equations

$$x = 2\cos t, \quad y = \sin t, \quad 0 \le t \le 2\pi.$$

Solution   The circumference of this curve is given by an integral which must be calculated numerically:

$$\text{Circumference} = \int_0^{2\pi} \sqrt{\left(\frac{dx}{dt}\right)^2 + \left(\frac{dy}{dt}\right)^2}\, dt = \int_0^{2\pi} \sqrt{(-2\sin t)^2 + (\cos t)^2}\, dt$$

$$= \int_0^{2\pi} \sqrt{4\sin^2 t + \cos^2 t}\, dt = 9.69.$$

Since the ellipse is inscribed in a circle of radius 2 and circumscribes a circle of radius 1, we would expect the length of the ellipse to be between $2\pi(2) \approx 12.57$ and $2\pi(1) \approx 6.28$, so the value of 9.69 is reasonable.

Exercises and Problems for Section 17.2

Exercises

In Exercises 1–6, find the velocity $\vec{v}(t)$ and speed $\|\vec{v}(t)\|$. Find any times at which the particle stops.

1. $x = t, y = t^2, z = t^3$

2. $x = \cos 3t, y = \sin 5t$

3. $x = 3t^2, y = t^3 + 1$

4. $x = (t-1)^2, y = 2, z = 2t^3 - 3t^2$

5. $x = 3\sin(t^2) - 1, y = 3\cos(t^2)$

6. $x = 3\sin^2 t, y = \cos t - 1, \quad z = t^2$

In Exercises 7–12, find the velocity and acceleration vectors.

7. $x = 2 + 3t, y = 4 + t, z = 1 - t$

8. $x = 2 + 3t^2, y = 4 + t^2, z = 1 - t^2$

9. $x = t, y = t^2, z = t^3$

10. $x = t, y = t^3 - t$

11. $x = 3\cos t, y = 4\sin t$

12. $x = 3\cos(t^2), y = 3\sin(t^2), z = t^2$

13. Find parametric equations for the tangent line at $t = 2$ for Exercise 4.

In Exercises 14–17, find the length of the curve.

14. $x = 3 + 5t, y = 1 + 4t, z = 3 - t$ for $1 \le t \le 2$. Check by calculating the length by another method.

15. $x = \cos 3t, y = \sin 5t$ for $0 \le t \le 2\pi$.

16. $x = \cos(e^t), y = \sin(e^t)$ for $0 \le t \le 1$. Check by calculating the length by another method.

17. $\vec{r}(t) = 2t\vec{i} + \ln t\vec{j} + t^2\vec{k}$ for $1 \le t \le 2$.

In Exercises 18–19, find the velocity and acceleration vectors of the uniform circular motion and check that they are perpendicular. Check that the speed and magnitude of the acceleration are constant.

18. $x = 3\cos(2\pi t), y = 3\sin(2\pi t), z = 0$

19. $x = 2\pi, y = 2\sin(3t), z = 2\cos(3t)$

In Exercises 20–21, find the velocity and acceleration vectors of the straight-line motion. Check that the acceleration vector points in the same direction as the velocity vector if the speed is increasing and in the opposite direction if the speed is decreasing.

20. $x = 2 + t^2, y = 3 - 2t^2, z = 5 - t^2$

21. $x = -2t^3 - 3t + 1, y = 4t^3 + 6t - 5, z = 6t^3 + 9t - 2$

Problems

22. The table gives x and y coordinates of a particle in the plane at time t. Assuming that the particle moves smoothly and that the points given show all the major features of the motion, estimate the following quantities:

 (a) The velocity vector and speed at time $t = 2$.
 (b) Any times when the particle is moving parallel to the y-axis.
 (c) Any times when the particle has come to a stop.

t	0	0.5	1.0	1.5	2.0	2.5	3.0	3.5	4.0
x	1	4	6	7	6	3	2	3	5
y	3	2	3	5	8	10	11	10	9

In Problems 23–24, find all values of t for which the particle is moving parallel to the x-axis and to the y-axis. Determine the end behavior and graph the particle's path.

23. $x = t^2 - 6t, \quad y = t^3 - 3t$

24. $x = t^3 - 12t, \quad y = t^2 + 10t$

25. A particle passes through the point $P = (5, 4, -2)$ at time $t = 4$, moving with constant velocity $\vec{v} = 2\vec{i} - 3\vec{j} + \vec{k}$. Find a parametric equation for its motion.

26. A particle starts at the point $P = (3, 2, -5)$ and moves along a straight line toward $Q = (5, 7, -2)$ at a speed of 5 cm/sec. Let x, y, z be measured in centimeters.

 (a) Find the particle's velocity vector.
 (b) Find parametric equations for the particle's motion.

27. A particle moves at a constant speed along a line from the point $P = (2, -1, 5)$ at time $t = 0$ to the point $Q = (5, 3, -1)$. Find parametric equations for the particle's motion if:

 (a) The particle takes 5 seconds to move from P to Q.
 (b) The speed of the particle is 5 units per second.

28. A particle travels along the line $x = 1 + t$, $y = 5 + 2t$, $z = -7 + t$, where t is in seconds and x, y, z are in meters.

 (a) When and where does the particle hit the plane $x + y + z = 1$?
 (b) How fast is the particle going when it hits the plane? Give units.

29. A stone is thrown from a rooftop at time $t = 0$ seconds. Its position at time t is given by

$$\vec{r}(t) = 10t\vec{i} - 5t\vec{j} + (6.4 - 4.9t^2)\vec{k}.$$

The origin is at the base of the building, which is standing on flat ground. Distance is measured in meters. The vector $\vec{i}$ points east, $\vec{j}$ points north, and $\vec{k}$ points up.

 (a) How high is the rooftop above the ground?
 (b) At what time does the stone hit the ground?
 (c) How fast is the stone moving when it hits the ground?
 (d) Where does the stone hit the ground?
 (e) What is the stone's acceleration when it hits the ground?

30. A child wanders slowly down a circular staircase from the top of a tower. With x, y, z in meters and the origin at the base of the tower, her position t minutes from the start is given by

$$x = 10 \cos t, \quad y = 10 \sin t, \quad z = 90 - 5t.$$

(a) How tall is the tower?
(b) When does the child reach the bottom?
(c) What is her speed at time t?
(d) What is her acceleration at time t?

31. The origin is on flat ground and the z-axis points upward. For time $0 \leq t \leq 10$ in seconds and distance in centimeters, a particle moves along a path given by

$$\vec{r} = 2t\vec{i} + 3t\vec{j} + (100 - (t - 5)^2)\vec{k}.$$

(a) When is the particle at the highest point? What is that point?
(b) When in the interval $0 \leq t \leq 10$ is the particle moving fastest? What is its speed at that moment?
(c) When in the interval $0 \leq t \leq 10$ is the particle moving slowest? What is its speed at that moment?

32. The function $w = f(x, y, z)$ has grad $f(7, 2, 5) = 4\vec{i} - 3\vec{j} + \vec{k}$. A particle moves along the curve $\vec{r}(t)$ arriving at the point $(7, 2, 5)$ with velocity $2\vec{i} + 3\vec{j} + 6\vec{k}$ when $t = 0$. Find the rate of change of w with respect to time at $t = 0$.

33. Suppose x measures horizontal distance in meters, and y measures distance above the ground in meters. At time $t = 0$ in seconds, a projectile starts from a point h meters above the origin with speed v meters/sec at an angle θ to the horizontal. Its path is given by

$$x = (v \cos \theta)t, \quad y = h + (v \sin \theta)t - \frac{1}{2}gt^2.$$

Using this information about a general projectile, analyze the motion of a ball which travels along the path

$$x = 20t, \quad y = 2 + 25t - 4.9t^2.$$

(a) When does the ball hit the ground?
(b) Where does the ball hit the ground?
(c) At what height above the ground does the ball start?
(d) What is the value of g, the acceleration due to gravity?
(e) What are the values of v and θ?

34. A particle is moving on a path in the xz-plane given by $x = 20t$, $z = 5t - 0.5t^2$, where z is the height of the particle above the ground in meters, x is the horizontal distance in meters, and t is time in seconds.

(a) What is the equation of the path in terms of x and z only?
(b) When is the particle at ground level?
(c) What is the velocity of the particle at time t?
(d) What is the speed of the particle at time t?
(e) Is the speed ever 0?
(f) When is the particle at the highest point?

35. The base of a 20-meter tower is at the origin; the base of a 20-meter tree is at $(0, 20, 0)$. The ground is flat and the z-axis points upward. The following parametric equations describe the motion of six projectiles each launched at time $t = 0$ in seconds.

(I) $\vec{r}(t) = (20 + t^2)\vec{k}$
(II) $\vec{r}(t) = 2t^2\vec{j} + 2t^2\vec{k}$
(III) $\vec{r}(t) = 20\vec{i} + 20\vec{j} + (20 - t^2)\vec{k}$
(IV) $\vec{r}(t) = 2t\vec{j} + (20 - t^2)\vec{k}$
(V) $\vec{r}(t) = (20 - 2t)\vec{i} + 2t\vec{j} + (20 - t)\vec{k}$
(VI) $\vec{r}(t) = t\vec{i} + t\vec{j} + t\vec{k}$

(a) Which projectile is launched from the top of the tower and goes downward? When and where does it hit the ground?
(b) Which projectile hits the top of the tree? When? From where is it launched?
(c) Which projectile is not launched from somewhere on the tower and hits the tree? Where and when does it hit the tree?

36. A particle moves on a circle of radius 5 cm, centered at the origin, in the xy-plane (x and y measured in centimeters). It starts at the point $(0, 5)$ and moves counterclockwise, going once around the circle in 8 seconds.

(a) Write a parameterization for the particle's motion.
(b) What is the particle's speed? Give units.

37. Determine the position vector $\vec{r}(t)$ for a rocket which is launched from the origin at time $t = 0$ seconds, reaches its highest point of $(x, y, z) = (1000, 3000, 10,000)$, where x, y, z are in meters, and after the launch is subject only to the acceleration due to gravity, 9.8 m/sec^2.

38. Suppose $\vec{r}(t) = \cos t\vec{i} + \sin t\vec{j} + 2t\vec{k}$ represents the position of a particle on a helix, where z is the height of the particle above the ground.

(a) Is the particle ever moving downward? When?
(b) When does the particle reach a point 10 units above the ground?
(c) What is the velocity of the particle when it is 10 units above the ground?
(d) When it is 10 units above the ground, the particle leaves the helix and moves along the tangent. Find parametric equations for this tangent line.

39. Your spaceship, the Overpriced, is locked on a trajectory given by $x(t) = \cos t; y(t) = \sin t; z(t) = t$ (starting at $t = -100$). You are planning to fire a torpedo, whose trajectory will be a straight line. The velocity of the torpedo will be constant and equal to the velocity of the Overpriced at firing time. What is the earliest time the torpedo can be fired in order to hit an enemy ship represented by a ball of center $(0, 0, 0)$ and radius 2?

40. Emily is standing on the outer edge of a merry-go-round, 10 meters from the center. The merry-go-round completes one full revolution every 20 seconds. As Emily passes over a point P on the ground, she drops a ball from 3 meters above the ground.

   (a) How fast is Emily going?

   (b) How far from P does the ball hit the ground? (The acceleration due to gravity is 9.8 m/sec².)

   (c) How far from Emily does the ball hit the ground?

41. A point P moves in a circle of radius a. Show that $\vec{r}(t)$, the position vector of P, and its velocity vector $\vec{r}\,'(t)$ are perpendicular.

42. A wheel of radius 1 meter rests on the x-axis with its center on the y-axis. There is a spot on the rim at the point $(1, 1)$. See Figure 17.15. At time $t = 0$ the wheel starts rolling on the x-axis in the direction shown at a rate of 1 radian per second.

   (a) Find parametric equations describing the motion of the center of the wheel.

   (b) Find parametric equations describing the motion of the spot on the rim. Plot its path.

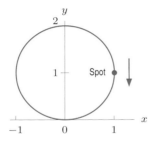

Figure 17.15

43. Show that the helix $\vec{r} = \alpha \cos t\vec{i} + \alpha \sin t\vec{j} + \beta t\vec{k}$ is parameterized with constant speed.

44. An ant crawls along the radius from the center to the edge of a circular disk of radius 1 meter, moving at a constant rate of 1 cm/sec. Meanwhile, the disk is turning counterclockwise about its center at 1 revolution per second.

   (a) Parameterize the motion of the ant.

   (b) Find the velocity and speed of the ant.

   (c) Determine the acceleration and magnitude of the acceleration of the ant.

45. The motion of a particle is given by $\vec{r}(t) = R\cos(\omega t)\vec{i} + R\sin(\omega t)\vec{j}$, with $R > 0$, $\omega > 0$.

   (a) Show that the particle moves on a circle and find the radius, direction, and period.

   (b) Determine the velocity vector of the particle and its direction and speed.

   (c) What are the direction and magnitude of the acceleration vector of the particle?

46. You bicycle along a straight flat road with a safety light attached to one foot. Your bike moves at a speed of 25 km/hr and your foot moves in a circle of radius 20 cm centered 30 cm above the ground, making one revolution per second.

   (a) Find parametric equations for x and y which describe the path traced out by the light, where y is distance (in cm) above the ground.

   (b) Sketch the light's path.

   (c) How fast (in revolutions/sec) would your foot have to be rotating if an observer standing at the side of the road sees the light moving backward?

47. How do the motions of objects A and B differ, if A has position vector $\vec{r}_A(t)$ and B has position vector $\vec{r}_B(t) = \vec{r}_A(2t)$ for $t \geq 0$. Illustrate your answer with $\vec{r}_A(t) = t\vec{i} + t^2\vec{j}$.

48. At time $t = 0$ an object is moving with velocity vector $\vec{v} = 2\vec{i} + \vec{j}$ and acceleration vector $\vec{a} = \vec{i} + \vec{j}$. Can it be in uniform circular motion about some point in the plane?

49. Figure 17.16 shows the velocity and acceleration vectors of an object in uniform circular motion about a point in the plane at a particular moment. Is it moving round the circle in the clockwise or counterclockwise direction?

Figure 17.16

50. The position of a particle at time t is given by $\vec{r}(t)$. Let $r = \|\vec{r}\|$ and $\vec{a}$ be a constant vector. Differentiate:

   (a) $\vec{r} \cdot \vec{r}$ **(b)** $\vec{a} \times \vec{r}$ **(c)** $r^3\vec{r}$

51. The function $f(x, y, z)$ is defined and smooth at every point in 3-space and grad $f(1, 7, 2) = \vec{i} - (\sqrt{6})\vec{j} + \vec{k}$. The curve C is $\vec{r} = (t + 1)^2\vec{i} + 7\cos t\vec{j} + 2e^t\vec{k}$.

   (a) Find an equation of the tangent plane to the level surface of f at the point $(1, 7, 2)$.

   (b) Find the angle between the normal to the level surface of f and the tangent to the curve C at $(1, 7, 2)$. (Note: There are two possible angles; give the smaller one. Your answer should be in radians.)

   (c) With x, y, z in centimeters, let f be the concentration of a pollutant in parts per million (ppm) at the point (x, y, z). A particle moves along the curve C with the given parameterization and t in seconds. Find how fast the concentration is changing at the time $t = 0$. Give units with your answer.

52. Let $\vec{v}(t)$ be the velocity of a particle moving in the plane. Let $s(t)$ be the magnitude of $\vec{v}$ and let $\theta(t)$ be the angle of $\vec{v}(t)$ with the positive x-axis at time t, so that $\vec{v} = s\cos\theta\,\vec{i} + s\sin\theta\,\vec{j}$.

Let $\vec{T}$ be the unit vector in the direction of $\vec{v}$, and let $\vec{N}$ be the unit vector in the direction of $\vec{k}\times\vec{v}$, perpendicular to $\vec{v}$. Show that the acceleration $\vec{a}(t)$ is given by

$$\vec{a} = \frac{ds}{dt}\vec{T} + s\frac{d\theta}{dt}\vec{N}.$$

This shows how to separate the acceleration into the sum of one component, $\dfrac{ds}{dt}\vec{T}$, due to changing speed and a perpendicular component, $s\dfrac{d\theta}{dt}\vec{N}$, due to changing direction of the motion.

53. A point particle P is acted on by a force, $\vec{F}$, which is directed toward a fixed point O; this is called a central force. Let $\vec{r}(t)$ be the position of the particle with respect to O and let $\vec{v}(t)$ be its velocity. Use Newton's second law $\vec{F} = m\vec{a}$, where m is mass and $\vec{a}$ is acceleration, to show that $\vec{r}(t)\times\vec{v}(t) = \vec{c}$, a constant vector. Explain why this tells us that the particle always moves in the same plane.

54. In this problem we justify the formula for the length of a curve given on page 950. Suppose the curve C is given by smooth parametric equations $x = x(t)$, $y = y(t)$, $z = z(t)$ for $a \leq t \leq b$. By dividing the parameter interval $a \leq t \leq b$ at points $t_1, \ldots, t_{n-1}$ into small segments

of length $\Delta t = t_{i+1} - t_i$, we get a corresponding division of the curve C into small pieces. See Figure 17.17, where the points $P_i = (x(t_i), y(t_i), z(t_i))$ on the curve C correspond to parameter values $t = t_i$. Let C_i be the portion of the curve C between P_i and P_{i+1}.

(a) Use local linearity to show that

$$\text{Length of } C_i \approx \sqrt{x'(t_i)^2 + y'(t_i)^2 + z'(t_i)^2}\,\Delta t.$$

(b) Use part (a) and a Riemann sum to explain why

$$\text{Length of } C = \int_a^b \sqrt{x'(t)^2 + y'(t)^2 + z'(t)^2}\,dt.$$

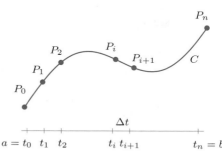

Figure 17.17: A subdivision of the parameter interval and the corresponding subdivision of the curve C

Strengthen Your Understanding

In Problems 55–57, explain what is wrong with the statement.

55. When a particle moves around a circle its velocity and acceleration are always orthogonal.

56. A particle with position $\vec{r}(t)$ at time t has acceleration equal to 3 m/sec^2 at time $t = 0$.

57. A parameterized curve $\vec{r}(t)$, $A \leq t \leq B$, has length $B - A$.

In Problems 58–59, give an example of:

58. A function $\vec{r}(t)$ such that the particle with position $\vec{r}(t)$ at time t has velocity $\vec{v} = \vec{i} + 2\vec{j}$ and acceleration $\vec{a} = 4\vec{i} + 6\vec{k}$ at $t = 0$.

59. An interval $a \leq t \leq b$ corresponding to a piece of the helix $\vec{r}(t) = \cos t\,\vec{i} + \sin t\,\vec{j} + t\,\vec{k}$ of length 10.

Are the statements in Problems 60–67 true or false? Give reasons for your answer.

60. A particle whose motion in the plane is given by $\vec{r}(t) = t^2\vec{i} + (1-t)\vec{j}$ has the same velocity at $t = 1$ and $t = -1$.

61. A particle whose motion in the plane is given by $\vec{r}(t) = t^2\vec{i} + (1-t)\vec{j}$ has the same speed at $t = 1$ and $t = -1$.

62. If a particle is moving along a parameterized curve $\vec{r}(t)$ then the acceleration vector at any point is always perpendicular to the velocity vector at that point.

63. If a particle is moving along a parameterized curve $\vec{r}(t)$ then the acceleration vector at a point cannot be parallel to the velocity vector at that point.

64. If $\vec{r}(t)$ for $a \leq t \leq b$ is a parameterized curve, then $\vec{r}(-t)$ for $a \leq t \leq b$ is the same curve traced backward.

65. If $\vec{r}(t)$ for $a \leq t \leq b$ is a parameterized curve C and the speed $\|\vec{v}(t)\| = 1$, then the length of C is $b - a$.

66. If a particle moves with motion $\vec{r}(t) = 3t\vec{i} + 2t\vec{j} + t\vec{k}$, then the particle stops at the origin.

67. If a particle moves with constant speed, the path of the particle must be a line.

For Problems 68–71, decide if the statement is true or false for all smooth parameterized curves $\vec{r}(t)$ and all values of t for which $\vec{r}'(t) \neq \vec{0}$.

68. The vector $\vec{r}'(t)$ is tangent to the curve at the point with position vector $\vec{r}(t)$.

69. $\vec{r}'(t) \times \vec{r}(t) = \vec{0}$

70. $\vec{r}'(t) \cdot \vec{r}(t) = 0$

71. $\vec{r}''(t) = -\omega^2 \vec{r}(t)$

17.3 VECTOR FIELDS

Introduction to Vector Fields

A *vector field* is a function that assigns a vector to each point in the plane or in 3-space. One example of a vector field is the gradient of a function $f(x, y)$; at each point (x, y) the vector grad $f(x, y)$ points in the direction of maximum rate of increase of f. In this section we look at other vector fields representing velocities and forces.

Velocity Vector Fields

Figure 17.18 shows the flow of a part of the Gulf Stream, a current in the Atlantic Ocean.[2] It is an example of a *velocity vector field*: each vector shows the velocity of the current at that point. The current is fastest where the velocity vectors are longest in the middle of the stream. Beside the stream are eddies where the water flows round and round in circles.

Force Fields

Another physical quantity represented by a vector is force. When we experience a force, sometimes it results from direct contact with the object that supplies the force (for example, a push). Many forces, however, can be felt at all points in space. For example, the earth exerts a gravitational pull on all other masses. Such forces can be represented by vector fields.

Figure 17.19 shows the gravitational force exerted by the earth on a mass of one kilogram at different points in space. This is a sketch of the vector field in 3-space. You can see that the vectors all point toward the earth (which is not shown in the diagram) and that the vectors farther from the earth are smaller in magnitude.

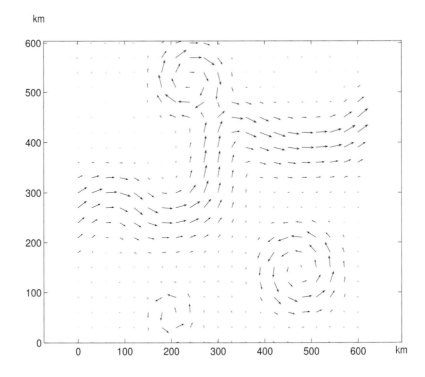

Figure 17.18: The velocity vector field of the Gulf Stream

[2]Based on data supplied by Avijit Gangopadhyay of the Jet Propulsion Laboratory.

Figure 17.19: The gravitational field of the earth

Definition of a Vector Field

Now that you have seen some examples of vector fields, we give a more formal definition.

> A **vector field** in 2-space is a function $\vec{F}(x, y)$ whose value at a point (x, y) is a 2-dimensional vector. Similarly, a vector field in 3-space is a function $\vec{F}(x, y, z)$ whose values are 3-dimensional vectors.

Notice the arrow over the function, $\vec{F}$, indicating that its value is a vector, not a scalar. We often represent the point (x, y) or (x, y, z) by its position vector $\vec{r}$ and write the vector field as $\vec{F}(\vec{r})$.

Visualising a Vector Field Given by a Formula

Since a vector field is a function that assigns a vector to each point, a vector field can often be given by a formula.

Example 1   Sketch the vector field in 2-space given by $\vec{F}(x, y) = -y\vec{i} + x\vec{j}$.

Solution   Table 17.1 shows the value of the vector field at a few points. Notice that each value is a vector. To plot the vector field, we plot $\vec{F}(x, y)$ with its tail at (x, y). (See Figure 17.20.)

Table 17.1   *Values of* $\vec{F}(x, y) = -y\vec{i} + x\vec{j}$

		y		
		-1	0	1
x	-1	$\vec{i} - \vec{j}$	$-\vec{j}$	$-\vec{i} - \vec{j}$
	0	$\vec{i}$	$\vec{0}$	$-\vec{i}$
	1	$\vec{i} + \vec{j}$	$\vec{j}$	$-\vec{i} + \vec{j}$

Now we look at the formula. The magnitude of the vector at (x, y) is the distance from (x, y) to the origin since

$$\|\vec{F}(x, y)\| = \| -y\vec{i} + x\vec{j}\| = \sqrt{x^2 + y^2}.$$

Therefore, all the vectors at a fixed distance from the origin (that is, on a circle centred at the origin) have the same magnitude. The magnitude gets larger as we move farther from the origin.

What about the direction? Figure 17.20 suggests that at each point (x, y) the vector $\vec{F}(x, y)$ is perpendicular to the position vector $\vec{r} = x\vec{i} + y\vec{j}$. We confirm this using the dot product:

$$\vec{r} \cdot \vec{F}(x, y) = (x\vec{i} + y\vec{j}) \cdot (-y\vec{i} + x\vec{j}) = 0.$$

Thus, the vectors are tangent to circles centred at the origin and get longer as we go out. In Figure 17.21, the vectors have been scaled so that they do not obscure each other.

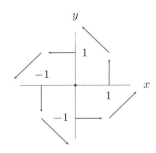

Figure 17.20: The value $\vec{F}(x, y)$ is placed at the point (x, y)

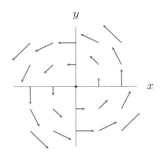

Figure 17.21: The vector field $\vec{F}(x, y) = -y\vec{i} + x\vec{j}$, vectors scaled smaller to fit in diagram

Example 2 Sketch the vector fields in 2-space given by (a) $\vec{F}(x, y) = x\vec{j}$ (b) $\vec{G}(x, y) = x\vec{i}$.

Solution (a) The vector $x\vec{j}$ is parallel to the y-direction, pointing up when x is positive and down when x is negative. Also, the larger $|x|$ is, the longer the vector. The vectors in the field are constant along vertical lines since the vector field does not depend on y. (See Figure 17.22.)

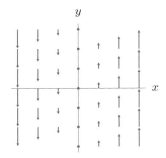

Figure 17.22: The vector field $\vec{F}(x, y) = x\vec{j}$

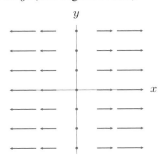

Figure 17.23: The vector field $\vec{F}(x, y) = x\vec{i}$

(b) This is similar to the previous example, except that the vector $x\vec{i}$ is parallel to the x-direction, pointing to the right when x is positive and to the left when x is negative. Again, the larger $|x|$ is the longer the vector, and the vectors are constant along vertical lines, since the vector field does not depend on y. (See Figure 17.23.)

Example 3 Describe the vector field in 3-space given by $\vec{F}(\vec{r}) = \vec{r}$, where $\vec{r} = x\vec{i} + y\vec{j} + z\vec{k}$.

Solution The notation $\vec{F}(\vec{r}) = \vec{r}$ means that the value of $\vec{F}$ at the point (x, y, z) with position vector $\vec{r}$ is the vector $\vec{r}$ with its tail at (x, y, z). Thus, the vector field points outward. See Figure 17.24. Note that the lengths of the vectors have been scaled down so as to fit into the diagram.

Figure 17.24: The vector field $\vec{F}(\vec{r}) = \vec{r}$

Figure 17.25: Force exerted on mass m by mass M

Finding a Formula for a Vector Field

Example 4   Newton's Law of Gravitation states that the magnitude of the gravitational force exerted by an object of mass M on an object of mass m is proportional to M and m and inversely proportional to the square of the distance between them. The direction of the force is from m to M along the line connecting them. (See Figure 17.25.) Find a formula for the vector field $\vec{F}(\vec{r})$ that represents the gravitational force, assuming M is located at the origin and m is located at the point with position vector $\vec{r}$.

Solution   Since the mass m is located at $\vec{r}$, Newton's law says that the magnitude of the force is given by

$$\|\vec{F}(\vec{r})\| = \frac{GMm}{\|\vec{r}\|^2},$$

where G is the universal gravitational constant. A unit vector in the direction of the force is $-\vec{r}/\|\vec{r}\|$, where the negative sign indicates that the direction of force is toward the origin (gravity is attractive). By taking the product of the magnitude of the force and a unit vector in the direction of the force, we obtain an expression for the force vector field:

$$\vec{F}(\vec{r}) = \frac{GMm}{\|\vec{r}\|^2}\left(-\frac{\vec{r}}{\|\vec{r}\|}\right) = \frac{-GMm\vec{r}}{\|\vec{r}\|^3}.$$

We have already seen a picture of this vector field in Figure 17.19.

Gradient Vector Fields

The gradient of a scalar function f is a function that assigns a vector to each point, and is therefore a vector field. It is called the *gradient field* of f. Many vector fields in physics are gradient fields.

Example 5   Sketch the gradient field of the functions in Figures 17.26–17.28.

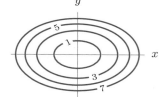

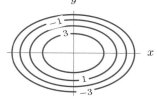

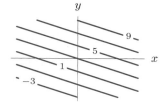

Figure 17.26: The contour map of $f(x, y) = x^2 + 2y^2$

Figure 17.27: The contour map of $g(x, y) = 5 - x^2 - 2y^2$

Figure 17.28: The contour map of $h(x, y) = x + 2y + 3$

Solution   See Figures 17.29–17.31. For a function $f(x, y)$, the gradient vector of f at a point is perpendicular to the contours in the direction of increasing f and its magnitude is the rate of change in that direction. The rate of change is large when the contours are close together and small when they are far apart. Notice that in Figure 17.29 the vectors all point outward, away from the local minimum of f, and in Figure 17.30 the vectors of grad g all point inward, toward the local maximum of g. Since h is a linear function, its gradient is constant, so grad h in Figure 17.31 is a constant vector field.

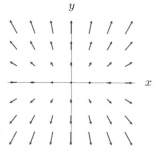

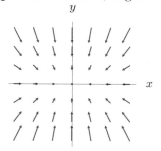

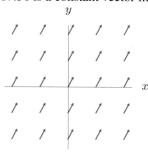

Figure 17.29: grad f

Figure 17.30: grad g

Figure 17.31: grad h

Exercises and Problems for Section 17.3

Exercises

In Exercises 1–4, assume $x, y > 0$ and decide if

(a) The vector field is parallel to the x-axis, parallel to the y-axis, or neither.

(b) As x increases, the length increases, decreases, or neither.

(c) As y increases, the length increases, decreases, or neither.

Assume $x, y > 0$.

1. $\vec{F} = x\vec{j}$ **2.** $\vec{F} = y\vec{i} + \vec{j}$

3. $\vec{F} = (x + e^{1-y})\vec{i}$ **4.** $\text{grad}(x^4 + e^{3y})$

For Exercises 5–10, find formulas for the vector fields. (There are many possible answers.)

5.

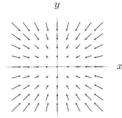

6.

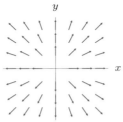

7.

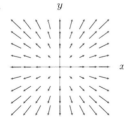

8.

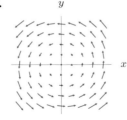

9.

10.

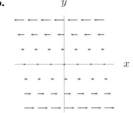

Sketch the vector fields in Exercises 11–20 in the xy-plane.

11. $\vec{F}(x, y) = 2\vec{i} + 3\vec{j}$ **12.** $\vec{F}(x, y) = y\vec{i}$

13. $\vec{F}(x, y) = -y\vec{j}$ **14.** $\vec{F}(\vec{r}) = 2\vec{r}$

15. $\vec{F}(\vec{r}) = \vec{r}/\|\vec{r}\|$ **16.** $\vec{F}(\vec{r}) = -\vec{r}/\|\vec{r}\|^3$

17. $\vec{F}(x, y) = -y\vec{i} + x\vec{j}$ **18.** $\vec{F}(x, y) = 2x\vec{i} + x\vec{j}$

19. $\vec{F}(x, y) = (x + y)\vec{i} + (x - y)\vec{j}$

20. $\vec{F} = \dfrac{y}{\sqrt{x^2 + y^2}}\vec{i} - \dfrac{x}{\sqrt{x^2 + y^2}}\vec{j}$

21. For each description of a vector field in (a)-(d), choose one or more of the vector fields I-IX.

(a) Pointing radially outward, increasing in length away from the origin.

(b) Pointing in a circular direction around the origin, remaining the same length.

(c) Pointing towards the origin, increasing in length farther from the origin.

(d) Pointing clockwise around the origin.

I. $\dfrac{x\vec{i} + y\vec{j}}{\sqrt{x^2 + y^2}}$ II. $\dfrac{-y\vec{i} + x\vec{j}}{\sqrt{x^2 + y^2}}$ III. $\vec{r}$

IV. $-\vec{r}$ V. $-y\vec{i} + x\vec{j}$ VI. $y\vec{i} - x\vec{j}$

VII. $y\vec{i} + x\vec{j}$ VIII. $\dfrac{\vec{r}}{\|\vec{r}\|^3}$ IX. $-\dfrac{\vec{r}}{\|\vec{r}\|^3}$

22. Each vector field in Figures (I)–(IV) represents the force on a particle at different points in space as a result of another particle at the origin. Match up the vector fields with the descriptions below.

(a) A repulsive force whose magnitude decreases as distance increases, such as between electric charges of the same sign.

(b) A repulsive force whose magnitude increases as distance increases.

(c) An attractive force whose magnitude decreases as distance increases, such as gravity.

(d) An attractive force whose magnitude increases as distance increases.

(I)

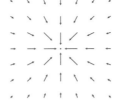

(II)

(III)

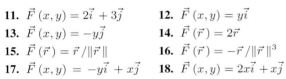

(IV)

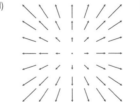

Problems

In Problems 23–27, give an example of a vector field $\vec{F}(x, y)$
in 2-space with the stated properties.

23. $\vec{F}$ is constant

24. $\vec{F}$ has a constant direction but $\|\vec{F}\|$ is not constant

25. $\|\vec{F}\|$ is constant but $\vec{F}$ is not constant

26. Neither $\|\vec{F}\|$ nor the direction of $\vec{F}$ is constant

27. $\vec{F}$ is perpendicular to $\vec{G} = (x + y)\vec{i} + (1 + y^2)\vec{j}$ at every point

28. Match the level curves in (I)–(IV) with the gradient fields in (A)–(D). All figures use the same square window.

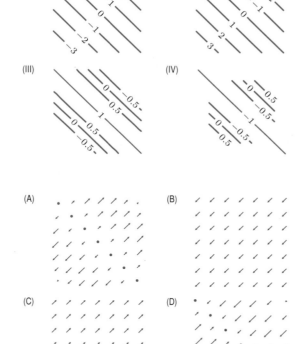

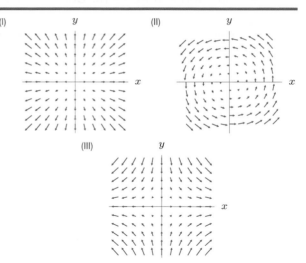

Problems 29–30 concern the vector fields $\vec{F} = x\vec{i} + y\vec{j}$, $\vec{G} = -y\vec{i} + x\vec{j}$, and $\vec{H} = x\vec{i} - y\vec{j}$.

29. Match $\vec{F}$, $\vec{G}$, $\vec{H}$ with their sketches in (I)–(III).

30. Match the vector fields with their sketches, (I)–(IV).

   (a) $\vec{F} + \vec{G}$ **(b)** $\vec{F} + \vec{H}$ **(c)** $\vec{G} + \vec{H}$ **(d)** $-\vec{F} + \vec{G}$

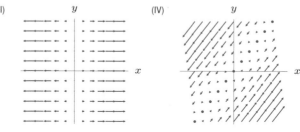

In Problems 31–33, write formulas for vector fields with the given properties.

31. All vectors are parallel to the x-axis; all vectors on a vertical line have the same magnitude.

32. All vectors point toward the origin and have constant length.

33. All vectors are of unit length and perpendicular to the position vector at that point.

34. (a) Let $\vec{F} = x\vec{i} + (x + y)\vec{j} + (x - y + z)\vec{k}$. Find a point at which $\vec{F}$ is parallel to l, the line $x = 5 + t$, $y = 6 - 2t$, $z = 7 - 3t$.

(b) Find a point at which $\vec{F}$ and l are perpendicular.

(c) Give an equation for and describe in words the set of all points at which $\vec{F}$ and l are perpendicular.

In Problems 35–36, let $\vec{F} = x\vec{i} + y\vec{j}$ and $\vec{G} = -y\vec{i} + x\vec{j}$.

35. Sketch the vector field $\vec{L} = a\vec{F} + \vec{G}$ if:

    (a) $a = 0$ **(b)** $a > 0$ **(c)** $a < 0$

36. Sketch the vector field $\vec{L} = \vec{F} + b\vec{G}$ if:

    (a) $b = 0$ **(b)** $b > 0$ **(c)** $b < 0$

37. In the middle of a wide, steadily flowing river there is a fountain that spouts water horizontally in all directions. The river flows in the $\vec{i}$-direction in the xy-plane and the fountain is at the origin.

    (a) If $A > 0$, $K > 0$, explain why the following expression could represent the velocity field for the combined flow of the river and the fountain:

$$\vec{v} = A\vec{i} + K(x^2 + y^2)^{-1}(x\vec{i} + y\vec{j}).$$

    (b) What is the significance of the constants A and K?

    (c) Using a computer, sketch the vector field $\vec{v}$ for $K = 1$ and $A = 1$ and $A = 2$, and for $A = 0.2$, $K = 2$.

38. Figures 17.32 and 17.33 show the gradient of the functions $z = f(x, y)$ and $z = g(x, y)$.

    (a) For each function, draw a rough sketch of the level curves, showing possible z-values.

    (b) The xz-plane cuts each of the surfaces $z = f(x, y)$ and $z = g(x, y)$ in a curve. Sketch each of these curves, making clear how they are similar and how they are different from one another.

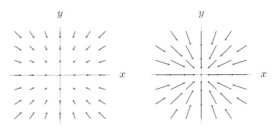

Figure 17.32: Gradient of $z = f(x, y)$ Figure 17.33: Gradient of $z = g(x, y)$

39. Let $\vec{F} = u\vec{i} + v\vec{j}$ be a vector field in 2-space with magnitude $F = \|\vec{F}\|$.

    (a) Let $\vec{T} = (1/F)\vec{F}$. Show that $\vec{T}$ is the unit vector in the direction of $\vec{F}$. See Figure 17.34.

    (b) Let $\vec{N} = (1/F)(\vec{k} \times \vec{F}) = (1/F)(-v\vec{i} + u\vec{j})$. Show that $\vec{N}$ is the unit vector pointing to the left of and at right angles to $\vec{F}$. See Figure 17.34.

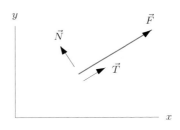

Figure 17.34

Strengthen Your Understanding

In Problems 40–41, explain what is wrong with the statement.

40. A plot of the vector field $\vec{G}(x, y, z) = \vec{F}(2x, 2y, 2z)$ can be obtained from a plot of the vector field $\vec{F}(x, y, z)$ by doubling the lengths of all the arrows.

41. A vector field $\vec{F}$ is defined by the formula $\vec{F}(x, y, z) = x^2 - yz$.

In Problems 42–43, give an example of:

42. A nonconstant vector field that is parallel to $\vec{i} + \vec{j} + \vec{k}$ at every point.

43. A nonconstant vector field with magnitude 1 at every point.

17.4 THE FLOW OF A VECTOR FIELD

When an iceberg is spotted in the North Atlantic, it is important to be able to predict where the iceberg is likely to be a day or a week later. To do this, one needs to know the velocity vector field of the ocean currents, that is, how fast and in what direction the water is moving at each point.

    In this section we use differential equations to find the path of an object in a fluid flow. This path is called a flow line. Figure 17.35 shows several flow lines for the Gulf Stream velocity vector field in Figure 17.18 on page 955. The arrows on each flow line indicate the direction of flow.

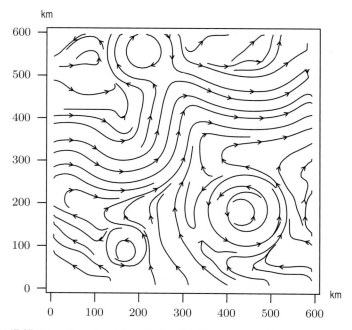

Figure 17.35: Flow lines for objects in the Gulf Stream with different starting points

How Do We Find a Flow Line?

Suppose that $\vec{F}$ is the velocity vector field of water on the surface of a creek and imagine a seed being carried along by the current. We want to know the position vector $\vec{r}(t)$ of the seed at time t. We know

$$\begin{array}{ccc} \text{Velocity of seed} & = & \text{Velocity of current at seed's position} \\ \text{at time } t & & \text{at time } t \end{array}$$

that is,

$$\vec{r}\,'(t) = \vec{F}\,(\vec{r}\,(t)).$$

We make the following definition:

A **flow line** of a vector field $\vec{v} = \vec{F}\,(\vec{r}\,)$ is a path $\vec{r}\,(t)$ whose velocity vector equals $\vec{v}$. Thus,

$$\vec{r}\,'(t) = \vec{v} = \vec{F}\,(\vec{r}\,(t)).$$

The **flow** of a vector field is the family of all of its flow lines.

A flow line is also called an *integral curve* or a *streamline*. We define flow lines for any vector field, as it turns out to be useful to study the flow of fields (for example, electric and magnetic) that are not velocity fields.

Resolving $\vec{F}$ and $\vec{r}$ into components, $\vec{F} = F_1\vec{i} + F_2\vec{j}$ and $\vec{r}\,(t) = x(t)\vec{i} + y(t)\vec{j}$, the definition of a flow line tells us that $x(t)$ and $y(t)$ satisfy the system of differential equations

$$x'(t) = F_1(x(t), y(t)) \quad \text{and} \quad y'(t) = F_2(x(t), y(t)).$$

Solving these differential equations gives a parameterisation of the flow line.

Example 1   Find the flow line of the constant velocity field $\vec{v} = 3\vec{i} + 4\vec{j}$ cm/sec that passes through the point $(1, 2)$ at time $t = 0$.

Solution   Let $\vec{r}(t) = x(t)\vec{i} + y(t)\vec{j}$ be the position in cm of a particle at time t, where t is in seconds. We have

$$x'(t) = 3 \quad \text{and} \quad y'(t) = 4.$$

Thus,

$$x(t) = 3t + x_0 \quad \text{and} \quad y(t) = 4t + y_0.$$

Since the path passes the point $(1, 2)$ at $t = 0$, we have $x_0 = 1$ and $y_0 = 2$ and so

$$x(t) = 3t + 1 \qquad \text{and} \qquad y(t) = 4t + 2.$$

Thus, the path is the line given parametrically by

$$\vec{r}(t) = (3t + 1)\vec{i} + (4t + 2)\vec{j}.$$

(See Figure 17.36.) To find an explicit equation for the path, eliminate t between these expressions to get

$$\frac{x - 1}{3} = \frac{y - 2}{4} \qquad \text{or} \qquad y = \frac{4}{3}x + \frac{2}{3}.$$

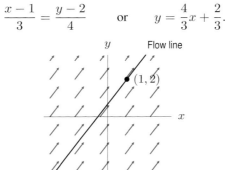

Figure 17.36: Vector field $\vec{F} = 3\vec{i} + 4\vec{j}$ with the flow line through $(1, 2)$

Example 2   The velocity of a flow at the point (x, y) is $\vec{F}(x, y) = \vec{i} + x\vec{j}$. Find the path of motion of an object in the flow that is at the point $(-2, 2)$ at time $t = 0$.

Solution   Figure 17.37 shows this field. Since $\vec{r}'(t) = \vec{F}(\vec{r}(t))$, we are looking for the flow line that satisfies the system of differential equations

$$x'(t) = 1, \quad y'(t) = x(t) \qquad \text{satisfying} \ x(0) = -2 \ \text{and} \ y(0) = 2.$$

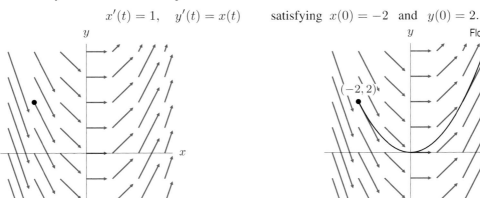

Figure 17.37: The velocity field $\vec{v} = \vec{i} + x\vec{j}$

Figure 17.38: A flow line of the velocity field $\vec{v} = \vec{i} + x\vec{j}$

Solving for $x(t)$ first, we get $x(t) = t + x_0$, where x_0 is a constant of integration. Thus, $y'(t) = t + x_0$, so $y(t) = \frac{1}{2}t^2 + x_0 t + y_0$, where y_0 is also a constant of integration. Since $x(0) = x_0 = -2$ and $y(0) = y_0 = 2$, the path of motion is given by

$$x(t) = t - 2, \quad y(t) = \tfrac{1}{2}t^2 - 2t + 2,$$

or, equivalently,

$$\vec{r}(t) = (t - 2)\vec{i} + (\tfrac{1}{2}t^2 - 2t + 2)\vec{j}.$$

The graph of this flow line in Figure 17.38 looks like a parabola. We check this by seeing that an explicit equation for the path is $y = \frac{1}{2}x^2$.

Example 3   Determine the flow of the vector field $\vec{v} = -y\vec{i} + x\vec{j}$.

Solution   Figure 17.39 suggests that the flow consists of concentric counterclockwise circles, centred at the origin. The system of differential equations for the flow is

$$x'(t) = -y(t) \qquad y'(t) = x(t).$$

The equations $(x(t), y(t)) = (a\cos t, \ a\sin t)$ parameterise a family of counterclockwise circles of radius a, centred at the origin. We check that this family satisfies the system of differential equations:

$$x'(t) = -a\sin t = -y(t) \quad \text{and} \quad y'(t) = a\cos t = x(t).$$

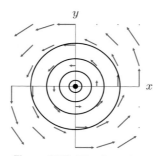

Figure 17.39: The flow of the
vector field $\vec{v} = -y\vec{i} + x\vec{j}$

Approximating Flow Lines Numerically

Often it is not possible to find formulas for the flow lines of a vector field. However, we can approximate them numerically by Euler's method for solving differential equations. Since the flow lines $\vec{r}(t) = x(t)\vec{i} + y(t)\vec{j}$ of a vector field $\vec{v} = \vec{F}(x, y)$ satisfy the differential equation $\vec{r}'(t) = \vec{F}(\vec{r}(t))$, we have

$$\vec{r}(t + \Delta t) \approx \vec{r}(t) + (\Delta t)\vec{r}'(t)$$
$$= \vec{r}(t) + (\Delta t)\vec{F}(\vec{r}(t)) \quad \text{for } \Delta t \text{ near } 0.$$

To approximate a flow line, we start at a point $\vec{r}_0 = \vec{r}(0)$ and estimate the position $\vec{r}_1$ of a particle at time Δt later:

$$\vec{r}_1 = \vec{r}(\Delta t) \approx \vec{r}(0) + (\Delta t)\vec{F}(\vec{r}(0))$$
$$= \vec{r}_0 + (\Delta t)\vec{F}(\vec{r}_0).$$

We then repeat the same procedure starting at $\vec{r}_1$, and so on. The general formula for getting from one point to the next is

$$\vec{r}_{n+1} = \vec{r}_n + (\Delta t)\vec{F}(\vec{r}_n).$$

The points with position vectors $\vec{r}_0, \vec{r}_1, \ldots$ trace out the path, as shown in the next example.

Example 4   Use Euler's method to approximate the flow line through $(1, 2)$ for the vector field $\vec{v} = y^2\vec{i} + 2x^2\vec{j}$.

Solution   The flow is determined by the differential equations $\vec{r}'(t) = \vec{v}$, or equivalently

$$x'(t) = y^2, \qquad y'(t) = 2x^2.$$

We use Euler's method with $\Delta t = 0.02$, giving

$$\begin{aligned} \vec{r}_{n+1} &= \vec{r}_n + 0.02\,\vec{v}(x_n, y_n) \\ &= x_n\vec{i} + y_n\vec{j} + 0.02(y_n^2\vec{i} + 2x_n^2\vec{j}), \end{aligned}$$

or equivalently,

$$x_{n+1} = x_n + 0.02y_n^2, \qquad y_{n+1} = y_n + 0.02 \cdot 2x_n^2.$$

When $t = 0$, we have $(x_0, y_0) = (1, 2)$. Then

$$\begin{aligned} x_1 &= x_0 + 0.02 \cdot y_0^2 = 1 + 0.02 \cdot 2^2 = 1.08, \\ y_1 &= y_0 + 0.02 \cdot 2x_0^2 = 2 + 0.02 \cdot 2 \cdot 1^2 = 2.04. \end{aligned}$$

So after one step $x(0.02) \approx 1.08$ and $y(0.02) \approx 2.04$. Similarly, $x(0.04) = x(2\Delta t) \approx 1.16$, $y(0.04) = y(2\Delta t) \approx 2.08$ and so on. Farther values along the flow line are given in Table 17.2 and plotted in Figure 17.40.

Table 17.2   *Approximated flow line starting at $(1, 2)$ for the vector field $\vec{v} = y^2\vec{i} + 2x^2\vec{j}$*

t	0	0.02	0.04	0.06	0.08	0.1	0.12	0.14	0.16	0.18
x	1	1.08	1.16	1.25	1.34	1.44	1.54	1.65	1.77	1.90
y	2	2.04	2.08	2.14	2.20	2.28	2.36	2.45	2.56	2.69

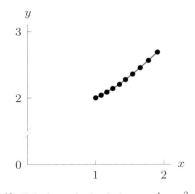

Figure 17.40: Euler's method solution to $x' = y^2$, $y' = 2x^2$

Exercises and Problems for Section 17.4

Exercises

In Exercises 1–3, sketch the vector field and its flow.

1. $\vec{v} = 3\vec{i}$ **2.** $\vec{v} = 2\vec{j}$ **3.** $\vec{v} = 3\vec{i} - 2\vec{j}$

In Exercises 4–9, sketch the vector field and the flow. Then find the system of differential equations associated with the vector field and check that the flow satisfies the system.

4. $\vec{v} = x\vec{i} + y\vec{j}$; $x(t) = ae^t, y(t) = be^t$

5. $\vec{v} = x\vec{i}$; $x(t) = ae^t, y(t) = b$

6. $\vec{v} = x\vec{j}$; $x(t) = a, y(t) = at + b$

7. $\vec{v} = x\vec{i} - y\vec{j}$; $x(t) = ae^t, y(t) = be^{-t}$

8. $\vec{v} = y\vec{i} - x\vec{j}$; $x(t) = a\sin t, y(t) = a\cos t$

9. $\vec{v} = y\vec{i} + x\vec{j}$; $x(t) = a(e^t + e^{-t}), y(t) = a(e^t - e^{-t})$

10. Use a computer or calculator with Euler's method to approximate the flow line through $(1, 2)$ for the vector field $\vec{v} = y^2\vec{i} + 2x^2\vec{j}$ using 5 steps with $\Delta t = 0.1$.

Problems

For Problems 11–14, find the region of the Gulf Stream velocity field in Figure 17.18 on page 955 represented by the given table of velocity vectors (in cm/sec).

11.

$-95\vec{i} - 60\vec{j}$	$18\vec{i} - 48\vec{j}$	$82\vec{i} - 22\vec{j}$
$-29\vec{i} + 48\vec{j}$	$76\vec{i} + 63\vec{j}$	$128\vec{i} - 16\vec{j}$
$26\vec{i} + 105\vec{j}$	$49\vec{i} + 119\vec{j}$	$88\vec{i} + 13\vec{j}$

12.

$35\vec{i} + 131\vec{j}$	$48\vec{i} + 92\vec{j}$	$47\vec{i} + \vec{j}$
$-32\vec{i} + 132\vec{j}$	$-44\vec{i} + 92\vec{j}$	$-42\vec{i} + \vec{j}$
$-51\vec{i} + 73\vec{j}$	$-119\vec{i} + 84\vec{j}$	$-128\vec{i} + 6\vec{j}$

13.

$10\vec{i} - 3\vec{j}$	$11\vec{i} + 16\vec{j}$	$20\vec{i} + 75\vec{j}$
$53\vec{i} - 7\vec{j}$	$58\vec{i} + 23\vec{j}$	$64\vec{i} + 80\vec{j}$
$119\vec{i} - 8\vec{j}$	$121\vec{i} + 31\vec{j}$	$114\vec{i} + 66\vec{j}$

14.

$97\vec{i} - 41\vec{j}$	$72\vec{i} - 24\vec{j}$	$54\vec{i} - 10\vec{j}$
$134\vec{i} - 49\vec{j}$	$131\vec{i} - 44\vec{j}$	$129\vec{i} - 18\vec{j}$
$103\vec{i} - 36\vec{j}$	$122\vec{i} - 30\vec{j}$	$131\vec{i} - 17\vec{j}$

15. $\vec{F}(x, y)$ and $\vec{G}(x, y) = 2\vec{F}(x, y)$ are two vector fields. Illustrating your answer with $\vec{F}(x, y) = -y\vec{i} + x\vec{j}$, describe the graphical difference between:

   (a) The vector fields   **(b)** Their flows

16. Show that the acceleration $\vec{a}$ of an object flowing in a velocity field $\vec{F}(x, y) = u(x, y)\vec{i} + v(x, y)\vec{j}$ is given by $\vec{a} = (u_x u + u_y v)\vec{i} + (v_x u + v_y v)\vec{j}$.

17. A velocity vector field $\vec{v} = -H_y\vec{i} + H_x\vec{j}$ is based on the partial derivatives of a smooth function $H(x, y)$. Explain why

   (a) $\vec{v}$ is perpendicular to grad H.

   (b) the flow lines of $\vec{v}$ are along the level curves of H.

18. Match the vector fields (a)–(f) with their flow lines (I)–(VI). Put arrows on the flow lines indicating the direction of flow.

   (a) $y\vec{i} + x\vec{j}$ **(b)** $-y\vec{i} + x\vec{j}$

   (c) $x\vec{i} + y\vec{j}$ **(d)** $-y\vec{i} + (x + y/10)\vec{j}$

   (e) $-y\vec{i} + (x - y/10)\vec{j}$ **(f)** $(x - y)\vec{i} + (x - y)\vec{j}$

(I)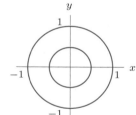

(II)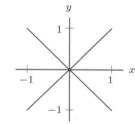

(III)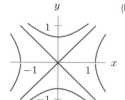

(IV)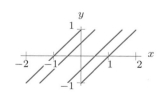

(V)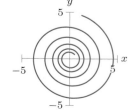

(VI)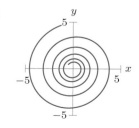

19. A solid metal ball has its center at the origin of a fixed set of axes. The ball rotates once every 24 hours around the z-axis. The direction of rotation is counterclockwise when viewed from above. Let $\vec{v}\,(x, y, z)$ be the velocity vector of the particle of metal at the point (x, y, z) inside the ball. Time is in hours and x, y, z are in meters.

    (a) Find a formula for the vector field $\vec{v}$. Give units for your answer.

    (b) Describe in words the flow lines of $\vec{v}$.

In Problems 20–22, show that every flow line of the vector field $\vec{v}$ lies on a level curve for the function $f(x, y)$.

20. $\vec{v} = x\vec{i} - y\vec{j}$, $f(x, y) = xy$

21. $\vec{v} = y\vec{i} + x\vec{j}$, $f(x, y) = x^2 - y^2$

22. $\vec{v} = ay\vec{i} + bx\vec{j}$, $f(x, y) = bx^2 - ay^2$

23. **(a)** Show that $h(t) = e^{-2at}(x^2 + y^2)$ is constant along any flow line of $\vec{v} = (ax - y)\vec{i} + (x + ay)\vec{j}$.

    (b) Show that points moving with the flow that are on the unit circle centered at the origin at time 0 are on the circle of radius e^{at} centered at the origin at time t.

Strengthen Your Understanding

In Problems 24–25, explain what is wrong with the statement.

24. The flow lines of a vector field whose components are linear functions are all straight lines.

25. If the flow lines of a vector field are all straight lines with the same slope pointing in the same direction, then the vector field is constant.

In Problems 26–27, give an example of:

26. A vector field $\vec{F}\,(x, y, z)$ such that the path $\vec{r}\,(t) = t\vec{i} + t^2\vec{j} + t^3\vec{k}$ is a flow line.

27. A vector field whose flow lines are rays from the origin.

Are the statements in Problems 28–37 true or false? Give reasons for your answer.

28. The flow lines for $\vec{F}\,(x, y) = x\vec{j}$ are parallel to the y-axis.

29. The flow lines of $\vec{F}\,(x, y) = y\vec{i} - x\vec{j}$ are hyperbolas.

30. The flow lines of $\vec{F}\,(x, y) = x\vec{i}$ are parabolas.

31. The vector field in Figure 17.41 has a flow line which lies in the first and third quadrants.

Figure 17.41

32. The vector field in Figure 17.41 has a flow line on which both x and y tend to infinity.

33. If $\vec{F}$ is a gradient vector field, $\vec{F}\,(x, y) = \nabla f(x, y)$, then the flow lines for $\vec{F}$ are the contours for f.

34. If the flow lines for the vector field $\vec{F}\,(\vec{r}\,)$ are all concentric circles centered at the origin, then $\vec{F}\,(\vec{r}\,) \cdot \vec{r} = 0$ for all $\vec{r}$.

35. If the flow lines for the vector field $\vec{F}\,(x, y)$ are all straight lines parallel to the constant vector $\vec{v} = 3\vec{i} + 5\vec{j}$, then $\vec{F}\,(x, y) = \vec{v}$.

36. No flow line for the vector field $\vec{F}\,(x, y) = x\vec{i} + 2\vec{j}$ has a point where the y-coordinate reaches a relative maximum.

37. The vector field $\vec{F}\,(x, y) = e^x\vec{i} + y\vec{j}$ has a flow line that crosses the x-axis.

CHAPTER SUMMARY (see also Ready Reference at the end of the book)

- **Parameterized Curves**
  Parameterizations representing motion in 2- and 3-space, change of parameter, vector form of parametric equations, parametric equation of a line.

- **Velocity and Acceleration Vectors**
  Computing velocity and acceleration, uniform circular motion, the length of a parametric curve.

- **Vector Fields**
  Definition of vector field, visualizing fields, gradient fields.

- **Flow Lines**
  Parametric equations of flow lines, approximating flow lines numerically.

REVIEW EXERCISES AND PROBLEMS FOR CHAPTER SEVENTEEN

Exercises

Write a parameterization for the curves in Exercises 1–13.

1. The equation of the line through $(2, -1, 3)$ and parallel to $5\vec{i} + 4\vec{j} - \vec{k}$.

2. The line passing through the points $(1, 2, 3)$ and $(3, 5, 7)$.

3. The horizontal line through the point $(0, 5)$.

4. The circle of radius 2 centered at the origin starting at the point $(0, 2)$ when $t = 0$.

5. The circle of radius 4 centered at the point $(4, 4)$ starting on the x-axis when $t = 0$.

6. The circle of radius 1 in the xy-plane centered at the origin, traversed counterclockwise when viewed from above.

7. The line through the points $(2, -1, 4)$ and $(1, 2, 5)$.

8. The line through the point $(1, 3, 2)$ perpendicular to the xz-plane.

9. The line through the point $(1, 1, 1)$ perpendicular to the plane $2x - 3y + 5z = 4$.

10. The circle of radius 3 in the xy-plane, centered at the origin, counterclockwise.

11. The circle of radius 3 parallel to the xz-plane, centered at the point $(0, 5, 0)$, and traversed counterclockwise when viewed from $(0, 10, 0)$.

12. The line of intersection of the planes $z = 4 + 2x + 5y$ and $z = 3 + x + 3y$.

13. The circle of radius 10 centered at the point $(0, 0, 7)$, lying horizontally, and traversed in a clockwise direction viewed from the point $(0, 0, 11)$. The parameterization should have period 30.

In Exercises 14–18, find the velocity vector.

14. $x = 3 \cos t, y = 4 \sin t$

15. $x = t, y = t^3 - t$

16. $x = 2 + 3t, y = 4 + t, z = 1 - t$

17. $x = 2 + 3t^2, y = 4 + t^2, z = 1 - t^2$

18. $x = t, y = t^2, z = t^3$

In Exercises 19–22, are the following quantities vectors or scalars? Find them.

19. The velocity of a particle moving, for $t \geq 0$, along the curve $x = 2 + 3\sin\sqrt{2t + 1}$, $y = 4 + 3\cos\sqrt{2t + 1}$, $z = 10 + \sqrt{2t + 1}$.

20. The speed of a particle moving along the curve $x = t^2$, $y = e^t$.

21. The velocity of a particle moving along the curve $x = 5 - \sqrt{3 + \sin t}, y = \sqrt{3 + \cos t}$.

22. The acceleration of a particle moving along the curve $x = te^t, y = e^{2t}$.

23. Are the lines $x = 3 + 2t, y = 5 - t, z = 7 + 3t$ and $x = 3 + t, y = 5 + 2t, z = 7 + 2t$ parallel?

24. Are the lines $x = 3 + 2t, y = 5 - t, z = 7 + 3t$ and $x = 5 + 4t, y = 3 - 2t, z = 1 + 6t$ parallel?

25. Explain how you know the following equations parameterize the same line:

$$\vec{r} = (3 - t)\vec{i} + (3 + 4t)\vec{j} - (1 + 2t)\vec{k}$$
$$\vec{r} = (1 + 2t)\vec{i} + (11 - 8t)\vec{j} + (4t - 5)\vec{k}$$

26. A line is parameterized by $\vec{r} = 10\vec{k} + t(\vec{i} + 2\vec{j} + 3\vec{k})$.

   (a) Suppose we restrict ourselves to $t < 0$. What part of the line do we get?
   (b) Suppose we restrict ourselves to $0 \leq t \leq 1$. What part of the line do we get?

Sketch the vector fields in Exercises 27–28.

27. $\vec{F} = y\vec{i} - x\vec{j}$

28. $\vec{F} = \dfrac{y}{x^2 + y^2}\vec{i} - \dfrac{x}{x^2 + y^2}\vec{j}$

Problems

29. Where does the line $x = 2t + 1, y = 3t - 2, z = -t + 3$ intersect the sphere $(x - 1)^2 + (y + 1)^2 + (z - 2)^2 = 2$?

30. A particle travels along a line, with position at time t given by $\vec{r}(t) = (2 + 5t)\vec{i} + (3 + t)\vec{j} + 2t\vec{k}$.

   (a) Where is the particle when $t = 0$?
   (b) When does the particle reach the point $(12, 5, 4)$?
   (c) Does the particle ever reach $(12, 4, 4)$? Explain.

31. Consider the parametric equations for $0 \leq t \leq \pi$:

   (I) $\vec{r} = \cos(2t)\vec{i} + \sin(2t)\vec{j}$
   (II) $\vec{r} = 2\cos t\vec{i} + 2\sin t\vec{j}$
   (III) $\vec{r} = \cos(t/2)\vec{i} + \sin(t/2)\vec{j}$
   (IV) $\vec{r} = 2\cos t\vec{i} - 2\sin t\vec{j}$

   (a) Match the equations above with four of the curves C_1, C_2, C_3, C_4, C_5 and C_6 in Figure 17.42. (Each curve is part of a circle.)
   (b) Give parametric equations for the curves which have not been matched, again assuming $0 \leq t \leq \pi$.

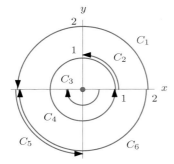

Figure 17.42

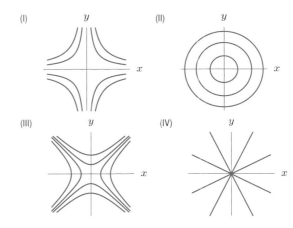

32. (a) What is meant by a vector field?
  (b) Suppose $\vec{a} = a_1\vec{i} + a_2\vec{j} + a_3\vec{k}$ is a constant vector. Which of the following are vector fields? Explain.

  (i) $\vec{r} + \vec{a}$ (ii) $\vec{r} \cdot \vec{a}$
  (iii) $x^2\vec{i} + y^2\vec{j} + z^2\vec{k}$ (iv) $x^2 + y^2 + z^2$

33. Match the level curves in (I)–(IV) with the gradient fields in (A)–(D). All figures have $-2 \le x \le 2, -2 \le y \le 2$.

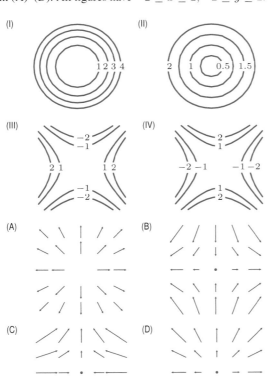

34. Each of the vector fields $\vec{E}, \vec{F}, \vec{G}, \vec{H}$ is tangent to one of the families of curves (I)–(IV). Match them.

$$\vec{E} = x\vec{i} + y\vec{j} \quad \vec{F} = x\vec{i} - y\vec{j}$$
$$\vec{G} = y\vec{i} - x\vec{j} \quad \vec{H} = y\vec{i} + x\vec{j}$$

35. A particle passes through the point $P = (5, 4, 3)$ at time $t = 7$, moving with constant velocity $\vec{v} = 3\vec{i} + \vec{j} + 2\vec{k}$. Find equations for its position at time t.

36. An object moving with constant velocity in 3-space, with coordinates in meters, passes through the point $(1, 1, 1)$, then through $(2, -1, 3)$ five seconds later. What is its velocity vector? What is its acceleration vector?

37. Find parametric equations for a particle moving along the line $y = -2x + 5$ with speed 3.

38. The temperature in °C at (x, y) in the plane is $H = f(x, y)$, where x, y are in centimeters. A particle moves along the curve $x = g(t)$, $y = k(t)$, with t in seconds.

  (a) What does the quantity $\| \operatorname{grad} f \|$ represent in this context? What is its units?
  (b) Write an expression for the speed of the particle. What is its units?
  (c) Write an expression for the rate of change of the particle's temperature with time. What is its units?

39. Find parametric equations for motion along the line $y = 3x + 7$ such that the x-coordinate decreases by 2 units for each unit of time.

40. The y-axis is vertical and the x-axis is horizontal; t represents time. The motion of a particle is given by

$$x = t^3 - 3t, \quad y = t^2 - 2t.$$

  (a) Does the particle ever come to a stop? If so, when and where?
  (b) Is the particle ever moving straight up or down? If so, when and where?
  (c) Is the particle ever moving straight horizontally right or left? If so, when and where?

41. The position of a particle at time t, is given by $\vec{r}(t) = \cos 4t\vec{i} + \sin 4t\vec{j} + 3t\vec{k}$.

    (a) Find the velocity and acceleration of the particle.
    (b) Find the speed of the particle.
    (c) Show that the particle moves with constant speed.
    (d) Find the angle between the particle's position and acceleration vector at time $t = 0$.

42. A stone is swung around on a string at a constant speed with period 2π seconds in a horizontal circle centered at the point $(0, 0, 8)$. When $t = 0$, the stone is at the point $(0, 5, 8)$; it travels clockwise when viewed from above. When the stone is at the point $(5, 0, 8)$, the string breaks and it moves under gravity.

    (a) Parameterize the stone's circular trajectory.
    (b) Find the velocity and acceleration of the stone at the moment before the string breaks.
    (c) Write, but do not solve, the differential equations (with initial conditions) satisfied by the coordinates x, y, z giving the position of the stone after it has left the circle.

43. The origin is on the surface of the earth, and the z-axis points upward. For $t \geq 0$, a particle moves according to

$$x(t) = 5t, \quad y(t) = 3t, \quad z(t) = 15 - t^2 + 2t.$$

    (a) What is the position, velocity, and acceleration at time $t = 0$?
    (b) When and where does the particle hit the ground? How fast is it moving then?

44. Let $f(x, y) = \dfrac{x^2 - y^2}{x^2 + y^2}$.

    (a) In which direction should you move from the point $(1, 1)$ to obtain the maximum rate of increase of f?
    (b) Find a direction in which the directional derivative at the point $(1, 1)$ is equal to zero.
    (c) Suppose you move along the curve $x = e^{2t}, y = 2t^3 + 6t + 1$. What is df/dt at $t = 0$?

45. An ant, starting at the origin, moves at 2 units/sec along the x-axis to the point $(1, 0)$. The ant then moves counterclockwise along the unit circle to $(0, 1)$ at a speed of $3\pi/2$ units/sec, then straight down to the origin at a speed of 2 units/sec along the y-axis.

    (a) Express the ant's coordinates as a function of time, t, in secs.
    (b) Express the reverse path as a function of time.

46. The temperature at the point (x, y) in the plane is given by $F(x, y) = 1/(x^2 + y^2)$. A ladybug moves along a parabola according to the parametric equations

$$x = t, \; y = t^2.$$

Assuming that the ladybug's temperature is the same as the plane at her current location, find the rate of change in the temperature of the ladybug at time t. Use the chain rule to show that for any temperature function $F(x, y)$ and any path of the ladybug $\vec{r} = x(t)\vec{i} + y(t)\vec{j}$, then writing $\vec{v}(t) = d\vec{r}/dt$ gives:

    Rate of change of temperature $= \nabla F(x, y) \cdot \vec{v}$.

47. The motion of the particle is given by the parametric equations

$$x = t^3 - 3t, \; y = t^2 - 2t.$$

Give parametric equations for the tangent line to the path of the particle at time $t = -2$.

48. At time $t = 0$ a particle in uniform circular motion in the plane has velocity $\vec{v} = 6\vec{i} - 4\vec{j}$ and acceleration $\vec{a} = 2\vec{i} + 3\vec{j}$. Find the radius and center of its orbit if at time $t = 0$ it is at the point

    (a) $P = (0, 0)$ **(b)** $P = (10, 50)$

49. Find parametric equations of the line passing through the points $(1, 2, 3)$, $(3, 5, 7)$ and calculate the shortest distance from the line to the origin.

50. On a calculator or a computer, plot $x = 2t/(t^2 + 1)$, $y = (t^2 - 1)/(t^2 + 1)$, first for $-50 \leq t \leq 50$ then for $-5 \leq t \leq 5$. Explain what you see. Is the curve really a circle?

51. A cheerleader has a 0.4 m long baton with a light on one end. She throws the baton in such a way that it moves entirely in a vertical plane. The origin is on the ground and the y-axis is vertical. The center of the baton moves along a parabola and the baton rotates counterclockwise around the center with a constant angular velocity. The baton is initially horizontal and 1.5 m above the ground; its initial velocity is 8 m/sec horizontally and 10 m/sec vertically, and its angular velocity is 2 revolutions per second. Find parametric equations describing the following motions:

    (a) The center of the baton relative to the ground.
    (b) The end of the baton relative to its center.
    (c) The path traced out by the end of the baton relative to the ground.
    (d) Sketch a graph of the motion of the end of the baton.

52. For a and ω positive constants and $t \geq 0$, the position vector of a particle moving in a spiral counterclockwise outward from the origin is given by

$$\vec{r}(t) = at\cos(\omega t)\vec{i} + at\sin(\omega t)\vec{j}.$$

What is the significance of the parameters ω and a?

53. An object is moving on a straight-line path. Can you conclude at all times that:

    (a) Its velocity vector is parallel to the line? Justify your answer.
    (b) Its acceleration vector is parallel to the line? Justify your answer.

54. If $\vec{F} = \vec{r}/\|\vec{r}\|^3$, find the following quantities in terms of x, y, z, or t.

(a) $\|\vec{F}\|$

(b) $\vec{F} \cdot \vec{r}$

(c) A unit vector parallel to $\vec{F}$ and pointing in the same direction

(d) A unit vector parallel to $\vec{F}$ and pointing in the opposite direction

(e) $\vec{F}$ if $\vec{r} = \cos t\vec{i} + \sin t\vec{j} + \vec{k}$

(f) $\vec{F} \cdot \vec{r}$ if $\vec{r} = \cos t\vec{i} + \sin t\vec{j} + \vec{k}$

55. Each of the following vector fields represents an ocean current. Sketch the vector field, and sketch the path of an iceberg in this current. Determine the location of an iceberg at time $t = 7$ if it is at the point $(1, 3)$ at time $t = 0$.

(a) The current everywhere is $\vec{i}$.

(b) The current at (x, y) is $2x\vec{i} + y\vec{j}$.

(c) The current at (x, y) is $-y\vec{i} + x\vec{j}$.

56. Wire is stretched taught from the point $P = (7, 12, -10)$ to the point $Q = (-2, -3, 2)$ and from the point $R = (-20, 17, 1)$ to the point $S = (37, 2, 25)$. Spherical beads of radius 8 cm slide along each wire through holes along an axis through their centers. Can the beads pass each other without touching, regardless of their position?

57. A particle moves with displacement vector $\vec{r}$ and constant speed. Show that the vector representing the velocity is perpendicular to the vector representing the acceleration.

CAS Challenge Problems

58. Let $\vec{r}_0 = x_0\vec{i} + y_0\vec{j} + z_0\vec{k}$, and let $\vec{e}_1$ and $\vec{e}_2$ be perpendicular unit vectors. A circle of radius R, centered at (x_0, y_0, z_0), and lying in the plane parallel to $\vec{e}_1$ and $\vec{e}_2$, is parameterized by $\vec{r}(t) = \vec{r}_0 + R\cos t\vec{e}_1 + R\sin t\vec{e}_2$. We want to parameterize a circle in 3-space with radius 5, centered at $(1, 2, 3)$, and lying in the plane $x + y + z = 6$.

(a) Let $\vec{e}_1 = a\vec{i} + b\vec{j}$ and $\vec{e}_2 = c\vec{i} + d\vec{j} + e\vec{k}$. Write down conditions on $\vec{e}_1$ and $\vec{e}_2$ that make them unit vectors, perpendicular to each other, and lying in the given plane.

(b) Solve the equations in part (a) for a, b, c, d, and e and write a parameterization of the circle.

59. Let $\vec{F}(x, y) = -y(1 - y^2)\vec{i} + x(1 - y^2)\vec{j}$.

(a) Show that $\vec{r} \cdot \vec{F} = 0$. What does this tell you about the shape of the flow lines?

(b) Show that $\vec{r}(t) = \cos t\vec{i} + \sin t\vec{j}$ has velocity vector parallel to $\vec{F}$ at every point, but is not a flow line.

(c) Show $\vec{r}(t) = (1/(\sqrt{1 + t^2}))\vec{i} + (t/(\sqrt{1 + t^2}))\vec{j}$

is a flow line for $\vec{F}$. What is the difference between this curve and the one in part (b)?

60. Let $\vec{F}(x, y) = (x + y)\vec{i} + (4x + y)\vec{j}$.

(a) Show that $\vec{r}(t) = (ae^{3t} + be^{-t})\vec{i} + (2ae^{3t} - 2be^{-t})\vec{j}$, for constant a, b, is a flow line for $\vec{F}$.

(b) Find the flow line passing through $(1, -2)$ at $t = 0$ and describe its behavior as $t \to \infty$. Do the same for the points $(1, -1.99)$ and $(1, -2.01)$. Compare the behavior of the three flow lines.

61. Two surfaces generally intersect in a curve. For each of the following pair of surfaces $f(x, y, z) = 0$ and $g(x, y, z) = 0$, find a parameterization for the curve of intersection by solving for two of the variables in terms of the third.

(a) $3x - 5y + z = 5$, $2x + y + z = 3$,

(b) $3x^2 - 5y + z = 5$, $2x + y + z = 3$,

(c) $x^2 + y^2 = 2$, $3x - y + z = 5$.

PROJECTS FOR CHAPTER SEVENTEEN

1. **Shooting a Basketball**

   A basketball player shoots the ball from 2 metres above the ground toward a basket that is 3 metres above the ground and 5 metres away horizontally.

   (a) Suppose she shoots the ball at an angle of A degrees above the horizontal $(0 < A < \pi/2)$ with an initial speed V. Give the x- and y-coordinates of the position of the basketball at time t. Assume the x-coordinate of the basket is 0 and that the x-coordinate of the shooter is -5. [Hint: There is an acceleration of -10 m/sec^2 in the y-direction; there is no acceleration in the x-direction. Ignore air resistance.]

   (b) Using the parametric equations you obtained in part (a), experiment with different values for V and A, plotting the path of the ball on a graphing calculator or computer to see how close the ball comes to the basket. (The tick marks on the y-axis can be used to locate the basket.) Find some values of V and A for which the shot goes in.

   (c) Find the angle A that minimizes the velocity needed for the ball to reach the basket. (This is a lengthy computation. First find an equation in V and A that holds if the path of the ball passes through the point 5 metres from the shooter and 3 metres above the ground. Then minimize V.)

2. **Kepler's Second Law**

   The planets do not orbit in circles with the sun at the center, nor does the moon orbit in a circle with the earth at the center. In fact, the moon's distance from the earth varies from 220,000 to 260,000 miles. In the last half of the 16th century the Danish astronomer Tycho Brahe (1546–1601) made measurements of the positions of the planets. Johann Kepler (1571–1630) studied this data and arrived at three laws now known as *Kepler's Laws*:

   I. The orbit of each planet is an ellipse with the sun at one focus. In particular, the orbit lies in a plane containing the sun.

   II. As a planet orbits around the sun, the line segment from the sun to the planet sweeps out equal areas in equal times. See Figure 17.43.

   III. The ratio p^2/d^3 is the same for every planet orbiting around the sun, where p is the period of the orbit (time to complete one revolution) and d is the mean distance of the orbit (average of the shortest and farthest distances from the sun).

   Kepler's Laws, impressive as they are, were purely descriptive; Newton's great achievement was to find an underlying cause for them. In this project, you will derive Kepler's Second Law from Newton's Law of Gravity.

   Consider a coordinate system centered at the sun.[3] Let $\vec{r}$ be the position vector of a planet and let $\vec{v}$ and $\vec{a}$ be the planet's velocity and acceleration, respectively. Define $\vec{L} = \vec{r} \times \vec{v}$. (This is a multiple of the planet's angular momentum.)

   (a) Show that $\dfrac{d\vec{L}}{dt} = \vec{r} \times \vec{a}$.

   (b) Consider the planet moving from $\vec{r}$ to $\vec{r} + \Delta\vec{r}$. Explain why the area ΔA about the origin swept out by the planet is approximately $\frac{1}{2}\|\Delta\vec{r} \times \vec{r}\|$.

   (c) Using part (b), explain why $\dfrac{dA}{dt} = \dfrac{1}{2}\|\vec{L}\|$.

   (d) Newton's Laws imply that the planet's gravitational acceleration, $\vec{a}$, is directed toward the sun. Using this fact and part (a), explain why $\vec{L}$ is constant.

   (e) Use parts (c) and (d) to explain Kepler's Second Law.

   (f) Using Kepler's Second Law, determine whether a planet is moving most quickly when it is closest to, or farthest from, the sun.

---

[3] We are assuming the center of the sun is the same as the center of mass of the planet/sun system. This is only approximately true.

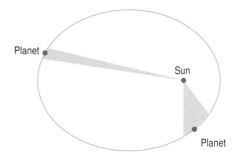

Figure 17.43: The line segment joining a planet to the sun sweeps out equal areas in equal times

3. **Flux Diagrams**

A *flux diagram* uses flow lines to represent a vector field. The arrows drawn on a flow line indicate the direction of the vector field. The flow lines are drawn in such a way that their density is proportional to the magnitude of the vector field at each point. (The density is the number of flow lines per unit length along a curve perpendicular to the vector field.)

Figure 17.44 is a flux diagram for the vector field in 2-space $\vec{F} = \vec{r}/\|\vec{r}\|^2$. Since the field points radially away from the origin, the flow lines are straight lines radiating from the origin. The number of flow lines passing through any circle centered at the origin is a constant k. Therefore, the flow lines passing through a small circle are more densely packed than those passing through a large circle, indicating that the magnitude of the vector field decreases as we move away from the origin. In fact,

$$\text{Density of lines} = \frac{\text{Number of lines passing through circle}}{\text{Circumference of circle}} = \frac{k}{2\pi r} = \frac{k}{2\pi} 1/r,$$

so that the density is proportional to $1/r$, the magnitude of the field.

Sometimes we have to start new lines to make the density proportional to the magnitude. For example, the flow lines of $\vec{v} = x\vec{i}$ are horizontal straight lines directed away from the y-axis. However, since the magnitude of $\vec{v}$ increases linearly with x, we have to make the density of lines increase linearly with x. We achieve this by starting new lines at regular intervals. (See Figure 17.45.)

Draw flux diagrams for the following vector fields:

(a) $\vec{v} = \vec{i}$ (b) $\vec{v} = -y\vec{i} + x\vec{j}$ (c) $\vec{v} = y\vec{i}$ (d) $\vec{v} = y\vec{j}$

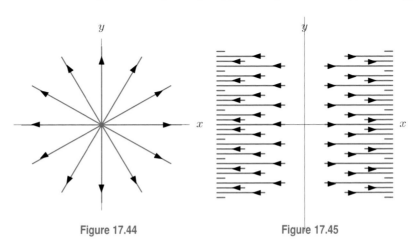

Figure 17.44   **Figure 17.45**

Chapter Eighteen

LINE INTEGRALS

Contents

18.1 THE IDEA OF A LINE INTEGRAL

Imagine that you are rowing on a river with a noticeable current. At times you may be working against the current and at other times you may be moving with it. At the end you have a sense of whether, overall, you were helped or hindered by the current. The line integral, defined in this section, measures the extent to which a curve in a vector field is, overall, going with the vector field or against it.

Orientation of a Curve

A curve can be traced out in two directions, as shown in Figure 18.1. We need to choose one direction before we can define a line integral.

> A curve is said to be **oriented** if we have chosen a direction of travel on it.

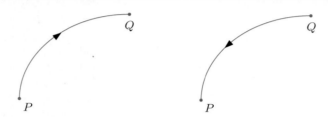

Figure 18.1: A curve with two different orientations represented by arrowheads

Definition of the Line Integral

Consider a vector field $\vec{F}$ and an oriented curve C. We begin by dividing C into n small, almost straight pieces along which $\vec{F}$ is approximately constant. Each piece can be represented by a displacement vector $\Delta\vec{r}_i = \vec{r}_{i+1} - \vec{r}_i$ and the value of $\vec{F}$ at each point of this small piece of C is approximately $\vec{F}(\vec{r}_i)$. See Figures 18.2 and 18.3.

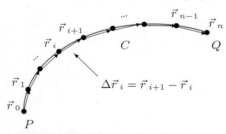

Figure 18.2: The curve C, oriented from P to Q, approximated by straight line segments represented by displacement vectors
$$\Delta\vec{r}_i = \vec{r}_{i+1} - \vec{r}_i$$

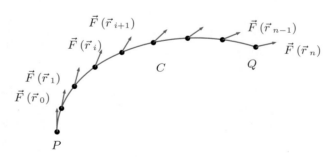

Figure 18.3: The vector field $\vec{F}$ evaluated at the points with position vector $\vec{r}_i$ on the curve C oriented from P to Q

Returning to our initial example, the vector field $\vec{F}$ represents the current and the oriented curve C is the path of the person rowing the boat. We wish to determine to what extent the vector field $\vec{F}$ helps or hinders motion along C. Since the dot product can be used to measure to what extent two vectors point in the same or opposing directions, we form the dot product $\vec{F}(\vec{r}_i) \cdot \Delta\vec{r}_i$ for each point with position vector $\vec{r}_i$ on C. Summing over all such pieces, we get a Riemann sum:

$$\sum_{i=0}^{n-1} \vec{F}(\vec{r}_i) \cdot \Delta\vec{r}_i.$$

We define the line integral, written $\int_C \vec{F} \cdot d\vec{r}$, by taking the limit as $\|\Delta\vec{r}_i\| \to 0$. Provided the limit exists, we make the following definition:

The **line integral** of a vector field $\vec{F}$ along an oriented curve C is

$$\int_C \vec{F} \cdot d\vec{r} = \lim_{\|\Delta \vec{r}_i\| \to 0} \sum_{i=0}^{n-1} \vec{F}\left(\vec{r}_i\right) \cdot \Delta \vec{r}_i.$$

How Does the Limit Defining a Line Integral Work?

The limit in the definition of a line integral exists if $\vec{F}$ is continuous on the curve C and if C is made by joining end to end a finite number of smooth curves. (A vector field $\vec{F} = F_1 \vec{i} + F_2 \vec{j} + F_3 \vec{k}$ is *continuous* if F_1, F_2, and F_3 are continuous, and a *smooth curve* is one that can be parameterised by smooth functions.) We subdivide the curve using a parameterisation which goes from one end of the curve to the other, in the forward direction, without retracing any portion of the curve. A subdivision of the parameter interval gives a subdivision of the curve. All the curves we consider in this book are *piecewise smooth* in this sense. Section 18.2 shows how to use a parameterisation to compute a line integral.

Example 1   Find the line integral of the constant vector field $\vec{F} = \vec{i} + 2\vec{j}$ along the path from $(1, 1)$ to $(10, 10)$ shown in Figure 18.4.

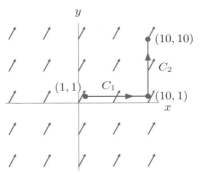

Figure 18.4: The constant vector field $\vec{F} = \vec{i} + 2\vec{j}$ and the path from $(1, 1)$ to $(10, 10)$

Solution   Let C_1 be the horizontal segment of the path going from $(1, 1)$ to $(10, 1)$. When we break this path into pieces, each piece $\Delta \vec{r}$ is horizontal, so $\Delta \vec{r} = \Delta x \vec{i}$ and $\vec{F} \cdot \Delta \vec{r} = (\vec{i} + 2\vec{j}) \cdot \Delta x \vec{i} = \Delta x$. Hence,

$$\int_{C_1} \vec{F} \cdot d\vec{r} = \int_{x=1}^{x=10} dx = 9.$$

Similarly, along the vertical segment C_2, we have $\Delta \vec{r} = \Delta y \vec{j}$ and $\vec{F} \cdot \Delta \vec{r} = (\vec{i} + 2\vec{j}) \cdot \Delta y \vec{j} = 2\Delta y$, so

$$\int_{C_2} \vec{F} \cdot d\vec{r} = \int_{y=1}^{y=10} 2 \, dy = 18.$$

Thus,

$$\int_C \vec{F} \cdot d\vec{r} = \int_{C_1} \vec{F} \cdot d\vec{r} + \int_{C_2} \vec{F} \cdot d\vec{r} = 9 + 18 = 27.$$

What Does the Line Integral Tell Us?

Remember that for any two vectors $\vec{u}$ and $\vec{v}$, the dot product $\vec{u} \cdot \vec{v}$ is positive if $\vec{u}$ and $\vec{v}$ point roughly in the same direction (that is, if the angle between them is less than $\pi/2$). The dot product is zero if $\vec{u}$ is perpendicular to $\vec{v}$ and is negative if they point roughly in opposite directions (that is, if the angle between them is greater than $\pi/2$).

The line integral of $\vec{F}$ adds the dot products of $\vec{F}$ and $\Delta\vec{r}$ along the path. If $||\vec{F}||$ is constant, the line integral gives a positive number if $\vec{F}$ is mostly pointing in the same direction as C, and a negative number if $\vec{F}$ is mostly pointing in the opposite direction to C. The line integral is zero if $\vec{F}$ is perpendicular to the path at all points or if the positive and negative contributions cancel out. In general, the line integral of a vector field $\vec{F}$ along a curve C measures the extent to which C is going with $\vec{F}$ or against it.

Example 2   The vector field $\vec{F}$ and the oriented curves C_1, C_2, C_3, C_4 are shown in Figure 18.5. The curves C_1 and C_3 are the same length. Which of the line integrals $\int_{C_i} \vec{F} \cdot d\vec{r}$, for $i = 1, 2, 3, 4$, are positive? Which are negative? Arrange these line integrals in ascending order.

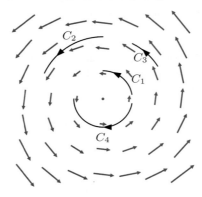

Figure 18.5: Vector field and paths C_1, C_2, C_3, C_4

Solution   The vector field $\vec{F}$ and the line segments $\Delta\vec{r}$ are approximately parallel and in the same direction for the curves C_1, C_2, and C_3. So the contributions of each term $\vec{F} \cdot \Delta\vec{r}$ are positive for these curves. Thus, $\int_{C_1} \vec{F} \cdot d\vec{r}$, $\int_{C_2} \vec{F} \cdot d\vec{r}$, and $\int_{C_3} \vec{F} \cdot d\vec{r}$ are each positive. For the curve C_4, the vector field and the line segments are in opposite directions, so each term $\vec{F} \cdot \Delta\vec{r}$ is negative, and therefore the integral $\int_{C_4} \vec{F} \cdot d\vec{r}$ is negative.

Since the magnitude of the vector field is smaller along C_1 than along C_3, and these two curves are the same length, we have

$$\int_{C_1} \vec{F} \cdot d\vec{r} < \int_{C_3} \vec{F} \cdot d\vec{r}.$$

In addition, the magnitude of the vector field is the same along C_2 and C_3, but the curve C_2 is longer than the curve C_3. Thus,

$$\int_{C_3} \vec{F} \cdot d\vec{r} < \int_{C_2} \vec{F} \cdot d\vec{r}.$$

Putting these results together with the fact that $\int_{C_4} \vec{F} \cdot d\vec{r}$ is negative, we have

$$\int_{C_4} \vec{F} \cdot d\vec{r} < \int_{C_1} \vec{F} \cdot d\vec{r} < \int_{C_3} \vec{F} \cdot d\vec{r} < \int_{C_2} \vec{F} \cdot d\vec{r}.$$

Interpretations of the Line Integral

Work

Recall from Section 13.3 that if a constant force $\vec{F}$ acts on an object while it moves along a straight line through a displacement $\vec{d}$, the work done by the force on the object is

$$\text{Work done} = \vec{F} \cdot \vec{d}.$$

Now suppose we want to find the work done by gravity on an object moving far above the surface of the earth. Since the force of gravity varies with distance from the earth and the path may not be straight, we can't use the formula $\vec{F} \cdot \vec{d}$. We approximate the path by line segments which are small enough that the force is approximately constant on each one. Suppose the force at a point with position vector $\vec{r}$ is $\vec{F}(\vec{r})$, as in Figures 18.2 and 18.3. Then

$$\begin{array}{l}\text{Work done by force } \vec{F}(\vec{r}_i) \\ \text{over small displacement } \Delta \vec{r}_i\end{array} \approx \vec{F}(\vec{r}_i) \cdot \Delta \vec{r}_i,$$

and so,

$$\begin{array}{l}\text{Total work done by force} \\ \text{along oriented curve } C\end{array} \approx \sum_i \vec{F}(\vec{r}_i) \cdot \Delta \vec{r}_i.$$

Taking the limit as $\|\Delta \vec{r}_i\| \to 0$, we get

$$\begin{array}{l}\text{Work done by force } \vec{F}(\vec{r}) \\ \text{along curve } C\end{array} = \lim_{\|\Delta \vec{r}_i\| \to 0} \sum_i \vec{F}(\vec{r}_i) \cdot \Delta \vec{r}_i = \int_C \vec{F} \cdot d\vec{r}.$$

Example 3 A mass lying on a flat table is attached to a spring whose other end is fastened to the wall. (See Figure 18.6.) The spring is extended 20 cm beyond its rest position and released. If the axes are as shown in Figure 18.6, when the spring is extended by a distance of x, the force exerted by the spring on the mass is given by

$$\vec{F}(x) = -kx\vec{i},$$

where k is a positive constant that depends on the strength of the spring.

Suppose the mass moves back to the rest position. How much work is done by the force exerted by the spring?

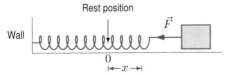

Figure 18.6: Force on mass due to an extended spring

Figure 18.7: Dividing up the interval $0 \leq x \leq 20$ in order to calculate the work done

Solution The path from $x = 20$ to $x = 0$ is divided as shown in Figure 18.7, with a typical segment represented by

$$\Delta \vec{r} = \Delta x \vec{i}.$$

Since we are moving from $x = 20$ to $x = 0$, the quantity Δx will be negative. The work done by the force as the mass moves through this segment is approximated by

$$\text{Work done} \approx \vec{F} \cdot \Delta \vec{r} = (-kx\vec{i}) \cdot (\Delta x \vec{i}) = -kx\,\Delta x.$$

Thus, we have

$$\text{Total work done} \approx \sum -kx\,\Delta x.$$

In the limit, as $\|\Delta x\| \to 0$, this sum becomes an ordinary definite integral. Since the path starts at $x = 20$, this is the lower limit of integration; $x = 0$ is the upper limit. Thus, we get

$$\text{Total work done} = \int_{x=20}^{x=0} -kx\,dx = -\frac{kx^2}{2}\Big|_{20}^{0} = \frac{k(20)^2}{2} = 200k.$$

Note that the work done is positive, since the force acts in the direction of motion.

Example 3 shows how a line integral over a path parallel to the x-axis reduces to a one-variable integral. Section 18.2 shows how to convert *any* line integral into a one-variable integral.

Example 4   A particle with position vector $\vec{r}$ is subject to a force, $\vec{F}$, due to gravity. What is the *sign* of the work done by $\vec{F}$ as the particle moves along the path C_1, a radial line through the centre of the earth, starting 8000 km from the centre and ending 10,000 km from the centre? (See Figure 18.8.)

Solution   We divide the path into small radial segments, $\Delta\vec{r}$, pointing away from the centre of the earth and parallel to the gravitational force. The vectors $\vec{F}$ and $\Delta\vec{r}$ point in opposite directions, so each term $\vec{F} \cdot \Delta\vec{r}$ is negative. Adding all these negative quantities and taking the limit results in a negative value for the total work. Thus, the work done by gravity is negative. The negative sign indicates that we would have to do work *against* gravity to move the particle along the path C_1.

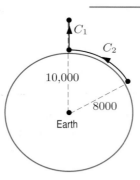

Figure 18.8: The earth

Example 5   Find the sign of the work done by gravity along the curve C_1 in Example 4, but with the opposite orientation.

Solution   Tracing a curve in the opposite direction changes the sign of the line integral because all the segments $\Delta\vec{r}$ change direction, and so every term $\vec{F} \cdot \Delta\vec{r}$ changes sign. Thus, the result will be the negative of the answer found in Example 4. Therefore, the work done by gravity as a particle moves along C_1 toward the centre of the earth is positive.

Example 6   Find the work done by gravity as a particle moves along C_2, an arc of a circle 8000 km long at a distance of 8000 km from the centre of the earth. (See Figure 18.8.)

Solution   Since C_2 is everywhere perpendicular to the gravitational force, $\vec{F} \cdot \Delta\vec{r} = 0$ for all $\Delta\vec{r}$ along C_2. Thus,

$$\text{Work done} = \int_{C_2} \vec{F} \cdot d\vec{r} = 0,$$

so the work done is zero. This is why satellites can remain in orbit without expending any fuel, once they have attained the correct altitude and velocity.

Circulation

The velocity vector field for the Gulf Stream on page 955 shows distinct eddies or regions where the water circulates. We can measure this circulation using a *closed curve*, that is, one that starts and ends at the same point.

> If C is an oriented closed curve, the line integral of a vector field $\vec{F}$ around C is called the **circulation** of $\vec{F}$ around C.

Circulation is a measure of the net tendency of the vector field to point around the curve C. To emphasise that C is closed, the circulation is sometimes denoted $\oint_C \vec{F} \cdot d\vec{r}$, with a small circle on the integral sign.

Example 7   Describe the rotation of the vector fields in Figures 18.9 and 18.10. Find the sign of the circulation of the vector fields around the indicated paths.

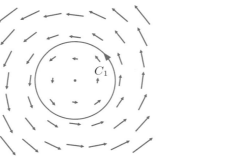

Figure 18.9: A circulating flow

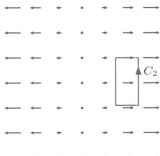

Figure 18.10: A flow with zero circulation

Solution   Consider the vector field in Figure 18.9. If you think of this as representing the velocity of water flowing in a pond, you see that the water is circulating. The line integral around C_1, measuring the circulation around C_1, is positive, because the vectors of the field are all pointing in the direction of the path. By way of contrast, look at the vector field in Figure 18.10. Here the line integral around C_2 is zero because the vertical portions of the path are perpendicular to the field and the contributions from the two horizontal portions cancel out. This means that there is no net tendency for the water to circulate around C_2.

It turns out that the vector field in Figure 18.10 has the property that its circulation around *any* closed path is zero. Water moving according to this vector field has no tendency to circulate around any point, and a leaf dropped into the water will not spin. We'll look at such special fields again later when we introduce the notion of the *curl* of a vector field.

Properties of Line Integrals

Line integrals share some basic properties with ordinary one-variable integrals:

For a scalar constant λ, vector fields $\vec{F}$ and $\vec{G}$, and oriented curves C, C_1, and C_2,

1. $\displaystyle\int_C \lambda \vec{F} \cdot d\vec{r} = \lambda \int_C \vec{F} \cdot d\vec{r}.$ **2.** $\displaystyle\int_C (\vec{F} + \vec{G}) \cdot d\vec{r} = \int_C \vec{F} \cdot d\vec{r} + \int_C \vec{G} \cdot d\vec{r}.$

3. $\displaystyle\int_{-C} \vec{F} \cdot d\vec{r} = -\int_C \vec{F} \cdot d\vec{r}.$ **4.** $\displaystyle\int_{C_1 + C_2} \vec{F} \cdot d\vec{r} = \int_{C_1} \vec{F} \cdot d\vec{r} + \int_{C_2} \vec{F} \cdot d\vec{r}.$

Properties 3 and 4 are concerned with the curve C over which the line integral is taken. If C is an oriented curve, then $-C$ is the same curve traversed in the opposite direction, that is, with the opposite orientation. (See Figure 18.11 on page 982.) Property 3 holds because if we integrate along $-C$, the vectors $\Delta \vec{r}$ point in the opposite direction and the dot products $\vec{F} \cdot \Delta \vec{r}$ are the negatives of what they were along C.

If C_1 and C_2 are oriented curves with C_1 ending where C_2 begins, we construct a new oriented curve, called $C_1 + C_2$, by joining them together. (See Figure 18.12.) Property 4 is the analogue for line integrals of the property for definite integrals which says that

$$\int_a^b f(x)\,dx = \int_a^c f(x)\,dx + \int_c^b f(x)\,dx.$$

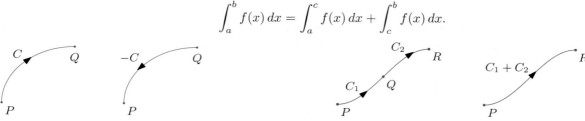

Figure 18.11: A curve, C, and its opposite, $-C$

Figure 18.12: Joining two curves, C_1, and C_2, to make a new one, $C_1 + C_2$

Exercises and Problems for Section 18.1

Exercises

In Exercises 1–6, say whether you expect the line integral of the pictured vector field over the given curve to be positive, negative, or zero.

1.

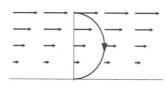

2.

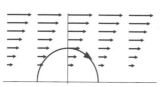

3.

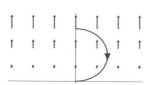

4.

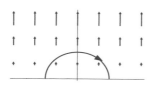

5.

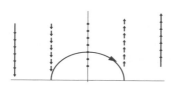

6.

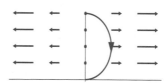

For the vector fields in Exercises 7–8, is the line integral positive, negative, or zero along

(a) A? **(b)** C_1, C_2, C_3, C_4?

(c) C, the closed curve consisting of all the Cs together?

7.

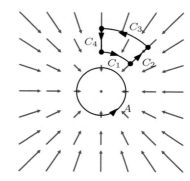

8.

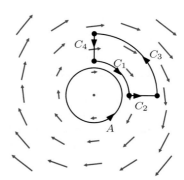

In Exercises 9–17, calculate the line integral of the vector field along the line between the given points.

9. $\vec{F} = x\vec{j}$, from $(1, 0)$ to $(3, 0)$

10. $\vec{F} = x\vec{j}$, from $(2, 0)$ to $(2, 5)$

11. $\vec{F} = 6\vec{i} - 7\vec{j}$, from $(0, 0)$ to $(7, 6)$

12. $\vec{F} = 6\vec{i} + y^2\vec{j}$, from $(3, 0)$ to $(7, 0)$

13. $\vec{F} = 3\vec{i} + 4\vec{j}$, from $(0, 6)$ to $(0, 13)$

14. $\vec{F} = x\vec{i}$, from $(2, 0)$ to $(6, 0)$

15. $\vec{F} = x\vec{i} + y\vec{j}$, from $(2, 0)$ to $(6, 0)$

16. $\vec{F} = \vec{r} = x\vec{i} + y\vec{j}$, from $(2, 2)$ to $(6, 6)$

17. $\vec{F} = x\vec{i} + 6\vec{j} - \vec{k}$, from $(0, -2, 0)$ to $(0, -10, 0)$

In Exercises 18–20, find $\int_C \vec{F} \cdot d\vec{r}$ for the given $\vec{F}$ and C.

18. $\vec{F} = 5\vec{i} + 7\vec{j}$, and C is the x-axis from $(-1, 0)$ to $(-9, 0)$.

19. $\vec{F} = x^2\vec{i} + y^2\vec{j}$, and C is the x-axis from $(2, 0)$ to $(3, 0)$.

20. $\vec{F} = 6x\vec{i} + (x + y^2)\vec{j}$; C is the y-axis from $(0, 3)$ to $(0, 5)$.

In Exercises 21–24, calculate the line integral.

21. $\int_C (2\vec{j} + 3\vec{k}) \cdot d\vec{r}$ where C is the y-axis from the origin to the point $(0, 10, 0)$.

22. $\int_C (2x\vec{i} + 3y\vec{j}) \cdot d\vec{r}$, where C is the line from $(1, 0, 0)$ to $(1, 0, 5)$.

23. $\int_C ((2y + 7)\vec{i} + 3x\vec{j}) \cdot d\vec{r}$, where C is the line from $(1, 0, 0)$ to $(5, 0, 0)$.

24. $\int_C (x\vec{i} + y\vec{j} + z\vec{k}) \cdot d\vec{r}$ where C is the unit circle in the xy-plane, oriented counterclockwise.

Problems

In Problems 25–28, let C_1 be the line from $(0, 0)$ to $(0, 1)$; let C_2 be the line from $(1, 0)$ to $(0, 1)$; let C_3 be the semicircle in the upper half plane from $(-1, 0)$ to $(1, 0)$. Do the line integrals of the vector field along each of the paths C_1, C_2, and C_3 appear to be positive, negative, or zero?

25.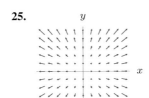

26.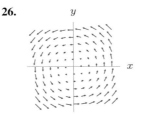

27.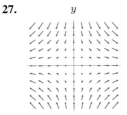

28.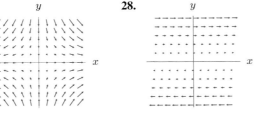

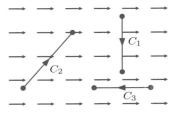

Figure 18.13

29. Consider the vector field $\vec{F}$ shown in Figure 18.13, together with the paths C_1, C_2, and C_3. Arrange the line integrals $\int_{C_1} \vec{F} \cdot d\vec{r}$, $\int_{C_2} \vec{F} \cdot d\vec{r}$ and $\int_{C_3} \vec{F} \cdot d\vec{r}$ in ascending order.

30. Compute $\int_C \vec{F} \cdot d\vec{r}$, where C is the oriented curve in Figure 18.14 and $\vec{F}$ is a vector field constant on each of the three straight segments of C:

$$\vec{F} = \begin{cases} \vec{i} & \text{on } PQ \\ 2\vec{i} - \vec{j} & \text{on } QR \\ 3\vec{i} + \vec{j} & \text{on } RS. \end{cases}$$

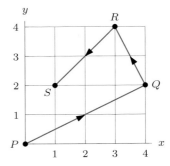

Figure 18.14

31. An object moves along the curve C in Figure 18.15 while being acted on by the force field $\vec{F}(x, y) = y\vec{i} + x^2\vec{j}$.

(a) Evaluate $\vec{F}$ at the points $(0, -1)$, $(1, -1)$, $(2, -1)$, $(3, -1)$, $(4, -1)$, $(4, 0)$, $(4, 1)$, $(4, 2)$, $(4, 3)$.

(b) Make a sketch showing the force field along C.
(c) Find the work done by $\vec{F}$ on the object.

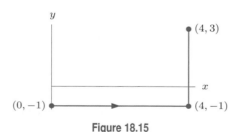

Figure 18.15

32. Let $\vec{F}$ be the constant force field $\vec{j}$ in Figure 18.16. On which of the paths C_1, C_2, C_3 is zero work done by $\vec{F}$? Explain.

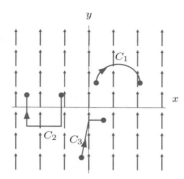

Figure 18.16

In Problems 33–37, give conditions on one or more of the constants a, b, c to ensure that the line integral $\int_C \vec{F} \cdot d\vec{r}$ has the given sign.

33. Positive for $\vec{F} = a\vec{i} + b\vec{j} + c\vec{k}$ and C is the line from the origin to $(10, 0, 0)$.

34. Positive for $\vec{F} = ay\vec{i} + c\vec{k}$ and C is the unit circle in the xy-plane, centered at the origin and oriented counterclockwise when viewed from above.

35. Negative for $\vec{F} = b\vec{j} + c\vec{k}$ and C is the parabola $y = x^2$ in the xy-plane from the origin to $(3, 9, 0)$.

36. Positive for $\vec{F} = ay\vec{i} - ax\vec{j} + (c-1)\vec{k}$ and C is the line segment from the origin to $(1, 1, 1)$.

37. Negative for $\vec{F} = a\vec{i} + b\vec{j} - \vec{k}$ and C is the line segment from $(1, 2, 3)$ to $(1, 2, c)$.

38. (a) For each of the vector fields, $\vec{F}$, shown in Figure 18.17, sketch a curve for which the integral $\int_C \vec{F} \cdot d\vec{r}$ is positive.
(b) For which of the vector fields is it possible to make your answer to part (a) a closed curve?

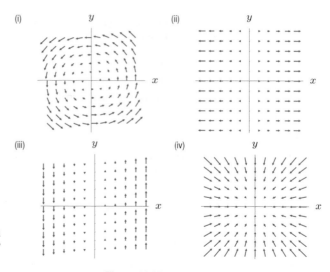

Figure 18.17

For Problems 39–43, say whether you expect the given vector field to have positive, negative, or zero circulation around the closed curve $C = C_1 + C_2 + C_3 + C_4$ in Figure 18.18. The segments C_1 and C_3 are circular arcs centered at the origin; C_2 and C_4 are radial line segments. You may find it helpful to sketch the vector field.

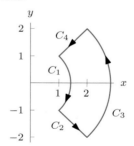

Figure 18.18

39. $\vec{F}(x, y) = x\vec{i} + y\vec{j}$

40. $\vec{F}(x, y) = -y\vec{i} + x\vec{j}$

41. $\vec{F}(x, y) = y\vec{i} - x\vec{j}$

42. $\vec{F}(x, y) = x^2\vec{i}$

43. $\vec{F}(x, y) = -\dfrac{y}{x^2 + y^2}\vec{i} + \dfrac{x}{x^2 + y^2}\vec{j}$

44. A horizontal square has sides of 1000 km running north-south and east-west. A wind blows from the east and decreases in magnitude toward the north at a rate of 6 meter/sec for every 500 km. Compute the circulation of the wind counterclockwise around the square.

45. Let $\vec{F} = x\vec{i} + y\vec{j}$ and let C_1 be the line joining $(1, 0)$ to $(0, 2)$ and let C_2 be the line joining $(0, 2)$ to $(-1, 0)$. Is $\int_{C_1} \vec{F} \cdot d\vec{r} = -\int_{C_2} \vec{F} \cdot d\vec{r}$? Explain.

46. The vector field $\vec{F}$ has $||\vec{F}|| \leq 7$ everywhere and C is the circle of radius 1 centered at the origin. What is the largest possible value of $\int_C \vec{F} \cdot d\vec{r}$? The smallest possible value? What conditions lead to these values?

47. Along a curve C, a vector field $\vec{F}$ is everywhere tangent to C in the direction of orientation and has constant magnitude $\|\vec{F}\| = m$. Use the definition of the line integral to explain why

$$\int_C \vec{F} \cdot d\vec{r} = m \cdot \text{Length of } C.$$

48. Explain why the following statement is true: Whenever the line integral of a vector field around every closed curve is zero, the line integral along a curve with fixed endpoints has a constant value independent of the path taken between the endpoints.

49. Explain why the converse to the statement in Problem 48 is also true: Whenever the line integral of a vector field depends only on endpoints and not on paths, the circulation around every closed curve is zero.

In Problems 50–51, use the fact that the force of gravity on a particle of mass m at the point with position vector $\vec{r}$ is

$$\vec{F} = -\frac{GMm\vec{r}}{\|\vec{r}\|^3}$$

where and G is a constant and M is the mass of the earth.

50. Calculate the work done by the force of gravity on a particle of mass m as it moves radially from 8000 km to 10,000 km from the center of the earth.

51. Calculate the work done by the force of gravity on a particle of mass m as it moves radially from 8000 km from the center of the earth to infinitely far away.

In Problems 52–55 we explore the notion of the electric potential of an electric field $\vec{E}$. We chose a point P_0 to be the ground, that is, the potential at P_0 is zero. Then the potential $\phi(P)$ at a point P is defined to be the work done moving a particle of charge 1 coulomb from P_0 to P. (The potential does not depend on the path chosen.) Q, E are constants.

52. Write a line integral for $\phi(P)$, using the fact that the electric field acts on a particle with force $\vec{E}$.

53. Let $\vec{E} = \left(\dfrac{Q}{4\pi\epsilon}\right)\dfrac{\vec{r}}{\|\vec{r}\|^3}$, and let P_0 be a point a units from the origin. Describe the set of points with zero potential.

54. An equipotential surface is a surface on which the potential is constant. Describe the equipotential surfaces of $\vec{E} = \dfrac{Q}{4\pi\epsilon}\dfrac{\vec{r}}{\|\vec{r}\|^3}$, where the ground P_0 is chosen to be at distance a from the origin.

55. Let $\vec{E} = \dfrac{Q}{4\pi\epsilon}\dfrac{\vec{r}}{\|\vec{r}\|^3}$ and let the ground P_0 be chosen to be a units from the origin.

    (a) Find a formula for ϕ.
    (b) Engineers often choose the ground point to be "at infinity." Why?

Strengthen Your Understanding

In Problems 56–57, explain what is wrong with the statement.

56. If $\vec{F}$ is a vector field and C is an oriented curve, then $\int_{-C} \vec{F} \cdot d\vec{r}$ must be less than zero.

57. It is possible that for a certain vector field $\vec{F}$ and oriented path C, we have $\int_C \vec{F} \cdot d\vec{r} = 2\vec{i} - 3\vec{j}$.

In Problems 58–59, give an example of:

58. A nonzero vector field $\vec{F}$ such that $\int_C \vec{F} \cdot d\vec{r} = 0$, where C is the straight line curve from $(0, 0)$ to $(1, 1)$.

59. Two oriented curves C_1 and C_2 in the plane such that, for $\vec{F}(x, y) = x\vec{j}$, we have $\int_{C_1} \vec{F} \cdot d\vec{r} > 0$ and $\int_{C_2} \vec{F} \cdot d\vec{r} < 0$.

Are the statements in Problems 60–62 true or false? Explain why or give a counterexample.

60. $\int_C \vec{F} \cdot d\vec{r}$ is a vector.

61. Suppose C_1 is the unit square joining the points $(0, 0)$, $(1, 0)$, $(1, 1)$, $(0, 1)$ oriented clockwise and C_2 is the same square but traversed twice in the opposite direction. If $\int_{C_1} \vec{F} \cdot d\vec{r} = 3$, then $\int_{C_2} \vec{F} \cdot d\vec{r} = -6$.

62. The line integral of $\vec{F} = x\vec{i} + y\vec{j} = \vec{r}$ along the semicircle $x^2 + y^2 = 1, y \geq 0$, oriented counterclockwise, is zero.

Are the statements in Problems 63–69 true or false? Give reasons for your answer.

63. The line integral $\int_C \vec{F} \cdot d\vec{r}$ is a scalar.

64. If C_1 and C_2 are oriented curves, and the length of C_1 is greater than the length of C_2, then $\int_{C_1} \vec{F} \cdot d\vec{r} > \int_{C_2} \vec{F} \cdot d\vec{r}$.

65. If C is an oriented curve and $\int_C \vec{F} \cdot d\vec{r} = 0$, then $\vec{F} = \vec{0}$.

66. If $\vec{F} = \vec{i}$ is a vector field in 2-space, then $\int_C \vec{F} \cdot d\vec{r} > 0$, where C is the oriented line from $(0, 0)$ to $(1, 0)$.

67. If $\vec{F} = \vec{i}$ is a vector field in 2-space, then $\int_C \vec{F} \cdot d\vec{r} > 0$, where C is the oriented line from $(0, 0)$ to $(0, 1)$.

68. If C_1 is the upper semicircle $x^2 + y^2 = 1, y \geq 0$ and C_2 is the lower semicircle $x^2 + y^2 = 1, y \leq 0$, both oriented counterclockwise, then for any vector field $\vec{F}$, we have $\int_{C_1} \vec{F} \cdot d\vec{r} = -\int_{C_2} \vec{F} \cdot d\vec{r}$.

69. The work done by the force $\vec{F} = -y\vec{i} + x\vec{j}$ on a particle moving clockwise around the boundary of the square $-1 \leq x \leq 1, -1 \leq y \leq 1$ is positive.

18.2   COMPUTING LINE INTEGRALS OVER PARAMETERIZED CURVES

The goal of this section is to show how to use a parameterisation of a curve to convert a line integral into an ordinary one-variable integral.

Using a Parameterisation to Evaluate a Line Integral

Recall the definition of the line integral,

$$\int_C \vec{F} \cdot d\vec{r} = \lim_{\|\Delta \vec{r}_i\| \to 0} \sum \vec{F}(\vec{r}_i) \cdot \Delta \vec{r}_i,$$

where the $\vec{r}_i$ are the position vectors of points subdividing the curve into short pieces. Now suppose we have a smooth parameterisation, $\vec{r}(t)$, of C for $a \le t \le b$, so that $\vec{r}(a)$ is the position vector of the starting point of the curve and $\vec{r}(b)$ is the position vector of the end. Then we can divide C into n pieces by dividing the interval $a \le t \le b$ into n pieces, each of size $\Delta t = (b - a)/n$. See Figures 18.19 and 18.20.

At each point $\vec{r}_i = \vec{r}(t_i)$ we want to compute

$$\vec{F}(\vec{r}_i) \cdot \Delta \vec{r}_i.$$

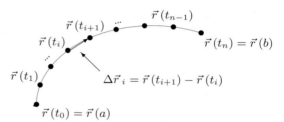

$\Delta \vec{r}_i = \vec{r}(t_{i+1}) - \vec{r}(t_i)$

Figure 18.19: Subdivision of the interval $a \le t \le b$

Figure 18.20: Corresponding subdivision of the parameterised path C

Since $t_{i+1} = t_i + \Delta t$, the displacement vectors $\Delta \vec{r}_i$ are given by

$$
\begin{aligned}
\Delta \vec{r}_i &= \vec{r}(t_{i+1}) - \vec{r}(t_i) \\
&= \vec{r}(t_i + \Delta t) - \vec{r}(t_i) \\
&= \frac{\vec{r}(t_i + \Delta t) - \vec{r}(t_i)}{\Delta t} \cdot \Delta t \\
&\approx \vec{r}\,'(t_i)\Delta t,
\end{aligned}
$$

where we use the facts that Δt is small and that $\vec{r}(t)$ is differentiable to obtain the last approximation.

Therefore,

$$\int_C \vec{F} \cdot d\vec{r} \approx \sum \vec{F}(\vec{r}_i) \cdot \Delta \vec{r}_i \approx \sum \vec{F}(\vec{r}(t_i)) \cdot \vec{r}\,'(t_i)\,\Delta t.$$

Notice that $\vec{F}(\vec{r}(t_i)) \cdot \vec{r}\,'(t_i)$ is the value at t_i of a one-variable function of t, so this last sum is really a one-variable Riemann sum. In the limit as $\Delta t \to 0$, we get a definite integral:

$$\lim_{\Delta t \to 0} \sum \vec{F}(\vec{r}(t_i)) \cdot \vec{r}\,'(t_i)\,\Delta t = \int_a^b \vec{F}(\vec{r}(t)) \cdot \vec{r}\,'(t)\,dt.$$

Thus, we have the following result:

If $\vec{r}(t)$, for $a \leq t \leq b$, is a smooth parameterisation of an oriented curve C and $\vec{F}$ is a vector field which is continuous on C, then

$$\int_C \vec{F} \cdot d\vec{r} = \int_a^b \vec{F}(\vec{r}(t)) \cdot \vec{r}'(t)\, dt.$$

In words: To compute the line integral of $\vec{F}$ over C, take the dot product of $\vec{F}$ evaluated on C with the velocity vector, $\vec{r}'(t)$, of the parameterisation of C, then integrate along the curve.

Even though we assumed that C is smooth, we can use the same formula to compute line integrals over curves which are *piecewise smooth*, such as the boundary of a rectangle. If C is piecewise smooth, we apply the formula to each one of the smooth pieces and add the results by applying property 4 on page 981.

Example 1 Compute $\int_C \vec{F} \cdot d\vec{r}$ where $\vec{F} = (x+y)\vec{i} + y\vec{j}$ and C is the quarter unit circle, oriented counterclockwise as shown in Figure 18.21.

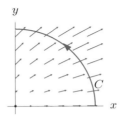

Figure 18.21: The vector field $\vec{F} = (x+y)\vec{i} + y\vec{j}$ and the quarter circle C

Solution Since most of the vectors in $\vec{F}$ along C point generally in a direction opposite to the orientation of C, we expect our answer to be negative. The first step is to parameterise C by

$$\vec{r}(t) = x(t)\vec{i} + y(t)\vec{j} = \cos t\,\vec{i} + \sin t\,\vec{j}, \quad 0 \leq t \leq \frac{\pi}{2}.$$

Substituting the parameterisation into $\vec{F}$, we get $\vec{F}(x(t), y(t)) = (\cos t + \sin t)\vec{i} + \sin t\,\vec{j}$. The vector $\vec{r}'(t) = x'(t)\vec{i} + y'(t)\vec{j} = -\sin t\,\vec{i} + \cos t\,\vec{j}$. Then

$$\int_C \vec{F} \cdot d\vec{r} = \int_0^{\pi/2} ((\cos t + \sin t)\vec{i} + \sin t\,\vec{j}) \cdot (-\sin t\,\vec{i} + \cos t\,\vec{j})\,dt$$

$$= \int_0^{\pi/2} (-\cos t \sin t - \sin^2 t + \sin t \cos t)\,dt$$

$$= \int_0^{\pi/2} -\sin^2 t\, dt = -\frac{\pi}{4} \approx -0.7854.$$

So the answer is negative, as expected.

Example 2 Consider the vector field $\vec{F} = x\vec{i} + y\vec{j}$.

(a) Suppose C_1 is the line segment joining $(1, 0)$ to $(0, 2)$ and C_2 is a part of a parabola with its vertex at $(0, 2)$, joining the same points in the same order. (See Figure 18.22.) Verify that

$$\int_{C_1} \vec{F} \cdot d\vec{r} = \int_{C_2} \vec{F} \cdot d\vec{r}.$$

(b) If C is the triangle shown in Figure 18.23, show that $\int_C \vec{F} \cdot d\vec{r} = 0$.

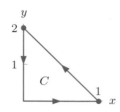

Figure 18.22   Figure 18.23

Solution   (a) We parameterise C_1 by $\vec{r}(t) = (1-t)\vec{i} + 2t\vec{j}$ with $0 \le t \le 1$. Then $\vec{r}'(t) = -\vec{i} + 2\vec{j}$, so

$$\int_{C_1} \vec{F} \cdot d\vec{r} = \int_0^1 \vec{F}(1-t, 2t) \cdot (-\vec{i} + 2\vec{j})\, dt = \int_0^1 ((1-t)\vec{i} + 2t\vec{j}) \cdot (-\vec{i} + 2\vec{j})\, dt$$

$$= \int_0^1 (5t - 1)\, dt = \frac{3}{2}.$$

To parameterise C_2, we use the fact that it is part of a parabola with vertex at $(0,2)$, so its equation is of the form $y = -kx^2 + 2$ for some k. Since the parabola crosses the x-axis at $(1,0)$, we find that $k = 2$ and $y = -2x^2 + 2$. Therefore, we use the parameterisation $\vec{r}(t) = t\vec{i} + (-2t^2 + 2)\vec{j}$ with $0 \le t \le 1$, which has $\vec{r}' = \vec{i} - 4t\vec{j}$. This traces out C_2 in reverse, since $t = 0$ gives $(0,2)$, and $t = 1$ gives $(1,0)$. Thus, we make $t = 0$ the upper limit of integration and $t = 1$ the lower limit:

$$\int_{C_2} \vec{F} \cdot d\vec{r} = \int_1^0 \vec{F}(t, -2t^2 + 2) \cdot (\vec{i} - 4t\vec{j})\, dt = -\int_0^1 (t\vec{i} + (-2t^2 + 2)\vec{j}) \cdot (\vec{i} - 4t\vec{j})\, dt$$

$$= -\int_0^1 (8t^3 - 7t)\, dt = \frac{3}{2}.$$

So the line integrals of $\vec{F}$ along C_1 and C_2 have the same value.

(b) We break $\int_C \vec{F} \cdot d\vec{r}$ into three pieces, one of which we have already computed (namely, the piece connecting $(1,0)$ to $(0,2)$, where the line integral has value $3/2$). The piece running from $(0,2)$ to $(0,0)$ can be parameterised by $\vec{r}(t) = (2-t)\vec{j}$ with $0 \le t \le 2$. The piece running from $(0,0)$ to $(1,0)$ can be parameterised by $\vec{r}(t) = t\vec{i}$ with $0 \le t \le 1$. Then

$$\int_C \vec{F} \cdot d\vec{r} = \frac{3}{2} + \int_0^2 \vec{F}(0, 2-t) \cdot (-\vec{j})\, dt + \int_0^1 \vec{F}(t, 0) \cdot \vec{i}\, dt$$

$$= \frac{3}{2} + \int_0^2 (2-t)\vec{j} \cdot (-\vec{j})\, dt + \int_0^1 t\vec{i} \cdot \vec{i}\, dt$$

$$= \frac{3}{2} + \int_0^2 (t-2)\, dt + \int_0^1 t\, dt = \frac{3}{2} + (-2) + \frac{1}{2} = 0.$$

Example 3   Let C be the closed curve consisting of the upper half-circle of radius 1 and the line forming its diameter along the x-axis, oriented counterclockwise. (See Figure 18.24.) Find $\int_C \vec{F} \cdot d\vec{r}$ where $\vec{F}(x, y) = -y\vec{i} + x\vec{j}$.

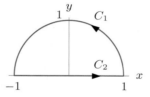

Figure 18.24: The curve $C = C_1 + C_2$ for Example 3

Solution We write $C = C_1 + C_2$ where C_1 is the half-circle and C_2 is the line, and compute $\int_{C_1} \vec{F} \cdot d\vec{r}$ and $\int_{C_2} \vec{F} \cdot d\vec{r}$ separately. We parameterise C_1 by $\vec{r}(t) = \cos t \vec{i} + \sin t \vec{j}$, with $0 \leq t \leq \pi$. Then

$$\int_{C_1} \vec{F} \cdot d\vec{r} = \int_0^\pi (-\sin t \vec{i} + \cos t \vec{j}) \cdot (-\sin t \vec{i} + \cos t \vec{j}) \, dt$$

$$= \int_0^\pi (\sin^2 t + \cos^2 t) \, dt = \int_0^\pi 1 \, dt = \pi.$$

For C_2, we have $\int_{C_2} \vec{F} \cdot d\vec{r} = 0$, since the vector field $\vec{F}$ has no $\vec{i}$ component along the x-axis (where $y = 0$) and is therefore perpendicular to C_2 at all points.

Finally, we can write

$$\int_C \vec{F} \cdot d\vec{r} = \int_{C_1} \vec{F} \cdot d\vec{r} + \int_{C_2} \vec{F} \cdot d\vec{r} = \pi + 0 = \pi.$$

It is no accident that the result for $\int_{C_1} \vec{F} \cdot d\vec{r}$ is the same as the length of the curve C_1. See Problem 47 on page 985 and Problem 34 on page 991.

The next example illustrates the computation of a line integral over a path in 3-space.

Example 4 A particle travels along the helix C given by $\vec{r}(t) = \cos t \vec{i} + \sin t \vec{j} + 2t\vec{k}$ and is subject to a force $\vec{F} = x\vec{i} + z\vec{j} - xy\vec{k}$. Find the total work done on the particle by the force for $0 \leq t \leq 3\pi$.

Solution The work done is given by a line integral, which we evaluate using the given parameterisation:

$$\text{Work done} = \int_C \vec{F} \cdot d\vec{r} = \int_0^{3\pi} \vec{F}(\vec{r}(t)) \cdot \vec{r}'(t) \, dt$$

$$= \int_0^{3\pi} (\cos t \vec{i} + 2t\vec{j} - \cos t \sin t \vec{k}) \cdot (-\sin t \vec{i} + \cos t \vec{j} + 2\vec{k}) \, dt$$

$$= \int_0^{3\pi} (-\cos t \sin t + 2t \cos t - 2\cos t \sin t) \, dt$$

$$= \int_0^{3\pi} (-3\cos t \sin t + 2t \cos t) \, dt = -4.$$

The Differential Notation $\int_C P \, dx + Q \, dy + R dz$

There is an alternative notation for line integrals that is often useful. For the vector field $\vec{F} = P(x, y, z)\vec{i} + Q(x, y, z)\vec{j} + R(x, y, z)\vec{k}$ and an oriented curve C, if we write $d\vec{r} = dx\vec{i} + dy\vec{j} + dz\vec{k}$ we have

$$\int_C \vec{F} \cdot d\vec{r} = \int_C P(x, y, z)dx + Q(x, y, z)dy + R(x, y, z)dz.$$

Example 5 Evaluate $\int_C xy \, dx - y^2 \, dy$ where C is the line segment from $(0, 0)$ to $(2, 6)$.

Solution We parameterise C by $x = t, y = 3t$, $0 \leq t \leq 2$. Thus, $dx = dt, dy = 3dt$, so

$$\int_C xy \, dx - y^2 \, dy = \int_0^2 t(3t)dt - (3t)^2(3dt) = \int_0^2 (-24t^2) \, dt = -64.$$

Line integrals can be expressed either using vectors or using differentials. If the independent variables are distances, then visualizing a line integral in terms of dot products can be useful. However, if the independent variables are, for example, temperature and volume, then the dot product does not have physical meaning, so differentials are more natural.

Independence of Parameterisation

Since there are many different ways of parameterising a given oriented curve, you may be wondering what happens to the value of a given line integral if you choose another parameterisation. The answer is that the choice of parameterisation makes no difference. Since we initially defined the line integral without reference to any particular parameterisation, this is exactly as we would expect.

Example 6   Consider the oriented path which is a straight-line segment L running from $(0,0)$ to $(1,1)$. Calculate the line integral of the vector field $\vec{F} = (3x-y)\vec{i} + x\vec{j}$ along L using each of the parameterisations

(a) $A(t) = (t,t), \quad 0 \le t \le 1,$

(b) $D(t) = (e^t - 1, e^t - 1), \quad 0 \le t \le \ln 2.$

Solution   The line L has equation $y = x$. Both $A(t)$ and $D(t)$ give a parameterisation of L: each has both coordinates equal and each begins at $(0,0)$ and ends at $(1,1)$. Now let's calculate the line integral of the vector field $\vec{F} = (3x - y)\vec{i} + x\vec{j}$ using each parameterisation.

(a) Using $A(t)$, we get

$$\int_L \vec{F} \cdot d\vec{r} = \int_0^1 ((3t - t)\vec{i} + t\vec{j}) \cdot (\vec{i} + \vec{j})\, dt = \int_0^1 3t\, dt = \frac{3t^2}{2}\Big|_0^1 = \frac{3}{2}.$$

(b) Using $D(t)$, we get

$$\int_L \vec{F} \cdot d\vec{r} = \int_0^{\ln 2} \left((3(e^t - 1) - (e^t - 1))\,\vec{i} + (e^t - 1)\vec{j}\right) \cdot (e^t\vec{i} + e^t\vec{j})\, dt$$
$$= \int_0^{\ln 2} 3(e^{2t} - e^t)\, dt = 3\left(\frac{e^{2t}}{2} - e^t\right)\Big|_0^{\ln 2} = \frac{3}{2}.$$

The fact that both answers are the same illustrates that the value of a line integral is independent of the parameterisation of the path. Problems 39–41 at the end of this section give another way of seeing this.

Exercises and Problems for Section 18.2

Exercises

In Exercises 1–3, write $\int_C \vec{F} \cdot d\vec{r}$ in the form $\int_a^b g(t)dt$. (Give a formula for g and numbers for a and b. You do not need to evaluate the integral.)

1. $\vec{F} = y\vec{i} + x\vec{j}$ and C is the semicircle from $(0,1)$ to $(0,-1)$ with $x > 0$.

2. $\vec{F} = x\vec{i} + z^2\vec{k}$ and C is the line from $(0,1,0)$ to $(2,3,2)$.

3. $\vec{F} = (\cos x)\vec{i} + (\cos y)\vec{j} + (\cos z)\vec{k}$ and C is the unit circle in the plane $z = 10$, centered on the z-axis and oriented counterclockwise when viewed from above.

In Exercises 4–18, find $\int_C \vec{F} \cdot d\vec{r}$ for the given $\vec{F}$ and C.

4. $\vec{F} = 2\vec{i} + \vec{j}$; C is the x-axis from $x = 10$ to $x = 7$.

5. $\vec{F} = 3\vec{j} - \vec{i}$; C is the line $y = x$ from $(1,1)$ to $(5,5)$.

6. $\vec{F} = x\vec{i} + y\vec{j}$ and C is the line from $(0,0)$ to $(3,3)$.

7. $\vec{F} = y\vec{i} - x\vec{j}$ and C is the right-hand side of the unit circle, starting at $(0,1)$.

8. $\vec{F} = x^2\vec{i} + y^2\vec{j}$ and C is the line from the point $(1,2)$ to the point $(3,4)$.

9. $\vec{F} = -y\sin x\vec{i} + \cos x\vec{j}$ and C is the parabola $y = x^2$ between $(0,0)$ and $(2,4)$.

10. $\vec{F} = y^3\vec{i} + x^2\vec{j}$ and C is the line from $(0,0)$ to $(3,2)$.

11. $\vec{F} = 2y\vec{i} - (\sin y)\vec{j}$ counterclockwise around the unit circle C starting at the point $(1,0)$.

12. $\vec{F} = \ln y\vec{i} + \ln x\vec{j}$ and C is the curve given parametrically by $(2t, t^3)$, for $2 \le t \le 4$.

13. $\vec{F} = x\vec{i} + 6\vec{j} - \vec{k}$, and C is the line $x = y = z$ from $(0,0,0)$ to $(2,2,2)$.

14. $\vec{F} = (2x - y + 4)\vec{i} + (5y + 3x - 6)\vec{j}$ and C is the triangle with vertices $(0,0)$, $(3,0)$, $(3,2)$ traversed counterclockwise.

15. $\vec{F} = x\vec{i} + 2zy\vec{j} + x\vec{k}$ and C is $\vec{r} = t\vec{i} + t^2\vec{j} + t^3\vec{k}$ for $1 \leq t \leq 2$.

16. $\vec{F} = x^3\vec{i} + y^2\vec{j} + z\vec{k}$ and C is the line from the origin to the point $(2,3,4)$.

17. $\vec{F} = -y\vec{i} + x\vec{j} + 5\vec{k}$ and C is the helix $x = \cos t, y = \sin t, z = t$, for $0 \leq t \leq 4\pi$.

18. $\vec{F} = e^y\vec{i} + \ln(x^2 + 1)\vec{j} + \vec{k}$ and C is the circle of radius 2 centered at the origin in the yz-plane in Figure 18.25.

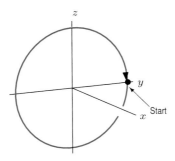

Figure 18.25

In Exercises 19–23, find the line integral.

19. $\displaystyle\int_C (3\vec{i} + (y+5)\vec{j}) \cdot d\vec{r}$ where C is the line from $(0,0)$ to $(0,3)$.

20. $\displaystyle\int_C (2x\vec{i} + 3y\vec{j}) \cdot d\vec{r}$ where C is the line from $(1,0,0)$ to $(5,0,0)$.

21. $\int_C (2y^2\vec{i} + x\vec{j}) \cdot d\vec{r}$ where C is the line segment from $(3,1)$ to $(0,0)$.

22. $\int_C (x\vec{i} + y\vec{j}) \cdot d\vec{r}$ where C is the semicircle with center at $(2,0)$ and going from $(3,0)$ to $(1,0)$ in the region $y > 0$.

23. Find $\int_C ((x^2 + y)\vec{i} + y^3\vec{j}) \cdot d\vec{r}$ where C consists of the three line segments from $(4,0,0)$ to $(4,3,0)$ to $(0,3,0)$ to $(0,3,5)$.

In Exercises 24–25, express the line integral $\int_C \vec{F} \cdot d\vec{r}$ in differential notation.

24. $\vec{F} = 3x\vec{i} - y\sin x\vec{j}$

25. $\vec{F} = y^2\vec{i} + z^2\vec{j} + (x^2 - 5)\vec{k}$

In Exercises 26–27, find $\vec{F}$ so that the line integral equals $\int_C \vec{F} \cdot d\vec{r}$.

26. $\int_C (x + 2y)dx + x^2 y\, dy$

27. $\int_C e^{-3y}dx - yz(\sin x)dy + (y + z)dy$

Evaluate the line integrals in Exercises 28–29.

28. $\int_C y\, dx + x\, dy$ where C is the parameterized path $x = t^2$, $y = t^3$, $1 \leq t \leq 5$.

29. $\int_C dx + y\, dy + z\, dz$ where C is one turn of the helix $x = \cos t, y = \sin t, z = 3t, 0 \leq t \leq 2\pi$.

Evaluate the line integrals in Exercises 30–31.

30. $\int_C 3y\, dx + 4x\, dy$ where C is the straight-line path from $(1,3)$ to $(5,9)$.

31. $\int_C x\, dx + z\, dy - y\, dz$ where C is the circle of radius 3 in the yz plane centered at the origin, oriented counterclockwise when viewed from the positive y-axis.

Problems

32. Curves C_1 and C_2 are parametrized as follows:

C_1 is $(x(t), y(t)) = (0, t)$ for $-1 \leq t \leq 1$

C_2 is $(x(t), y(t)) = (\cos t, \sin t)$ for $\dfrac{\pi}{2} \leq t \leq \dfrac{3\pi}{2}$.

   (a) Sketch C_1 and C_2 with arrows showing their orientation.

   (b) Suppose $\vec{F} = (x+3y)\vec{i} + y\vec{j}$. Calculate $\int_C \vec{F} \cdot d\vec{r}$, where C is the curve given by $C = C_1 + C_2$.

33. In Example 6 on page 990 we integrated $\vec{F} = (3x - y)\vec{i} + x\vec{j}$ over two parameterizations of the line from $(0,0)$ to $(1,1)$, getting $3/2$ each time. Now compute the line integral along two different paths with the same endpoints, and show that the answers are different.

   (a) The path (t, t^2), with $0 \leq t \leq 1$

   (b) The path (t^2, t), with $0 \leq t \leq 1$

34. Let $\vec{F} = -y\vec{i} + x\vec{j}$ and let C be the unit circle oriented counterclockwise.

   (a) Show that $\vec{F}$ has a constant magnitude of 1 on C.

   (b) Show that $\vec{F}$ is always tangent to the circle C.

   (c) Show that $\int_C \vec{F} \cdot d\vec{r} = $ Length of C.

35. A spiral staircase in a building is in the shape of a helix of radius 5 meters. Between two floors of the building, the stairs make one full revolution and climb by 4 meters. A person carries a bag of groceries up two floors. The combined mass of the person and the groceries is 70 kg and the gravitational force is $70g$ downward, where g is the acceleration due to gravity. Calculate the work done by the person against gravity.

36. Suppose C is the line segment from the point $(0,0)$ to the point $(4,12)$ and $\vec{F} = xy\vec{i} + x\vec{j}$.

   (a) Is $\int_C \vec{F} \cdot d\vec{r}$ greater than, less than, or equal to zero? Give a geometric explanation.

(b) A parameterization of C is $(x(t), y(t)) = (t, 3t)$ for $0 \le t \le 4$. Use this to compute $\int_C \vec{F} \cdot d\vec{r}$.

(c) Suppose a particle leaves the point $(0, 0)$, moves along the line toward the point $(4, 12)$, stops before reaching it and backs up, stops again and reverses direction, then completes its journey to the endpoint. All travel takes place along the line segment joining the point $(0, 0)$ to the point $(4, 12)$. If we call this path C', explain why $\int_{C'} \vec{F} \cdot d\vec{r} = \int_C \vec{F} \cdot d\vec{r}$.

(d) A parameterization for a path like C' is given, for $0 \le t \le 4$, by

$$(x(t), y(t)) = \left(\frac{t^3 - 6t^2 + 11t}{3}, (t^3 - 6t^2 + 11t) \right)$$

Check that this parameterization begins at the point $(0, 0)$ and ends at the point $(4, 12)$. Check also that all points of C' lie on the line segment connecting the point $(0, 0)$ to the point $(4, 12)$. What are the values of t at which the particle changes direction?

(e) Find $\int_{C'} \vec{F} \cdot d\vec{r}$ using the parameterization in part (d). Do you get the same answer as in part (b)?

37. Calculate the line integral of $\vec{F} = (3x - y)\vec{i} + x\vec{j}$ along the line segment L from $(0, 0)$ to $(1, 1)$ using each of the parameterizations

(a) $B(t) = (2t, 2t), \quad 0 \le t \le 1/2$

(b) $C(t) = \left(\dfrac{t^2 - 1}{3}, \dfrac{t^2 - 1}{3} \right), \quad 1 \le t \le 2$

38. If C is $\vec{r} = (t + 1)\vec{i} + 2t\vec{j} + 3t\vec{k}$ for $0 \le t \le 1$, we know $\int_C \vec{F}(\vec{r}) \cdot d\vec{r} = 5$. Find the value of the integrals:

(a) $\int_1^0 \vec{F}((t + 1)\vec{i} + 2t\vec{j} + 3t\vec{k}) \cdot (\vec{i} + 2\vec{j} + 3\vec{k})dt$

(b) $\int_0^1 \vec{F}((t^2 + 1)\vec{i} + 2t^2\vec{j} + 3t^2\vec{k}) \cdot (2t\vec{i} + 4t\vec{j} + 6t\vec{k})dt$

(c) $\int_{-1}^1 \vec{F}((t^2 + 1)\vec{i} + 2t^2\vec{j} + 3t^2\vec{k}) \cdot (2t\vec{i} + 4t\vec{j} + 6t\vec{k})dt$

In Example 6 on page 990 two parameterizations, $A(t)$, and $D(t)$, are used to convert a line integral into a definite integral. In Problem 37, two other parameterizations, $B(t)$ and $C(t)$, are used on the same line integral. In Problems 39–41 show that two definite integrals corresponding to two of the given parameterizations are equal by finding a substitution which converts one integral to the other. This gives us another way of seeing why changing the parameterization of the curve does not change the value of the line integral.

39. $A(t)$ and $B(t)$

40. $A(t)$ and $C(t)$

41. $A(t)$ and $D(t)$

Strengthen Your Understanding

In Problems 42–43, explain what is wrong with the statement.

42. For the vector field $\vec{F} = x\vec{i} - y\vec{j}$ and oriented path C parameterized by $x = \cos t, y = \sin t, 0 \le t \le \pi/2$, we have

$$\int_C \vec{F} \cdot d\vec{r} = \int_0^{\pi/2} (\cos t\vec{i} - \sin t\vec{j}) \cdot (\cos t\vec{i} + \sin t\vec{j}) \, dt.$$

43. $\int_C 3 \, dx + 4 \, dy > 0$

In Problems 44–45, give an example of:

44. A vector field $\vec{F}$ such that, for the parameterized path $\vec{r}(t) = 3 \cos t\vec{i} + 3 \sin t\vec{j}, -\pi/2 \le t \le \pi/2$, the integral $\int_C \vec{F} \cdot d\vec{r}$ can be computed geometrically, without using the parameterization.

45. A parameterized path C such that, for the vector field $\vec{F}(x, y) = \sin y\vec{i}$, the integral $\int_C \vec{F} \cdot d\vec{r}$ is nonzero and can be computed geometrically, without using the parameterization.

Are the statements in Problems 46–54 true or false? Give reasons for your answer.

46. If C_1 and C_2 are oriented curves with C_2 beginning where C_1 ends, then $\int_{C_1 + C_2} \vec{F} \cdot d\vec{r} > \int_{C_1} \vec{F} \cdot d\vec{r}$.

47. The line integral $\int_C 4\vec{i} \cdot d\vec{r}$ over the curve C parameterized by $\vec{r}(t) = t\vec{i} + t^2\vec{j}$, for $0 \le t \le 2$, is positive.

48. If C_1 is the curve parameterized by $\vec{r}_1(t) = \cos t\vec{i} + \sin t\vec{j}$, with $0 \le t \le \pi$, and C_2 is the curve parameterized by $\vec{r}_2(t) = \cos t\vec{i} - \sin t\vec{j}, 0 \le t \le \pi$, then for any vector field $\vec{F}$ we have $\int_{C_1} \vec{F} \cdot d\vec{r} = \int_{C_2} \vec{F} \cdot d\vec{r}$.

49. If C_1 is the curve parameterized by $\vec{r}_1(t) = \cos t\vec{i} + \sin t\vec{j}$, with $0 \le t \le \pi$, and C_2 is the curve parameterized by $\vec{r}_2(t) = \cos(2t)\vec{i} + \sin(2t)\vec{j}, 0 \le t \le \frac{\pi}{2}$, then for any vector field $\vec{F}$ we have $\int_{C_1} \vec{F} \cdot d\vec{r} = \int_{C_2} \vec{F} \cdot d\vec{r}$.

50. If C is the curve parameterized by $\vec{r}(t)$, for $a \le t \le b$ with $\vec{r}(a) = \vec{r}(b)$, then $\int_C \vec{F} \cdot d\vec{r} = 0$ for any vector field $\vec{F}$. (Note that C starts and ends at the same place.)

51. If C_1 is the line segment from $(0, 0)$ to $(1, 0)$ and C_2 is the line segment from $(0, 0)$ to $(2, 0)$, then for any vector field $\vec{F}$, we have $\int_{C_2} \vec{F} \cdot d\vec{r} = 2 \int_{C_1} \vec{F} \cdot d\vec{r}$.

52. If C is a circle of radius a, centered at the origin and oriented counterclockwise, then $\int_C 2x\vec{i} + y\vec{j} \cdot d\vec{r} = 0$.

53. If C is a circle of radius a, centered at the origin and oriented counterclockwise, then $\int_C 2y\vec{i} + x\vec{j} \cdot d\vec{r} = 0$.

54. If C_1 is the curve parameterized by $\vec{r}_1(t) = t\vec{i} + t^2\vec{j}$, with $0 \le t \le 2$, and C_2 is the curve parameterized by $\vec{r}_2(t) = (2 - t)\vec{i} + (2 - t)^2\vec{j}, 0 \le t \le 2$, then for any vector field $\vec{F}$ we have $\int_{C_1} \vec{F} \cdot d\vec{r} = -\int_{C_2} \vec{F} \cdot d\vec{r}$.

55. If C_1 is the path parameterized by $\vec{r}_1(t) = (t, t)$, $0 \le t \le 1$, and if C_2 is the path parameterized by $\vec{r}_2(t) = (1-t, 1-t)$, $0 \le t \le 1$, and if $\vec{F} = x\vec{i} + y\vec{j}$, which of the following is true?

(a) $\int_{C_1} \vec{F} \cdot d\vec{r} > \int_{C_2} \vec{F} \cdot d\vec{r}$

(b) $\int_{C_1} \vec{F} \cdot d\vec{r} < \int_{C_2} \vec{F} \cdot d\vec{r}$

(c) $\int_{C_1} \vec{F} \cdot d\vec{r} = \int_{C_2} \vec{F} \cdot d\vec{r}$

56. If C_1 is the path parameterized by $\vec{r}_1(t) = (t, t)$, for $0 \le t \le 1$, and if C_2 is the path parameterized by $\vec{r}_2(t) = (\sin t, \sin t)$, for $0 \le t \le 1$, and if $\vec{F} = x\vec{i} + y\vec{j}$, which of the following is true?

(a) $\int_{C_1} \vec{F} \cdot d\vec{r} > \int_{C_2} \vec{F} \cdot d\vec{r}$

(b) $\int_{C_1} \vec{F} \cdot d\vec{r} < \int_{C_2} \vec{F} \cdot d\vec{r}$

(c) $\int_{C_1} \vec{F} \cdot d\vec{r} = \int_{C_2} \vec{F} \cdot d\vec{r}$

18.3 GRADIENT FIELDS AND PATH-INDEPENDENT FIELDS

For a function, f, of one variable, the Fundamental Theorem of Calculus tells us that the definite integral of a rate of change, f', gives the total change in f:

$$\int_a^b f'(t)\, dt = f(b) - f(a).$$

What about functions of two or more variables? The quantity that describes the rate of change is the gradient vector field. If we know the gradient of a function f, can we compute the total change in f between two points? The answer is yes, using a line integral.

Finding the Total Change in f from grad f: The Fundamental Theorem

To find the change in f between two points P and Q, we choose a smooth path C from P to Q, then divide the path into many small pieces. See Figure 18.26.

First we estimate the change in f as we move through a displacement $\Delta \vec{r}_i$ from $\vec{r}_i$ to $\vec{r}_{i+1}$. Suppose $\vec{u}$ is a unit vector in the direction of $\Delta \vec{r}_i$. Then the change in f is given by

$$f(\vec{r}_{i+1}) - f(\vec{r}_i) \approx \text{Rate of change of } f \times \text{Distance moved in direction of } \vec{u}$$
$$= f_{\vec{u}}(\vec{r}_i)\|\Delta \vec{r}_i\|$$
$$= \text{grad } f \cdot \vec{u}\,\|\Delta \vec{r}_i\|$$
$$= \text{grad } f \cdot \Delta \vec{r}_i. \qquad \text{since } \Delta \vec{r}_i = \|\Delta \vec{r}_i\|\vec{u}$$

Therefore, summing over all pieces of the path, the total change in f is given by

$$\text{Total change} = f(Q) - f(P) \approx \sum_{i=0}^{n-1} \text{grad } f(\vec{r}_i) \cdot \Delta \vec{r}_i.$$

In the limit as $\|\Delta \vec{r}_i\|$ approaches zero, this suggests the following result:

Theorem 18.1: The Fundamental Theorem of Calculus for Line Integrals

Suppose C is a piecewise smooth oriented path with starting point P and ending point Q. If f is a function whose gradient is continuous on the path C, then

$$\int_C \text{grad } f \cdot d\vec{r} = f(Q) - f(P).$$

Notice that there are many different paths from P to Q. (See Figure 18.27.) However, the value of the line integral $\int_C \text{grad } f \cdot d\vec{r}$ depends only on the endpoints of C; it does not depend on where C goes in between. Problem 62 on page 1003 shows how the Fundamental Theorem for Line Integrals can be derived from the one-variable Fundamental Theorem of Calculus.

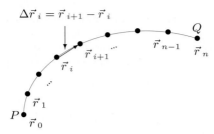

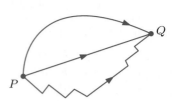

Figure 18.26: Subdivision of the path from P to Q. We estimate the change in f along $\Delta\vec{r}_i$

Figure 18.27: There are many different paths from P to Q: all give the same value of $\int_C \text{grad } f \cdot d\vec{r}$

Example 1   Suppose that $\text{grad } f$ is everywhere perpendicular to the curve joining P and Q shown in Figure 18.28.

(a) Explain why you expect the path joining P and Q to be a contour.
(b) Using a line integral, show that $f(P) = f(Q)$.

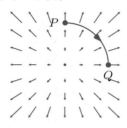

Figure 18.28: The gradient vector field of the function f

Solution   (a) The gradient of f is everywhere perpendicular to the path from P to Q, as you expect along a contour.

(b) Consider the path from P to Q shown in Figure 18.28 and evaluate the line integral

$$\int_C \text{grad } f \cdot d\vec{r} = f(Q) - f(P).$$

Since $\text{grad } f$ is everywhere perpendicular to the path, the line integral is 0. Thus, $f(Q) = f(P)$.

Example 2   Consider the vector field $\vec{F} = x\vec{i} + y\vec{j}$. In Example 2 on page 987 we calculated $\int_{C_1} \vec{F} \cdot d\vec{r}$ and $\int_{C_2} \vec{F} \cdot d\vec{r}$ over the oriented curves shown in Figure 18.29 and found they were the same. Find a scalar function f with $\text{grad } f = \vec{F}$. Hence, find an easy way to calculate the line integrals, and explain how we could have expected them to be the same.

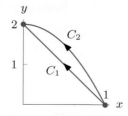

Figure 18.29: Find the line integral of $\vec{F} = x\vec{i} + y\vec{j}$ over the curves C_1 and C_2

Solution   One possibility for f is

$$f(x, y) = \frac{x^2}{2} + \frac{y^2}{2}.$$

You can check that $\operatorname{grad} f = x\vec{i} + y\vec{j}$. Now we can use the Fundamental Theorem to compute the line integral. Since $\vec{F} = \operatorname{grad} f$ we have

$$\int_{C_1} \vec{F} \cdot d\vec{r} = \int_{C_1} \operatorname{grad} f \cdot d\vec{r} = f(0, 2) - f(1, 0) = \frac{3}{2}.$$

Notice that the calculation looks exactly the same for C_2. Since the value of the integral depends only on the values of f at the endpoints, it is the same no matter what path we choose.

Path-Independent, or Conservative, Vector Fields

In the previous example, the line integral was independent of the path taken between the two (fixed) endpoints. We give vector fields whose line integrals have this property a special name.

> A vector field $\vec{F}$ is said to be **path-independent**, or **conservative**, if for any two points P and Q, the line integral $\int_C \vec{F} \cdot d\vec{r}$ has the same value along any piecewise smooth path C from P to Q lying in the domain of $\vec{F}$.

If, on the other hand, the line integral $\int_C \vec{F} \cdot d\vec{r}$ does depend on the path C joining P to Q, then $\vec{F}$ is said to be a *path-dependent* vector field.

Now suppose that $\vec{F}$ is any continuous gradient field, so $\vec{F} = \operatorname{grad} f$. If C is a path from P to Q, the Fundamental Theorem for Line Integrals tells us that

$$\int_C \vec{F} \cdot d\vec{r} = f(Q) - f(P).$$

Since the right-hand side of this equation does not depend on the path, but only on the endpoints of the path, the vector field $\vec{F}$ is path-independent. Thus, we have the following important result:

> If $\vec{F}$ is a continuous gradient vector field, then $\vec{F}$ is path-independent.

Why Do We Care About Path-Independent, or Conservative, Vector Fields?

Many of the fundamental vector fields of nature are path-independent—for example, the gravitational field and the electric field of particles at rest. The fact that the gravitational field is path-independent means that the work done by gravity when an object moves depends only on the starting and ending points and not on the path taken. For example, the work done by gravity (computed by the line integral) on a bicycle being carried to a sixth floor apartment is the same whether it is carried up the stairs in a zig-zag path or taken straight up in an elevator.

When a vector field is path-independent, we can define the *potential energy* of a body. When the body moves to another position, the potential energy changes by an amount equal to the work done by the vector field, which depends only on the starting and ending positions. If the work done had not been path-independent, the potential energy would depend both on the body's current position *and* on how it got there, making it impossible to define a useful potential energy.

Project 1 on page 1021 explains why path-independent force vector fields are also called *conservative* vector fields: When a particle moves under the influence of a conservative vector field, the *total energy* of the particle is *conserved*. It turns out that the force field is obtained from the gradient of the potential energy function.

Path-Independent Fields and Gradient Fields

We have seen that every gradient field is path-independent. What about the converse? That is, given a path-independent vector field $\vec{F}$, can we find a function f such that $\vec{F} = \text{grad } f$? The answer is yes, provided that $\vec{F}$ is continuous.

How to Construct f from $\vec{F}$

First, notice that there are many different choices for f, since we can add a constant to f without changing grad f. If we pick a fixed starting point P, then by adding or subtracting a constant to f, we can ensure that $f(P) = 0$. For any other point Q, we define $f(Q)$ by the formula

$$f(Q) = \int_C \vec{F} \cdot d\vec{r}, \quad \text{where } C \text{ is any path from } P \text{ to } Q.$$

Since $\vec{F}$ is path-independent, it does not matter which path we choose from P to Q. On the other hand, if $\vec{F}$ is not path-independent, then different choices might give different values for $f(Q)$, so f would not be a function (a function has to have a single value at each point).

We still have to show that the gradient of the function f really is $\vec{F}$; we do this on page 998. However, by constructing a function f in this manner, we have the following result:

Theorem 18.2: Path-independent Fields Are Gradient Fields

If $\vec{F}$ is a continuous path-independent vector field on an open region R, then $\vec{F} = \text{grad } f$ for some f defined on R.

Combining Theorems 18.1 and 18.2, we have

A continuous vector field $\vec{F}$ defined on an open region is path-independent if and only if $\vec{F}$ is a gradient vector field.

The function f is sufficiently important that it is given a special name:

If a vector field $\vec{F}$ is of the form $\vec{F} = \text{grad } f$ for some scalar function f, then f is called a **potential function** for the vector field $\vec{F}$.

Warning

Physicists use the convention that a function ϕ is a potential function for a vector field $\vec{F}$ if $\vec{F} = -\text{grad }\phi$. See Problem 63 on page 1003.

Example 3   Show that the vector field $\vec{F}(x, y) = y \cos x \vec{i} + (\sin x + y)\vec{j}$ is path-independent.

Solution   Let's suppose $\vec{F}$ does have a potential function f, so that $\vec{F} = \operatorname{grad} f$. This means

$$\frac{\partial f}{\partial x} = y \cos x \qquad \text{and} \qquad \frac{\partial f}{\partial y} = \sin x + y.$$

Integrating the expression for $\partial f / \partial x$ with respect to x shows that

$$f(x, y) = y \sin x + C(y) \qquad \text{where } C(y) \text{ is a function of } y \text{ only.}$$

The constant of integration here is an arbitrary function $C(y)$ of y, since $\partial(C(y))/\partial x = 0$. Differentiating this expression for $f(x, y)$ with respect to y and using $\partial f / \partial y = \sin x + y$ gives

$$\frac{\partial f}{\partial y} = \sin x + C'(y) = \sin x + y.$$

Thus, we must have $C'(y) = y$, so $g(y) = y^2/2 + A$, where A is some constant. Thus,

$$f(x, y) = y \sin x + \frac{y^2}{2} + A$$

is a potential function for $\vec{F}$. Therefore, $\vec{F}$ is path-independent.

Example 4   The gravitational field, $\vec{F}$, of an object of mass M is given by

$$\vec{F} = -\frac{GM}{r^3}\vec{r}.$$

Show that $\vec{F}$ is a gradient field by finding a potential function for $\vec{F}$.

Solution   The vector $\vec{F}$ points directly in toward the origin. If $\vec{F} = \operatorname{grad} f$, then $\vec{F}$ must be perpendicular to the level surfaces of f, so the level surfaces of f must be spheres. Also, if $\operatorname{grad} f = \vec{F}$, then $\|\operatorname{grad} f\| = \|\vec{F}\| = GM/r^2$ is the rate of change of f in the direction toward the origin. Now, differentiating with respect to r gives the rate of change in a radially outward direction. Thus, if $w = f(x, y, z)$ we have

$$\frac{dw}{dr} = -\frac{GM}{r^2} = GM\left(-\frac{1}{r^2}\right) = GM\frac{d}{dr}\left(\frac{1}{r}\right).$$

So let's try

$$w = \frac{GM}{r} \qquad \text{or} \qquad f(x, y, z) = \frac{GM}{\sqrt{x^2 + y^2 + z^2}}.$$

We calculate

$$f_x = \frac{\partial}{\partial x}\frac{GM}{\sqrt{x^2 + y^2 + z^2}} = \frac{-GMx}{(x^2 + y^2 + z^2)^{3/2}},$$

$$f_y = \frac{\partial}{\partial y}\frac{GM}{\sqrt{x^2 + y^2 + z^2}} = \frac{-GMy}{(x^2 + y^2 + z^2)^{3/2}},$$

$$f_z = \frac{\partial}{\partial z}\frac{GM}{\sqrt{x^2 + y^2 + z^2}} = \frac{-GMz}{(x^2 + y^2 + z^2)^{3/2}}.$$

So

$$\operatorname{grad} f = f_x\vec{i} + f_y\vec{j} + f_z\vec{k} = \frac{-GM}{(x^2 + y^2 + z^2)^{3/2}}(x\vec{i} + y\vec{j} + z\vec{k}) = \frac{-GM}{r^3}\vec{r} = \vec{F}.$$

Our computations show that $\vec{F}$ is a gradient field and that $f = GM/r$ is a potential function for $\vec{F}$.

Path-independent vector fields are rare, but often important. Section 18.4 gives a method for determining whether a vector field has the property.

Why Path-Independent Vector Fields Are Gradient Fields: Showing grad $f = \vec{F}$

Suppose $\vec{F}$ is a path-independent vector field. On page 996 we defined the function f, which we hope will satisfy grad $f = \vec{F}$, as follows:

$$f(x_0, y_0) = \int_C \vec{F} \cdot d\vec{r},$$

where C is a path from a fixed starting point P to a point $Q = (x_0, y_0)$. This integral has the same value for any path from P to Q because $\vec{F}$ is path-independent. Now we show why grad $f = \vec{F}$. We consider vector fields in 2-space; the argument in 3-space is essentially the same.

First, we write the line integral in terms of the components $\vec{F}(x, y) = F_1(x, y)\vec{i} + F_2(x, y)\vec{j}$ and the components $d\vec{r} = dx\vec{i} + dy\vec{j}$:

$$f(x_0, y_0) = \int_C F_1(x, y)dx + F_2(x, y)dy.$$

We want to compute the partial derivatives of f, that is, the rate of change of f at (x_0, y_0) parallel to the axes. To do this easily, we choose a path which reaches the point (x_0, y_0) on a horizontal or vertical line segment. Let C' be a path from P which stops short of Q at a fixed point (a, b) and let L_x and L_y be the paths shown in Figure 18.30. Then we can split the line integral into three pieces. Since $d\vec{r} = \vec{j}\, dy$ on L_y and $d\vec{r} = \vec{i}\, dx$ on L_x, we have:

$$f(x_0, y_0) = \int_{C'} \vec{F} \cdot d\vec{r} + \int_{L_y} \vec{F} \cdot d\vec{r} + \int_{L_x} \vec{F} \cdot d\vec{r} = \int_{C'} \vec{F} \cdot d\vec{r} + \int_b^{y_0} F_2(a, y)dy + \int_a^{x_0} F_1(x, y_0)dx.$$

The first two integrals do not involve x_0. Thinking of x_0 as a variable and differentiating with respect to it gives

$$f_{x_0}(x_0, y_0) = \frac{\partial}{\partial x_0} \int_{C'} \vec{F} \cdot d\vec{r} + \frac{\partial}{\partial x_0} \int_b^{y_0} F_2(a, y)dy + \frac{\partial}{\partial x_0} \int_a^{x_0} F_1(x, y_0)dx$$
$$= 0 + 0 + F_1(x_0, y_0) = F_1(x_0, y_0),$$

and thus

$$f_x(x, y) = F_1(x, y).$$

A similar calculation for y using the path from P to Q shown in Figure 18.31 gives

$$f_{y_0}(x_0, y_0) = F_2(x_0, y_0).$$

Therefore, as we claimed,

$$\text{grad}\, f = f_x\vec{i} + f_y\vec{j} = F_1\vec{i} + F_2\vec{j} = \vec{F}.$$

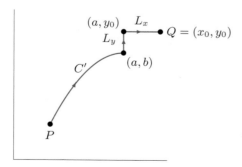

Figure 18.30: The path $C' + L_y + L_x$ is used to show $f_x = F_1$

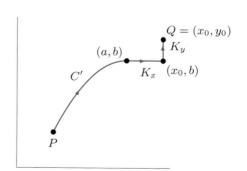

Figure 18.31: The path $C' + K_x + K_y$ is used to show $f_y = F_2$

Exercises and Problems for Section 18.3

Exercises

In Exercises 1–6, does the vector field appear to be path-independent (conservative)?

1.

2.

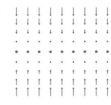

3.

4.

5.

6.
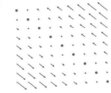

7. If $\vec{F} = \text{grad}(x^2 + y^4)$, find $\int_C \vec{F} \cdot d\vec{r}$ where C is the quarter of the circle $x^2 + y^2 = 4$ in the first quadrant, oriented counterclockwise.

8. If $\vec{F} = \text{grad}(\sin(xy) + e^z)$, find $\int_C \vec{F} \cdot d\vec{r}$ where C consists of a line from $(0, 0, 0)$ to $(0, 0, 1)$ followed by a line to $(0, \sqrt{2}, 3)$, followed by a line to $(\sqrt{2}, \sqrt{5}, 2)$.

In Exercises 9–12, let C be the curve consisting of a square of side 2, centered at the origin with sides on the lines $x = \pm 1$, $y = \pm 1$ and traversed counterclockwise. What is the sign of the line integrals of the vector fields around the curve C? Indicate whether each vector field is path-independent.

9.
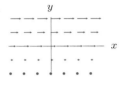

10.
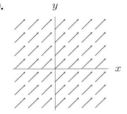

11.

12.

13. Find f if $\text{grad } f = 2xy\vec{i} + x^2\vec{j}$.

14. Find f if $\text{grad } f = 2xy\vec{i} + (x^2 + 8y^3)\vec{j}$.

15. Find f if $\text{grad } f = (yze^{xyz} + z^2\cos(xz^2))\vec{i} + xze^{xyz}\vec{j} + (xye^{xyz} + 2xz\cos(xz^2))\vec{k}$.

16. Let $f(x, y, z) = x^2 + 2y^3 + 3z^4$ and $\vec{F} = \text{grad } f$. Find $\int_C \vec{F} \cdot d\vec{r}$ where C consists of four line segments from $(4, 0, 0)$ to $(4, 3, 0)$ to $(0, 3, 0)$ to $(0, 3, 5)$ to $(0, 0, 5)$.

In Exercises 17–25, use the Fundamental Theorem of Line Integrals to calculate $\int_C \vec{F} \cdot d\vec{r}$ exactly.

17. $\vec{F} = 3x^2\vec{i} + 4y^3\vec{j}$ around the top of the unit circle from $(1, 0)$ to $(-1, 0)$.

18. $\vec{F} = (x + 2)\vec{i} + (2y + 3)\vec{j}$ and C is the line from $(1, 0)$ to $(3, 1)$.

19. $\vec{F} = 2x\vec{i} - 4y\vec{j} + (2z - 3)\vec{k}$ and C is the line from $(1, 1, 1)$ to $(2, 3, -1)$.

20. $\vec{F} = 2\sin(2x + y)\vec{i} + \sin(2x + y)\vec{j}$ along the path consisting of a line from $(\pi, 0)$ to $(2, 5)$ followed by a line to $(5\pi, 0)$ followed by a quarter circle to $(0, 5\pi)$.

21. $\vec{F} = y\sin(xy)\vec{i} + x\sin(xy)\vec{j}$ and C is the parabola $y = 2x^2$ from $(1, 2)$ to $(3, 18)$.

22. $\vec{F} = x^{2/3}\vec{i} + e^{7y}\vec{j}$, and C is the unit circle oriented clockwise.

23. $\vec{F} = x^{2/3}\vec{i} + e^{7y}\vec{j}$, and C is the quarter of the unit circle in the first quadrant, traced counterclockwise from $(1, 0)$ to $(0, 1)$.

24. $\vec{F} = ye^{xy}\vec{i} + xe^{xy}\vec{j} + (\cos z)\vec{k}$ along the curve consisting of a line from $(0, 0, \pi)$ to $(1, 1, \pi)$ followed by the parabola $z = \pi x^2$ in the plane $y = 1$ to the point $(3, 1, 9\pi)$.

25. $\vec{F} = 2xy^2ze^{x^2y^2z}\vec{i} + 2x^2yze^{x^2y^2z}\vec{j} + x^2y^2e^{x^2y^2z}\vec{k}$ and C is the circle of radius 1 in the plane $z = 1$, centered on the z-axis, starting at $(1, 0, 1)$ and oriented counterclockwise viewed from above.

Problems

26. Let $\vec{v} = \text{grad}(x^2 + y^2)$. Consider the path C which is a line between any two of the following points: $(0,0)$; $(5,0)$; $(-5,0)$; $(0,6)$; $(0,-6)$; $(5,4)$; $(-3,-5)$. Suppose you want to choose the path C in order to maximize $\int_C \vec{v} \cdot d\vec{r}$. What point should be the start of C? What point should be the end of C? Explain your answer.

27. Let $\vec{F} = \text{grad}(2x^2 + 3y^2)$. Which one of the three paths PQ, QR, and RS in Figure 18.32 should you choose as C in order to maximize $\int_C \vec{F} \cdot d\vec{r}$?

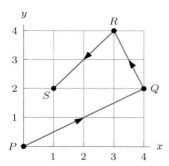

Figure 18.32

28. Compute
$$\int_C \left(\cos(xy)e^{\sin(xy)}(y\vec{i} + x\vec{j}) + \vec{k} \right) \cdot d\vec{r}$$
where C is the line from $(\pi, 2, 5)$ to $(0.5, \pi, 7)$.

29. The vector field $\vec{F}(x,y) = x\vec{i} + y\vec{j}$ is path-independent. Compute algebraically the line integrals over the three paths A, B, and C shown in Figure 18.33 from $(0,0)$ to $(1,1)$ and check that they are equal. Here A is a line segment, B is part of the graph of $f(x) = x^2$, and C consists of two line segments meeting at a right angle.

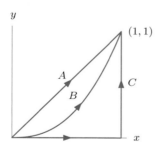

Figure 18.33

30. The vector field $\vec{F}(x,y) = x\vec{i} + y\vec{j}$ is path-independent. Compute geometrically the line integrals over the three paths A, B, and C shown in Figure 18.34 from $(1,0)$ to $(0,1)$ and check that they are equal. Here A is a portion of a circle, B is a line, and C consists of two line segments meeting at a right angle.

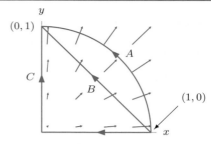

Figure 18.34

In Problems 31–34, decide whether the vector field could be a gradient vector field. Justify your answer.

31. $\vec{F}(x,y) = x\vec{i}$

32. $\vec{G}(x,y) = (x^2 - y^2)\vec{i} - 2xy\vec{j}$

33. $\vec{F}(x,y,z) = \dfrac{-z}{\sqrt{x^2 + z^2}}\vec{i} + \dfrac{y}{\sqrt{x^2 + z^2}}\vec{j} + \dfrac{x}{\sqrt{x^2 + z^2}}\vec{k}$

34. $\vec{F}(\vec{r}) = \vec{r}/\|\vec{r}\|^3$, where $\vec{r} = x\vec{i} + y\vec{j} + z\vec{k}$

35. If $\vec{F}(x,y,z) = 2xe^{x^2+yz}\vec{i} + ze^{x^2+yz}\vec{j} + ye^{x^2+yz}\vec{k}$, find exactly the line integral of $\vec{F}$ along the curve consisting of the two half circles in the plane $z = 0$ in Figure 18.35.

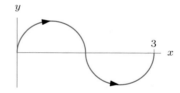

Figure 18.35

36. Let $\text{grad} f = 2xe^{x^2} \sin y\vec{i} + e^{x^2} \cos y\vec{j}$. Find the change in f between $(0,0)$ and $(1, \pi/2)$:

(a) By computing a line integral.
(b) By computing f.

37. In the Sea of Fixed Thermal Fingerprints, Prince Humperdink tries to track the Dread Pirate Roberts. The temperature of the water is given by $T(x,y) = e^{y-(x-y^2)^2}$ and Prince Humperdink follows the route C given by $x(t) = t^2$; $y(t) = t$ where $0 \le t \le 1$.

(a) Find the line integral of ∇T along the curve C. What does it mean? Do you really have to calculate ∇T to do that?

(b) Is C a flowline for the vector field $\vec{F} = \nabla T$?

38. Let C be the quarter of the unit circle centered at the origin, traversed counterclockwise starting on the negative x-axis. Find the exact values of

(a) $\displaystyle\int_C (2\pi x\vec{i} + y^2\vec{j}) \cdot d\vec{r}$ **(b)** $\displaystyle\int_C (-2y\vec{i} + x\vec{j}) \cdot d\vec{r}$

For the vector fields in Problems 39–42, find the line integral along the curve C from the origin along the x-axis to the point $(3, 0)$ and then counterclockwise around the circumference of the circle $x^2 + y^2 = 9$ to the point $(3/\sqrt{2}, 3/\sqrt{2})$.

39. $\vec{F} = x\vec{i} + y\vec{j}$

40. $\vec{H} = -y\vec{i} + x\vec{j}$

41. $\vec{F} = y(x+1)^{-1}\vec{i} + \ln(x+1)\vec{j}$

42. $\vec{G} = (ye^{xy} + \cos(x+y))\vec{i} + (xe^{xy} + \cos(x+y))\vec{j}$

43. Let C be the helix $x = \cos t$, $y = \sin t$, $z = t$ for $0 \le t \le 1.25\pi$. Find $\int_C \vec{F} \cdot d\vec{r}$ exactly for

$$\vec{F} = yz^2 e^{xyz^2}\vec{i} + xz^2 e^{xyz^2}\vec{j} + 2xyze^{xyz^2}\vec{k}.$$

44. Let $\vec{F} = 2x\vec{i} + 2y\vec{j} + 2z\vec{k}$ and $\vec{G} = (2x+y)\vec{i} + 2y\vec{j} + 2z\vec{k}$. Let C be the line from the origin to the point $(1, 5, 9)$. Find $\int_C \vec{F} \cdot d\vec{r}$ and use the result to find $\int_C \vec{G} \cdot d\vec{r}$.

45. **(a)** If $\vec{F} = ye^x\vec{i} + e^x\vec{j}$, explain how the Fundamental Theorem of Line Integrals enables you to calculate $\int_C \vec{F} \cdot d\vec{r}$ where C is any curve going from the point $(1, 2)$ to the point $(3, 7)$. Explain why it does not matter how the curve goes.

(b) If C is the line from the point $(1, 2)$ to the point $(3, 7)$, calculate the line integral in part (a) without using the Fundamental Theorem.

46. Calculate the line integral $\int_C \vec{F} \cdot d\vec{r}$ exactly, where C is the curve from P to Q in Figure 18.36 and

$$\vec{F} = \sin\left(\frac{x}{2}\right)\sin\left(\frac{y}{2}\right)\vec{i} - \cos\left(\frac{x}{2}\right)\cos\left(\frac{y}{2}\right)\vec{j}.$$

The curves PR, RS and SQ are trigonometric functions of period 2π and amplitude 1.

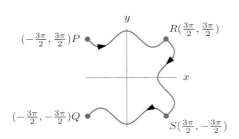

Figure 18.36

47. The domain of $f(x, y)$ is the xy-plane; values of f are in Table 18.1. Find $\int_C \text{grad } f \cdot d\vec{r}$, where C is

(a) A line from $(0, 2)$ to $(3, 4)$.

(b) A circle of radius 1 centered at $(1, 2)$ traversed counterclockwise.

Table 18.1

$y \backslash x$	0	1	2	3	4
0	53	57	59	58	56
1	56	58	59	59	57
2	57	58	59	60	59
3	59	60	61	62	61
4	62	63	65	66	69

48. Figure 18.37 shows the vector field $\vec{F}(x, y) = x\vec{j}$.

(a) Find paths C_1, C_2, and C_3 from P to Q such that

$$\int_{C_1} \vec{F} \cdot d\vec{r} = 0, \quad \int_{C_2} \vec{F} \cdot d\vec{r} > 0, \quad \int_{C_3} \vec{F} \cdot d\vec{r} < 0.$$

(b) Is $\vec{F}$ a gradient field? Explain.

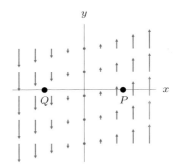

Figure 18.37

49. **(a)** Figure 18.38 shows level curves of $f(x, y)$. Sketch a vector at P in the direction of grad f.

(b) Is the length of grad f at P longer, shorter, or the same length as the length of grad f at Q?

(c) If C is a curve going from P to Q, evaluate $\int_C \text{grad } f \cdot d\vec{r}$.

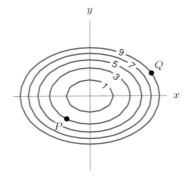

Figure 18.38

50. Consider the line integrals, $\int_{C_i} \vec{F} \cdot d\vec{r}$, for $i = 1, 2, 3, 4$, where C_i is the path from P_i to Q_i shown in Figure 18.39 and $\vec{F} = \text{grad } f$. Level curves of f are also shown in Figure 18.39.

(a) Which of the line integral(s) is (are) zero?

(b) Arrange the four line integrals in ascending order (from least to greatest).

(c) Two of the nonzero line integrals have equal and opposite values. Which are they? Which is negative and which is positive?

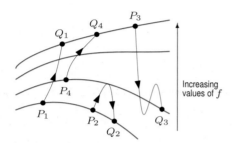

Figure 18.39

51. Consider the vector field $\vec{F}$ shown in Figure 18.40.

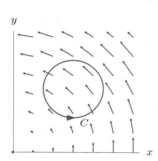

Figure 18.40

(a) Is $\int_C \vec{F} \cdot d\vec{r}$ positive, negative, or zero?

(b) From your answer to part (a), can you determine whether or not $\vec{F} = \text{grad } f$ for some function f?

(c) Which of the following formulas best fits $\vec{F}$?

$$\vec{F}_1 = \frac{x}{x^2 + y^2}\vec{i} + \frac{y}{x^2 + y^2}\vec{j},$$

$$\vec{F}_2 = -y\vec{i} + x\vec{j},$$

$$\vec{F}_3 = \frac{-y}{(x^2 + y^2)^2}\vec{i} + \frac{x}{(x^2 + y^2)^2}\vec{j}.$$

52. If $\vec{F}$ is a path-independent vector field, with $\int_{(0,0)}^{(1,0)} \vec{F} \cdot d\vec{r} = 5.1$ and $\int_{(1,0)}^{(1,1)} \vec{F} \cdot d\vec{r} = 3.2$ and $\int_{(0,1)}^{(1,1)} \vec{F} \cdot d\vec{r} = 4.7$, find

$$\int_{(0,1)}^{(0,0)} \vec{F} \cdot d\vec{r}.$$

53. The path C is a line segment of length 10 in the plane starting at $(2, 1)$. For $f(x, y) = 3x + 4y$, consider

$$\int_C \text{grad } f \cdot d\vec{r}.$$

(a) Where should the other end of the line segment C be placed to maximize the value of the integral?

(b) What is the maximum value of the integral?

54. Let $\vec{r} = x\vec{i} + y\vec{j} + z\vec{k}$ and $\vec{a} = a_1\vec{i} + a_2\vec{j} + a_3\vec{k}$, a constant vector.

(a) Find $\text{grad}(\vec{r} \cdot \vec{a})$

(b) Let C be a path from the origin to the point with position vector $\vec{r}_0$. Find $\int_C \text{grad}(\vec{r} \cdot \vec{a}) \cdot d\vec{r}$.

(c) If $||\vec{r}_0|| = 10$, what is the maximum possible value of $\int_C \text{grad}(\vec{r} \cdot \vec{a}) \cdot d\vec{r}$? Explain.

55. The force exerted by gravity on a refrigerator of mass m is $\vec{F} = -mg\vec{k}$.

(a) Find the work done against this force in moving from the point $(1, 0, 0)$ to the point $(1, 0, 2\pi)$ along the curve $x = \cos t, y = \sin t, z = t$ by calculating a line integral.

(b) Is $\vec{F}$ conservative (that is, path independent)? Give a reason for your answer.

56. A particle subject to a force $\vec{F}(x, y) = y\vec{i} - x\vec{j}$ moves clockwise along the arc of the unit circle, centered at the origin, that begins at $(-1, 0)$ and ends at $(0, 1)$.

(a) Find the work done by $\vec{F}$. Explain the sign of your answer.

(b) Is $\vec{F}$ path-independent? Explain.

57. Suppose $\vec{F}(x, y) - \vec{G}(x, y)$ is parallel to $\text{grad } h(x, y)$ at every point, and that C is an oriented path from P to Q lying entirely on a contour of h.

(a) Show that $\int_C \vec{F} \cdot d\vec{r} = \int_C \vec{G} \cdot d\vec{r}$.

(b) If $\vec{G} = \text{grad } \phi$, show that $\int_C \vec{F} \cdot d\vec{r} = \phi(Q) - \phi(P)$. This result can be useful when $\vec{F}$ is not a gradient field.

In Problems 58–61, let $\vec{F} = y\vec{i} + 2x\vec{j}$.

(a) Show that $\vec{F} - \text{grad } \phi$ is parallel to $\text{grad } h$.

(b) Use ϕ and the Fundamental Theorem of Calculus for Line Integrals to evaluate $\int_C \vec{F} \cdot d\vec{r}$, where C is the oriented path on a contour of h from P to Q.

58. $\phi = xy, h = y, P = (3, 10), Q = (8, 10)$

59. $\phi = 2xy, h = x, P = (3, 5), Q = (3, 10)$

60. $\phi = 2x^3/3 + xy, h = y - x^2, P = (0, 4), Q = (6, 40)$

61. $\phi = x^2/2 + 3xy + y^2, h = x + y, P = (10, 30), Q = (20, 20)$

62. In this problem, we see how the Fundamental Theorem for Line Integrals can be derived from the Fundamental Theorem for ordinary definite integrals. Suppose that $(x(t), y(t))$, for $a \leq t \leq b$, is a parameterization of C, with endpoints $P = (x(a), y(a))$ and $Q = (x(b), y(b))$. The values of f along C are given by the single variable function $h(t) = f(x(t), y(t))$.

   (a) Use the chain rule to show that
$$h'(t) = f_x(x(t), y(t))x'(t) + f_y(x(t), y(t))y'(t).$$
   (b) Use the Fundamental Theorem of Calculus applied to $h(t)$ to show
$$\int_C \operatorname{grad} f \cdot d\vec{r} = f(Q) - f(P).$$

63. Let $\vec{F}$ be a path-independent vector field. In physics, the potential function ϕ is usually required to satisfy the equation $\vec{F} = -\nabla\phi$. This problem illustrates the significance of the negative sign.[1]

   (a) Let the xy-plane represent part of the earth's surface with the z-axis pointing upward. (The scale is small enough that a flat plane is a good approximation to the earth's surface.) Let $\vec{r} = x\vec{i} + y\vec{j} + z\vec{k}$, with $z \geq 0$ and x, y, z in meters, be the position vector of a rock of unit mass. The gravitational potential energy function for the rock is $\phi(x, y, z) = gz$, where $g \approx 9.8$ m/sec^2. Describe in words the level surfaces of ϕ. Does the potential energy increase or decrease with height above the earth?

   (b) What is the relation between the gravitational vector, $\vec{F}$, and the vector $\nabla\phi$? Explain the significance of the negative sign in the equation $\vec{F} = -\nabla\phi$.

64. An *ideal electric dipole* consists of two equal and opposite charges separated by a small distance and is represented by a dipole moment vector $\vec{p}$. The electric field $\vec{D}$, at the point with position vector $\vec{r}$, due to an ideal electric dipole located at the origin is given by
$$\vec{D}(\vec{r}) = 3\frac{(\vec{r} \cdot \vec{p})\vec{r}}{||\vec{r}||^5} - \frac{\vec{p}}{||\vec{r}||^3}.$$

   (a) Check that φ is a potential function for $\vec{D}$, in the sense that $\vec{D} = -\operatorname{grad}\varphi$, where
$$\varphi(\vec{r}) = \frac{\vec{p} \cdot \vec{r}}{||\vec{r}||^3}.$$

   (b) Is $\vec{D}$ a path-independent vector field?

Strengthen Your Understanding

In Problems 65–67, explain what is wrong with the statement.

65. If $\vec{F}$ is a gradient field and C is an oriented path from point P to point Q, then $\int_C \vec{F} \cdot d\vec{r} = \vec{F}(Q) - \vec{F}(P)$.

66. Given any vector field $\vec{F}$ and a point P, the function $f(Q) = \int_C \vec{F} \cdot d\vec{r}$, where C is a path from P to Q, is a potential function for $\vec{F}$.

67. If a vector field $\vec{F}$ is not a gradient vector field, then $\int_C \vec{F} \cdot d\vec{r}$ can't be evaluated.

In Problems 68–69, give an example of:

68. A vector field $\vec{F}$ such that $\int_C \vec{F} \cdot d\vec{r} = 100$, for every oriented path C from $(0, 0)$ to $(1, 2)$.

69. A path-independent vector field.

In Problems 70–73, each of the statements is *false*. Explain why or give a counterexample.

70. If $\int_C \vec{F} \cdot d\vec{r} = 0$ for one particular closed path C, then $\vec{F}$ is path-independent.

71. $\int_C \vec{F} \cdot d\vec{r}$ is the total change in $\vec{F}$ along C.

72. If the vector fields $\vec{F}$ and $\vec{G}$ have $\int_C \vec{F} \cdot d\vec{r} = \int_C \vec{G} \cdot d\vec{r}$ for a particular path C, then $\vec{F} = \vec{G}$.

73. If the total change of a function f along a curve C is zero, then C must be a contour of f.

Are the statements in Problems 74–75 true or false? Explain why or give a counterexample.

74. The fact that the line integral of a vector field $\vec{F}$ is zero around the unit circle $x^2 + y^2 = 1$ means that $\vec{F}$ must be a gradient vector field.

75. If C is the line segment that starts at $(0, 0)$ and ends at (a, b) then $\int_C (x\vec{i} + y\vec{j}) \cdot d\vec{r} = \frac{1}{2}(a^2 + b^2)$.

Are the statements in Problems 76–84 true or false? Give reasons for your answer.

76. The circulation of any vector field $\vec{F}$ around any closed curve C is zero.

77. If $\vec{F} = \operatorname{grad} f$, then $\vec{F}$ is path-independent.

78. If $\vec{F}$ is path-independent, then $\int_{C_1} \vec{F} \cdot d\vec{r} = \int_{C_2} \vec{F} \cdot d\vec{r}$, where C_1 and C_2 are any paths.

79. The line integral $\int_C \vec{F} \cdot d\vec{r}$ is the total change of $\vec{F}$ along C.

80. If $\vec{F}$ is path-independent, then there is a potential function for $\vec{F}$.

81. If $f(x, y) = e^{\cos(xy)}$, and C_1 is the upper semicircle $x^2 + y^2 = 1$ from $(-1, 0)$ to $(1, 0)$, and C_2 is the line from $(-1, 0)$ to $(1, 0)$, then $\int_{C_1} \operatorname{grad} f \cdot d\vec{r} = \int_{C_2} \operatorname{grad} f \cdot d\vec{r}$.

---

[1] Adapted from V.I. Arnold, *Mathematical Methods of Classical Mechanics*, 2nd edition, Graduate Texts in Mathematics, Springer, 1989. Potential energy is also discussed in Project 1 on page 1021.

82. If $\vec{F}$ is path-independent, and C is any closed curve, then $\int_C \vec{F} \cdot d\vec{r} = 0$.

83. The vector field $\vec{F}(x, y) = y^2 \vec{i} + k\vec{j}$, where k is con-

stant, is a gradient field.

84. If $\int_C \vec{F} \cdot d\vec{r} = 0$, where C is any circle of the form $x^2 + y^2 = a^2$, then $\vec{F}$ is path-independent.

18.4 PATH-DEPENDENT VECTOR FIELDS AND GREEN'S THEOREM

Suppose we are given a vector field but are not told whether it is path-independent. How can we tell if it has a potential function, that is, if it is a gradient field?

How to Tell If a Vector Field Is Path-Dependent Using Line Integrals

One way to decide if a vector field is path-dependent is to find two paths with the same endpoints such that the line integrals of the vector field along the two paths have different values.

Example 1   Is the vector field $\vec{G}$ shown in Figure 18.41 path-independent? At any point $\vec{G}$ has magnitude equal to the distance from the origin and direction perpendicular to the line joining the point to the origin.

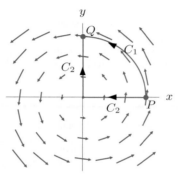

Figure 18.41: Is this vector field
path-independent?

Solution   We choose $P = (1, 0)$ and $Q = (0, 1)$ and two paths between them: C_1, a quarter circle of radius 1, and C_2, formed by parts of the x- and y-axes. (See Figure 18.41.)
Along C_1, the line integral $\int_{C_1} \vec{G} \cdot d\vec{r} > 0$, since $\vec{G}$ points in the direction of the curve.
Along C_2, however, we have $\int_{C_2} \vec{G} \cdot d\vec{r} = 0$, since $\vec{G}$ is perpendicular to C_2 everywhere.
Thus, $\vec{G}$ is not path-independent.

Path-Dependent Fields and Circulation

Notice that the vector field in the previous example has nonzero circulation around the origin. What can we say about the circulation of a general path-independent vector field $\vec{F}$ around a closed curve, C? Suppose C is a *simple* closed curve, that is, a closed curve that does not cross itself. If P and Q are any two points on the path, then we can think of C (oriented as shown in Figure 18.42) as made up of the path C_1 followed by $-C_2$. Since $\vec{F}$ is path-independent, we know that

$$\int_{C_1} \vec{F} \cdot d\vec{r} = \int_{C_2} \vec{F} \cdot d\vec{r}.$$

Thus, we see that the circulation around C is zero:

$$\int_C \vec{F} \cdot d\vec{r} = \int_{C_1} \vec{F} \cdot d\vec{r} + \int_{-C_2} \vec{F} \cdot d\vec{r} = \int_{C_1} \vec{F} \cdot d\vec{r} - \int_{C_2} \vec{F} \cdot d\vec{r} = 0.$$

If the closed curve C does cross itself, we break it into simple closed curves as shown in Figure 18.43 and apply the same argument to each one.

Now suppose we know that the line integral around any closed curve is zero. For any two points, P and Q, with two paths, C_1 and C_2, between them, create a closed curve, C, as in Figure 18.42. Since the circulation around this closed curve, C, is zero, the line integrals along the two paths, C_1 and C_2, are equal.[2] Thus, $\vec{F}$ is path-independent.

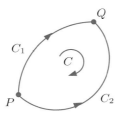

Figure 18.42: A simple closed curve C broken into two pieces, C_1 and C_2

Figure 18.43: A curve C which crosses itself can be broken into simple closed curves

Thus, we have the following result:

A vector field is path-independent if and only if $\int_C \vec{F} \cdot d\vec{r} = 0$ for every closed curve C.

Hence, to see if a field is *path-dependent*, we look for a closed path with nonzero circulation. For instance, the vector field in Example 1 has nonzero circulation around a circle around the origin, showing it is path-dependent.

How to Tell If a Vector Field Is Path-Dependent Algebraically: The Curl

Example 2   Does the vector field $\vec{F} = 2xy\vec{i} + xy\vec{j}$ have a potential function? If so, find it.

Solution   Let's suppose $\vec{F}$ does have a potential function, f, so $\vec{F} = \text{grad } f$. This means that

$$\frac{\partial f}{\partial x} = 2xy \quad \text{and} \quad \frac{\partial f}{\partial y} = xy.$$

Integrating the expression for $\partial f/\partial x$ shows that we must have

$$f(x, y) = x^2 y + C(y) \qquad \text{where } C(y) \text{ is a function of } y.$$

Differentiating this expression for $f(x, y)$ with respect to y and using the fact that $\partial f/\partial y = xy$, we get

$$\frac{\partial f}{\partial y} = x^2 + C'(y) = xy.$$

Thus, we must have

$$C'(y) = xy - x^2.$$

But this expression for $C'(y)$ is impossible because $C'(y)$ is a function of y alone. This argument shows that there is no potential function for the vector field $\vec{F}$.

---

[2] A similar argument is used in Problems 48 and 49 on page 985.

Is there an easier way to see that a vector field has no potential function, other than by trying to find the potential function and failing? The answer is yes. First we look at a 2-dimensional vector field $\vec{F} = F_1\vec{i} + F_2\vec{j}$. If $\vec{F}$ is a gradient field, then there is a potential function f such that

$$\vec{F} = F_1\vec{i} + F_2\vec{j} = \frac{\partial f}{\partial x}\vec{i} + \frac{\partial f}{\partial y}\vec{j}.$$

Thus,

$$F_1 = \frac{\partial f}{\partial x} \quad \text{and} \quad F_2 = \frac{\partial f}{\partial y}.$$

Let us assume that f has continuous second partial derivatives. Then, by the equality of mixed partial derivatives,

$$\frac{\partial F_1}{\partial y} = \frac{\partial^2 f}{\partial y \partial x} = \frac{\partial^2 f}{\partial x \partial y} = \frac{\partial F_2}{\partial x}.$$

Thus, we have the following result:

If $\vec{F}(x, y) = F_1\vec{i} + F_2\vec{j}$ is a gradient vector field with continuous partial derivatives, then

$$\frac{\partial F_2}{\partial x} - \frac{\partial F_1}{\partial y} = 0.$$

If $\vec{F}(x, y) = F_1\vec{i} + F_2\vec{j}$ is an arbitrary vector field, then we define the 2-dimensional or scalar **curl** of the vector field $\vec{F}$ to be

$$\frac{\partial F_2}{\partial x} - \frac{\partial F_1}{\partial y}.$$

Notice that we now know that if $\vec{F}$ is a gradient field, then its curl is 0. We do not (yet) know whether the converse is true. (That is: If the curl is 0, does $\vec{F}$ have to be a gradient field?) However, the curl already enables us to show that a vector field is *not* a gradient field.

Example 3   Show that $\vec{F} = 2xy\vec{i} + xy\vec{j}$ cannot be a gradient vector field.

Solution   We have $F_1 = 2xy$ and $F_2 = xy$. Since $\partial F_1/\partial y = 2x$ and $\partial F_2/\partial x = y$, in this case

$$\partial F_2/\partial x - \partial F_1/\partial y \neq 0$$

so $\vec{F}$ cannot be a gradient field.

Green's Theorem

We now have two ways of seeing that a vector field $\vec{F}$ in the plane is path-dependent. We can evaluate $\int_C \vec{F} \cdot d\vec{r}$ for some closed curve and find it is not zero, or we can show that $\partial F_2/\partial x - \partial F_1/\partial y \neq 0$. It's natural to think that

$$\int_C \vec{F} \cdot d\vec{r} \quad \text{and} \quad \frac{\partial F_2}{\partial x} - \frac{\partial F_1}{\partial y}$$

might be related. The relation is called Green's Theorem.

Theorem 18.3: Green's Theorem

Suppose C is a piecewise smooth simple closed curve that is the boundary of a region R in the plane and oriented so that the region is on the left as we move around the curve. See Figure 18.44. Suppose $\vec{F} = F_1\vec{i} + F_2\vec{j}$ is a smooth vector field on an open region containing R and C. Then

$$\int_C \vec{F} \cdot d\vec{r} = \int_R \left(\frac{\partial F_2}{\partial x} - \frac{\partial F_1}{\partial y} \right) dx\,dy.$$

The online supplement at www.wiley.com/college/hughes-hallett contains a proof of Green's Theorem with different, but equivalent, conditions on the region R.

We first prove Green's Theorem in the case where the region R is the rectangle $a \leq x \leq b, c \leq y \leq d$. Figure 18.45 shows the boundary of R divided into four curves.

On C_1, where $y = c$ and $dy = 0$, we have $d\vec{r} = dx\vec{i}$ and thus

$$\int_{C_1} \vec{F} \cdot d\vec{r} = \int_a^b F_1(x, c)\,dx.$$

Similarly, on C_3 where $y = d$ we have

$$\int_{C_3} \vec{F} \cdot d\vec{r} = \int_b^a F_1(x, d)\,dx = -\int_a^b F_1(x, d)\,dx.$$

Hence

$$\int_{C_1+C_3} \vec{F} \cdot d\vec{r} = \int_a^b F_1(x, c)\,dx - \int_a^b F_1(x, d)\,dx = -\int_a^b (F_1(x, d) - F_1(x, c))\,dx.$$

By the Fundamental Theorem of Calculus,

$$F_1(x, d) - F_1(x, c) = \int_c^d \frac{\partial F_1}{\partial y}\,dy$$

and therefore

$$\int_{C_1+C_3} \vec{F} \cdot d\vec{r} = -\int_a^b \int_c^d \frac{\partial F_1}{\partial y}\,dy\,dx = -\int_c^d \int_a^b \frac{\partial F_1}{\partial y}\,dx\,dy.$$

Along the curve C_2, where $x = b$, and the curve C_4, where $x = a$, we get, by a similar argument,

$$\int_{C_2+C_4} \vec{F} \cdot d\vec{r} = \int_c^d (F_2(b, y) - F_2(a, y))\,dy = \int_c^d \int_a^b \frac{\partial F_2}{\partial x}\,dx\,dy.$$

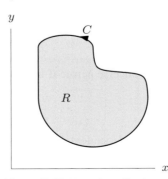

Figure 18.44: Boundary C oriented
with R on the left

Adding the line integrals over $C_1 + C_3$ and $C_2 + C_4$, we get

$$\int_C \vec{F} \cdot d\vec{r} = \int_R \left(\frac{\partial F_2}{\partial x} - \frac{\partial F_1}{\partial y} \right) dx \, dy.$$

If R is not a rectangle, we subdivide it into small rectangular pieces as shown in Figure 18.46. The contribution to the integral of the non-rectangular pieces can be made as small as we like by making the subdivision fine enough. The double integrals over each piece add up to the double integral over the whole region R. Figure 18.47 shows how the circulations around adjacent pieces cancel along the common edge, so the circulations around all the pieces add up to the circulation around the boundary C. Since Green's Theorem holds for the rectangular pieces, it holds for the whole region R.

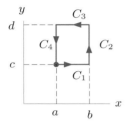

Figure 18.45: The boundary of a rectangle broken into C_1, C_2, C_3, C_4

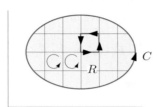

Figure 18.46: Region R bounded by a closed curve C and split into many small regions, ΔR

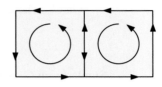

Figure 18.47: Two adjacent small closed curves

Example 4   Use Green's Theorem to evaluate $\int_C \left(y^2 \vec{i} + x \vec{j} \right) \cdot d\vec{r}$ where C is the counterclockwise path around the perimeter of the rectangle $0 \le x \le 2, 0 \le y \le 3$.

Solution   We have $F_1 = y^2$ and $F_2 = x$. By Green's Theorem

$$\int_C \left(y^2 \vec{i} + x \vec{j} \right) \cdot d\vec{r} = \int_R \left(\frac{\partial F_2}{\partial x} - \frac{\partial F_1}{\partial y} \right) dx \, dy = \int_0^3 \int_0^2 (1 - 2y) \, dx \, dy = -12.$$

The Curl Test for Vector Fields in the Plane

We already know that if $\vec{F} = F_1 \vec{i} + F_2 \vec{j}$ is a gradient field with continuous partial derivatives, then

$$\frac{\partial F_2}{\partial x} - \frac{\partial F_1}{\partial y} = 0.$$

Now we show that the converse is true if the domain of $\vec{F}$ has no holes in it. This means that we assume that

$$\frac{\partial F_2}{\partial x} - \frac{\partial F_1}{\partial y} = 0$$

and show that $\vec{F}$ is path-independent. If C is any oriented simple closed curve in the domain of $\vec{F}$ and R is the region inside C, then

$$\int_R \left(\frac{\partial F_2}{\partial x} - \frac{\partial F_1}{\partial y} \right) dx \, dy = 0$$

since the integrand is identically 0. Therefore, by Green's Theorem,

$$\int_C \vec{F} \cdot d\vec{r} = \int_R \left(\frac{\partial F_2}{\partial x} - \frac{\partial F_1}{\partial y} \right) dx\,dy = 0.$$

Thus, $\vec{F}$ is path-independent and therefore a gradient field. This argument is valid for every closed curve, C, provided the region R is entirely in the domain of $\vec{F}$. Thus, we have the following result:

The Curl Test for Vector Fields in 2-Space

Suppose $\vec{F} = F_1\vec{i} + F_2\vec{j}$ is a vector field with continuous partial derivatives such that
- The domain of $\vec{F}$ has the property that every closed curve in it encircles a region that lies entirely within the domain. In particular, the domain of $\vec{F}$ has no holes.
- $\dfrac{\partial F_2}{\partial x} - \dfrac{\partial F_1}{\partial y} = 0.$

Then $\vec{F}$ is path-independent, so $\vec{F}$ is a gradient field and has a potential function.

Why Are Holes in the Domain of the Vector Field Important?

The reason for assuming that the domain of the vector field $\vec{F}$ has no holes is to ensure that the region R inside C is actually contained in the domain of $\vec{F}$. Otherwise, we cannot apply Green's Theorem. The next two examples show that if $\partial F_2/\partial x - \partial F_1/\partial y = 0$ but the domain of $\vec{F}$ contains a hole, then $\vec{F}$ can either be path-independent or path-dependent.

Example 5   Let $\vec{F}$ be the vector field given by $\vec{F}(x, y) = \dfrac{-y\vec{i} + x\vec{j}}{x^2 + y^2}$.

(a) Calculate $\dfrac{\partial F_2}{\partial x} - \dfrac{\partial F_1}{\partial y}$. Does the curl test imply that $\vec{F}$ is path-independent?

(b) Calculate $\displaystyle\int_C \vec{F} \cdot d\vec{r}$, where C is the unit circle centred at the origin and oriented counterclockwise. Is $\vec{F}$ a path-independent vector field?

(c) Explain why the answers to parts (a) and (b) do not contradict Green's Theorem.

Solution   (a) Taking partial derivatives, we have

$$\frac{\partial F_2}{\partial x} = \frac{\partial}{\partial x}\left(\frac{x}{x^2+y^2}\right) = \frac{1}{x^2+y^2} - \frac{x \cdot 2x}{(x^2+y^2)^2} = \frac{y^2 - x^2}{(x^2+y^2)^2}.$$

Similarly,

$$\frac{\partial F_1}{\partial y} = \frac{\partial}{\partial y}\left(\frac{-y}{x^2+y^2}\right) = \frac{-1}{x^2+y^2} + \frac{y \cdot 2y}{(x^2+y^2)^2} = \frac{y^2 - x^2}{(x^2+y^2)^2}.$$

Thus,

$$\frac{\partial F_2}{\partial x} - \frac{\partial F_1}{\partial y} = 0.$$

Since $\vec{F}$ is undefined at the origin, the domain of $\vec{F}$ contains a hole. Therefore, the curl test does not apply.

(b) On the unit circle, $\vec{F}$ is tangent to the circle and $||\vec{F}|| = 1$. Thus,[3]

$$\int_C \vec{F} \cdot d\vec{r} = ||\vec{F}|| \cdot \text{Length of curve} = 1 \cdot 2\pi = 2\pi.$$

Since the line integral around the closed curve C is nonzero, $\vec{F}$ is not path-independent. We observe that $\vec{F} = \text{grad}(\arctan(y/x))$ and $\arctan(y/x)$ is θ from polar coordinates, for $-\pi/2 < \theta < \pi/2$. The fact that θ increases by 2π each time we wind once around the origin counterclockwise explains why $\vec{F}$ is not path-independent.

(c) The domain of $\vec{F}$ is the "punctured plane," as shown in Figure 18.48. Since $\vec{F}$ is not defined at the origin, which is inside C, Green's Theorem does not apply. In this case

$$2\pi = \int_C \vec{F} \cdot d\vec{r} \neq \int_R \left(\frac{\partial F_2}{\partial x} - \frac{\partial F_1}{\partial y} \right) dx\, dy = 0.$$

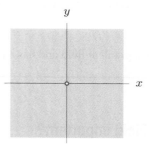

Figure 18.48: The domain of $\vec{F}(x,y) = \frac{-y\vec{i} + x\vec{j}}{x^2+y^2}$ is the plane minus the origin

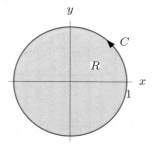

Figure 18.49: The region R is *not* contained in the domain of $\vec{F}(x,y) = \frac{-y\vec{i} + x\vec{j}}{x^2+y^2}$

Although the vector field $\vec{F}$ in the last example was not defined at the origin, this by itself does not prevent the vector field from being path-independent, as we see in the following example.

Example 6   Consider the vector field $\vec{F}$ given by $\vec{F}(x,y) = \dfrac{x\vec{i} + y\vec{j}}{x^2 + y^2}$.

(a) Calculate $\dfrac{\partial F_2}{\partial x} - \dfrac{\partial F_1}{\partial y}$. Does the curl test imply that $\vec{F}$ is path-independent?

(b) Explain how we know that $\displaystyle\int_C \vec{F} \cdot d\vec{r} = 0$, where C is the unit circle centred at the origin and oriented counterclockwise. Does this imply that $\vec{F}$ is path-independent?

(c) Check that $f(x,y) = \frac{1}{2}\ln(x^2 + y^2)$ is a potential function for $\vec{F}$. Does this imply that $\vec{F}$ is path-independent?

Solution   (a) Taking partial derivatives, we have

$$\frac{\partial F_2}{\partial x} = \frac{\partial}{\partial x}\left(\frac{y}{x^2 + y^2} \right) = \frac{-2xy}{(x^2 + y^2)^2}, \quad \text{and} \quad \frac{\partial F_1}{\partial y} = \frac{\partial}{\partial y}\left(\frac{x}{x^2 + y^2} \right) = \frac{-2xy}{(x^2 + y^2)^2}.$$

Therefore,

$$\frac{\partial F_2}{\partial x} - \frac{\partial F_1}{\partial y} = 0.$$

This does *not* imply that $\vec{F}$ is path-independent: The domain of $\vec{F}$ contains a hole since $\vec{F}$ is undefined at the origin. Thus, the curl test does not apply.

---

[3] See Problem 47 on page 985.

(b) Since $\vec{F}(x, y) = x\vec{i} + y\vec{j} = \vec{r}$ on the unit circle C, the field $\vec{F}$ is everywhere perpendicular to C. Thus

$$\int_C \vec{F} \cdot d\vec{r} = 0.$$

The fact that $\int_C \vec{F} \cdot d\vec{r} = 0$ when C is the unit circle does *not* imply that $\vec{F}$ is path-independent. To be sure that $\vec{F}$ is path-independent, we would have to show that $\int_C \vec{F} \cdot d\vec{r} = 0$ for *every* closed curve C in the domain of $\vec{F}$, not just the unit circle.

(c) To check that $\text{grad } f = \vec{F}$, we differentiate f:

$$f_x = \frac{1}{2} \frac{\partial}{\partial x} \ln(x^2 + y^2) = \frac{1}{2} \frac{2x}{x^2 + y^2} = \frac{x}{x^2 + y^2},$$

and

$$f_y = \frac{1}{2} \frac{\partial}{\partial y} \ln(x^2 + y^2) = \frac{1}{2} \frac{2y}{x^2 + y^2} = \frac{y}{x^2 + y^2},$$

so that

$$\text{grad } f = \frac{x\vec{i} + y\vec{j}}{x^2 + y^2} = \vec{F}.$$

Thus, $\vec{F}$ is a gradient field and therefore is path-independent—even though $\vec{F}$ is undefined at the origin.

The Curl Test for Vector Fields in 3-Space

The curl test is a convenient way of deciding whether a 2-dimensional vector field is path-independent. Fortunately, there is an analogous test for 3-dimensional vector fields, although we cannot justify it until Chapter 20.

If $\vec{F}(x, y, z) = F_1\vec{i} + F_2\vec{j} + F_3\vec{k}$ is a vector field on 3-space we define a new vector field, $\text{curl } \vec{F}$, on 3-space by

$$\text{curl } \vec{F} = \left(\frac{\partial F_3}{\partial y} - \frac{\partial F_2}{\partial z} \right) \vec{i} + \left(\frac{\partial F_1}{\partial z} - \frac{\partial F_3}{\partial x} \right) \vec{j} + \left(\frac{\partial F_2}{\partial x} - \frac{\partial F_1}{\partial y} \right) \vec{k}.$$

The vector field $\text{curl } \vec{F}$ can be used to determine whether the vector field $\vec{F}$ is path-independent.

The Curl Test for Vector Fields in 3-Space

Suppose $\vec{F}$ is a vector field on 3-space with continuous partial derivatives such that
- The domain of $\vec{F}$ has the property that every closed curve in it can be contracted to a point in a smooth way, staying at all times within the domain.
- $\text{curl } \vec{F} = \vec{0}$.

Then $\vec{F}$ is path-independent, so $\vec{F}$ is a gradient field and has a potential function.

For the 2-dimensional curl test, the domain of $\vec{F}$ must have no holes. This meant that if $\vec{F}$ was defined on a simple closed curve C, then it was also defined at all points inside C. One way to test for holes is to try to "lasso" them with a closed curve. If every closed curve in the domain can be pulled to a point without hitting a hole, that is, without straying outside the domain, then the domain has no holes. In 3-space, we need the same condition to be satisfied: we must be able to pull every closed curve to a point, like a lasso, without straying outside the domain.

Example 7   Decide if the following vector fields are path-independent and whether or not the curl test applies.

(a) $\vec{F} = \dfrac{x\vec{i} + y\vec{j} + z\vec{k}}{(x^2 + y^2 + z^2)^{3/2}}$

(b) $\vec{G} = \dfrac{-y\vec{i} + x\vec{j}}{x^2 + y^2} + z^2\vec{k}$

Solution

(a) Suppose $f = -(x^2 + y^2 + z^2)^{-1/2}$. Then $f_x = x(x^2 + y^2 + z^2)^{-3/2}$ and f_y and f_z are similar, so grad $f = \vec{F}$. Thus, $\vec{F}$ is a gradient field and therefore path-independent. Calculations show curl $\vec{F} = \vec{0}$. The domain of $\vec{F}$ is all of 3-space minus the origin, and any closed curve in the domain can be pulled to a point without leaving the domain. Thus, the curl test applies.

(b) Let C be the circle $x^2 + y^2 = 1, z = 0$ traversed counterclockwise when viewed from the positive z-axis. Since $z = 0$ on the curve C, the vector field $\vec{G}$ reduces to the vector field in Example 5 and is everywhere tangent to C and of magnitude 1, so

$$\int_C \vec{G} \cdot d\vec{r} = \|\vec{G}\| \cdot \text{Length of curve} = 1 \cdot 2\pi = 2\pi.$$

Since the line integral around this closed curve is nonzero, $\vec{G}$ is path-dependent. Computations show curl $\vec{G} = \vec{0}$. However, the domain of $\vec{G}$ is all of 3-space minus the z-axis, and it does not satisfy the curl test domain criterion. For example, the circle, C, is lassoed around the z-axis, and cannot be pulled to a point without hitting the z-axis. Thus, the curl test does not apply.

Exercises and Problems for Section 18.4

Exercises

In Exercises 1–10, decide if the given vector field is the gradient of a function f. If so, find f. If not, explain why not.

1. $y\vec{i} + y\vec{j}$

2. $y\vec{i} - x\vec{j}$

3. $2xy\vec{i} + 2xy\vec{j}$

4. $2xy\vec{i} + x^2\vec{j}$

5. $(x^2 + y^2)\vec{i} + 2xy\vec{j}$

6. $(2xy^3 + y)\vec{i} + (3x^2y^2 + x)\vec{j}$

7. $\dfrac{\vec{i}}{x} + \dfrac{\vec{j}}{y} + \dfrac{\vec{k}}{z}$

8. $\dfrac{\vec{i}}{x} + \dfrac{\vec{j}}{y} + \dfrac{\vec{k}}{xy}$

9. $2x\cos(x^2 + z^2)\vec{i} + \sin(x^2 + z^2)\vec{j} + 2z\cos(x^2 + z^2)\vec{k}$

10. $\dfrac{y}{x^2 + y^2}\vec{i} - \dfrac{x}{x^2 + y^2}\vec{j}$

In Exercises 11–14, use Green's Theorem to calculate the circulation of $\vec{F}$ around the curve, oriented counterclockwise.

11. $\vec{F} = y\vec{i} - x\vec{j}$ around the unit circle.

12. $\vec{F} = xy\vec{j}$ around the square $0 \le x \le 1, 0 \le y \le 1$.

13. $\vec{F} = (2x^2 + 3y)\vec{i} + (2x + 3y^2)\vec{j}$ around the triangle with vertices $(2, 0), (0, 3), (-2, 0)$.

14. $\vec{F} = 3y\vec{i} + xy\vec{j}$ around the unit circle.

15. Calculate $\int_C ((3x + 5y)\vec{i} + (2x + 7y)\vec{j}) \cdot d\vec{r}$ where C is the circular path with center (a, b) and radius m, oriented counterclockwise. Use Green's Theorem.

16. (a) Sketch $\vec{F} = y\vec{i}$ and determine the sign of the circulation of $\vec{F}$ around the unit circle centered at the origin and traversed counterclockwise.

(b) Use Green's Theorem to compute the circulation in part (a) exactly.

Problems

17. Use Green's Theorem to evaluate $\int_C (y^2\vec{i} + x\vec{j}) \cdot d\vec{r}$ where C is the counterclockwise path around the perimeter of the rectangle $0 \le x \le 2, 0 \le y \le 3$.

18. Let $\vec{F} = (\sin x)\vec{i} + (x + y)\vec{j}$. Find the line integral of $\vec{F}$ around the perimeter of the rectangle with corners $(3, 0)$, $(3, 5), (-1, 5), (-1, 0)$, traversed in that order.

19. Find $\int_C (\sin(x^2)\cos y)\vec{i} + (\sin(y^2) + e^x)\vec{j} \cdot d\vec{r}$ where C is the square of side 1 in the first quadrant of the xy-plane, with one vertex at the origin and sides along the axes, and oriented counterclockwise when viewed from above.

20. Find the line integral of $\vec{F} = (x - y)\vec{i} + x\vec{j}$ around the closed curve in Figure 18.50. (The arc is part of a circle.)

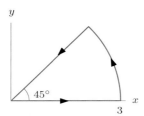

Figure 18.50

21. Find the line integral of $\vec{F} = (x+y)\vec{i} + \sin y\vec{j}$ around the closed curve in Figure 18.50. (The arc is part of a circle.)

22. Let $\vec{F} = 2xe^y\vec{i} + x^2e^y\vec{j}$ and $\vec{G} = (x-y)\vec{i} + (x+y)\vec{j}$. Let C be the line from $(0,0)$ to $(2,4)$. Find exactly:

(a) $\int_C \vec{F} \cdot d\vec{r}$ **(b)** $\int_C \vec{G} \cdot d\vec{r}$

23. Let $\vec{F} = y\vec{i} + x\vec{j}$ and $\vec{G} = 3y\vec{i} - 3x\vec{j}$. In Figure 18.51, the curve C_2 is the semicircle centered at the origin from $(-1,1)$ to $(1,-1)$ and C_1 is the line segment from $(-1,1)$ to $(1,-1)$, and $C = C_2 - C_1$. Find the following line integrals:

(a) $\int_{C_1} \vec{F} \cdot d\vec{r}$ **(b)** $\int_C \vec{F} \cdot d\vec{r}$

(c) $\int_{C_2} \vec{F} \cdot d\vec{r}$ **(d)** $\int_{C_2} \vec{G} \cdot d\vec{r}$

(e) $\int_C \vec{G} \cdot d\vec{r}$ **(f)** $\int_{C_1} \vec{G} \cdot d\vec{r}$

(g) $\int_C (\vec{F} + \vec{G}) \cdot d\vec{r}$

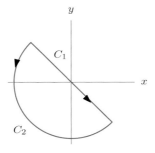

Figure 18.51

24. Show that the line integral of $\vec{F} = x\vec{j}$ around a closed curve in the xy-plane, oriented as in Green's Theorem, measures the area of the region enclosed by the curve.

In Problems 25–27, use the result of Problem 24 to calculate the area of the region within the parameterized curves. In each case, sketch the curve.

25. The ellipse $x^2/a^2 + y^2/b^2 = 1$ parameterized by $x = a\cos t$, $y = b\sin t$, for $0 \le t \le 2\pi$.

26. The hypocycloid $x^{2/3} + y^{2/3} = a^{2/3}$ parameterized by $x = a\cos^3 t$, $y = a\sin^3 t$, $0 \le t \le 2\pi$.

27. The folium of Descartes, $x^3 + y^3 = 3xy$, parameterized by $x = \dfrac{3t}{1+t^3}$, $y = \dfrac{3t^2}{1+t^3}$, for $0 \le t < \infty$.

28. Calculate $\int_C \left((x^2 - y)\vec{i} + (y^2 + x)\vec{j}\right) \cdot d\vec{r}$ if:

(a) C is the circle $(x-5)^2 + (y-4)^2 = 9$ oriented counterclockwise.

(b) C is the circle $(x-a)^2 + (y-b)^2 = R^2$ in the xy-plane oriented counterclockwise.

29. Consider the following parametric equations:

$$C_1 : \vec{r}(t) = t\cos(2\pi t)\,\vec{i} + t\sin(2\pi t)\,\vec{k}, 0 \le t \le 2$$
$$C_2 : \vec{r}(t) = t\cos(2\pi t)\,\vec{i} + t\vec{j} + t\sin(2\pi t)\,\vec{k}, 0 \le t \le 2$$

(a) Describe, in words, the motion of a particle moving through each of the paths.

(b) Evaluate $\int_{C_2} \vec{F} \cdot d\vec{r}$, for the vector field $\vec{F} = yz\,\vec{i} + z(x+1)\,\vec{j} + (xy + y + 1)\,\vec{k}$.

(c) Find a non-zero vector field $\vec{G}$ such that:

$$\int_{C_1} \vec{G} \cdot d\vec{r} = \int_{C_2} \vec{G} \cdot d\vec{r}.$$

Explain how you reasoned to find $\vec{G}$.

(d) Find two different, non-zero vector fields $\vec{H}_1$, $\vec{H}_2$ such that:

$$\int_{C_1} \vec{H}_1 \cdot d\vec{r} = \int_{C_1} \vec{H}_2 \cdot d\vec{r}.$$

Explain how you reasoned to find the two fields.

30. The vector field $\vec{F}$ is defined on the disk D of radius 5 centered at the origin in the plane:

$$\vec{F} = (-y^3 + y\sin(xy))\vec{i} + (4x(1-y^2) + x\sin(xy))\vec{j}.$$

Consider the line integral $\int_C \vec{F} \cdot d\vec{r}$, where C is some closed curve contained in D. For which C is the value of this integral the largest? [Hint: Assume C is a closed curve, made up of smooth pieces and never crossing itself, and oriented counterclockwise.]

31. Example 1 on page 1004 showed that the vector field in Figure 18.52 could not be a gradient field by showing that it is not path-independent. Here is another way to see the same thing. Suppose that the vector field were the gradient of a function f. Draw and label a diagram showing what the contours of f would have to look like, and explain why it would not be possible for f to have a single value at any given point.

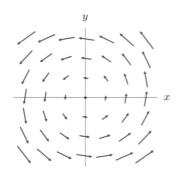

Figure 18.52

32. Repeat Problem 31 for the vector field in Problem 48 on page 1001.

33. (a) By finding potential functions, show that each of the vector fields $\vec{F}, \vec{G}, \vec{H}$ is a gradient field on some domain (not necessarily the whole plane).
(b) Find the line integrals of $\vec{F}, \vec{G}, \vec{H}$ around the unit circle in the xy-plane, centered at the origin, and traversed counterclockwise.
(c) For which of the three vector fields can Green's Theorem be used to calculate the line integral in part (b)? Why or why not?

$$\vec{F} = y\vec{i} + x\vec{j}, \quad \vec{G} = \frac{y\vec{i} - x\vec{j}}{x^2 + y^2}, \quad \vec{H} = \frac{x\vec{i} + y\vec{j}}{(x^2 + y^2)^{1/2}}$$

34. (a) For which of the following can you use Green's Theorem to evaluate the integral? Explain.

I $\displaystyle\int_C (x^2 + y^2)\,dx + (x^2 + y^2)\,dy$ where C is the curve defined by $y = x$, $y = x^2$, $0 \leq x \leq 1$ with counterclockwise orientation.

II $\displaystyle\int_C \frac{1}{\sqrt{x^2 + y^2}}\,dx - \frac{1}{\sqrt{x^2 + y^2}}\,dy$ where C is the unit circle centered at the origin, oriented counterclockwise.

III $\displaystyle\int_C \vec{F} \cdot d\vec{r}$ where $\vec{F} = x\vec{i} + y\vec{j}$ and C is the line segment from the origin to $(1, 1)$.

(b) Use Green's Theorem to evaluate the integrals in part (a) that can be done that way.

35. Arrange the line integrals L_1, L_2, L_3 in ascending order, where

$$L_i = \int_{C_i} (-x^2 y\vec{i} + (xy^2 - x)\vec{j}) \cdot d\vec{r}.$$

The points A, B, D lie on the unit circle and C_i is one of the curves shown in Figure 18.53.

C_1: Line segment A to B

C_2: Line segment A to D followed by line segment D to B

C_3: Semicircle ADB

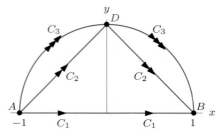

Figure 18.53

36. For all x, y, let $\vec{F} = F_1(x, y)\vec{i} + F_2(x, y)\vec{j}$ satisfy

$$\frac{\partial F_2}{\partial x} - \frac{\partial F_1}{\partial y} = 3.$$

(a) Calculate $\int_{C_1} \vec{F} \cdot d\vec{r}$ where C_1 is the unit circle in the xy-plane centered at the origin, oriented counterclockwise.
(b) Calculate $\int_{C_2} \vec{F} \cdot d\vec{r}$ where C_2 is the boundary of the rectangle of $4 \leq x \leq 7$, $5 \leq y \leq 7$, oriented counterclockwise.
(c) Let C_3 be the circle of radius 7 centered at the point $(10, 2)$; let C_4 be the circle of radius 8 centered at the origin; let C_5 be the square of side 14 centered at $(7, 7)$ with sides parallel to the axes; C_3, C_4, C_5 are all oriented counterclockwise. Arrange the integrals $\int_{C_3} \vec{F} \cdot d\vec{r}, \int_{C_4} \vec{F} \cdot d\vec{r}, \int_{C_5} \vec{F} \cdot d\vec{r}$ in increasing order.

37. Let $\vec{F} = (3x^2 y + y^3 + e^x)\vec{i} + (e^{y^2} + 12x)\vec{j}$. Consider the line integral of $\vec{F}$ around the circle of radius a, centered at the origin and traversed counterclockwise.

(a) Find the line integral for $a = 1$.
(b) For which value of a is the line integral a maximum? Explain.

38. A volume of water swirls down according to the velocity field

$$\vec{F} = -\frac{y + xz}{(z^2 + 1)^2}\vec{i} - \frac{yz - x}{(z^2 + 1)^2}\vec{j} - \frac{1}{z^2 + 1}\vec{k}$$

Let C be a circle on the xy plane, with center at the origin and radius 1.

(a) Find the line integral of $\vec{F}$ along the border of the circle (oriented any way you wish).
(b) Is $\vec{F}$ a gradient vector field? How can you tell?

39. Let

$$\vec{F}(x, y) = \frac{-y\vec{i} + x\vec{j}}{x^2 + y^2}$$

and let oriented curves C_1 and C_2 be as in Figure 18.54. The curve C_2 is an arc of the unit circle centered at the origin. Show that

(a) The curl of $\vec{F}$ is zero.
(b) $\int_{C_1} \vec{F} \cdot d\vec{r} = \int_{C_2} \vec{F} \cdot d\vec{r}$.

(c) $\int_{C_1} \vec{F} \cdot d\vec{r} = \theta$, the angle at the origin subtended by the oriented curve C_1.

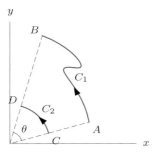

Figure 18.54

40. The electric field $\vec{E}$, at the point with position vector $\vec{r}$ in 3-space, due to a charge q at the origin is given by

$$\vec{E}(\vec{r}) = q\frac{\vec{r}}{||\vec{r}||^3}.$$

(a) Compute curl $\vec{E}$. Is $\vec{E}$ a path-independent vector field? Explain.

(b) Find a potential function φ for $\vec{E}$, if possible.

Strengthen Your Understanding

In Problems 41–42, explain what is wrong with the statement.

41. If $\int_C \vec{F} \cdot d\vec{r} = 0$ for a specific closed path C, then $\vec{F}$ must be path-independent.

42. Let $\vec{F} = F_1(x, y)\vec{i} + F_2(x, y)\vec{j}$ with

$$\frac{\partial F_2}{\partial x} - \frac{\partial F_1}{\partial y} = 3$$

and let C be the path consisting of line segments from $(0, 0)$ to $(1, 1)$ to $(2, 0)$. Then

$$\int_C \vec{F} \cdot d\vec{r} = 3.$$

In Problems 43–45, give an example of:

43. A function $Q(x, y)$ such that $\vec{F} = xy\vec{i} + Q(x, y)\vec{j}$ is a gradient field.

44. Two oriented curves, C_1 and C_2, from $(1, 0)$ to $(0, 1)$ such that if

$$\vec{F}(x, y) = \frac{-y\vec{i} + x\vec{j}}{x^2 + y^2},$$

then

$$\int_{C_1} \vec{F} \cdot d\vec{r} \neq \int_{C_2} \vec{F} \cdot d\vec{r}.$$

[Note that the scalar curl of $\vec{F}$ is 0 where $\vec{F}$ is defined.]

45. A vector field that is not a gradient field.

Are the statements in Problems 46–53 true or false? Give reasons for your answer.

46. If $f(x)$ and $g(y)$ are continuous one-variable functions, then the vector field $\vec{F} = f(x)\vec{i} + g(y)\vec{j}$ is path-independent.

47. If $\vec{F} = \text{grad } f$, and C is the perimeter of a square of side length a oriented counterclockwise and surrounding the region R, then

$$\int_C \vec{F} \cdot d\vec{r} = \int_R f \, dA.$$

48. If $\vec{F}$ and $\vec{G}$ are both path-independent vector fields, then $\vec{F} + \vec{G}$ is path-independent.

49. If $\vec{F}$ and $\vec{G}$ are both path-dependent vector fields, then $\vec{F} + \vec{G}$ is path-dependent.

50. The vector field $\vec{F}(\vec{r}) = \vec{r}$ in 3-space is path-independent.

51. A constant vector field $\vec{F} = a\vec{i} + b\vec{j}$ is path-independent.

52. If $\vec{F}$ is path-independent and k is a constant, then the vector field $k\vec{F}$ is path-independent.

53. If $\vec{F}$ is path-independent and $h(x, y)$ is a scalar function, then the vector field $h(x, y)\vec{F}$ is path-independent.

CHAPTER SUMMARY (see also Ready Reference at the end of the book)

- **Line Integrals**
  Oriented curves, definition as a limit of a Riemann sum, work interpretation, circulation, algebraic properties, computing line integrals over parameterized curves, independence of parameterization, differential notation.

- **Gradient Fields**

Fundamental Theorem for line integrals, path-independent (conservative) fields and their relation to gradient fields, potential functions.

- **Green's Theorem**
  Statement of the theorem, curl test for path-independence, holes in the domain.

REVIEW EXERCISES AND PROBLEMS FOR CHAPTER EIGHTEEN

Exercises

The figures in Exercises 1–2 show a vector field $\vec{F}$ and a curve C. Decide if $\int_C \vec{F} \cdot d\vec{r}$ is positive, zero, or negative.

1.

2.

3. Is $\int_C (3\vec{i} + 4\vec{j}) \cdot d\vec{r}$, where C is the line from $(5, 2)$ to $(1, 8)$, a vector or a scalar? Calculate it.

4. Is $\int_C (x\vec{i} + y\vec{j}) \cdot d\vec{r}$, where C is the line from $(0, 2)$ to $(0, 6)$, a vector or a scalar? Calculate it.

In Exercises 5–10, find $\int_C \vec{F} \cdot d\vec{r}$ for the given $\vec{F}$ and C.

5. $\vec{F} = 6\vec{i} - 7\vec{j}$, and C is an oriented curve from $(2, -6)$ to $(4, 4)$.

6. $\vec{F} = x\vec{i} + y\vec{j}$ and C is the unit circle in xy-plane oriented counterclockwise.

7. $\vec{F} = x\vec{i} + y\vec{j}$ and C is the y-axis from the origin to $(0, 10)$.

8. $\vec{F} = (x^2 - y)\vec{i} + (y^2 + x)\vec{j}$ and C is the parabola $y = x^2 + 1$ traversed from $(0, 1)$ to $(1, 2)$.

9. $\vec{F} = x\vec{i} + y\vec{j} + z\vec{k}$ and C is the path consisting of a line from $(2, 3, 0)$ to $(4, 5, 0)$, followed by a line from $(4, 5, 0)$ to $(0, 0, 7)$.

10. $\vec{F} = xy\vec{i} + (x - y)\vec{j}$ and C is the triangle joining $(1, 0)$, $(0, 1)$ and $(-1, 0)$ in the clockwise direction.

In Exercises 11–12, evaluate the line integrals.

11. $\int_C 3x^2 \, dx + 4y \, dy$ where C is the path $y = x^2$ from $(1, 1)$ to $(5, 25)$.

12. $\int_C y \, dx + x \, dy$ where C is the path $y = \sin x$ from $(0, 0)$ to $(\pi/2, 1)$.

In Exercises 13–19, which of the vector fields are path-independent on all of 3-space?

13. $y\vec{i}$

14. $y\vec{j}$

15. $z\vec{k}$

16. $z\vec{j} + z\vec{k}$

17. $y\vec{i} + x\vec{j}$

18. $(x + y)\vec{i}$

19. $yz\vec{i} + zx\vec{j} + xy\vec{k}$

In Exercises 20–25, find the line integral of $\vec{F} = 5x\vec{i} + 3x\vec{j}$ along the path C.

20. C is the line from $(2, 3)$ to $(2, 8)$.

21. C is the line from $(2, 3)$ to $(12, 3)$.

22. C is the curve $y = x^2$ from $(1, 1)$ to $(2, 4)$.

23. C is the semicircle of radius 3 from $(3, 0)$ to $(-3, 0)$ in the upper half plane.

24. C is the path in Figure 18.55.

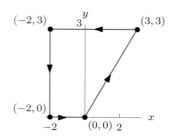

Figure 18.55

25. C is the path in Figure 18.56.

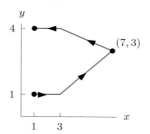

Figure 18.56

In Exercises 26–28, find the line integral of $\vec{F}$ around C_1 and C_2, where C_1 is the circle of radius 3, centered at $(5, 4)$ and oriented clockwise and C_2 is the semicircle starting at $(3, 4)$ and ending at $(7, 4)$, passing above the line $y = 4$.

26. $\vec{F} = 5\vec{i} + 4\vec{j}$

27. $\vec{F} = 5x\vec{i} + 4y\vec{j}$

28. $\vec{F} = 5y\vec{i} + 4x\vec{j}$

Problems

29. Which two of the vector fields (i)-(iv) could represent gradient vector fields on the whole plane? Give reasons for your answer.

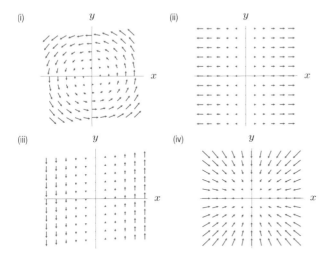

30. Let $\vec{F}(x, y)$ be the path-independent vector field in Figure 18.57. The vector field $\vec{F}$ associates with each point a unit vector pointing radially outward. The curves $C_1, C_2, \ldots, C_7$ have the directions shown. Consider the line integrals $\int_{C_i} \vec{F} \cdot d\vec{r}$, $i = 1, \ldots, 7$. Without computing any integrals,

(a) List all the line integrals which you expect to be zero.

(b) List all the line integrals which you expect to be negative.

(c) Arrange the positive line integrals in ascending order.

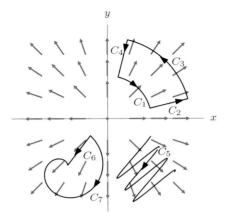

Figure 18.57

31. If C is $\vec{r} = (\cos t)\vec{i} + (\sin t)\vec{j}$ for $0 \le t \le 2\pi$, we know $\int_C \vec{F}(\vec{r}) \cdot d\vec{r} = 12$. Find the value of the integrals:

(a) $\int_0^{4\pi} \vec{F}\left((\cos t)\vec{i} + (\sin t)\vec{j}\right) \cdot \left((-\sin t)\vec{i} + (\cos t)\vec{j}\right) dt$

(b) $\int_{2\pi}^0 \vec{F}\left((\cos t)\vec{i} + (\sin t)\vec{j}\right) \cdot \left((\sin t)\vec{i} - (\cos t)\vec{j}\right) dt$

(c) $\int_0^{2\pi} \vec{F}\left((\sin t)\vec{i} + (\cos t)\vec{j}\right) \cdot \left((-\cos t)\vec{i} - (\sin t)\vec{j}\right) dt$

32. Let C be the straight path from $(0, 0)$ to $(5, 5)$ and let $\vec{F} = (y - x + 2)\vec{i} + (\sin(y - x) + 2)\vec{j}$.

(a) At each point of C, what angle does $\vec{F}$ make with a tangent vector to C?

(b) Find the magnitude $\|\vec{F}\|$ at each point of C.

(c) Evaluate $\int_C \vec{F} \cdot d\vec{r}$.

33. The line integral of $\vec{F} = (x + y)\vec{i} + x\vec{j}$ along each of the following paths is $3/2$:

(i) The path (t, t^2), with $0 \le t \le 1$

(ii) The path (t^2, t), with $0 \le t \le 1$

(iii) The path (t, t^n), with $n > 0$ and $0 \le t \le 1$

Show this

(a) Using the given parameterization to compute the line integral.

(b) Using the Fundamental Theorem of Calculus for Line Integrals.

Problems 34–37 refer to the star-shaped region R in Figure 18.58.

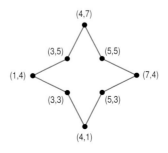

Figure 18.58

34. Let C be the path from $(5, 5)$ to $(4, 7)$. Find

$$\int_C (2\vec{i} + 13\vec{j}) \cdot d\vec{r}.$$

35. Let C be the path from $(5, 5)$ to $(4, 7)$. Find

$$\int_C (4x\vec{i} + 3y\vec{j}) \cdot d\vec{r}.$$

36. Let C be the path from $(5, 5)$ to $(4, 7)$. Find

$$\int_C ((4x + 5y)\vec{i} + (2x + 3y)\vec{j}) \cdot d\vec{r}.$$

37. Let C be the path around the outside of the star, traced counterclockwise. Find $\int_C ((4x + 5y)\vec{i} + (2x + 3y)\vec{j}) \cdot d\vec{r}$.

38. Let $\vec{F} = 2y\vec{i} + 5x\vec{j}$. Let C be the M-shaped closed curve consisting of line segments starting at $(-5, 1)$, going along the line $y = 1$ to $(15, 1)$, then to $(10, 9)$, then to $(5, 1)$, and then to $(0, 9)$, and then back to $(-5, 1)$. Let C_1 be the part of C along the line $y = 1$; let C_2 be the rest of C and $C = C_1 + C_2$.

(a) Find $\int_C \vec{F} \cdot d\vec{r}$.
(b) Find $\int_{C_1} \vec{F} \cdot d\vec{r}$.
(c) Find $\int_{C_2} \vec{F} \cdot d\vec{r}$.

39. Let $\vec{F} = 2xe^y\vec{i} + x^2 e^y\vec{j}$ and $\vec{G} = (x-y)\vec{i} + (x+y)\vec{j}$. Let C be the path consisting of lines from $(0, 0)$ to $(3, 0)$ to $(3, 8)$ to $(0, 0)$. Find exactly:

(a) $\int_C \vec{F} \cdot d\vec{r}$ **(b)** $\int_C \vec{G} \cdot d\vec{r}$

40. Let $\vec{F} = (x^2 + 3x^2y^4)\vec{i} + 4x^3y^3\vec{j}$ and $\vec{G} = (x^4 + x^3y^2)\vec{i} + x^2y^3\vec{j}$. Let C_1 be the path along the x-axis from $(2, 0)$ to $(-2, 0)$; let C_2 be the semi-circle in the upper half plane from $(2, 0)$ to $(-2, 0)$. Find exactly:

(a) $\int_{C_1} \vec{F} \cdot d\vec{r}$ **(b)** $\int_{C_2} \vec{F} \cdot d\vec{r}$
(c) $\int_{C_1} \vec{G} \cdot d\vec{r}$ **(d)** $\int_{C_2} \vec{G} \cdot d\vec{r}$

41. Calculate the line integral of $\vec{F} = -y\vec{i} + x\vec{j}$ along the following paths in the xy-plane.

(a) Line from the origin to the point $(2, 3)$.
(b) Line from $(2, 3)$ to $(0, 3)$.
(c) Counterclockwise around a circle of radius 5 centered at the origin, starting from $(5, 0)$ to $(0, -5)$.
(d) Counterclockwise around the perimeter of a triangle of area 7.

42. Let C_1 and C_2 be the curves in Figure 18.59. Let $\vec{F} = (6x + y^2)\vec{i} + 2xy\vec{j}$ and $\vec{G} = (x - y)\vec{i} + (x + y)\vec{j}$. [Note C_1 is made up of line segments and C_2 is part of a circle.] Compute the following line integrals.

(a) $\int_{C_1} \vec{F} \cdot d\vec{r}$ **(b)** $\int_{C_1} \vec{G} \cdot d\vec{r}$
(c) $\int_{C_2} \vec{F} \cdot d\vec{r}$ **(d)** $\int_{C_2} \vec{G} \cdot d\vec{r}$

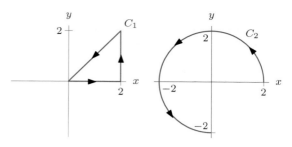

Figure 18.59

43. Let $\vec{F} = x\vec{i} + y\vec{j}$. Find the line integral of $\vec{F}$:

(a) Along the x-axis from the origin to the point $(3, 0)$.
(b) Around the path from A to B to O in Figure 18.60. (The curve is part of a circle centered at the origin.)

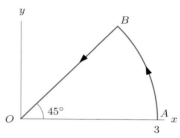

Figure 18.60

44. (a) Sketch the curves C_1 and C_2:

C_1 is $(x, y) = (0, t)$ for $-1 \leq t \leq 1$
C_2 is $(x, y) = (\cos t, \sin t)$ for $\pi/2 \leq t \leq 3/\pi/2$.

(b) Find $\int_C ((x+3y)\vec{i} + y\vec{j}) \cdot d\vec{r}$, where $C = C_1 + C_2$.

45. Draw an oriented curve C and a vector field $\vec{F}$ along C that is not always perpendicular to C, but for which $\int_C \vec{F} \cdot d\vec{r} = 0$.

46. (a) Sketch the curve, C, consisting of three parts, $C = C_1 + C_2 + C_3$, where C_1 is $\vec{r} = t\vec{i}$ for $0 \leq t \leq 1$, and C_2 is $x = 1 - t, y = t$ for $0 \leq t \leq 1$, and C_3 is $\vec{r} = (1 - t)\vec{j}$ for $0 \leq t \leq 1$. Label the coordinates of the points where C_1, C_2, C_3 meet. Each curve is oriented in the direction of increasing t.

(b) Sketch the vector field $\vec{F} = -\vec{i} + \vec{j}$
(c) Find

(i) $\int_{C_1} \vec{F} \cdot d\vec{r}$ (ii) $\int_{C_2} \vec{F} \cdot d\vec{r}$
(iii) $\int_{C_3} \vec{F} \cdot d\vec{r}$ (iv) $\int_C \vec{F} \cdot d\vec{r}$

47. For each of the following vector fields in the plane, use Green's Theorem to sketch a closed curve, C, in the plane with $\int_C \vec{F} \cdot d\vec{r} > 0$. Show the orientation of your curve.

(a) $\vec{F} = (x^3 - y)\vec{i} + (y^5 + x)\vec{j}$
(b) $\vec{F} = x^3\vec{i} + (y^5 - xy)\vec{j}$

48. Suppose P and Q both lie on the same contour of f. What can you say about the total change in f from P to Q? Explain your answer in terms of $\int_C \text{grad } f \cdot d\vec{r}$ where C is a part of the contour that goes from P to Q.

49. Figure 18.61 shows level curves of the function $f(x, y)$.

   (a) Sketch ∇f at P.
   (b) Is the vector ∇f at P longer than, shorter than, or the same length as, ∇f at Q?
   (c) If C is a curve from P to Q, evaluate $\int_C \nabla f \cdot d\vec{r}$.

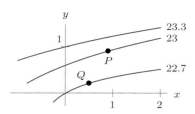

Figure 18.61

50. **(a)** Compute $\int_C \vec{v} \cdot d\vec{r}$ where $\vec{v} = y\vec{i} + 2x\vec{j}$ and C is
   (i) The line joining $(0, 1)$ to $(1, 0)$
   (ii) The arc of the unit circle joining $(0, 1)$ to $(1, 0)$
   (b) What can you conclude about $\vec{v}$?

51. Let C be the straight path from $(0, 0)$ to $(5, 5)$ and let $\vec{F} = (y - x + 2)\vec{i} + (\sin(y - x) - 2)\vec{j}$.

   (a) At each point of C, what angle does $\vec{F}$ make with a tangent vector to C?
   (b) Evaluate $\int_C \vec{F} \cdot d\vec{r}$.

52. Let $\vec{F} = F_1\vec{i} + F_2\vec{j}$ and

$$\frac{\partial F_2}{\partial x} - \frac{\partial F_1}{\partial y} = 3(x^2 + y^2) - (x^2 + y^2)^{3/2}.$$

Let C_a be the circle of radius a in the xy-plane, centered at the origin and oriented counterclockwise. For what value of a is the line integral $\int_{C_a} \vec{F} \cdot d\vec{r}$ largest? What is the largest value?

53. The fact that an electric current gives rise to a magnetic field is the basis for some electric motors. Ampère's Law relates the magnetic field $\vec{B}$ to a steady current I. It says

$$\int_C \vec{B} \cdot d\vec{r} = kI$$

where I is the current[4] flowing through a closed curve C and k is a constant. Figure 18.62 shows a rod carrying a current and the magnetic field induced around the rod. If the rod is very long and thin, experiments show that the magnetic field $\vec{B}$ is tangent to every circle that is perpendicular to the rod and has center on the axis of the rod (like C in Figure 18.62). The magnitude of $\vec{B}$ is constant along every such circle. Use Ampère's Law to show that around a circle of radius r, the magnetic field due to a current I has magnitude given by

$$\|\vec{B}\| = \frac{kI}{2\pi r}.$$

(In other words, the strength of the field is inversely proportional to the radial distance from the rod.)

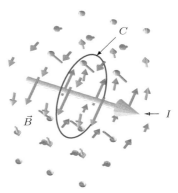

Figure 18.62

54. A *central vector field* is a vector field whose direction is always toward (or away from) a fixed point O (the center) and whose magnitude at a point P is a function only of the distance from P to O. In two dimensions this means that the vector field has constant magnitude on circles centered at O. The gravitational and electrical fields of spherically symmetric sources are both central fields.

   (a) Sketch an example of a central vector field.
   (b) Suppose that the central field $\vec{F}$ is a gradient field, that is, $\vec{F} = \operatorname{grad} f$. What must be the shape of the contours of f? Sketch some contours for this case.
   (c) Is every gradient field a central vector field? Explain.
   (d) In Figure 18.63, two paths are shown between the points Q and P. Assuming that the three circles C_1, C_2, and C_3 are centered at O, explain why the work done by a central vector field $\vec{F}$ is the same for either path.
   (e) It is in fact true that every central vector field is a gradient field. Use an argument suggested by Figure 18.63 to explain why any central vector field must be path-independent.

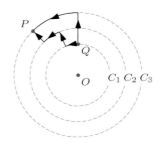

Figure 18.63

---

[4]More precisely, I is the net current through any surface that has C as its boundary.

55. A *free vortex* circulating about the origin in the xy-plane (or about the z-axis in 3-space) has vector field $\vec{v} = K(x^2 + y^2)^{-1}(-y\vec{i} + x\vec{j})$ where K is a constant. The Rankine model of a tornado hypothesizes an inner core that rotates at constant angular velocity, surrounded by a free vortex. Suppose that the inner core has radius 100 meters and that $\|\vec{v}\| = 3 \cdot 10^5$ meters/hr at a distance of 100 meters from the center.

(a) Assuming that the tornado rotates counterclockwise (viewed from above the xy-plane) and that $\vec{v}$ is continuous, determine ω and K such that

$$\vec{v} = \begin{cases} \omega(-y\vec{i} + x\vec{j}) & \text{if } \sqrt{x^2 + y^2} < 100 \\ K(x^2 + y^2)^{-1}(-y\vec{i} + x\vec{j}) \\ \quad\quad \text{if } \sqrt{x^2 + y^2} \geq 100. \end{cases}$$

(b) Sketch the vector field $\vec{v}$.

(c) Find the circulation of $\vec{v}$ around the circle of radius r centered at the origin, traversed counterclockwise.

56. Figure 18.64 shows the tangential velocity as a function of radius for the tornado that hit Dallas on April 2, 1957. Use it and Problem 55 to estimate K and ω for the Rankine model of this tornado.[5]

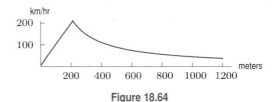

Figure 18.64

CAS Challenge Problems

57. Let C_a be the circle of radius a, centered at the origin, oriented in the counterclockwise direction, and let

$$\vec{F} = (-y + \frac{2}{3}y^3)\vec{i} + (2x - \frac{x^3}{3} + xy^2)\vec{j}.$$

(a) Evaluate $\int_{C_a} \vec{F} \cdot d\vec{r}$. For what positive value of a does the integral take its maximum value?

(b) Use Green's Theorem to convert the integral to a double integral. Without evaluating the double integral, give a geometric explanation of the value of a you found in part (a).

58. If f is a potential function for the two-dimensional vector field $\vec{F}$ then the Fundamental Theorem of Calculus for Line Integrals says that $\int_C \vec{F} \cdot d\vec{r} = f(x, y) - f(0, 0)$

where C is any path from $(0, 0)$ to (x, y). Using this fact and choosing C to be a straight line, find potential functions for the following conservative fields (where a, b, c are constants):

(a) $\vec{F} = ay\vec{i} + ax\vec{j}$
(b) $\vec{F} = abye^{bxy}\vec{i} + (c + abxe^{bxy})\vec{j}$

59. Let $\vec{F} = (ax + by)\vec{i} + (cx + dy)\vec{j}$. Evaluate the line integral of $\vec{F}$ along the paths

$$C_1 : \vec{r}(t) = 2t\vec{i} + t^2\vec{j}, \quad 0 \leq t \leq 3$$
$$C_2 : \vec{r}(t) = 2(3 - t)\vec{i} + (3 - t)^2\vec{j}, \quad 0 \leq t \leq 3$$

Describe and explain the relationship between the two integrals.

PROJECTS FOR CHAPTER EIGHTEEN

1. Conservation of Energy

(a) A particle moves with position vector $\vec{r}(t) = x(t)\vec{i} + y(t)\vec{j} + z(t)\vec{k}$. Let $\vec{v}(t)$ and $\vec{a}(t)$ be its velocity and acceleration vectors. Show that

$$\frac{1}{2}\frac{d}{dt}\|\vec{v}(t)\|^2 = \vec{a}(t) \cdot \vec{v}(t).$$

(b) We now derive the principle of Conservation of Energy. The kinetic energy of a particle of mass m moving with speed v is $(1/2)mv^2$. Suppose the particle has potential energy $f(\vec{r})$ at the position $\vec{r}$ due to a force field $\vec{F} = -\nabla f$. If the particle moves with position vector $\vec{r}(t)$ and velocity $\vec{v}(t)$, then the Conservation of Energy principle says that

$$\text{Total energy} = \text{Kinetic energy} + \text{Potential energy} = \frac{1}{2}m\|\vec{v}(t)\|^2 + f(\vec{r}(t)) = \text{Constant}.$$

[5]Adapted from *Encyclopædia Britannica, Macropædia*, Vol. 16, page 477, "Climate and the Weather," Tornados and Waterspouts, 1991.

Let P and Q be two points in space and let C be a path from P to Q parameterized by $\vec{r}(t)$ for $t_0 \le t \le t_1$, where $\vec{r}(t_0) = P$ and $\vec{r}(t_1) = Q$.

(i) Using part (a) and Newton's law $\vec{F} = m\vec{a}$, show

$$\begin{array}{c} \text{Work done by } \vec{F} \\ \text{as particle moves along } C \end{array} = \text{Kinetic energy at } Q - \text{Kinetic energy at } P.$$

(ii) Use the Fundamental Theorem of Calculus for Line Integrals to show that

$$\begin{array}{c} \text{Work done by } \vec{F} \\ \text{as particle moves along } C \end{array} = \text{Potential energy at } P - \text{Potential energy at } Q.$$

(iii) Use parts (a) and (b) to show that the total energy at P is the same as at Q.

This problem explains why force vector fields which are *path-independent* are usually called *conservative* (force) vector fields.

2. Planimeters

A planimeter is a mechanical device that exploits Green's Theorem to find the area of a planar region by tracing out its boundary.

A *linear planimeter* is a rod with one end, the foot, that moves along a straight track as a wheel at the other end rolls and slides along a curve being traced. The rod acts as an axle for the wheel so that the wheel rolls without sliding for motion perpendicular to the rod, and slides without rolling for motion parallel to the rod. The wheel vector is defined by rotating the displacement vector from wheel to foot clockwise $90°$. For motion in a direction making an angle θ with the wheel vector, the ratio of roll to distance traveled is $\cos\theta$. (This counts rolling as positive for one direction of rotation and negative for the other.) A meter on the planimeter measures the amount the wheel rolls. When the user has finished tracing out a closed curve, the total roll of the wheel determines the area of the enclosed region. In Figure 18.65 the foot of a planimeter of length L slides up and down the x-axis as the wheel moves along C from P to Q.

The key fact that makes this work is that the planimeter computes the line integral of the vector field of unit vectors $\vec{F}$ in the direction of the wheel vector along the oriented curve C:

$$\int_C \vec{F} \cdot d\vec{r} = \text{Total roll of wheel.}$$

To see this, divide C into n small, almost straight pieces along which $\vec{F}$ has approximately constant direction. Each piece can be represented by a displacement vector $\Delta\vec{r}_i = \vec{r}_{i+1} - \vec{r}_i$, and $\vec{F}(\vec{r}_i)$ is the unit vector in the direction of the wheel vector at the point of C with position vector $\vec{r}_i$. We have

$$\vec{F}(\vec{r}_i)\cdot\Delta\vec{r}_i = \|\vec{F}(\vec{r}_i)\|\|\Delta\vec{r}_i\|\cos\theta_i \approx 1\cdot(\text{Length of } i\text{th piece of } C)\cdot\cos\theta_i \approx \text{Roll on } i\text{th piece of } C.$$

Summing over all the pieces and taking the limit as $\|\Delta\vec{r}_i\| \to 0$ gives the result.

In the problems, we find a formula for $\vec{F}$ and use it to show how area is computed.

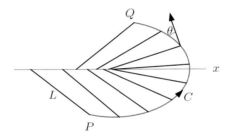

Figure 18.65

(a) Let $L > 0$, and define $\vec{F}(x, y)$ in the horizontal strip $-L < y < L$ as follows. Let $\vec{m}$ be the displacement vector of magnitude L from a point $(a, 0)$ on the x-axis to the point (x, y) in the strip, where $a < x$. Let $\vec{F}(x, y)$ be the unit vector in the direction of $\vec{k} \times \vec{m}$, perpendicular to $\vec{m}$. The flow lines of $\vec{F}$ are arcs of circles of radius L with center on the x-axis, oriented counterclockwise. See Figure 18.66. Show that

$$\vec{F}(x, y) = \frac{-y}{L}\vec{i} + \frac{1}{L}\sqrt{L^2 - y^2}\vec{j}.$$

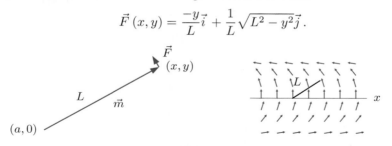

Figure 18.66

(b) Show that $\operatorname{curl} \vec{F} = 1/L$, for $\vec{F}$ as in part (a).

(c) Let C be a simple closed curve oriented counterclockwise in the strip $-L < y < L$, and let R be the enclosed region. Use Green's Theorem to show that

$$\text{Area of } R = L \int_C \vec{F} \cdot d\vec{r} = L \cdot (\text{Total roll of planimeter wheel}).$$

3. **Ampère's Law**

Ampère's Law, introduced in Problem 53 on page 1019, relates the net current, I, flowing through a surface to the magnetic field around the boundary, C, of the surface. The law says that

$$\int_C \vec{B} \cdot d\vec{r} = kI, \quad \text{for some constant } k.$$

The orientation of C is given by the right-hand rule: if the fingers of your right hand lie against the surface and curl in the direction of the oriented curve C, then your thumb points in the direction of positive current.

(a) We consider an infinitely long cylindrical wire having a radius r_0, where $r_0 > 0$. Suppose the wire is centered on the z-axis and carries a constant current, I, uniformly distributed across a cross-section of the wire. The magnitude of $\vec{B}$ is constant along any circle which is centered on and perpendicular to the z-axis. The direction of $\vec{B}$ is tangent to such circles. Show that the magnitude of $\vec{B}$, at a distance r from the z-axis, is given by

$$\|\vec{B}\| = \begin{cases} \dfrac{aI}{2\pi r} & \text{for } r \geq r_0 \\[2mm] \dfrac{aIr}{2\pi r_0^2} & \text{for } r < r_0. \end{cases}$$

(b) A *torus* is a doughnut-shaped surface obtained by rotating around the z-axis the circle $(x - \beta)^2 + z^2 = \alpha^2$ of radius α and center $(\beta, 0, 0)$, where $0 < \alpha < \beta$. A *toroidal solenoid* is constructed by wrapping a thin wire a large number, say N, times around the torus. Experiments show that if a constant current I flows though the wire, then the magnitude of the magnetic field is constant on circles inside the torus which are centered on and perpendicular to the z-axis. The direction of $\vec{B}$ is tangent to such circles. Explain why, at all points inside the torus,

$$\|\vec{B}\| = \frac{aNI}{2\pi r},$$

and $\|\vec{B}\| = 0$ otherwise. [Hint: Apply Ampère's Law to a suitably chosen surface S with boundary curve C. What is the net current passing through S?]

4. **Conservative Forces, Friction, and Gravity**

The work done by a conservative force on an object does not depend on the path taken but only on the endpoints—in other words, the vector field representing the force is path-independent. Only for a conservative force is the work done stored as energy; for a non-conservative force, the work may be dissipated.

(a) The frictional drag force $\vec{F}$ on an airplane flying through the air with velocity $\vec{v}$ is given by

$$\vec{F} = -c\|\vec{v}\|\vec{v}.$$

The constant c is positive and depends on the shape of the airplane. One airplane takes off from an airport located at the point $(2, 0, 0)$ and follows the path

$$\vec{r}(t) = (2\cos t)\vec{i} + (2\sin t)\vec{j} + 3t\vec{k}, \quad \text{from } t = 0 \text{ to } t = \pi.$$

An identical airplane takes off from the same point and follows the path

$$\vec{r}(t) = \frac{2 - 4t}{\pi}t\vec{i} + 3t\vec{k} \quad \text{also from } t = 0 \text{ to } t = \pi.$$

(i) Compute the total work done by the drag force $\vec{F}$ on each airplane.

(ii) Is the drag force conservative? Explain.

(b) Suppose the two airplanes of the preceding problem each have mass m, while the earth has mass M. The gravitational force $\vec{F}$ acting on either airplane at a point with position vector $\vec{r}$ is given by

$$\vec{F}(\vec{r}) = -GMm\frac{\vec{r}}{\|\vec{r}\|^3}, \quad \text{where } G \text{ is the gravitational constant.}$$

(i) Compute the total work done by the gravitational force $\vec{F}$ on each airplane as it moves from the point $(2, 0, 0)$ to the point $(-2, 0, 3\pi)$.

(ii) Is the gravitational force conservative? Explain.

Chapter Nineteen

FLUX INTEGRALS AND DIVERGENCE

Contents

19.1 THE IDEA OF A FLUX INTEGRAL

Flow Through a Surface

Imagine water flowing through a fishing net stretched across a stream. Suppose we want to measure the flow rate of water through the net, that is, the volume of fluid that passes through the surface per unit time.

Example 1   A flat square surface of area A, in m^2, is immersed in a fluid. The fluid flows with constant velocity $\vec{v}$, in m/sec, perpendicular to the square. Write an expression for the rate of flow in m^3/sec.

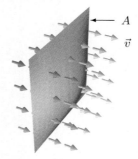

A

$\vec{v}$

Figure 19.1: Fluid flowing perpendicular
to a surface

Solution   In one second a given particle of water moves a distance of $\|\vec{v}\|$ in the direction perpendicular to the square. Thus, the entire body of water moving through the square in one second is a box of length $\|\vec{v}\|$ and cross-sectional area A. So the box has volume $\|\vec{v}\|A$ m^3, and

$$\text{Flow rate} = \|\vec{v}\|A \text{ m}^3/\text{sec.}$$

This flow rate is called the *flux* of the fluid through the surface. We can also compute the flux of vector fields, such as electric and magnetic fields, where no flow is actually taking place. If the vector field is constant and perpendicular to the surface, and if the surface is flat, as in Example 1, the flux is obtained by multiplying the speed by the area.

Next we find the flux of a constant vector field through a flat surface that is not perpendicular to the vector field, using a dot product. In general, we break a surface into small pieces which are approximately flat and where the vector field is approximately constant, leading to a flux integral.

Orientation of a Surface

Before computing the flux of a vector field through a surface, we need to decide which direction of flow through the surface is the positive direction; this is described as choosing an orientation.[1]

> At each point on a smooth surface there are two unit normals, one in each direction. **Choosing an orientation** means picking one of these normals at every point of the surface in a continuous way. The unit normal vector in the direction of the orientation is denoted by $\vec{n}$. For a closed surface (that is, the boundary of a solid region), we choose the **outward orientation** unless otherwise specified.

We say the flux through a piece of surface is positive if the flow is in the direction of the orientation and negative if it is in the opposite direction. (See Figure 19.2.)

[1] Although we will not study them, there are a few surfaces for which this cannot be done. See page 1032.

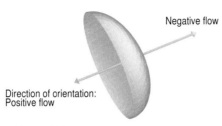

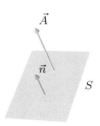

Figure 19.2: An oriented surface showing directions of positive and negative flow

Figure 19.3: Area vector $\vec{A} = \vec{n}\,A$ of flat surface with area A and orientation $\vec{n}$

The Area Vector

The flux through a flat surface depends both on the area of the surface and its orientation. Thus, it is useful to represent its area by a vector as shown in Figure 19.3.

> The **area vector** of a flat, oriented surface is a vector $\vec{A}$ such that
> - The magnitude of $\vec{A}$ is the area of the surface.
> - The direction of $\vec{A}$ is the direction of the orientation vector $\vec{n}$.

The Flux of a Constant Vector Field Through a Flat Surface

Suppose the velocity vector field, $\vec{v}$, of a fluid is constant and $\vec{A}$ is the area vector of a flat surface. The flux through this surface is the volume of fluid that flows through in one unit of time. The skewed box in Figure 19.4 has cross-sectional area $\|\vec{A}\|$ and height $\|\vec{v}\|\cos\theta$, so its volume is $(\|\vec{v}\|\cos\theta)\|\vec{A}\| = \vec{v}\cdot\vec{A}$. Thus, we have the following result:

> If $\vec{v}$ is constant and $\vec{A}$ is the area vector of a flat surface, then
> $$\text{Flux through surface} = \vec{v}\cdot\vec{A}.$$

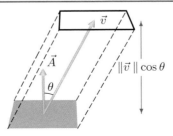

Figure 19.4: Flux of $\vec{v}$ through a surface with area vector $\vec{A}$ is the volume of this skewed box

Example 2   Water is flowing down a cylindrical pipe 2 cm in radius with a velocity of 3 cm/sec. Find the flux of the velocity vector field through the ellipse-shaped region shown in Figure 19.5. The normal to the ellipse makes an angle of θ with the direction of flow and the area of the ellipse is $4\pi/(\cos\theta)$ cm^2.

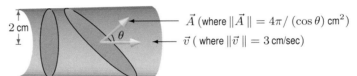

Figure 19.5: Flux through ellipse-shaped region across a cylindrical pipe

Solution   There are two ways to approach this problem. One is to use the formula we just derived, which gives

$$\text{Flux through ellipse} = \vec{v} \cdot \vec{A} = \|\vec{v}\|\|\vec{A}\| \cos\theta = 3(\text{Area of ellipse}) \cos\theta$$

$$= 3\left(\frac{4\pi}{\cos\theta}\right)\cos\theta = 12\pi \ \text{cm}^3/\text{sec}.$$

The second way is to notice that the flux through the ellipse is equal to the flux through the circle perpendicular to the pipe in Figure 19.5. Since the flux is the rate at which water is flowing down the pipe, we have

$$\text{Flux through circle} = \begin{matrix}\text{Velocity} \\ \text{of water}\end{matrix} \times \begin{matrix}\text{Area of} \\ \text{circle}\end{matrix} = \left(3\ \frac{\text{cm}}{\text{sec}}\right)(\pi 2^2 \ \text{cm}^2) = 12\pi \ \text{cm}^3/\text{sec}.$$

The Flux Integral

If the vector field, $\vec{F}$, is not constant or the surface, S, is not flat, we divide the surface into a patchwork of small, almost flat pieces. (See Figure 19.6.) For a particular patch with area ΔA, we pick a unit orientation vector $\vec{n}$ at a point on the patch and define the area vector of the patch, $\Delta \vec{A}$, as

$$\Delta \vec{A} = \vec{n}\,\Delta A.$$

(See Figure 19.7.) If the patches are small enough, we can assume that $\vec{F}$ is approximately constant on each piece. Then we know that

$$\text{Flux through patch} \approx \vec{F} \cdot \Delta \vec{A},$$

so, adding the fluxes through all the small pieces, we have

$$\text{Flux through whole surface} \approx \sum \vec{F} \cdot \Delta \vec{A},$$

As each patch becomes smaller and $\|\Delta \vec{A}\| \to 0$, the approximation gets better and we get

$$\text{Flux through } S = \lim_{\|\Delta \vec{A}\| \to 0} \sum \vec{F} \cdot \Delta \vec{A}.$$

Thus, provided the limit exists, we make the following definition:

> The **flux integral** of the vector field $\vec{F}$ through the oriented surface S is
>
> $$\int_S \vec{F} \cdot d\vec{A} = \lim_{\|\Delta \vec{A}\| \to 0} \sum \vec{F} \cdot \Delta \vec{A}.$$
>
> If S is a closed surface oriented outward, we describe the flux through S as the flux out of S.

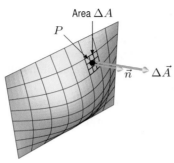

Figure 19.6: Surface S divided into small, almost flat pieces, showing a typical orientation vector $\vec{n}$ and area vector $\Delta \vec{A}$

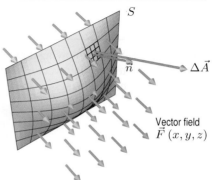

Figure 19.7: Flux of a vector field through a curved surface S

In computing a flux integral, we have to divide the surface up in a reasonable way, or the limit might not exist. In practice this problem seldom arises; however, one way to avoid it is to define flux integrals by the method used to compute them shown in Section 21.3.

Flux and Fluid Flow

If $\vec{v}$ is the velocity vector field of a fluid, we have

$$
\begin{array}{ccccc}
\text{Rate fluid flows} & = & \text{Flux of } \vec{v} & = & \displaystyle\int_S \vec{v} \cdot d\vec{A} \\
\text{through surface } S & & \text{through } S & &
\end{array}
$$

The rate of fluid flow is measured in units of volume per unit time.

Example 3   Find the flux of the vector field $\vec{B}(x, y, z)$ shown in Figure 19.8 through the square S of side 2 shown in Figure 19.9, oriented in the $\vec{j}$ direction, where

$$
\vec{B}(x, y, z) = \frac{-y\vec{i} + x\vec{j}}{x^2 + y^2}.
$$

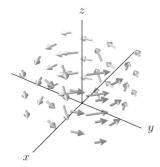

Figure 19.8: The vector field
$\vec{B}(x, y, z) = \frac{-y\vec{i} + x\vec{j}}{x^2 + y^2}$

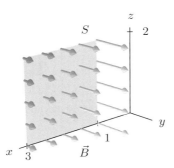

Figure 19.9: Flux of $\vec{B}$ through the square S of side 2 in xy-plane and oriented in $\vec{j}$ direction

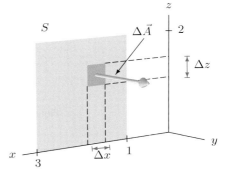

Figure 19.10: A small patch of surface with area $\|\Delta\vec{A}\| = \Delta x \Delta z$

Solution   Consider a small rectangular patch with area vector $\Delta\vec{A}$ in S, with sides Δx and Δz so that $\|\Delta\vec{A}\| = \Delta x \Delta z$. Since $\Delta\vec{A}$ points in the $\vec{j}$ direction, we have $\Delta\vec{A} = \vec{j}\,\Delta x \Delta z$. (See Figure 19.10.)

At the point $(x, 0, z)$ in S, substituting $y = 0$ into $\vec{B}$ gives $\vec{B}(x, 0, z) = (1/x)\vec{j}$. Thus, we have

$$
\text{Flux through small patch} \approx \vec{B} \cdot \Delta\vec{A} = \left(\frac{1}{x}\vec{j}\right) \cdot (\vec{j}\,\Delta x \Delta z) = \frac{1}{x}\,\Delta x\,\Delta z.
$$

Therefore,

$$
\text{Flux through surface} = \int_S \vec{B} \cdot d\vec{A} = \lim_{\|\Delta\vec{A}\| \to 0} \sum \vec{B} \cdot \Delta\vec{A} = \lim_{\substack{\Delta x \to 0 \\ \Delta z \to 0}} \sum \frac{1}{x}\,\Delta x\,\Delta z.
$$

This last expression is a Riemann sum for the double integral $\int_R \frac{1}{x}\,dA$, where R is the square $1 \le x \le 3$, $0 \le z \le 2$. Thus,

$$
\text{Flux through surface} = \int_S \vec{B} \cdot d\vec{A} = \int_R \frac{1}{x}\,dA = \int_0^2 \int_1^3 \frac{1}{x}\,dx\,dz = 2\ln 3.
$$

The result is positive since the vector field is passing through the surface in the positive direction.

Example 4   Each of the vector fields in Figure 19.11 consists entirely of vectors parallel to the xy-plane, and is constant in the z direction (that is, the vector field looks the same in any plane parallel to the xy-plane). For each one, say whether you expect the flux through a closed surface surrounding the origin to be positive, negative, or zero. In part (a) the surface is a closed cube with faces perpendicular to the axes; in parts (b) and (c) the surface is a closed cylinder. In each case we choose the outward orientation. (See Figure 19.12.)

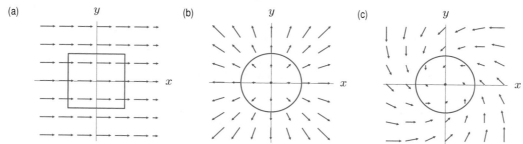

Figure 19.11: Flux of a vector field through the closed surfaces whose cross-sections are shown in the xy-plane

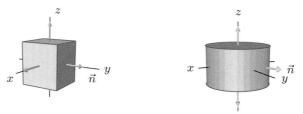

Figure 19.12: The closed cube and closed cylinder, both oriented outward

Solution   (a) Since the vector field appears to be parallel to the faces of the cube which are perpendicular to the y- and z-axes, we expect the flux through these faces to be zero. The fluxes through the two faces perpendicular to the x-axis appear to be equal in magnitude and opposite in sign, so we expect the net flux to be zero.

(b) Since the top and bottom of the cylinder are parallel to the flow, the flux through them is zero. On the curved surface of the cylinder, $\vec{v}$ and $\Delta\vec{A}$ appear to be everywhere parallel and in the same direction, so we expect each term $\vec{v} \cdot \Delta\vec{A}$ to be positive, and therefore the flux integral $\int_S \vec{v} \cdot d\vec{A}$ to be positive.

(c) As in part (b), the flux through the top and bottom of the cylinder is zero. In this case $\vec{v}$ and $\Delta\vec{A}$ are not parallel on the round surface of the cylinder, but since the fluid appears to be flowing inward as well as swirling, we expect each term $\vec{v} \cdot \Delta\vec{A}$ to be negative, and therefore the flux integral to be negative.

Calculating Flux Integrals Using $d\vec{A} = \vec{n}\,d\,A$

For a small patch of surface ΔS with unit normal $\vec{n}$ and area ΔA, the area vector is $\Delta\vec{A} = \vec{n}\,\Delta A$. The next example shows how we can use this relationship to compute a flux integral.

Example 5   An electric charge q is placed at the origin in 3-space. The resulting electric field $\vec{E}\,(\vec{r}\,)$ at the point with position vector $\vec{r}$ is given by

$$\vec{E}\,(\vec{r}\,) = q\frac{\vec{r}}{\|\vec{r}\,\|^3}, \qquad \vec{r} \neq \vec{0}\,.$$

Find the flux of $\vec{E}$ out of the sphere of radius R centred at the origin. (See Figure 19.13.)

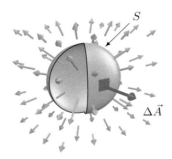

Figure 19.13: Flux of $\vec{E} = q\vec{r}/\|\vec{r}\|^3$ through the surface of a sphere of radius R centred at the origin

Solution

This vector field points radially outward from the origin in the same direction as $\vec{n}$. Thus, since $\vec{n}$ is a unit vector,

$$\vec{E} \cdot \Delta\vec{A} = \vec{E} \cdot \vec{n}\, \Delta A = \|\vec{E}\|\, \Delta A.$$

On the sphere, $\|\vec{E}\| = q/R^2$, so

$$\int_S \vec{E} \cdot d\vec{A} = \lim_{\|\Delta\vec{A}\| \to 0} \sum \vec{E} \cdot \Delta\vec{A} = \lim_{\Delta A \to 0} \sum \frac{q}{R^2}\, \Delta A = \frac{q}{R^2} \lim_{\Delta A \to 0} \sum \Delta A.$$

The last sum approximates the surface area of the sphere. In the limit as the subdivisions get finer we have

$$\lim_{\Delta A \to 0} \sum \Delta A = \text{Surface area of sphere}.$$

Thus, the flux is given by

$$\int_S \vec{E} \cdot d\vec{A} = \frac{q}{R^2} \lim_{\Delta A \to 0} \sum \Delta A = \frac{q}{R^2} \cdot (\text{Surface area of sphere}) = \frac{q}{R^2}(4\pi R^2) = 4\pi q.$$

This result is known as Gauss's law.

To compute a flux with an integral instead of Riemann sums, we often write $d\vec{A} = \vec{n}\, dA$, as in the next example.

Example 6

Suppose S is the surface of the cube bounded by the six planes $x = \pm 1$, $y = \pm 1$, and $z = \pm 1$. Compute the flux of the electric field $\vec{E}$ of the previous example outward through S.

Solution

It is enough to compute the flux of $\vec{E}$ through a single face, say the top face S_1 defined by $z = 1$, where $-1 \leq x \leq 1$ and $-1 \leq y \leq 1$. By symmetry, the flux of $\vec{E}$ through the other five faces of S must be the same.

On the top face, S_1, we have $d\vec{A} = \vec{n}\, dA = \vec{k}\, dx\, dy$ and

$$\vec{E}(x, y, 1) = q\frac{x\vec{i} + y\vec{j} + \vec{k}}{(x^2 + y^2 + 1)^{3/2}}.$$

The corresponding flux integral is given by

$$\int_{S_1} \vec{E} \cdot d\vec{A} = q \int_{-1}^{1} \int_{-1}^{1} \frac{x\vec{i} + y\vec{j} + \vec{k}}{(x^2 + y^2 + 1)^{3/2}} \cdot \vec{k}\, dx\, dy = q \int_{-1}^{1} \int_{-1}^{1} \frac{1}{(x^2 + y^2 + 1)^{3/2}}\, dx\, dy.$$

Computing this integral numerically shows that

$$\text{Flux through top face} = \int_{S_1} \vec{E} \cdot d\vec{A} \approx 2.0944q.$$

Thus,

$$\text{Total flux of } \vec{E} \text{ out of cube} = \int_S \vec{E} \cdot d\vec{A} \approx 6(2.0944q) = 12.5664q.$$

Example 5 on page 1030 showed that the flux of $\vec{E}$ through a sphere of radius R centred at the origin is $4\pi q$. Since $4\pi \approx 12.5664$, Example 6 suggests that

$$\text{Total flux of } \vec{E} \text{ out of cube} = 4\pi q.$$

By computing the flux integral in Example 6 exactly, it is possible to verify that the flux of $\vec{E}$ through the cube and the sphere are exactly equal. When we encounter the Divergence Theorem in Chapter 20 we will see why this is so.

Notes on Orientation

Two difficulties can occur in choosing an orientation. The first is that if the surface is not smooth, it may not have a normal vector at every point. For example, a cube does not have a normal vector along its edges. When we have a surface, such as a cube, which is made of a finite number of smooth pieces, we choose an orientation for each piece separately. The best way to do this is usually clear. For example, on the cube we choose the outward orientation on each face. (See Figure 19.14.)

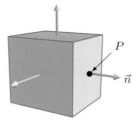

Figure 19.14: The orientation vector field $\vec{n}$ on the cube surface determined by the choice of unit normal vector at the point P

Figure 19.15: The Möbius strip is an example of a non-orientable surface

The second difficulty is that there are some surfaces which cannot be oriented at all, such as the *Möbius strip* in Figure 19.15.

Exercises and Problems for Section 19.1

Exercises

1. Let S be the disk of radius 5 perpendicular to the y-axis, centered at $(0, 7, 0)$ and oriented toward the origin. Is
$$\int_S (3\vec{i} + 4\vec{j}) \cdot d\vec{A}$$
a vector or a scalar? Calculate it.

In Exercises 2–5, find the area vector of the oriented flat surface.

2. The triangle with vertices $(0, 0, 0)$, $(0, 2, 0)$, $(0, 0, 3)$ oriented in the negative x direction.

3. Circular disc of radius 5 in the xy-plane, oriented upward.

4. $y = 10$, $0 \leq x \leq 5$, $0 \leq z \leq 3$, oriented away from the xz-plane.

5. $y = -10$, $0 \leq x \leq 5$, $0 \leq z \leq 3$, oriented away from the xz-plane.

6. Find an oriented flat surface with area vector $150\vec{j}$.

7. Compute $\int_S (4\vec{i} + 5\vec{k}) \cdot d\vec{A}$, where S is the square of side length 3 perpendicular to the z-axis, centered at $(0, 0, -2)$ and oriented

(a) Toward the origin. (b) Away from the origin.

8. Compute $\int_S (2\vec{i} + 3\vec{k}) \cdot d\vec{A}$, where S is the disk of radius 4 perpendicular to the x-axis, centered at $(5, 0, 0)$ and oriented

(a) Toward the origin. (b) Away from the origin.

9. Let $\vec{F}(x, y, z) = z\vec{i}$. For each of the surfaces in (a)–(e), say whether the flux of $\vec{F}$ through the surface is positive, negative, or zero. The orientation of the surface is indicated by a normal vector.

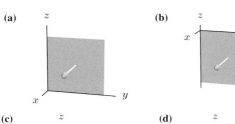

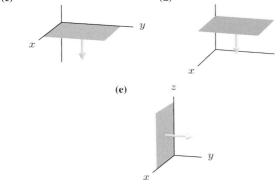

10. Repeat Exercise 9 with $\vec{F}(x, y, z) = -z\vec{i} + x\vec{k}$.

11. Repeat Exercise 9 with the vector field $\vec{F}(\vec{r}) = \vec{r}$.

For Exercises 12–15 find the flux of the constant vector field $\vec{v} = \vec{i} - \vec{j} + 3\vec{k}$ through the given surface.

12. A disk of radius 2 in the xy-plane oriented upward.

13. A triangular plate of area 4 in the yz-plane oriented in the positive x-direction.

14. A square plate of area 4 in the yz-plane oriented in the positive x-direction.

15. The triangular plate with vertices $(1, 0, 0)$, $(0, 1, 0)$, $(0, 0, 1)$, oriented away from the origin.

In Exercises 16–18, find the flux of $\vec{H} = 2\vec{i} + 3\vec{j} + 5\vec{k}$ through the surface S.

16. S is the cylinder $x^2 + y^2 = 1$, closed at the ends by the planes $z = 0$ and $z = 1$ and oriented outward.

17. S is the disk of radius 1 in the plane $x = 2$ oriented in the positive x-direction.

18. S is the disk of radius 1 in the plane $x + y + z = 1$ oriented in upward.

Find the flux of the vector fields in Exercises 19–21 out of the closed box $0 \le x \le 1, 0 \le y \le 2, 0 \le z \le 3$.

19. $\vec{F} = 3\vec{i} + 2\vec{j} + \vec{k}$ **20.** $\vec{G} = x\vec{i}$

21. $\vec{H} = zx\vec{k}$

In Exercises 22–25, calculate the flux integral.

22. $\int_S (x\vec{i} + 4\vec{j}) \cdot d\vec{A}$ where S is the disk of radius 5 perpendicular to the x-axis, centered at $(3, 0, 0)$ and oriented toward the origin.

23. $\int_S \vec{r} \cdot d\vec{A}$ where S is the sphere of radius 3 centered at the origin.

24. $\int_S (\sin x \, \vec{i} + (y^2 + z^2)\vec{j} + y^2\vec{k}) \cdot d\vec{A}$ where S is a disk of radius π in the plane $x = 3\pi/2$, oriented in the positive x-direction.

25. $\int_S (5\vec{i} + 5\vec{j} + 5\vec{k}) \cdot d\vec{A}$ where S is a disk of radius 3 in the plane $x + y + z = 1$, oriented upward.

In Exercises 26–48, calculate the flux of the vector field through the surface.

26. $\vec{F} = 2\vec{i} + 3\vec{j}$ through the square of side π in the xy-plane, oriented upward.

27. $\vec{F} = 2\vec{i} + 3\vec{j}$ through the unit disk in the yz-plane, centered at the origin and oriented in the positive x-direction.

28. $\vec{F} = x\vec{i} + y\vec{j} + z\vec{k}$ through the square of side 1.6 centered at $(2, 5, 8)$, parallel to the xz-plane and oriented away from the origin.

29. $\vec{F} = z\vec{k}$ through a square of side $\sqrt{14}$ in a horizontal plane 2 units below the xy-plane and oriented downward.

30. $\vec{F} = -y\vec{i} + x\vec{j}$ and S is the square plate in the yz-plane with corners at $(0, 1, 1)$, $(0, -1, 1)$, $(0, 1, -1)$, and $(0, -1, -1)$, oriented in the positive x-direction.

31. $\vec{F} = 7\vec{i} + 6\vec{j} + 5\vec{k}$ and S is a disk of radius 2 in the yz-plane, centered at the origin and oriented in the positive x-direction.

32. $\vec{F} = x\vec{i} + 2y\vec{j} + 3z\vec{k}$ and S is a square of side 2 in the plane $y = 3$, oriented in the positive y-direction.

33. $\vec{F} = 7\vec{i} + 6\vec{j} + 5\vec{k}$ and S is a sphere of radius π centered at the origin.

34. $\vec{F} = -5\vec{r}$ through the sphere of radius 2 centered at the origin.

35. $\vec{F} = x\vec{i} + y\vec{j} + (z^2 + 3)\vec{k}$ and S is the rectangle $z = 4$, $0 \le x \le 2, 0 \le y \le 3$, oriented in the positive z-direction.

36. $\vec{F} = 6\vec{i} + 7\vec{j}$ through triangle of area 10 in the plane $x + y = 5$, oriented in the positive x-direction.

37. $\vec{F} = 6\vec{i} + x^2\vec{j} - \vec{k}$, through the square of side 4 in the plane $y = 3$, centered on the y-axis, with sides parallel to the x and z axes, and oriented in the positive y-direction.

38. $\vec{F} = (x + 3)\vec{i} + (y + 5)\vec{j} + (z + 7)\vec{k}$ through the rectangle $x = 4$, $0 \le y \le 2, 0 \le z \le 3$, oriented in the positive x-direction.

39. $\vec{F} = 7\vec{r}$ through the sphere of radius 3 centered at the origin.

40. $\vec{F} = -3\vec{r}$ through the sphere of radius 2 centered at the origin.

41. $\vec{F} = 2z\vec{i} + x\vec{j} + x\vec{k}$ through the rectangle $x = 4$, $0 \le y \le 2, 0 \le z \le 3$, oriented in the positive x-direction.

42. $\vec{F} = \vec{i} + 2\vec{j}$ through a square of side 2 lying in the plane $x + y + z = 1$, oriented away from the origin.

43. $\vec{F} = (x^2 + y^2)\vec{k}$ through the disk of radius 3 in the xy-plane and centered at the origin and oriented upward.

44. $\vec{F} = \cos(x^2 + y^2)\vec{k}$ through the disk $x^2 + y^2 \le 9$ oriented upward in the plane $z = 1$.

45. $\vec{F} = e^{y^2 + z^2}\vec{i}$ through the disk of radius 2 in the yz-plane, centered at the origin and oriented in the positive x-direction.

46. $\vec{F} = -y\vec{i} + x\vec{j}$ through the disk in the xy-plane with radius 2, oriented upward and centered at the origin.

47. $\vec{F} = \vec{r}$ through the disk of radius 2 parallel to the xy-plane oriented upward and centered at $(0, 0, 2)$.

48. $\vec{F} = (2 - x)\vec{i}$ through the cube whose vertices include the points $(0, 0, 0)$, $(3, 0, 0)$, $(0, 3, 0)$, $(0, 0, 3)$, and oriented outward.

Problems

49. Let B be the surface of a box centered at the origin, with edges parallel to the axes and in the planes $x = \pm 1$, $y = \pm 1$, $z = \pm 1$, and let S be the sphere of radius 1 centered at origin.

(a) Indicate whether the following flux integrals are positive, negative, or zero. No reasons needed.

   (a) $\int_B x\vec{i} \cdot d\vec{A}$ **(b)** $\int_B y\vec{i} \cdot d\vec{A}$
   (c) $\int_S |x|\vec{i} \cdot d\vec{A}$ **(d)** $\int_S (y - x)\vec{i} \cdot d\vec{A}$

(b) Explain with reasons how you know which flux integral is greater:

$$\int_S x\vec{i} \cdot d\vec{A} \quad \text{or} \quad \int_B x\vec{i} \cdot d\vec{A} ?$$

50. Suppose that $\vec{E}$ is a uniform electric field on 3-space, so $\vec{E}(x, y, z) = a\vec{i} + b\vec{j} + c\vec{k}$, for all points (x, y, z), where a, b, c are constants. Show, with the aid of symmetry, that the flux of $\vec{E}$ through each of the following closed surfaces S is zero:

(a) S is the cube bounded by the planes $x = \pm 1$, $y = \pm 1$, and $z = \pm 1$.
(b) S is the sphere $x^2 + y^2 + z^2 = 1$.
(c) S is the cylinder bounded by $x^2 + y^2 = 1$, $z = 0$, and $z = 2$.

51. Water is flowing down a cylindrical pipe of radius 2 cm; its speed is $(3 - (3/4)r^2)$ cm/sec at a distance r cm from the center of the pipe. Find the flux through the circular cross-section of the pipe, oriented so that the flux is positive.

52. **(a)** What do you think will be the electric flux through the cylindrical surface that is placed as shown in the constant electric field in Figure 19.16? Why?
(b) What if the cylinder is placed upright, as shown in Figure 19.17? Explain.

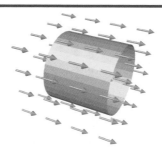

Figure 19.16

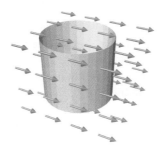

Figure 19.17

53. Let S be part of a cylinder centered on the y-axis. Explain why the three vectors fields $\vec{F}$, $\vec{G}$, and $\vec{H}$ have the same flux through S. Do not compute the flux.
$$\vec{F} = x\vec{i} + 2yz\vec{k}$$
$$\vec{G} = x\vec{i} + y\sin x\vec{j} + 2yz\vec{k}$$
$$\vec{H} = x\vec{i} + \cos(x^2 + z)\vec{j} + 2yz\vec{k}$$

54. Find the flux of $\vec{F} = \vec{r}/\|\vec{r}\|^3$ out of the sphere of radius R centered at the origin.

55. Find the flux of $\vec{F} = \vec{r}/\|r\|^2$ out of the sphere of radius R centered at the origin.

56. Consider the flux of the vector field $\vec{F} = \vec{r}/\|\vec{r}\|^p$ for $p \ge 0$ out of the sphere of radius 2 centered at the origin.

(a) For what value of p is the flux a maximum?
(b) What is that maximum value?

57. Let S be the cube with side length 2, faces parallel to the coordinate planes, and centered at the origin.

(a) Calculate the total flux of the constant vector field $\vec{v} = -\vec{i} + 2\vec{j} + \vec{k}$ out of S by computing the flux through each face separately.

(b) Calculate the flux out of S for any constant vector field $\vec{v} = a\vec{i} + b\vec{j} + c\vec{k}$.

(c) Explain why the answers to parts (a) and (b) make sense.

58. Let S be the tetrahedron with vertices at the origin and at $(1, 0, 0)$, $(0, 1, 0)$ and $(0, 0, 1)$.

(a) Calculate the total flux of the constant vector field $\vec{v} = -\vec{i} + 2\vec{j} + \vec{k}$ out of S by computing the flux through each face separately.

(b) Calculate the flux out of S in part (a) for any constant vector field $\vec{v}$.

(c) Explain why the answers to parts (a) and (b) make sense.

59. A volume of water swirls down according to the velocity field

$$\vec{F} = -\frac{y + xz}{(z^2 + 1)^2}\vec{i} - \frac{yz - x}{(z^2 + 1)^2}\vec{j} - \frac{1}{z^2 + 1}\vec{k}$$

Let C be a circle on the xy plane, with center at the origin and radius 1.

(a) Find the Flux Integral of $\vec{F}$ through the circle going downwards.

(b) Consider S, the patch of the surface $x^2 + y^2 - z^2 = 1$ between $z = -5$ and $z = 5$ (this is a hyperbola rotated around the z-axis), oriented outwards. Find the coordinates of a normal vector to this surface at the point (x, y, z).

(c) Find the Flux Integral of $\vec{F}$ through S.

60. Let $P(x, y, z)$ be the pressure at the point (x, y, z) in a fluid. Let $\vec{F}(x, y, z) = P(x, y, z)\vec{k}$. Let S be the surface of a body submerged in the fluid. If S is oriented inward, show that $\int_S \vec{F} \cdot d\vec{A}$ is the buoyant force on the body, that is, the force upward on the body due to the pressure of the fluid surrounding it. [Hint: $\vec{F} \cdot d\vec{A} = P(x, y, z)\vec{k} \cdot d\vec{A} = (P(x, y, z) \, d\vec{A}) \cdot \vec{k}$.]

61. A region of 3-space has a temperature which varies from point to point. Let $T(x, y, z)$ be the temperature at a point (x, y, z). Newton's law of cooling says that $\operatorname{grad} T$ is proportional to the heat flow vector field, $\vec{F}$, where $\vec{F}$ points in the direction in which heat is flowing and has magnitude equal to the rate of flow of heat.

(a) Suppose $\vec{F} = k \operatorname{grad} T$ for some constant k. What is the sign of k?

(b) Explain why this form of Newton's law of cooling makes sense.

(c) Let W be a region of space bounded by the surface S. Explain why

$$\begin{array}{l} \text{Rate of heat} \\ \text{loss from } W \end{array} = k \int_S (\operatorname{grad} T) \cdot d\vec{A}.$$

62. The z-axis carries a constant electric charge density of λ units of charge per unit length, with $\lambda > 0$. The resulting electric field is $\vec{E}$.

(a) Sketch the electric field, $\vec{E}$, in the xy-plane, given

$$\vec{E}(x, y, z) = 2\lambda \frac{x\vec{i} + y\vec{j}}{x^2 + y^2}.$$

(b) Compute the flux of $\vec{E}$ outward through the cylinder $x^2 + y^2 = R^2$, for $0 \le z \le h$.

63. An infinitely long straight wire lying along the z-axis carries an electric current I flowing in the $\vec{k}$ direction. Ampère's Law in magnetostatics says that the current gives rise to a magnetic field $\vec{B}$ given by

$$\vec{B}(x, y, z) = \frac{I}{2\pi} \frac{-y\vec{i} + x\vec{j}}{x^2 + y^2}.$$

(a) Sketch the field $\vec{B}$ in the xy-plane.

(b) Suppose S_1 is a disk with center at $(0, 0, h)$, radius a, and parallel to the xy-plane, oriented in the $\vec{k}$ direction. What is the flux of $\vec{B}$ through S_1? Is your answer reasonable?

(c) Suppose S_2 is the rectangle given by $x = 0$, $a \le y \le b$, $0 \le z \le h$, and oriented in the $-\vec{i}$ direction. What is the flux of $\vec{B}$ through S_2? Does your answer seem reasonable?

64. An ideal electric dipole in electrostatics is characterized by its position in 3-space and its dipole moment vector $\vec{p}$. The electric field $\vec{D}$, at the point with position vector $\vec{r}$, of an ideal electric dipole located at the origin with dipole moment $\vec{p}$ is given by

$$\vec{D}(\vec{r}) = 3\frac{(\vec{r} \cdot \vec{p})\vec{r}}{\|\vec{r}\|^5} - \frac{\vec{p}}{\|\vec{r}\|^3}.$$

Assume $\vec{p} = p\vec{k}$, so the dipole points in the $\vec{k}$ direction and has magnitude p.

(a) What is the flux of $\vec{D}$ through a sphere S with center at the origin and radius $a > 0$?

(b) The field $\vec{D}$ is a useful approximation to the electric field $\vec{E}$ produced by two "equal and opposite" charges, q at $\vec{r}_2$ and $-q$ at $\vec{r}_1$, where the distance $\|\vec{r}_2 - \vec{r}_1\|$ is small. The dipole moment of this configuration of charges is defined to be $q(\vec{r}_2 - \vec{r}_1)$. Gauss's Law in electrostatics says that the flux of $\vec{E}$ through S is equal to 4π times the total charge enclosed by S. What is the flux of $\vec{E}$ through S if the charges at $\vec{r}_1$ and $\vec{r}_2$ are enclosed by S? How does this compare with your answer for the flux of $\vec{D}$ through S if $\vec{p} = q(\vec{r}_2 - \vec{r}_1)$?

Strengthen Your Understanding

In Problems 65–66, explain what is wrong with the statement.

65. For a certain vector field $\vec{F}$ and oriented surface S, we have $\int_S \vec{F} \cdot d\vec{A} = 2\vec{i} - 3\vec{j} + \vec{k}$.

66. If S is a region in the xy-plane oriented upwards then $\int_S \vec{F} \cdot d\vec{A} > 0$.

In Problems 67–68, give an example of:

67. A nonzero vector field $\vec{F}$ such that $\int_S \vec{F} \cdot d\vec{A} = 0$, where S is the triangular surface with corners $(1,0,0)$, $(0,1,0)$, $(0,0,1)$, oriented away from the origin.

68. A nonconstant vector field $\vec{F}(x,y,z)$ and an oriented surface S such that $\int_S \vec{F} \cdot d\vec{A} = 1$.

Are the statements in Problems 69–78 true or false? Give reasons for your answer.

69. The value of a flux integral is a scalar.

70. The area vector $\vec{A}$ of a flat, oriented surface is parallel to the surface.

71. If S is the unit sphere centered at the origin, oriented outward and the flux integral $\int_S \vec{F} \cdot d\vec{A}$ is zero, then $\vec{F} = \vec{0}$.

72. The flux of the vector field $\vec{F} = \vec{i}$ through the plane $x = 0$, with $0 \le y \le 1, 0 \le z \le 1$, oriented in the $\vec{i}$ direction is positive.

73. If S is the unit sphere centered at the origin, oriented outward and $\vec{F} = x\vec{i} + y\vec{j} + z\vec{k} = \vec{r}$, then the flux integral $\int_S \vec{F} \cdot d\vec{A}$ is positive.

74. If S is the cube bounded by the six planes $x = \pm 1, y = \pm 1, z = \pm 1$, oriented outward, and $\vec{F} = \vec{k}$, then $\int_S \vec{F} \cdot d\vec{A} = 0$.

75. If S is an oriented surface in 3-space, and $-S$ is the same surface, but with the opposite orientation, then $\int_S \vec{F} \cdot d\vec{A} = -\int_{-S} \vec{F} \cdot d\vec{A}$.

76. If S_1 is a rectangle with area 1 and S_2 is a rectangle with area 2, then $2\int_{S_1} \vec{F} \cdot d\vec{A} = \int_{S_2} \vec{F} \cdot d\vec{A}$.

77. If $\vec{F} = 2\vec{G}$, then $\int_S \vec{F} \cdot d\vec{A} = 2\int_S \vec{G} \cdot d\vec{A}$.

78. If $\int_S \vec{F} \cdot d\vec{A} > \int_S \vec{G} \cdot d\vec{A}$ then $||\vec{F}|| > ||\vec{G}||$ at all points on the surface S.

79. For each of the surfaces in (a)–(e), pick the vector field $\vec{F}_1, \vec{F}_2, \vec{F}_3, \vec{F}_4, \vec{F}_5$, with the largest flux through the surface. The surfaces are all squares of the same size. Note that the orientation is shown.

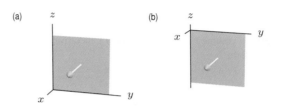

$$\vec{F}_1 = 2\vec{i} - 3\vec{j} - 4\vec{k}$$
$$\vec{F}_2 = \vec{i} - 2\vec{j} + 7\vec{k}$$
$$\vec{F}_3 = -7\vec{i} + 5\vec{j} + 6\vec{k}$$
$$\vec{F}_4 = -11\vec{i} + 4\vec{j} - 5\vec{k}$$
$$\vec{F}_5 = -5\vec{i} + 3\vec{j} + 5\vec{k}$$

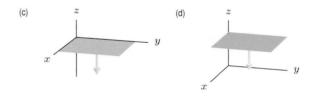

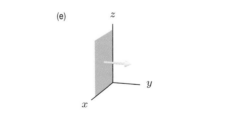

19.2 FLUX INTEGRALS FOR GRAPHS, CYLINDERS, AND SPHERES

In Section 19.1 we computed flux integrals in certain simple cases. In this section we see how to compute flux through surfaces that are graphs of functions, through cylinders, and through spheres.

Flux of a Vector Field Through the Graph of $z = f(x, y)$

Suppose S is the graph of the differentiable function $z = f(x, y)$, oriented upward, and that $\vec{F}$ is a smooth vector field. In Section 19.1 we subdivided the surface into small pieces with area vector

$\Delta \vec{A}$ and defined the flux of $\vec{F}$ through S as follows:

$$\int_S \vec{F} \cdot d\vec{A} = \lim_{\|\Delta \vec{A}\| \to 0} \sum \vec{F} \cdot \Delta \vec{A}.$$

How do we divide S into small pieces? One way is to use the cross-sections of f with x or y constant and take the patches in a wire frame representation of the surface. So we must calculate the area vector of one of these patches, which is approximately a parallelogram.

The Area Vector of a Coordinate Patch

According to the geometric definition of the cross product on page 763, the vector $\vec{v} \times \vec{w}$ has magnitude equal to the area of the parallelogram formed by $\vec{v}$ and $\vec{w}$ and direction perpendicular to this parallelogram and determined by the right-hand rule. Thus, we have

$$\boxed{\text{Area vector of parallelogram} = \vec{A} = \vec{v} \times \vec{w}.}$$

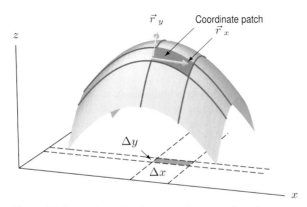

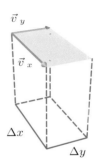

Figure 19.18: Surface showing coordinate patch and tangent vectors $\vec{r}_x$ and $\vec{r}_y$

Figure 19.19: Parallelogram-shaped patch in the tangent plane to the surface

Consider the patch of surface above the rectangular region with sides Δx and Δy in the xy-plane shown in Figure 19.18. We approximate the area vector, $\Delta \vec{A}$, of this patch by the area vector of the corresponding patch on the tangent plane to the surface. See Figure 19.19. This patch is the parallelogram determined by the vectors $\vec{v}_x$ and $\vec{v}_y$, so its area vector is given by

$$\Delta \vec{A} \approx \vec{v}_x \times \vec{v}_y.$$

To find $\vec{v}_x$ and $\vec{v}_y$, notice that a point on the surface has position vector $\vec{r} = x\vec{i} + y\vec{j} + f(x,y)\vec{k}$. Thus, a cross-section of S with y constant has tangent vector

$$\vec{r}_x = \frac{\partial \vec{r}}{\partial x} = \vec{i} + f_x \vec{k},$$

and a cross-section with x constant has tangent vector

$$\vec{r}_y = \frac{\partial \vec{r}}{\partial y} = \vec{j} + f_y \vec{k}.$$

The vectors $\vec{r}_x$ and $\vec{v}_x$ are parallel because they are both tangent to the surface and parallel to the xz-plane. Since the x-component of $\vec{r}_x$ is $\vec{i}$ and the x-component of $\vec{v}_x$ is $(\Delta x)\vec{i}$, we have

$\vec{v}_x = (\Delta x)\vec{r}_x$. Similarly, we have $\vec{v}_y = (\Delta y)\vec{r}_y$. So the upward-pointing area vector of the parallelogram is

$$\Delta \vec{A} \approx \vec{v}_x \times \vec{v}_y = (\vec{r}_x \times \vec{r}_y) \, \Delta x \, \Delta y = \left(-f_x \vec{i} - f_y \vec{j} + \vec{k} \right) \Delta x \, \Delta y.$$

This is our approximation for the area vector $\Delta \vec{A}$ on the surface. Replacing $\Delta \vec{A}$, Δx, and Δy by $d\vec{A}$, dx and dy, we write

$$d\vec{A} = \left(-f_x \vec{i} - f_y \vec{j} + \vec{k} \right) dx \, dy.$$

The Flux of $\vec{F}$ Through a Surface Given by a Graph of $z = f(x, y)$

Suppose the surface S is the part of the graph of $z = f(x, y)$ above a region R in the xy-plane, and suppose S is oriented upward. The flux of $\vec{F}$ through S is

$$\int_S \vec{F} \cdot d\vec{A} = \int_R \vec{F}(x, y, f(x, y)) \cdot \left(-f_x \vec{i} - f_y \vec{j} + \vec{k} \right) dx \, dy.$$

Example 1   Compute $\int_S \vec{F} \cdot d\vec{A}$ where $\vec{F}(x, y, z) = z\vec{k}$ and S is the rectangular plate with corners $(0, 0, 0)$, $(1, 0, 0)$, $(0, 1, 3)$, $(1, 1, 3)$, oriented upward. See Figure 19.20.

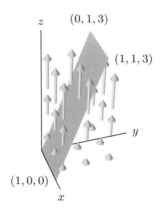

Figure 19.20: The vector field $\vec{F} = z\vec{k}$ on the rectangular surface S

Solution   We find the equation for the plane S in the form $z = f(x, y)$. Since f is linear, with x-slope equal to 0 and y-slope equal to 3, and $f(0, 0) = 0$, we have

$$z = f(x, y) = 0 + 0x + 3y = 3y.$$

Thus, we have

$$d\vec{A} = (-f_x \vec{i} - f_y \vec{j} + \vec{k}) \, dx \, dy = (0\vec{i} - 3\vec{j} + \vec{k}) \, dx \, dy = (-3\vec{j} + \vec{k}) \, dx \, dy.$$

The flux integral is therefore

$$\int_S \vec{F} \cdot d\vec{A} = \int_0^1 \int_0^1 3y\vec{k} \cdot (-3\vec{j} + \vec{k}) \, dx \, dy = \int_0^1 \int_0^1 3y \, dx \, dy = 1.5.$$

Flux of a Vector Field Through a Cylindrical Surface

Consider the cylinder of radius R centred on the z-axis illustrated in Figure 19.21 and oriented away from the z-axis. The coordinate patch in Figure 19.22 has surface area given by

$$\Delta A \approx R \, \Delta \theta \, \Delta z.$$

The outward unit normal $\vec{n}$ points in the direction of $x\vec{i} + y\vec{j}$, so

$$\vec{n} = \frac{x\vec{i} + y\vec{j}}{\|x\vec{i} + y\vec{j}\|} = \frac{R\cos\theta\vec{i} + R\sin\theta\vec{j}}{R} = \cos\theta\vec{i} + \sin\theta\vec{j}.$$

Therefore, the area vector of the coordinate patch is approximated by

$$\Delta\vec{A} = \vec{n}\,\Delta A \approx \left(\cos\theta\vec{i} + \sin\theta\vec{j}\right) R\,\Delta z\,\Delta\theta.$$

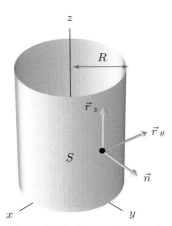

Figure 19.21: Outward-oriented cylinder

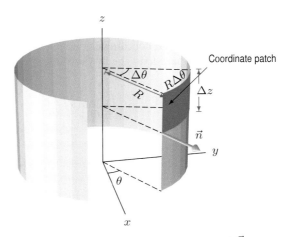

Figure 19.22: Coordinate patch with area $\Delta\vec{A}$ on surface of a cylinder

Replacing $\Delta\vec{A}$, Δz, and $\Delta\theta$ by $d\vec{A}$, dz, and $d\theta$, we write

$$d\vec{A} = \left(\cos\theta\vec{i} + \sin\theta\vec{j}\right) R\,dz\,d\theta.$$

This gives the following result:

The Flux of a Vector Field Through a Cylinder

The flux of $\vec{F}$ through the cylindrical surface S, of radius R and oriented away from the z-axis, is given by

$$\int_S \vec{F} \cdot d\vec{A} = \int_T \vec{F}(R,\theta,z) \cdot \left(\cos\theta\vec{i} + \sin\theta\vec{j}\right) R\,dz\,d\theta,$$

where T is the θz-region corresponding to S.

Example 2   Compute $\int_S \vec{F} \cdot d\vec{A}$ where $\vec{F}(x, y, z) = y\vec{j}$ and S is the part of the cylinder of radius 2 centred on the z-axis with $x \geq 0, y \geq 0$, and $0 \leq z \leq 3$. The surface is oriented toward the z-axis.

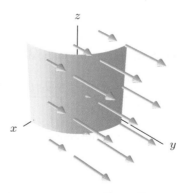

Figure 19.23: The vector field $\vec{F} = y\vec{j}$ on the surface S

Solution   In cylindrical coordinates, we have $R = 2$ and $\vec{F} = y\vec{j} = 2\sin\theta\vec{j}$. Since the orientation of S is toward the z-axis, the flux through S is given by

$$\int_S \vec{F} \cdot d\vec{A} = -\int_T 2\sin\theta\vec{j} \cdot (\cos\theta\vec{i} + \sin\theta\vec{j})2\,dz\,d\theta = -4\int_0^{\pi/2}\int_0^3 \sin^2\theta\,dz\,d\theta = -3\pi.$$

Flux of a Vector Field Through a Spherical Surface

Consider the piece of the sphere of radius R centred at the origin, oriented outward, as illustrated in Figure 19.24. The coordinate patch in Figure 19.24 has surface area given by

$$\Delta A \approx R^2 \sin\phi\,\Delta\phi\,\Delta\theta.$$

The outward unit normal $\vec{n}$ points in the direction of $\vec{r} = x\vec{i} + y\vec{j} + z\vec{k}$, so

$$\vec{n} = \frac{\vec{r}}{\|\vec{r}\|} = \sin\phi\cos\theta\vec{i} + \sin\phi\sin\theta\vec{j} + \cos\phi\vec{k}.$$

Therefore, the area vector of the coordinate patch is approximated by

$$\Delta\vec{A} \approx \vec{n}\,\Delta A = \frac{\vec{r}}{\|\vec{r}\|}\Delta A = \left(\sin\phi\cos\theta\vec{i} + \sin\phi\sin\theta\vec{j} + \cos\phi\vec{k}\right)R^2\sin\phi\,\Delta\phi\,\Delta\theta.$$

Replacing $\Delta\vec{A}$, $\Delta\phi$, and $\Delta\theta$ by $d\vec{A}$, $d\phi$, and $d\theta$, we write

$$d\vec{A} = \frac{\vec{r}}{\|\vec{r}\|}dA = \left(\sin\phi\cos\theta\vec{i} + \sin\phi\sin\theta\vec{j} + \cos\phi\vec{k}\right)R^2\sin\phi\,d\phi\,d\theta.$$

Thus, we obtain the following result:

The Flux of a Vector Field Through a Sphere

The flux of $\vec{F}$ through the spherical surface S, with radius R and oriented away from the origin, is given by

$$\int_S \vec{F} \cdot d\vec{A} = \int_S \vec{F} \cdot \frac{\vec{r}}{\|\vec{r}\|} \, dA$$

$$= \int_T \vec{F}(R, \theta, \phi) \cdot \left(\sin\phi\cos\theta\vec{i} + \sin\phi\sin\theta\vec{j} + \cos\phi\vec{k} \right) R^2 \sin\phi \, d\phi \, d\theta,$$

where T is the $\theta\phi$-region corresponding to S.

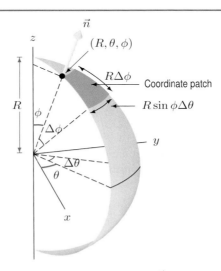

Figure 19.24: Coordinate patch with area $\Delta\vec{A}$ on surface of a sphere

Example 3   Find the flux of $\vec{F} = z\vec{k}$ through S, the upper hemisphere of radius 2 centred at the origin, oriented outward.

Solution   The hemisphere S is parameterised by spherical coordinates θ and ϕ, with $0 \le \theta \le 2\pi$ and $0 \le \phi \le \pi/2$. Since $R = 2$ and $\vec{F} = z\vec{k} = 2\cos\phi\vec{k}$, the flux is

$$\int_S \vec{F} \cdot d\vec{A} = \int_S 2\cos\phi\vec{k} \cdot (\sin\phi\cos\theta\vec{i} + \sin\phi\sin\theta\vec{j} + \cos\phi\vec{k})4\sin\phi \, d\phi \, d\theta$$

$$= \int_0^{2\pi} \int_0^{\pi/2} 8\sin\phi\cos^2\phi \, d\phi \, d\theta = 2\pi \left(8 \left(\frac{-\cos^3\phi}{3} \right) \Big|_{\phi=0}^{\pi/2} \right) = \frac{16\pi}{3}.$$

Example 4   The magnetic field $\vec{B}$ due to an *ideal magnetic dipole*, $\vec{\mu}$, located at the origin is defined to be

$$\vec{B}(\vec{r}) = -\frac{\vec{\mu}}{\|\vec{r}\|^3} + \frac{3(\vec{\mu} \cdot \vec{r})\vec{r}}{\|\vec{r}\|^5}.$$

Figure 19.25 shows a sketch of $\vec{B}$ in the plane $z = 0$ for the dipole $\vec{\mu} = \vec{i}$. Notice that $\vec{B}$ is similar to the magnetic field of a bar magnet with its north pole at the tip of the vector $\vec{i}$ and its south pole at the tail of the vector $\vec{i}$.

Compute the flux of $\vec{B}$ outward through the sphere S with centre at the origin and radius R.

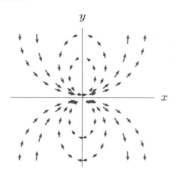

Figure 19.25: The magnetic field of a dipole, $\vec{i}$, at the origin: $\vec{B} = \dfrac{-\vec{i}}{\|\vec{r}\|^3} + \dfrac{3(\vec{i} \cdot \vec{r})\vec{r}}{\|\vec{r}\|^5}$

Solution   Since $\vec{i} \cdot \vec{r} = x$ and $\|\vec{r}\| = R$ on the sphere of radius R, we have

$$\int_S \vec{B} \cdot d\vec{A} = \int_S \left(-\frac{\vec{i}}{\|\vec{r}\|^3} + \frac{3(\vec{i} \cdot \vec{r})\vec{r}}{\|\vec{r}\|^5} \right) \cdot \frac{\vec{r}}{\|\vec{r}\|} \, dA = \int_S \left(-\frac{\vec{i} \cdot \vec{r}}{\|\vec{r}\|^4} + \frac{3(\vec{i} \cdot \vec{r})\|\vec{r}\|^2}{\|\vec{r}\|^6} \right) dA$$

$$= \int_S \frac{2\vec{i} \cdot \vec{r}}{\|\vec{r}\|^4} \, dA = \int_S \frac{2x}{\|\vec{r}\|^4} \, dA = \frac{2}{R^4} \int_S x \, dA.$$

But the sphere S is centred at the origin. Thus, the contribution to the integral from each positive x-value is cancelled by the contribution from the corresponding negative x-value; so $\int_S x \, dA = 0$. Therefore,

$$\int_S \vec{B} \cdot d\vec{A} = \frac{2}{R^4} \int_S x \, dA = 0.$$

Exercises and Problems for Section 19.2

Exercises

In Exercises 1–4, find the area vector $d\vec{A}$ for the surface $z = f(x, y)$, oriented upward.

1. $f(x, y) = 8x + 7y$ **2.** $f(x, y) = 3x - 5y$

3. $f(x, y) = xy + y^2$ **4.** $f(x, y) = 2x^2 - 3y^2$

In Exercises 5–8, write an iterated integral for the flux of $\vec{F}$ through the surface S, which is the part of the graph of $z = f(x, y)$ corresponding to the region R, oriented upward. Do not evaluate the integral.

5. $\vec{F}(x, y, z) = z\vec{i} + x\vec{j} + y\vec{k}$
   $f(x, y) = 50 - 4x + 10y$
   $R: 0 \leq x \leq 4, 0 \leq y \leq 8$

6. $\vec{F}(x, y, z) = 10\vec{i} + 20\vec{j} + 30\vec{k}$
   $f(x, y) = 2x - 3y$
   $R: -2 \leq x \leq 3, 0 \leq y \leq 5$

7. $\vec{F}(x, y, z) = \cos(x + 2y)\vec{j}$
   $f(x, y) = xe^{3y}$
   $R:$ Quarter disk of radius 5 centered at the origin, in quadrant I

8. $\vec{F}(x, y, z) = yz\vec{i} + xy\vec{j} + xy\vec{k}$
   $f(x, y) = \cos x + \sin 2y$
   $R:$ Triangle with vertices $(0, 0)$, $(0, 5)$, $(5, 0)$

In Exercises 9–10, compute the flux of $\vec{v} = z\vec{k}$ through the rectangular region with the orientation shown.

9.

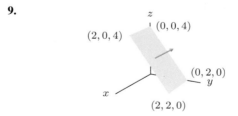

10.

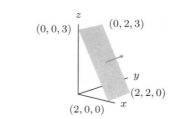

In Exercises 11–14, compute the flux of $\vec{F}$ through the surface S, which is the part of the graph of $z = f(x, y)$ corresponding to region R, oriented upward.

11. $\vec{F}(x, y, z) = 3\vec{i} - 2\vec{j} + 6\vec{k}$
$f(x, y) = 4x - 2y$
$R: 0 \leq x \leq 5, 0 \leq y \leq 10$

12. $\vec{F}(x, y, z) = \vec{i} - 2\vec{j} + z\vec{k}$
$f(x, y) = xy$
$R: 0 \leq x \leq 10, 0 \leq y \leq 10$

13. $\vec{F}(x, y, z) = \cos y\vec{i} + z\vec{j} + \vec{k}$
$f(x, y) = x^2 + 2y$
$R: 0 \leq x \leq 1, 0 \leq y \leq 1$

14. $\vec{F}(x, y, z) = x\vec{i} + z\vec{k}$
$f(x, y) = x + y + 2$
$R:$ Triangle with vertices $(-1, 0), (1, 0), (0, 1)$

In Exercises 15–18, write an iterated integral for the flux of $\vec{F}$ through the cylindrical surface S centered on the z-axis, oriented away from the z-axis. Do not evaluate the integral.

15. $\vec{F}(x, y, z) = \vec{i} + 2\vec{j} + 3\vec{k}$
$S:$ radius 10, $x \geq 0, y \geq 0, 0 \leq z \leq 5$

16. $\vec{F}(x, y, z) = x\vec{i} + 2y\vec{j} + 3z\vec{k}$
$S:$ radius 10, $0 \leq z \leq 5$

17. $\vec{F}(x, y, z) = z^2\vec{i} + e^x\vec{j} + \vec{k}$
$S:$ radius 6, inside sphere of radius 10

18. $\vec{F}(x, y, z) = x^2yz\vec{j} + z^3\vec{k}$
$S:$ radius 2, between the xy-plane and the paraboloid $z = x^2 + y^2$

In Exercises 19–22, write an iterated integral for the flux of $\vec{F}$ through the spherical surface S centered at the origin, oriented away from the origin. Do not evaluate the integral.

19. $\vec{F}(x, y, z) = \vec{i} + 2\vec{j} + 3\vec{k}$
$S:$ radius 10, $z \geq 0$

20. $\vec{F}(x, y, z) = x\vec{i} + 2y\vec{j} + 3z\vec{k}$
$S:$ radius 5, entire sphere

21. $\vec{F}(x, y, z) = z^2\vec{i}$
$S:$ radius 2, $x \geq 0$

22. $\vec{F}(x, y, z) = e^x\vec{k}$
$S:$ radius 3, $y \geq 0, z \leq 0$

In Exercises 23–26, compute the flux of $\vec{F}$ through the cylindrical surface S centered on the z-axis, oriented away from the z-axis.

23. $\vec{F}(x, y, z) = z\vec{j} + 6x\vec{k}$
$S:$ radius 5, $y \geq 0, 0 \leq z \leq 20$

24. $\vec{F}(x, y, z) = y\vec{i} + xz\vec{k}$
$S:$ radius 10, $x \geq 0, y \geq 0, 0 \leq z \leq 3$

25. $\vec{F}(x, y, z) = xyz\vec{j} + xe^z\vec{k}$
$S:$ radius 2, $0 \leq y \leq x, 0 \leq z \leq 10$

26. $\vec{F}(x, y, z) = xy\vec{i} + 2z\vec{j}$
$S:$ radius 1, $x \geq 0, 0 \leq y \leq 1/2, 0 \leq z \leq 2$

In Exercises 27–29, compute the flux of $\vec{F}$ through the spherical surface S centered at the origin, oriented away from the origin.

27. $\vec{F}(x, y, z) = z\vec{i}$
$S:$ radius 20, $x \geq 0, y \geq 0, z \geq 0$

28. $\vec{F}(x, y, z) = y\vec{i} - x\vec{j} + z\vec{k}$
$S:$ radius 4, entire sphere

29. $\vec{F}(x, y, z) = x\vec{i} + y\vec{j}$
$S:$ radius 1, above the cone $\phi = \pi/4$.

Problems

In Problems 30–31, compute the flux of $\vec{F}$ through the cylindrical surface in Figure 19.26, oriented away from the z-axis.

Figure 19.26

30. $\vec{F} = x\vec{i} + y\vec{j}$

31. $\vec{F} = xz\vec{i} + yz\vec{j} + z^3\vec{k}$

In Problems 32–47 compute the flux of the vector field $\vec{F}$ through the surface S.

32. $\vec{F} = (x - y)\vec{i} + z\vec{j} + 3x\vec{k}$ and S is the part of the plane $z = x + y$ above the rectangle $0 \leq x \leq 2, 0 \leq y \leq 3$, oriented upward.

33. $\vec{F} = z\vec{k}$ and S is the portion of the plane $x + y + z = 1$ that lies in the first octant, oriented upward.

34. $\vec{F} = -y\vec{j} + z\vec{k}$ and S is the part of the surface $z = y^2 + 5$ over the rectangle $-2 \leq x \leq 1, 0 \leq y \leq 1$, oriented upward.

35. $\vec{F} = 2x\vec{j} + y\vec{k}$ and S is the part of the surface $z = -y + 1$ above the square $0 \leq x \leq 1, 0 \leq y \leq 1$, oriented upward.

36. $\vec{F} = 5\vec{i} + 7\vec{j} + z\vec{k}$ and S is a closed cylinder of radius 3 centered on the z-axis, with $-2 \leq z \leq 2$, and oriented outward.

37. $\vec{F} = \ln(x^2)\vec{i} + e^x\vec{j} + \cos(1 - z)\vec{k}$ and S is the part of the surface $z = -y + 1$ above the square $0 \leq x \leq 1$, $0 \leq y \leq 1$, oriented upward.

38. $\vec{F} = x\vec{i} + y\vec{j} + z\vec{k}$ and S is a closed cylinder of radius 2 centered on the y-axis, with $-3 \leq y \leq 3$, and oriented outward.

39. $\vec{F} = 3x\vec{i} + y\vec{j} + z\vec{k}$ and S is the part of the surface $z = -2x - 4y + 1$, oriented upward, with (x, y) in the triangle R with vertices $(0, 0)$, $(0, 2)$, $(1, 0)$.

40. $\vec{F} = \cos(x^2 + y^2)\vec{k}$ and S is as in Exercise 41.

41. $\vec{F} = x\vec{i} + y\vec{j}$ and S is the part of the surface $z = 25 - (x^2 + y^2)$ above the disk of radius 5 centered at the origin, oriented upward.

42. $\vec{F} = \vec{r}$ and S is the part of the plane $x + y + z = 1$ above the rectangle $0 \leq x \leq 2$, $0 \leq y \leq 3$, oriented downward.

43. $\vec{F} = xz\vec{i} + y\vec{k}$ and S is the hemisphere $x^2 + y^2 + z^2 = 9$, $z \geq 0$, oriented upward.

44. $\vec{F} = \vec{r}$ and S is the part of the surface $z = x^2 + y^2$ above the disk $x^2 + y^2 \leq 1$, oriented downward.

45. $\vec{F} = -xz\vec{i} - yz\vec{j} + z^2\vec{k}$ and S is the cone $z = \sqrt{x^2 + y^2}$ for $0 \leq z \leq 6$, oriented upward.

46. $\vec{F} = y\vec{i} + \vec{j} - xz\vec{k}$ and S is the surface $y = x^2 + z^2$, with $x^2 + z^2 \leq 1$, oriented in the positive y-direction.

47. $\vec{F} = x^2\vec{i} + y^2\vec{j} + z^2\vec{k}$ and S is the oriented triangular surface shown in Figure 19.27.

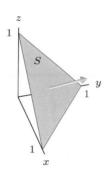

Figure 19.27

In Problems 48–51, compute the flux of $\vec{F}$ through the spherical surface, S.

48. $\vec{F} = z\vec{k}$ and S is the upper hemisphere of radius 2 centered at the origin, oriented outward.

49. $\vec{F} = y\vec{i} - x\vec{j} + z\vec{k}$ and S is the spherical cap given by $x^2 + y^2 + z^2 = 1$, $z \geq 0$, oriented upward.

50. $\vec{F} = z^2\vec{k}$ and S is the upper hemisphere of the sphere $x^2 + y^2 + z^2 = 25$, oriented away from the origin.

51. $\vec{F} = x\vec{i} + y\vec{j} + z\vec{k}$ and S is the surface of the sphere $x^2 + y^2 + z^2 = a^2$, oriented outward.

52. Compute the flux of $\vec{F} = x\vec{i} + y\vec{j} + z\vec{k}$ over the quarter cylinder S given by $x^2 + y^2 = 1$, $0 \leq x \leq 1$, $0 \leq y \leq 1$, $0 \leq z \leq 1$, oriented outward.

53. Compute the flux of $\vec{F} = x\vec{i} + \vec{j} + \vec{k}$ through the surface S given by $x = \sin y \sin z$, with $0 \leq y \leq \pi/2$, $0 \leq z \leq \pi/2$, oriented in the direction of increasing x.

54. Compute the flux of $\vec{F} = (x + z)\vec{i} + \vec{j} + z\vec{k}$ through the surface S given by $y = x^2 + z^2$, with $0 \leq y \leq 1$, $x \geq 0$, $z \geq 0$, oriented toward the xz-plane.

55. Let $\vec{F} = (xze^{yz})\vec{i} + xz\vec{j} + (5 + x^2 + y^2)\vec{k}$. Calculate the flux of $\vec{F}$ through the disk $x^2 + y^2 \leq 1$ in the xy-plane, oriented upward.

56. Let $\vec{H} = (e^{xy} + 3z + 5)\vec{i} + (e^{xy} + 5z + 3)\vec{j} + (3z + e^{xy})\vec{k}$. Calculate the flux of $\vec{H}$ through the square of side 2 with one vertex at the origin, one edge along the positive y-axis, one edge in the xz-plane with $x > 0$, $z > 0$, and the normal $\vec{n} = \vec{i} - \vec{k}$.

57. Electric charge is distributed in space with density (in coulomb/m^3) given in spherical coordinates by

$$\delta(\rho, \phi, \theta) = \begin{cases} \delta_0 \text{ (a constant)} & \rho \leq a \\ 0 & \rho > a. \end{cases}$$

(a) Describe the charge distribution in words.
(b) Find the electric field $\vec{E}$ due to δ. Assume that $\vec{E}$ can be written in spherical coordinates as $\vec{E} = E(\rho)\vec{e}_\rho$, where $\vec{e}_\rho$ is the unit outward normal to the sphere of radius ρ. In addition, $\vec{E}$ satisfies Gauss's Law for any simple closed surface S enclosing a volume W:

$$\int_S \vec{E} \cdot d\vec{A} = k\int_W \delta\, dV, \quad k \text{ a constant.}$$

58. Electric charge is distributed in space with density (in coulomb/m^3) given in cylindrical coordinates by

$$\delta(r, \theta, z) = \begin{cases} \delta_0 \text{ (a constant)} & \text{if } r \leq a \\ 0 & \text{if } r > a \end{cases}$$

(a) Describe the charge distribution in words.
(b) Find the electric field $\vec{E}$ due to δ. Assume that $\vec{E}$ can be written in cylindrical coordinates as $\vec{E} = E(r)\vec{e}_r$, where $\vec{e}_r$ is the unit outward vector to the cylinder of radius r, and that $\vec{E}$ satisfies Gauss's Law (see Problem 57).

Strengthen Your Understanding

In Problems 59–60, explain what is wrong with the statement.

59. Flux outward through the cone, given in cylindrical co-ordinates by $z = r$, can be computed using the formula $d\vec{A} = \left(\cos\theta\vec{i} + \sin\theta\vec{j}\right) R\, dz\, d\theta$.

60. For the surface $z = f(x, y)$ oriented upward, the formula

$$d\vec{A} = \vec{n}\, dA = \left(-f_x\vec{i} - f_y\vec{j} + \vec{k}\right) dx\, dy$$

gives $\vec{n} = -f_x\vec{i} - f_y\vec{j} + \vec{k}$ and $dA = dx\, dy$.

In Problems 61–62, give an example of:

61. A function $f(x, y)$ such that, for the surface $z = f(x, y)$ oriented upwards, we have $d\vec{A} = (\vec{i} + \vec{j} + \vec{k})\, dx\, dy$.

62. An oriented surface S on the cylinder of radius 10 centered on the z-axis such that $\int_S \vec{F} \cdot d\vec{A} = 600$, where $\vec{F} = x\vec{i} + y\vec{j}$.

Are the statements in Problems 63–65 true or false? Give reasons for your answer.

63. If S is the part of the graph of f lying above $a \le x \le b, c \le y \le d$, then S has surface area $\int_a^b \int_c^d \sqrt{f_x^2 + f_y^2 + 1}\, dx\, dy$.

64. If $\vec{A}(x, y)$ is the area vector for $z = f(x, y)$ oriented upward and $\vec{B}(x, y)$ is the area vector for $z = -f(x, y)$ oriented upward, then $\vec{A}(x, y) = -\vec{B}(x, y)$.

65. If S is the sphere $x^2 + y^2 + z^2 = 1$ oriented outward and $\int_S \vec{F} \cdot d\vec{A} = 0$, then $\vec{F}(x, y, z)$ is perpendicular to $x\vec{i} + y\vec{j} + z\vec{k}$ at every point of S.

66. The vector field, $\vec{F}$, in Figure 19.28 depends only on z; that is, it is of the form $g(z)\vec{k}$, where g is an increasing function. The integral $\int_S \vec{F} \cdot d\vec{A}$ represents the flux of $\vec{F}$ through this rectangle, S, oriented upward. In each of the following cases, how does the flux change?

(a) The rectangle is twice as wide in the x-direction, with new corners at the origin, $(2, 0, 0)$, $(2, 1, 3)$, $(0, 1, 3)$.

(b) The rectangle is moved so that its corners are at $(1, 0, 0)$, $(2, 0, 0)$, $(2, 1, 3)$, $(1, 1, 3)$.

(c) The orientation is changed to downward.

(d) The rectangle is tripled in size, so that its new corners are at the origin, $(3, 0, 0)$, $(3, 3, 9)$, $(0, 3, 9)$.

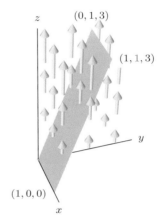

Figure 19.28

19.3 THE DIVERGENCE OF A VECTOR FIELD

Imagine that the vector fields in Figures 19.29 and 19.30 are velocity vector fields describing the flow of a fluid.[2] Figure 19.29 suggests outflow from the origin; for example, it could represent the expanding cloud of matter in the big-bang theory of the origin of the universe. We say that the origin is a *source*. Figure 19.30 suggests flow into the origin; in this case we say that the origin is a *sink*.

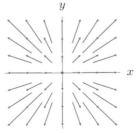

Figure 19.29: Vector field showing a source

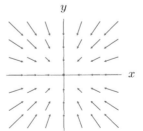

Figure 19.30: Vector field showing a sink

---

[2]Although not all vector fields represent physically realistic fluid flows, it is useful to think of them in this way.

In this section we use the flux out of a closed surface surrounding a point to measure the outflow per unit volume there, also called the *divergence*, or *flux density*.

Definition of Divergence

To measure the outflow per unit volume of a vector field at a point, we calculate the flux out of a small sphere centred at the point, divide by the volume enclosed by the sphere, then take the limit of this flux-to-volume ratio as the sphere contracts around the point.

Geometric Definition of Divergence

The **divergence**, or **flux density**, of a smooth vector field $\vec{F}$, written **div** $\vec{F}$, is a scalar-valued function defined by

$$\operatorname{div} \vec{F}(x, y, z) = \lim_{\text{Volume} \to 0} \frac{\int_S \vec{F} \cdot d\vec{A}}{\text{Volume of } S}.$$

Here S is a sphere centred at (x, y, z), oriented outward, that contracts down to (x, y, z) in the limit.

The limit can be computed using other shapes as well, such as the cubes in Example 2.

Cartesian Coordinate Definition of Divergence

If $\vec{F} = F_1 \vec{i} + F_2 \vec{j} + F_3 \vec{k}$, then

$$\operatorname{div} \vec{F} = \frac{\partial F_1}{\partial x} + \frac{\partial F_2}{\partial y} + \frac{\partial F_3}{\partial z}.$$

The dot product formula gives an easy way to remember the Cartesian coordinate definition, and suggests another common notation for div $\vec{F}$, namely $\nabla \cdot \vec{F}$. Using $\nabla = \frac{\partial}{\partial x}\vec{i} + \frac{\partial}{\partial y}\vec{j} + \frac{\partial}{\partial z}\vec{k}$, we can write

$$\operatorname{div} \vec{F} = \nabla \cdot \vec{F} = \left(\frac{\partial}{\partial x}\vec{i} + \frac{\partial}{\partial y}\vec{j} + \frac{\partial}{\partial z}\vec{k} \right) \cdot (F_1\vec{i} + F_2\vec{j} + F_3\vec{k}) = \frac{\partial F_1}{\partial x} + \frac{\partial F_2}{\partial y} + \frac{\partial F_3}{\partial z}.$$

Example 1   Calculate the divergence of $\vec{F}(\vec{r}) = \vec{r}$ at the origin

(a) Using the geometric definition.

(b) Using the Cartesian coordinate definition.

Solution   (a) Using the method of Example 5 on page 1030, you can calculate the flux of $\vec{F}$ out of the sphere of radius a, centred at the origin; it is $4\pi a^3$. So we have

$$\operatorname{div} \vec{F}(0, 0, 0) = \lim_{a \to 0} \frac{\text{Flux}}{\text{Volume}} = \lim_{a \to 0} \frac{4\pi a^3}{\frac{4}{3}\pi a^3} = \lim_{a \to 0} 3 = 3.$$

(b) In Cartesian coordinates, $\vec{F}(x, y, z) = x\vec{i} + y\vec{j} + z\vec{k}$, so

$$\operatorname{div} \vec{F} = \frac{\partial}{\partial x}(x) + \frac{\partial}{\partial y}(y) + \frac{\partial}{\partial z}(z) = 1 + 1 + 1 = 3.$$

The next example shows that the divergence can be negative if there is net inflow to a point.

Example 2 (a) Using the geometric definition, find the divergence of $\vec{v} = -x\vec{i}$ at: (i) $(0, 0, 0)$ (ii) $(2, 2, 0)$.
(b) Confirm that the coordinate definition gives the same results.

Solution (a) (i) The vector field $\vec{v} = -x\vec{i}$ is parallel to the x-axis and is shown in the xy-plane in Figure 19.31. To compute the flux density at $(0, 0, 0)$, we use a cube S_1, centred at the origin with edges parallel to the axes, of length $2c$. Then the flux through the faces perpendicular to the y- and z-axes is zero (because the vector field is parallel to these faces). On the faces perpendicular to the x-axis, the vector field and the outward normal are parallel but point in opposite directions. On the face at $x = c$, where $\vec{v} = -c\vec{i}$ and $\Delta\vec{A} = \|\vec{A}\|\vec{i}$, we have

$$\vec{v} \cdot \Delta\vec{A} = -c\,\|\Delta\vec{A}\|.$$

On the face at $x = -c$, where $\vec{v} = c\vec{i}$ and $\Delta\vec{A} = -\|\vec{A}\|\vec{i}$, the dot product is still negative:

$$\vec{v} \cdot \Delta\vec{A} = -c\,\|\Delta\vec{A}\|.$$

Therefore, the flux through the cube is given by

$$\int_{S_1} \vec{v} \cdot d\vec{A} = \int_{\text{Face } x=-c} \vec{v} \cdot d\vec{A} + \int_{\text{Face } x=c} \vec{v} \cdot d\vec{A}$$
$$= -c \cdot \text{Area of one face} + (-c) \cdot \text{Area of other face} = -2c(2c)^2 = -8c^3.$$

Thus,

$$\text{div}\,\vec{v}\,(0, 0, 0) = \lim_{\text{Volume}\to 0} \frac{\int_S \vec{v} \cdot d\vec{A}}{\text{Volume of cube}} = \lim_{c\to 0} \left(\frac{-8c^3}{(2c)^3}\right) = -1.$$

Since the vector field points inward toward the yz-plane, it makes sense that the divergence is negative at the origin.

(ii) Take S_2 to be a cube as before, but centred this time at the point $(2, 2, 0)$. See Figure 19.31. As before, the flux through the faces perpendicular to the y- and z-axes is zero. On the face at $x = 2 + c$,

$$\vec{v} \cdot \Delta\vec{A} = -(2 + c)\,\|\Delta\vec{A}\|.$$

On the face at $x = 2 - c$ with outward normal, the dot product is positive, and

$$\vec{v} \cdot \Delta\vec{A} = (2 - c)\,\|\Delta\vec{A}\|.$$

Therefore, the flux through the cube is given by

$$\int_{S_2} \vec{v} \cdot d\vec{A} = \int_{\text{Face } x=2-c} \vec{v} \cdot d\vec{A} + \int_{\text{Face } x=2+c} \vec{v} \cdot d\vec{A}$$
$$= (2 - c) \cdot \text{Area of one face} - (2 + c) \cdot \text{Area of other face} = -2c(2c)^2 = -8c^3.$$

Then, as before,

$$\text{div}\,\vec{v}\,(2, 2, 0) = \lim_{\text{Volume}\to 0} \frac{\int_S \vec{v} \cdot d\vec{A}}{\text{Volume of cube}} = \lim_{c\to 0} \left(\frac{-8c^3}{(2c)^3}\right) = -1.$$

Although the vector field is flowing away from the point $(2, 2, 0)$ on the left, this outflow is smaller in magnitude than the inflow on the right, so the net outflow is negative.

(b) Since $\vec{v} = -x\vec{i} + 0\vec{j} + 0\vec{k}$, the formula gives

$$\text{div}\,\vec{v} = \frac{\partial}{\partial x}(-x) + \frac{\partial}{\partial y}(0) + \frac{\partial}{\partial z}(0) = -1 + 0 + 0 = -1.$$

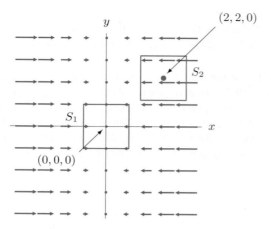

Figure 19.31: Vector field $\vec{v} = -x\vec{i}$ in the xy-plane

Why Do the Two Definitions of Divergence Give the Same Result?

The geometric definition defines div $\vec{F}$ as the flux density of $\vec{F}$. To see why the coordinate definition is also the flux density, imagine computing the flux out of a small box-shaped surface S at (x_0, y_0, z_0), with sides of length Δx, Δy, and Δz parallel to the axes. On S_1 (the back face of the box shown in Figure 19.32, where $x = x_0$), the outward normal is in the negative x-direction, so $d\vec{A} = -dy\,dz\,\vec{i}$. Assuming $\vec{F}$ is approximately constant on S_1, we have

$$\int_{S_1} \vec{F} \cdot d\vec{A} = \int_{S_1} \vec{F} \cdot (-\vec{i})\,dy\,dz \approx -F_1(x_0, y_0, z_0)\int_{S_1} dy\,dz$$
$$= -F_1(x_0, y_0, z_0) \cdot \text{Area of } S_1 = -F_1(x_0, y_0, z_0)\,\Delta y\,\Delta z.$$

On S_2, the face where $x = x_0 + \Delta x$, the outward normal points in the positive x-direction, so $d\vec{A} = dy\,dz\,\vec{i}$. Therefore,

$$\int_{S_2} \vec{F} \cdot d\vec{A} = \int_{S_2} \vec{F} \cdot \vec{i}\,dy\,dz \approx F_1(x_0 + \Delta x, y_0, z_0)\int_{S_2} dy\,dz$$
$$= F_1(x_0 + \Delta x, y_0, z_0) \cdot \text{Area of } S_2 = F_1(x_0 + \Delta x, y_0, z_0)\,\Delta y\,\Delta z.$$

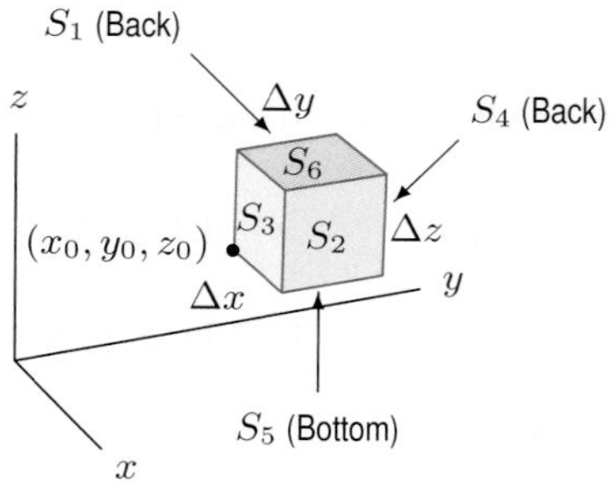

Figure 19.32: Box used to find div $\vec{F}$ at (x_0, y_0, z_0)

Thus,

$$\int_{S_1} \vec{F} \cdot d\vec{A} + \int_{S_2} \vec{F} \cdot d\vec{A} \approx F_1(x_0 + \Delta x, y_0, z_0)\Delta y\Delta z - F_1(x_0, y_0, z_0)\Delta y\Delta z$$
$$= \frac{F_1(x_0 + \Delta x, y_0, z_0) - F_1(x_0, y_0, z_0)}{\Delta x}\Delta x\Delta y\Delta z$$
$$\approx \frac{\partial F_1}{\partial x}\Delta x\Delta y\Delta z.$$

By an analogous argument, the contribution to the flux from S_3 and S_4 (the surfaces perpendicular to the y-axis) is approximately

$$\frac{\partial F_2}{\partial y} \Delta x \, \Delta y \, \Delta z,$$

and the contribution to the flux from S_5 and S_6 is approximately

$$\frac{\partial F_3}{\partial z} \Delta x \, \Delta y \, \Delta z.$$

Thus, adding these contributions, we have

$$\text{Total flux through } S \approx \frac{\partial F_1}{\partial x} \Delta x \, \Delta y \, \Delta z + \frac{\partial F_2}{\partial y} \Delta x \, \Delta y \, \Delta z + \frac{\partial F_3}{\partial z} \Delta x \, \Delta y \, \Delta z.$$

Since the volume of the box is $\Delta x \, \Delta y \, \Delta z$, the flux density is

$$\frac{\text{Total flux through } S}{\text{Volume of box}} \approx \frac{\dfrac{\partial F_1}{\partial x}\Delta x \Delta y \Delta z + \dfrac{\partial F_2}{\partial y}\Delta x \Delta y \Delta z + \dfrac{\partial F_3}{\partial z}\Delta x \Delta y \Delta z}{\Delta x \Delta y \Delta z}$$

$$= \frac{\partial F_1}{\partial x} + \frac{\partial F_2}{\partial y} + \frac{\partial F_3}{\partial z}.$$

Divergence-Free Vector Fields

A vector field $\vec{F}$ is said to be *divergence free* or *solenoidal* if $\text{div} \vec{F} = 0$ everywhere that $\vec{F}$ is defined.

Example 3   Figure 19.33 shows, for three values of the constant p, the vector field

$$\vec{E} = \frac{\vec{r}}{\|\vec{r}\|^p} \qquad \vec{r} = x\vec{i} + y\vec{j} + z\vec{k}, \ \vec{r} \neq \vec{0}.$$

(a) Find a formula for $\text{div} \, \vec{E}$.
(b) Is there a value of p for which $\vec{E}$ is divergence-free? If so, find it.

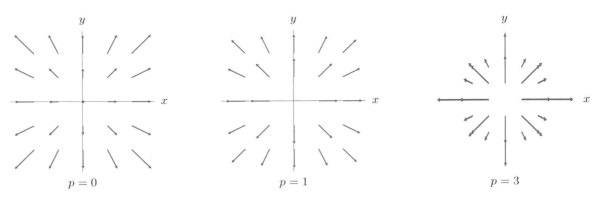

Figure 19.33: The vector field $\vec{E}(\vec{r}) = \vec{r}/\|\vec{r}\|^p$ for $p = 0, 1,$ and 3 in the xy-plane

Solution   (a) The components of $\vec{E}$ are

$$\vec{E} = \frac{x}{(x^2 + y^2 + z^2)^{p/2}}\vec{i} + \frac{y}{(x^2 + y^2 + z^2)^{p/2}}\vec{j} + \frac{z}{(x^2 + y^2 + z^2)^{p/2}}\vec{k}.$$

We compute the partial derivatives

$$\frac{\partial}{\partial x}\left(\frac{x}{(x^2+y^2+z^2)^{p/2}}\right) = \frac{1}{(x^2+y^2+z^2)^{p/2}} - \frac{px^2}{(x^2+y^2+z^2)^{(p/2)+1}}$$

$$\frac{\partial}{\partial y}\left(\frac{y}{(x^2+y^2+z^2)^{p/2}}\right) = \frac{1}{(x^2+y^2+z^2)^{p/2}} - \frac{py^2}{(x^2+y^2+z^2)^{(p/2)+1}}$$

$$\frac{\partial}{\partial z}\left(\frac{z}{(x^2+y^2+z^2)^{p/2}}\right) = \frac{1}{(x^2+y^2+z^2)^{p/2}} - \frac{pz^2}{(x^2+y^2+z^2)^{(p/2)+1}}.$$

So

$$\operatorname{div}\vec{E} = \frac{3}{(x^2+y^2+z^2)^{p/2}} - \frac{p(x^2+y^2+z^2)}{(x^2+y^2+z^2)^{(p/2)+1}}$$

$$= \frac{3-p}{(x^2+y^2+z^2)^{p/2}} = \frac{3-p}{\|\vec{r}\|^p}.$$

(b) The divergence is zero when $p = 3$, so $\vec{F}(\vec{r}) = \vec{r}/\|\vec{r}\|^3$ is a divergence-free vector field. Notice that the divergence is zero even though the vectors point outward from the origin.

Magnetic Fields

An important class of divergence-free vector fields is the magnetic fields. One of Maxwell's Laws of Electromagnetism is that the magnetic field $\vec{B}$ satisfies

$$\operatorname{div}\vec{B} = 0.$$

Example 4   An infinitesimal current loop, similar to that shown in Figure 19.34, is called a *magnetic dipole*. Its magnitude is described by a constant vector $\vec{\mu}$, called the dipole moment. The magnetic field due to a magnetic dipole with moment $\vec{\mu}$ is

$$\vec{B} = -\frac{\vec{\mu}}{\|\vec{r}\|^3} + \frac{3(\vec{\mu}\cdot\vec{r})\vec{r}}{\|\vec{r}\|^5}, \qquad \vec{r} \neq \vec{0}.$$

Show that $\operatorname{div}\vec{B} = 0$.

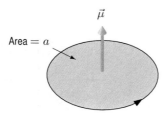

Area $= a$

$\vec{\mu}$

Figure 19.34: A current loop

Solution   To show that $\operatorname{div}\vec{B} = 0$ we can use the following version of the product rule for the divergence: if g is a scalar function and $\vec{F}$ is a vector field, then

$$\operatorname{div}(g\vec{F}) = (\operatorname{grad} g)\cdot\vec{F} + g\operatorname{div}\vec{F}.$$

(See Problem 35 on page 1053.) Thus, since $\operatorname{div}\vec{\mu} = 0$, we have

$$\operatorname{div}\left(\frac{\vec{\mu}}{\|\vec{r}\|^3}\right) = \operatorname{div}\left(\frac{1}{\|\vec{r}\|^3}\vec{\mu}\right) = \operatorname{grad}\left(\frac{1}{\|\vec{r}\|^3}\right)\cdot\vec{\mu} + \left(\frac{1}{\|\vec{r}\|^3}\right)0$$

and

$$\operatorname{div}\left(\frac{(\vec{\mu}\cdot\vec{r})\vec{r}}{\|\vec{r}\|^5}\right) = \operatorname{div}\left(\vec{\mu}\cdot\vec{r}\,\frac{\vec{r}}{\|\vec{r}\|^5}\right) = \operatorname{grad}(\vec{\mu}\cdot\vec{r})\cdot\frac{\vec{r}}{\|\vec{r}\|^5} + (\vec{\mu}\cdot\vec{r})\operatorname{div}\left(\frac{\vec{r}}{\|\vec{r}\|^5}\right).$$

From Problems 67 and 68 on page 814 and Example 3 on page 1049, we have

$$\operatorname{grad}\left(\frac{1}{\|\vec{r}\|^3}\right) = \frac{-3\vec{r}}{\|\vec{r}\|^5}, \qquad \operatorname{grad}(\vec{\mu}\cdot\vec{r}) = \vec{\mu}, \qquad \operatorname{div}\left(\frac{\vec{r}}{\|\vec{r}\|^5}\right) = \frac{-2}{\|\vec{r}\|^5}.$$

Putting these results together gives

$$\operatorname{div}\vec{B} = -\operatorname{grad}\left(\frac{1}{\|\vec{r}\|^3}\right)\cdot\vec{\mu} + 3\operatorname{grad}(\vec{\mu}\cdot\vec{r})\cdot\frac{\vec{r}}{\|\vec{r}\|^5} + 3(\vec{\mu}\cdot\vec{r})\operatorname{div}\left(\frac{\vec{r}}{\|\vec{r}\|^5}\right)$$

$$= \frac{3\vec{r}\cdot\vec{\mu}}{\|\vec{r}\|^5} + \frac{3\vec{\mu}\cdot\vec{r}}{\|\vec{r}\|^5} - \frac{6\vec{\mu}\cdot\vec{r}}{\|\vec{r}\|^5}$$

$$= 0.$$

Exercises and Problems for Section 19.3

Exercises

1. Is $\operatorname{div}\left(\dfrac{y\vec{i} - x\vec{j}}{x^2 + y^2}\right)$ a vector or a scalar? Calculate it.

2. Which of the following two vector fields, sketched in the xy-plane, appears to have the greater divergence at the origin? The scales are the same on each.

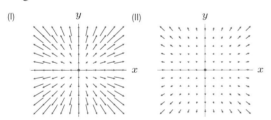

In Exercises 3–11, find the divergence of the vector field. (Note: $\vec{r} = x\vec{i} + y\vec{j} + z\vec{k}$.)

3. $\vec{F}(x, y) = -y\vec{i} + x\vec{j}$

4. $\vec{F}(x, y) = -x\vec{i} + y\vec{j}$

5. $\vec{F}(x, y, z) = (-x + y)\vec{i} + (y + z)\vec{j} + (-z + x)\vec{k}$

6. $\vec{F}(x, y) = (x^2 - y^2)\vec{i} + 2xy\vec{j}$

7. $\vec{F}(x, y, z) = 3x^2\vec{i} - \sin(xz)(\vec{i} + \vec{k})$

8. $\vec{F} = \left(\ln\left(x^2 + 1\right)\vec{i} + (\cos y)\vec{j} + (xye^z)\vec{k}\right)$

9. $\vec{F}(\vec{r}) = \vec{a} \times \vec{r}$

10. $\vec{F}(x, y) = \dfrac{-y\vec{i} + x\vec{j}}{x^2 + y^2}$

11. $\vec{F}(\vec{r}) = \dfrac{\vec{r} - \vec{r}_0}{\|\vec{r} - \vec{r}_0\|}, \quad \vec{r} \neq \vec{r}_0$

12. For each of the following vector fields, sketched in the xy-plane, decide if the divergence is positive, zero, or negative at the indicated point.

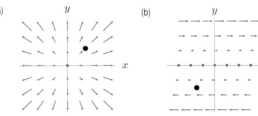

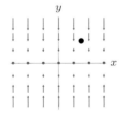

Problems

13. Draw two vector fields that have positive divergence everywhere.

14. Draw two vector fields that have negative divergence everywhere.

15. Draw two vector fields that have zero divergence everywhere.

16. A small sphere of radius 0.1 surrounds the point $(2, 3, -1)$. The flux of a vector field $\vec{G}$ into this sphere is 0.00004π. Estimate div $\vec{G}$ at the point $(2, 3, -1)$.

17. A smooth vector field $\vec{F}$ has div $\vec{F}(1, 2, 3) = 5$. Estimate the flux of $\vec{F}$ out of a small sphere of radius 0.01 centered at the point $(1, 2, 3)$.

18. Let $\vec{F}$ be a vector field with div $\vec{F} = x^2 + y^2 - z$. Estimate $\int_S \vec{F} \cdot d\vec{A}$ where S is

   (a)   (i) A sphere of radius 0.1 centered at $(2, 0, 0)$.

        (ii) A box of side 0.2 with edges parallel to the axes and centered at $(0, 0, 10)$.

   (b) The point $(2, 0, 0)$ is called a *source* for the vector field $\vec{F}$; the point $(0, 0, 10)$ is called a *sink*. Explain the reason for these names using your answer to part (a).

19. The flux of $\vec{F}$ out of a small sphere of radius 0.1 centered at $(4, 5, 2)$, is 0.0125. Estimate:

   (a) div $\vec{F}$ at $(4, 5, 2)$

   (b) The flux of $\vec{F}$ out of a sphere of radius 0.2 centered at $(4, 5, 2)$.

20. **(a)** Find the flux of $\vec{F} = 2x\vec{i} - 3y\vec{j} + 5z\vec{k}$ through a box with four of its corners at the points $(a, b, c), (a+w, b, c), (a, b+w, c), (a, b, c+w)$ and edge length w. See Figure 19.35.

   (b) Use the geometric definition and part (a) to find div $\vec{F}$ at the point (a, b, c).

   (c) Find div $\vec{F}$ using partial derivatives.

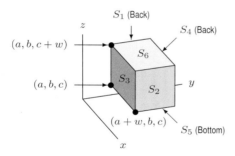

Figure 19.35

21. Suppose $\vec{F} = (3x+2)\vec{i} + 4x\vec{j} + (5x+1)\vec{k}$. Use the method of Exercise 20 to find div $\vec{F}$ at the point (a, b, c) by two different methods.

22. Use the geometric definition of divergence to find div $\vec{v}$ at the origin, where $\vec{v} = -2\vec{r}$. Check that you get the same result using the definition in Cartesian coordinates.

23. **(a)** Let $f(x, y) = axy + ax^2y + y^3$. Find div grad f.

   (b) If possible, choose a so that div grad $f = 0$ for all x, y.

24. Let

$$\vec{F} = (9a^2x + 10ay^2)\vec{i} + (10z^3 - 6ay)\vec{j} - (3z + 10x^2 + 10y^2)\vec{k}.$$

Find the value(s) of a making div $\vec{F}$

   (a)   0   **(b)**   A minimum

25. The vector field $\vec{F}(\vec{r}) = \vec{r}/\|\vec{r}\|^3$ is not defined at the origin. Nevertheless, we can attempt to use the flux definition to compute div $\vec{F}$ at the origin. What is the result?

26. The divergence of a magnetic vector field $\vec{B}$ must be zero everywhere. Which of the following vector fields cannot be a magnetic vector field?

   (a) $\vec{B}(x, y, z) = -y\vec{i} + x\vec{j} + (x+y)\vec{k}$

   (b) $\vec{B}(x, y, z) = -z\vec{i} + y\vec{j} + x\vec{k}$

   (c) $\vec{B}(x, y, z) = (x^2 - y^2 - x)\vec{i} + (y - 2xy)\vec{j}$

27. Let $\vec{F}(x, y, z) = z\vec{k}$.

   (a) Calculate div $\vec{F}$.

   (b) Sketch $\vec{F}$. Does it appear to be diverging? Does this agree with your answer to part (a)?

28. Let $\vec{F}(\vec{r}) = \vec{r}/\|\vec{r}\|^3$ (in 3-space), $\vec{r} \neq \vec{0}$.

   (a) Calculate div $\vec{F}$.

   (b) Sketch $\vec{F}$. Does it appear to be diverging? Does this agree with your answer to part (a)?

Problems 29–30 involve electric fields. Electric charge produces a vector field $\vec{E}$, called the electric field, which represents the force on a unit positive charge placed at the point. Two positive or two negative charges repel one another, whereas two charges of opposite sign attract one another. The divergence of $\vec{E}$ is proportional to the density of the electric charge (that is, the charge per unit volume), with a positive constant of proportionality.

29. A certain distribution of electric charge produces the electric field shown in Figure 19.36. Where are the charges that produced this electric field concentrated? Which concentrations are positive and which are negative?

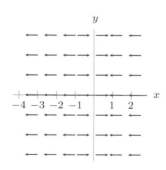

Figure 19.36

30. The electric field at the point $\vec{r}$ as a result of a point charge at the origin is $\vec{E}\,(\vec{r}) = k\vec{r}/\|\vec{r}\|^3$.

    (a) Calculate div $\vec{E}$ for $\vec{r} \neq \vec{0}$.

    (b) Calculate the limit suggested by the geometric definition of div $\vec{E}$ at the point $(0,0,0)$.

    (c) Explain what your answers mean in terms of charge density.

31. Due to roadwork ahead, the traffic on a highway slows linearly from 80 km/hr to 20 km/hr over a 600-meter stretch of road, then crawls along at 20 km/hr for 1800 meters, then speeds back up linearly to 80 km/hr in the next 300 meters, after which it moves steadily at 80 km/hr.

    (a) Sketch a velocity vector field for the traffic flow.

    (b) Write a formula for the velocity vector field $\vec{v}$ (km/hr) as a function of the distance x meters from the initial point of slowdown. (Take the direction of motion to be $\vec{i}$ and consider the various sections of the road separately.)

    (c) Compute div $\vec{v}$ at $x = 1300, 1800, 2500, 3000$. Be sure to include the proper units.

32. The velocity field $\vec{v}$ in Problem 31 does not give a complete description of the traffic flow, for it takes no account of the spacing between vehicles. Let ρ be the density (cars/km) of highway, where we assume that ρ depends only on x.

    (a) Using your highway experience, arrange in ascending order: $\rho(0), \rho(300), \rho(1800)$.

    (b) What are the units and interpretation of the vector field $\rho\vec{v}$?

    (c) Would you expect $\rho\vec{v}$ to be constant? Why? What does this mean for $\mathrm{div}(\rho\vec{v})$?

    (d) Determine $\rho(x)$ if $\rho(0) = 50$ cars/km and $\rho\vec{v}$ is constant.

    (e) If the highway has two lanes, find the approximate number of meters between cars at $x = 0, 300$, and 1800.

33. Let $\vec{r} = x\vec{i} + y\vec{j} + z\vec{k}$ and $\vec{c} = c_1\vec{i} + c_2\vec{j} + c_3\vec{k}$, a constant vector; let S be a sphere of radius R centered at the origin. Find

    (a) $\mathrm{div}(\vec{r} \times \vec{c})$ **(b)** $\int_S (\vec{r} \times \vec{c}) \cdot d\vec{A}$

34. Show that if $\vec{a}$ is a constant vector and $f(x,y,z)$ is a function, then $\mathrm{div}(f\vec{a}) = (\mathrm{grad}\,f) \cdot \vec{a}$.

35. Show that if $g(x,y,z)$ is a scalar-valued function and $\vec{F}\,(x,y,z)$ is a vector field, then

$$\mathrm{div}(g\vec{F}) = (\mathrm{grad}\,g) \cdot \vec{F} + g\,\mathrm{div}\,\vec{F}.$$

36. If $f(x,y,z)$ and $g(x,y,z)$ are functions with continuous second partial derivatives, show that

$$\mathrm{div}(\mathrm{grad}\,f \times \mathrm{grad}\,g) = 0.$$

In Problems 37–39, use Problems 35 and 36 to find the divergence of the vector field. The vectors $\vec{a}$ and $\vec{b}$ are constant.

37. $\vec{F} = \dfrac{1}{\|\vec{r}\|^p}\,\vec{a} \times \vec{r}$ **38.** $\vec{B} = \dfrac{1}{x^a}\vec{r}$

39. $\vec{G} = (\vec{b} \cdot \vec{r})\vec{a} \times \vec{r}$

40. Let $\vec{F}\,(x,y) = u(x,y)\vec{i} + v(x,y)\vec{j}$ be a 2-dimensional vector field. Let $F(x,y)$ be the magnitude of $\vec{F}$ and let $\theta(x,y)$ be the angle of $\vec{F}$ with the positive x-axis at the point (x,y), so that $u = F\cos\theta$ and $v = F\sin\theta$. Let $\vec{T}$ be the unit vector in the direction of $\vec{F}$, and let $\vec{N}$ be the unit vector in the direction of $\vec{k} \times \vec{F}$, perpendicular to $\vec{F}$. Show that

$$\mathrm{div}\,\vec{F} = F\theta_{\vec{N}} + F_{\vec{T}}.$$

This problem shows that the divergence of the vector field $\vec{F}$ is the sum of two terms. The first term, $F\theta_{\vec{N}}$, is due to changes in the direction of $\vec{F}$ perpendicular to the flow lines, so it reflects the extent to which flow lines of $\vec{F}$ fail to be parallel. The second term, $F_{\vec{T}}$, is due to changes in the magnitude of $\vec{F}$ along flow lines of $\vec{F}$.

41. In Problem 61 on page 1035 it was shown that the rate of heat loss from a volume V in a region of non-uniform temperature equals $k\int_S (\mathrm{grad}\,T) \cdot d\vec{A}$, where k is a constant, S is the surface bounding V, and $T(x,y,z)$ is the temperature at the point (x,y,z) in space. By taking the limit as V contracts to a point, show that, at that point,

$$\frac{\partial T}{\partial t} = B\,\mathrm{div}\,\mathrm{grad}\,T$$

where B is a constant with respect to x, y, z, but may depend on time, t.

42. A vector field, $\vec{v}$, in the plane is a *point source* at the origin if its direction is away from the origin at every point, its magnitude depends only on the distance from the origin, and its divergence is zero away from the origin.

    (a) Explain why a point source at the origin must be of the form $\vec{v} = \left[f(x^2 + y^2)\right](x\vec{i} + y\vec{j})$ for some positive function f.

    (b) Show that $\vec{v} = K(x^2 + y^2)^{-1}(x\vec{i} + y\vec{j})$ is a point source at the origin if $K > 0$.

    (c) What is the magnitude $\|\vec{v}\|$ of the source in part (b) as a function of the distance from its center?

    (d) Sketch the vector field $\vec{v} = (x^2 + y^2)^{-1}(x\vec{i} + y\vec{j})$.

    (e) Show that $\phi = \frac{K}{2}\log(x^2 + y^2)$ is a potential function for the source in part (b).

43. A vector field, $\vec{v}$, in the plane is a *point sink* at the origin if its direction is toward the origin at every point, its magnitude depends only on the distance from the origin, and its divergence is zero away from the origin.

    (a) Explain why a point sink at the origin must be of the form $\vec{v} = \left[f(x^2 + y^2)\right](x\vec{i} + y\vec{j})$ for some negative function f.

    (b) Show that $\vec{v} = K(x^2 + y^2)^{-1}(x\vec{i} + y\vec{j})$ is a point sink at the origin if $K < 0$.

    (c) Determine the magnitude $\|\vec{v}\|$ of the sink in part (b) as a function of the distance from its center.

    (d) Sketch $\vec{v} = -(x^2 + y^2)^{-1}(x\vec{i} + y\vec{j})$.

    (e) Show that $\phi = \frac{K}{2}\log(x^2 + y^2)$ is a potential function for the sink in part (b).

Strengthen Your Understanding

In Problems 44–46, explain what is wrong with the statement.

44. $\operatorname{div}(2x\vec{i}) = 2\vec{i}$.

45. For $\vec{F}(x, y, z) = (x^2 + y)\vec{i} + (2y + z)\vec{j} - z^2\vec{k}$ we have $\operatorname{div}\vec{F} = 2x\vec{i} + 2\vec{j} - 2z\vec{k}$.

46. The divergence of $f(x, y, z) = x^2 + yz$ is given by $\operatorname{div} f(x, y, z) = 2x + z + y$.

In Problems 47–49, give an example of:

47. A vector field $\vec{F}(x, y, z)$ whose divergence is a nonzero constant.

48. A nonzero vector field $\vec{F}(x, y, z)$ whose divergence is zero.

49. A vector field that is not divergence free.

Are the statements in Problems 50–62 true or false? Give reasons for your answer.

50. $\operatorname{div}(\vec{F} + \vec{G}) = \operatorname{div}\vec{F} + \operatorname{div}\vec{G}$

51. $\operatorname{grad}(\vec{F} \cdot \vec{G}) = \vec{F}(\operatorname{div}\vec{G}) + (\operatorname{div}\vec{F})\vec{G}$

52. $\operatorname{div}\vec{F}$ is a scalar whose value can vary from point to point.

53. If $\vec{F}$ is a vector field in 3-space, then $\operatorname{div}\vec{F}$ is also a vector field.

54. A constant vector field $\vec{F} = a\vec{i} + b\vec{j} + c\vec{k}$ has zero divergence.

55. If a vector field $\vec{F}$ in 3-space has zero divergence then $\vec{F} = a\vec{i} + b\vec{j} + c\vec{k}$ where a, b and c are constants.

56. If $\vec{F}$ is a vector field in 3-space, and f is a scalar function, then $\operatorname{div}(f\vec{F}) = f\operatorname{div}\vec{F}$.

57. If $\vec{F}$ is a vector field in 3-space, and $\vec{F} = \operatorname{grad} f$, then $\operatorname{div}\vec{F} = 0$.

58. If $\vec{F}$ is a vector field in 3-space, then $\operatorname{grad}(\operatorname{div}\vec{F}) = \vec{0}$.

59. The field $\vec{F}(\vec{r}) = \vec{r}$ is divergence free.

60. If $f(x, y, z)$ is any given continuous scalar function, then there is at least one vector field $\vec{F}$ such that $\operatorname{div}\vec{F} = f$.

61. If $\vec{F}$ and $\vec{G}$ are vector fields satisfying $\operatorname{div}\vec{F} = \operatorname{div}\vec{G}$ then $\vec{F} = \vec{G}$.

62. There exist a scalar function f and a vector field $\vec{F}$ satisfying $\operatorname{div}(\operatorname{grad} f) = \operatorname{grad}(\operatorname{div}\vec{F})$.

63. For $\vec{r} = x\vec{i} + y\vec{j} + z\vec{k}$, an arbitrary function $f(x, y, z)$ and an arbitrary vector field, $\vec{F}(x, y, z)$, which of the following is a vector field, and which is a constant vector field?

    (a) $\operatorname{grad} f$ **(b)** $(\operatorname{div}\vec{F})\vec{i}$ **(c)** $(\operatorname{div}\vec{r})\vec{i}$

    (d) $(\operatorname{div}\vec{i})\vec{F}$ **(e)** $\operatorname{grad}(\operatorname{div}\vec{F})$

19.4 THE DIVERGENCE THEOREM

The Divergence Theorem is a multivariable analogue of the Fundamental Theorem of Calculus; it says that the integral of the flux density over a solid region equals the flux integral through the boundary of the region.

The Boundary of a Solid Region

A solid region is an open region in 3-space. The boundary of a solid region may be thought of as the skin between the interior of the region and the space around it. For example, the boundary of a solid ball is a spherical surface, the boundary of a solid cube is its six faces, and the boundary of a solid cylinder is a tube sealed at both ends by disks. (See Figure 19.37). A surface which is the boundary of a solid region is called a *closed surface*.

W = Ball
S = Sphere

W = Solid Cylinder
S = Tube and two disks

W = Solid cube
S = 6 square faces

Figure 19.37: Several solid regions and their boundaries

Calculating the Flux from the Flux Density

Consider a solid region W in 3-space whose boundary is the closed surface S. There are two ways to find the total flux of a vector field $\vec{F}$ out of W. One is to calculate the flux of $\vec{F}$ through S:

$$\text{Flux out of } W = \int_S \vec{F} \cdot d\vec{A}.$$

Another way is to use $\operatorname{div} \vec{F}$, which gives the flux density at any point in W. We subdivide W into small boxes, as shown in Figure 19.38. Then, for a small box of volume ΔV,

$$\text{Flux out of box} \approx \text{Flux density} \cdot \text{Volume} = \operatorname{div} \vec{F} \, \Delta V.$$

What happens when we add the fluxes out of all the boxes? Consider two adjacent boxes, as shown in Figure 19.39. The flux through the shared wall is counted twice, once out of the box on each side. When we add the fluxes, these two contributions cancel, so we get the flux out of the solid region formed by joining the two boxes. Continuing in this way, we find that

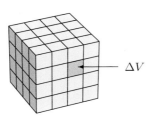

Figure 19.38: Subdivision of region into small boxes

ΔV

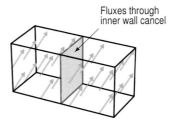

Fluxes through inner wall cancel

Figure 19.39: Adding the flux out of adjacent boxes

$$\text{Flux out of } W = \sum \text{Flux out of small boxes} \approx \sum \operatorname{div} \vec{F} \, \Delta V.$$

We have approximated the flux by a Riemann sum. As the subdivision gets finer, the sum approaches an integral, so

$$\text{Flux out of } W = \int_W \operatorname{div} \vec{F} \, dV.$$

We have calculated the flux in two ways, as a flux integral and as a volume integral. Therefore, these two integrals must be equal. This result holds even if W is not a rectangular solid. Thus, we have the following result.[3]

---

[3] A proof of the Divergence Theorem using the coordinate definition of the divergence can be found in the online supplement at www.wiley.com/college/hughes-hallett.

Theorem 19.1: The Divergence Theorem

If W is a solid region whose boundary S is a piecewise smooth surface, and if $\vec{F}$ is a smooth vector field on an open region containing W and S, then

$$\int_S \vec{F} \cdot d\vec{A} = \int_W \text{div} \, \vec{F} \, dV,$$

where S is given the outward orientation.

Example 1   Use the Divergence Theorem to calculate the flux of the vector field $\vec{F}(\vec{r}) = \vec{r}$ through the sphere of radius a centred at the origin.

Solution   In Example 5 on page 1030 we computed the flux using the definition of a flux integral, giving

$$\int_S \vec{r} \cdot d\vec{A} = 4\pi a^3.$$

Now we use div $\vec{F} = \text{div}(x\vec{i} + y\vec{j} + z\vec{k}) = 3$ and the Divergence Theorem:

$$\int_S \vec{r} \cdot d\vec{A} = \int_W \text{div} \, \vec{F} \, dV = \int_W 3 \, dV = 3 \cdot \frac{4}{3}\pi a^3 = 4\pi a^3.$$

Example 2   Use the Divergence Theorem to calculate the flux of the vector field

$$\vec{F}(x, y, z) = (x^2 + y^2)\vec{i} + (y^2 + z^2)\vec{j} + (x^2 + z^2)\vec{k}$$

through the cube in Figure 19.40.

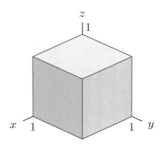

Figure 19.40

Solution   The divergence of $\vec{F}$ is div $\vec{F} = 2x + 2y + 2z$. Since div $\vec{F}$ is positive everywhere in the first quadrant, the flux through S is positive. By the Divergence Theorem,

$$\int_S \vec{F} \cdot d\vec{A} = \int_0^1 \int_0^1 \int_0^1 2(x + y + z) \, dx \, dy \, dz = \int_0^1 \int_0^1 (x^2 + 2x(y + z)) \Big|_0^1 \, dy \, dz$$

$$= \int_0^1 \int_0^1 1 + 2(y + z) \, dy \, dz = \int_0^1 (y + y^2 + 2yz) \Big|_0^1 \, dz$$

$$= \int_0^1 (2 + 2z) \, dz = (2z + z^2) \Big|_0^1 = 3.$$

The Divergence Theorem and Divergence-Free Vector Fields

An important application of the Divergence Theorem is the study of divergence-free vector fields.

Example 3   In Example 3 on page 1049 we saw that the following vector field is divergence free:

$$\vec{F}(\vec{r}) = \frac{\vec{r}}{\|\vec{r}\|^3}, \qquad \vec{r} \neq \vec{0}.$$

Calculate $\int_S \vec{F} \cdot d\vec{A}$, using the Divergence Theorem if possible, for the following surfaces:
(a) S_1 is the sphere of radius a centred at the origin.
(b) S_2 is the sphere of radius a centred at the point $(2a, 0, 0)$.

Solution   (a) We cannot use the Divergence Theorem directly because $\vec{F}$ is not defined everywhere inside the sphere (it is not defined at the origin). Since $\vec{F}$ points outward everywhere on S_1, the flux out of S_1 is positive. On S_1,

$$\vec{F} \cdot d\vec{A} = \|\vec{F}\| dA = \frac{a}{a^3} dA,$$

so

$$\int_{S_1} \vec{F} \cdot d\vec{A} = \frac{1}{a^2} \int_{S_1} dA = \frac{1}{a^2}(\text{Area of } S_1) = \frac{1}{a^2} 4\pi a^2 = 4\pi.$$

Notice that the flux is not zero, although $\text{div}\,\vec{F}$ is zero everywhere it is defined.
(b) Suppose W is the solid region enclosed by S_2. Since $\text{div}\,\vec{F} = 0$ everywhere in W, we can use the Divergence Theorem in this case, giving

$$\int_{S_2} \vec{F} \cdot d\vec{A} = \int_W \text{div}\,\vec{F}\, dV = \int_W 0\, dV = 0.$$

The Divergence Theorem applies to any solid region W and its boundary S, even in cases where the boundary consists of two or more surfaces. For example, if W is the solid region between the sphere S_1 of radius 1 and the sphere S_2 of radius 2, both centred at the same point, then the boundary of W consists of both S_1 and S_2. The Divergence Theorem requires the outward orientation, which on S_2 points away from the centre and on S_1 points toward the centre. (See Figure 19.41.)

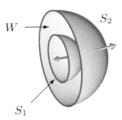

Figure 19.41: Cut-away view of the region W between two spheres, showing orientation vectors

Example 4   Let S_1 be the sphere of radius 1 centred at the origin and let S_2 be the ellipsoid $x^2 + y^2 + 4z^2 = 16$, both oriented outward. For

$$\vec{F}(\vec{r}) = \frac{\vec{r}}{\|\vec{r}\|^3}, \qquad \vec{r} \neq \vec{0},$$

show that

$$\int_{S_1} \vec{F} \cdot d\vec{A} = \int_{S_2} \vec{F} \cdot d\vec{A}.$$

Solution   The ellipsoid contains the sphere; let W be the solid region between them. Since W does not contain the origin, $\operatorname{div} \vec{F}$ is defined and equal to zero everywhere in W. Thus, if S is the boundary of W, then

$$\int_S \vec{F} \cdot d\vec{A} = \int_W \operatorname{div} \vec{F} \, dV = 0.$$

But S consists of S_2 oriented outward and S_1 oriented inward, so

$$0 = \int_S \vec{F} \cdot d\vec{A} = \int_{S_2} \vec{F} \cdot d\vec{A} - \int_{S_1} \vec{F} \cdot d\vec{A},$$

and thus

$$\int_{S_2} \vec{F} \cdot d\vec{A} = \int_{S_1} \vec{F} \cdot d\vec{A}.$$

In Example 3 we showed that $\int_{S_1} \vec{F} \cdot d\vec{A} = 4\pi$, so $\int_{S_2} \vec{F} \cdot d\vec{A} = 4\pi$ also. Note that it would have been more difficult to compute the integral over the ellipsoid directly.

Electric Fields

The electric field produced by a positive point charge q placed at the origin is

$$\vec{E} = q \frac{\vec{r}}{\|\vec{r}\|^3}.$$

Using Example 3, we see that the flux of the electric field through any sphere centred at the origin is $4\pi q$. In fact, using the idea of Example 4, we can show that the flux of $\vec{E}$ through any closed surface containing the origin is $4\pi q$. See Problems 39 and 40 on page 1060. This is a special case of Gauss's Law, which states that the flux of an electric field through any closed surface is proportional to the total charge enclosed by the surface. Carl Friedrich Gauss (1777–1855) also discovered the Divergence Theorem, which is sometimes called Gauss's Theorem.

Exercises and Problems for Section 19.4

Exercises

For Exercises 1–5, compute the flux integral $\int_S \vec{F} \cdot d\vec{A}$ in two ways, if possible, directly and using the Divergence Theorem. In each case, S is closed and oriented outward.

1. $\vec{F}(\vec{r}) = \vec{r}$ and S is the cube enclosing the volume $0 \le x \le 2, 0 \le y \le 2$, and $0 \le z \le 2$.

2. $\vec{F} = x^2\vec{i} + 2y^2\vec{j} + 3z^2\vec{k}$ and S is the surface of the box with faces $x = 1, x = 2, y = 0, y = 1, z = 0, z = 1$.

3. $\vec{F} = (z^2 + x)\vec{i} + (x^2 + y)\vec{j} + (y^2 + z)\vec{k}$ and S is the closed cylinder $x^2 + z^2 = 1$, with $0 \le y \le 1$, oriented outward.

4. $\vec{F} = -z\vec{i} + x\vec{k}$ and S is a square pyramid with height 3 and base on the xy-plane of side length 1.

5. $\vec{F} = y\vec{j}$ and S is a closed vertical cylinder of height 2, with its base a circle of radius 1 on the xy-plane, centered at the origin.

In Exercises 6–7, use the Divergence Theorem to calculate the flux of the vector field through the surface.

6. $\vec{F} = x^2\vec{i} + (y - 2xy)\vec{j} + 10z\vec{k}$ through a sphere of radius 5 centered at the origin, oriented outward.

7. $\vec{F} = -z\vec{i} + x\vec{k}$ through the sphere of radius a centered at the origin. Explain your answer geometrically.

Find the flux of the vector fields in Exercises 8–14 out of the closed box $0 \le x \le 2, 0 \le y \le 3, 0 \le z \le 4$.

8. $\vec{F} = 4\vec{i} + 7\vec{j} - \vec{k}$

9. $\vec{G} = y\vec{i} + z\vec{k}$

10. $\vec{H} = xy\vec{i} + z\vec{j} + y\vec{k}$

11. $\vec{J} = xy^2\vec{j} + x\vec{k}$

12. $\vec{N} = e^z\vec{i} + \sin(xy)\vec{k}$

13. $\vec{M} = ((3x + 4y)\vec{i} + (4y + 5z)\vec{j} + (5z + 3x)\vec{k}$

14. $\vec{M}$ where $\operatorname{div} \vec{M} = xy + 5$

Problems

15. Find the flux of $\vec{F} = z\vec{i} + y\vec{j} + x\vec{k}$ out of a sphere of radius 3 centered at the origin.

16. Find the flux of $\vec{F} = xy\vec{i} + yz\vec{j} + zx\vec{k}$ out of a sphere of radius 1 centered at the origin.

17. Find the flux of $\vec{F} = x^3\vec{i} + y^3\vec{j} + z^3\vec{k}$ through the closed surface bounding the solid region $x^2 + y^2 \leq 4$, $0 \leq z \leq 5$, oriented outward.

18. The region W lies between the spheres $x^2 + y^2 + z^2 = 4$ and $x^2 + y^2 + z^2 = 9$ and within the cone $z = \sqrt{x^2 + y^2}$ with $z \geq 0$; its boundary is the closed surface, S, oriented outward. Find the flux of $\vec{F} = x^3\vec{i} + y^3\vec{j} + z^3\vec{k}$ out of S.

19. Find the flux of $\vec{F}$ through the closed cylinder of radius 2, centered on the z-axis, with $3 \leq z \leq 7$, if $\vec{F} = (x + 3e^{yz})\vec{i} + (\ln(x^2z^2 + 1) + y)\vec{j} + z\vec{k}$.

20. Find the flux of $\vec{F} = e^{y^2z^2}\vec{i} + (\tan(0.001x^2z^2) + y^2)\vec{j} + (\ln(1 + x^2y^2) + z^2)\vec{k}$ out of the closed box $0 \leq x \leq 5, 0 \leq y \leq 4, 0 \leq z \leq 3$.

21. Find the flux of $\vec{F} = x^2\vec{i} + z\vec{j} + y\vec{k}$ out of the closed cone $x = \sqrt{y^2 + z^2}$, with $0 \leq x \leq 1$.

22. Suppose $\vec{F}$ is a vector field with $\operatorname{div} \vec{F} = 10$. Find the flux of $\vec{F}$ out of a cylinder of height a and radius a, centered on the z-axis and with base in the xy-plane.

23. A cone has its tip at the point $(0, 0, 5)$ and its base the disk D, $x^2 + y^2 \leq 1$, in the xy-plane. The surface of the cone is the curved and slanted face, S, oriented upward, and the flat base, D, oriented downward. The flux of the constant vector field $\vec{F} = a\vec{i} + b\vec{j} + c\vec{k}$ through S is given by

$$\int_S \vec{F} \cdot d\vec{A} = 3.22.$$

Is it possible to calculate $\int_D \vec{F} \cdot d\vec{A}$? If so, give the answer. If not, explain what additional information you would need to be able to make this calculation.

24. If V is a volume surrounded by a closed surface S, show that $\frac{1}{3} \int_S \vec{r} \cdot d\vec{A} = V$.

25. A vector field $\vec{F}$ satisfies $\operatorname{div} \vec{F} = 0$ everywhere. Show that $\int_S \vec{F} \cdot d\vec{A} = 0$ for every closed surface S.

26. Let $\vec{r} = x\vec{i} + y\vec{j} + z\vec{k}$ and let $\vec{F}$ be the vector field given by

$$\vec{F} = \frac{\vec{r}}{\|\vec{r}\|^3}.$$

(a) Calculate the flux of $\vec{F}$ out of the unit sphere $x^2 + y^2 + z^2 = 1$ oriented outward.
(b) Calculate $\operatorname{div} \vec{F}$. Show your work and simplify your answer completely.
(c) Use your answers to parts (a) and (b) to calculate the flux out of a box of side 10 centered at the origin and with sides parallel to the coordinate planes. (The box is also oriented outward.)

27. (a) Find $\operatorname{div}(\vec{r}/\|\vec{r}\|^2)$ where $\vec{r} = x\vec{i} + y\vec{j}$ for $\vec{r} \neq \vec{0}$.
(b) Can you use the Divergence Theorem to compute the flux of $\vec{r}/\|\vec{r}\|^2$ out of a closed cylinder of radius 1, length 2, centered at the origin, and with its axis along the z-axis?
(c) Compute the flux of $\vec{r}/\|\vec{r}\|^2$ out of the cylinder in part (b).
(d) Find the flux of $\vec{r}/\|\vec{r}\|^2$ out of a closed cylinder of radius 2, length 2, centered at the origin, and with its axis along the z-axis.

28. Let S be the cube in the first quadrant with side 2, one corner at the origin and edges parallel to the axes. Let

$$\vec{F}_1 = (xy^2 + 3xz^2)\vec{i} + (3x^2y + 2yz^2)\vec{j} + 3zy^2\vec{k}$$
$$\vec{F}_2 = (xy^2 + 5e^{yz})\vec{i} + (yz^2 + 7\sin(xz))\vec{j} + (x^2z + \cos(xy))\vec{k}$$
$$\vec{F}_3 = \left(xz^2 + \frac{x^3}{3}\right)\vec{i} + \left(yz^2 + \frac{y^3}{3}\right)\vec{j} + \left(zy^2 + \frac{z^3}{3}\right)\vec{k}.$$

Arrange the flux integrals of $\vec{F}_1$, $\vec{F}_2$, $\vec{F}_3$ out of S in increasing order.

29. Let $\operatorname{div} \vec{F} = 2(6 - x)$ and $0 \leq a, b, c \leq 10$.

(a) Find the flux of $\vec{F}$ out of the rectangular solid $0 \leq x \leq a, 0 \leq y \leq b, 0 \leq z \leq c$.
(b) For what values of a, b, c is the flux largest? What is that largest flux?

30. Assume $\vec{r} \neq \vec{0}$. Let $\vec{F} = \vec{r}/\|\vec{r}\|^3$.

(a) Calculate the flux of $\vec{F}$ out of the unit sphere $x^2 + y^2 + z^2 = 1$ oriented outward.
(b) Calculate $\operatorname{div} \vec{F}$. Simplify your answer completely.
(c) Use your answer to parts (a) and (b) to calculate the flux out of a box of side 10 centered at the origin and with sides parallel to the coordinate planes. (The box is also oriented outward.)

In Problems 31–32, find the flux of $\vec{F} = \vec{r}/\|\vec{r}\|^3$ through the surface. [Hint: Use the method of Problem 30.]

31. S is the ellipsoid $x^2 + 2y^2 + 3z^2 = 6$.

32. S is the closed cylinder $y^2 + z^2 = 4$, $-2 \leq x \leq 2$.

33. (a) Let $\operatorname{div} \vec{F} = x^2 + y^2 + z^2 + 3$. Calculate $\int_{S_1} \vec{F} \cdot d\vec{A}$ where S_1 is the sphere of radius 1 centered at the origin.
(b) Let S_2 be the sphere of radius 2 centered at the origin; let S_3 be the sphere of radius 3 centered at the origin; let S_4 be the box of side 6 centered at the origin with edges parallel to the axes. Without calculating them, arrange the following integrals in increasing order:

$$\int_{S_2} \vec{F} \cdot d\vec{A}, \quad \int_{S_3} \vec{F} \cdot d\vec{A}, \quad \int_{S_4} \vec{F} \cdot d\vec{A}.$$

34. Suppose div $\vec{F} = xyz^2$.

(a) Find div $\vec{F}$ at the point $(1, 2, 1)$. [Note: You are given div $\vec{F}$, not $\vec{F}$.]

(b) Using your answer to part (a), but no other information about the vector field $\vec{F}$, estimate the flux out of a small box of side 0.2 centered at the point $(1, 2, 1)$ and with edges parallel to the axes.

(c) Without computing the vector field $\vec{F}$, calculate the exact flux out of the box.

35. Suppose div $\vec{F} = x^2 + y^2 + 3$. Find a surface S such that $\int_S \vec{F} \cdot d\vec{A}$ is negative, or explain why no such surface exists.

36. A volume of water swirls down according to the velocity field

$$\vec{F} = -\frac{y + xz}{(z^2 + 1)^2}\vec{i} - \frac{yz - x}{(z^2 + 1)^2}\vec{j} - \frac{1}{z^2 + 1}\vec{k}$$

Let C be a circle on the xy plane, with center at the origin and radius 1. The flux of this vector field through the drain oriented downward is

$$\int_C \vec{F} \cdot d\vec{A} = \pi$$

(a) Calculate the divergence of $\vec{F}$.

(b) Calculate the flux of $\vec{F}$ through the surface defined by $z \leq 0$ and $x^2 + y^2 + z^2 = 1$ (a hemisphere, center at the origin, radius 1) oriented downward.

37. As a result of radioactive decay, heat is generated uniformly throughout the interior of the earth at a rate of 30 watts per cubic kilometer. (A watt is a rate of heat production.) The heat then flows to the earth's surface where it is lost to space. Let $\vec{F}(x, y, z)$ denote the rate of flow of heat measured in watts per square kilometer. By definition, the flux of $\vec{F}$ across a surface is the quantity of heat flowing through the surface per unit of time.

(a) What is the value of div $\vec{F}$? Include units.

(b) Assume the heat flows outward symmetrically. Verify that $\vec{F} = \alpha \vec{r}$, where $\vec{r} = x\vec{i} + y\vec{j} + z\vec{k}$ and α is a suitable constant, satisfies the given conditions. Find α.

(c) Let $T(x, y, z)$ denote the temperature inside the earth. Heat flows according to the equation $\vec{F} = -k \operatorname{grad} T$, where k is a constant. Explain why this makes sense physically.

(d) If T is in $°C$, then $k = 30{,}000$ watts/km$°$C. Assuming the earth is a sphere with radius 6400 km and surface temperature $20°C$, what is the temperature at the center?

38. If a surface S is submerged in an incompressible fluid, a force $\vec{F}$ is exerted on one side of the surface by the pressure in the fluid. If the z-axis is vertical, with the positive direction upward and the fluid level at $z = 0$, then the component of force in the direction of a unit vector $\vec{u}$ is given by the following:

$$\vec{F} \cdot \vec{u} = -\int_S z\delta g\vec{u} \cdot d\vec{A},$$

where δ is the density of the fluid (mass/volume), g is the acceleration due to gravity, and the surface is oriented away from the side on which the force is exerted. In this problem we consider a totally submerged closed surface enclosing a volume V. We are interested in the force of the liquid on the external surface, so S is oriented inward. Use the Divergence Theorem to show that:

(a) The force in the $\vec{i}$ and $\vec{j}$ directions is zero.

(b) The force in the $\vec{k}$ direction is $\delta g V$, the weight of the volume of fluid with the same volume as V. This is *Archimedes' Principle*.

39. According to Coulomb's Law, the electrostatic field $\vec{E}$ at the point $\vec{r}$ due to a charge q at the origin is given by

$$\vec{E}(\vec{r}) = q\frac{\vec{r}}{\|\vec{r}\|^3}.$$

(a) Compute div $\vec{E}$.

(b) Let S_a be the sphere of radius a centered at the origin and oriented outward. Show that the flux of $\vec{E}$ through S_a is $4\pi q$.

(c) Could you have used the Divergence Theorem in part (b)? Explain why or why not.

(d) Let S be an arbitrary, closed, outward-oriented surface surrounding the origin. Show that the flux of $\vec{E}$ through S is again $4\pi q$. [Hint: Apply the Divergence Theorem to the solid region lying between a small sphere S_a and the surface S.]

40. According to Coulomb's Law, the electric field $\vec{E}$ at the point $\vec{r}$ due to a charge q at the point $\vec{r}_0$ is given by

$$\vec{E}(\vec{r}) = q\frac{(\vec{r} - \vec{r}_0)}{\|\vec{r} - \vec{r}_0\|^3}.$$

Suppose S is a closed, outward-oriented surface and that $\vec{r}_0$ does not lie on S. Use Problem 39 to show that

$$\int_S \vec{E} \cdot d\vec{A} = \begin{cases} 4\pi q & \text{if } q \text{ lies inside } S, \\ 0 & \text{if } q \text{ lies outside } S. \end{cases}$$

Strengthen Your Understanding

In Problems 41–42, explain what is wrong with the statement.

41. The flux integral $\int_S \vec{F} \cdot d\vec{A}$ can be evaluated using the

Divergence Theorem, where $\vec{F} = 2x\vec{i} - 3\vec{j}$ and S is the triangular surface with corners $(1, 0, 0)$, $(0, 1, 0)$, $(0, 0, 1)$ oriented away from the origin.

42. If S is the boundary of a solid region W, where S is oriented outward, and $\vec{F}$ is a vector field, then

$$\int_S \operatorname{div} \vec{F} \, d\vec{A} = \int_W \vec{F} \, dV.$$

In Problems 43–44, give an example of:

43. A surface S that is the boundary of a solid region such that $\int_S \vec{F} \cdot d\vec{A} = 0$ if $\vec{F}(x, y, z) = y\vec{i} + xz\vec{j} + y^2\vec{k}$.

44. A vector field $\vec{F}$ such that the flux of $\vec{F}$ out of a sphere of radius 1 centered at the origin is 3.

Are the statements in Problems 45–49 true or false? The smooth vector field $\vec{F}$ is defined everywhere in 3-space and has constant divergence equal to 4.

45. The field $\vec{F}$ has a net inflow per unit volume at the point $(-3, 4, 0)$.

46. The vector field $\vec{F}$ could be the field $\vec{F} = x\vec{i} + (3y)\vec{j} + (y - 5x)\vec{k}$.

47. The vector field $\vec{F}$ could be a constant field.

48. The flux of $\vec{F}$ through a circle of radius 5 lying anywhere on the xy-plane and oriented upward is $4(\pi 5^2)$.

49. The flux of $\vec{F}$ through a closed cylinder of radius 1 centered along the y-axis, $0 \le y \le 3$ and oriented outward is $4(3\pi)$.

Are the statements in Problems 50–58 true or false? Give reasons for your answer.

50. $\int_S \vec{F} \cdot d\vec{A} = \operatorname{div} \vec{F}$.

51. If $\vec{F}$ is a vector field in 3-space, and W is a solid region with boundary surface S, then $\int_S \operatorname{div} \vec{F} \cdot d\vec{A} = \int_W \vec{F} \, dV$.

52. If $\vec{F}$ is a divergence-free vector field in 3-space, and S is a closed surface oriented inward, then $\int_S \vec{F} \cdot d\vec{A} = 0$.

53. If $\vec{F}$ is a vector field in 3-space satisfying $\operatorname{div} \vec{F} = 1$, and S is a closed surface oriented outward, then $\int_S \vec{F} \cdot d\vec{A}$ is equal to the volume enclosed by S.

54. Let W be the solid region between the sphere S_1 of radius 1 and S_2 of radius 2, both centered at the origin. If $\vec{F}$ is a vector field in 3-space, then $\int_W \operatorname{div} \vec{F} \, dV = \int_{S_2} \vec{F} \cdot d\vec{A} - \int_{S_1} \vec{F} \cdot d\vec{A}$, where both S_1 and S_2 are oriented outward.

55. Let S_1 be the square $0 \le x \le 1, 0 \le y \le 1, z = 0$ oriented downward and let S_2 be the square $0 \le x \le 1, 0 \le y \le 1, z = 1$ oriented upward. If $\vec{F}$ is a vector field, then $\int_W \operatorname{div} \vec{F} \, dV = \int_{S_2} \vec{F} \cdot d\vec{A} + \int_{S_1} \vec{F} \cdot d\vec{A}$, where W is the solid cube $0 \le x \le 1, 0 \le y \le 1, 0 \le z \le 1$.

56. Let S_1 be the square $0 \le x \le 1, 0 \le y \le 1, z = 0$ oriented downward and let S_2 be the square $0 \le x \le 1, 0 \le y \le 1, z = 1$ oriented upward. If $\vec{F} = \cos(xyz)\vec{k}$, then $\int_W \operatorname{div} \vec{F} \, dV = \int_{S_2} \vec{F} \cdot d\vec{A} + \int_{S_1} \vec{F} \cdot d\vec{A}$, where W is the solid cube $0 \le x \le 1, 0 \le y \le 1, 0 \le z \le 1$.

57. If S is a sphere of radius 1, centered at the origin, oriented outward, and $\vec{F}$ is a vector field satisfying $\int_S \vec{F} \cdot d\vec{A} = 0$, then $\operatorname{div} \vec{F} = 0$ at all points inside S.

58. Let S_h be the surface consisting of a cylinder of height h, closed at the top. The curved sides are $x^2 + y^2 = 1$, for $0 \le z \le h$, and the top $x^2 + y^2 \le 1$, for $z = h$, oriented outward. If $\vec{F}$ is divergence free, then $\int_{S_h} \vec{F} \cdot d\vec{A}$ is independent of the height h.

59. Let $\vec{F} = (5x + 7y)\vec{i} + (7y + 9z)\vec{j} + (9z + 11x)\vec{k}$, and let Q_i be the flux of $\vec{F}$ through the surfaces S_i for $i = 1$–4. Arrange Q_i in ascending order, where

(a) S_1 is the sphere of radius 2 centered at the origin
(b) S_2 is the cube of side 2 centered at the origin and with sides parallel to the axes
(c) S_3 is the sphere of radius 1 centered at the origin
(d) S_4 is a pyramid with all four corners lying on S_3

CHAPTER SUMMARY (see also Ready Reference at the end of the book)

- **Flux Integrals**
  Oriented surfaces, definition of flux through a surface, definition of flux integral as a limit of a Riemann sum

- **Calculating flux integrals** over surfaces
  Graphs: $d\vec{A} = (-f_x\vec{i} - f_y\vec{j} + \vec{k}) \, dx \, dy$
  Cylinders: $d\vec{A} = (\cos\theta\vec{i} + \sin\theta\vec{j})R \, dz \, d\theta$
  Spheres:
  $d\vec{A} = (\sin\phi\cos\theta\vec{i} + \sin\phi\sin\theta\vec{j} + \cos\phi\vec{k})R^2 \, d\phi \, d\theta$

- **Divergence**
  Geometric and coordinate definition of divergence, calculating divergence, interpretation in terms of outflow per unit volume.

- **The Divergence Theorem**
  Statement of the theorem, divergence-free fields, harmonic functions.

REVIEW EXERCISES AND PROBLEMS FOR CHAPTER NINETEEN

Exercises

1. Let S be the disk of radius 3 perpendicular the the y-axis, centered at $(0, 6, 0)$ and oriented away from the origin. Is
$$\int_S (x\vec{i} + y\vec{j}) \cdot d\vec{A} \text{ a vector or a scalar?}$$

In Exercises 2–5, compute the flux of $\vec{v} = \vec{i} + 2\vec{j} - 3\vec{k}$ through the rectangular region with the orientation shown.

2. **3.**

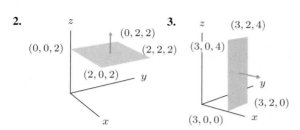

4. **5.**

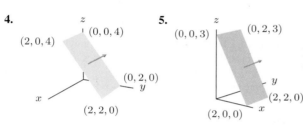

6. Let $\vec{F}(x, y, z) = \vec{i} + 2\vec{j} + \vec{k}$. Each of the surfaces in (a)–(e) are squares of side length 4, with orientation given by the normal vector shown. Compute the flux of $\vec{F}$ through these surfaces.

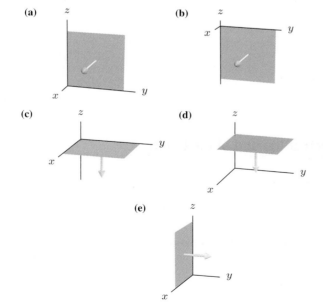

Are the quantities in Exercises 7–8 vectors or scalars? Calculate them.

7. div $\left((x^2 + y)\vec{i} + (xye^z)\vec{j} - \ln(x^2 + y^2)\vec{k}\right)$

8. div $\left((2\sin(xy) + \tan z)\vec{i} + (\tan y)\vec{j} + (e^{x^2+y^2})\vec{k}\right)$

In Exercises 9–12, are the flux integrals positive, negative, or zero? Let S be a disk of radius 3 in the plane $x = 7$, centered on the x-axis and oriented toward the origin.

9. $\int_S (x\vec{j} + y\vec{k}) \cdot d\vec{A}$ **10.** $\int_S (x\vec{i} + y\vec{k}) \cdot d\vec{A}$

11. $\int_S (y\vec{i} + x\vec{k}) \cdot d\vec{A}$

12. $\int_S ((x - 10)\vec{i} + (x + 10)\vec{j}) \cdot d\vec{A}$

In Exercises 13–28, find the flux of the vector field through the surface S.

13. $\vec{F} = 2\vec{i} + 3\vec{j}$ through a disk of radius 5 in the plane $y = 2$ oriented in the direction of increasing y.

14. $\vec{F} = \vec{i} + 2\vec{j}$ through a sphere of radius 3 at the origin.

15. $\vec{F} = y\vec{j}$ through the square of side 4 in the plane $y = 5$. The square is centered on the y-axis, has sides parallel to the axes, and is oriented in the positive y-direction.

16. $\vec{F} = x\vec{i}$ through a square of side 3 in the plane $x = -5$. The square is centered on the x-axis, has sides parallel to the axes, and is oriented in the positive x-direction.

17. $\vec{F} = (y+3)\vec{j}$ through a square of side 2 in the xz-plane, oriented in the negative y-direction.

18. $\vec{F} = x\vec{k}$ through the square $0 \leq x \leq 3, 0 \leq y \leq 3$ in the xy-plane, with sides parallel to the axes, and oriented upward.

19. $\vec{F} = \vec{i} - \vec{j} - \vec{k}$ through a cube of side 2 with sides parallel to the axes.

20. $\vec{F} = 6\vec{i} + x^2\vec{j} - \vec{k}$, through a square of side 2 in the plane $z = 3$, oriented upward.

21. $\vec{F} = (x^2 + y^2)\vec{i} + xy\vec{j}$ and S is the square in the xy-plane with corners at $(1, 1, 0), (-1, 1, 0), (1, -1, 0), (-1, -1, 0)$, and oriented upward.

22. $\vec{F} = z\vec{i} + y\vec{j} + 2x\vec{k}$ and S is the rectangle $z = 4$, $0 \leq x \leq 2, 0 \leq y \leq 3$, oriented in the positive z-direction.

23. $\vec{F} = (x + \cos z)\vec{i} + y\vec{j} + 2x\vec{k}$ and S is the rectangle $x = 2, 0 \leq y \leq 3, 0 \leq z \leq 4$, oriented in the positive x-direction.

24. $\vec{F} = x^2\vec{i} + (x + e^y)\vec{j} - \vec{k}$, and S is the rectangle $y = -1$, $0 \leq x \leq 2$, $0 \leq z \leq 4$, oriented in the negative y-direction.

25. $\vec{F} = (5 + xy)\vec{i} + z\vec{j} + yz\vec{k}$ and S is the 2×2 square plate in the yz-plane centered at the origin, oriented in the positive x-direction.

26. $\vec{F} = x\vec{i} + y\vec{j}$ and S is the surface of a closed cylinder of radius 2 and height 3 centered on the z-axis with its base in the xy-plane.

27. $\vec{F} = -y\vec{i} + x\vec{j} + z\vec{k}$ and S is the surface of a closed cylinder of radius 1 centered on the z-axis with base in the plane $z = -1$ and top in the plane $z = 1$.

28. $\vec{F} = x^2\vec{i} + y^2\vec{j} + z\vec{k}$ and S is the cone $z = \sqrt{x^2 + y^2}$, oriented upward with $x^2 + y^2 \leq 1$, $x \geq 0$, $y \geq 0$.

In Problems 29–32, give conditions on one or more of the constants a, b, c to ensure that the flux integral $\int_S \vec{F} \cdot d\vec{A}$ has the given sign.

29. Positive for $\vec{F} = a\vec{i} + b\vec{j} + c\vec{k}$ and S is a disk perpendicular to the y-axis, through $(0, 5, 0)$ and oriented away from the origin.

30. Negative for $\vec{F} = a\vec{i} + b\vec{j} + c\vec{k}$ and S is the upper half of the unit sphere centered at the origin and oriented downward.

31. Positive for $\vec{F} = ax\vec{i} + ay\vec{j} + az\vec{k}$ and S is a sphere centered at the origin.

32. Positive for $\vec{F} = a\vec{i} + b\vec{j} + c\vec{k}$ and S is the triangle cut off by $x + y + z = 1$ in the first quadrant, oriented upward.

33. Calculate the flux of $\vec{F} = xy\vec{i} + yz\vec{j} + zx\vec{k}$ out of the closed box $0 \leq x \leq 1$, $0 \leq y \leq 1$, $0 \leq z \leq 1$
  (a) Directly **(b)** Using the Divergence Theorem.

34. Compute the flux integral $\int_S (x^3\vec{i} + 2y\vec{j} + 3\vec{k}) \cdot d\vec{A}$, where S is the $2 \times 2 \times 2$ rectangular surface centered at the origin, oriented outward. Do this in two ways:
  (a) Directly **(b)** Using the Divergence Theorem

In Exercises 35–40, use the Divergence Theorem to calculate the flux of the vector field out of the surface.

35. $\vec{F} = -x\vec{i} + 2y\vec{j} + (3 + 2z)\vec{k}$ out of the sphere of radius 2 centered at $(1, 2, 3)$.

36. $\vec{F} = (2x + y)\vec{i} + (3y + z)\vec{j} + (4z + x)\vec{k}$ and S is the sphere of radius 5 centered at the origin.

37. $\vec{F} = x^2\vec{i} + y^2\vec{j} + e^{xy}\vec{k}$ and S is the cube of side 3 in the first octant with one corner at the origin, and sides parallel to the axes.

38. $\vec{F} = (e^z + x)\vec{i} + (2y + \sin x)\vec{j} + \left(e^{x^2} - z\right)\vec{k}$ and S is the sphere of radius 1 centered at $(2, 1, 0)$.

39. $\vec{F} = x^3\vec{i} + y^3\vec{j} + \left(x^2 + y^2\right)\vec{k}$ and S is the closed cylinder $x^2 + y^2 = 4$ with $0 \leq z \leq 5$.

40. $\vec{F} = x\vec{i} + y\vec{j} + z\vec{k}$ and S is the open cylinder (ends not included) $x^2 + y^2 \leq 1$, with $0 \leq z \leq 1$, oriented outward.

Problems

41. Arrange the following flux integrals, $\int_{S_i} \vec{F} \cdot d\vec{A}$, with $i = 1, 2, 3, 4$, in ascending order if $\vec{F} = -\vec{i} - \vec{j} + \vec{k}$ and if the S_i are the following surfaces:

- S_1 is a horizontal square of side 1 with one corner at $(0, 0, 2)$, above the first quadrant of the xy-plane, oriented upward.
- S_2 is a horizontal square of side 1 with one corner at $(0, 0, 3)$, above the third quadrant of the xy-plane, oriented upward.
- S_3 is a square of side $\sqrt{2}$ in the xz-plane with one corner at the origin, one edge along the positive x-axis, one along the negative z-axis, oriented in the negative y-direction.
- S_4 is a square of side $\sqrt{2}$ with one corner at the origin, one edge along the positive y-axis, one corner at $(1, 0, 1)$, oriented upward.

42. Let $f(x, y, z) = xy + e^{xyz}$. Find
  (a) grad f
  (b) $\int_C$ grad $f \cdot d\vec{r}$, where C is the line from the point $(1, 1, 1)$ to $(2, 3, 4)$.
  (c) $\int_S$ grad $f \cdot d\vec{A}$, where S is the quarter disk in the xy-plane $x^2 + y^2 \leq 4$, $x \geq 0$, $y \geq 0$, oriented up-

ward.

43. The flux of the constant vector field $a\vec{i} + b\vec{j} + c\vec{k}$ through the square of side 2 in the plane $x = 5$, oriented in the positive x-direction, is 24. Which of the constants a, b, c can be determined from the information given? Give the value(s).

44. **(a)** Let $\vec{F} = (x^2 + 4)\vec{i} + y\vec{j}$. Which of the following flux integrals is the largest? Explain.
  $\int_{S_1} \vec{F} \cdot d\vec{A}$, where S_1 is the disk of radius 1 in the plane $x = 2$ centered on the x-axis and oriented away from the origin.
  $\int_{S_2} \vec{F} \cdot d\vec{A}$, where S_2 is the disk of radius 1 in the plane $y = 4$ centered on the y-axis and oriented away from the origin.
  $\int_{S_3} \vec{F} \cdot d\vec{A}$, where S_3 is the disk of radius 1 in the plane $x = -3$ centered on the x-axis and oriented toward the origin.
  (b) Calculate the integral you chose in part (a).

45. Figure 19.42 shows a cross-section of the earth's magnetic field. Assume that the earth's magnetic and geographic poles coincide. Is the magnetic flux through a horizontal plate, oriented skyward, positive, negative, or zero if the plate is

(a) At the north pole?   (b) At the south pole?

(c) On the equator?

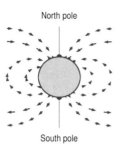

North pole

South pole

Figure 19.42

46. (a) Let $\operatorname{div}(\vec{F}) = x^2 + y^2 - z^2$. Estimate the flux out of a small sphere of radius 0.1 centered at each of the following points

(i) $(2, 1, 1)$ (ii) $(0, 0, 1)$

(b) What do the signs of your answers to part (a) tell you about the vector field near each of these two points?

For Problems 47–49,

(a) Find the flux of the given vector field out of a cube in the first octant with edge length c, one corner at the origin and edges along the axes.

(b) Use your answer to part (a) to find $\operatorname{div} \vec{F}$ at the origin using the geometric definition.

(c) Compute $\operatorname{div} \vec{F}$ at the origin using partial derivatives.

47. $\vec{F} = x\vec{i}$

48. $\vec{F} = 2\vec{i} + y\vec{j} + 3\vec{k}$

49. $\vec{F} = x\vec{i} + y\vec{j}$

In Problems 50–54, calculate the flux of $\vec{F}$ through the cylinder $x^2 + y^2 = 2$, $-3 \le z \le 3$ and its base, oriented outward. The cylinder is open at the top.

50. $\vec{F} = z^2\vec{i} + x^2\vec{j} + 5\vec{k}$

51. $\vec{F} = y^2\vec{i} + z^2\vec{j} + (x^2 + y^2)\vec{k}$

52. $\vec{F} = z\vec{i} + x\vec{j} + y\vec{k}$

53. $\vec{F} = y^2\vec{i} + x^2\vec{j} + 7z\vec{k}$

54. $\vec{F} = x^3\vec{i} + y^3\vec{j} + \vec{k}$

55. Find the constant vector field $\vec{F}$ parallel to $\vec{i} + \vec{k}$ and giving a flux of -7 through S, the cone $z = r$, with $0 \le z \le 4$, oriented upward.

56. (a) Find $\operatorname{div}(\vec{r}/\|\vec{r}\|^2)$ where $\vec{r} = x\vec{i} + y\vec{j}$ for $\vec{r} \ne \vec{0}$.

(b) Where in 3-space is $\operatorname{div}(\vec{r}/\|\vec{r}\|^2)$ undefined?

(c) Find the flux of $\vec{r}/\|\vec{r}\|^2$ through the closed cylinder of radius 2, length 3, centered at $(5, 0, 0)$, with its axis parallel to the z-axis.

(d) Find the flux of $\vec{r}/\|\vec{r}\|^2$ through the curved sides of the open cylinder in part (c).

57. (a) Let $\vec{F}$ be a smooth vector field defined throughout 3-space. What must be true of $\vec{F}$ if the flux of $\vec{F}$ through any closed surface is zero?

(b) What value of a ensures that the flux of $\vec{F} = a(e^x + y^2 - x)\vec{i} + 12y(1 - e^x)\vec{j}$ through any closed surface is zero?

58. Let $\vec{a} = a_1\vec{i} + a_2\vec{j} + a_3\vec{k}$ be a constant vector and let $\vec{r} = x\vec{i} + y\vec{j} + z\vec{k}$.

(a) Calculate $\operatorname{div}(\vec{r} \times \vec{a})$.

(b) Calculate the flux of $\vec{r} \times \vec{a}$ out of a cube of side 5 centered at the origin with edges parallel to the axes.

59. Find the flux of $\vec{F}$ out of the closed surface S given by $x^2 + y^2 + z^2 = 100$. You are given that $\vec{F}$ is continuous, with $\operatorname{div} \vec{F} = 3$ inside the cube $-2 \le x \le 2$, $-2 \le y \le 2$, $-2 \le z \le 2$ and $\operatorname{div} \vec{F} = 5$ outside the cube.

60. The closed surface S consists of S_1, the cone $x = \sqrt{y^2 + z^2}$ for $0 \le x \le 2$, and a disk S_2. Let $\vec{F} = 3x\vec{i} + 4y\vec{j} + 5z\vec{k}$.

(a) In what plane does the disk S_2 lie? How is it oriented?

(b) Find the flux of $\vec{F}$ through

(i) S_2 (ii) S_1

61. Find the flux integral, using $\vec{r} = x\vec{i} + y\vec{j} + z\vec{k}$.

(a) $\int_{S_1} \vec{r} \cdot d\vec{A}$, where S_1 is the disk $x = 5$, $y^2 + z^2 \le 7$, oriented away from the origin.

(b) $\int_{S_2} \vec{r} \cdot d\vec{A}$, where S_2 is the closed cylinder $y^2 + z^2 = 7$, $0 \le x \le 5$.

(c) $\int_{S_3} \vec{r} \cdot d\vec{A}$, where S_3 is the curved side of the open cylinder $y^2 + z^2 = 7$, $0 \le x \le 5$, oriented away from the x-axis.

62. Let $\vec{F}(x, y, z) = f_1(x, y, z)\vec{i} + f_2(x, y, z)\vec{j} + \vec{k}$ be a vector field with the property that $\operatorname{div} \vec{F} = 5$ everywhere. Let S be the hemisphere $z = -\sqrt{9 - x^2 - y^2}$, with its boundary in the xy-plane and oriented downward. Find $\int_S \vec{F} \cdot d\vec{A}$.

63. Let $\vec{F} = \vec{r}/\|\vec{r}\|^3$.

(a) Calculate $\operatorname{div} \vec{F}$. Where is $\operatorname{div} \vec{F}$ undefined?

(b) Find $\int_S \vec{F} \cdot d\vec{A}$ where S is a sphere of radius 10 centered at the origin.

(c) Find $\int_{B_1} \vec{F} \cdot d\vec{A}$ where B_1 is a box of side 1 centered at the point $(3, 0, 0)$ with sides parallel to the axes.

(d) Find $\int_{B_2} \vec{F} \cdot d\vec{A}$ where B_2 is a box of side 1 centered at the origin with sides parallel to the axes.

(e) Using your results to parts (c) and (d), explain how, with no further calculation, you can find the flux of this vector field through any closed surface, provided the origin does not lie on the surface.

64. The gravitational field, $\vec{F}$, of a planet of mass m at the origin is given by

$$\vec{F} = -Gm\frac{\vec{r}}{\|\vec{r}\|^3}.$$

Use the Divergence Theorem to show that the flux of the gravitational field through the sphere of radius a is independent of a. [Hint: Consider the region bounded by two concentric spheres.]

65. A basic property of the electric field $\vec{E}$ is that its divergence is zero at points where there is no charge. Suppose that the only charge is along the z-axis, and that the electric field $\vec{E}$ points radially out from the z-axis and its magnitude depends only on the distance r from the z-axis. Use the Divergence Theorem to show that the magnitude of the field is proportional to $1/r$. [Hint: Consider a solid region consisting of a cylinder of finite length whose axis is the z-axis, and with a smaller concentric cylinder removed.]

66. A fluid is flowing along a cylindrical pipe of radius a in the $\vec{i}$ direction. The velocity of the fluid at a radial distance r from the center of the pipe is $\vec{v} = u(1-r^2/a^2)\vec{i}$.

(a) What is the significance of the constant u?

(b) What is the velocity of the fluid at the wall of the pipe?

(c) Find the flux through a circular cross-section of the pipe.

67. A closed surface S encloses a volume W. The function $\rho(x, y, z)$ gives the electrical charge density at points in space. The vector field $\vec{J}(x, y, z)$ gives the electric current density at any point in space and is defined so that the current through a small area $d\vec{A}$ is given by

Current through small area $\approx \vec{J} \cdot d\vec{A}$.

(a) What do the following integrals represent, in terms of electricity?

(i) $\displaystyle\int_W \rho\, dV$ (ii) $\displaystyle\int_S \vec{J} \cdot d\vec{A}$

(b) Using the fact that an electric current through a surface is the rate at which electric charge passes through the surface per unit time, explain why

$$\int_S \vec{J} \cdot d\vec{A} = -\frac{\partial}{\partial t}\left(\int_W \rho\, dV\right).$$

68. (a) A river flows across the xy-plane in the positive x-direction and around a circular rock of radius 1 centered at the origin. The velocity of the river can be modeled using the potential function $\phi = x + (x/(x^2 + y^2))$. Compute the velocity vector field, $\vec{v} = \text{grad}\,\phi$.

(b) Show that $\text{div}\,\vec{v} = 0$.

(c) Show that the flow of $\vec{v}$ is tangent to the circle $x^2 + y^2 = 1$. This means that no water crosses the circle. The water on the outside must therefore all flow around the circle.

(d) Use a computer to sketch the vector field $\vec{v}$ in the region outside the unit circle.

69. A vector field is a *point source* at the origin in 3-space if its direction is away from the origin at every point, its magnitude depends only on the distance from the origin, and its divergence is zero except at the origin. (Such a vector field might be used to model the photon flow out of a star or the neutrino flow out of a supernova.)

(a) Show that $\vec{v} = K(x^2+y^2+z^2)^{-3/2}(x\vec{i}+y\vec{j}+z\vec{k})$ is a point source at the origin if $K > 0$.

(b) Determine the magnitude $\|\vec{v}\|$ of the source in part (a) as a function of the distance from its center.

(c) Compute the flux of $\vec{v}$ through a sphere of radius r centered at the origin.

(d) Compute the flux of $\vec{v}$ through a closed surface that does not contain the origin.

CAS Challenge Problems

70. Let S be the part of the ellipsoid $x^2 + y^2 + 2z^2 = 1$ lying above the rectangle $-1/2 \le x \le 1/2$, $-1/2 \le y \le 1/2$, oriented upward. For each vector field (a)–(c), say whether you expect $\int_S \vec{F} \cdot d\vec{A}$ to be positive, negative, or zero. Then evaluate the integral exactly using a computer algebra system and find numerical approximations for your answers. Describe and explain what you notice.

(a) $\vec{F} = x\vec{i}$

(b) $\vec{F} = (x+1)\vec{i}$

(c) $\vec{F} = y\vec{j}$

71. Let $\vec{F} = (z+4)\vec{k}$, and let S be the surface with normal pointing in the direction of the negative y-axis parameterized, for $0 \le s \le 2, 0 \le t \le 2\pi$, by

$$\vec{r}(s, t) = s^2 \cos t\,\vec{i} + s\vec{j} + s^2 \sin t\,\vec{k}.$$

(a) Sketch S and, without evaluating the integral, say whether $\int_S \vec{F} \cdot d\vec{A}$ is positive, negative, or zero. Give a geometric explanation for your answer.

(b) Evaluate the flux integral. Does it agree with your answer to part (a)?

PROJECTS FOR CHAPTER NINETEEN

1. **Solid Angle**

   Let

   $$\vec{F}(x, y) = \frac{x\vec{i} + y\vec{j} + z\vec{k}}{(x^2 + y^2 + z^2)^{3/2}}$$

   and let oriented surfaces S_1 and S_2 be as in Figure 19.43. The surface S_2 is on the unit sphere centered at the origin. Lines from the origin to the boundary of S_1 intersect the unit sphere at the boundary of S_2. Both surfaces are oriented away from the origin. Show that

   (a) The divergence of $\vec{F}$ is zero.

   (b) $\int_{S_1} \vec{F} \cdot d\vec{A} = \int_{S_2} \vec{F} \cdot d\vec{A}$

   (c) $\int_{S_1} \vec{F} \cdot d\vec{r} = \Omega$, the area of S_2. The value Ω is called the *solid angle* subtended by S_1 at the origin.

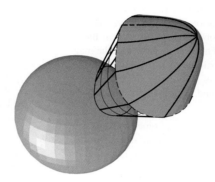

Figure 19.43

2. **Divergence of Spherically Symmetric Vector Fields**

   A vector field is spherically symmetric about the origin if, on every sphere centered at the origin, it has constant magnitude and points either away from or toward the origin. A vector field that is spherically symmetric about the origin can be written in terms of the spherical coordinate $\rho = \|\vec{r}\|$ as $\vec{F} = f(\rho)\vec{e}_\rho$ where f is a function of the distance ρ from the origin, $f(0) = 0$, and $\vec{e}_\rho$ is a unit vector pointing away from the origin.

   (a) Show that

   $$\text{div}\,\vec{F} = \frac{1}{\rho^2}\frac{d}{d\rho}\left(\rho^2 f(\rho)\right), \quad \rho \neq 0.$$

   (b) Use part (a) to show that if $\vec{F}$ is a spherically symmetric vector field such that $\text{div}\,\vec{F} = 0$ away from the origin then, for some constant k,

   $$\vec{F} = k\frac{1}{\rho^2}\vec{e}_\rho, \quad \rho \neq 0.$$

   (c) Use part (a) to confirm the Divergence Theorem for the flux of a spherically symmetric vector field through a sphere centered at the origin.

   (d) A form of Gauss's Law for an electric field $\vec{E}$ states that $\text{div}\,\vec{E}(\vec{r}) = \delta(\vec{r})$, where δ is a scalar-valued function giving the density of electric charge at every point. Use Gauss's Law and part (a) to find the electric field when the charge density is

   $$\delta(\vec{r}) = \begin{cases} \delta_0 & \|\vec{r}\| \leq a \\ 0 & \|\vec{r}\| > a, \end{cases}$$

   for some nonnegative a. Assume that $\vec{E}$ is spherically symmetric and continuous everywhere.

3. **Gauss's Law Applied to a Charged Wire and a Charged Sheet**
   Gauss's Law states that the flux of an electric field through a closed surface, S, is proportional to the quantity of charge, q, enclosed within S. That is,

   $$\int_S \vec{E} \cdot d\vec{A} = kq.$$

   In this project we use Gauss's Law to calculate the electric field of a uniformly charged wire in part (a) and a flat sheet of charge in part (b).

   (a) Consider the electric field due to an infinitely long, straight, uniformly charged wire. (There is no current running through the wire—all charges are fixed.) Assuming that the wire is infinitely long means that we can assume that the electric field is perpendicular to any cylinder that has the wire as an axis and that the magnitude of the field is constant on any such cylinder. Denote by E_r the magnitude of the electric field due to the wire on a cylinder of radius r. (See Figure 19.44.)

   Imagine a closed surface S made up of two cylinders, one of radius a and one of larger radius b, both coaxial with the wire, and the two washers that cap the ends. (See Figure 19.45.) The outward orientation of S means that a normal on the outer cylinder points away from the wire and a normal on the inner cylinder points toward the wire.

   (i) Explain why the flux of $\vec{E}$, the electric field, through the washers is 0.

   (ii) Explain why Gauss's Law implies that the flux through the inner cylinder is the same as the flux through the outer cylinder. [Hint: The charge on the wire is not inside the surface S].

   (iii) Use part (ii) to show that $E_b/E_a = a/b$.

   (iv) Explain why part (iii) shows that the strength of the field due to an infinitely long uniformly charged wire is proportional to $1/r$.

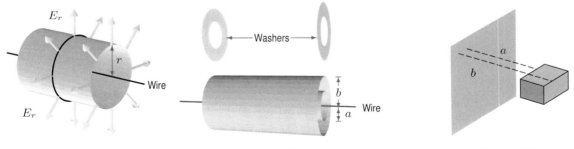

| Figure 19.44 | Figure 19.45 | Figure 19.46 |

   (b) Now consider an infinite flat sheet uniformly covered with charge. As in part (a), symmetry shows that the electric field $\vec{E}$ is perpendicular to the sheet and has the same magnitude at all points that are the same distance from the sheet. Use Gauss's Law to explain why, on any one side of the sheet, the electric field is the same at all points in space off the sheet. [Hint: Consider the flux through the box with sides parallel to the sheet shown in Figure 19.46.]

4. **Flux Across a Cylinder: Obtaining Gauss's Law from Coulomb's Law**
   An electric charge q is placed at the origin in 3-space. The induced electric field $\vec{E}(\vec{r})$ at the point with position vector $\vec{r}$ is given by Coulomb's Law, which says

   $$\vec{E}(\vec{r}) = q\frac{\vec{r}}{\|\vec{r}\|^3}, \qquad \vec{r} \neq \vec{0}.$$

   In this project, Gauss's Law is obtained for a cylinder enclosing a point charge by direct calculation from Coulomb's Law.

(a) Let S be the open cylinder of height $2H$ and radius R given by $x^2 + y^2 = R^2$, $-H \leq z \leq H$, oriented outward.

   (i) Show that the flux of $\vec{E}$, the electric field, through S is given by

$$\int_S \vec{E} \cdot d\vec{A} = 4\pi q \frac{H}{\sqrt{H^2 + R^2}}.$$

   (ii) What are the limits of the flux $\int_S \vec{E} \cdot d\vec{A}$ if
   - $H \to 0$ or $H \to \infty$ when R is fixed?
   - $R \to 0$ or $R \to \infty$ when H is fixed?

(b) Let T be the outward-oriented, closed cylinder of height $2H$ and radius R whose curved side is given by $x^2 + y^2 = R^2$, $-H \leq z \leq H$, whose top is given by $z = H$, $x^2 + y^2 \leq R^2$, and bottom by $z = -H$, $x^2 + y^2 \leq R^2$. Use part (a) to show that the flux of the electric field, $\vec{E}$, through T is given by

$$\int_T \vec{E} \cdot d\vec{A} = 4\pi q.$$

Notice that this is Gauss's Law. In particular, the flux is independent of both the height, H, and radius, R, of the cylinder.

Chapter Twenty

THE CURL AND STOKES' THEOREM

Contents

20.1 THE CURL OF A VECTOR FIELD

The divergence is a scalar derivative which measures the outflow of a vector field per unit volume. Now we introduce a vector derivative, the curl, which measures the circulation of a vector field.

Imagine holding the paddle-wheel in Figure 20.1 in the flow shown by Figure 20.2. The speed at which the paddle-wheel spins measures the strength of circulation. Notice that the angular velocity depends on the direction in which the stick is pointing. If the stick is pointing horizontally the paddle-wheel does not spin; if the stick is vertical, the paddle wheel spins.

Figure 20.1: A device for measuring circulation

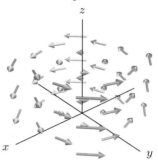

Figure 20.2: A vector field with circulation about the z-axis

Circulation Density

We measure the strength of the circulation using a closed curve. Suppose C is a circle with centre $P = (x, y, z)$ in the plane perpendicular to $\vec{n}$, traversed in the direction determined from $\vec{n}$ by the right-hand rule. (See Figures 20.3 and 20.4.)

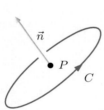

Figure 20.3: Direction of C relates to direction of $\vec{n}$ by the right-hand rule

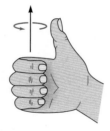

Figure 20.4: When the thumb points in the direction of $\vec{n}$, the fingers curl in the forward direction around C

We make the following definition:

The **circulation density** of a smooth vector field $\vec{F}$ at (x, y, z) around the direction of the unit vector $\vec{n}$ is defined, provided the limit exists, to be

$$\text{circ}_{\vec{n}}\, \vec{F}\,(x, y, z) = \lim_{\text{Area} \to 0} \frac{\text{Circulation around } C}{\text{Area inside } C} = \lim_{\text{Area} \to 0} \frac{\int_C \vec{F} \cdot d\vec{r}}{\text{Area inside } C},$$

The circle C is in the plane perpendicular to $\vec{n}$ and oriented by the right-hand rule.

We can use other closed curves for C, such as rectangles, that lie in a plane perpendicular to $\vec{n}$ and include (x, y, z).

The circulation density determines the angular velocity of the paddle-wheel in Figure 20.1 provided you could make one sufficiently small and light and insert it without disturbing the flow.

Example 1   Consider the vector field $\vec{F}$ in Figure 20.2. Suppose that $\vec{F}$ is parallel to the xy-plane and that at a distance r from the z-axis it has magnitude $2r$. Calculate $\mathrm{circ}_{\vec{n}}\ \vec{F}$ at the origin for
(a) $\vec{n} = \vec{k}$ (b) $\vec{n} = -\vec{k}$ (c) $\vec{n} = \vec{i}$.

Solution   (a) Take a circle C of radius a in the xy-plane, centred at the origin, traversed in a direction determined from $\vec{k}$ by the right hand rule. Then, since $\vec{F}$ is tangent to C everywhere and points in the forward direction around C, we have

$$\text{Circulation around } C = \int_C \vec{F} \cdot d\vec{r} = \|\vec{F}\| \cdot \text{Circumference of } C = 2a(2\pi a) = 4\pi a^2.$$

Thus, the circulation density is

$$\mathrm{circ}_{\vec{k}}\ \vec{F} = \lim_{a \to 0} \frac{\text{Circulation around } C}{\text{Area inside } C} = \lim_{a \to 0} \frac{4\pi a^2}{\pi a^2} = 4.$$

(b) If $\vec{n} = -\vec{k}$ the circle is traversed in the opposite direction, so the line integral changes sign. Thus,

$$\mathrm{circ}_{-\vec{k}}\ \vec{F} = -4.$$

(c) The circulation around $\vec{i}$ is calculated using circles in the yz-plane. Since $\vec{F}$ is everywhere perpendicular to such a circle C,

$$\int_C \vec{F} \cdot d\vec{r} = 0.$$

Thus, we have

$$\mathrm{circ}_{\vec{i}}\ \vec{F} = \lim_{a \to 0} \frac{\int_C \vec{F} \cdot d\vec{r}}{\pi a^2} = \lim_{a \to 0} \frac{0}{\pi a^2} = 0.$$

Definition of the Curl

Example 1 shows that the circulation density of a vector field can be positive, negative, or zero, depending on the direction. We assume that there is one direction in which the circulation density is greatest and define a single vector quantity that incorporates all these different circulation densities. We give two definitions, one geometric and one algebraic, which turn out to lead to the same result.

Geometric Definition of Curl

The curl of a smooth vector field $\vec{F}$, written $\mathrm{curl}\,\vec{F}$, is the vector field with the following properties
- The direction of $\mathrm{curl}\,\vec{F}(x, y, z)$ is the direction $\vec{n}$ for which $\mathrm{circ}_{\vec{n}}\ \vec{F}(x, y, z)$ is the greatest.
- The magnitude of $\mathrm{curl}\,\vec{F}(x, y, z)$ is the circulation density of $\vec{F}$ around that direction.

If the circulation density is zero around every direction, then we define the curl to be $\vec{0}$.

Cartesian Coordinate Definition of Curl

If $\vec{F} = F_1\vec{i} + F_2\vec{j} + F_3\vec{k}$, then

$$\mathrm{curl}\,\vec{F} = \left(\frac{\partial F_3}{\partial y} - \frac{\partial F_2}{\partial z}\right)\vec{i} + \left(\frac{\partial F_1}{\partial z} - \frac{\partial F_3}{\partial x}\right)\vec{j} + \left(\frac{\partial F_2}{\partial x} - \frac{\partial F_1}{\partial y}\right)\vec{k}.$$

The cross-product formula gives an easy way to remember the Cartesian coordinate definition, and suggests another common notation for curl $\vec{F}$, namely $\nabla \times \vec{F}$. Using $\nabla = \frac{\partial}{\partial x}\vec{i} + \frac{\partial}{\partial y}\vec{j} + \frac{\partial}{\partial z}\vec{k}$, we can write

$$\text{curl}\,\vec{F} = \nabla \times \vec{F} = \begin{vmatrix} \vec{i} & \vec{j} & \vec{k} \\ \frac{\partial}{\partial x} & \frac{\partial}{\partial y} & \frac{\partial}{\partial z} \\ F_1 & F_2 & F_3 \end{vmatrix}.$$

Example 2   For each field in Figure 20.5, use the sketch and the geometric definition to decide whether the curl at the origin appears to point up, down, or to be the zero vector. Then check your answer using the coordinate definition of curl. Note that the vector fields have no z-components and are independent of z.

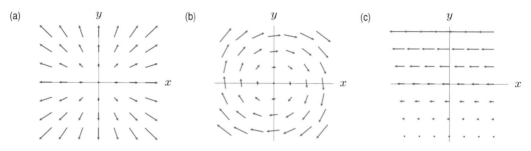

Figure 20.5: Sketches in the xy-plane of (a) $\vec{F} = x\vec{i} + y\vec{j}$ (b) $\vec{F} = y\vec{i} - x\vec{j}$ (c) $\vec{F} = -(y+1)\vec{i}$

Solution

(a) This vector field shows no rotation, and the circulation around any circle in the xy-plane centered at the origin appears to be zero, so we suspect that the circulation density around $\vec{k}$ is zero. The coordinate definition of curl gives

$$\text{curl}\,\vec{F} = \left(\frac{\partial(0)}{\partial y} - \frac{\partial y}{\partial z}\right)\vec{i} + \left(\frac{\partial x}{\partial z} - \frac{\partial(0)}{\partial x}\right)\vec{j} + \left(\frac{\partial y}{\partial x} - \frac{\partial x}{\partial y}\right)\vec{k} = \vec{0}.$$

(b) This vector field appears to be rotating around the z-axis. By the right-hand rule, the circulation density around $\vec{k}$ is negative, so we expect the z-component of the curl points down. The coordinate definition gives

$$\text{curl}\,\vec{F} = \left(\frac{\partial(0)}{\partial y} - \frac{\partial(-x)}{\partial z}\right)\vec{i} + \left(\frac{\partial y}{\partial z} - \frac{\partial(0)}{\partial x}\right)\vec{j} + \left(\frac{\partial(-x)}{\partial x} - \frac{\partial y}{\partial y}\right)\vec{k} = -2\vec{k}.$$

(c) At first glance, you might expect this vector field to have zero curl, as all the vectors are parallel to the x-axis. However, if you find the circulation around the curve C in Figure 20.6, the sides contribute nothing (they are perpendicular to the vector field), the bottom contributes a negative quantity (the curve is in the opposite direction to the vector field), and the top contributes a larger positive quantity (the curve is in the same direction as the vector field and the magnitude of the vector field is larger at the top than at the bottom). Thus, the circulation around C is positive and hence we expect the curl to be nonzero and point up. The coordinate definition gives

$$\text{curl}\,\vec{F} = \left(\frac{\partial(0)}{\partial y} - \frac{\partial(0)}{\partial z}\right)\vec{i} + \left(\frac{\partial(-(y+1))}{\partial z} - \frac{\partial(0)}{\partial x}\right)\vec{j} + \left(\frac{\partial(0)}{\partial x} - \frac{\partial(-(y+1))}{\partial y}\right)\vec{k} = \vec{k}.$$

Another way to see that the curl is nonzero in this case is to imagine the vector field representing the velocity of moving water. A boat sitting in the water tends to rotate, as the water moves faster on one side than the other.

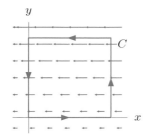

Figure 20.6: Rectangular curve in xy-plane

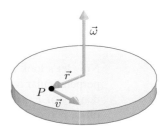

Figure 20.7: Rotating flywheel

Example 3 A flywheel is rotating with angular velocity $\vec{\omega}$ and the velocity of a point P with position vector $\vec{r}$ is given by $\vec{v} = \vec{\omega} \times \vec{r}$. (See Figure 20.7.) Calculate $\operatorname{curl} \vec{v}$.

Solution If $\vec{\omega} = \omega_1 \vec{i} + \omega_2 \vec{j} + \omega_3 \vec{k}$, using the determinant notation introduced in Section 13.4, we have

$$\vec{v} = \vec{\omega} \times \vec{r} = \begin{vmatrix} \vec{i} & \vec{j} & \vec{k} \\ \omega_1 & \omega_2 & \omega_3 \\ x & y & z \end{vmatrix} = (\omega_2 z - \omega_3 y)\vec{i} + (\omega_3 x - \omega_1 z)\vec{j} + (\omega_1 y - \omega_2 x)\vec{k}.$$

The curl formula can also be written using a determinant:

$$\operatorname{curl} \vec{v} = \begin{vmatrix} \vec{i} & \vec{j} & \vec{k} \\ \frac{\partial}{\partial x} & \frac{\partial}{\partial y} & \frac{\partial}{\partial z} \\ \omega_2 z - \omega_3 y & \omega_3 x - \omega_1 z & \omega_1 y - \omega_2 x \end{vmatrix}$$

$$= \left(\frac{\partial}{\partial y}(\omega_1 y - \omega_2 x) - \frac{\partial}{\partial z}(\omega_3 x - \omega_1 z) \right)\vec{i} + \left(\frac{\partial}{\partial z}(\omega_2 z - \omega_3 y) - \frac{\partial}{\partial x}(\omega_1 y - \omega_2 x) \right)\vec{j}$$

$$+ \left(\frac{\partial}{\partial x}(\omega_3 x - \omega_1 z) - \frac{\partial}{\partial y}(\omega_2 z - \omega_3 y) \right)\vec{k}$$

$$= 2\omega_1 \vec{i} + 2\omega_2 \vec{j} + 2\omega_3 \vec{k} = 2\vec{\omega}.$$

Thus, as we would expect, $\operatorname{curl} \vec{v}$ is parallel to the axis of rotation of the flywheel (namely, the direction of $\vec{\omega}$) and the magnitude of $\operatorname{curl} \vec{v}$ is larger the faster the flywheel is rotating (that is, the larger the magnitude of $\vec{\omega}$).

Why Do the Two Definitions of Curl Give the Same Result?

Using Green's Theorem in Cartesian coordinates, we can show that for $\operatorname{curl} \vec{F}$ defined in Cartesian coordinates

$$\boxed{\operatorname{curl} \vec{F} \cdot \vec{n} = \operatorname{circ}_{\vec{n}} \vec{F}.}$$

This shows that $\operatorname{curl} \vec{F}$ defined in Cartesian coordinates satisfies the geometric definition, since the left-hand side takes its maximum value when $\vec{n}$ points in the same direction as $\operatorname{curl} \vec{F}$, and in that case its value is $\| \operatorname{curl} \vec{F} \|$.

The following example justifies this formula in a specific case.

Example 4 Use the definition of curl in Cartesian coordinates and Green's Theorem to show that

$$\left(\operatorname{curl} \vec{F} \right) \cdot \vec{k} = \operatorname{circ}_{\vec{k}} \vec{F}.$$

Solution
Using the definition of curl in Cartesian coordinates, the left-hand side of the formula is

$$\left(\text{curl}\,\vec{F}\right)\cdot\vec{k} = \frac{\partial F_2}{\partial x} - \frac{\partial F_1}{\partial y}.$$

Now let's look at the right hand side. The circulation density around $\vec{k}$ is calculated using circles perpendicular to $\vec{k}$; hence, the $\vec{k}$-component of $\vec{F}$ does not contribute to it; that is, the circulation density of $\vec{F}$ around $\vec{k}$ is the same as the circulation density of $F_1\vec{i} + F_2\vec{j}$ around $\vec{k}$. But in any plane perpendicular to $\vec{k}$, z is constant, so in that plane F_1 and F_2 are functions of x and y alone. Thus, $F_1\vec{i} + F_2\vec{j}$ can be thought of as a two-dimensional vector field on the horizontal plane through the point (x, y, z) where the circulation density is being calculated. Let C be a circle in this plane, with radius a and centred at (x, y, z), and let R be the region enclosed by C. Green's Theorem says that

$$\int_C (F_1\vec{i} + F_2\vec{j})\cdot d\vec{r} = \int_R \left(\frac{\partial F_2}{\partial x} - \frac{\partial F_1}{\partial y}\right) dA.$$

When the circle is small, $\partial F_2/\partial x - \partial F_1/\partial y$ is approximately constant on R, so

$$\int_R \left(\frac{\partial F_2}{\partial x} - \frac{\partial F_1}{\partial y}\right) dA \approx \left(\frac{\partial F_2}{\partial x} - \frac{\partial F_1}{\partial y}\right)\cdot \text{Area of } R = \left(\frac{\partial F_2}{\partial x} - \frac{\partial F_1}{\partial y}\right)\pi a^2.$$

Thus, taking a limit as the radius of the circle goes to zero, we have

$$\text{circ}_{\vec{k}}\,\vec{F}(x, y, z) = \lim_{a\to 0}\frac{\displaystyle\int_C (F_1\vec{i} + F_2\vec{j})\cdot d\vec{r}}{\pi a^2} = \lim_{a\to 0}\frac{\displaystyle\int_R \left(\frac{\partial F_2}{\partial x} - \frac{\partial F_1}{\partial y}\right) dA}{\pi a^2} = \frac{\partial F_2}{\partial x} - \frac{\partial F_1}{\partial y}.$$

Curl-Free Vector Fields

A vector field is said to be *curl free* or *irrotational* if $\text{curl}\,\vec{F} = \vec{0}$ everywhere that $\vec{F}$ is defined.

Example 5
Figure 20.8 shows the vector field $\vec{B}$ for three values of the constant p, where $\vec{B}$ is defined on 3-space by

$$\vec{B} = \frac{-y\vec{i} + x\vec{j}}{(x^2 + y^2)^{p/2}}.$$

(a) Find a formula for $\text{curl}\,\vec{B}$.
(b) Is there a value of p for which $\vec{B}$ is curl free? If so, find it.

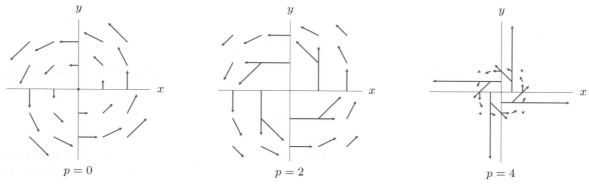

$p = 0$ $p = 2$ $p = 4$

Figure 20.8: The vector field $\vec{B}\,(\vec{r}) = (-y\vec{i} + x\vec{j})/(x^2 + y^2)^{p/2}$ for $p = 0, 2$, and 4

Solution
(a) We can use the following version of the product rule for curl. If ϕ is a scalar function and $\vec{F}$ is a vector field, then

$$\text{curl}(\phi\vec{F}) = \phi\,\text{curl}\,\vec{F} + (\text{grad}\,\phi)\times\vec{F}.$$

(See Problem 28 on page 1076.) We write $\vec{B} = \phi\vec{F} = \dfrac{1}{(x^2 + y^2)^{p/2}}(-y\vec{i} + x\vec{j})$. Then

$$\text{curl } \vec{F} = \text{curl}(-y\vec{i} + x\vec{j}) = 2\vec{k}$$

$$\text{grad } \phi = \text{grad}\left(\frac{1}{(x^2 + y^2)^{p/2}}\right) = \frac{-p}{(x^2 + y^2)^{(p/2)+1}}(x\vec{i} + y\vec{j}).$$

Thus, we have

$$
\begin{aligned}
\text{curl } \vec{B} &= \frac{1}{(x^2 + y^2)^{p/2}} \text{curl}(-y\vec{i} + x\vec{j}) + \text{grad}\left(\frac{1}{(x^2 + y^2)^{p/2}}\right) \times (-y\vec{i} + x\vec{j}) \\
&= \frac{1}{(x^2 + y^2)^{p/2}} 2\vec{k} + \frac{-p}{(x^2 + y^2)^{(p/2)+1}}(x\vec{i} + y\vec{j}) \times (-y\vec{i} + x\vec{j}) \\
&= \frac{1}{(x^2 + y^2)^{p/2}} 2\vec{k} + \frac{-p}{(x^2 + y^2)^{(p/2)+1}}(x^2 + y^2)\vec{k} \\
&= \frac{2 - p}{(x^2 + y^2)^{p/2}} \vec{k}.
\end{aligned}
$$

(b) The curl is zero when $p = 2$. Thus, when $p = 2$ the vector field is curl free:

$$\vec{B} = \frac{-y\vec{i} + x\vec{j}}{x^2 + y^2}.$$

Exercises and Problems for Section 20.1

Exercises

1. Is $\text{curl}(z\vec{i} - x\vec{j} + y\vec{k})$ a vector or a scalar? Calculate it.

In Exercises 2–9, compute the curl of the vector field.

2. $\vec{F} = 3x\vec{i} - 5z\vec{j} + y\vec{k}$

3. $\vec{F} = (x^2 - y^2)\vec{i} + 2xy\vec{j}$

4. $\vec{F} = (-x + y)\vec{i} + (y + z)\vec{j} + (-z + x)\vec{k}$

5. $\vec{F} = 2yz\vec{i} + 3xz\vec{j} + 7xy\vec{k}$

6. $\vec{F} = x^2\vec{i} + y^3\vec{j} + z^4\vec{k}$

7. $\vec{F} = e^x\vec{i} + \cos y\vec{j} + e^{z^2}\vec{k}$

8. $\vec{F} = (x + yz)\vec{i} + (y^2 + xzy)\vec{j} + (zx^3y^2 + x^7y^6)\vec{k}$

9. $\vec{F}(\vec{r}) = \vec{r}/\|\vec{r}\|$

In Exercises 10–13, decide whether the vector field appears to have nonzero curl at the origin. The vector field is shown in the xy-plane; it has no z-component and is independent of z.

10.

11.

12.

13.

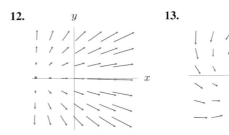

Problems

14. Let $\vec{F}$ be the vector field in Figure 20.2 on page 1070. It is rotating counterclockwise around the z-axis when viewed from above. At a distance r from the z-axis, $\vec{F}$ has magnitude $2r$.

(a) Find a formula for $\vec{F}$.

(b) Find $\text{curl } \vec{F}$ using the coordinate definition and relate your answer to circulation density.

15. Use the geometric definition to find the curl of the vector field $\vec{F}(\vec{r}) = \vec{r}$. Check your answer using the coordinate definition.

16. A smooth vector field $\vec{G}$ has curl $\vec{G}(0,0,0) = 2\vec{i} - 3\vec{j} + 5\vec{k}$. Estimate the circulation around a circle of radius 0.01 centered at the origin in each of the following planes:

(a) xy-plane, oriented counterclockwise when viewed from the positive z-axis.

(b) yz-plane, oriented counterclockwise when viewed from the positive x-axis.

(c) xz-plane, oriented counterclockwise when viewed from the positive y-axis.

17. Three small circles, C_1, C_2, and C_3, each with radius 0.1 and centered at the origin are in the xy-, yz-, and xz-planes, respectively. The circles are oriented counterclockwise when viewed from the positive z-, x-, and y-axes, respectively. A vector field, $\vec{F}$, has circulation around C_1 of 0.02π, around C_2 of 0.5π, and around C_3 of 3π. Estimate curl $\vec{F}$ at the origin.

18. Using your answers to Exercises 6–7, make a conjecture about a particular form of the vector field $\vec{F} \neq \vec{0}$ that has curl $\vec{F} = \vec{0}$. What form? Show why your conjecture is true.

19. **(a)** Find curl $\vec{G}$ if $\vec{G} = (ay^3 + be^z)\vec{i} + (cz + dx^2)\vec{j} + (e\sin x + fy)\vec{k}$ and a, b, c, d, e, f are constants.

(b) If curl $\vec{G}$ is everywhere parallel to the yz-plane, what can you say about the constants a–f?

(c) If curl $\vec{G}$ is everywhere parallel to the z-axis, what can you say about the constants a–f?

20. Figure 20.9 gives a sketch of the velocity vector field $\vec{F} = y\vec{i} + x\vec{j}$ in the xy-plane.

(a) What is the direction of rotation of a thin twig placed at the origin along the x-axis?

(b) What is the direction of rotation of a thin twig placed at the origin along the y-axis?

(c) Compute curl $\vec{F}$.

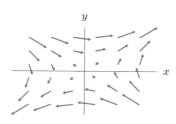

Figure 20.9

21. A tornado is formed when a tube of air circling a horizontal axis is tilted up vertically by the updraft from a thunderstorm. If t is time, this process can be modeled by the wind velocity field

$$\vec{F}(t, x, y, z) = (\cos t\vec{j} + \sin t\vec{k}) \times \vec{r} \quad \text{and} \quad 0 \leq t \leq \frac{\pi}{2}.$$

Determine the direction of curl $\vec{F}$:

(a) At $t = 0$ **(b)** At $t = \pi/2$

(c) For $0 < t < \pi/2$

22. A large fire becomes a fire-storm when the nearby air acquires a circulatory motion. The associated updraft has the effect of bringing more air to the fire, causing it to burn faster. Records show that a fire-storm developed during the Chicago Fire of 1871 and during the Second World War bombing of Hamburg, Germany, but there was no fire-storm during the Great Fire of London in 1666. Explain how a fire-storm could be identified using the curl of a vector field.

23. A vortex that rotates at constant angular velocity ω about the z-axis has velocity vector field $\vec{v} = \omega(-y\vec{i} + x\vec{j})$.

(a) Sketch the vector field with $\omega = 1$ and the vector field with $\omega = -1$.

(b) Determine the speed $\|\vec{v}\|$ of the vortex as a function of the distance from its center.

(c) Compute div $\vec{v}$ and curl $\vec{v}$.

(d) Compute the circulation of $\vec{v}$ counterclockwise about the circle of radius R in the xy-plane, centered at the origin.

24. A central vector field is one of the form $\vec{F} = f(r)\vec{r}$ where f is any function of $r = \|\vec{r}\|$. Show that any central vector field is irrotational.

25. Show that curl $(\vec{F} + \vec{C}) = $ curl $\vec{F}$ for a constant vector field $\vec{C}$.

26. If $\vec{F}$ is any vector field whose components have continuous second partial derivatives, show div curl $\vec{F} = 0$.

27. We have seen that the Fundamental Theorem of Calculus for Line Integrals implies $\int_C \text{grad } f \cdot d\vec{r} = 0$ for any smooth closed path C and any smooth function f.

(a) Use the geometric definition of curl to deduce that curl grad $f = \vec{0}$.

(b) Show that curl grad $f = \vec{0}$ using the coordinate definition.

28. Show that curl $(\phi\vec{F}) = \phi$ curl $\vec{F} + (\text{grad } \phi) \times \vec{F}$ for a scalar function ϕ and a vector field $\vec{F}$.

29. Show that if $\vec{F} = f \text{ grad } g$ for some scalar functions f and g, then curl $\vec{F}$ is everywhere perpendicular to $\vec{F}$.

30. Let $\vec{F}$ be a smooth vector field and let $\vec{u}$ and $\vec{v}$ be constant vectors. Using the definition of curl $\vec{F}$ in Cartesian coordinates, show that

$$\text{grad}(\vec{F} \cdot \vec{v}) \cdot \vec{u} - \text{grad}(\vec{F} \cdot \vec{u}) \cdot \vec{v} = (\text{curl } \vec{F}) \cdot \vec{u} \times \vec{v}.$$

31. Let $\vec{T} = a\vec{i} + b\vec{j}$ be a fixed unit vector, and let $\vec{F} = F(x, y)\vec{T}$ be a vector field everywhere parallel to $\vec{T}$, but of varying magnitude F. Show that curl $\vec{F}$ equals $\vec{k}$ times the directional derivative of F in the direction of $\vec{F} \times \vec{k}$. Do this in two ways:

(a) Graphically, using line integrals

(b) Algebraically

(Note: The direction of $\vec{F} \times \vec{k}$ is obtained by rotating $\vec{F}$ through $90°$ clockwise as viewed from above the xy-plane.)

32. Let $r = (x^2 + y^2)^{1/2}$. Figure 20.10 shows the vector field $r^A(-y\vec{i} + x\vec{j})$ for $r \neq 0$ and $A = -1, -2$, and -3. The vector fields are shown in the xy-plane; they have no z-component and are independent of z.

(a) Show that $\text{curl}(r^A(-y\vec{i} + x\vec{j})) = (2+A)r^A\vec{k}$ for any constant A.

(b) Using your answer to part (a), find the direction of the curl of the three vector fields for $A = -1, -2, -3$.

(c) For each value of A, what (if anything) does your answer to part (b) tell you about the sign of the circulation around a small circle oriented counterclockwise when viewed from above, and centered at $(1, 1, 1)$? Centered at $(0, 0, 0)$?

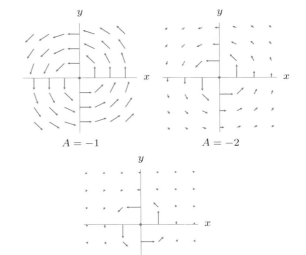

Figure 20.10: The vector field $r^A(-y\vec{i} + x\vec{j})$ for three values of A

Strengthen Your Understanding

In Problems 33–34, explain what is wrong with the statement.

33. A vector field $\vec{F}$ has curl given by curl $\vec{F} = 2x - 3y$.

34. If all the vectors of a vector field $\vec{F}$ are parallel, then curl $\vec{F} = \vec{0}$.

In Problems 35–36, give an example of:

35. A vector field $\vec{F}(x, y, z)$ such that curl $\vec{F} = \vec{0}$.

36. A vector field $\vec{F}(x, y, z)$ such that curl $\vec{F} = \vec{j}$.

In Problems 37–45, is the statement true or false? Assume $\vec{F}$ and $\vec{G}$ are smooth vector fields and f is a smooth function on 3-space. Explain.

37. The circulation density, $\text{circ}_{\vec{n}} \vec{F}(x, y, z)$, is a scalar.

38. curl grad $f = 0$

39. If $\vec{F}$ is a vector field with div$\vec{F} = 0$ and curl$\vec{F} = \vec{0}$, then $\vec{F} = \vec{0}$.

40. If $\vec{F}$ and $\vec{G}$ are vector fields, then curl$(\vec{F} + \vec{G}) = \text{curl}\vec{F} + \text{curl}\vec{G}$.

41. If $\vec{F}$ and $\vec{G}$ are vector fields, then curl$(\vec{F} \cdot \vec{G}) = \text{curl}\vec{F} \cdot \text{curl}\vec{G}$.

42. If $\vec{F}$ and $\vec{G}$ are vector fields, then curl$(\vec{F} \times \vec{G}) = (\text{curl}\vec{F}) \times (\text{curl}\vec{G})$.

43. curl$(f\vec{G}) = (\text{grad } f) \times \vec{G} + f(\text{curl }\vec{G})$

44. For any vector field $\vec{F}$, the curl of $\vec{F}$ is perpendicular at every point to $\vec{F}$.

45. If $\vec{F}$ is as shown in Figure 20.11, then curl $\vec{F} \cdot \vec{j} > 0$.

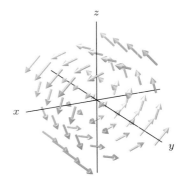

Figure 20.11

46. Of the following vector fields, which ones have a curl which is parallel to one of the axes? Which axis?

(a) $y\vec{i} - x\vec{j} + z\vec{k}$ **(b)** $y\vec{i} + z\vec{j} + x\vec{k}$ **(c)** $-z\vec{i} + y\vec{j} + x\vec{k}$
(d) $x\vec{i} + z\vec{j} - y\vec{k}$ **(e)** $z\vec{i} + x\vec{j} + y\vec{k}$

20.2 STOKES' THEOREM

The Divergence Theorem says that the integral of the flux density over a solid region is equal to the flux through the surface bounding the region. Similarly, Stokes' Theorem says that the integral of the circulation density over a surface is equal to the circulation around the boundary of the surface.

The Boundary of a Surface

The *boundary* of a surface S is the curve or curves running around the edge of S (like the hem around the edge of a piece of cloth). An orientation of S determines an orientation for its boundary, C, as follows. Pick a positive normal vector $\vec{n}$ on S, near C, and use the right-hand rule to determine a direction of travel around $\vec{n}$. This in turn determines a direction of travel around the boundary C. See Figure 20.12. Another way of describing the orientation on C is that someone walking along C in the forward direction, body upright in the direction of the positive normal on S, would have the surface on their left. Notice that the boundary can consist of two or more curves, as the surface on the right in Figure 20.12 shows.

Figure 20.12: Two oriented surfaces and their boundaries

Calculating the Circulation from the Circulation Density

Consider a closed, oriented curve C in 3-space. We can find the circulation of a vector field $\vec{F}$ around C by calculating the line integral:

$$\begin{array}{c} \text{Circulation} \\ \text{around } C \end{array} = \int_C \vec{F} \cdot d\vec{r}.$$

If C is the boundary of an oriented surface S, there is another way to calculate the circulation using $\operatorname{curl} \vec{F}$. We subdivide S into pieces as shown on the surface on the left in Figure 20.12. If $\vec{n}$ is a positive unit normal vector to a piece of surface with area ΔA, then $\Delta \vec{A} = \vec{n} \, \Delta A$. In addition, $\operatorname{circ}_{\vec{n}} \vec{F}$ is the circulation density of $\vec{F}$ around $\vec{n}$, so

$$\begin{array}{c} \text{Circulation of } \vec{F} \text{ around} \\ \text{boundary of the piece} \end{array} \approx \left(\operatorname{circ}_{\vec{n}} \vec{F} \right) \Delta A = ((\operatorname{curl} \vec{F}) \cdot \vec{n}) \Delta A = (\operatorname{curl} \vec{F}) \cdot \Delta \vec{A}.$$

Next we add up the circulations around all the small pieces. The line integral along the common edge of a pair of adjacent pieces appears with opposite sign in each piece, so it cancels out. (See Figure 20.13.) When we add up all the pieces the internal edges cancel and we are left with the circulation around C, the boundary of the entire surface. Thus,

$$\begin{array}{c} \text{Circulation} \\ \text{around } C \end{array} = \sum \begin{array}{c} \text{Circulation around} \\ \text{boundary of pieces} \end{array} \approx \sum \operatorname{curl} \vec{F} \cdot \Delta \vec{A}.$$

Taking the limit as $\Delta A \to 0$, we get

$$\begin{array}{c}\text{Circulation}\\\text{around } C\end{array} = \int_S \text{curl}\, \vec{F} \cdot d\vec{A}\,.$$

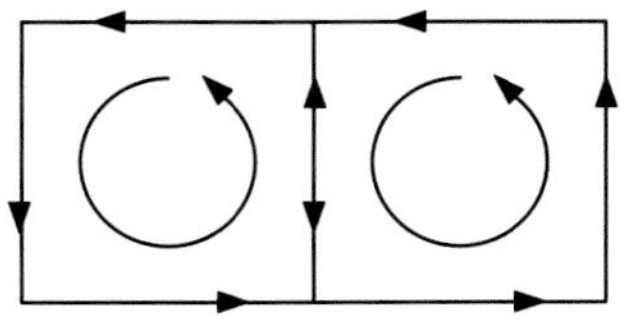

Figure 20.13: Two adjacent pieces of the surface

We have expressed the circulation as a line integral around C and as a flux integral over S; thus, the two integrals must be equal. Hence we have[1]

Theorem 20.1: Stokes' Theorem

If S is a smooth oriented surface with piecewise smooth, oriented boundary C, and if $\vec{F}$ is a smooth vector field on an open region containing S and C, then

$$\int_C \vec{F} \cdot d\vec{r} = \int_S \text{curl}\, \vec{F} \cdot d\vec{A}\,.$$

The orientation of C is determined from the orientation of S according to the right-hand rule.

Example 1   Let $\vec{F}(x, y, z) = -2y\vec{i} + 2x\vec{j}$. Use Stokes' Theorem to find $\int_C \vec{F} \cdot d\vec{r}$, where C is a circle

(a) Parallel to the yz-plane, of radius a, centred at a point on the x-axis, with either orientation.
(b) Parallel to the xy-plane, of radius a, centred at a point on the z-axis, oriented counterclockwise as viewed from a point on the z-axis above the circle.

Solution   We have $\text{curl}\, \vec{F} = 4\vec{k}$. Figure 20.14 shows sketches of $\vec{F}$ and $\text{curl}\, \vec{F}$.

(a) Let S be the disk enclosed by C. Since S lies in a vertical plane and $\text{curl}\, \vec{F}$ points vertically everywhere, the flux of $\text{curl}\, \vec{F}$ through S is zero. Hence, by Stokes' Theorem,

$$\int_C \vec{F} \cdot d\vec{r} = \int_S \text{curl}\, \vec{F} \cdot d\vec{A} = 0.$$

It makes sense that the line integral is zero. If C is parallel to the yz-plane (even if it is not lying in the plane), the symmetry of the vector field means that the line integral of $\vec{F}$ over the top half of the circle cancels the line integral over the bottom half.

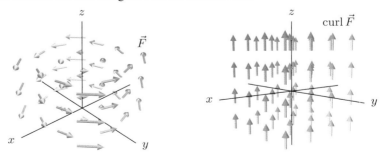

Figure 20.14: The vector fields $\vec{F}$ and $\text{curl}\, \vec{F}$

---

[1] A proof of Stokes' Theorem using the coordinate definition of curl can be found in the online supplement at www.wiley.com/college/hughes-hallett.

(b) Let S be the horizontal disk enclosed by C. Since curl $\vec{F}$ is a constant vector field pointing in the direction of $\vec{k}$, we have, by Stokes' Theorem,

$$\int_C \vec{F} \cdot d\vec{r} = \int_S \text{curl}\,\vec{F} \cdot d\vec{A} = \|\,\text{curl}\,\vec{F}\,\| \cdot \text{Area of } S = 4\pi a^2.$$

Since $\vec{F}$ is circling around the z-axis in the same direction as C, we expect the line integral to be positive. In fact, in Example 1 on page 1071, we computed this line integral directly.

Curl Free Vector Fields

Stokes' Theorem applies to any oriented surface S and its boundary C, even in cases where the boundary consists of two or more curves. This is useful in studying curl-free vector fields.

Example 2   A current I flows along the z-axis in the $\vec{k}$ direction. The induced magnetic field $\vec{B}(x, y, z)$ is

$$\vec{B}(x, y, z) = \frac{2I}{c}\left(\frac{-y\vec{i} + x\vec{j}}{x^2 + y^2}\right),$$

where c is the speed of light. In Example 5 on page 1074 we showed that curl $\vec{B} = \vec{0}$.

(a) Compute the circulation of $\vec{B}$ around the circle C_1 in the xy-plane of radius a, centred at the origin, and oriented counterclockwise when viewed from above.

(b) Use part (a) and Stokes' Theorem to compute $\int_{C_2} \vec{B} \cdot d\vec{r}$, where C_2 is the ellipse $x^2 + 9y^2 = 9$ in the plane $z = 2$, oriented counterclockwise when viewed from above.

Solution   (a) On the circle C_1, we have $\|\vec{B}\| = 2I/(ca)$. Since $\vec{B}$ is tangent to C_1 everywhere and points in the forward direction around C_1,

$$\int_{C_1} \vec{B} \cdot d\vec{r} = \|\vec{B}\| \cdot \text{Length of } C_1 = \frac{2I}{ca} \cdot 2\pi a = \frac{4\pi I}{c}.$$

(b) Let S be the conical surface extending from C_1 to C_2 in Figure 20.15. The boundary of this surface has two pieces, $-C_2$ and C_1. The orientation of C_1 leads to the outward normal on S, which forces us to choose the clockwise orientation on C_2. By Stokes' Theorem,

$$\int_S \text{curl}\,\vec{B} \cdot d\vec{A} = \int_{-C_2} \vec{B} \cdot d\vec{r} + \int_{C_1} \vec{B} \cdot d\vec{r} = -\int_{C_2} \vec{B} \cdot d\vec{r} + \int_{C_1} \vec{B} \cdot d\vec{r}.$$

Since curl $\vec{B} = \vec{0}$, we have $\int_S \text{curl}\,\vec{B} \cdot d\vec{A} = 0$, so the two line integrals must be equal:

$$\int_{C_2} \vec{B} \cdot d\vec{r} = \int_{C_1} \vec{B} \cdot d\vec{r} = \frac{4\pi I}{c}.$$

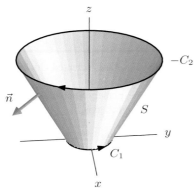

Figure 20.15: Surface joining C_1 to C_2,
oriented to satisfy the conditions of
Stokes' Theorem

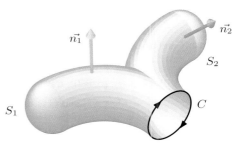

Figure 20.16: The flux of a curl is the same
through the two surfaces S_1 and S_2 if they
determine the same orientation on the boundary,
C

Curl Fields

A vector field $\vec{F}$ is called a *curl field* if $\vec{F} = \operatorname{curl} \vec{G}$ for some vector field $\vec{G}$. Recall that if $\vec{F} = \operatorname{grad} f$, then f is called a potential function. By analogy, if a vector field $\vec{F} = \operatorname{curl} \vec{G}$, then $\vec{G}$ is called a *vector potential* for $\vec{F}$. The following example shows that the flux of a curl field through a surface depends only on the boundary of the surface. This is analogous to the fact that the line integral of a gradient field depends only on the endpoints of the path.

Example 3   Suppose $\vec{F} = \operatorname{curl} \vec{G}$. Suppose that S_1 and S_2 are two oriented surfaces with the same boundary C. Show that, if S_1 and S_2 determine the same orientation on C (as in Figure 20.16), then

$$\int_{S_1} \vec{F} \cdot d\vec{A} = \int_{S_2} \vec{F} \cdot d\vec{A}.$$

If S_1 and S_2 determine opposite orientations on C, then

$$\int_{S_1} \vec{F} \cdot d\vec{A} = -\int_{S_2} \vec{F} \cdot d\vec{A}.$$

Solution   Since $\vec{F} = \operatorname{curl} \vec{G}$, by Stokes' Theorem we have

$$\int_{S_1} \vec{F} \cdot d\vec{A} = \int_{S_1} \operatorname{curl} \vec{G} \cdot d\vec{A} = \int_{C} \vec{G} \cdot d\vec{r}$$

and

$$\int_{S_2} \vec{F} \cdot d\vec{A} = \int_{S_2} \operatorname{curl} \vec{G} \cdot d\vec{A} = \int_{C} \vec{G} \cdot d\vec{r}.$$

In each case the line integral on the right must be computed using the orientation determined by the surface. Thus, the two flux integrals of $\vec{F}$ are the same if the orientations are the same and they are opposite if the orientations are opposite.

Exercises and Problems for Section 20.2

Exercises

In Exercises 1–5, calculate the circulation, $\int_C \vec{F} \cdot d\vec{r}$, in two ways, directly and using Stokes' Theorem.

1. $\vec{F} = (x - y + z)(\vec{i} + \vec{j})$ and C is the triangle with vertices $(0, 0, 0)$, $(5, 0, 0)$, $(5, 5, 0)$, traversed in that order.

2. $\vec{F} = (x + z)\vec{i} + x\vec{j} + y\vec{k}$ and C is the upper half of the circle $x^2 + z^2 = 9$ in the plane $y = 0$, together with the x-axis from $(3, 0, 0)$ to $(-3, 0, 0)$, traversed counterclockwise when viewed from the positive y-axis.

3. $\vec{F} = y\vec{i} - x\vec{j}$ and C is the boundary of S, the part of the surface $z = 4 - x^2 - y^2$ above the xy-plane, oriented upward.

4. $\vec{F} = xy\vec{i} + yz\vec{j} + xz\vec{k}$ and C is the boundary of S, the surface $z = 1 - x^2$ for $0 \leq x \leq 1$ and $-2 \leq y \leq 2$, oriented upward. Sketch S and C.

5. $\vec{F} = y\vec{i} + z\vec{j} + x\vec{k}$ and C is the boundary of S, the paraboloid $z = 1 - (x^2 + y^2)$, $z \geq 0$ oriented upward. [Hint: Use polar coordinates.]

In Exercises 6–9, use Stokes' Theorem to find the circulation of the vector field around the given paths.

6. $\vec{F} = (z - 2y)\vec{i} + (3x - 4y)\vec{j} + (z + 3y)\vec{k}$ and C is the circle $x^2 + y^2 = 4$, $z = 1$, oriented counterclockwise when viewed from above.

7. $\vec{F} = (2x - y)\vec{i} + (x + 4y)\vec{j}$ and C is a circle of radius 10, centered at the origin.

(a) In the xy-plane, oriented clockwise as viewed from the positive z-axis.

(b) In the yz-plane, oriented clockwise as viewed from the positive x-axis.

8. $\vec{F} = \vec{r}/\|\vec{r}\|^3$ and C is the path consisting of line segments from $(1, 0, 1)$ to $(1, 0, 0)$ to $(0, 0, 1)$ to $(1, 0, 1)$.

9. $\vec{F} = xz\vec{i} + (x + yz)\vec{j} + x^2\vec{k}$ and C is the circle $x^2 + y^2 = 1$, $z = 2$, oriented counterclockwise when viewed from above.

10. Let $\vec{F} = y\vec{i} - x\vec{j}$ and let C be the unit circle in the xy-plane centered at the origin and oriented counterclockwise when viewed from above.

(a) Calculate $\int_C \vec{F} \cdot d\vec{r}$ by parameterizing the circle.

(b) Calculate curl $\vec{F}$.

(c) Calculate $\int_C \vec{F} \cdot d\vec{r}$ using your result from part (b).

(d) What theorem did you use in part (c)?

11. (a) If $\vec{F} = (\cos x)\vec{i} + e^y\vec{j} + (x - y - z)\vec{k}$, find curl $\vec{F}$.

(b) Find $\int_C \vec{F} \cdot d\vec{r}$ where C is the circle of radius 3 in the plane $x + y + z = 1$, centered at $(1, 0, 0)$ oriented counterclockwise when viewed from above.

12. Can you use Stokes' Theorem to compute the line integral $\int_C (2x\vec{i} + 2y\vec{j} + 2z\vec{k}) \cdot d\vec{r}$ where C is the straight line from the point $(1, 2, 3)$ to the point $(4, 5, 6)$? Why or why not?

Problems

In Problems 13–18, find $\int_C \vec{F} \cdot d\vec{r}$ where C is a circle of radius 2 in the plane $x + y + z = 3$, centered at $(1, 1, 1)$ and oriented clockwise when viewed from the origin.

13. $\vec{F} = \vec{i} + \vec{j} + 3\vec{k}$

14. $\vec{F} = -y\vec{i} + x\vec{j} + z\vec{k}$

15. $\vec{F} = y\vec{i} - x\vec{j} + (y - x)\vec{k}$

16. $\vec{F} = (2y + e^x)\vec{i} + ((\sin y) - x)\vec{j} + (2y - x + \cos z^2)\vec{k}$

17. $\vec{F} = -z\vec{j} + y\vec{k}$

18. $\vec{F} = (z - y)\vec{i} + (x - z)\vec{j} + (y - x)\vec{k}$

19. For positive constants a, b, and c, let

$$f(x, y, z) = \ln(1 + ax^2 + by^2 + cz^2).$$

(a) What is the domain of f?

(b) Find grad f.

(c) Find curl(grad f).

(d) Find $\int_C \vec{F} \cdot d\vec{r}$ where C is the helix $x = \cos t$, $y = \sin t$, $z = t$ for $0 \leq t \leq 13\pi/2$ and

$$\vec{F} = \frac{2x\vec{i} + 4y\vec{j} + 6z\vec{k}}{1 + x^2 + 2y^2 + 3z^2}.$$

20. Figure 20.17 shows an open cylindrical can, S, standing on the xy-plane. (S has a bottom and sides, but no top.)

(a) Give equation(s) for the rim, C.

(b) If S is oriented outward and downward, find $\int_S \text{curl}(-y\vec{i} + x\vec{j} + z\vec{k}) \cdot d\vec{A}$.

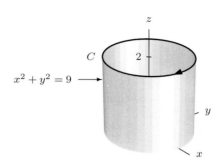

$x^2 + y^2 = 9$

Figure 20.17

21. At all points in 3-space curl $\vec{F}$ points in the direction of $\vec{i} - \vec{j} - \vec{k}$. Let C be a circle in the yz-plane, oriented clockwise when viewed from the positive x-axis. Is the circulation of $\vec{F}$ around C positive, zero, or negative?

22. If curl $\vec{F} = (x^2 + z^2)\vec{j} + 5\vec{k}$, find $\int_C \vec{F} \cdot d\vec{r}$, where C is a circle of radius 3, centered at the origin, with

   (a) C in the xy-plane, oriented counterclockwise when viewed from above.

   (b) C in the xz-plane, oriented counterclockwise, when viewed from the positive y-axis.

23. (a) Find curl$(y\vec{i} + z\vec{j} + x\vec{k})$.

   (b) Find $\int_C (y\vec{i} + z\vec{j} + x\vec{k}) \cdot d\vec{r}$ where C is the boundary of the triangle with vertices $(2, 0, 0)$, $(0, 3, 0)$, $(-2, 0, 0)$, traversed in that order.

24. (a) Let $\vec{F} = y\vec{i} + z\vec{j} + x\vec{k}$. Find curl $\vec{F}$.

   (b) Calculate $\int_C \vec{F} \cdot d\vec{r}$ where C is

     (i) A circle of radius 2 centered at $(1, 1, 3)$ in the plane $z = 3$, oriented counterclockwise when viewed from above.

     (ii) The triangle obtained by tracing out the path $(2, 0, 0)$ to $(2, 0, 5)$ to $(2, 3, 5)$ to $(2, 0, 0)$.

25. (a) Find curl$(z\vec{i} + x\vec{j} + y\vec{k})$.

   (b) Find $\int_C (z\vec{i} + x\vec{j} + y\vec{k}) \cdot d\vec{r}$ where C is a square of side 2 lying in the plane $x + y + z = 5$, oriented counterclockwise when viewed from the origin.

26. Evaluate $\int_C (-z\vec{i} + y\vec{j} + x\vec{k}) \cdot d\vec{r}$, where C is a circle of radius 2, parallel to the xz-plane and around the y-axis with the orientation shown in Figure 20.18.

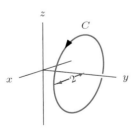

Figure 20.18

27. Evaluate the circulation of $\vec{G} = xy\vec{i} + z\vec{j} + 3y\vec{k}$ around a square of side 6, centered at the origin, lying in the yz-plane, and oriented counterclockwise viewed from the positive x-axis.

28. Find the flux of $\vec{F} = \text{curl}((x^3 + \cos(z^2))\vec{i} + (x + \sin(y^2))\vec{j} + (y^2\sin(x^2))\vec{k})$ through the upper half of the sphere of radius 2, with center at the origin and oriented upward.

29. Suppose that C is a closed curve in the xy-plane, oriented counterclockwise when viewed from above. Show that $\frac{1}{2}\int_C (-y\vec{i} + x\vec{j}) \cdot d\vec{r}$ equals the area of the region R in the xy-plane enclosed by C.

30. In the region between the circles $C_1 : x^2 + y^2 = 4$ and $C_2 : x^2 + y^2 = 25$ in the xy-plane, the vector field $\vec{F}$ has curl $\vec{F} = 3\vec{k}$. If C_1 and C_2 are both oriented counterclockwise when viewed from above, find the value of

$$\int_{C_2} \vec{F} \cdot d\vec{r} - \int_{C_1} \vec{F} \cdot d\vec{r}.$$

31. Let curl $\vec{F} = 3x\vec{i} + 3y\vec{j} - 6z\vec{k}$ and let C_1 and C_2 be the closed curves in the planes $z = 0$ and $z = 5$ in Figure 20.19. Find

$$\int_{C_1} \vec{F} \cdot d\vec{r} + \int_{C_2} \vec{F} \cdot d\vec{r}.$$

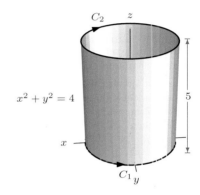

Figure 20.19

32. (a) Find curl$(x^3\vec{i} + \sin(y^3)\vec{j} + e^{z^3}\vec{k})$.

   (b) What does your answer to part (a) tell you about $\int_C (x^3\vec{i} + \sin(y^3)\vec{j} + e^{z^3}\vec{k}) \cdot d\vec{r}$ where C is the circle $(x - 10)^2 + (y - 20)^2 = 1$ in the xy-plane, oriented clockwise?

   (c) If C is any closed curve, what can you say about $\int_C (x^3\vec{i} + \sin(y^3)\vec{j} + e^{z^3}\vec{k}) \cdot d\vec{r}$?

33. Let $\vec{F}(x, y, z) = F_1(x, y)\vec{i} + F_2(x, y)\vec{j}$, where F_1 and F_2 are continuously differentiable for all x, y.

   (a) Describe in words how $\vec{F}$ varies through space.

   (b) Find an expression for curl $\vec{F}$ in terms of F_1 and F_2.

   (c) Let C be a closed curve in the xy-plane, oriented counterclockwise when viewed from above, and let S be the region enclosed by C. Use your answer to part (b) to simplify the statement of Stokes' Theorem for this $\vec{F}$ and C.

   (d) The result in part (c) is usually known by another name. What is it?

34. Water in a bathtub has velocity vector field near the drain given, for x, y, z in cm, by

$$\vec{F} = -\frac{y + xz}{(z^2 + 1)^2}\vec{i} - \frac{yz - x}{(z^2 + 1)^2}\vec{j} - \frac{1}{z^2 + 1}\vec{k} \text{ cm/sec.}$$

(a) Rewriting $\vec{F}$ as follows, describe in words how the water is moving:

$$\vec{F} = \frac{-y\vec{i} + x\vec{j}}{(z^2 + 1)^2} + \frac{-z(x\vec{i} + y\vec{j})}{(z^2 + 1)^2} - \frac{\vec{k}}{z^2 + 1}.$$

(b) The drain in the bathtub is a disk in the xy-plane with center at the origin and radius 1 cm. Find the rate at which the water is leaving the bathtub. (That is, find the rate at which water is flowing through the disk.) Give units for your answer.

(c) Find the divergence of $\vec{F}$.
(d) Find the flux of the water through the hemisphere of radius 1, centered at the origin, lying below the xy-plane and oriented downward.
(e) Find $\int_C \vec{G} \cdot d\vec{r}$ where C is the edge of the drain, oriented clockwise when viewed from above, and where

$$\vec{G} = \frac{1}{2}\left(\frac{y}{z^2 + 1}\vec{i} - \frac{x}{z^2 + 1}\vec{j} - \frac{x^2 + y^2}{(z^2 + 1)^2}\vec{k}\right).$$

(f) Calculate curl $\vec{G}$.
(g) Explain why your answers to parts (d) and (e) are equal.

Strengthen Your Understanding

In Problems 35–36, explain what is wrong with the statement.

35. The line integral $\int_C \vec{F} \cdot d\vec{r}$ can be evaluated using Stokes' Theorem, where $\vec{F} = 2x\vec{i} - 3\vec{j} + \vec{k}$ and C is an oriented curve from $(0, 0, 0)$ to $(3, 4, 5)$.

36. If S is the unit circular disc $x^2 + y^2 \leq 1$, $z = 0$, in the xy-plane, oriented downward, and C is the unit circle in the xy-plane oriented counterclockwise, and $\vec{F}$ is a vector field, then

$$\int_C \vec{F} \cdot d\vec{r} = \int_S \text{curl}\,\vec{F} \cdot d\vec{A}.$$

In Problems 37–38, give an example of:

37. An oriented closed curve C such that $\int_C \vec{F} \cdot d\vec{r} = 0$, where $\vec{F}(x, y, z) = x\vec{i} + y^2\vec{j} + z^3\vec{k}$.

38. A surface S, oriented appropriately to use Stokes' Theorem, which has as its boundary the circle C of radius 1 centered at the origin, lying in the xy-plane, and oriented counterclockwise, when viewed from above.

In Problems 39–47, is the statement true or false? Give a reason for your answer.

39. If curl $\vec{F}$ everywhere perpendicular to the z-axis, and if C is a circle in the xy-plane, then the circulation of $\vec{F}$ around C is zero.

40. If S is the upper unit hemisphere $x^2 + y^2 + z^2 = 1$, $z \geq 0$, oriented upward, then the boundary of S used in Stokes' Theorem is the circle $x^2 + y^2 = 1$, $z = 0$, with orientation counterclockwise when viewed from the positive z-axis.

41. Let S be the cylinder $x^2 + z^2 = 1$, $0 \leq y \leq 2$, oriented with inward pointing normal. Then the boundary of S consists of two circles C_1 ($x^2 + z^2 = 1$, $y = 0$) and C_2 ($x^2 + z^2 = 1$, $y = 2$), both oriented clockwise when viewed from the positive y-axis.

42. If C is the boundary of an oriented surface S, oriented by the right-hand rule, then $\int_C \text{curl}\,\vec{F} \cdot d\vec{r} = \int_S \vec{F} \cdot d\vec{A}$.

43. Let S_1 be the disk $x^2 + y^2 \leq 1$, $z = 0$ and let S_2 be the upper unit hemisphere $x^2 + y^2 + z^2 = 1$, $z \geq 0$, both oriented upward. If $\vec{F}$ is a vector field then $\int_{S_1} \text{curl}\vec{F} \cdot d\vec{A} = \int_{S_2} \text{curl}\vec{F} \cdot d\vec{A}$.

44. Let S be the closed unit sphere $x^2 + y^2 + z^2 = 1$, oriented outward. If $\vec{F}$ is a vector field, then $\int_S \text{curl}\vec{F} \cdot d\vec{A} = 0$.

45. If $\vec{F}$ and $\vec{G}$ are vector fields satisfying curl$\vec{F} = $ curl$\vec{G}$, then $\int_C \vec{F} \cdot d\vec{r} = \int_C \vec{G} \cdot d\vec{r}$, where C is any oriented circle in 3-space.

46. If $\vec{F}$ is a vector field satisfying curl$\vec{F} = \vec{0}$, then $\int_C \vec{F} \cdot d\vec{r} = 0$, where C is any oriented path around a rectangle in 3-space.

47. Let S be an oriented surface, with oriented boundary C, and suppose that $\vec{F}$ is a vector field such that $\int_C \vec{F} \cdot d\vec{r} = 0$. Then curl $\vec{F} = \vec{0}$ everywhere on S.

48. The circle C has radius 3 and lies in a plane through the origin. Let $\vec{F} = (2z + 3y)\vec{i} + (x - z)\vec{j} + (6y - 7x)\vec{k}$. What is the equation of the plane and what is the orientation of the circle that make the circulation, $\int_C \vec{F} \cdot d\vec{r}$, a maximum? [Note: You should specify the orientation of the circle by saying that it is clockwise or counterclockwise when viewed from the positive or negative x- or y- or z-axis.]

20.3 THE THREE FUNDAMENTAL THEOREMS

We have now seen three multivariable versions of the Fundamental Theorem of Calculus. In this section we will examine some consequences of these theorems.

Fundamental Theorem of Calculus for Line Integrals

$$\int_C \operatorname{grad} f \cdot d\vec{r} = f(Q) - f(P).$$

Stokes' Theorem

$$\int_S \operatorname{curl} \vec{F} \cdot d\vec{A} = \int_C \vec{F} \cdot d\vec{r}.$$

Divergence Theorem

$$\int_W \operatorname{div} \vec{F} \, dV = \int_S \vec{F} \cdot d\vec{A}.$$

Notice that, in each case, the region of integration on the right is the boundary of the region on the left (except that for the first theorem we simply evaluate f at the boundary points); the integrand on the left is a sort of derivative of the integrand on the right; see Figure 20.20.

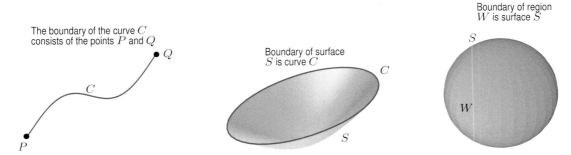

The boundary of the curve C consists of the points P and Q

Boundary of surface S is curve C

Boundary of region W is surface S

Figure 20.20: Regions and their boundaries for the three fundamental theorems

The Gradient and the Curl

Suppose that $\vec{F}$ is a smooth gradient field, so $\vec{F} = \operatorname{grad} f$ for some function f. Using the Fundamental Theorem for Line Integrals, we saw in Chapter 18 that

$$\int_C \vec{F} \cdot d\vec{r} = 0$$

for any closed curve C. Thus, for any unit vector $\vec{n}$

$$\operatorname{circ}_{\vec{n}} \vec{F} = \lim_{\text{Area} \to 0} \frac{\int_C \vec{F} \cdot d\vec{r}}{\text{Area of } C} = \lim_{\text{Area} \to 0} \frac{0}{\text{Area}} = 0,$$

where the limit is taken over circles C in a plane perpendicular to $\vec{n}$, and oriented by the right-hand rule. Thus, the circulation density of $\vec{F}$ is zero in every direction, so $\operatorname{curl} \vec{F} = \vec{0}$, that is,

$$\boxed{\operatorname{curl} \operatorname{grad} f = \vec{0}.}$$

(This formula can also be verified using the coordinate definition of curl. See Problem 27 on page 1076.)

Is the converse true? Is any vector field whose curl is zero a gradient field? Suppose that $\operatorname{curl} \vec{F} = \vec{0}$ and let us consider the line integral $\int_C \vec{F} \cdot d\vec{r}$ for a closed curve C contained in the domain of $\vec{F}$. If C is the boundary curve of an oriented surface S that lies wholly in the domain of $\operatorname{curl} \vec{F}$, then Stokes' Theorem asserts that

$$\int_C \vec{F} \cdot d\vec{r} = \int_S \operatorname{curl} \vec{F} \cdot d\vec{A} = \int_S \vec{0} \cdot d\vec{A} = 0.$$

If we knew that $\int_C \vec{F} \cdot d\vec{r} = 0$ for every closed curve C, then $\vec{F}$ would be path-independent, and hence a gradient field. Thus, we need to know whether every closed curve in the domain of $\vec{F}$ is the boundary of an oriented surface contained in the domain. It can be quite difficult to determine if a given curve is the boundary of a surface (suppose, for example, that the curve is knotted in a complicated way). However, if the curve can be contracted smoothly to a point, remaining all the time in the domain of $\vec{F}$, then it is the boundary of a surface, namely, the surface it sweeps through as it contracts. Thus, we have proved the test for a gradient field that we stated in Chapter 18.

The Curl Test for Vector Fields in 3-Space

Suppose $\vec{F}$ is a smooth vector field on 3-space such that
- The domain of $\vec{F}$ has the property that every closed curve in it can be contracted to a point in a smooth way, staying at all times within the domain.
- curl $\vec{F} = \vec{0}$.

Then $\vec{F}$ is path-independent, and thus is a gradient field.

Example 7 on page 1012 shows how the curl test is applied.

The Curl and the Divergence

In this section we will use the second two fundamental theorems to get a test for a vector field to be a curl field, that is, a field of the form $\vec{F} = \text{curl}\,\vec{G}$ for some $\vec{G}$.

Example 1   Suppose that $\vec{F}$ is a smooth curl field. Use Stokes' Theorem to show that for any closed surface, S, contained in the domain of $\vec{F}$

$$\int_S \vec{F} \cdot d\vec{A} = 0.$$

Solution   Suppose $\vec{F} = \text{curl}\,\vec{G}$. Draw a closed curve C on the surface S, thus dividing S into two surfaces S_1 and S_2 as shown in Figure 20.21. Pick the orientation for C corresponding to S_1; then the orientation of C corresponding to S_2 is the opposite. Thus, using Stokes' Theorem,

$$\int_{S_1} \vec{F} \cdot d\vec{A} = \int_{S_1} \text{curl}\,\vec{G} \cdot d\vec{A} = \int_C \vec{G} \cdot d\vec{r} = -\int_{S_2} \text{curl}\,\vec{G} \cdot d\vec{A} = -\int_{S_2} \vec{F} \cdot d\vec{A}.$$

Thus, for any closed surface S, we have

$$\int_S \vec{F} \cdot d\vec{A} = \int_{S_1} \vec{F} \cdot d\vec{A} + \int_{S_2} \vec{F} \cdot d\vec{A} = 0.$$

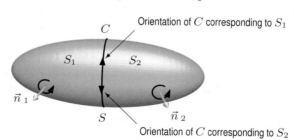

Figure 20.21: The closed surface S divided into two surfaces S_1 and S_2

Thus, if $\vec{F} = \text{curl}\,\vec{G}$, we use the result of Example 1 to see that

$$\text{div}\,\vec{F} = \lim_{\text{Volume}\to 0} \frac{\displaystyle\int_S \vec{F} \cdot d\vec{A}}{\text{Volume enclosed by } S} = \lim_{\text{Volume}\to 0} \frac{0}{\text{Volume}} = 0,$$

where the limit is taken over spheres S contracting down to a point. So we conclude that:

$$\boxed{\text{div}\,\text{curl}\,\vec{G} = 0.}$$

(This formula can also be verified using coordinates. See Problem 26 on page 1076.)

Is every vector field whose divergence is zero a curl field? It turns out that we have the following analogue of the curl test, though we will not prove it.

The Divergence Test for Vector Fields in 3-Space

Suppose $\vec{F}$ is a smooth vector field on 3-space such that
- The domain of $\vec{F}$ has the property that every closed surface in it is the boundary of a solid region completely contained in the domain.
- $\text{div}\,\vec{F} = 0$.

Then $\vec{F}$ is a curl field.

Example 2   Consider the vector fields $\vec{E} = q\dfrac{\vec{r}}{\|\vec{r}\|^3}$ and $\vec{B} = \dfrac{2I}{c}\left(\dfrac{-y\vec{i} + x\vec{j}}{x^2 + y^2}\right)$.

(a) Calculate $\text{div}\,\vec{E}$ and $\text{div}\,\vec{B}$.
(b) Do $\vec{E}$ and $\vec{B}$ satisfy the divergence test?
(c) Is either $\vec{E}$ or $\vec{B}$ a curl field?

Solution   (a) Example 3 on page 1049 shows that $\text{div}\,\vec{E} = 0$. The following calculation shows $\text{div}\,\vec{B} = 0$ also:

$$\text{div}\,\vec{B} = \frac{2I}{c}\left(\frac{\partial}{\partial x}\left(\frac{-y}{x^2 + y^2}\right) + \frac{\partial}{\partial y}\left(\frac{x}{x^2 + y^2}\right) + \frac{\partial}{\partial z}(0)\right)$$

$$= \frac{2I}{c}\left(\frac{2xy}{(x^2 + y^2)^2} + \frac{-2yx}{(x^2 + y^2)^2}\right) = 0.$$

(b) The domain of $\vec{E}$ is 3-space minus the origin, so a region is contained in the domain if it misses the origin. Thus, the surface of a sphere centred at the origin is contained in the domain of E, but the solid ball inside is not. Hence, $\vec{E}$ does not satisfy the divergence test.

The domain of $\vec{B}$ is 3-space minus the z-axis, so a region is contained in the domain if it avoids the z-axis. If S is a surface bounding a solid region W, then the z-axis cannot pierce W without piercing S as well. Hence, if S avoids the z-axis, so does W. Thus, $\vec{B}$ satisfies the divergence test.

(c) In Example 3 on page 1057 we computed the flux of $\vec{r}/\|\vec{r}\|^3$ through a sphere centred at the origin, and found it was 4π, so the flux of $\vec{E}$ through this sphere is $4\pi q$. Thus, $\vec{E}$ cannot be a curl field, because by Example 1, the flux of a curl field through a closed surface is zero.

On the other hand, $\vec{B}$ satisfies the divergence test, so it must be a curl field. In fact, Problem 21 shows that

$$\vec{B} = \text{curl}\left(\frac{-I}{c}\ln(x^2 + y^2)\vec{k}\right).$$

Exercises and Problems for Section 20.3

Exercises

In Exercises 1–6, is the vector field a curl field?

1. $\vec{F} = z\vec{i} + x\vec{j} + y\vec{k}$
2. $\vec{F} = z\vec{i} + y\vec{j} + x\vec{k}$
3. $\vec{F} = 2x\vec{i} - y\vec{j} - z\vec{k}$
4. $\vec{F} = (x+y)\vec{i} + (y+z)\vec{j} + (x+z)\vec{k}$
5. $\vec{F} = (-xy)\vec{i} + (2yz)\vec{j} + (yz - z^2))\vec{k}$
6. $\vec{F} = (xy)\vec{i} + (xy)\vec{j} + (xy)\vec{k}$

In Exercises 7–12, is the vector field a gradient field?

7. $\vec{F} = yz\vec{i} + (xz + z^2)\vec{j} + (xy + 2yz)\vec{k}$
8. $\vec{G} = -y\vec{i} + x\vec{j}$

9. $\vec{F} = 2x\vec{i} + z\vec{j} + y\vec{k}$
10. $\vec{F} = y\vec{i} + z\vec{j} + x\vec{k}$
11. $\vec{F} = (y + 2z)\vec{i} + (x + z)\vec{j} + (2x + y)\vec{k}$
12. $\vec{F} = (y - 2z)\vec{i} + (x - z)\vec{j} + (2x - y)\vec{k}$

In Exercises 13–16, can the curl test and the divergence test be applied to a vector field whose domain is the given region?

13. All points (x, y, z) not on the y-axis.
14. All points (x, y, z) such that $z > 0$.
15. All points (x, y, z) except the x-axis with $0 \leq x \leq 1$.
16. All points (x, y, z) not on the positive z-axis.

Problems

17. Let $\vec{B} = b\vec{k}$, for some constant b. Show that the following are all possible vector potentials for $\vec{B}$:
    (a) $\vec{A} = -by\vec{i}$ (b) $\vec{A} = bx\vec{j}$
    (c) $\vec{A} = \frac{1}{2}\vec{B} \times \vec{r}$.

18. Express $(3x + 2y)\vec{i} + (4x + 9y)\vec{j}$ as the sum of a curl-free vector field and a divergence-free vector field.

19. Find a vector field $\vec{F}$ such that curl $\vec{F} = 2\vec{i} - 3\vec{j} + 4\vec{k}$. [Hint: Try $\vec{F} = \vec{v} \times \vec{r}$ for some vector $\vec{v}$.]

20. Find a vector potential for the constant vector field $\vec{B}$ whose value at every point is $\vec{b}$.

21. Show that $\vec{A} = \dfrac{-I}{c}\ln(x^2 + y^2)\vec{k}$ is a vector potential for
$$\vec{B} = \frac{2I}{c}\left(\frac{-y\vec{i} + x\vec{j}}{x^2 + y^2}\right).$$

In Problems 22–23, does a vector potential exist for the vector field given? If so, find one.

22. $\vec{F} = 2x\vec{i} + (3y - z^2)\vec{j} + (x - 5z)\vec{k}$
23. $\vec{G} = x^2\vec{i} + y^2\vec{j} + z^2\vec{k}$

24. Use Stokes' Theorem to show that if $u(x, y)$ and $v(x, y)$ are two functions of x and y and C is a closed curve in the xy-plane oriented counterclockwise, then
$$\int_C (u\vec{i} + v\vec{j}) \cdot d\vec{r} = \int_R \left(\frac{\partial v}{\partial x} - \frac{\partial u}{\partial y}\right) dx dy$$
where R is the region in the xy-plane enclosed by C. This is Green's Theorem.

25. An electric charge q at the origin produces an electric field $\vec{E} = q\vec{r}/\|\vec{r}\|^3$.
    (a) Does curl $\vec{E} = \vec{0}$?
    (b) Does $\vec{E}$ satisfy the curl test?
    (c) Is $\vec{E}$ a gradient field?

26. Suppose c is the speed of light. A thin wire along the z-axis carrying a current I produces a magnetic field
$$\vec{B} = \frac{2I}{c}\left(\frac{-y\vec{i} + x\vec{j}}{x^2 + y^2}\right).$$
    (a) Does curl $\vec{B} = \vec{0}$?
    (b) Does $\vec{B}$ satisfy the curl test?
    (c) Is $\vec{B}$ a gradient field?

27. For constant p, consider the vector field $\vec{E} = \dfrac{\vec{r}}{\|\vec{r}\|^p}$.
    (a) Find curl $\vec{E}$.
    (b) Find the domain of $\vec{E}$.
    (c) For which values of p does $\vec{E}$ satisfy the curl test? For those values of p, find a potential function for $\vec{E}$.

28. The magnetic field, $\vec{B}$, due to a magnetic dipole with moment $\vec{\mu}$ satisfies div $\vec{B} = 0$ and is given by
$$\vec{B} = -\frac{\vec{\mu}}{\|\vec{r}\|^3} + \frac{3(\vec{\mu} \cdot \vec{r})\vec{r}}{\|\vec{r}\|^5}, \qquad \vec{r} \neq \vec{0}.$$
    (a) Does $\vec{B}$ satisfy the divergence test?
    (b) Show that $\vec{A} = \dfrac{\vec{\mu} \times \vec{r}}{\|\vec{r}\|^3}$ is a vector potential for $\vec{B}$. [Hint: Use Problem 28 on page 1076. The identities in Example 3 on page 1073, Problem 68 on page 814, and Problem 49 on page 768 may also be useful.]
    (c) Does your answer to part (a) contradict your answer to part (b)? Explain.

29. Suppose that $\vec{A}$ is a vector potential for $\vec{B}$.

    (a) Show that $\vec{A} + \operatorname{grad} \psi$ is also a vector potential for $\vec{B}$, for any function ψ with continuous second-order partial derivatives. (The vector potentials $\vec{A}$ and $\vec{A} + \operatorname{grad} \psi$ are called *gauge equivalent* and the transformation, for any ψ, from $\vec{A}$ to $\vec{A} + \operatorname{grad} \psi$ is called a *gauge transformation*.)

    (b) What is the divergence of $\vec{A} + \operatorname{grad} \psi$? How should ψ be chosen such that $\vec{A} + \operatorname{grad} \psi$ has zero divergence? (If $\operatorname{div} \vec{A} = 0$, the magnetic vector potential $\vec{A}$ is said to be in *Coulomb gauge*.)

Strengthen Your Understanding

In Problems 30–31, explain what is wrong with the statement.

30. The curl of a vector field $\vec{F}$ is given by $\operatorname{curl} \vec{F} = x\vec{i}$.

31. For a certain vector field $\vec{F}$, we have $\operatorname{curl} \operatorname{div} \vec{F} = y\vec{i}$.

In Problems 32–33, give an example of:

32. A vector field $\vec{F}$ that is not the curl of another vector field.

33. A function f such that $\operatorname{div} \operatorname{grad} f \neq 0$.

In Problems 34–37, is the statement true or false? Give a reason for your answer.

34. There exists a vector field $\vec{F}$ with $\operatorname{curl} \vec{F} = \vec{i}$.

35. There exists a vector field $\vec{F}$ (whose components have continuous second partial derivatives) satisfying $\operatorname{curl} \vec{F} = x\vec{i}$.

36. Let S be an oriented surface, with oriented boundary C, and suppose that $\vec{F}$ is a vector field such that $\int_S \operatorname{curl} \vec{F} \cdot d\vec{A} = 0$. Then $\vec{F}$ is a gradient field.

37. If $\vec{F}$ is a gradient field, then $\int_S \operatorname{curl} \vec{F} \cdot d\vec{A} = 0$, for any smooth oriented surface, S, in 3-space.

38. Let $f(x, y, z)$ be a scalar function with continuous second partial derivatives. Let $\vec{F}(x, y, z)$ be a vector field with continuous second partial derivatives. Which of the following quantities are identically zero?

    (a) $\operatorname{curl} \operatorname{grad} f$ **(b)** $\vec{F} \times \operatorname{curl} \vec{F}$

    (c) $\operatorname{grad} \operatorname{div} \vec{F}$ **(d)** $\operatorname{div} \operatorname{curl} \vec{F}$

    (e) $\operatorname{div} \operatorname{grad} f$

CHAPTER SUMMARY (see also Ready Reference at the end of the book)

- **Curl**
  Geometric and coordinate definition of curl, calculating curl, interpretation in terms of circulation per unit area.
- **Stokes' Theorem**
  Statement of the theorem, curl-free and curl fields.

- **Three Fundamental Theorems**
  Combining the fundamental theorem of line integrals, Stokes' theorem and the Divergence theorem to show $\operatorname{curl} \operatorname{grad} f = \vec{0}$ and $\operatorname{div} \operatorname{curl} \vec{G} = 0$.
  Curl test for gradient field in 3-space.
  Divergence test for curl field in 3-space.

REVIEW EXERCISES AND PROBLEMS FOR CHAPTER TWENTY

Exercises

1. Find $\text{curl}((x + y)\vec{i} - (y + z)\vec{j} + (x + z)\vec{k})$

2. Find $(\text{curl}\,\vec{n}) \cdot \vec{j}$
where $\vec{n} = (2x + 3y)\vec{i} + (4y + 5z)\vec{j} + (6z + 7x)\vec{k}$

Exercises 3–5 concern the vector fields in Figure 20.22. In each case, assume that the cross-section is the same in all other planes parallel to the given cross-section.

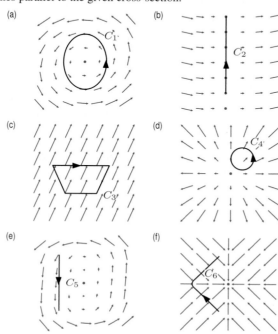

Figure 20.22

3. Three of the vector fields have zero curl at each point shown. Which are they? How do you know?

4. Three of the vector fields have zero divergence at each point shown. Which are they? How do you know?

5. Four of the line integrals $\int_{C_i} \vec{F} \cdot d\vec{r}$ are zero. Which are they? How do you know?

In Exercises 6–9, are the quantities defined? For those that are, is the quantity a vector or scalar? Let $f(x, y, z)$ be a smooth function and let $\vec{F}(\vec{r})$ and $\vec{G}(\vec{r})$ be smooth vector fields.

6. $\displaystyle\int_C (\text{grad}\,\vec{F}) \cdot d\vec{r}$

7. $\displaystyle\int_S (\vec{F}(\vec{r}) \times \vec{G}(\vec{r})) \cdot d\vec{A}$

8. $\text{div}((\text{grad}\,f) \times \vec{r})$

9. $(\text{curl}\,\vec{F}) \times \vec{F}$

In Exercises 10–11 decide whether the vector fields appear to have nonzero curl at the point marked. The vector field is shown in the xy-plane; it has no z-component and is independent of z.

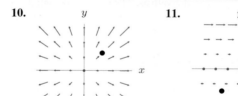

12. Is $\text{curl}(-2z\vec{i} - z\vec{j} + xy\vec{k})$ a vector or scalar? Calculate it.

In Exercises 13–16, calculate $\text{div}\vec{F}$ and $\text{curl}\vec{F}$. Is $\vec{F}$ solenoidal or irrotational?

13. $\vec{F} = x^2\vec{i} + y^3\vec{j} + z^4\vec{k}$

14. $\vec{F} = xy\vec{i} + yz\vec{j} + zx\vec{k}$

15. $\vec{F} = (\cos x)\,\vec{i} + e^y\vec{j} + (x + y + z)\,\vec{k}$

16. $\vec{F} = e^{y+z}\vec{i} + \sin(x + z)\vec{j} + (x^2 + y^2)\,\vec{k}$

17. Let S be the curved side of the cylinder $y^2 + z^2 = 5$, for $0 \le x \le 3$, oriented outward. Let $\vec{F} = xz\vec{j} - xy\vec{k}$. Find $\int_S \text{curl}\,\vec{F} \cdot d\vec{A}$

    (a) Directly   **(b)** Using Stokes' Theorem

18. Compute $\int_C ((yz^2 - y)\vec{i} + (xz^2 + x)\vec{j} + 2xyz\vec{k}) \cdot d\vec{r}$, where the line integral is around C, the circle of radius 3 in the xy-plane, centered at the origin, oriented counterclockwise as viewed from the positive z-axis. Do this in two ways:

    (a) Directly   **(b)** Using Stokes' Theorem

19. Find the flux of $\text{curl}(e^{x^2}\vec{i} + (x + y)\vec{k})$ through the disk $y^2 + z^2 \le 1$, $x = 0$, oriented toward the positive x-axis using:

    (a) Stokes' Theorem   **(b)** Direct calculation

In Exercises 20–23, use Stokes' Theorem to calculate the integral.

20. $\int_C \vec{F} \cdot d\vec{r}$ where $\vec{F} = x^2\vec{i} + y^2\vec{j} + z^2\vec{k}$ and C is the unit circle in the xz-plane, oriented counterclockwise when viewed from the positive y-axis.

21. $\int_C \vec{F} \cdot d\vec{r}$ where $\vec{F} = (y - x)\vec{i} + (z - y)\vec{j} + (x - z)\vec{k}$ and C is the circle $x^2 + y^2 = 5$ in the xy-plane, oriented counterclockwise when viewed from above.

22. $\int_S \text{curl}\,\vec{F} \cdot d\vec{A}$ where $\vec{F} = -y\vec{i} + x\vec{j} + (xy + \cos z)\vec{k}$ and S is the disk $x^2 + y^2 \le 9$, oriented upward in the xy-plane.

23. $\int_S \text{curl}\,\vec{F} \cdot d\vec{A}$ where $\vec{F} = (x + 7)\vec{j} + e^{x+y+z}\vec{k}$ and S is the rectangle $0 \le x \le 3$, $0 \le y \le 2$, $z = 0$, oriented counterclockwise when viewed from above.

Problems

24. Let $\vec{r} = x\vec{i} + y\vec{j} + z\vec{k}$ and $\vec{a}$ be a constant vector. For each of the quantities in (a)–(f), choose one of the statements in (I)–(V).

(a) $\operatorname{div}(\vec{r} + \vec{a})$ (b) $\operatorname{div}(\vec{r} \times \vec{a})$ (c) $\operatorname{div}(\vec{r} \cdot \vec{a})$

(d) $\operatorname{curl}(\vec{r} + \vec{a})$ (e) $\operatorname{curl}(\vec{r} \times \vec{a})$ (f) $\operatorname{curl}(\vec{r} \cdot \vec{a})$

(I) Scalar, independent of $\vec{a}$.

(II) Scalar, depends on $\vec{a}$.

(III) Vector, independent of $\vec{a}$.

(IV) Vector, depends on $\vec{a}$.

(V) Not defined.

25. Calculate the following quantities or say why it is impossible. Let $\vec{r} = x\vec{i} + y\vec{j} + z\vec{k}$ and let $\vec{a} = a_1\vec{i} + a_2\vec{j} + a_3\vec{k}$ be a constant vector.

(a) $\operatorname{grad}(\vec{r} \cdot \vec{a})$ (b) $\operatorname{div}(\vec{r} \cdot \vec{a})$

(c) $\operatorname{curl}(\vec{r} \cdot \vec{a})$ (d) $\operatorname{grad}(\vec{r} \times \vec{a})$

(e) $\operatorname{div}(\vec{r} \times \vec{a})$ (f) $\operatorname{curl}(\vec{r} \times \vec{a})$

26. Calculate each of the following integrals or say why it cannot be done with the methods and theorems in this book. Let $\vec{F} = x^3\vec{i} + y^3\vec{j} + z^3\vec{k}$.

(a) $\int_S \vec{F} \cdot d\vec{A}$ where S is the disk of radius 3 in the plane $y = 5$, oriented toward the origin.

(b) $\int_W \vec{F}\, dV$ where W is the solid sphere of radius 2 centered at the origin.

(c) $\int_S \operatorname{curl} \vec{F} \cdot d\vec{A}$ where S is the disk of radius 3 in the plane $y = 5$, oriented toward the origin.

(d) $\int_C \operatorname{grad} \vec{F} \cdot d\vec{r}$ where C is the line from the origin to $(2, 3, 4)$.

(e) $\int_W \operatorname{div} \vec{F}\, dV$ where W is the solid sphere of radius 2 centered at the origin.

(f) $\int_C \vec{F} \cdot d\vec{r}$ where C is the line from the origin to $(2, 3, 4)$.

(g) $\int_W \operatorname{curl} \vec{F}\, dV$ where W is the box $0 \le x \le 1, 0 \le y \le 2, 0 \le z \le 3$.

(h) $\int_W \vec{F} \cdot (\vec{i} + \vec{j} + \vec{k})\, dV$ where W is the box $0 \le x \le 1, 0 \le y \le 2, 0 \le z \le 3$.

27. Let $\vec{F}$ be a vector field with continuous partial derivatives at all points in 3-space. Let S_1 be the upper half of the sphere of radius 1 centered at the origin, oriented upward. Let S_2 be the disk of radius 1 in the xy-plane centered at the origin and oriented upward. Let C be the unit circle in the xy-plane, oriented counterclockwise when viewed from above. For each of the following integrals, say whether or not it is defined. If it is defined, list which of the other integrals it must equal (if any) and name the theorem.

(a) $\int_C \vec{F} \cdot d\vec{r}$ (b) $\int_C \vec{F} \cdot d\vec{A}$

(c) $\int_{S_1} \vec{F} \cdot d\vec{r}$ (d) $\int_{S_2} \vec{F} \cdot d\vec{A}$

(e) $\int_{S_1} \operatorname{curl} \vec{F} \cdot d\vec{A}$ (f) $\int_{S_2} \operatorname{curl} \vec{F} \cdot d\vec{A}$

(g) $\int_C \operatorname{curl} \vec{F} \cdot d\vec{r}$

28. Let $\operatorname{curl} \vec{F} = 2x\vec{i} + 5\vec{j} - 2z\vec{k}$, let $P = (3, 2, 4)$, and let C be the circle of radius 0.01 centered at P in the plane $x + y + z = 9$, oriented clockwise when viewed from the origin.

(a) Find $\operatorname{curl} \vec{F} \cdot (\vec{i} + \vec{j} + \vec{k})$ at P.

(b) What does your answer to part (a) tell you about $\int_C \vec{F} \cdot d\vec{r}$?

29. Three small squares, S_1, S_2, and S_3, each with side 0.1 and centered at the point $(4, 5, 7)$, lie parallel to the xy-, yz- and xz-planes, respectively. The squares are oriented counterclockwise when viewed from the positive z-, x-, and y-axes, respectively. A vector field $\vec{G}$ has circulation around S_1 of -0.02, around S_2 of 6, and around S_3 of -5. Estimate $\operatorname{curl} \vec{G}$ at the point $(4, 5, 7)$.

30. Figures 20.23 and 20.24 show the vector fields $\vec{F}$ and $\vec{G}$. Each vector field has no z-component and is independent of z. All the axes have the same scales.

(a) What can you say about $\operatorname{div} \vec{F}$ and $\operatorname{div} \vec{G}$ at the origin?

(b) What can you say about $\operatorname{curl} \vec{F}$ and $\operatorname{curl} \vec{G}$ at the origin?

(c) Is there a closed surface around the origin such that $\vec{F}$ has a nonzero flux through it?

(d) Repeat part (c) for $\vec{G}$.

(e) Is there a closed curve around the origin such that $\vec{F}$ has a nonzero circulation around it?

(f) Repeat part (e) for $\vec{G}$.

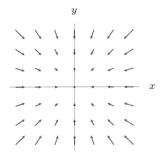

Figure 20.23: Cross-section of $\vec{F}$

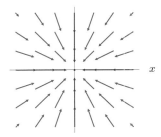

Figure 20.24: Cross-section of $\vec{G}$

31. Let $\vec{F} = y\vec{i} + \pi x\vec{j} + z\vec{k}$. Find the following integrals:

(a) $\int_S \vec{F} \cdot d\vec{A}$, where S is the disk $x^2 + y^2 \leq 10$, $z = \sqrt{5}$, oriented downward.

(b) $\int_C \vec{F} \cdot d\vec{r}$, where C is the line from $(0, \sqrt{3}, 0)$ to $(2, \sqrt{3}, 0)$.

(c) $\int_S \text{curl } \vec{F} \cdot d\vec{A}$, where S is the rectangle with corners at $(0, 0, 0)$, $(0, \sqrt{3}, 0)$, $(2, \sqrt{3}, 0)$, $(2, 0, 0)$, oriented upward.

In Problems 32–43, calculate the integral.

32. $\int_C \left(x^2\vec{i} + y^2\vec{j} + (x + y + z)\vec{k} \right) \cdot d\vec{r}$ where C is the circle $(x - 1)^2 + (y - 2)^2 = 4$ in the xy-plane, oriented counterclockwise when viewed from the positive z-axis.

33. $\int_C (-y^3\vec{i} + x^3\vec{j} + e^z\vec{k}) \cdot d\vec{r}$ where C is $x^2 + y^2 = 3$, $z = 4$, oriented counterclockwise when viewed from above.

34. $\int_C \left(\sin\left(x^2\right)\vec{i} + \cos\left(y^2\right)\vec{j} + (x + y)\vec{k} \right) \cdot d\vec{r}$ where C is the circle $(y - 1)^2 + (z - 2)^2 = 4$ in the yz-plane, oriented counterclockwise when viewed from the positive x-axis.

35. $\int_S \text{curl}\vec{F} \cdot d\vec{A}$ where $\vec{F} = (z + y)\vec{i} - (z + x)\vec{j} + (y + x)\vec{k}$. and S is the disk $y^2 + z^2 \leq 3$, $x = 0$, oriented in the positive x-direction.

36. $\int_S \vec{F} \cdot d\vec{A}$, where $\vec{F} = 3x\vec{i} + 4y\vec{j} + xy\vec{k}$ and S is the closed rectangular box whose top face has corners $(0, 0, 0)$, $(3, 0, 0)$, $(3, 5, 0)$, $(0, 5, 0)$, and whose bottom face contains the corner $(0, 0, -2)$.

37. $\int_S \vec{F} \cdot d\vec{A}$ where $\vec{F} = \left(y^2 + 3x \right)\vec{i} + \left(x^2 - y \right)\vec{j} + 2z\vec{k}$ and S is the unit sphere centered at the origin.

38. $\int_S \vec{F} \cdot d\vec{A}$ where $\vec{F} = x^3\vec{i} + y^3\vec{j} + z^3\vec{k}$ and S is the sphere of radius 1 centered at the origin.

39. $\int_C \vec{F} \cdot d\vec{r}$ where $\vec{F} = (x + y)\vec{i} + (y + 2z)\vec{j} + (z + 3x)\vec{k}$ and C is a square of side 7 in the xz-plane, oriented counterclockwise when viewed from the positive y-axis.

40. $\int_C \vec{F} \cdot d\vec{r}$ where $\vec{F} = (x - y^3 + z)\vec{i} + (x^3 + y + z)\vec{j} + (x + y + z^3)\vec{k}$ and C is the circle $x^2 + y^2 = 10$, oriented counterclockwise when viewed from above.

41. $\int_S \vec{F} \cdot d\vec{A}$ where $\vec{F} = (y^3z^3)\vec{i} + y^3\vec{j} + z^3\vec{k}$ and S is the cylinder $y^2 + z^2 = 16$, $-1 \leq x \leq 1$.

42. $\int_S \text{curl } \vec{F} \cdot d\vec{A}$ where $\vec{F} = -xe^y\vec{i} + ye^x\vec{j} + x^2y^2z\vec{k}$ and S is the top and sides of the cube $0 \leq x \leq 1$, $0 \leq y \leq 1$, $0 \leq z \leq 1$, oriented outward.

43. $\int_C \vec{F} \cdot d\vec{r}$ if $\text{curl } \vec{F} = 4\vec{k}$ and C is a unit circle in the xy-plane, oriented counterclockwise when viewed from above.

44. A box of side 1 in the first octant has faces in the planes $x = 0$, $x = 3$, $y = 0$, $y = 3$, $z = 0$, $z = 3$. Remove the face in the yz-plane to have an open surface, S, with a square boundary C. Let $\vec{F} = (x + y)\vec{i} - z\vec{j} + y\vec{k}$.

(a) Calculate $\int_C \vec{F} \cdot d\vec{r}$, where C is oriented counterclockwise when viewed from the positive x-axis.

(b) Calculate $\int_S \vec{F} \cdot d\vec{A}$, when S is oriented in the direction of the positive x-axis.

45. Suppose $\text{div } \vec{F}(x, y, z) = 4$ everywhere. Which of the following quantities can be computed from this information? Give the value of those that can be computed.

(a) $\int_S \vec{F} \cdot d\vec{A}$, where S is a sphere of radius 2 centered at the origin and oriented outward.

(b) $\int_C \vec{F} \cdot d\vec{r}$, where C is the unit circle in the xy-plane, oriented counterclockwise viewed from above.

(c) $\int_S \text{curl } \vec{F} \cdot d\vec{A}$, where S is a sphere of radius 2 centered at the origin and oriented outward.

46. Are the following vector fields conservative?

(a) $\vec{F}(x, y, z) = y^2z\vec{i} + 2xyz\vec{j} + xy\vec{k}$
(b) The vector field in Figure 20.25.

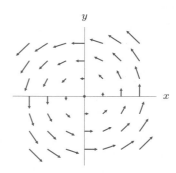

Figure 20.25

47. Let $\vec{F} = \dfrac{-y\vec{i} + x\vec{j}}{x^2 + y^2}$.

(a) Calculate $\text{curl } \vec{F}$. What is the domain of $\text{curl } \vec{F}$?
(b) Find the circulation of $\vec{F}$ around the unit circle C_1 in the xy-plane, oriented counterclockwise when viewed from above.
(c) Find the circulation of $\vec{F}$ around the circle C_2 in the plane $z = 4$ and with equation $(x - 3)^2 + y^2 = 1$, oriented counterclockwise when viewed from above.
(d) Find the circulation of $\vec{F}$ around the square S with corners $(2, 2, 0)$, $(-2, 2, 0)$, $(-2, -2, 0)$, $(2, -2, 0)$, oriented counterclockwise when viewed from above.
(e) Using your results to parts (b)–(d) explain how, with no additional calculation, you can find the circulation of $\vec{F}$ around any simple closed curve in the xy-plane, provided it does not intersect the z-axis. (A simple closed curve does not cross itself.)

48. Let C_1 be the circle of radius 3 in the xy-plane oriented counterclockwise and centered at the origin. Let

$$\vec{F} = \frac{-y\vec{i} + x\vec{j}}{x^2 + y^2}.$$

(a) Find $\int_{C_1} \vec{F} \cdot d\vec{r}$ by direct computation.

(b) Calculate curl $\vec{F}$.

(c) If possible, use Stokes' Theorem to calculate $\int_{C_1} \vec{F} \cdot d\vec{r}$? If it cannot be used, explain why not.

(d) Let C_2 be the circle of radius 3 in the xy-plane centered at $(5, 0)$. If possible, use Stokes' Theorem to calculate $\int_{C_2} \vec{F} \cdot d\vec{r}$? If it cannot be used, explain why not.

(e) Is $\vec{F}$ a gradient field?

49. Consider the circulation of the vector fields in parts (a)–(c) around the sets of closed curves in (I)–(V). For which of the sets, (I)–(V), is the circulation zero on every curve?

(I) All closed curves in the xy-plane.

(II) All closed curves in the yz-plane.

(III) All closed curves in the xz-plane.

(IV) All closed curves in the plane $x + y + z = 0$.

(V) All closed curves in all planes of the form $mx + ny = d$, where m, n, d are constants.

(a) $\vec{F} = -y\vec{i} + x\vec{j}$ (b) $\vec{G} = y\vec{i} + x\vec{j}$

(c) $\vec{H} = z\vec{j}$

50. Let $\vec{F} = y\vec{i} - x\vec{j} + z\vec{k}$. Evaluate:

(a) $\int_C \vec{F} \cdot d\vec{r}$ where C is the z-axis from the origin to $(0, 0, 10)$.

(b) $\int_S \vec{F} \cdot d\vec{A}$ where S is the disk $x^2 + y^2 \leq 3$, $z = 10$.

(c) $\int_S \vec{F} \cdot d\vec{A}$ where S is the closed box with edges of length 2 in the first octant, with one corner at the origin and edges along the axes.

(d) $\int_C \vec{F} \cdot d\vec{r}$ where C is the circle of radius 3, centered on the z-axis in the plane $z = 4$, and oriented counterclockwise when viewed from above.

CAS Challenge Problems

51. (a) Let $\vec{F} = x^3 y\vec{i} + 2xz^3\vec{j} + (z^3 + 4x^2)\vec{k}$. Compute curl $\vec{F}(1, 2, 1)$.

(b) Consider the family of curves C_a given, for $0 \leq t \leq 2\pi$, by

$$\vec{r}(t) = \vec{i} + (2 + a\cos t)\vec{j} + (1 + a\sin t)\vec{k}.$$

Evaluate the line integral $\int_{C_a} \vec{F} \cdot d\vec{r}$ and compute the limit

$$\lim_{a \to 0} \frac{\int_{C_a} \vec{F} \cdot d\vec{r}}{\pi a^2}.$$

(c) Repeat part (b) for the family D_a given, for $0 \leq t \leq 2\pi$, by

$$\vec{r}(t) = (1 + a\sin t)\vec{i} + 2\vec{j} + (1 + a\cos t)\vec{k}.$$

(d) Repeat part (b) for the family E_a given, for $0 \leq t \leq 2\pi$, by

$$\vec{r}(t) = (1 + a\cos t)\vec{i} + (2 + a\sin t)\vec{j} + \vec{k}.$$

(e) Compare your answers to parts (b)-(d) with part (a) and explain using the geometric definition of curl.

52. Let S be the sphere of radius R centered at the origin with outward orientation and let

$$\vec{F} = (ax^2 + bxz)\vec{i} + (cy^2 + py)\vec{j} + (qz + rx^3)\vec{k}.$$

(a) Use the Divergence Theorem to express the flux integral $\int_S \vec{F} \cdot d\vec{A}$ as a triple integral. Then use symmetry and the volume formula for a sphere to evaluate the triple integral.

(b) Check your answer in part (a) by computing the flux integral directly.

53. Let $\vec{F}(x, y, z) = x^2 y\vec{i} + 2xz\vec{j} + (z^3 + 4x^2)\vec{k}$ and let S_a be the sphere of radius a centered at $(1, 1, 1)$, oriented outward, parameterized by $\vec{r}(\phi, \theta) = (1 + a\sin\phi\cos\theta)\vec{i} + (1 + a\sin\phi\sin\theta)\vec{j} + (1 + a\cos\phi)\vec{k}$, $0 \leq \phi \leq \pi$, $0 \leq \theta \leq 2\pi$.

(a) Compute div $\vec{F}(1, 1, 1)$.

(b) Use the geometric definition of divergence to estimate $\int_{S_a} \vec{F} \cdot d\vec{A}$ for $a = 0.1$.

(c) Evaluate the flux integral $\int_{S_a} \vec{F} \cdot d\vec{A}$. Compare its value for $a = 0.1$ with your answer to part (b). Then compute the limit

$$\lim_{a \to 0} \frac{\int_{S_a} \vec{F} \cdot d\vec{A}}{\text{Volume inside } S_a}$$

and compare your result with part (a). Explain your answer in terms of the geometric definition of the divergence.

PROJECTS FOR CHAPTER TWENTY

1. **Magnetic Field Generated by a Current in a Wire**

    Under steady-state conditions, a magnetic field $\vec{B}$ has $\operatorname{curl}\vec{B} = \vec{0}$ in a region where there is no current. We study the steady-state magnetic field $\vec{B}$ due to a constant current in an infinitely long straight thin wire. The magnitude of the magnetic field at a point depends only on the distance from the wire and its direction is tangent to the circle around the wire and determined by the right-hand rule. Suppose the current is flowing upward along the z-axis. Then $\vec{B}$ is parallel to the xy-plane and, by the right-hand rule, points counterclockwise around a circle centered on the z-axis. (See Figure 20.26.)

    (a) Find the flux of $\operatorname{curl}\vec{B}$ through the surface S between two concentric circles in the xy-plane of radius R_1 and R_2 centered on the wire.

    (b) Calculate the circulation of $\vec{B}$ around each of the two boundary pieces of S, where their orientations are determined by the upward orientation of S.

    (c) Use Stokes' Theorem to deduce that the magnitude of the magnetic field $\vec{B}$ is proportional to the reciprocal of the distance from the wire.

    (d) Compare the magnitude $\|\vec{B}\|$ at two points P and Q if Q is twice as far from the wire as P.

    (e) To decrease the magnitude of $\vec{B}$ by 20%, by what factor does the distance from the wire have to increase?

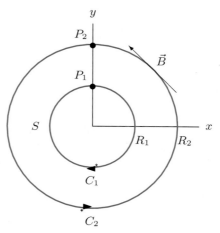

Figure 20.26: Current along positive z-axis
(out of page)

2. **Curl in Natural Coordinates**

    Let $\vec{F}(x, y) = u(x, y)\vec{i} + v(x, y)\vec{j}$ be a 2-dimensional vector field. Let $F(x, y)$ be the magnitude of $\vec{F}$ and let $\theta(x, y)$ be the angle of $\vec{F}$ with the positive x-axis at the point (x, y), so that $u = F\cos\theta$ and $v = F\sin\theta$. Let $\vec{T}$ be the unit vector in the direction of $\vec{F}$, and let $\vec{N}$ be the unit vector in the direction of $\vec{k} \times \vec{F}$, perpendicular to $\vec{F}$. Show that

$$\operatorname{curl}\vec{F} = c\vec{k} \qquad \text{where} \qquad c = F\theta_{\vec{T}} - F_{\vec{N}}.$$

The scalar c is called the *scalar curl* or the *vorticity* of the vector field $\vec{F}$. This problem shows that the vorticity is the difference of two terms, the *curvature vorticity*, $F\theta_{\vec{T}}$, due to turning of the flow lines of $\vec{F}$, and the *shear vorticity*, $F_{\vec{N}}$, due to changes in the magnitude of $\vec{F}$ in a direction normal to $\vec{F}$.

Chapter Twenty-One

PARAMETERS, COORDINATES, AND INTEGRALS

Contents

21.1 COORDINATES AND PARAMETERIZED SURFACES

In Chapter 17 we parameterized curves in 2- and 3-space, and in Chapter 16 we used polar, cylindrical, and spherical coordinates to simplify iterated integrals. We now take a second look at parameterizations and coordinate systems, and see that they are the same thing in different disguises: functions from one space to another.

We have already seen this with parameterized curves, which we view as a function from an interval $a \leq t \leq b$ to a curve in xyz-space. See Figure 21.1.

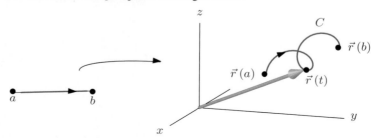

Figure 21.1: The parameterisation is a function from the interval, $a \leq t \leq b$, to 3-space, whose image is the curve, C

Polar, Cylindrical, and Spherical Coordinates Revisited

The equations for polar coordinates,

$$x = r \cos \theta$$
$$y = r \sin \theta,$$

can also be viewed as defining a function from the $r\theta$-plane into the xy-plane. This function transforms the rectangle on the left of Figure 21.2 into the quarter disk on the right. We need two parameters to describe this disk because it is a two-dimensional object.

Polar Coordinates as Families of Parameterized Curves

Polar coordinates give two families of parameterized curves, which form the polar coordinate grid. The lines $r = $ Constant in the $r\theta$-plane correspond to circles in the xy-plane, each circle parameterized by θ; the lines $\theta = $ Constant correspond to rays in the xy-plane, each ray parameterized by r.

Cylindrical and Spherical Coordinates

Similarly, cylindrical and spherical coordinates may be viewed as functions from 3-space to 3-space. Cylindrical coordinates take rectangular boxes in $r\theta z$-space and map them to cylindrical regions in xyz-space; spherical coordinates take rectangular boxes in $\rho\theta\phi$-space and map them to spherical regions in xyz-space.

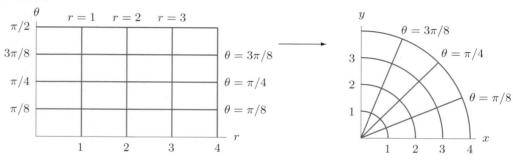

Figure 21.2: A grid in the $r\theta$-plane and the corresponding curved grid in the xy-plane

General Parameterizations

In general, a parameterization or coordinate system provides a way of representing a curved object by means of a simple region in the *parameter space* (an interval, rectangle, or rectangular box), along with a function mapping that region into the curved object. In the next section, we use this idea to parameterize curved surfaces in 3-space.

How Do We Parameterise a Surface?

In Section 17.1 we parameterised a circle in 2-space using the equations

$$x = \cos t, \quad y = \sin t.$$

In 3-space, the same circle in the xy-plane has parametric equations

$$x = \cos t, \quad y = \sin t, \quad z = 0.$$

We add the equation $z = 0$ to specify that the circle is in the xy-plane. If we wanted a circle in the plane $z = 3$, we would use the equations

$$x = \cos t, \quad y = \sin t, \quad z = 3.$$

Suppose now we let z vary freely, as well as t. We get circles in every horizontal plane, forming a cylinder as in the left of Figure 21.3. Thus, we need two parameters, t and z, to parameterise the cylinder.

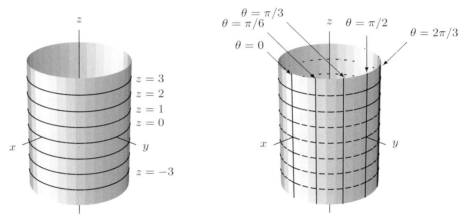

Figure 21.3: The cylinder $x = \cos t$, $y = \sin t$, $z = z$

We can contrast curves and surfaces. A curve, though it may live in two or three dimensions, is itself one-dimensional; if we move along it we can only move backward and forward in one direction. Thus, it only requires one parameter to trace out a curve.

A surface is 2-dimensional; at any given point there are two independent directions we can move. For example, on the cylinder we can move vertically, or we can circle around the z-axis horizontally. So we need *two* parameters to describe it. We can think of the parameters as map coordinates, like longitude and latitude on the surface of the earth. Just as polar coordinates give a polar grid on a circular region, so the parameters for a surface give a grid on the surface. See Figure 21.3 on the right.

In the case of the cylinder our parameters are t and z, so

$$x = \cos t, \quad y = \sin t, \quad z = z, \quad 0 \le t < 2\pi, \quad -\infty < z < \infty.$$

The last equation, $z = z$, looks strange, but it reminds us that we are in three dimensions, not two, and that the z-coordinate on our surface is allowed to vary freely.

In general, we express the coordinates, (x, y, z), of a point on a surface S in terms of two parameters, s and t:

$$x = f_1(s, t), \quad y = f_2(s, t), \quad z = f_3(s, t).$$

As the values of s and t vary, the corresponding point (x, y, z) sweeps out the surface, S. (See Figure 21.4.) The function which sends the point (s, t) to the point (x, y, z) is called the *parameterisation of the surface*.

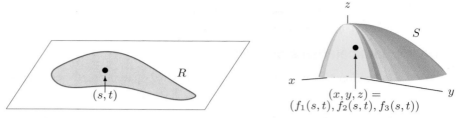

Figure 21.4: The parameterisation sends each point (s, t) in the parameter region, R, to a point $(x, y, z) = (f_1(s, t), f_2(s, t), f_3(s, t))$ in the surface, S

Using Position Vectors

We can use the position vector $\vec{r} = x\vec{i} + y\vec{j} + z\vec{k}$ to combine the three parametric equations for a surface into a single vector equation. For example, the parameterisation of the cylinder $x = \cos t, y = \sin t, z = z$ can be written as

$$\vec{r}(t, z) = \cos t\,\vec{i} + \sin t\,\vec{j} + z\vec{k} \qquad 0 \le t < 2\pi, \quad -\infty < z < \infty.$$

For a general parameterised surface S, we write

$$\vec{r}(s, t) = f_1(s, t)\vec{i} + f_2(s, t)\vec{j} + f_3(s, t)\vec{k}.$$

Parameterising a Surface of the Form $z = f(x, y)$

The graph of a function $z = f(x, y)$ can be given parametrically simply by letting the parameters s and t be x and y:

$$x = s, \quad y = t, \quad z = f(s, t).$$

Example 1   Give a parametric description of the lower hemisphere of the sphere $x^2 + y^2 + z^2 = 1$.

Solution   The surface is the graph of the function $z = -\sqrt{1 - x^2 - y^2}$ over the region $x^2 + y^2 \le 1$ in the plane. Then parametric equations are $x = s, y = t, z = -\sqrt{1 - s^2 - t^2}$, where the parameters s and t vary inside the unit circle.

In practice we often think of x and y as parameters rather than introduce new parameters s and t. Thus, we may write $x = x, y = y, z = f(x, y)$.

Parameterising Planes

Consider a plane containing two nonparallel vectors $\vec{v}_1$ and $\vec{v}_2$ and a point P_0 with position vector $\vec{r}_0$. We can get to any point on the plane by starting at P_0 and moving parallel to $\vec{v}_1$ or $\vec{v}_2$, adding multiples of them to $\vec{r}_0$. (See Figure 21.5.)

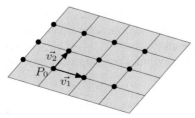

Figure 21.5: The plane $\vec{r}(s, t) = \vec{r}_0 + s\vec{v}_1 + t\vec{v}_2$ and some points corresponding to various choices of s and t

Since $s\vec{v}_1$ is parallel to $\vec{v}_1$ and $t\vec{v}_2$ is parallel to $\vec{v}_2$, we have the following result:

> ### Parameterising a Plane
>
> The plane through the point with position vector $\vec{r}_0$ and containing the two nonparallel vectors $\vec{v}_1$ and $\vec{v}_2$ has parameterization
>
> $$\vec{r}(s,t) = \vec{r}_0 + s\vec{v}_1 + t\vec{v}_2.$$

If $\vec{r}_0 = x_0\vec{i} + y_0\vec{j} + z_0\vec{k}$, and $\vec{v}_1 = a_1\vec{i} + a_2\vec{j} + a_3\vec{k}$, and $\vec{v}_2 = b_1\vec{i} + b_2\vec{j} + b_3\vec{k}$, then the parameterization of the plane can be expressed with the parametric equations

$$x = x_0 + sa_1 + tb_1, \quad y = y_0 + sa_2 + tb_2, \quad z = z_0 + sa_3 + tb_3.$$

Notice that the parameterisation of the plane expresses the coordinates $x, y,$ and z as linear functions of the parameters s and t.

Example 2
Write a parameterization for the plane through the point $(2, -1, 3)$ and containing the vectors $\vec{v}_1 = 2\vec{i} + 3\vec{j} - \vec{k}$ and $\vec{v}_2 = \vec{i} - 4\vec{j} + 5\vec{k}$.

Solution
A possible parameterization is

$$\begin{aligned}
\vec{r}(s,t) = \vec{r}_0 + s\vec{v}_1 + t\vec{v}_2 &= 2\vec{i} - \vec{j} + 3\vec{k} + s(2\vec{i} + 3\vec{j} - \vec{k}) + t(\vec{i} - 4\vec{j} + 5\vec{k}) \\
&= (2 + 2s + t)\vec{i} + (-1 + 3s - 4t)\vec{j} + (3 - s + 5t)\vec{k},
\end{aligned}$$

or equivalently,

$$x = 2 + 2s + t, \quad y = -1 + 3s - 4t, \quad z = 3 - s + 5t.$$

Parameterisations Using Spherical Coordinates

Recall the spherical coordinates ρ, ϕ, and θ introduced on page 916 of Chapter 16. On a sphere of radius $\rho = a$ we can use ϕ and θ as coordinates, similar to latitude and longitude on the surface of the earth. (See Figure 21.6.) The latitude, however, is measured from the equator, whereas ϕ is measured from the north pole. If the positive x-axis passes through the Greenwich meridian, the longitude and θ are equal for $0 \le \theta \le \pi$.

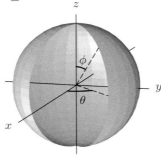

Figure 21.6: Parameterising the sphere by ϕ and θ

Example 3
You are at a point on a sphere with $\phi = 3\pi/4$. Are you in the northern or southern hemisphere? If ϕ decreases, do you move closer to or farther from the equator?

Solution
The equator has $\phi = \pi/2$. Since $3\pi/4 > \pi/2$, you are in the southern hemisphere. If ϕ decreases, you move closer to the equator.

Example 4
On a sphere, you are standing at a point with coordinates θ_0 and ϕ_0. Your *antipodal* point is the point on the other side of the sphere on a line through you and the centre. What are the θ, ϕ coordinates of your antipodal point?

Solution   Figure 21.7 shows that the coordinates are $\theta = \theta_0 + \pi$ if $\theta_0 < \pi$ or $\theta = \theta_0 - \pi$ if $\pi \leq \theta_0 \leq 2\pi$, and $\phi = \pi - \phi_0$. Notice that if you are on the equator, then so is your antipodal point.

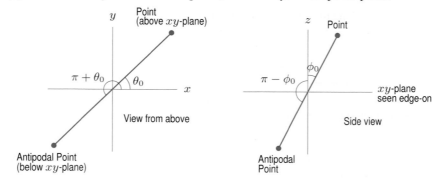

Figure 21.7: Two views of the xyz-coordinate system showing coordinates of antipodal points

Parameterising a Sphere Using Spherical Coordinates

The sphere with radius 1 centred at the origin is parameterised by

$$x = \sin \phi \cos \theta, \qquad y = \sin \phi \sin \theta, \qquad z = \cos \phi,$$

where $0 \leq \theta \leq 2\pi$ and $0 \leq \phi \leq \pi$. (See Figure 21.8.)

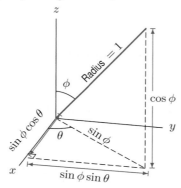

Figure 21.8: The relationship between x, y, z and ϕ, θ on a sphere of radius 1

We can also write these equations in vector form:

$$\vec{r}(\theta, \phi) = \sin \phi \cos \theta \, \vec{i} + \sin \phi \sin \theta \, \vec{j} + \cos \phi \, \vec{k}.$$

Since $x^2 + y^2 + z^2 = \sin^2 \phi (\cos^2 \theta + \sin^2 \theta) + \cos^2 \phi = \sin^2 \phi + \cos^2 \phi = 1$, this verifies that the point with position vector $\vec{r}(\theta, \phi)$ does lie on the sphere of radius 1. Notice that the z-coordinate depends only on the parameter ϕ. Geometrically, this means that all points on the same latitude have the same z-coordinate.

Example 5   Find parametric equations for the following spheres:

(a) Centre at the origin and radius 2.
(b) Centre at the point $(2, -1, 3)$ and radius 2.

Solution   (a) We must scale the distance from the origin by 2. Thus, we have

$$x = 2 \sin \phi \cos \theta, \qquad y = 2 \sin \phi \sin \theta, \qquad z = 2 \cos \phi,$$

where $0 \leq \theta \leq 2\pi$ and $0 \leq \phi \leq \pi$. In vector form, this is written

$$\vec{r}(\theta, \phi) = 2 \sin \phi \cos \theta \vec{i} + 2 \sin \phi \sin \theta \vec{j} + 2 \cos \phi \vec{k}.$$

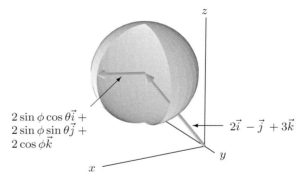

$$2\sin\phi\cos\theta\vec{i} +$$
$$2\sin\phi\sin\theta\vec{j} +$$
$$2\cos\phi\vec{k}$$

$$2\vec{i} - \vec{j} + 3\vec{k}$$

Figure 21.9: Sphere with centre at the point $(2, -1, 3)$ and radius 2

(b) To shift the centre of the sphere from the origin to the point $(2, -1, 3)$, we add the vector parameterisation we found in part (a) to the position vector of $(2, -1, 3)$. (See Figure 21.9.) This gives

$$\vec{r}(\theta, \phi) = 2\vec{i} - \vec{j} + 3\vec{k} + (2\sin\phi\cos\theta\vec{i} + 2\sin\phi\sin\theta\vec{j} + 2\cos\phi\vec{k})$$
$$= (2 + 2\sin\phi\cos\theta)\vec{i} + (-1 + 2\sin\phi\sin\theta)\vec{j} + (3 + 2\cos\phi)\vec{k},$$

where $0 \leq \theta \leq 2\pi$ and $0 \leq \phi \leq \pi$. Alternatively,

$$x = 2 + 2\sin\phi\cos\theta, \qquad y = -1 + 2\sin\phi\sin\theta, \qquad z = 3 + 2\cos\phi.$$

Note that the same point can have more than one value for θ or ϕ. For example, points with $\theta = 0$ also have $\theta = 2\pi$, unless we restrict θ to the range $0 \leq \theta < 2\pi$. Also, the north pole, at $\phi = 0$, and the south pole, at $\phi = \pi$, can have any value of θ.

Parameterising Surfaces of Revolution

Many surfaces have an axis of rotational symmetry and circular cross-sections perpendicular to that axis. These surfaces are referred to as *surfaces of revolution.*

Example 6   Find a parameterisation of the cone whose base is the circle $x^2 + y^2 = a^2$ in the xy-plane and whose vertex is at height h above the xy-plane. (See Figure 21.10.)

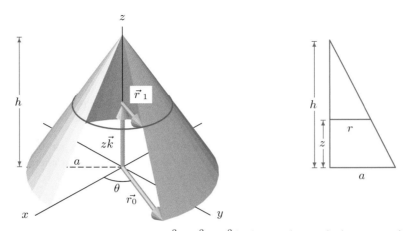

Figure 21.10: The cone whose base is the circle $x^2 + y^2 = a^2$ in the xy-plane and whose vertex is at the point $(0, 0, h)$ and the vertical cross-section through the cone

Solution   We use cylindrical coordinates, r, θ, z. (See Figure 21.10.) In the xy-plane, the radius vector, $\vec{r}_0$, from the z-axis to a point on the cone in the xy-plane is

$$\vec{r}_0 = a \cos \theta \vec{i} + a \sin \theta \vec{j}.$$

Above the xy-plane, the radius of the circular cross-section, r, decreases linearly from $r = a$ when $z = 0$ to $r = 0$ when $z = h$. From the similar triangles in Figure 21.10,

$$\frac{a}{h} = \frac{r}{h - z}.$$

Solving for r, we have

$$r = \left(1 - \frac{z}{h}\right) a.$$

The horizontal radius vector, $\vec{r}_1$, at height z has components similar to $\vec{r}_0$, but with a replaced by r:

$$\vec{r}_1 = r \cos \theta \vec{i} + r \sin \theta \vec{j} = \left(1 - \frac{z}{h}\right) a \cos \theta \vec{i} + \left(1 - \frac{z}{h}\right) a \sin \theta \vec{j}.$$

As θ goes from 0 to 2π, the vector $\vec{r}_1$ traces out the horizontal circle in Figure 21.10. We get the position vector, $\vec{r}$, of a point on the cone by adding the vector $z\vec{k}$, so

$$\vec{r} = \vec{r}_1 + z\vec{k} = a \left(1 - \frac{z}{h}\right) \cos \theta \vec{i} + a \left(1 - \frac{z}{h}\right) \sin \theta \vec{j} + z\vec{k}, \quad \text{for } 0 \le z \le h \text{ and } 0 \le \theta \le 2\pi.$$

These equations can be written as

$$x = \left(1 - \frac{z}{h}\right) a \cos \theta, \quad y = \left(1 - \frac{z}{h}\right) a \sin \theta, \quad z = z.$$

The parameters are θ and z.

Example 7   Consider the bell of a trumpet. A model for the radius $z = f(x)$ of the horn (in cm) at a distance x cm from the large open end is given by the function

$$f(x) = \frac{6}{(x + 1)^{0.7}}.$$

The bell is obtained by rotating the graph of f about the x-axis. Find a parameterisation for the first 24 cm of the bell. (See Figure 21.11.)

Figure 21.11: The bell of a trumpet obtained by rotating the
graph of $z = f(x)$ about the x-axis

Solution   At distance x from the large open end of the horn, the cross-section parallel to the yz-plane is a circle of radius $f(x)$, with centre on the x-axis. Such a circle can be parameterised by $y = f(x) \cos \theta$, $z = f(x) \sin \theta$. Thus, we have the parameterisation

$$x = x, \quad y = \left(\frac{6}{(x + 1)^{0.7}}\right) \cos \theta, \quad z = \left(\frac{6}{(x + 1)^{0.7}}\right) \sin \theta, \quad 0 \le x \le 24, \quad 0 \le \theta \le 2\pi.$$

The parameters are x and θ.

Parameter Curves

On a parameterised surface, the curve obtained by setting one of the parameters equal to a constant and letting the other vary is called a *parameter curve*. If the surface is parameterised by

$$\vec{r}(s,t) = f_1(s,t)\vec{i} + f_2(s,t)\vec{j} + f_3(s,t)\vec{k},$$

there are two families of parameter curves on the surface, one family with t constant and the other with s constant.

Example 8   Consider the vertical cylinder

$$x = \cos t, \quad y = \sin t, \quad z = z.$$

(a) Describe the two parameter curves through the point $(0, 1, 1)$.
(b) Describe the family of parameter curves with t constant and the family with z constant.

Solution   (a) Since the point $(0, 1, 1)$ corresponds to the parameter values $t = \pi/2$ and $z = 1$, there are two parameter curves, one with $t = \pi/2$ and the other with $z = 1$. The parameter curve with $t = \pi/2$ has the parametric equations

$$x = \cos\left(\frac{\pi}{2}\right) = 0, \quad y = \sin\left(\frac{\pi}{2}\right) = 1, \quad z = z,$$

with parameter z. This is a line through the point $(0, 1, 1)$ parallel to the z-axis.
    The parameter curve with $z = 1$ has the parametric equations

$$x = \cos t, \quad y = \sin t, \quad z = 1,$$

with parameter t. This is a unit circle parallel to and one unit above the xy-plane centred on the z-axis.

(b) First, fix $t = t_0$ for t and let z vary. The curves parameterised by z have equations

$$x = \cos t_0, \quad y = \sin t_0, \quad z = z.$$

These are vertical lines on the cylinder parallel to the z-axis. (See Figure 21.12.)
    The other family is obtained by fixing $z = z_0$ and varying t. Curves in this family are parameterised by t and have equations

$$x = \cos t, \quad y = \sin t, \quad z = z_0.$$

They are circles of radius 1 parallel to the xy-plane centred on the z-axis. (See Figure 21.13.)

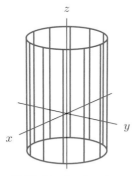

Figure 21.12: The family of parameter curves with $t = t_0$ for the cylinder $x = \cos t, y = \sin t, z = z$

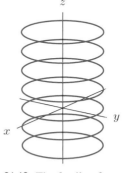

Figure 21.13: The family of parameter curves with $z = z_0$ for the cylinder $x = \cos t, y = \sin t, z = z$

Example 9   Describe the families of parameter curves with $\theta = \theta_0$ and $\phi = \phi_0$ for the sphere

$$x = \sin\phi\cos\theta, \quad y = \sin\phi\sin\theta, \quad z = \cos\phi,$$

where $0 \le \theta \le 2\pi, 0 \le \phi \le \pi$.

Solution   Since ϕ measures latitude, the family with ϕ constant consists of the circles of constant latitude. (See Figure 21.14.) Similarly, the family with θ constant consists of the meridians (semicircles) running between the north and south poles. (See Figure 21.15.)

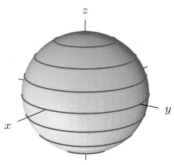
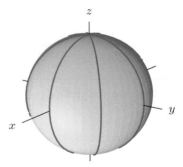

Figure 21.14: The family of parameter curves with $\phi = \phi_0$ for the sphere parameterised by (θ, ϕ)

Figure 21.15: The family of parameter curves with $\theta = \theta_0$ for the sphere parameterised by (θ, ϕ)

We have seen parameter curves before on pages 693-694 of Section 12.2: The cross-sections with $x = a$ or $y = b$ on a surface $z = f(x, y)$ are examples of parameter curves. So are the grid lines on a computer sketch of a surface. The small regions shaped like parallelograms surrounded by nearby pairs of parameter curves are called *parameter rectangles*. See Figure 21.16.

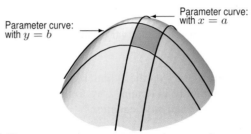

Parameter curve: with $y = b$

Parameter curve: with $x = a$

Figure 21.16: Parameter curves $x = a$ or $y = b$ on a surface $z = f(x, y)$; the darker region is a parameter rectangle

Exercises and Problems for Section 21.1

Exercises

Describe in words the objects parameterized by the equations in Exercises 1–4. (Note: r and θ are cylindrical coordinates.)

1. $x = r \cos \theta \qquad y = r \sin \theta \qquad z = 7$
$\quad\ 0 \le r \le 5 \qquad 0 \le \theta \le 2\pi$

2. $x = 5 \cos \theta \qquad y = 5 \sin \theta \qquad z = z$
$\quad\ 0 \le \theta \le 2\pi \qquad 0 \le z \le 7$

3. $x = 5 \cos \theta \qquad y = 5 \sin \theta \qquad z = 50$
$\quad\ 0 \le \theta \le 2\pi$

4. $x = r \cos \theta \qquad y = r \sin \theta \qquad z = r$
$\quad\ 0 \le r \le 5 \qquad 0 \le \theta \le 2\pi$

In Exercises 5–8, for a sphere parameterized using the spherical coordinates θ and ϕ, describe in words the part of the sphere given by the restrictions.

5. $0 \le \theta < 2\pi, \quad 0 \le \phi \le \pi/2$

6. $\pi \le \theta < 2\pi, \quad 0 \le \phi \le \pi$

7. $\pi/4 \le \theta < \pi/3, \quad 0 \le \phi \le \pi$

8. $0 \le \theta \le \pi, \quad \pi/4 \le \phi < \pi/3$

In Exercises 9–12 decide if the parameterization describes a curve or a surface.

9. $\vec{r}(s) = s\vec{i} + (3 - s)\vec{j} + s^2\vec{k}$

10. $\vec{r}(s, t) = (s + t)\vec{i} + (3 - s)\vec{j}$

11. $\vec{r}(s, t) = \cos s\, \vec{i} + \sin s\, \vec{j} + t^2\vec{k}$

12. $\vec{r}(s) = \cos s\, \vec{i} + \sin s\, \vec{j} + s^2\vec{k}$

Problems

In Problems 13–16, describe the families of parameter curves with $s = s_0$ and $t = t_0$ for the parameterized surface.

13. $x = s$, $y = t$, $z = 1$ for $-\infty < s < \infty$, $-\infty < t < \infty$

14. $x = s$, $y = \cos t$, $z = \sin t$ for $-\infty < s < \infty$, $0 \le t \le 2\pi$

15. $x = s$ $y = t$, $z = s^2 + t^2$ for $-\infty < s < \infty$, $-\infty < t < \infty$

16. $x = \cos s \sin t$ $y = \sin s \sin t$, $z = \cos t$ for $0 \le s \le 2\pi$, $0 \le t \le \pi$

In Problems 17–18, parameterize the plane that contains the three points.

17. $(0, 0, 0)$, $(1, 2, 3)$, $(2, 1, 0)$

18. $(1, 2, 3)$, $(2, 5, 8)$, $(5, 2, 0)$

In Problems 19–20, parameterize the plane through the point with the given normal vector.

19. $(3, 5, 7)$, $\vec{i} + \vec{j} + \vec{k}$

20. $(5, 1, 4)$, $\vec{i} + 2\vec{j} + 3\vec{k}$

21. Does the plane $\vec{r}(s, t) = (2 + s)\vec{i} + (3 + s + t)\vec{j} + 4t\vec{k}$ contain the following points?
(a) $(4, 8, 12)$ (b) $(1, 2, 3)$

22. Are the following two planes parallel?

$$x = 2 + s + t, \quad y = 4 + s - t, \quad z = 1 + 2s, \quad \text{and}$$

$$x = 2 + s + 2t, \quad y = t, \quad z = s - t.$$

23. A city is described parametrically by the equation

$$\vec{r} = (x_0\vec{i} + y_0\vec{j} + z_0\vec{k}) + s\vec{v_1} + t\vec{v_2}$$

where $\vec{v}_1 = 2\vec{i} - 3\vec{j} + 2\vec{k}$ and $\vec{v}_2 = \vec{i} + 4\vec{j} + 5\vec{k}$. A city block is a rectangle determined by $\vec{v}_1$ and $\vec{v}_2$. East is in the direction of $\vec{v}_1$ and north is in the direction of $\vec{v}_2$. Starting at the point (x_0, y_0, z_0), you walk 5 blocks east, 4 blocks north, 1 block west and 2 blocks south. What are the parameters of the point where you end up? What are your x, y and z coordinates at that point?

24. You are at a point on the earth with longitude $80°$ West of Greenwich, England, and latitude $40°$ North of the equator.

(a) If your latitude decreases, have you moved nearer to or farther from the equator?
(b) If your latitude decreases, have you moved nearer to or farther from the north pole?
(c) If your longitude increases (say, to $90°$ West), have you moved nearer to or farther from Greenwich?

25. Describe in words the curve $\phi = \pi/4$ on the surface of the globe.

26. Describe in words the curve $\theta = \pi/4$ on the surface of the globe.

27. A decorative oak post is $48cm$ long and is turned on a lathe so that its profile is sinusoidal, as shown in Figure 21.17.

(a) Describe the surface of the post parametrically using cylindrical coordinates.
(b) Find the volume of the post.

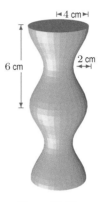

$\mapsto$4 cm$\mapsto$

2 cm

6 cm

Figure 21.17

28. Find parametric equations for the sphere $(x - a)^2 + (y - b)^2 + (z - c)^2 = d^2$.

29. Suppose you are standing at a point on the equator of a sphere, parameterized by spherical coordinates θ_0 and ϕ_0. If you go halfway around the equator and halfway up toward the north pole along a longitude, what are your new θ and ϕ coordinates?

30. Find parametric equations for the cone $x^2 + y^2 = z^2$.

31. Parameterize the cone in Example 6 on page 1101 in terms of r and θ.

32. Give a parameterization of the circle of radius a centered at the point (x_0, y_0, z_0) and in the plane parallel to two given unit vectors $\vec{u}$ and $\vec{v}$ such that $\vec{u} \cdot \vec{v} = 0$.

For Problems 33–35,

(a) Write an equation in x, y, z and identify the parametric surface.
(b) Draw a picture of the surface.

33. $x = 2s$ $y = s + t$ $z = 1 + s - t$
    $0 \le s \le 1$ $0 \le t \le 1$

34. $x = s$ $y = t$ $z = \sqrt{1 - s^2 - t^2}$
    $s^2 + t^2 \le 1$ $s, t \ge 0$

35. $x = s + t$ $y = s - t$ $z = s^2 + t^2$
    $0 \le s \le 1$ $0 \le t \le 1$

Strengthen Your Understanding

In Problems 36–37, explain what is wrong with the statement.

36. The parameter curves of a parameterized surface intersect at right angles.

37. The parameter curves for constant ϕ on the sphere $\vec{r}(\theta, \phi) = R \sin \phi \cos \theta \vec{i} + R \sin \phi \sin \theta \vec{j} + R \cos \phi \vec{k}$ are circles of radius R.

In Problems 38–40, give an example of:

38. A parameterization $\vec{r}(s, t)$ of the plane tangent to the unit sphere at the point where $\theta = \pi/4$ and $\phi = \pi/4$.

39. An equation of the form $f(x, y, z) = 0$ for the plane

$$\vec{r}(s, t) = (s + 1)\vec{i} + (t + 2)\vec{j} + (s + t)\vec{k}.$$

40. A parameterized curve on the sphere $\vec{r}(\theta, \phi) = \sin \phi \cos \theta \vec{i} + \sin \phi \sin \theta \vec{j} + \cos \phi \vec{k}$ that is not a parameter curve.

Are the statements in Problems 41–47 true or false? Give reasons for your answer.

41. The equations $x = s + 1, y = t - 2, z = 3$ parameterize a plane.

42. The equations $x = 2s - 1, y = -s + 3, z = 4 + s$ parameterize a plane.

43. If $\vec{r} = \vec{r}(s, t)$ parameterizes the upper hemisphere $x^2 + y^2 + z^2 = 1, z \geq 0$, then $\vec{r} = -\vec{r}(s, t)$ parameterizes the lower hemisphere $x^2 + y^2 + z^2 = 1, z \leq 0$.

44. If $\vec{r} = \vec{r}(s, t)$ parameterizes the upper hemisphere $x^2 + y^2 + z^2 = 1, z \geq 0$, then $\vec{r} = \vec{r}(-s, -t)$ parameterizes the lower hemisphere $x^2 + y^2 + z^2 = 1, z \leq 0$.

45. If $\vec{r_1}(s, t)$ parameterizes a plane then $\vec{r_2}(s, t) = \vec{r}_1(s, t) + 2\vec{i} - 3\vec{j} + \vec{k}$ parameterizes a parallel plane.

46. Every point on a parameterized surface has a parameter curve passing through it.

47. If $s_0 \neq s_1$, then the parameter curves $\vec{r}(s_0, t)$ and $\vec{r}(s_1, t)$ do not intersect.

48. Match the parameterizations (I)–(IV) with the surfaces (a)–(d). In all cases $0 \leq s \leq \pi/2, 0 \leq t \leq \pi/2$. Note that only part of the surface may be described by the given parameterization.

    (a) Cylinder
    (b) Plane
    (c) Sphere
    (d) Cone

I. $x = \cos s, \quad y = \sin t, \quad z = \cos s + \sin t$
II. $x = \cos s, \quad y = \sin s, \quad z = \cos t$
III. $x = \sin s \cos t, \quad y = \sin s \sin t, \quad z = \cos s$
IV. $x = \cos s, \quad y = \sin t, \quad z = \sqrt{\cos^2 s + \sin^2 t}$

21.2 CHANGE OF COORDINATES IN A MULTIPLE INTEGRAL

In Chapter 16 we used polar, cylindrical, and spherical coordinates to simplify iterated integrals. In this section, we discuss more general changes of coordinate. In the process, we see where the factors r and $\rho^2 \sin \phi$ come from when we convert to polar, cylindrical, or spherical coordinates (see pages 910, 915, and 918).

Polar Change of Coordinates Revisited

Consider the integral $\int_R (x + y) \, dA$ where R is the region in the first quadrant bounded by the circle $x^2 + y^2 = 16$ and the x and y-axes. Writing the integral in Cartesian and polar coordinates, we have

$$\int_R (x + y) \, dA = \int_0^4 \int_0^{\sqrt{16-x^2}} (x + y) \, dy \, dx = \int_0^{\pi/2} \int_0^4 (r \cos \theta + r \sin \theta) r \, dr \, d\theta.$$

The integral on the right is over the rectangle in the $r\theta$-plane given by $0 \leq r \leq 4, 0 \leq \theta \leq \pi/2$. The conversion from polar to Cartesian coordinates changes this rectangle into a quarter-disk. Figure 21.18 shows how a typical rectangle (shaded) in the $r\theta$-plane with sides of length Δr and $\Delta \theta$ corresponds to a curved rectangle in the xy-plane with sides of length Δr and $r\Delta \theta$. The extra r is needed because the correspondence between r, θ and x, y not only curves the lines $r = 1, 2, 3 \ldots$ into circles, it also stretches those lines around larger and larger circles.

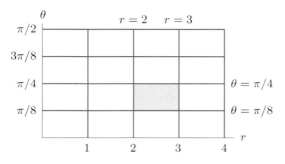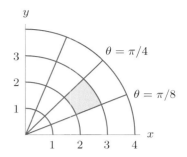

Figure 21.18: A grid in the $r\theta$-plane and the corresponding curved grid in the xy-plane

General Change of Coordinates

We now consider a general change of coordinates, where x, y coordinates are related to s, t coordinates by the differentiable functions

$$x = x(s, t) \quad y = y(s, t).$$

Just as a rectangular region in the $r\theta$-plane corresponds to a region in the xy-plane, a rectangular region, T, in the st-plane corresponds to a region, R, in the xy-plane. We assume that the change of coordinates is one-to-one, that is, that each point in R corresponds to only one point in T.

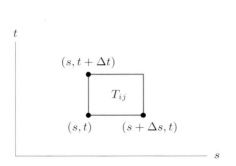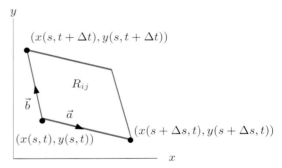

Figure 21.19: A small rectangle T_{ij} in the st-plane and the corresponding region R_{ij} of the xy-plane

We divide T into small rectangles T_{ij} with sides of length Δs and Δt. (See Figure 21.19.) The corresponding piece R_{ij} of the xy-plane is a quadrilateral with curved sides. If we choose Δs and Δt small, then by local linearity of $x(s, t)$ and $y(s, t)$, we know R_{ij} is approximately a parallelogram.

Recall from Chapter 13 that the area of the parallelogram with sides $\vec{a}$ and $\vec{b}$ is $\|\vec{a} \times \vec{b}\|$. Thus, we need to find the sides of R_{ij} as vectors. The side of R_{ij} corresponding to the bottom side of T_{ij} has endpoints $(x(s, t), y(s, t))$ and $(x(s + \Delta s, t), y(s + \Delta s, t))$, so in vector form that side is

$$\vec{a} = (x(s + \Delta s, t) - x(s, t))\vec{i} + (y(s + \Delta s, t) - y(s, t))\vec{j} \approx \left(\frac{\partial x}{\partial s}\Delta s\right)\vec{i} + \left(\frac{\partial y}{\partial s}\Delta s\right)\vec{j}.$$

Similarly, the side of R_{ij} corresponding to the left edge of T_{ij} is given by

$$\vec{b} \approx \left(\frac{\partial x}{\partial t}\Delta t\right)\vec{i} + \left(\frac{\partial y}{\partial t}\Delta t\right)\vec{j}.$$

Computing the cross product, we get

$$\text{Area } R_{ij} \approx \|\vec{a} \times \vec{b}\| \approx \left|\left(\frac{\partial x}{\partial s}\Delta s\right)\left(\frac{\partial y}{\partial t}\Delta t\right) - \left(\frac{\partial x}{\partial t}\Delta t\right)\left(\frac{\partial y}{\partial s}\Delta s\right)\right|$$

$$= \left|\frac{\partial x}{\partial s} \cdot \frac{\partial y}{\partial t} - \frac{\partial x}{\partial t} \cdot \frac{\partial y}{\partial s}\right|\Delta s \Delta t.$$

Using determinant notation,[1] we define the *Jacobian*, $\dfrac{\partial(x, y)}{\partial(s, t)}$, as follows:

$$\frac{\partial(x, y)}{\partial(s, t)} = \frac{\partial x}{\partial s} \cdot \frac{\partial y}{\partial t} - \frac{\partial x}{\partial t} \cdot \frac{\partial y}{\partial s} = \begin{vmatrix} \frac{\partial x}{\partial s} & \frac{\partial x}{\partial t} \\[2mm] \frac{\partial y}{\partial s} & \frac{\partial y}{\partial t} \end{vmatrix}.$$

Thus, we can write

$$\text{Area } R_{ij} \approx \left| \frac{\partial(x, y)}{\partial(s, t)} \right| \Delta s \, \Delta t.$$

To compute $\int_R f(x, y) \, dA$, where f is a continuous function, we look at the Riemann sum obtained by dividing the region R into the small curved regions R_{ij}, giving

$$\int_R f(x, y) \, dA \approx \sum_{i,j} f(u_{ij}, v_{ij}) \cdot \text{Area of } R_{ij} \approx \sum_{i,j} f(u_{ij}, v_{ij}) \left| \frac{\partial(x, y)}{\partial(s, t)} \right| \Delta s \, \Delta t.$$

Each point (u_{ij}, v_{ij}) in R_{ij} corresponds to a point (s_{ij}, t_{ij}) in T_{ij}, so the sum can be written in terms of s and t:

$$\sum_{i,j} f(x(s_{ij}, t_{ij}), y(s_{ij}, t_{ij})) \left| \frac{\partial(x, y)}{\partial(s, t)} \right| \Delta s \, \Delta t.$$

This is a Riemann sum in terms of s and t, so as Δs and Δt approach 0, we get

$$\int_R f(x, y) \, dA = \int_T f(x(s, t), y(s, t)) \left| \frac{\partial(x, y)}{\partial(s, t)} \right| ds \, dt.$$

To convert an integral from x, y to s, t coordinates we make three changes:
1. Substitute for x and y in the integrand in terms of s and t.
2. Change the xy region R into an st region T.
3. Use the absolute value of the Jacobian to change the area element by making the substi-
   tution $dx \, dy = \left| \dfrac{\partial(x, y)}{\partial(s, t)} \right| ds \, dt.$

Example 1   Check that the Jacobian $\dfrac{\partial(x, y)}{\partial(r, \theta)} = r$ for polar coordinates $x = r \cos \theta$, $y = r \sin \theta$.

Solution   We have $\dfrac{\partial(x, y)}{\partial(r, \theta)} = \begin{vmatrix} \frac{\partial x}{\partial r} & \frac{\partial x}{\partial \theta} \\[2mm] \frac{\partial y}{\partial r} & \frac{\partial y}{\partial \theta} \end{vmatrix} = \begin{vmatrix} \cos \theta & -r \sin \theta \\ \sin \theta & r \cos \theta \end{vmatrix} = r \cos^2 \theta + r \sin^2 \theta = r.$

Example 2   Find the area of the ellipse $\dfrac{x^2}{a^2} + \dfrac{y^2}{b^2} = 1$.

Solution   Let $x = as$, $y = bt$. Then the ellipse $x^2/a^2 + y^2/b^2 = 1$ in the xy-plane corresponds to the circle $s^2 + t^2 = 1$ in the st-plane. The Jacobian is $\begin{vmatrix} a & 0 \\ 0 & b \end{vmatrix} = ab$. Thus, if R is the ellipse in the xy-plane and T is the unit circle in the st-plane, we get

$$\text{Area of } xy\text{-ellipse} = \int_R 1 \, dA = \int_T 1 \, ab \, ds \, dt = ab \int_T ds \, dt = ab \cdot \text{Area of } st\text{-circle} = \pi ab.$$

---

[1] See Appendix E.

Change of Coordinates in Triple Integrals

For triple integrals, there is a similar formula. Suppose the differentiable functions

$$x = x(s, t, u), \quad y = y(s, t, u), \quad z = z(s, t, u)$$

define a one-to-one change of coordinates from a region S in stu-space to a region W in xyz-space. Then, the Jacobian of this change of coordinates is given by the determinant[2]

$$\frac{\partial(x, y, z)}{\partial(s, t, u)} = \begin{vmatrix} \frac{\partial x}{\partial s} & \frac{\partial x}{\partial t} & \frac{\partial x}{\partial u} \\ \frac{\partial y}{\partial s} & \frac{\partial y}{\partial t} & \frac{\partial y}{\partial u} \\ \frac{\partial z}{\partial s} & \frac{\partial z}{\partial t} & \frac{\partial z}{\partial u} \end{vmatrix}.$$

Just as the Jacobian in two dimensions gives us the change in the area element, the Jacobian in three dimensions represents the change in the volume element. Thus, we have

$$\int_W f(x, y, z) \, dx \, dy \, dz = \int_S f(x(s,t,u), y(s,t,u), z(s,t,u)) \left| \frac{\partial(x, y, z)}{\partial(s, t, u)} \right| ds \, dt \, du.$$

Problem 13 at the end of this section asks you to check that the Jacobian for the change of coordinates to spherical coordinates is $\rho^2 \sin \phi$. The next example generalises Example 2 to ellipsoids.

Example 3   Find the volume of the ellipsoid $\dfrac{x^2}{a^2} + \dfrac{y^2}{b^2} + \dfrac{z^2}{c^2} = 1$.

Solution   Let $x = as$, $y = bt$, $z = cu$. The Jacobian is computed to be abc. The xyz-ellipsoid corresponds to the stu-sphere $s^2 + t^2 + u^2 = 1$. Thus, as in Example 2,

$$\text{Volume of } xyz\text{-ellipsoid} = abc \cdot \text{Volume of } stu\text{-sphere} = abc \frac{4}{3}\pi = \frac{4}{3}\pi abc.$$

Exercises and Problems for Section 21.2

Exercises

In Exercises 1–4, find the absolute value of the Jacobian, $\left| \dfrac{\partial(x,y)}{\partial(s,t)} \right|$, for the given change of coordinates.

1. $x = 5s + 2t, y = 3s + t$

2. $x = s^2 - t^2, y = 2st$

3. $x = e^s \cos t, y = e^s \sin t$

4. $x = s^3 - 3st^2, y = 3s^2 t - t^3$

In Exercises 5–6, find the Jacobian.

5. $\dfrac{\partial(x, y, z)}{\partial(s, t, u)}$, where $x = 3s + t + 2u, y = s + 5t - u, z = 2s - t + u$.

6. $\dfrac{\partial(x, y, z)}{\partial(r, \theta, z)}$, where $x = r \cos \theta, y = r \sin \theta, z = z$.

In Exercises 7–9, find positive numbers a and b so that the change of coordinates $s = ax, t = by$ transforms the integral $\int \int_R dx \, dy$ into

$$\int \int_T \left| \frac{\partial(x, y)}{\partial(s, t)} \right| ds \, dt$$

for the given regions R and T.

7. R is the rectangle $0 \le x \le 10, 0 \le y \le 1$ and T is the square $0 \le s, t \le 1$.

8. R is the rectangle $0 \le x \le 1, 0 \le y \le 1/4$ and T is the square $0 \le s, t \le 1$.

9. R is the rectangle $0 \le x \le 50, 0 \le y \le 10$ and T is the square $0 \le s, t \le 1$.

---

[2] See Appendix E.

In Exercises 10–11, find a number a so that the change of coordinates $s = x + ay, t = y$ transforms the integral $\int \int_R dx \, dy$ over the parallelogram R in the xy-plane into an integral

$$\int \int_T \left| \frac{\partial(x, y)}{\partial(s, t)} \right| ds \, dt$$

over a rectangle T in the st-plane.

10. R has vertices $(0, 0)$, $(10, 0)$, $(12, 3)$, $(22, 3)$

11. R has vertices $(0, 0)$, $(10, 0)$, $(-15, 5)$, $(-5, 5)$

Problems

12. Find the region R in the xy-plane corresponding to the region $T = \{(s, t) \mid 0 \le s \le 3, 0 \le t \le 2\}$ under the change of coordinates $x = 2s - 3t$, $y = s - 2t$. Check that

$$\int_R dx \, dy = \int_T \left| \frac{\partial(x, y)}{\partial(s, t)} \right| ds \, dt.$$

13. Compute the Jacobian for the change of coordinates into spherical coordinates:

$$x = \rho \sin \phi \cos \theta, \quad y = \rho \sin \phi \sin \theta, \quad z = \rho \cos \phi.$$

14. For the change of coordinates $x = 3s - 4t$, $y = 5s + 2t$, show that

$$\frac{\partial(x, y)}{\partial(s, t)} \cdot \frac{\partial(s, t)}{\partial(x, y)} = 1$$

15. Use the change of coordinates $x = 2s + t$, $y = s - t$ to compute the integral $\int_R (x + y) \, dA$, where R is the parallelogram formed by $(0, 0)$, $(3, -3)$, $(5, -2)$, and $(2, 1)$.

16. Use the change of coordinates $s = x + y$, $t = y$ to find the area of the ellipse $x^2 + 2xy + 2y^2 \le 1$.

17. Use the change of coordinates $s = y$, $t = y - x^2$ to evaluate $\int \int_R x \, dx \, dy$ over the region R in the first quadrant bounded by $y = 0$, $y = 16$, $y = x^2$, and $y = x^2 - 9$.

18. If R is the triangle bounded by $x + y = 1$, $x = 0$, and $y = 0$, evaluate the integral $\int_R \cos\left(\frac{x - y}{x + y}\right) dx \, dy$.

19. Two independent random numbers x and y from a normal distribution with mean 0 and standard deviation σ have joint density function $p(x, y) = (1/(2\pi\sigma^2))e^{-(x^2 + y^2)/(2\sigma^2)}$. The average $z = (x + y)/2$ has a one-variable probability density function of its own.

(a) Give a double integral expression for $F(t)$, the probability that $z \le t$.

(b) Give a single integral expression for $F(t)$. To do this, make the change of coordinates: $u = (x + y)/2$, $v = (x - y)/2$ and then do the integral on dv. Use the fact that $\int_{-\infty}^{\infty} e^{-x^2/a^2} dx = a\sqrt{\pi}$.

(c) Find the probability density function $F'(t)$ of z.

(d) What is the name of the distribution of z?

20. A river follows the path $y = f(x)$ where x, y are in kilometers. Near the sea, it widens into a lagoon, then narrows again at its mouth. See Figure 21.20. At the point (x, y), the depth, $d(x, y)$, of the lagoon is given by

$$d(x, y) = 40 - 160(y - f(x))^2 - 40x^2 \text{ meters.}$$

The lagoon itself is described by $d(x, y) \ge 0$. What is the volume of the lagoon in cubic meters? [Hint: Use new coordinates $u = x/2$, $v = y - f(x)$ and Jacobians.]

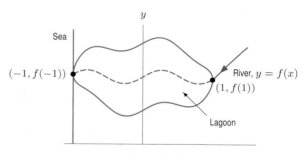

Figure 21.20

Strengthen Your Understanding

In Problems 21–22, explain what is wrong with the statement.

21. If R is the region $0 \le x \le 1$, $0 \le y \le 4$ and T is the region $0 \le s \le 1$, $-2 \le t \le 2$, using the formulas $x = s$, $y = t^2$, we have

$$\int_R f(x, y) \, dx \, dy = \int_T f(s, t^2) \left| \frac{\partial(x, y)}{\partial(s, t)} \right| ds \, dt.$$

22. If R and T are corresponding regions of the xy- and st-planes, the change of coordinates $x = t^3$, $y = s$ leads to

the formula

$$\int_R (x + 2y) \, dx \, dy = \int_T \left(t^3 + 2s\right)\left(-3t^2\right) ds \, dt.$$

In Problems 23–24, give an example of:

23. A change of coordinates $x = x(s, t)$, $y = y(s, t)$ where the rectangle $0 \le s \le 1$, $0 \le t \le 1$ in the st-plane corresponds to a different rectangle in the xy-plane.

24. A change of coordinates $x = x(s, t)$, $y = y(s, t)$ where every region in the st-plane corresponds to a region in the xy-plane with twice the area.

In Problems 25–26, consider a change of variable in the integral $\int_R f(x, y) \, dA$ from x, y to s, t. Are the following statements true or false?

25. If the Jacobian $\left| \dfrac{\partial(x, y)}{\partial(s, t)} \right| > 1$, the value of the s, t-integral is greater than the original x, y-integral.

26. The Jacobian cannot be negative.

21.3 FLUX INTEGRALS OVER PARAMETERIZED SURFACES

Most of the flux integrals we are likely to encounter can be computed using the methods of Sections 19.1 and 19.2. In this section, we briefly consider the general case: how to compute the flux of a smooth vector field $\vec{F}$ through a smooth oriented surface, S, parameterised by

$$\vec{r} = \vec{r}(s, t),$$

for (s, t) in some region R of the parameter space. The method is similar to the one used for graphs in Section 19.2. We consider a parameter rectangle on the surface S corresponding to a rectangular region with sides Δs and Δt in the parameter space. (See Figure 21.21.)

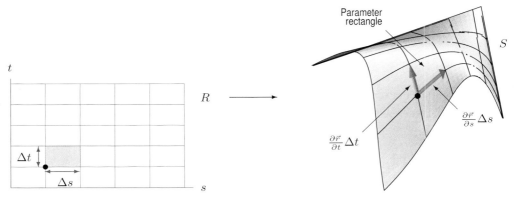

Figure 21.21: Parameter rectangle on the surface S corresponding to a small rectangular region in the parameter space, R

If Δs and Δt are small, the area vector, $\Delta \vec{A}$, of the patch is approximately the area vector of the parallelogram defined by the vectors

$$\vec{r}(s + \Delta s, t) - \vec{r}(s, t) \approx \frac{\partial \vec{r}}{\partial s} \Delta s, \qquad \text{and} \qquad \vec{r}(s, t + \Delta t) - \vec{r}(s, t) \approx \frac{\partial \vec{r}}{\partial t} \Delta t.$$

Thus,

$$\Delta \vec{A} \approx \frac{\partial \vec{r}}{\partial s} \times \frac{\partial \vec{r}}{\partial t} \Delta s \, \Delta t.$$

We assume that the vector $\partial \vec{r} / \partial s \times \partial \vec{r} / \partial t$ is never zero and points in the direction of the unit normal orientation vector $\vec{n}$. If the vector $\partial \vec{r} / \partial s \times \partial \vec{r} / \partial t$ points in the opposite direction to $\vec{n}$, we reverse the order of the cross product. Replacing $\Delta \vec{A}$, Δs, and Δt by $d\vec{A}$, ds, and dt, we write

$$d\vec{A} = \left(\frac{\partial \vec{r}}{\partial s} \times \frac{\partial \vec{r}}{\partial t} \right) ds \, dt.$$

The Flux of a Vector Field Through a Parameterised Surface

The flux of a smooth vector field $\vec{F}$ through a smooth oriented surface S parameterised by $\vec{r} = \vec{r}(s, t)$, where (s, t) varies in a parameter region R, is given by

$$\int_S \vec{F} \cdot d\vec{A} = \int_R \vec{F}(\vec{r}(s, t)) \cdot \left(\frac{\partial \vec{r}}{\partial s} \times \frac{\partial \vec{r}}{\partial t} \right) ds\, dt.$$

We choose the parameterisation so that $\partial \vec{r}/\partial s \times \partial \vec{r}/\partial t$ is never zero and points in the direction of $\vec{n}$ everywhere.

Example 1   Find the flux of the vector field $\vec{F} = x\vec{i} + y\vec{j}$ through the surface S, oriented downward and given by

$$x = 2s, \quad y = s + t, \quad z = 1 + s - t, \qquad \text{where } 0 \le s \le 1, \quad 0 \le t \le 1.$$

Solution   Since S is parameterised by

$$\vec{r}(s, t) = 2s\vec{i} + (s + t)\vec{j} + (1 + s - t)\vec{k},$$

we have

$$\frac{\partial \vec{r}}{\partial s} = 2\vec{i} + \vec{j} + \vec{k} \quad \text{and} \quad \frac{\partial r}{\partial t} = \vec{j} - \vec{k},$$

so

$$\frac{\partial \vec{r}}{\partial s} \times \frac{\partial \vec{r}}{\partial t} = \begin{vmatrix} \vec{i} & \vec{j} & \vec{k} \\ 2 & 1 & 1 \\ 0 & 1 & -1 \end{vmatrix} = -2\vec{i} + 2\vec{j} + 2\vec{k}.$$

Since the vector $-2\vec{i} + 2\vec{j} + 2\vec{k}$ points upward, we use $2\vec{i} - 2\vec{j} - 2\vec{k}$ for downward orientation. Thus, the flux integral is given by

$$\int_S \vec{F} \cdot d\vec{A} = \int_0^1 \int_0^1 (2s\vec{i} + (s + t)\vec{j}) \cdot (2\vec{i} - 2\vec{j} - 2\vec{k})\, ds\, dt$$

$$= \int_0^1 \int_0^1 (4s - 2s - 2t)\, ds\, dt = \int_0^1 \int_0^1 (2s - 2t)\, ds\, dt$$

$$= \int_0^1 \left(s^2 - 2st \Big|_{s=0}^{s=1} \right) dt = \int_0^1 (1 - 2t)\, dt = t - t^2 \Big|_0^1 = 0.$$

Area of a Parameterised Surface

The area ΔA of a small parameter rectangle is the magnitude of its area vector $\Delta \vec{A}$. Therefore,

$$\text{Area of } S = \sum \Delta A = \sum \|\Delta \vec{A}\| \approx \sum \left\| \frac{\partial \vec{r}}{\partial s} \times \frac{\partial \vec{r}}{\partial t} \right\| \Delta s\, \Delta t.$$

Taking the limit as the area of the parameter rectangles tends to zero, we are led to the following expression for the area of S.

The Area of a Parameterised Surface

The area of a surface S which is parameterised by $\vec{r} = \vec{r}(s, t)$, where (s, t) varies in a parameter region R, is given by

$$\int_S dA = \int_R \left\| \frac{\partial \vec{r}}{\partial s} \times \frac{\partial \vec{r}}{\partial t} \right\| ds\, dt.$$

Example 2   Compute the surface area of a sphere of radius a.

Solution   We take the sphere S of radius a centred at the origin and parameterise it with the spherical coordinates ϕ and θ. The parameterisation is

$$x = a \sin \phi \cos \theta, \quad y = a \sin \phi \sin \theta, \quad z = a \cos \phi, \quad \text{for } 0 \le \theta \le 2\pi, \quad 0 \le \phi \le \pi.$$

We compute

$$\frac{\partial \vec{r}}{\partial \phi} \times \frac{\partial \vec{r}}{\partial \theta} = (a \cos \phi \cos \theta \vec{i} + a \cos \phi \sin \theta \vec{j} - a \sin \phi \vec{k}) \times (-a \sin \phi \sin \theta \vec{i} + a \sin \phi \cos \theta \vec{j})$$

$$= a^2 (\sin^2 \phi \cos \theta \vec{i} + \sin^2 \phi \sin \theta \vec{j} + \sin \phi \cos \phi \vec{k})$$

and so

$$\left\| \frac{\partial \vec{r}}{\partial \phi} \times \frac{\partial \vec{r}}{\partial \theta} \right\| = a^2 \sin \phi.$$

Thus, we see that the surface area of the sphere S is given by

$$\text{Surface area} = \int_S dA = \int_R \left\| \frac{\partial \vec{r}}{\partial \phi} \times \frac{\partial \vec{r}}{\partial \theta} \right\| d\phi d\theta = \int_{\phi=0}^{\pi} \int_{\theta=0}^{2\pi} a^2 \sin \phi \, d\theta \, d\phi = 4\pi a^2.$$

Exercises and Problems for Section 21.3

Exercises

In Exercises 1–3 compute $d\vec{A}$ for the given parameterization for one of the two orientations.

1. $x = s + t, \quad y = s - t, \quad z = st$

2. $x = \sin t, \quad y = \cos t, \quad z = s + t$

3. $x = e^s, \quad y = \cos t, \quad z = \sin t$

In Exercises 4–8 compute the flux of the vector field $\vec{F}$ through the parameterized surface S.

4. $\vec{F} = z\vec{k}$ and S is oriented upward and given, for $0 \le s \le 1, \ 0 \le t \le 1$, by

$$x = s + t, \quad y = s - t, \quad z = s^2 + t^2.$$

5. $\vec{F} = x\vec{i} + y\vec{j}$ and S is oriented downward and given, for $0 \le s \le 1, 0 \le t \le 1$, by

$$x = 2s, \quad y = s + t, \quad z = 1 + s - t.$$

6. $\vec{F} = x\vec{i}$ through the surface S oriented downward and parameterized for $0 \le s \le 4, 0 \le t \le \pi/6$ by

$$x = e^s, \quad y = \cos(3t), z = 6s.$$

7. $\vec{F} = y\vec{i} + x\vec{j}$ and S is oriented away from the z-axis and given, for $0 \le s \le \pi, \ 0 \le t \le 1$, by

$$x = 3 \sin s, \quad y = 3 \cos s, \quad z = t + 1.$$

8. $\vec{F} = x^2 y^2 z\vec{k}$ and S is the cone $\sqrt{x^2 + y^2} = z$, with $0 \le z \le R$, oriented downward. Parameterize the cone using cylindrical coordinates. (See Figure 21.22.)

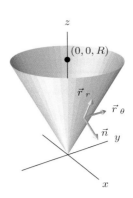

Figure 21.22

In Exercises 9–10, find the surface area.

9. A cylinder of radius a and length L.

10. The region S in the plane $z = 3x + 2y$ such that $0 \le x \le 10$ and $0 \le y \le 20$.

Problems

11. Compute the flux of the vector field $\vec{F} = (x + z)\vec{i} + \vec{j} + z\vec{k}$ through the surface S given by $y = x^2 + z^2$, $1/4 \leq x^2 + z^2 \leq 1$ oriented away from the y-axis.

12. Find the area of the ellipse S on the plane $2x + y + z = 2$ cut out by the circular cylinder $x^2 + y^2 = 2x$. (See Figure 21.23.)

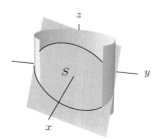

Figure 21.23

13. Consider the surface S formed by rotating the graph of $y = f(x)$ around the x-axis between $x = a$ and $x = b$. Assume that $f(x) \geq 0$ for $a \leq x \leq b$. Show that the surface area of S is $2\pi \int_a^b f(x)\sqrt{1 + f'(x)^2}\, dx$.

14. A rectangular channel of width w and depth h meters lies in the $\vec{j}$ direction. At a point d_1 meters from one side and d_2 meters from the other side, the velocity vector of fluid in the channel is $\vec{v} = kd_1d_2\vec{j}$ meters/sec. Find the flux through a rectangle stretching the full width and depth of the channel, and perpendicular to the flow.

15. The base of a cone is the unit circle centered at the origin in the xy-plane and vertex $P = (a, b, c)$, where $c > 0$.

(a) Parameterize the cone.

(b) Express the surface area of the cone as an integral.

(c) Use a numerical method to find the surface area of the cone with vertex $P = (2, 0, 1)$.

As we remarked in Section 19.1, the limit defining a flux integral might not exist if we subdivide the surface in the wrong way. One way to get around this is to take the formula for a flux integral over a parameterized surface that we have developed in this section and to use it as the *definition* of the flux integral. In Problems 16–19 we explore how this works.

16. Use a parameterization to verify the formula for a flux integral over a surface graph on page 1038.

17. Use a parameterization to verify the formula for a flux integral over a cylindrical surface on page 1039.

18. Use a parameterization to verify the formula for a flux integral over a spherical surface on page 1040.

19. One problem with defining the flux integral using a parameterization is that the integral appears to depend on the choice of parameterization. However, the flux through a surface ought not to depend on how the surface is parameterized. Suppose that the surface S has two parameterizations, $\vec{r} = \vec{r}(s, t)$ for (s, t) in the region R of st-space, and also $\vec{r} = r(u, v)$ for (u, v) in the region T in uv-space, and suppose that the two parameterizations are related by the change of coordinates

$$u = u(s, t) \quad v = v(s, t).$$

Suppose that the Jacobian determinant $\partial(u, v)/\partial(s, t)$ is positive at every point (s, t) in R. Use the change of coordinates formula for double integrals on page 1108 to show that computing the flux integral using either parameterization gives the same result.

Strengthen Your Understanding

In Problems 20–21, explain what is wrong with the statement.

20. The area of the surface parameterized by $x = s, y = t, z = f(s, t)$ above the square $0 \leq x \leq 1, 0 \leq y \leq 1$ is given by the integral

$$\text{Area} = \int_0^1 \int_0^1 f(s, t)\, ds\, dt.$$

21. The surface S parameterized by $x = f(s, t), y = g(s, t)$, $z = h(s, t)$, where $0 \leq s \leq 2, 0 \leq t \leq 3$, has area 6.

In Problems 22–23, give an example of:

22. A parameterization $\vec{r} = \vec{r}(s, t)$ of the xy-plane such that $dA = 2\, ds\, dt$.

23. A vector field $\vec{F}$ such that $\int_S \vec{F} \cdot d\vec{A} > 0$, where S is

the surface $\vec{r} = (s - t)\vec{i} + t^2\vec{j} + (s + t)\vec{k}, 0 \leq s \leq 1$, $0 \leq t \leq 1$, oriented in the direction of $\dfrac{\partial \vec{r}}{\partial s} \times \dfrac{\partial \vec{r}}{\partial t}$.

Are the statements in Problems 24–26 true or false? Give reasons for your answer.

24. If $\vec{r}(s, t), 0 \leq s \leq 1, 0 \leq t \leq 1$ is an oriented parameterized surface S, and $\vec{F}$ is a vector field that is everywhere tangent to S, then the flux of $\vec{F}$ through S is zero.

25. For any parameterization of the surface $x^2 - y^2 + z^2 = 6$, $d\vec{A}$ at $(1, 2, 3)$ is a multiple of $(2\vec{i} - 4\vec{j} + 6\vec{k})dx\, dy$.

26. If you parameterize the plane $3x + 4y + 5z = 7$, then there is a constant c such that, at any point (x, y, z), $d\vec{A} = c(3\vec{i} + 4\vec{j} + 5\vec{k})dx\, dy$.

27. Let S be the hemisphere $x^2 + y^2 + z^2 = 1$ with $x \leq 0$, oriented away from the origin. Which of the following integrals represents the flux of $\vec{F}(x, y, z)$ through S?

(a) $\displaystyle \int_R \vec{F}(x, y, z(x, y)) \cdot \frac{\partial \vec{r}}{\partial x} \times \frac{\partial \vec{r}}{\partial y} \, dx \, dy$

(b) $\displaystyle \int_R \vec{F}(x, y, z(x, y)) \cdot \frac{\partial \vec{r}}{\partial y} \times \frac{\partial \vec{r}}{\partial x} \, dy \, dx$

(c) $\displaystyle \int_R \vec{F}(x, y(x, z), z) \cdot \frac{\partial \vec{r}}{\partial x} \times \frac{\partial \vec{r}}{\partial z} \, dx \, dz$

(d) $\displaystyle \int_R \vec{F}(x, y(x, z), z) \cdot \frac{\partial \vec{r}}{\partial z} \times \frac{\partial \vec{r}}{\partial x} \, dz \, dx$

(e) $\displaystyle \int_R \vec{F}(x(y, z), y, z) \cdot \frac{\partial \vec{r}}{\partial y} \times \frac{\partial \vec{r}}{\partial z} \, dy \, dz$

(f) $\displaystyle \int_R \vec{F}(x(y, z), y, z) \cdot \frac{\partial \vec{r}}{\partial z} \times \frac{\partial \vec{r}}{\partial y} \, dz \, dy$

CHAPTER SUMMARY (see also Ready Reference at the end of the book)

- **Parameterized Curves and Surfaces**
  Parametric equations for curves in 2- and 3-space, planes, graphs of functions, spheres, and cylinders; parameter curves.

- **Change of Variable**
  Polar coordinates, spherical and cylindrical coordinates, general change of variables and Jacobians.

- **Calculating flux integrals over parameterized surfaces**
  Surfaces parameterized by $\vec{r}(s, t)$:
  $d\vec{A} = (\partial \vec{r}/\partial s) \times (\partial \vec{r}/\partial t) \, ds \, dt$

- **Area of parameterized surface**
  $dA = \|(\partial \vec{r}/\partial s) \times (\partial \vec{r}/\partial t)\| \, ds \, dt$

REVIEW EXERCISES AND PROBLEMS FOR CHAPTER TWENTY-ONE

Exercises

In Exercises 1–2, find positive numbers a and b so that the change of coordinates $s = ax, t = by$ transforms the integral $\int \int_R dx \, dy$ into

$$\int \int_T \left| \frac{\partial(x, y)}{\partial(s, t)} \right| \, ds \, dt$$

for the given regions R and T.

1. R is the circular disc of radius 15 centered at the origin and T is the circular disc $s^2 + t^2 \leq 1$.

2. R is the elliptical region $x^2/4 + y^2/9 \leq 1$ and T is the circular disc $s^2 + t^2 \leq 1$.

Describe in words the objects parameterized by the equations in Exercises 3–4.

3. $x = 2z \cos\theta \quad y = 2z \sin\theta \quad z = z$
$\quad 0 \leq z \leq 7 \quad\quad 0 \leq \theta \leq 2\pi$

4. $x = x \quad y = x^2 \quad z = z$
$\quad -5 \leq x \leq 5 \quad 0 \leq z \leq 7$

5. Find a number a so that the change of coordinates $s = x + ay, t = y$ transforms the integral $\int \int_R dx \, dy$ over the parallelogram R with vertices $(10, 15)$, $(30, 15)$,

$(20, 35)$, $(40, 35)$ in the xy-plane into an integral

$$\int \int_T \left| \frac{\partial(x, y)}{\partial(s, t)} \right| \, ds \, dt$$

over a rectangle T in the st-plane.

In Exercises 6–7 compute the flux of the vector field $\vec{F}$ through the parameterized surface S.

6. $\vec{F} = z\vec{i} + x\vec{j}$ and S is oriented upward and given, for $0 \leq s \leq 1, \ 1 \leq t \leq 3$, by

$$x = s^2, \quad y = 2s + t^2, \quad z = 5t.$$

7. $\vec{F} = -\dfrac{2}{x}\vec{i} + \dfrac{2}{y}\vec{j}$ and S is oriented upward and parameterized by a and θ, where, for $1 \leq a \leq 3, \ 0 \leq \theta \leq \pi$,

$$x = a\cos\theta, \quad y = a\sin\theta, \quad z = \sin a^2.$$

8. Parameterize the plane containing the three points $(5, 5, 5)$, $(10, -10, 10)$, $(0, 20, 40)$.

9. Find parametric equations for the sphere centered at the point $(2, -1, 3)$ and with radius 5.

Problems

10. Parameterize a cone of height h and maximum radius a with vertex at the origin and opening upward. Do this in two ways, giving the range of values for each parameter in each case: **(a)** Use r and θ. **(b)** Use z and θ.

11. (a) Describe the surface given parametrically by the

equations

$$x = \cos(s - t), \quad y = \sin(s - t), \quad z = s + t.$$

(b) Describe the two families of parameter curves on the surface.

12. Find a parameterization for the plane through $(1, 3, 4)$ and orthogonal to $\vec{n} = 2\vec{i} + \vec{j} - \vec{k}$.

13. Adapt the parameterization for the sphere to find a parameterization for the ellipsoid

$$\frac{x^2}{a^2} + \frac{y^2}{b^2} + \frac{z^2}{c^2} = 1.$$

14. Parameterize a vase formed by rotating the curve $z = 10\sqrt{x-1}$, $1 \le x \le 2$, around the z-axis. Sketch the vase.

15. For the surface given parametrically by

$$x = 3\sin s \quad y = 3\cos s \quad z = t + 1,$$

where $0 \le s \le \pi$ and $0 \le t \le 1$

(a) Write an equation in x, y, z and identify the parametric surface.

(b) Draw a picture of the surface.

16. Find the region R in the xy-plane corresponding to the region $T = \{(s,t) \,|\, 0 \le s \le 2,\ s \le t \le 2\}$ under the

change of coordinates $x = s^2$, $y = t$. Check that

$$\int_R dx\,dy = \int_T \left| \frac{\partial(x,y)}{\partial(s,t)} \right| ds\,dt.$$

17. Use the change of coordinates $s = x - y$, $t = x + y$ to evaluate $\int \int_R \sin(x+y)\,dx\,dy$ over the disc $x^2 + y^2 \le 1$.

18. Use the change of coordinates $s = xy$, $t = xy^2$ to compute $\int_R xy^2\,dA$, where R is the region bounded by $xy = 1$, $xy = 4$, $xy^2 = 1$, $xy^2 = 4$.

19. Evaluate $\int_S \vec{F} \cdot d\vec{A}$, where $\vec{F} = (bx/a)\vec{i} + (ay/b)\vec{j}$ and S is the elliptic cylinder oriented away from the z-axis, and given by $x^2/a^2 + y^2/b^2 = 1$, $|z| \le c$, where a, b, c are positive constants.

20. Find $\int_S (x^2\vec{i} + y^2\vec{j} + z^2\vec{k}) \cdot d\vec{A}$ where S is the surface of the sphere $(x-a)^2 + (y-b)^2 + (z-c)^2 = d^2$, oriented outward.

PROJECTS FOR CHAPTER TWENTY-ONE

1. Stereographic Projection

We parameterize the sphere $x^2 + y^2 + z^2 = 1$ by a famous method called stereographic projection. Draw a line from a point (x, y) in the xy-plane to the north pole $(0, 0, 1)$. This line intersects the sphere in a point (x, y, z). This gives a parameterization of the sphere by points in the plane.

(a) Which point corresponds to the south pole?

(b) Which points correspond to the equator?

(c) Do we get all the points of the sphere by this parameterization?

(d) Which points correspond to the upper hemisphere?

(e) Which points correspond to the lower hemisphere?

2. Parameterizing a Torus

A torus (doughnut) is constructed by rotating a small circle of radius a in a large circle of radius b about the origin. The small circle is in a (rotating) vertical plane through the z-axis and the large circle is in the xy-plane. See Figure 21.24. Parameterize the torus as follows.

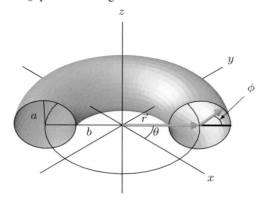

Figure 21.24

(a) Parameterize the large circle.

(b) For a typical point on the large circle, find two unit vectors which are perpendicular to one another and in the plane of the small circle at that point. Use these vectors to parameterize the small circle relative to its center.

(c) Combine your answers to parts (b) and (c) to parameterize the torus.

APPENDICES

A   ROOTS, ACCURACY, AND BOUNDS

It is often necessary to find the zeros of a polynomial or the points of intersection of two curves. So far, you have probably used algebraic methods, such as the quadratic formula, to solve such problems. Unfortunately, however, mathematicians' search for similar solutions to more complicated equations has not been all that successful. The formulas for the solutions to third- and fourth-degree equations are so complicated that you'd never want to use them. Early in the nineteenth century, it was proved that there is no algebraic formula for the solutions to equations of degree 5 and higher. Most non-polynomial equations cannot be solved using a formula either.

However, we can still find roots of equations, provided we use approximation methods, not formulas. In this section we will discuss three ways to find roots: algebraic, graphical, and numerical. Of these, only the algebraic method gives exact solutions.

First, let's get some terminology straight. Given the equation $x^2 = 4$, we call $x = -2$ and $x = 2$ the *roots*, or *solutions of the equation*. If we are given the function $f(x) = x^2 - 4$, then -2 and 2 are called the *zeros of the function*; that is, the zeros of the function f are the roots of the equation $f(x) = 0$.

The Algebraic Viewpoint: Roots by Factoring

If the product of two numbers is zero, then one or the other or both must be zero, that is, if $AB = 0$, then $A = 0$ or $B = 0$. This observation lies behind finding roots by factoring. You may have spent a lot of time factoring polynomials. Here you will also factor expressions involving trigonometric and exponential functions.

Example 1   Find the roots of $x^2 - 7x = 8$.

Solution   Rewrite the equation as $x^2 - 7x - 8 = 0$. Then factor the left side: $(x + 1)(x - 8) = 0$. By our observation about products, either $x + 1 = 0$ or $x - 8 = 0$, so the roots are $x = -1$ and $x = 8$.

Example 2   Find the roots of $\dfrac{1}{x} - \dfrac{x}{(x + 2)} = 0$.

Solution   Rewrite the left side with a common denominator:
$$\frac{x + 2 - x^2}{x(x + 2)} = 0.$$

Whenever a fraction is zero, the numerator must be zero. Therefore we must have
$$x + 2 - x^2 = (-1)(x^2 - x - 2) = (-1)(x - 2)(x + 1) = 0.$$

We conclude that $x - 2 = 0$ or $x + 1 = 0$, so 2 and -1 are the roots. They can be checked by substitution.

Example 3   Find the roots of $e^{-x} \sin x - e^{-x} \cos x = 0$.

Solution   Factor the left side: $e^{-x}(\sin x - \cos x) = 0$. The factor e^{-x} is never zero; it is impossible to raise e to a power and get zero. Therefore, the only possibility is that $\sin x - \cos x = 0$. This equation is equivalent to $\sin x = \cos x$. If we divide both sides by $\cos x$, we get
$$\frac{\sin x}{\cos x} = \frac{\cos x}{\cos x} \quad \text{so} \quad \tan x = 1.$$

The roots of this equation are
$$\ldots, \frac{-7\pi}{4}, \frac{-3\pi}{4}, \frac{\pi}{4}, \frac{5\pi}{4}, \frac{9\pi}{4}, \frac{13\pi}{4}, \ldots.$$

Warning: Using factoring to solve an equation only works when one side of the equation is 0. It is not true that if, say, $AB = 7$ then $A = 7$ or $B = 7$. For example, you *cannot* solve $x^2 - 4x = 2$ by factoring $x(x - 4) = 2$ and then assuming that either x or $x - 4$ equals 2.

The problem with factoring is that factors are not easy to find. For example, the left side of the quadratic equation $x^2 - 4x - 2 = 0$ does not factor, at least not into "nice" factors with integer coefficients. For the general quadratic equation

$$ax^2 + bx + c = 0,$$

there is the quadratic formula for the roots:

$$x = \frac{-b \pm \sqrt{b^2 - 4ac}}{2a}.$$

Thus the roots of $x^2 - 4x - 2 = 0$ are $(4 \pm \sqrt{24})/2$, or $2 + \sqrt{6}$ and $2 - \sqrt{6}$.

Notice that in each of these examples, we have found the roots exactly.

The Graphical Viewpoint: Roots by Zooming

To find the roots of an equation $f(x) = 0$, it helps to draw the graph of f. The roots of the equation, that is the zeros of f, are *the values of x where the graph of f crosses the x-axis.* Even a very rough sketch of the graph can be useful in determining how many zeros there are and their approximate values. If you have a computer or graphing calculator, then finding solutions by graphing is the easiest method, especially if you use the zoom feature. However, a graph can never tell you the exact value of a root, only an approximate one.

Example 4   Find the roots of $x^3 - 4x - 2 = 0$.

Solution   Attempting to factor the left side with integer coefficients will convince you it cannot be done, so we cannot easily find the roots by algebra. We know the graph of $f(x) = x^3 - 4x - 2$ will have the usual cubic shape; see Figure A.1.

There are clearly three roots: one between $x = -2$ and $x = -1$, another between $x = -1$ and $x = 0$, and a third between $x = 2$ and $x = 3$. Zooming in on the largest root with a graphing calculator or computer shows that it lies in the following interval:

$$2.213 < x < 2.215.$$

Thus, the root is $x = 2.21$, accurate to two decimal places. Zooming in on the other two roots shows them to be $x = -1.68$ and $x = -0.54$, accurate to two decimal places.

Useful trick: Suppose you want to solve the equation $\sin x - \cos x = 0$ graphically. Instead of graphing $f(x) = \sin x - \cos x$ and looking for zeros, you may find it easier to rewrite the equation as $\sin x = \cos x$ and graph $g(x) = \sin x$ and $h(x) = \cos x$. (After all, you already know what these two graphs look like. See Figure A.2.) The roots of the original equation are then precisely the x coordinates of the points of intersection of the graphs of $g(x)$ and $h(x)$.

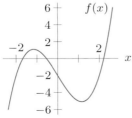

Figure A.1: The cubic
$f(x) = x^3 - 4x - 2$

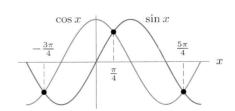

Figure A.2: Finding roots of
$\sin x - \cos x = 0$

Example 5   Find the roots of $2\sin x - x = 0$.

Solution   Rewrite the equation as $2\sin x = x$, and graph both sides. Since $g(x) = 2\sin x$ is always between -2 and 2, there are no roots of $2\sin x = x$ for $x > 2$ or for $x < -2$. We need only consider the graphs between -2 and 2 (or between $-\pi$ and π, which makes graphing the sine function easier). Figure A.3 shows the graphs. There are three points of intersection: one appears to be at $x = 0$, one between $x = \pi/2$ and $x = \pi$, and one between $x = -\pi/2$ and $x = -\pi$. You can tell that $x = 0$ is the exact value of one root because it satisfies the original equation exactly. Zooming in shows that there is a second root $x \approx 1.9$, and the third root is $x \approx -1.9$ by symmetry.

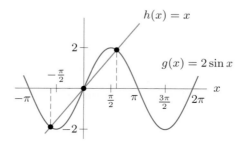

Figure A.3: Finding roots of $2\sin x - x = 0$

The Numerical Viewpoint: Roots by Bisection

We now look at a numerical method of approximating the solutions to an equation. This method depends on the idea that if the value of a function $f(x)$ changes sign in an interval, and if we believe there is no break in the graph of the function there, then there is a root of the equation $f(x) = 0$ in that interval.

Let's go back to the problem of finding the root of $f(x) = x^3 - 4x - 2 = 0$ between 2 and 3. To locate the root, we close in on it by evaluating the function at the midpoint of the interval, $x = 2.5$. Since $f(2) = -2$, $f(2.5) = 3.625$, and $f(3) = 13$, the function changes sign between $x = 2$ and $x = 2.5$, so the root is between these points. Now we look at $x = 2.25$.

Since $f(2.25) = 0.39$, the function is negative at $x = 2$ and positive at $x = 2.25$, so there is a root between 2 and 2.25. Now we look at 2.125. We find $f(2.125) = -0.90$, so there is a root between 2.125 and 2.25, ... and so on. (You may want to round the decimals as you work.) See Figure A.4. The intervals containing the root are listed in Table A.1 and show that the root is $x = 2.21$ to two decimal places.

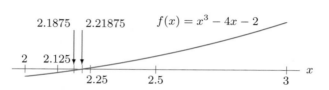

Figure A.4: Locating a root of $x^3 - 4x - 2 = 0$

Table A.1   *Intervals containing root of*
$x^3 - 4x - 2 = 0$ *(Note: $[2, 3]$ means $2 \le x \le 3$)*

$[2, 3]$
$[2, 2.5]$
$[2, 2.25]$
$[2.125, 2.25]$
$[2.1875, 2.25]$ So $x = 2.2$ rounded to one decimal place
$[2.1875, 2.21875]$
$[2.203125, 2.21875]$
$[2.2109375, 2.21875]$
$[2.2109375, 2.2148438]$ So $x = 2.21$ rounded to two decimal places

This method of estimating roots is called the **Bisection Method**:
- To solve an equation $f(x) = 0$ using the bisection method, we need two starting values for x, say, $x = a$ and $x = b$, such that $f(a)$ and $f(b)$ have opposite signs and f is continuous on $[a, b]$.
- Evaluate f at the midpoint of the interval $[a, b]$, and decide in which half-interval the root lies.
- Repeat, using the new half-interval instead of $[a, b]$.

There are some problems with the bisection method:
- The function may not change signs near the root. For example, $f(x) = x^2 - 2x + 1 = 0$ has a root at $x = 1$, but $f(x)$ is never negative because $f(x) = (x - 1)^2$, and a square cannot be negative. (See Figure A.5.)
- The function f must be continuous between the starting values $x = a$ and $x = b$.
- If there is more than one root between the starting values $x = a$ and $x = b$, the method will find only one of the roots. For example, if we had tried to solve $x^3 - 4x - 2 = 0$ starting at $x = -12$ and $x = 10$, the bisection method would zero in on the root between $x = -2$ and $x = -1$, not the root between $x = 2$ and $x = 3$ that we found earlier. (Try it! Then see what happens if you use $x = -10$ instead of $x = -12$.)
- The bisection method is slow and not very efficient. Applying bisection three times in a row only traps the root in an interval $\left(\frac{1}{2}\right)^3 = \frac{1}{8}$ as large as the starting interval. Thus, if we initially know that a root is between, say, 2 and 3, then we would need to apply the bisection method at least four times to know the first digit after the decimal point.

There are much more powerful methods available for finding roots, such as Newton's method, which are more complicated but which avoid some of these difficulties.

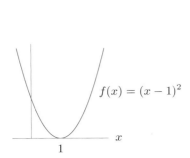

Figure A.5: f does not change sign at the root

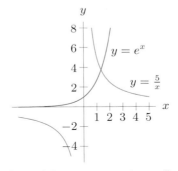

Figure A.6: Intersection of $y = e^x$ and $y = 5/x$

Table A.2   *Bisection method for* $f(x) = xe^x - 5 = 0$ *(Note that* $[1, 2]$ *means the interval* $1 \leq x \leq 2$*)*

Interval Containing Root
$[1, 2]$
$[1, 1.5]$
$[1.25, 1.5]$
$[1.25, 1.375]$
$[1.3125, 1.375]$
$[1.3125, 1.34375]$

Example 6   Find all the roots of $xe^x = 5$ to at least one decimal place.

Solution   If we rewrite the equation as $e^x = 5/x$ and graph both sides, as in Figure A.6, it is clear that there is exactly one root, and it is somewhere between 1 and 2. Table A.2 shows the intervals obtained by the bisection method. After five iterations, we have the root trapped between 1.3125 and 1.34375, so we can say the root is $x = 1.3$ to one decimal place.

Iteration

Both zooming in and bisection as discussed here are examples of *iterative* methods, in which a sequence of steps is repeated over and over again, using the results of one step as the input for the next. We can use such methods to locate a root to any degree of accuracy. In bisection, each iteration traps the root in an interval that is half the length of the previous one. Each time you zoom in on a calculator, you trap the root in a smaller interval; how much smaller depends on the settings on the calculator.

Accuracy and Error

In the previous discussion, we used the phrase "accurate to 2 decimal places." For an iterative process where we get closer and closer estimates of some quantity, we take a common-sense approach to accuracy: we watch the numbers carefully, and when a digit stays the same for several iterations, we assume it has stabilized and is correct, especially if the digits to the right of that digit also stay the same. For example, suppose 2.21429 and 2.21431 are two successive estimates for a zero of $f(x) = x^3 - 4x - 2$. Since these two estimates agree to the third digit after the decimal point, we probably have at least 3 decimal places correct.

There is a problem with this, however. Suppose we are estimating a root whose true value is 1, and the estimates are converging to the value from below—say, 0.985, 0.991, 0.997 and so on. In this case, not even the first decimal place is "correct," even though the difference between the estimates and the actual answer is very small—much less than 0.1. To avoid this difficulty, we say that an estimate a for some quantity r is *accurate to p decimal places* if the error, which is the absolute value of the difference between a and r, or $|r - a|$, is as follows:

Accuracy to p decimal places	means	Error less than
$p = 1$		0.05
2		0.005
3		0.0005
$\vdots$		$\vdots$
n		$0.\underbrace{000\ldots0}_{n}5$

This is the same as saying that r must lie in an interval of length twice the maximum error, centered on a. For example, if a is accurate to 1 decimal place, r must lie in the following interval:

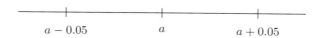

$$a - 0.05 \qquad\qquad a \qquad\qquad a + 0.05$$

Since both the graphing calculator and the bisection method give us an interval in which the root is trapped, this definition of decimal accuracy is a natural one for these processes.

Example 7 Suppose the numbers $\sqrt{10}$, $22/7$, and 3.14 are given as approximations to $\pi = 3.1415\ldots$. To how many decimal places is each approximation accurate?

Solution Using $\sqrt{10} = 3.1622\ldots$,

$$|\sqrt{10} - \pi| = |3.1622\ldots - 3.1415\ldots| = 0.0206\ldots < 0.05,$$

so $\sqrt{10}$ is accurate to one decimal place. Similarly, using $22/7 = 3.1428\ldots$,

$$\left|\frac{22}{7} - \pi\right| = |3.1428\ldots - 3.1415\ldots| = 0.0013\ldots < 0.005,$$

so 22/7 is accurate to two decimal places. Finally,

$$|3.14 - 3.1415\ldots| = 0.0015\ldots < 0.005,$$

so 3.14 is accurate to two decimal places.

Warning:

- Saying that an approximation is accurate to, say, 2 decimal places does *not* guarantee that its first two decimal places are "correct," that is, that the two digits of the approximation are the same as the corresponding two digits in the true value. For example, an approximate value of 5.997 is accurate to 2 decimal places if the true value is 6.001, but neither of the 9s in the approximation agrees with the 0s in the true value (nor does the digit 5 agree with the digit 6).

- When finding a root r of an equation, the number of decimal places of accuracy refers to the number of digits that have stabilized in the root. It does *not* refer to the number of digits of $f(r)$ that are zero. For example, Table A.1 on page 1120 shows that $x = 2.2$ is a root of $f(x) = x^3 - 4x - 2 = 0$, accurate to one decimal place. Yet, $f(2.2) = -0.152$, so $f(2.2)$ does not have one zero after the decimal point. Similarly, $x = 2.21$ is the root accurate to two decimal places, but $f(2.21) = -0.046$ does not have two zeros after the decimal point.

Example 8   Is $x = 2.2143$ a zero of $f(x) = x^3 - 4x - 2$ accurate to four decimal places?

Solution   We want to know whether r, the exact value of the zero, lies in the interval

$$2.2143 - 0.00005 < r < 2.2143 + 0.00005$$

which is the same as

$$2.21425 < r < 2.21435.$$

Since $f(2.21425) < 0$ and $f(2.21435) > 0$, the zero does lie in this interval, and so $r = 2.2143$ is accurate to four decimal places.

How to Write a Decimal Answer

The graphing calculator and bisection method naturally give an interval for a root or a zero. However, other numerical techniques do not give a pair of numbers bounding the true value, but rather a single number near the true value. What should you do if you want a single number, rather than an interval, for an answer? In general, averaging the endpoint of the interval is the best solution.

    When giving a single number as an answer and interpreting it, be careful about giving rounded answers. For example, suppose you know a root lies in the interval between 0.81 and 0.87. Averaging gives 0.84 as a single number estimating the root. But it would be wrong to round 0.84 to 0.8 and say that the answer is 0.8 accurate to one decimal place; the true value could be 0.86, which is not within 0.05 of 0.8. The right thing to say is that the answer is 0.84 accurate to one decimal place. Similarly, to give an answer accurate to, say, 2 decimal places, you may have to show 3 decimal places in your answer.

Bounds of a Function

Knowing how big or how small a function gets can sometimes be useful, especially when you can't easily find exact values of the function. You can say, for example, that $\sin x$ always stays between -1 and 1 and that $2\sin x + 10$ always stays between 8 and 12. But 2^x is not confined between any two numbers, because 2^x will exceed any number you can name if x is large enough. We say that $\sin x$ and $2\sin x + 10$ are *bounded* functions, and that 2^x is an *unbounded* function.

A function f is **bounded** on an interval if there are numbers L and U such that

$$L \leq f(x) \leq U$$

for all x in the interval. Otherwise, f is **unbounded** on the interval.

We say that L is a **lower bound** for f on the interval, and that U is an **upper bound** for f on the interval.

Example 9   Use Figures A.7 and A.8 to decide which of the following functions are bounded.

(a) x^3 on $-\infty < x < \infty$; on $0 \leq x \leq 100$.

(b) $2/x$ on $0 < x < \infty$; on $1 \leq x < \infty$.

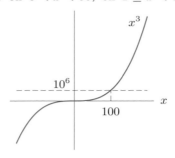

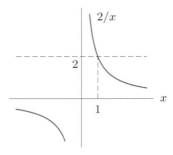

Figure A.7: Is x^3 bounded?   **Figure A.8**: Is $2/x$ bounded?

Solution   (a) The graph of x^3 in Figure A.7 shows that x^3 will exceed any number, no matter how large, if x is big enough, so x^3 does not have an upper bound on $-\infty < x < \infty$. Therefore, x^3 is unbounded on $-\infty < x < \infty$. But on the interval $0 \leq x \leq 100$, x^3 stays between 0 (a lower bound) and $100^3 = 1{,}000{,}000$ (an upper bound). Therefore, x^3 is bounded on the interval $0 \leq x \leq 100$. Notice that upper and lower bounds, when they exist, are not unique. For example, -100 is another lower bound and $2{,}000{,}000$ another upper bound for x^3 on $0 \leq x \leq 100$.

(b) $2/x$ is unbounded on $0 < x < \infty$, since it has no upper bound on that interval. But $0 \leq 2/x \leq 2$ for $1 \leq x < \infty$, so $2/x$ is bounded, with lower bound 0 and upper bound 2, on $1 \leq x < \infty$. (See Figure A.8.)

Best Possible Bounds

Consider a group of people whose height in feet, h, ranges from 5 feet to 6 feet. Then 5 feet is a lower bound for the people in the group and 6 feet is an upper bound:

$$5 \leq h \leq 6.$$

But the people in this group are also all between 4 feet and 7 feet, so it is also true that

$$4 \leq h \leq 7.$$

So, there are many lower bounds and many upper bounds. However, the 5 and the 6 are considered the best bounds because they are the closest together of all the possible pairs of bounds.

The **best possible bounds** for a function, f, over an interval are numbers A and B such that, for all x in the interval,

$$A \leq f(x) \leq B$$

and where A and B are as close together as possible. A is called the **greatest lower bound** and B is the **least upper bound**.

What Do Bounds Mean Graphically?

Upper and lower bounds can be represented on a graph by horizontal lines. See Figure A.9.

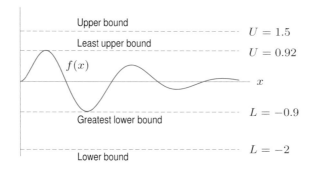

Figure A.9: Upper and lower bounds for the function f

Exercises for Appendix A

1. Use a calculator or computer graph of $f(x) = 13 - 20x - x^2 - 3x^4$ to determine:

   (a) The range of this function;
   (b) The number of zeros of this function.

For Problems 2–12, determine the roots or points of intersection to an accuracy of one decimal place.

2. (a) The root of $x^3 - 3x + 1 = 0$ between 0 and 1
   (b) The root of $x^3 - 3x + 1 = 0$ between 1 and 2
   (c) The smallest root of $x^3 - 3x + 1 = 0$

3. The root of $x^4 - 5x^3 + 2x - 5 = 0$ between -2 and -1

4. The root of $x^5 + x^2 - 9x - 3 = 0$ between -2 and -1

5. The largest real root of $2x^3 - 4x^2 - 3x + 1 = 0$

6. All real roots of $x^4 - x - 2 = 0$

7. All real roots of $x^5 - 2x^2 + 4 = 0$

8. The smallest positive root of $x \sin x - \cos x = 0$

9. The left-most point of intersection between $y = 2x$ and $y = \cos x$

10. The left-most point of intersection between $y = 1/2^x$ and $y = \sin x$

11. The point of intersection between $y = e^{-x}$ and $y = \ln x$

12. All roots of $\cos t = t^2$

13. Estimate all real zeros of the following polynomials, accurate to 2 decimal places:

    (a) $f(x) = x^3 - 2x^2 - x + 3$
    (b) $f(x) = x^3 - x^2 - 2x + 2$

14. Find the largest zero of

$$f(x) = 10xe^{-x} - 1$$

to two decimal places, using the bisection method. Make sure to demonstrate that your approximation is as good as you claim.

15. (a) Find the smallest positive value of x where the graphs of $f(x) = \sin x$ and $g(x) = 2^{-x}$ intersect.
    (b) Repeat with $f(x) = \sin 2x$ and $g(x) = 2^{-x}$.

16. Use a graphing calculator to sketch $y = 2 \cos x$ and $y = x^3 + x^2 + 1$ on the same set of axes. Find the positive zero of $f(x) = 2 \cos x - x^3 - x^2 - 1$. A friend claims there is one more real zero. Is your friend correct? Explain.

17. Use the table below to investigate the zeros of the function

$$f(\theta) = (\sin 3\theta)(\cos 4\theta) + 0.8$$

in the interval $0 \le \theta \le 1.8$.

θ	0	0.2	0.4	0.6	0.8	1.0	1.2	1.4	1.6	1.8
$f(\theta)$	0.80	1.19	0.77	0.08	0.13	0.71	0.76	0.12	-0.19	0.33

   (a) Decide how many zeros the function has in the interval $0 \le \theta \le 1.8$.
   (b) Locate each zero, or a small interval containing each zero.
   (c) Are you sure you have found all the zeros in the interval $0 \le \theta \le 1.8$? Graph the function on a calculator or computer to decide.

18. (a) Use Table A.3 to locate approximate solution(s) to

$$(\sin 3x)(\cos 4x) = \frac{x^3}{\pi^3}$$

in the interval $1.07 \leq x \leq 1.15$. Give an interval of length 0.01 in which each solution lies.

Table A.3

x	x^3/π^3	$(\sin 3x)(\cos 4x)$
1.07	0.0395	0.0286
1.08	0.0406	0.0376
1.09	0.0418	0.0442
1.10	0.0429	0.0485
1.11	0.0441	0.0504
1.12	0.0453	0.0499
1.13	0.0465	0.0470
1.14	0.0478	0.0417
1.15	0.0491	0.0340

(b) Make an estimate for each solution accurate to two decimal places.

19. (a) With your calculator in radian mode, take the arctangent of 1 and multiply that number by 4. Now, take the arctangent of the result and multiply it by 4. Continue this process 10 times or so and record each result as in the accompanying table. At each step, you get 4 times the arctangent of the result of the previous step.

1
3.14159...
5.05050...
5.50129...
⋮

(b) Your table allows you to find a solution of the equation

$$4\arctan x = x.$$

Why? What is that solution?

(c) What does your table in part (a) have to do with Figure A.10?

[Hint: The coordinates of P_0 are $(1, 1)$. Find the coordinates of $P_1, P_2, P_3, \ldots$]

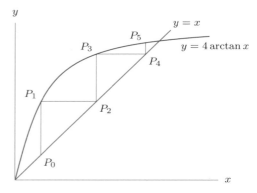

Figure A.10

(d) In part (a), what happens if you start with an initial guess of 10? Of -10? What types of behavior do you observe? (That is, for which initial guesses is the sequence increasing, and for which is it decreasing; does the sequence approach a limit?) Explain your answers graphically, as in part (c).

20. Using radians, apply the iteration method of Problem 19 to the equation

$$\cos x = x.$$

Represent your results graphically, as in Figure A.10.

For Problems 21–23, draw a graph to decide if the function is bounded on the interval given. Give the best possible upper and lower bounds for any function which is bounded.

21. $f(x) = 4x - x^2$ on $[-1, 4]$

22. $h(\theta) = 5 + 3\sin\theta$ on $[-2\pi, 2\pi]$

23. $f(t) = \dfrac{\sin t}{t^2}$ on $[-10, 10]$

B COMPLEX NUMBERS

The quadratic equation

$$x^2 - 2x + 2 = 0$$

is not satisfied by any real number x. If you try applying the quadratic formula, you get

$$x = \frac{2 \pm \sqrt{4 - 8}}{2} = 1 \pm \frac{\sqrt{-4}}{2}.$$

Apparently, you need to take a square root of -4. But -4 does not have a square root, at least, not one which is a real number. Let's give it a square root.

We define the imaginary number i to be a number such that

$$i^2 = -1.$$

Using this i, we see that $(2i)^2 = -4$, so

$$x = 1 \pm \frac{\sqrt{-4}}{2} = 1 \pm \frac{2i}{2} = 1 \pm i.$$

This solves our quadratic equation. The numbers $1 + i$ and $1 - i$ are examples of complex numbers.

A **complex number** is defined as any number that can be written in the form

$$z = a + bi,$$

where a and b are real numbers and $i^2 = -1$, so we say $i = \sqrt{-1}$.
The *real part* of z is the number a; the *imaginary part* is the number b.

Calling the number i imaginary makes it sound as if i does not exist in the same way that real numbers exist. In some cases, it is useful to make such a distinction between real and imaginary numbers. For example, if we measure mass or position, we want our answers to be real numbers. But the imaginary numbers are just as legitimate mathematically as the real numbers are.

As an analogy, consider the distinction between positive and negative numbers. Originally, people thought of numbers only as tools to count with; their concept of "five" or "ten" was not far removed from "five arrows" or "ten stones." They were unaware that negative numbers existed at all. When negative numbers were introduced, they were viewed only as a device for solving equations like $x + 2 = 1$. They were considered "false numbers," or, in Latin, "negative numbers." Thus, even though people started to use negative numbers, they did not view them as existing in the same way that positive numbers did. An early mathematician might have reasoned: "The number 5 exists because I can have 5 dollars in my hand. But how can I have -5 dollars in my hand?" Today we have an answer: "I have -5 dollars" means I owe somebody 5 dollars. We have realized that negative numbers are just as useful as positive ones, and it turns out that complex numbers are useful too. For example, they are used in studying wave motion in electric circuits.

Algebra of Complex Numbers

Numbers such as 0, 1, $\frac{1}{2}$, π, and $\sqrt{2}$ are called *purely real* because they contain no imaginary components. Numbers such as i, $2i$, and $\sqrt{2}i$ are called *purely imaginary* because they contain only the number i multiplied by a nonzero real coefficient.

Two complex numbers are called *conjugates* if their real parts are equal and if their imaginary parts are opposites. The complex conjugate of the complex number $z = a + bi$ is denoted $\overline{z}$, so we have

$$\overline{z} = a - bi.$$

(Note that z is real if and only if $z = \overline{z}$.) Complex conjugates have the following remarkable property: if $f(x)$ is any polynomial with real coefficients ($x^3 + 1$, say) and $f(z) = 0$, then $f(\overline{z}) = 0$. This means that if z is the solution to a polynomial equation with real coefficients, then so is $\overline{z}$.

- Two complex numbers are equal if and only if their real parts are equal and their imaginary parts are equal. Consequently, if $a + bi = c + di$, then $a = c$ and $b = d$.

- Adding two complex numbers is done by adding real and imaginary parts separately:

$$(a + bi) + (c + di) = (a + c) + (b + d)i.$$

- Subtracting is similar:

$$(a + bi) - (c + di) = (a - c) + (b - d)i.$$

- Multiplication works just as for polynomials, using $i^2 = -1$:

$$(a + bi)(c + di) = a(c + di) + bi(c + di)$$
$$= ac + adi + bci + bdi^2$$
$$= ac + adi + bci - bd = (ac - bd) + (ad + bc)i.$$

- Powers of i: We know that $i^2 = -1$; then $i^3 = i \cdot i^2 = -i$, and $i^4 = (i^2)^2 = (-1)^2 = 1$. Then $i^5 = i \cdot i^4 = i$, and so on. Thus we have

$$i^n = \begin{cases} i & \text{for } n = 1, 5, 9, 13, \ldots \\ -1 & \text{for } n = 2, 6, 10, 14, \ldots \\ -i & \text{for } n = 3, 7, 11, 15, \ldots \\ 1 & \text{for } n = 0, 4, 8, 12, 16, \ldots \end{cases}$$

- The product of a number and its conjugate is always real and nonnegative:

$$z \cdot \bar{z} = (a + bi)(a - bi) = a^2 - abi + abi - b^2 i^2 = a^2 + b^2.$$

- Dividing by a nonzero complex number is done by multiplying the denominator by its conjugate, thereby making the denominator real:

$$\frac{a + bi}{c + di} = \frac{a + bi}{c + di} \cdot \frac{c - di}{c - di} = \frac{ac - adi + bci - bdi^2}{c^2 + d^2} = \frac{ac + bd}{c^2 + d^2} + \frac{bc - ad}{c^2 + d^2} i.$$

Example 1   Compute $(2 + 7i)(4 - 6i) - i$.

Solution $(2 + 7i)(4 - 6i) - i = 8 + 28i - 12i - 42i^2 - i = 8 + 15i + 42 = 50 + 15i.$

Example 2   Compute $\dfrac{2 + 7i}{4 - 6i}$.

Solution

$$\frac{2 + 7i}{4 - 6i} = \frac{2 + 7i}{4 - 6i} \cdot \frac{4 + 6i}{4 + 6i} = \frac{8 + 12i + 28i + 42i^2}{4^2 + 6^2} = \frac{-34 + 40i}{52} = \frac{-17}{26} + \frac{10}{13} i.$$

You can check by multiplying out that $(-17/26 + 10i/13)(4 - 6i) = 2 + 7i$.

The Complex Plane and Polar Coordinates

It is often useful to picture a complex number $z = x + iy$ in the plane, with x along the horizontal axis and y along the vertical. The xy-plane is then called the *complex plane*. Figure B.11 shows the complex numbers $-2i$, $1 + i$, and $-2 + 3i$.

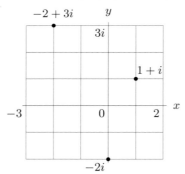

Figure B.11: Points in the complex plane

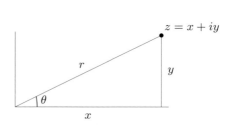

Figure B.12: The point $z = x + iy$ in the complex plane, showing polar coordinates

The triangle in Figure B.12 shows that a complex number can be written using polar coordinates as follows:

$$z = x + iy = r \cos \theta + ir \sin \theta.$$

Example 3   Express $z = -2i$ and $z = -2 + 3i$ using polar coordinates. (See Figure B.11.)

Solution   For $z = -2i$, the distance of z from the origin is 2, so $r = 2$. Also, one value for θ is $\theta = 3\pi/2$. Using polar coordinates, $-2i = 2\cos(3\pi/2) + i\,2(\sin 3\pi/2)$.
      For $z = -2 + 3i$, we have $x = -2$, $y = 3$. So $r = \sqrt{(-2)^2 + 3^2} \approx 3.61$, and one solution of $\tan\theta = 3/(-2)$ with θ in quadrant II is $\theta \approx 2.16$. So $-2 + 3i \approx 3.61\cos(2.16) + i\,3.61\sin(2.16)$.

Example 4   Consider the point with polar coordinates $r = 5$ and $\theta = 3\pi/4$. What complex number does this point represent?

Solution   Since $x = r\cos\theta$ and $y = r\sin\theta$ we see that $x = 5\cos 3\pi/4 = -5/\sqrt{2}$, and $y = 5\sin 3\pi/4 = 5/\sqrt{2}$, so $z = -5/\sqrt{2} + i\,5/\sqrt{2}$.

Derivatives and Integrals of Complex-Valued Functions

Suppose $z(t) = x(t) + iy(t)$, where t is real, then we define $z'(t)$ and $\int z(t)\,dt$ by treating i like any other constant:

$$z'(t) = x'(t) + iy'(t)$$

$$\int z(t)\,dt = \int x(t)\,dt + i\int y(t)\,dt.$$

With these definitions, all the usual properties of differentiation and integration hold, such as

$$\int z'(t)\,dt = z(t) + C, \qquad \text{for } C \text{ is a complex constant.}$$

Euler's Formula

Consider the complex number z lying on the unit circle in Figure B.13. Writing z in polar coordinates, and using the fact that $r = 1$, we have

$$z = f(\theta) = \cos\theta + i\sin\theta.$$

It turns out that there is a particularly beautiful and compact way of rewriting $f(\theta)$ using complex exponentials. We take the derivative of f using the fact that $i^2 = -1$:

$$f'(\theta) = -\sin\theta + i\cos\theta = i\cos\theta + i^2\sin\theta.$$

Factoring out an i gives

$$f'(\theta) = i(\cos\theta + i\sin\theta) = i \cdot f(\theta).$$

As you know from Chapter 11, page 628, the only real-valued function whose derivative is proportional to the function itself is the exponential function. In other words, we know that if

$$g'(x) = k \cdot g(x), \quad \text{then} \quad g(x) = Ce^{kx}$$

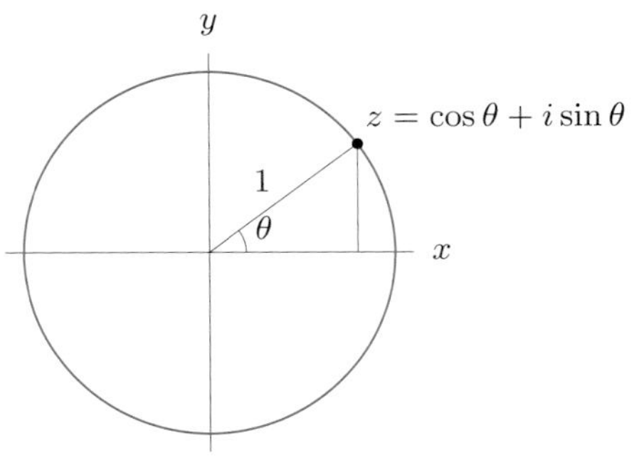

Figure B.13: Complex number represented by a point on the unit circle for some constant C. If we assume that a similar result holds for complex-valued functions, then we have

$$f'(\theta) = i \cdot f(\theta), \quad \text{so} \quad f(\theta) = Ce^{i\theta}$$

for some constant C. To find C we substitute $\theta = 0$. Now $f(0) = Ce^{i \cdot 0} = C$, and since $f(0) = \cos 0 + i \sin 0 = 1$, we must have $C = 1$. Therefore $f(\theta) = e^{i\theta}$. Thus we have

---

Euler's formula

$$e^{i\theta} = \cos\theta + i\sin\theta.$$

This elegant and surprising relationship was discovered by the Swiss mathematician Leonhard Euler in the eighteenth century, and it is particularly useful in solving second-order differential equations. Another way of obtaining Euler's formula (using Taylor series) is given in Problem 46 on page 569. It allows us to write the complex number represented by the point with polar coordinates (r, θ) in the following form:

$$z = r(\cos\theta + i\sin\theta) = re^{i\theta}.$$

Similarly, since $\cos(-\theta) = \cos\theta$ and $\sin(-\theta) = -\sin\theta$, we have

$$re^{-i\theta} = r\left(\cos(-\theta) + i\sin(-\theta)\right) = r(\cos\theta - i\sin\theta).$$

Example 5   Evaluate $e^{i\pi}$.

Solution   Using Euler's formula, $e^{i\pi} = \cos\pi + i\sin\pi = -1$.

Example 6   Express the complex number represented by the point $r = 8$, $\theta = 3\pi/4$ in Cartesian form and polar form, $z = re^{i\theta}$.

Solution   Using Cartesian coordinates, the complex number is

$$z = 8\left(\cos\left(\frac{3\pi}{4}\right) + i\sin\left(\frac{3\pi}{4}\right)\right) = \frac{-8}{\sqrt{2}} + i\frac{8}{\sqrt{2}}.$$

Using polar coordinates, we have

$$z = 8e^{i\,3\pi/4}.$$

The polar form of complex numbers makes finding powers and roots of complex numbers much easier. Writing $z = re^{i\theta}$, we find any power of z as follows:

$$z^p = (re^{i\theta})^p = r^p e^{ip\theta}.$$

To find roots, we let p be a fraction, as in the following example.

Example 7   Find a cube root of the complex number represented by the point with polar coordinates $(8, 3\pi/4)$.

Solution   In Example 6, we saw that this complex number could be written as $z = 8e^{i3\pi/4}$. So,

$$\sqrt[3]{z} = \left(8e^{i\,3\pi/4}\right)^{1/3} = 8^{1/3}e^{i(3\pi/4)\cdot(1/3)} = 2e^{\pi i/4} = 2\left(\cos(\pi/4) + i\sin(\pi/4)\right)$$

$$= 2\left(1/\sqrt{2} + i/\sqrt{2}\right) = \sqrt{2}(1 + i).$$

You can check by multiplying out that $(\sqrt{2}(1 + i))^3 = -(8/\sqrt{2}) + i(8/\sqrt{2}) = z$.

Using Complex Exponentials

Euler's formula, together with the fact that exponential functions are simple to manipulate, allows us to obtain many results about trigonometric functions easily.

The following example uses the fact that for complex z, the function e^z has all the usual algebraic properties of exponents.

Example 8   Use Euler's formula to obtain the double-angle identities

$$\cos 2\theta = \cos^2 \theta - \sin^2 \theta \qquad \text{and} \qquad \sin 2\theta = 2 \cos \theta \sin \theta.$$

Solution   We use the fact that $e^{2i\theta} = e^{i\theta} \cdot e^{i\theta}$. This can be rewritten as

$$\cos 2\theta + i \sin 2\theta = (\cos \theta + i \sin \theta)^2.$$

Multiplying out $(\cos \theta + i \sin \theta)^2$, using the fact that $i^2 = -1$ gives

$$\cos 2\theta + i \sin 2\theta = \cos^2 \theta - \sin^2 \theta + i(2 \cos \theta \sin \theta).$$

Since two complex numbers are equal only if the real and imaginary parts are equal, we must have

$$\cos 2\theta = \cos^2 \theta - \sin^2 \theta \qquad \text{and} \qquad \sin 2\theta = 2 \cos \theta \sin \theta.$$

If we solve $e^{i\theta} = \cos \theta + i \sin \theta$ and $e^{-i\theta} = \cos \theta - i \sin \theta$ for $\sin \theta$ and $\cos \theta$, we obtain

$$\sin \theta = \frac{e^{i\theta} - e^{-i\theta}}{2i} \qquad \text{and} \qquad \cos \theta = \frac{e^{i\theta} + e^{-i\theta}}{2}.$$

By differentiating the formula $e^{ik\theta} = \cos(k\theta) + i \sin(k\theta)$, for θ real and k a real constant, it can be shown that

$$\frac{d}{d\theta}\left(e^{ik\theta}\right) = ik e^{ik\theta} \qquad \text{and} \qquad \int e^{ik\theta}\, d\theta = \frac{1}{ik} e^{ik\theta} + C.$$

Thus complex exponentials are differentiated and integrated just like real exponentials.

Example 9   Use $\cos \theta = \left(e^{i\theta} + e^{-i\theta}\right)/2$ to obtain the derivative formula for $\cos \theta$.

Solution   Differentiating gives

$$\frac{d}{d\theta}(\cos \theta) = \frac{d}{d\theta}\left(\frac{e^{i\theta} + e^{-i\theta}}{2}\right) = \frac{ie^{i\theta} - ie^{-i\theta}}{2} = \frac{i(e^{i\theta} - e^{-i\theta})}{2}$$

$$= -\frac{e^{i\theta} - e^{-i\theta}}{2i} = -\sin \theta.$$

The facts that e^z has all the usual properties when z is complex leads to

$$\frac{d}{d\theta}(e^{(a+ib)\theta}) = (a + ib)e^{(a+ib)\theta} \qquad \text{and} \qquad \int e^{(a+ib)\theta}\, d\theta = \frac{1}{a + ib} e^{(a+ib)\theta} + C.$$

Example 10   Use the formula for $\int e^{(a+ib)\theta}\, d\theta$ to obtain formulas for $\int e^{ax}\cos bx\, dx$ and $\int e^{ax}\sin bx\, dx$.

Solution   The formula for $\int e^{(a+ib)\theta}\, d\theta$ allows us to write

$$\int e^{ax}e^{ibx}\, dx = \int e^{(a+ib)x}\, dx = \frac{1}{a+ib}e^{(a+ib)x} + C = \frac{a-ib}{a^2+b^2}e^{ax}e^{ibx} + C.$$

The left-hand side of this equation can be rewritten as

$$\int e^{ax}e^{ibx}\, dx = \int e^{ax}\cos bx\, dx + i\int e^{ax}\sin bx\, dx.$$

The right-hand side can be rewritten as

$$\frac{a-ib}{a^2+b^2}e^{ax}e^{ibx} = \frac{e^{ax}}{a^2+b^2}(a-ib)(\cos bx + i\sin bx),$$

$$= \frac{e^{ax}}{a^2+b^2}\left(a\cos bx + b\sin bx + i\left(a\sin bx - b\cos bx\right)\right).$$

Equating real parts gives

$$\int e^{ax}\cos bx\, dx = \frac{e^{ax}}{a^2+b^2}\left(a\cos bx + b\sin bx\right) + C,$$

and equating imaginary parts gives

$$\int e^{ax}\sin bx\, dx = \frac{e^{ax}}{a^2+b^2}\left(a\sin bx - b\cos bx\right) + C.$$

These two formulas are usually obtained by integrating by parts twice.

Example 11   Using complex exponentials, find a formula for $\int \sin 2x \sin 3x\, dx$.

Solution   Replacing $\sin 2x$ and $\sin 3x$ by their exponential form, we have

$$\int \sin 2x \sin 3x\, dx = \int \frac{\left(e^{2ix} - e^{-2ix}\right)}{2i}\frac{\left(e^{3ix} - e^{-3ix}\right)}{2i}\, dx$$

$$= \frac{1}{(2i)^2}\int \left(e^{5ix} - e^{-ix} - e^{ix} + e^{-5ix}\right)\, dx$$

$$= -\frac{1}{4}\left(\frac{1}{5i}e^{5ix} + \frac{1}{i}e^{-ix} - \frac{1}{i}e^{ix} - \frac{1}{5i}e^{-5ix}\right) + C$$

$$= -\frac{1}{4}\left(\frac{e^{5ix} - e^{-5ix}}{5i} - \frac{e^{ix} - e^{-ix}}{i}\right) + C$$

$$= -\frac{1}{4}\left(\frac{2}{5}\sin 5x - 2\sin x\right) + C$$

$$= -\frac{1}{10}\sin 5x + \frac{1}{2}\sin x + C.$$

This result is usually obtained by using a trigonometric identity.

Exercises for Appendix B

For Problems 1–8, express the given complex number in polar form, $z = re^{i\theta}$.

For Problems 9–18, perform the indicated calculations. Give your answer in Cartesian form, $z = x + iy$.

1. $2i$

2. -5

3. $1 + i$

4. $-3 - 4i$

9. $(2 + 3i) + (-5 - 7i)$

10. $(2 + 3i)(5 + 7i)$

5. 0

6. $-i$

7. $-1 + 3i$

8. $5 - 12i$

11. $(2 + 3i)^2$

12. $(1 + i)^2 + (1 + i)$

13. $(0.5 - i)(1 - i/4)$ **14.** $(2i)^3 - (2i)^2 + 2i - 1$

15. $(e^{i\pi/3})^2$ **16.** $\sqrt{e^{i\pi/3}}$

17. $(5e^{i7\pi/6})^3$ **18.** $\sqrt[4]{10e^{i\pi/2}}$

By writing the complex numbers in polar form, $z = re^{i\theta}$, find a value for the quantities in Problems 19–28. Give your answer in Cartesian form, $z = x + iy$.

19. $\sqrt{i}$ **20.** $\sqrt{-i}$ **21.** $\sqrt[3]{i}$

22. $\sqrt{7i}$ **23.** $(1 + i)^{100}$ **24.** $(1 + i)^{2/3}$

25. $(-4 + 4i)^{2/3}$ **26.** $(\sqrt{3} + i)^{1/2}$ **27.** $(\sqrt{3}+i)^{-1/2}$

28. $(\sqrt{5} + 2i)^{\sqrt{2}}$

29. Calculate i^n for $n = -1, -2, -3, -4$. What pattern do you observe? What is the value of i^{-36}? Of i^{-41}?

Solve the simultaneous equations in Problems 30–31 for A_1 and A_2.

30. $A_1 + A_2 = 2$
$(1 - i)A_1 + (1 + i)A_2 = 3$

31. $A_1 + A_2 = 2$
$(i - 1)A_1 + (1 + i)A_2 = 0$

32. **(a)** Calculate a and b if $\dfrac{3 - 4i}{1 + 2i} = a + bi$.
(b) Check your answer by calculating $(1 + 2i)(a + bi)$.

33. Check that $z = \dfrac{ac + bd}{c^2 + d^2} + \dfrac{bc - ad}{c^2 + d^2}i$ is the quotient $\dfrac{a + bi}{c + di}$ by showing that the product $z \cdot (c + di)$ is $a + bi$.

34. Let $z_1 = -3 - i\sqrt{3}$ and $z_2 = -1 + i\sqrt{3}$.

    (a) Find $z_1 z_2$ and z_1/z_2. Give your answer in Cartesian form, $z = x + iy$.

    (b) Put z_1 and z_2 into polar form, $z = re^{i\theta}$. Find $z_1 z_2$ and z_1/z_2 using the polar form, and verify that you get the same answer as in part (a).

35. Let $z_1 = a_1 + b_1 i$ and $z_2 = a_2 + b_2 i$. Show that $\overline{z_1 z_2} = \bar{z}_1 \bar{z}_2$.

36. If the roots of the equation $x^2 + 2bx + c = 0$ are the complex numbers $p \pm iq$, find expressions for p and q in terms of b and c.

Are the statements in Problems 37–42 true or false? Explain your answer.

37. Every nonnegative real number has a real square root.

38. For any complex number z, the product $z \cdot \bar{z}$ is a real number.

39. The square of any complex number is a real number.

40. If f is a polynomial, and $f(z) = i$, then $f(\bar{z}) = i$.

41. Every nonzero complex number z can be written in the form $z = e^w$, where w is another complex number.

42. If $z = x+iy$, where x and y are positive, then $z^2 = a+ib$ has a and b positive.

For Problems 43–47, use Euler's formula to derive the following relationships. (Note that if a, b, c, d are real numbers, $a + bi = c + di$ means that $a = c$ and $b = d$.)

43. $\sin^2\theta + \cos^2\theta = 1$ **44.** $\sin 2\theta = 2\sin\theta\cos\theta$

45. $\cos 2\theta = \cos^2\theta - \sin^2\theta$ **46.** $\dfrac{d}{d\theta}\sin\theta = \cos\theta$

47. $\dfrac{d^2}{d\theta^2}\cos\theta = -\cos\theta$

48. Use complex exponentials to show that

$$\sin(-x) = -\sin x.$$

49. Use complex exponentials to show that

$$\sin(x + y) = \sin x \cos y + \cos x \sin y.$$

50. For real t, show that if $z_1(t) = x_1(t) + iy_1(t)$ and $z_2(t) = x_2(t) + iy_2(t)$ then

$$(z_1 + z_2)' = z_1' + z_2' \quad \text{and} \quad (z_1 z_2)' = z_1' z_2 + z_1 z_2'.$$

C   NEWTON'S METHOD

Many problems in mathematics involve finding the root of an equation. For example, we might have to locate the zeros of a polynomial, or determine the point of intersection of two curves. Here we will see a numerical method for approximating solutions which cannot be calculated exactly.

    One such method, bisection, is described in Appendix A. Although it is very simple, the bisection method has two major drawbacks. First, it cannot locate a root where the curve is tangent to, but does not cross, the x-axis. Second, it is relatively slow in the sense that it requires a considerable number of iterations to achieve a desired level of accuracy. Although speed may not be important in solving a single equation, a practical problem may involve solving thousands of equations as a parameter changes. In such a case, any reduction in the number of steps can be important.

Using Newton's Method

We now consider a powerful root-finding method developed by Newton. Suppose we have a function $y = f(x)$. The equation $f(x) = 0$ has a root at $x = r$, as shown in Figure C.14. We begin with an initial estimate, x_0, for this root. (This can be a guess.) We will now obtain a better estimate x_1. To do this, construct the tangent line to the graph of f at the point $x = x_0$, and extend it until it crosses the x-axis, as shown in Figure C.14. The point where it crosses the axis is usually much closer to r, and we use that point as the next estimate, x_1. Having found x_1, we now repeat the process starting with x_1 instead of x_0. We construct a tangent line to the curve at $x = x_1$, extend it until it crosses the x-axis, use that x-intercept as the next approximation, x_2, and so on. The resulting sequence of x-intercepts usually converges rapidly to the root r.

Let's see how this looks algebraically. We know that the slope of the tangent line at the initial estimate x_0 is $f'(x_0)$, and so the equation of the tangent line is

$$y - f(x_0) = f'(x_0)(x - x_0).$$

At the point where this tangent line crosses the x-axis, we have $y = 0$ and $x = x_1$, so that

$$0 - f(x_0) = f'(x_0)(x_1 - x_0).$$

Solving for x_1, we obtain

$$x_1 = x_0 - \frac{f(x_0)}{f'(x_0)}$$

provided that $f'(x_0)$ is not zero. We now repeat this argument and find that the next approximation is

$$x_2 = x_1 - \frac{f(x_1)}{f'(x_1)}.$$

Summarizing, for any $n = 0, 1, 2, \ldots$, we obtain the following result.

Newton's Method to Solve the Equation $f(x) = 0$

Choose x_0 near a solution and compute the sequence $x_1, x_2, x_3 \ldots$ using the rule

$$x_{n+1} = x_n - \frac{f(x_n)}{f'(x_n)}$$

provided that $f'(x_n)$ is not zero. For large n, the solution is well approximated by x_n.

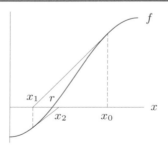

Figure C.14: Newton's method: successive approximations $x_0, x_1, x_2, \ldots$ to the root, r

---

Example 1   Use Newton's method to find the fifth root of 23. (By calculator, this is 1.872171231, correct to nine decimal places.)

Solution   To use Newton's method, we need an equation of the form $f(x) = 0$ having $23^{1/5}$ as a root. Since $23^{1/5}$ is a root of $x^5 = 23$ or $x^5 - 23 = 0$, we take $f(x) = x^5 - 23$. The root of this equation is between 1 and 2 (since $1^5 = 1$ and $2^5 = 32$), so we will choose $x_0 = 2$ as our initial estimate. Now $f'(x) = 5x^4$, so we can set up Newton's method as

$$x_{n+1} = x_n - \frac{x_n^5 - 23}{5x_n^4}.$$

In this case, we can simplify using a common denominator, to obtain

$$x_{n+1} = \frac{4x_n^5 + 23}{5x_n^4}.$$

Therefore, starting with $x_0 = 2$, we find that $x_1 = 1.8875$. This leads to $x_2 = 1.872418193$ and $x_3 = 1.872171296$. These values are in Table C.4. Since we have $f(1.872171231) > 0$ and $f(1.872171230) < 0$, the root lies between 1.872171230 and 1.872171231. Therefore, in just four iterations of Newton's method, we have achieved eight-decimal accuracy.

Table C.4 *Newton's method: $x_0 = 2$*

n	x_n	$f(x_n)$
0	2	9
1	1.8875	0.957130661
2	1.872418193	0.015173919
3	1.872171296	0.000004020
4	1.872171231	0.000000027

Table C.5 *Newton's method: $x_0 = 10$*

n	x_n	n	x_n
0	10	6	2.679422313
1	8.000460000	7	2.232784753
2	6.401419079	8	1.971312452
3	5.123931891	9	1.881654220
4	4.105818871	10	1.872266333
5	3.300841811	11	1.872171240

As a general guideline for Newton's method, once the first correct decimal place is found, each successive iteration approximately doubles the number of correct digits.

What happens if we select a very poor initial estimate? In the preceding example, suppose x_0 were 10 instead of 2. The results are in Table C.5. Notice that even with $x_0 = 10$, the sequence of values moves reasonably quickly toward the solution: We achieve six-decimal place accuracy by the eleventh iteration.

Example 2   Find the first point of intersection of the curves given by $f(x) = \sin x$ and $g(x) = e^{-x}$.

Solution   The graphs in Figure C.15 make it clear that there are an infinite number of points of intersection, all with $x > 0$. In order to find the first one numerically, we consider the function

$$F(x) = f(x) - g(x) = \sin x - e^{-x}$$

whose derivative is $F'(x) = \cos x + e^{-x}$. From the graph, we see that the point we want is fairly close to $x = 0$, so we start with $x_0 = 0$. The values in Table C.6 are approximations to the root. Since $F(0.588532744) > 0$ and $F(0.588532743) < 0$, the root lies between 0.588532743 and 0.588532744. (Remember, your calculator must be set in radians.)

Table C.6 *Successive approximations to root of $\sin x = e^{-x}$*

n	x_n
0	0
1	0.5
2	0.585643817
3	0.588529413
4	0.588532744
5	0.588532744

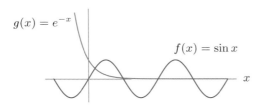

Figure C.15: Root of $\sin x = e^{-x}$

When Does Newton's Method Fail?

In most practical situations, Newton's method works well. Occasionally, however, the sequence x_0, x_1, x_2, ... fails to converge or fails to converge to the root you want. Sometimes, for example, the sequence can jump from one root to another. This is particularly likely to happen if the magnitude of the derivative $f'(x_n)$ is small for some x_n. In this case, the tangent line is nearly horizontal and so x_{n+1} will be far from x_n. (See Figure C.16.)

If the equation $f(x) = 0$ has *no* root, then the sequence will not converge. In fact, the sequence obtained by applying Newton's method to $f(x) = 1 + x^2$ is one of the best known examples of *chaotic behavior* and has attracted considerable research interest recently. (See Figure C.17.)

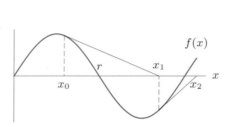

Figure C.16: Problems with Newton's method: Converges to wrong root

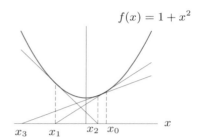

Figure C.17: Problems with Newton's method: Chaotic behavior

Exercises for Appendix C

1. Suppose you want to find a solution of the equation

$$x^3 + 3x^2 + 3x - 6 = 0.$$

Consider $f(x) = x^3 + 3x^2 + 3x - 6$.

   (a) Find $f'(x)$, and use it to show that $f(x)$ increases everywhere.
   (b) How many roots does the original equation have?
   (c) For each root, find an interval which contains it.
   (d) Find each root to two decimal places, using Newton's method.

For Problems 2–4, use Newton's method to find the given quantities to two decimal places:

2. $\sqrt[3]{50}$ 3. $\sqrt[4]{100}$ 4. $10^{-1/3}$

For Problems 5–8, solve each equation and give each answer to two decimal places:

5. $\sin x = 1 - x$ 6. $\cos x = x$

7. $e^{-x} = \ln x$

8. $e^x \cos x = 1$, for $0 < x < \pi$

9. Find, to two decimal places, all solutions of $\ln x = 1/x$.

10. How many zeros do the following functions have? For each zero, find an upper and a lower bound which differ by no more than 0.1.
    (a) $f(x) = x^3 + x - 1$ (b) $f(x) = \sin x - \frac{2}{3}x$
    (c) $f(x) = 10xe^{-x} - 1$

11. Find the largest zero of

$$f(x) = x^3 + x - 1$$

to six decimal places, using Newton's method. How do you know your approximation is as good as you claim?

12. For any positive number, a, the problem of calculating the square root, $\sqrt{a}$, is often done by applying Newton's method to the function $f(x) = x^2 - a$. Apply the method to obtain an expression for x_{n+1} in terms of x_n. Use this to approximate $\sqrt{a}$ for $a = 2, 10, 1000$, and π, correct to four decimal places, starting at $x_0 = a/2$ in each case.

D   VECTORS IN THE PLANE

Position Vectors

Consider a point (a, b) lying on a curve C in the plane (see Figure D.18). The arrow from the origin to the point (a, b) is called the *position vector* of the point, written $\vec{r}$. As the point moves

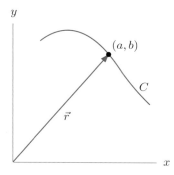

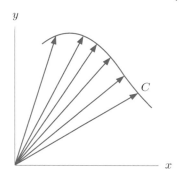

Figure D.18: A position vector $\vec{r}$ **Figure D.19**: Position vectors of points on curve C

along the curve, the position vector sweeps across the plane, the arrowhead touching the curve (see Figure D.19).

A position vector is defined by its magnitude (or length) and its direction. Figure D.20 shows two position vectors with the same magnitude but different directions. Figure D.21 shows two position vectors with the same direction but different magnitudes. An object that possesses both magnitude and direction is called a *vector*, and a position vector is one example. Other physical quantities (such as force, electric and magnetic fields, velocity and acceleration) that have both magnitude and direction can be represented by vectors. To distinguish them from vectors, real numbers (which have magnitude but no direction) are sometimes called *scalars*.

Vectors can be written in several ways. One is to write $\langle a, b \rangle$ for the position vector with tip at (a, b)— the use of the angle brackets signifies that we're talking about a vector, not a point. Another notation uses the special vectors $\vec{i}$ and $\vec{j}$ along the axes. The position vector $\vec{i}$ points to $(1, 0)$ and $\vec{j}$ points to $(0, 1)$; both have magnitude 1. The position vector $\vec{r}$ pointing to (a, b) can be written

$$\vec{r} = a\vec{i} + b\vec{j} .$$

The terms $a\vec{i}$ and $b\vec{j}$ are called the *components* of the vector.

Other special vectors include the zero vector, $\vec{0} = 0\vec{i} + 0\vec{j}$. Any vector with magnitude 1 is called a *unit vector*.

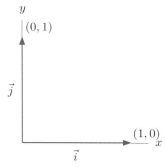

Figure D.20: Position vectors with same magnitude, different direction

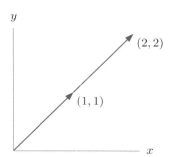

Figure D.21: Position vectors with same direction, different magnitude

Example 1   What are the components of the position vector in Figure D.22?

Solution   Since the vector points to $(3, -\sqrt{3})$, we have

$$\vec{r} = 3\vec{i} - \sqrt{3}\,\vec{j}.$$

Thus, the components of the vector are $3\vec{i}$ and $-\sqrt{3}\,\vec{j}$.

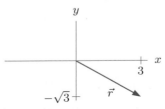

Figure D.22: Find the components
of this position vector

Magnitude and Direction

If $\vec{r}$ is the position vector $a\vec{i} + b\vec{j}$, then the Pythagorean Theorem gives the magnitude of $\vec{r}$, written $\|\vec{r}\|$. From Figure D.23, we see

$$\|\vec{r}\| = \|a\vec{i} + b\vec{j}\| = \sqrt{a^2 + b^2}.$$

The direction of a position vector $\vec{r} = a\vec{i} + b\vec{j}$ is given by the angle θ between the vector and the positive x-axis, measured counterclockwise. This angle satisfies

$$\tan\theta = \left(\frac{b}{a}\right).$$

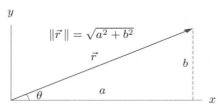

Figure D.23: Magnitude $\|\vec{r}\|$ and
direction of the position vector $\vec{r}$

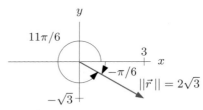

Figure D.24: Magnitude and direction of
the vector $3\vec{i} - \sqrt{3}\vec{j}$

Example 2   Find the magnitude and direction of the position vector $\vec{r} = 3\vec{i} - \sqrt{3}\vec{j}$ in Figure D.22.

Solution   The magnitude is $\|\vec{r}\| = \sqrt{3^2 + (-\sqrt{3})^2} = \sqrt{12} = 2\sqrt{3}$. For the direction, we find $\arctan(-\sqrt{3}/3) = -\pi/6$. Thus, the angle with the positive x-axis is $\theta = 2\pi - \pi/6 = 11\pi/6$. See Figure D.24.

Describing Motion with Position Vectors

The motion given by the parametric equations

$$x = f(t), y = g(t)$$

can be represented by a changing position vector

$$\vec{r}(t) = f(t)\vec{i} + g(t)\vec{j}.$$

For example, $\vec{r}(t) = \cos t\vec{i} + \sin t\vec{j}$ represents the motion $x = \cos t, y = \sin t$ around the unit circle.

Displacement Vectors

Position vectors are vectors that begin at the origin. More general vectors can start at any point in the plane. We view such an arrow as an instruction to move from one point to another and call it a displacement vector. Figure D.25 shows the same displacement vector starting at two different points; we say they are the same vector since they have the same direction and magnitude. Thus, a position vector $\vec{r}$ is a displacement vector beginning at the origin. The zero vector $\vec{0} = 0\vec{i} + 0\vec{j}$ represents no displacement at all.

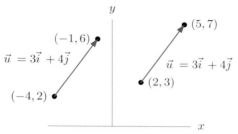

Figure D.25: Two equal displacement vectors:
Same magnitude and direction

Vector Operations

The sum $\vec{u}_1 + \vec{u}_2$ of two displacement vectors is the result of displacing an object first by $\vec{u}_1$ and then by $\vec{u}_2$; see Figure D.26. In terms of components:

If $\vec{u}_1 = a_1\vec{i} + b_1\vec{j}$ and $\vec{u}_2 = a_2\vec{i} + b_2\vec{j}$, then the sum is

$$\vec{u}_1 + \vec{u}_2 = (a_1 + a_2)\vec{i} + (b_1 + b_2)\vec{j}.$$

In other words, to add vectors, add their components separately.

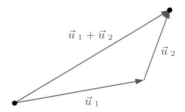

Figure D.26: Vector addition

Example 3   Find the sum of the following pairs of vectors:

(a) $3\vec{i} + 2\vec{j}$ and $-\vec{i} + \vec{j}$ (b) $\vec{i}$ and $3\vec{i} + \vec{j}$ (c) $\vec{i}$ and $\vec{j}$.

Solution   (a) $(3\vec{i} + 2\vec{j}) + (-\vec{i} + \vec{j}) = 2\vec{i} + 3\vec{j}$
(b) $(\vec{i} + 0\vec{j}) + (3\vec{i} + \vec{j}) = 4\vec{i} + \vec{j}$
(c) $(\vec{i} + 0\vec{j}) + (0\vec{i} + \vec{j}) = \vec{i} + \vec{j}$.

Vectors can be multiplied by a number. This operation is called scalar multiplication because it represents changing ("scaling") the magnitude of a vector while keeping its direction the same or reversing it. See Figure D.27.

If c is a real number and $\vec{u} = a\vec{i} + b\vec{j}$, then the *scalar multiple of $\vec{u}$ by c, $c\vec{u}$* , is

$$c\vec{u} = ca\vec{i} + cb\vec{j} .$$

In other words, to multiply a vector by a scalar c, multiply each component by c.

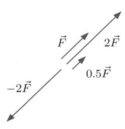

Figure D.27: Scalar multiplication

Example 4   If $\vec{u}_1 = 2\vec{i}$ and $\vec{u}_2 = \vec{i} + 3\vec{j}$, evaluate $6\vec{u}_2, (-2)\vec{u}_1$, and $2\vec{u}_1 + 5\vec{u}_2$.

Solution   We have

$$6\vec{u}_2 = 6\vec{i} + 18\vec{j} ,$$
$$(-2)\vec{u}_1 = -4\vec{i} ,$$
$$2\vec{u}_1 + 5\vec{u}_2 = (4\vec{i}) + (5\vec{i} + 15\vec{j}) = 9\vec{i} + 15\vec{j} .$$

Velocity Vectors

For a particle moving along a line with position $s(t)$, the instantaneous velocity is ds/dt. For a particle moving in the plane, the velocity is a vector. If the position vector is $\vec{r}(t) = x(t)\vec{i} + y(t)\vec{j}$, the particle's displacement during a time interval Δt is

$$\Delta \vec{r}(t) = \Delta x\vec{i} + \Delta y\vec{j} .$$

Dividing by Δt and letting $\Delta t \to 0$, we get the following result:

For motion in the plane with position vector $\vec{r}(t) = x(t)\vec{i} + y(t)\vec{j}$, the **velocity vector** is

$$\vec{v}(t) = \frac{dx}{dt}\vec{i} + \frac{dy}{dt}\vec{j}.$$

The direction of $\vec{v}(t)$ is tangent to the curve. The magnitude $\|\vec{v}(t)\|$ is the **speed**.

Notice the vector viewpoint agrees with the formulas for speed, v_x and v_y, given in Section 4.8, so we write

$$\vec{v}(t) = v_x\vec{i} + v_y\vec{j}.$$

Recall that for motion on a line, the acceleration is $a = dv/dt = d^2s/dt^2$. For motion in the plane, we have the following:

If the position vector is $\vec{r}(t) = x(t)\vec{i} + y(t)\vec{j}$, the **acceleration vector** is

$$\vec{a}(t) = \frac{d^2x}{dt^2}\vec{i} + \frac{d^2y}{dt^2}\vec{j}.$$

The acceleration measures both change in speed and change in direction of the velocity vector.

Example 5   Let $\vec{r}(t) = \cos(2t)\vec{i} + \sin(2t)\vec{j}$. Find the

(a)   Velocity   (b)   Speed   (c)   Acceleration

Solution   (a) Differentiating $\vec{r}(t)$ gives the velocity vector

$$\text{Velocity} = \vec{v}(t) = -2\sin(2t)\vec{i} + 2\cos(2t)\vec{j}.$$

(b) Finding the magnitude of $\vec{v}(t)$, we have

$$\text{Speed} = \|\vec{v}(t)\| = \sqrt{(-2\sin(2t))^2 + (2\cos(2t))^2} = 2.$$

Notice that the speed is constant.
(c) Differentiating $\vec{v}(t)$ gives the acceleration vector

$$\text{Acceleration} = \vec{a}(t) = -4\cos(2t)\vec{i} - 4\sin(2t)\vec{j}.$$

Notice that even though the speed is constant, the acceleration vector is not $\vec{0}$, since the velocity vector is changing direction.

Exercises for Appendix D

Exercises

In Exercises 1–3, find the magnitude of the vector and the angle between the vector and the positive x-axis.

1. $3\vec{i}$

2. $2\vec{i} + \vec{j}$

3. $-\sqrt{2}\,\vec{i} + \sqrt{2}\,\vec{j}$

In Exercises 4–6, perform the indicated operations on the following vectors

$$\vec{u} = 2\vec{j} \qquad \vec{v} = \vec{i} + 2\vec{j} \qquad \vec{w} = -2\vec{i} + 3\vec{j}.$$

4. $\vec{v} + \vec{w}$

5. $2\vec{v} + \vec{w}$

6. $\vec{w} + (-2)\vec{u}$

Exercises 7–9 concern the following vectors:

$$3\vec{i} + 4\vec{j}, \quad \vec{i} + \vec{j}, \quad -5\vec{i}, \quad 5\vec{j}, \quad \sqrt{2}\vec{j}, \quad 2\vec{i} + 2\vec{j}, \quad -6\vec{j}$$

.

7. Which vectors have the same magnitude?

8. Which vectors have the same direction?

9. Which vectors have opposite direction?

10. If k is any real number and $\vec{r} = a\vec{i} + b\vec{j}$ is any vector, show that $\|k\vec{r}\| = |k| \|\vec{r}\|$.

11. Find a unit vector (that is, with magnitude 1) that is

   (a) In the same direction as the vector $-3\vec{i} + 4\vec{j}$.
   (b) In the direction opposite to the vector $-3\vec{i} + 4\vec{j}$.

In Exercises 12–15, express the vector in components.

12. The vector of magnitude 5 making an angle of $90°$ with the positive x-axis.

13. The vector in the same direction as $4\vec{i} - 3\vec{j}$ but with twice the magnitude.

14. The vector with the same magnitude as $4\vec{i} - 3\vec{j}$ and in the opposite direction.

15. The vector from $(3, 2)$ to $(4, 4)$.

In Exercises 16–19, determine whether the vectors are equal.

16. $6\vec{i} - 6\vec{j}$ and the vector from $(6, 6)$ to $(-6, -6)$.

17. The vector from $(7, 9)$ to $(9, 11)$ and the vector from $(8, 10)$ to $(10, 12)$.

18. $-\vec{i} + \vec{j}$ and the vector of length $\sqrt{2}$ making an angle of $\pi/4$ with the positive x-axis.

19. $5\vec{i} - 2\vec{j}$ and the vector from $(1, 12)$ to $(6, 10)$.

In Exercises 20–22, find the velocity vector and the speed, and acceleration.

20. $\vec{r}(t) = t\vec{i} + t^2\vec{j}, \quad t = 1$

21. $\vec{r}(t) = e^t\vec{i} + \ln(1 + t)\vec{j}, \quad t = 0$

22. $\vec{r}(t) = 5\cos t\,\vec{i} + 5\sin t\,\vec{j}, \quad t = \pi/2$

23. A particle is moving along the curve $\vec{r}(t) = \cos t\,\vec{i} + \sin t\,\vec{j}$. Find the particle's position and velocity vectors and its speed when $t = \pi/4$.

E   DETERMINANTS

We introduce the determinant of an array of numbers. Each 2 by 2 array of numbers has another number associated with it, called its determinant, which is given by

$$\begin{vmatrix} a_1 & a_2 \\ b_1 & b_2 \end{vmatrix} = a_1 b_2 - a_2 b_1.$$

For example

$$\begin{vmatrix} 2 & 5 \\ -4 & -6 \end{vmatrix} = 2(-6) - 5(-4) = 8.$$

Each 3 by 3 array of numbers also has a number associated with it, also called a determinant, which is defined in terms of 2 by 2 determinants as follows:

$$\begin{vmatrix} a_1 & a_2 & a_3 \\ b_1 & b_2 & b_3 \\ c_1 & c_2 & c_3 \end{vmatrix} = a_1 \begin{vmatrix} b_2 & b_3 \\ c_2 & c_3 \end{vmatrix} - a_2 \begin{vmatrix} b_1 & b_3 \\ c_1 & c_3 \end{vmatrix} + a_3 \begin{vmatrix} b_1 & b_2 \\ c_1 & c_2 \end{vmatrix}.$$

Notice that the determinant of the 2 by 2 array multiplied by a_i is the determinant of the array found by removing the row and column containing a_i. Also, note the minus sign in the second term. An example is given by

$$\begin{vmatrix} 2 & 1 & -3 \\ 0 & 3 & -1 \\ 4 & 0 & 5 \end{vmatrix} = 2 \begin{vmatrix} 3 & -1 \\ 0 & 5 \end{vmatrix} - 1 \begin{vmatrix} 0 & -1 \\ 4 & 5 \end{vmatrix} + (-3) \begin{vmatrix} 0 & 3 \\ 4 & 0 \end{vmatrix} = 2(15 + 0) - 1(0 - (-4)) + (-3)(0 - 12) = 62.$$

Suppose the vectors $\vec{a}$ and $\vec{b}$ have components $\vec{a} = a_1 \vec{i} + a_2 \vec{j} + a_3 \vec{k}$ and $\vec{b} = b_1 \vec{i} + b_2 \vec{j} + b_3 \vec{k}$. Recall that the cross product $\vec{a} \times \vec{b}$ is given by the expression

$$\vec{a} \times \vec{b} = (a_2 b_3 - a_3 b_2)\vec{i} + (a_3 b_1 - a_1 b_3)\vec{j} + (a_1 b_2 - a_2 b_1)\vec{k}.$$

Notice that if we expand the following determinant, we get the cross product:

$$\begin{vmatrix} \vec{i} & \vec{j} & \vec{k} \\ a_1 & a_2 & a_3 \\ b_1 & b_2 & b_3 \end{vmatrix} = \vec{i}(a_2 b_3 - a_3 b_2) - \vec{j}(a_1 b_3 - a_3 b_1) + \vec{k}(a_1 b_2 - a_2 b_1) = \vec{a} \times \vec{b}.$$

Determinants give a useful way of computing cross products.

READY REFERENCE

This section contains a concise summary of the main definitions, theorems, and ideas presented throughout the book.

READY REFERENCE

Multivariable Functions

Points in 3-space are represented by a system of **Cartesian coordinates** (p. 686). The **distance** between (x, y, z) and (a, b, c) is $\sqrt{(x-a)^2 + (y-b)^2 + (z-c)^2}$ (p. 689).

Functions of two variables can be represented by **graphs** (p. 692), **contour diagrams** (p. 699), **cross-sections** (p. 694), and **tables** (p. 685).

Functions of three variables can be represented by the family of **level surfaces** $f(x, y, z) = c$ for various values of the constant c (p. 718).

A **linear function** $f(x, y)$ has equation

$$f(x, y) = z_0 + m(x - x_0) + n(y - y_0) \text{ (p. 712)}$$

$$= c + mx + ny, \text{ where } c = z_0 - mx_0 - ny_0.$$

Its **graph** is a plane with slope m in the x-direction, slope n in the y-direction, through (x_0, y_0, z_0) (p. 712). Its **table of values** has linear rows (of same slope) and linear columns (of same slope) (p. 713). Its **contour diagram** is equally spaced parallel straight lines (p. 713).

The **limit** of f at the point (a, b), written $\lim_{(x,y) \to (a,b)} f(x, y)$, is the number L, if one exists, such that $f(x, y)$ is as close to L as we please whenever the distance from the point (x, y) to the point (a, b) is sufficiently small, but not zero. (p. 725).

A function f is **continuous at the point** (a, b) if $\lim_{(x,y) \to (a,b)} f(x, y) = f(a, b)$. A function is **continuous on a region** R if it is continuous at each point of R (p. 725).

Vectors

A **vector** $\vec{v}$ has **magnitude** (denoted $\|\vec{v}\|$) and **direction**. Examples are **displacement vectors** (p. 736), **velocity and acceleration vectors** (pp. 745, 746). and **force** (p. 746). We can add vectors, and multiply a vector by a scalar (p. 737). Two non-zero vectors, $\vec{v}$ and $\vec{w}$, are **parallel** if one is a scalar multiple of the other (p. 738).

A **unit vector** has magnitude 1. The vectors $\vec{i}$, $\vec{j}$, and $\vec{k}$ are unit vectors in the directions of the coordinate axes. A unit vector in the direction of any nonzero vector $\vec{v}$ is $\vec{u} = \vec{v} / \|\vec{v}\|$ (p. 742). We **resolve** $\vec{v}$ **into components** by writing $\vec{v} = v_1 \vec{i} + v_2 \vec{j} + v_3 \vec{k}$ (p. 739).

If $\vec{v} = v_1 \vec{i} + v_2 \vec{j} + v_3 \vec{k}$ and $\vec{w} = w_1 \vec{i} + w_2 \vec{j} + w_3 \vec{k}$ then

$$\|\vec{v}\| = \sqrt{v_1^2 + v_2^2 + v_3^2} \text{ (p. 740)}$$

$$\vec{v} + \vec{w} = (v_1 + w_1)\vec{i} + (v_2 + w_2)\vec{j} + (v_3 + w_3)\vec{k} \text{ (p. 741)},$$

$$\lambda\vec{v} = \lambda v_1 \vec{i} + \lambda v_2 \vec{j} + \lambda v_3 \vec{k} \text{ (p. 741)}.$$

The **displacement vector** from $P_1 = (x_1, y_1, z_1)$ to $P_2 = (x_2, y_2, z_2)$ is

$$\overrightarrow{P_1 P_2} = (x_2 - x_1)\vec{i} + (y_2 - y_1)\vec{j} + (z_2 - z_1)\vec{k} \quad \text{(p. 740)}.$$

The **position vector** of $P = (x, y, z)$ is $\overrightarrow{OP}$ (p. 740). A **vector in** n **dimensions** is a string of numbers $\vec{v} = (v_1, v_2, \ldots, v_n)$ (p. 748).

Dot Product (Scalar Product) (p. 752).
Geometric definition: $\vec{v} \cdot \vec{w} = \|\vec{v}\|\|\vec{w}\| \cos\theta$ where θ is the angle between $\vec{v}$ and $\vec{w}$ and $0 \le \theta \le \pi$.
Algebraic definition: $\vec{v} \cdot \vec{w} = v_1 w_1 + v_2 w_2 + v_3 w_3$.

Two nonzero vectors $\vec{v}$ and $\vec{w}$ are **perpendicular** if and only if $\vec{v} \cdot \vec{w} = 0$ (p. 754). Magnitude and dot product are related by $\vec{v} \cdot \vec{v} = \|\vec{v}\|^2$ (p. 754). If $\vec{u} = (u_1, \ldots, u_n)$ and $\vec{v} = (v_1, \ldots, v_n)$ then the dot product of $\vec{u}$ and $\vec{v}$ is $\vec{u} \cdot \vec{v} = u_1 v_1 + \ldots + u_n v_n$ (p. 756).

The **equation of the plane** with **normal vector** $\vec{n} = a\vec{i} + b\vec{j} + c\vec{k}$ and containing the point $P_0 = (x_0, y_0, z_0)$ is $\vec{n} \cdot (\vec{r} - \vec{r}_0) = a(x - x_0) + b(y - y_0) + c(z - z_0) = 0$ or $ax + by + cz = d$, where $d = ax_0 + by_0 + cz_0$ (p. 755).

If $\vec{v}_{\text{parallel}}$ and $\vec{v}_{\text{perp}}$ are components of $\vec{v}$ which are parallel and perpendicular, respectively, to a unit vector $\vec{u}$, then $\vec{v}_{\text{parallel}} = (\vec{v} \cdot \vec{u})\vec{u}$ and $\vec{v}_{\text{perp}} = \vec{v} - \vec{v}_{\text{parallel}}$ (p. 756).

The **work**, W, done by a force $\vec{F}$ acting on an object through a displacement $\vec{d}$ is $W = \vec{F} \cdot \vec{d}$ (p. 758).

Cross Product (Vector Product) (p. 762, 763)
Geometric definition

$$\vec{v} \times \vec{w} = \begin{pmatrix} \text{Area of parallelogram} \\ \text{with edges } \vec{v} \text{ and } \vec{w} \end{pmatrix} \vec{n}$$

$$= (\|\vec{v}\|\|\vec{w}\| \sin\theta)\vec{n},$$

where $0 \le \theta \le \pi$ is the angle between $\vec{v}$ and $\vec{w}$ and $\vec{n}$ is the unit vector perpendicular to $\vec{v}$ and $\vec{w}$ pointing in the direction given by the **right-hand rule**.
Algebraic definition

$$\vec{v} \times \vec{w} = (v_2 w_3 - v_3 w_2)\vec{i} + (v_3 w_1 - v_1 w_3)\vec{j} + (v_1 w_2 - v_2 w_1)\vec{k}$$

$$\vec{v} = v_1 \vec{i} + v_2 \vec{j} + v_3 \vec{k}, \vec{w} = w_1 \vec{i} + w_2 \vec{j} + w_3 \vec{k}.$$

To find the **equation of a plane through three points** that do not lie on a line, determine two vectors in the plane and then find a normal vector using the cross product (p. 765). The **area of a parallelogram** with edges $\vec{v}$ and $\vec{w}$ is $\|\vec{v} \times \vec{w}\|$. The **volume of a parallelepiped** with edges $\vec{a}$, $\vec{b}$, $\vec{c}$ is $\left| (\vec{b} \times \vec{c}) \cdot \vec{a} \right|$ (p. 766).

Differentiation of Multivariable Functions

Partial derivatives of f (p. 777).

$$f_x(a,b) = \begin{array}{c}\text{Rate of change of } f \text{ with respect to } x \\ \text{at the point } (a,b)\end{array}$$

$$= \lim_{h \to 0} \frac{f(a+h,b) - f(a,b)}{h},$$

$$f_y(a,b) = \begin{array}{c}\text{Rate of change of } f \text{ with respect to } y \\ \text{at the point } (a,b)\end{array}$$

$$= \lim_{h \to 0} \frac{f(a,b+h) - f(a,b)}{h}.$$

On the graph of f, the partial derivatives $f_x(a,b)$ and $f_y(a,b)$ give the slope in the x and y directions, respectively (p. 778). The **tangent plane** to $z = f(x,y)$ at (a,b) is

$$z = f(a,b) + f_x(a,b)(x-a) + f_y(a,b)(y-b) \quad \text{(p. 790)}.$$

Partial derivatives can be estimated from a contour diagram or table of values using difference quotients (p. 778), and can be computed algebraically using the same rules of differentiation as for one-variable calculus (p. 784). Partial derivatives for functions of three or more variables are defined and computed in the same way (p 785).

The gradient vector $\operatorname{grad} f$ of f is $\operatorname{grad} f(a,b) = f_x(a,b)\vec{i} + f_y(a,b)\vec{j}$ (2 variables) (p. 800) or $\operatorname{grad} f(a,b,c) = f_x(a,b,c)\vec{i} + f_y(a,b,c)\vec{j} + f_z(a,b,c)\vec{k}$ (3 variables) (p. 808). The gradient vector at P: Points in the direction of increasing f; is perpendicular to the level curve or level surface of f through P; and has magnitude $\|\operatorname{grad} f\|$ equal to the maximum rate of change of f at P (pp. 801, 808). The magnitude is large when the level curves or surfaces are close together and small when they are far apart.

The **directional derivative** of f at P in the direction of a unit vector $\vec{u}$ is (pp. 798, 800)

$$f_{\vec{u}}(P) = \begin{array}{c}\text{Rate of change} \\ \text{of } f \text{ in direction} \\ \text{of } \vec{u} \text{ at } P\end{array} = \operatorname{grad} f(P) \cdot \vec{u}$$

The **tangent plane approximation** to $f(x,y)$ for (x,y) near the point (a,b) is

$$f(x,y) \approx f(a,b) + f_x(a,b)(x-a) + f_y(a,b)(y-b).$$

The right-hand side is the **local linearization** (p. 791). The **differential of** $z = f(x,y)$ at (a,b) is the linear function of dx and dy

$$df = f_x(a,b)\,dx + f_y(a,b)\,dy \quad \text{(p. 793)}.$$

Local linearity with three or more variables follows the same pattern as for functions of two variables (p. 792).

The **tangent plane to a level surface** of a function of three-variables f at (a,b,c) is (p. 811)

$$f_x(a,b,c)(x-a) + f_y(a,b,c)(y-b) + f_z(a,b,c)(z-c) = 0.$$

The Chain Rule for the partial derivative of one variable with respect to another in a chain of composed functions (p. 817):

- Draw a diagram expressing the relationship between the variables, and label each link in the diagram with the derivative relating the variables at its ends.
- For each path between the two variables, multiply together the derivatives from each step along the path.
- Add the contributions from each path.

If $z = f(x,y)$, and $x = g(t)$, and $y = h(t)$, then

$$\frac{dz}{dt} = \frac{\partial z}{\partial x}\frac{dx}{dt} + \frac{\partial z}{\partial y}\frac{dy}{dt} \quad \text{(p. 816)}.$$

If $z = f(x,y)$, with $x = g(u,v)$ and $y = h(u,v)$, then

$$\frac{\partial z}{\partial u} = \frac{\partial z}{\partial x}\frac{\partial x}{\partial u} + \frac{\partial z}{\partial y}\frac{\partial y}{\partial u},$$

$$\frac{\partial z}{\partial v} = \frac{\partial z}{\partial x}\frac{\partial x}{\partial v} + \frac{\partial z}{\partial y}\frac{\partial y}{\partial v} \quad \text{(p. 818)}.$$

Second-order partial derivatives (p. 825)

$$\frac{\partial^2 z}{\partial x^2} = f_{xx} = (f_x)_x, \qquad \frac{\partial^2 z}{\partial x \partial y} = f_{yx} = (f_y)_x,$$

$$\frac{\partial^2 z}{\partial y \partial x} = f_{xy} = (f_x)_y, \qquad \frac{\partial^2 z}{\partial y^2} = f_{yy} = (f_y)_y.$$

Theorem: Equality of Mixed Partial Derivatives. If f_{xy} and f_{yx} are continuous at (a,b), an interior point of their domain, then $f_{xy}(a,b) = f_{yx}(a,b)$ (p. 826).

Taylor Polynomial of Degree 1 Approximating $f(x,y)$ for (x,y) near (a,b) (p. 830)

$$f(x,y) \approx L(x,y) = f(a,b) + f_x(a,b)(x-a) + f_y(a,b)(y-b).$$

Taylor Polynomial of Degree 2 (p. 830)

$$f(x,y) \approx Q(x,y)$$
$$= f(a,b) + f_x(a,b)(x-a) + f_y(a,b)(y-b)$$
$$+ \frac{f_{xx}(a,b)}{2}(x-a)^2 + f_{xy}(a,b)(x-a)(y-b)$$
$$+ \frac{f_{yy}(a,b)}{2}(y-b)^2.$$

Definition of Differentiability (p. 834). A function $f(x,y)$ is **differentiable at the point** (a,b) if there is a linear function $L(x,y) = f(a,b) + m(x-a) + n(y-b)$ such that if the **error** $E(x,y)$ is defined by

$$f(x,y) = L(x,y) + E(x,y),$$

and if $h = x - a, k = y - b$, then the **relative error** $E(a+h, b+k)/\sqrt{h^2 + k^2}$ satisfies

$$\lim_{\substack{h \to 0 \\ k \to 0}} \frac{E(a+h, b+k)}{\sqrt{h^2 + k^2}} = 0.$$

Theorem: Continuity of Partial Derivatives Implies Differentiability (p. 838). If the partial derivatives, f_x and f_y, of a function f exist and are continuous on a small disk centered at the point (a,b), then f is differentiable at (a,b).

Optimization

A function f has a **local maximum** at the point P_0 if $f(P_0) \geq f(P)$ for all points P near P_0, and a **local minimum** at the point P_0 if $f(P_0) \leq f(P)$ for all points P near P_0 (p. 848). A **critical point** of a function f is a point where grad f is either $\vec{0}$ or undefined. If f has a local maximum or minimum at a point P_0, not on the boundary of its domain, then P_0 is a critical point (p. 848). A **quadratic function** $f(x, y) = ax^2 + bxy + cz^2$ generally has one critical point, which can be a local maximum, a local minimum, or a **saddle point** (p. 851).

Second derivative test for functions of two variables (p. 853). Suppose grad $f(x_0, y_0) = \vec{0}$. Let $D = f_{xx}(x_0, y_0)f_{yy}(x_0, y_0) - (f_{xy}(x_0, y_0))^2$.

- If $D > 0$ and $f_{xx}(x_0, y_0) > 0$, then f has a local minimum at (x_0, y_0).

- If $D > 0$ and $f_{xx}(x_0, y_0) < 0$, then f has a local maximum at (x_0, y_0).

- If $D < 0$, then f has a saddle point at (x_0, y_0).

- If $D = 0$, anything can happen.

Unconstrained optimization

A function f defined on a region R has a **global maximum on** R at the point P_0 if $f(P_0) \geq f(P)$ for all points P in R, and a **global minimum on** R at the point P_0 if $f(P_0) \leq f(P)$ for all points P in R (p. 857). For an **unconstrained optimization problem**, find the critical points and investigate whether the critical points give global maxima or minima (p. 858).

A **closed** region is one which contains its boundary; a **bounded** region is one which does not stretch to infinity in any direction (p. 862).

Extreme Value Theorem for Multivariable Functions. If f is a continuous function on a closed and bounded region R, then f has a global maximum at some point (x_0, y_0) in R and a global minimum at some point (x_1, y_1) in R (p. 863).

Constrained optimization

Suppose P_0 is a point satisfying the constraint $g(x, y) = c$. A function f has a **local maximum** at P_0 **subject to the constraint** if $f(P_0) \geq f(P)$ for all points P near P_0 satisfying the constraint (p. 868). It has a **global maximum** at P_0 **subject to the constraint** if $f(P_0) \geq f(P)$ for all points P satisfying the constraint (p. 868). Local and global minima are defined similarly (p. 868). A local maximum or minimum of $f(x, y)$ subject to a constraint $g(x, y) = c$ occurs at a point where the constraint is tangent to a level curve of f, and thus where grad g is parallel to grad f (p. 868).

To optimize f subject to the constraint $g = c$ (p. 869), find the points satisfying the equations

$$\text{grad } f = \lambda \text{ grad } g \quad \text{and} \quad g = c.$$

Then compare values of f at these points, at points on the constraint where grad $g = \vec{0}$, and at the endpoints of the constraint. The number λ is called the **Lagrange multiplier**.

To optimize f subject to the constraint $g \leq c$ (p. 870), find all points in the interior $g(x, y) < c$ where grad f is zero

or undefined; then use Lagrange multipliers to find the local extrema of f on the boundary $g(x, y) = c$. Evaluate f at the points found and compare the values.

The value of λ is the rate of change of the optimum value of f as c increases (where $g(x, y) = c$) (p. 872). The **Lagrangian function** $\mathcal{L}(x, y, \lambda) = f(x, y) - \lambda(g(x, y) - c)$ can be used to convert a constrained optimization problem for f subject the constraint $g = c$ into an unconstrained problem for $\mathcal{L}$ (p. 873).

Multivariable Integration

The **definite integral** of f, a continuous function of two variables, over R, the rectangle $a \leq x \leq b$, $c \leq y \leq d$, is called a **double integral**, and is a limit of **Riemann sums**

$$\int_R f \, dA = \lim_{\Delta x, \Delta y \to 0} \sum_{i,j} f(u_{ij}, v_{ij}) \Delta x \Delta y \quad \text{(p. 887)}.$$

The Riemann sum is constructed by subdividing R into subrectangles of width Δx and height Δy, and choosing a point (u_{ij}, v_{ij}) in the ij-th rectangle.

A **triple integral** of f, a continuous function of three variables, over W, the box $a \leq x \leq b$, $c \leq y \leq d$, $p \leq z \leq q$ in 3-space, is defined in a similar way using three-variable Riemann sums (p. 903).

Interpretations

If $f(x, y)$ is positive, $\int_R f \, dA$ is the **volume** under graph of f above the region R (p. 888). If $f(x, y) = 1$ for all x and y, then the area of R is $\int_R 1 \, dA = \int_R dA$ (p. 889). If $f(x, y)$ is a **density**, then $\int_R f \, dA$ is the **total quantity** in the region R (p. 886). The **average value** of $f(x, y)$ on the region R is $\frac{1}{\text{Area of } R} \int_R f \, dA$ (p. 890). In probability, if $p(x, y)$ is a **joint density function** then $\int_a^b \int_c^d p(x, y) \, dy \, dx$ is the fraction of population with $a \leq x \leq b$ and $c \leq y \leq d$ (p. 925).

Iterated integrals

Double and triple integrals can be written as **iterated integrals**

$$\int_R f \, dA = \int_c^d \int_a^b f(x, y) \, dx \, dy \quad \text{(p. 894)}$$

$$\int_W f \, dV = \int_p^q \int_c^d \int_a^b f(x, y, z) \, dx \, dy \, dz \quad \text{(p. 903)}$$

Other orders of integration are possible. For iterated integrals over **non-rectangular regions** (p. 895), limits on outer integral are constants and limits on inner integrals involve only the variables in the integrals further out (pp. 897, 905).

Integrals in other coordinate systems

When computing double integrals in polar coordinates, put $dA = r \, dr \, d\theta$ or $dA = r \, d\theta \, dr$ (p. 909). Cylindrical coordinates are given by $x = r \cos \theta$, $y = r \sin \theta$, $z = z$, for $0 \leq r < \infty$, $0 \leq \theta \leq 2\pi$, $-\infty < z < \infty$ (p. 914). Spherical coordinates are given by $x = \rho \sin \phi \cos \theta$, $y = \rho \sin \phi \sin \theta$, $z = \rho \cos \phi$, for $0 \leq \rho < \infty$, $0 \leq \phi \leq \pi$, $0 \leq \theta \leq 2\pi$ (p. 917). When computing triple integrals in cylindrical or spherical coordinates, put $dV = r \, dr \, d\theta \, dz$ for cylindrical coordinates (p. 915), $dV = \rho^2 \sin\phi \, d\rho \, d\phi \, d\theta$ for spherical coordinates (p. 918). Other orders of integration are also possible.

For a **change of variables** $x = x(s,t)$, $y = y(s,t)$, the **Jacobian** is

$$\frac{\partial(x,y)}{\partial(s,t)} = \frac{\partial x}{\partial s} \cdot \frac{\partial y}{\partial t} - \frac{\partial x}{\partial t} \cdot \frac{\partial y}{\partial s} = \begin{vmatrix} \frac{\partial x}{\partial s} & \frac{\partial x}{\partial t} \\ \frac{\partial y}{\partial s} & \frac{\partial y}{\partial t} \end{vmatrix} \quad \text{(p. 1108)}.$$

To convert an integral from x, y to s, t coordinates (p. 1108): Substitute for x and y in terms of s and t, change the xy region R into an st region T, and change the area element by making the substitution $dxdy = \left| \frac{\partial(x,y)}{\partial(s,t)} \right| dsdt$. For triple integrals, there is a similar formula (p. 1109).

Parameterizations and Vector Fields

Parameterized curves

The motion of a particle is described by **parametric equations** $x = f(t), y = g(t)$ (2-space) or $x = f(t), y = g(t), z = h(t)$ (3-space). The path of the particle is a **parameterized curve** (p. 936). Parameterizations are also written in **vector form** $\vec{r}(t) = f(t)\vec{i} + g(t)\vec{j} + h(t)\vec{k}$ (p. 937). For a **curve segment** we restrict the parameter to to a closed interval $a \leq t \leq b$ (p. 939). **Parametric equations for the graph** of $y = f(x)$ are $x = t, y = f(t)$.

Parametric equations for a line through (x_0, y_0) in the direction of $\vec{v} = a\vec{i} + b\vec{j}$ are $x = x_0 + at$, $y = y_0 + bt$. In 3-space, the line through (x_0, y_0, z_0) in the direction of $\vec{v} = a\vec{i} + b\vec{j} + c\vec{k}$ is $x = x_0 + at$, $y = y_0 + bt$, $z = z_0 + ct$ (p. 937). In vector form, the equation for a line is $\vec{r}(t) = \vec{r}_0 + t\vec{v}$, where $\vec{r}_0 = x_0\vec{i} + y_0\vec{j} + z_0\vec{k}$ (p. 939).

Parametric equations for a circle of radius R in the plane, centered at the origin are $x = R\cos t$, $y = R\sin t$ (counterclockwise), $x = R\cos t, y = -R\sin t$ (clockwise).

To find the **intersection points** of a curve $\vec{r}(t) = f(t)\vec{i} + g(t)\vec{j} + h(t)\vec{k}$ with a surface $F(x,y,z) = c$, solve $F(f(t), g(t), h(t)) = c$ for t (p. 939). To find the intersection points of two curves $\vec{r}_1(t)$ and $\vec{r}_2(t)$, solve $\vec{r}_1(t_1) = \vec{r}_2(t_2)$ for t_1 and t_2 (p. 939).

The **length of a curve segment** C given parametrically for $a \leq t \leq b$ with velocity vector $\vec{v}$ is $\int_a^b \|\vec{v}\| dt$ if $\vec{v} \neq \vec{0}$ for $a < t < b$ (p. 950).

The **velocity** and **acceleration** of a moving object with position vector $\vec{r}(t)$ at time t are

$$\vec{v}(t) = \lim_{\Delta t \to 0} \frac{\Delta \vec{r}}{\Delta t} \quad \text{(p. 946)}$$

$$\vec{a}(t) = \lim_{\Delta t \to 0} \frac{\Delta \vec{v}}{\Delta t} \quad \text{(p. 948)}$$

We write $\vec{v} = \frac{d\vec{r}}{dt} = \vec{r}'(t)$ and $\vec{a} = \frac{d\vec{v}}{dt} = \frac{d^2\vec{r}}{dt^2} = \vec{r}''(t)$.

The **components of the velocity and acceleration vectors** are

$$\vec{v}(t) = \frac{dx}{dt}\vec{i} + \frac{dy}{dt}\vec{j} + \frac{dz}{dt}\vec{k} \quad \text{(p. 946)}$$

$$\vec{a}(t) = \frac{d^2x}{dt^2}\vec{i} + \frac{d^2y}{dt^2}\vec{j} + \frac{d^2z}{dt^2}\vec{k} \quad \text{(p. 948)}$$

The **speed** is $\|\vec{v}\| = \sqrt{(dx/dt)^2 + (dy/dt)^2 + (dz/dt)^2}$ (p. 950). Analogous formulas for velocity, speed, and acceleration hold in 2-space.

Uniform Circular Motion (p. 949) For a particle $\vec{r}(t) = R\cos(\omega t)\vec{i} + R\sin(\omega t)\vec{j}$: motion is in a circle of radius R with period $2\pi/\omega$; velocity, $\vec{v}$, is tangent to the circle and speed is constant $\|\vec{v}\| = \omega R$; acceleration, $\vec{a}$, points toward the center of the circle with $\|\vec{a}\| = \|\vec{v}\|^2/R$.

Motion in a Straight Line (p. 950) For a particle $\vec{r}(t) = \vec{r}_0 + f(t)\vec{v}_0$: Motion is along a straight line through the point with position vector $\vec{r}_0$ parallel to $\vec{v}_0$; velocity, $\vec{v}$, and acceleration, $\vec{a}$, are parallel to the line.

Vector fields

A **vector field** in 2-space is a function $\vec{F}(x,y)$ whose value at a point (x,y) is a 2-dimensional vector (p. 955). Similarly, a vector field in 3-space is a function $\vec{F}(x,y,z)$ whose values are 3-dimensional vectors (p. 955). Examples are the **gradient** of a differentiable function f, the **velocity field** of a fluid flow, and **force fields** (p. 955). A **flow line** of a vector field $\vec{v} = \vec{F}(\vec{r})$ is a path $\vec{r}(t)$ whose velocity vector equals $\vec{v}$, thus $\vec{r}'(t) = \vec{v} = \vec{F}(\vec{r}(t))$ (p. 962). The **flow** of a vector field is the family of all of its flow line (p. 962). Flow lines can be approximated numerically using Euler's method (p. 964).

Parameterized surfaces

We **parameterize a surface** with two parameters, $x = f_1(s,t)$, $y = f_2(s,t)$, $z = f_3(s,t)$ (p. 1098). We also use the vector form $\vec{r}(s,t) = f_1(s,t)\vec{i} + f_2(s,t)\vec{j} + f_3(s,t)\vec{k}$ (p. 1098). **Parametric equations for the graph** of $z = f(x,y)$ are $x = s$, $y = t$, and $z = f(s,t)$ (p. 1098). **Parametric equation for a plane** through the point with position vector $\vec{r}_0$ and containing the two nonparallel vectors $\vec{v}_1$ and $\vec{v}_2$ is $\vec{r}(s,t) = \vec{r}_0 + s\vec{v}_1 + t\vec{v}_2$ (p. 1099). **Parametric equation for a sphere** of radius R centered at the origin is $\vec{r}(\theta, \phi) = R\sin\phi\cos\theta\vec{i} + R\sin\phi\sin\theta\vec{j} + \cos\phi\vec{k}$, $0 \leq \theta \leq 2\pi, 0 \leq \phi \leq \pi$ (p. 1099). **Parametric equation for a cylinder** of radius R along the z-axis is $\vec{r}(\theta, z) = R\cos\theta\vec{i} + R\sin\theta\vec{j} + z\vec{k}$, $0 \leq \theta \leq 2\pi, -\infty < z < \infty$ (p. 1097). A **parameter curve** is the curve obtained by holding one of the parameters constant and letting the other vary (p. 1103).

Line Integrals

The **line integral** of a vector field $\vec{F}$ along an **oriented curve** C (p. 976) is

$$\int_C \vec{F} \cdot d\vec{r} = \lim_{\|\Delta\vec{r}_i\| \to 0} \sum_{i=0}^{n-1} \vec{F}(\vec{r}_i) \cdot \Delta\vec{r}_i,$$

where the direction of $\Delta\vec{r}_i$ is the direction of the orientation (p. 977).

The line integral measures the extent to which C is going with $\vec{F}$ or against it (p. 977). For oriented curves C, C_1, and C_2, $\int_{-C} \vec{F} \cdot d\vec{r} = -\int_C \vec{F} \cdot d\vec{r}$, where $-C$ is the curve C parameterized in the opposite direction, and $\int_{C_1+C_2} \vec{F} \cdot d\vec{r} = \int_{C_1} \vec{F} \cdot d\vec{r} + \int_{C_2} \vec{F} \cdot d\vec{r}$, where $C_1 + C_2$ is the curve obtained by joining the endpoint of C_1 to the starting point of C_2 (p. 981).

The **work done by a force** $\vec{F}$ along a curve C is $\int_C \vec{F} \cdot d\vec{r}$ (p. 978). The **circulation** of $\vec{F}$ around an oriented closed curve is $\int_C \vec{F} \cdot d\vec{r}$ (p. 980).

Given a parameterization of C, $\vec{r}(t)$, for $a \leq t \leq b$, the line integral can be calculated as

$$\int_C \vec{F} \cdot d\vec{r} = \int_a^b \vec{F}(\vec{r}(t)) \cdot \vec{r}'(t)\, dt \quad \text{(p. 987)}.$$

Fundamental Theorem for Line Integrals (p. 993):

Suppose C is a piecewise smooth oriented path with starting point P and endpoint Q. If f is a function whose gradient is continuous on the path C, then

$$\int_C \operatorname{grad} f \cdot d\vec{r} = f(Q) - f(P).$$

Path-independent fields and gradient fields

A vector field $\vec{F}$ is said to be **path-independent**, or **conservative**, if for any two points P and Q, the line integral $\int_C \vec{F} \cdot d\vec{r}$ has the same value along any piecewise smooth path C from P to Q lying in the domain of $\vec{F}$ (p. 995). A **gradient field** is a vector field of the form $\vec{F} = \operatorname{grad} f$ for some scalar function f, and f is called a **potential function** for the vector field $\vec{F}$ (p. 996). A vector field $\vec{F}$ is path-independent if and only if $\vec{F}$ is a gradient vector field (p. 996). A vector field $\vec{F}$ is path-independent if and only if $\int_C \vec{F} \cdot d\vec{r} = 0$ for every closed curve C (p. 1005). If $\vec{F}$ is a gradient field, then $\frac{\partial F_2}{\partial x} - \frac{\partial F_1}{\partial y} = 0$ (p. 1006). The quantity $\frac{\partial F_2}{\partial x} - \frac{\partial F_1}{\partial y}$ is called the 2-dimensional or scalar **curl** of $\vec{F}$.

Green's Theorem (p. 1007):

Suppose C is a piecewise smooth simple closed curve that is the boundary of an open region R in the plane and oriented so that the region is on the left as we move around the curve. Suppose $\vec{F} = F_1 \vec{i} + F_2 \vec{j}$ is a smooth vector field defined at every point of the region R and boundary C. Then

$$\int_C \vec{F} \cdot d\vec{r} = \int_R \left(\frac{\partial F_2}{\partial x} - \frac{\partial F_1}{\partial y} \right) dx\, dy.$$

Curl test for vector fields in 2-space: If $\frac{\partial F_2}{\partial x} - \frac{\partial F_1}{\partial y} = 0$ and the domain of $\vec{F}$ has no holes, then $\vec{F}$ is path-independent, and hence a gradient field (p. 1009). The condition that the domain have no holes is important. It is not always true that if the scalar curl of $\vec{F}$ is zero then $\vec{F}$ is a gradient field (p. 1009).

Surface Integrals

A surface is **oriented** if a unit normal vector $\vec{n}$ has been chosen at every point on it in a continuous way (p. 1026). For a closed surface, we usually choose the outward orientation (p. 1026). The **area vector** of a flat, oriented surface is a vector $\vec{A}$ whose magnitude is the area of the surface, and whose direction is the direction of the orientation vector $\vec{n}$ (p. 1027). If $\vec{v}$ is the velocity vector of a constant fluid flow and $\vec{A}$ is the area vector of a flat surface, then the total flow through the surface in units of volume per unit time is called the **flux** of $\vec{v}$ through the surface and is given by $\vec{v} \cdot \vec{A}$ (p. 1027).

The **surface integral** or **flux integral** of the vector field $\vec{F}$ through the oriented surface S is

$$\int_S \vec{F} \cdot d\vec{A} = \lim_{\|\Delta \vec{A}\| \to 0} \sum \vec{F} \cdot \Delta \vec{A},$$

where the direction of $\Delta \vec{A}$ is the direction of the orientation (p. 1028). If $\vec{v}$ is a variable vector field and then $\int_S \vec{v} \cdot d\vec{A}$ is the flux through the surface S (p. 1029).

Simple flux integrals can be calculated by putting $d\vec{A} = \vec{n}\, dA$ and using geometry or converting to a double integral (p. 1030).

The **flux through a graph** of $z = f(x, y)$ above a region R in the xy-plane, oriented upward, is

$$\int_R \vec{F}(x, y, f(x, y)) \cdot \left(-f_x \vec{i} - f_y \vec{j} + \vec{k} \right) dx\, dy \quad \text{(p. 1038)}.$$

The **flux through a cylindrical surface** S of radius R and oriented away from the z-axis is

$$\int_T \vec{F}(R, \theta, z) \cdot \left(\cos\theta \vec{i} + \sin\theta \vec{j} \right) R\, dz\, d\theta \quad \text{(p. 1039)},$$

where T is the θz-region corresponding to S.

The **flux through a spherical surface** S of radius R and oriented away from the origin is

$$\int_T \vec{F}(R, \theta, \phi) \cdot \left(\sin\phi\cos\theta \vec{i} + \sin\phi\sin\theta \vec{j} + \cos\phi \vec{k} \right)$$
$$R^2 \sin\phi\, d\phi\, d\theta, \quad \text{(p. 1040)}$$

where T is the $\theta\phi$-region corresponding to S.

The **flux through a parameterized surface** S, parameterized by $\vec{r} = \vec{r}(s, t)$, where (s, t) varies in a parameter region R, is

$$\int_R \vec{F}(\vec{r}(s, t)) \cdot \left(\frac{\partial \vec{r}}{\partial s} \times \frac{\partial \vec{r}}{\partial t} \right) ds\, dt \quad \text{(p. 1112)}.$$

We choose the parameterization so that $\partial \vec{r}/\partial s \times \partial \vec{r}/\partial t$ is never zero and points in the direction of $\vec{n}$ everywhere.

The **area of a parameterized surface** S, parameterized by $\vec{r} = \vec{r}(s, t)$, where (s, t) varies in a parameter region R, is

$$\int_S dA = \int_R \left\| \frac{\partial \vec{r}}{\partial s} \times \frac{\partial \vec{r}}{\partial t} \right\| ds\, dt \quad \text{(p. 1112)}.$$

Divergence and Curl

Divergence

Definition of Divergence (p. 1046).
Geometric definition: The **divergence** of $\vec{F}$ is

$$\operatorname{div} \vec{F}(x,y,z) = \lim_{\text{Volume}\to 0} \frac{\int_S \vec{F} \cdot d\vec{A}}{\text{Volume of } S}.$$

Here S is a sphere centered at (x,y,z), oriented outwards, that contracts down to (x,y,z) in the limit.

Cartesian coordinate definition: If $\vec{F} = F_1 \vec{i} + F_2 \vec{j} + F_3 \vec{k}$, then

$$\operatorname{div} \vec{F} = \frac{\partial F_1}{\partial x} + \frac{\partial F_2}{\partial y} + \frac{\partial F_3}{\partial z}.$$

The divergence can be thought of as the outflow per unit volume of the vector field. A vector field $\vec{F}$ is said to be **divergence free** or **solenoidal** if $\operatorname{div} \vec{F} = 0$ everywhere that $\vec{F}$ is defined. Magnetic fields are divergence free (p. 1049).

The Divergence Theorem (p. 1056). If W is a solid region whose boundary S is a piecewise smooth surface, and if $\vec{F}$ is a smooth vector field which is defined everywhere in W and on S, then

$$\int_S \vec{F} \cdot d\vec{A} = \int_W \operatorname{div} \vec{F} \, dV,$$

where S is given the outward orientation. In words, the Divergence Theorem says that the total flux out of a closed surface is the integral of the flux density over the volume it encloses.

Curl

The **circulation density** of a smooth vector field $\vec{F}$ at (x,y,z) around the direction of the unit vector $\vec{n}$ is defined to be

$$\operatorname{circ}_{\vec{n}} \vec{F}(x,y,z) = \lim_{\text{Area}\to 0} \frac{\text{Circulation around } C}{\text{Area inside } C}$$

$$= \lim_{\text{Area}\to 0} \frac{\int_C \vec{F} \cdot d\vec{r}}{\text{Area inside } C} \quad (\text{p. 1070}).$$

Circulation density is calculated using the **right-hand rule** (p. 1070).

Definition of curl (p. 1071).
Geometric definition The curl of $\vec{F}$, written $\operatorname{curl} \vec{F}$, is the vector field with the following properties

- The direction of $\operatorname{curl} \vec{F}(x,y,z)$ is the direction $\vec{n}$ for which $\operatorname{circ}_{\vec{n}}(x,y,z)$ is greatest.
- The magnitude of $\operatorname{curl} \vec{F}(x,y,z)$ is the circulation density of $\vec{F}$ around that direction.

Cartesian coordinate definition If $\vec{F} = F_1 \vec{i} + F_2 \vec{j} + F_3 \vec{k}$, then

$$\operatorname{curl} \vec{F} = \left(\frac{\partial F_3}{\partial y} - \frac{\partial F_2}{\partial z} \right) \vec{i} + \left(\frac{\partial F_1}{\partial z} - \frac{\partial F_3}{\partial x} \right) \vec{j}$$

$$+ \left(\frac{\partial F_2}{\partial x} - \frac{\partial F_1}{\partial y} \right) \vec{k}.$$

Curl and circulation density are related by $\operatorname{circ}_{\vec{n}} \vec{F} = \operatorname{curl} \vec{F} \cdot \vec{n}$ (p. 1073). A vector field is said to be **curl free** or **irrotational** if $\operatorname{curl} \vec{F} = \vec{0}$ everywhere that $\vec{F}$ is defined (p. 1074).

Given an oriented surface S with a boundary curve C we use the right-hand rule to determine the orientation of C (p. 1078).

Stokes' Theorem (p. 1079). If S is a smooth oriented surface with piecewise smooth, oriented boundary C, and if $\vec{F}$ is a smooth vector field which is defined on S and C, then

$$\int_C \vec{F} \cdot d\vec{r} = \int_S \operatorname{curl} \vec{F} \cdot d\vec{A}.$$

Stokes' Theorem says that the total circulation around C is the integral over S of the circulation density. A **curl field** is a vector field $\vec{F}$ that can be written as $\vec{F} = \operatorname{curl} \vec{G}$ for some vector field $\vec{G}$, called a **vector potential** for $\vec{F}$ (p. 1081).

Relation between divergence, gradient, and curl
The curl and gradient are related by $\operatorname{curl} \operatorname{grad} f = 0$ (p. 1085). Divergence and curl are related by $\operatorname{div} \operatorname{curl} \vec{F} = 0$ (p. 1087).

The curl test for vector fields in 3-space (p. 1086) Suppose that $\operatorname{curl} \vec{F} = \vec{0}$, and that the domain of $\vec{F}$ has the property that every closed curve in it can be contracted to a point in a smooth way, staying at all times within the domain. Then $\vec{F}$ is path-independent, so $\vec{F}$ is a gradient field and has a potential function.

The divergence test for vector fields in 3-space (p. 1087) Suppose that $\operatorname{div} \vec{F} = 0$, and that the domain of $\vec{F}$ has the property that every closed surface in it is the boundary of a solid region completely contained in the domain. Then $\vec{F}$ is a curl field.

ANSWERS TO ODD NUMBERED PROBLEMS

Section 12.1

1 A, B, C

3 Q

5 $(1, -1, 1)$; Front, left, above

7 $(2, 4.5, 3)$

9

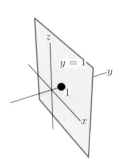

11

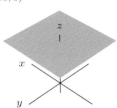

13 $(2, 3, 15)$

15 $(x - 1)^2 + y^2 + z^2 = 4$

17 (a) 20-30°C
   (b) 0-12°C
   (c) 0-40°C

19

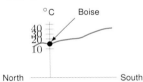

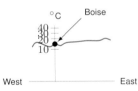

23 (a) −28°C
   (b) 16 km/hr
   (c) About 12 km/hr
   (d) About −14°C

29 $(1.5, 0.5, -0.5)$

31 Cone, tip at origin, along x-axis with slope of 1

33 (a) $z = 7, z = -1$
   (b) $x = 6, x = -2$
   (c) $y = 7, y = -1$

35 (a) $(12, 7, 2); (5, 7, 2); (12, 1, 2)$
   (b) $(5, 1, 4); (5, 7, 4); (12, 1, 4)$

37 (a) yz-plane: circle $(y + 3)^2 + (z - 2)^2 = 3$
       xz-plane: none
       xy-plane: point $(1, -3, 0)$

(b) Does not intersect

39 $(8, 0, \sqrt{3})$

41 $y = 1$ is a plane, not a line

43 Distance is 5

45 $(-2, -1, -5)$

47 True

49 False

51 True

53 True

55 True

57 False

59 False

Section 12.2

1 (a) (IV)
  (b) (II)
  (c) (I)
  (d) (V)
  (e) (III)

3 Horizontal plane 3 units above the xy-plane

5 Bowl opening up, vertex $(0, 0, 4)$

7 Parabolic cylinder extended along x-axis

9 Circular cylinder of radius 2 in the z-direction

11 $x^2 + z^2 = 7$

13 $x = 1 - (y - 3)^2 - (z - 5)^2$

15 $f(3, 2) = 0.037$ mg/liter

17

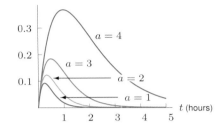

19 (a) is (IV)
   (b) is (IX)
   (c) is (VII)
   (d) is (I)
   (e) is (VIII)
   (f) is (II)
   (g) is (VI)
   (h) is (III)
   (i) is (V)

21 (a) (i)

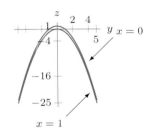

(ii)

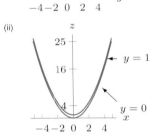

(b) (i)

(ii)

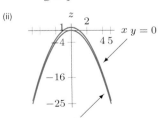

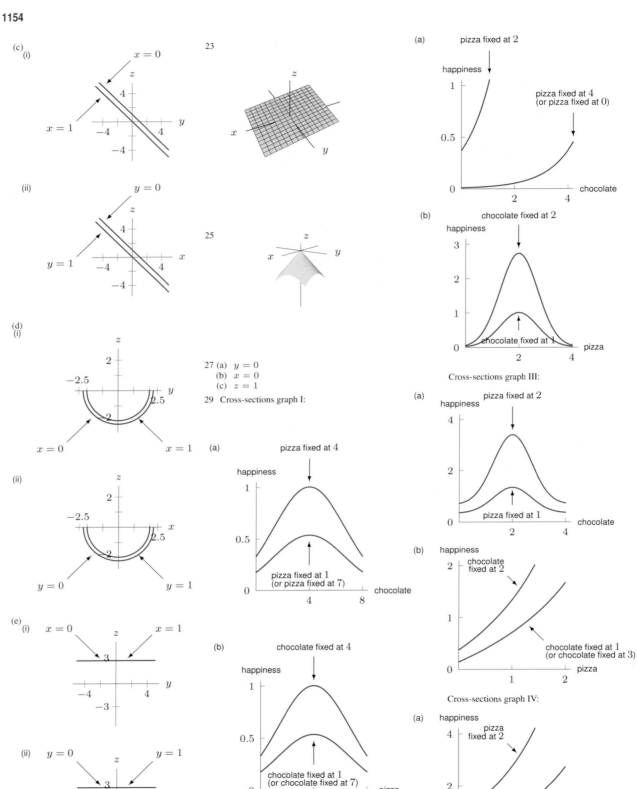

(c)
(i) $x = 0$
$x = 1$

(ii) $y = 0$
$y = 1$

(d)
(i)
$x = 0$
$x = 1$

(ii)
$y = 0$
$y = 1$

(e)
(i) $x = 0$ $x = 1$

(ii) $y = 0$ $y = 1$

23

25

27 (a) $y = 0$
   (b) $x = 0$
   (c) $z = 1$

29 Cross-sections graph I:

(a) pizza fixed at 4

(b) chocolate fixed at 4

Cross-sections graph II:

Cross-sections graph III:

(a) pizza fixed at 2
    pizza fixed at 4
    (or pizza fixed at 0)

(b) chocolate fixed at 2
    chocolate fixed at 1

(a) pizza fixed at 2
    pizza fixed at 1

(b) chocolate fixed at 2
    chocolate fixed at 1
    (or chocolate fixed at 3)

Cross-sections graph IV:

(a) pizza fixed at 2
    pizza fixed at 1

(b)

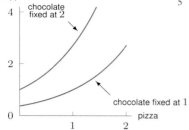

happiness

5

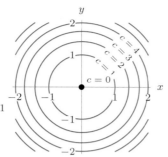

17 $3x^2y + 7x + 20 = 805$

19

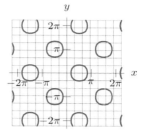

31 (a)

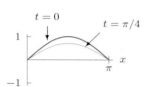

$t = 0$ $t = \pi/4$

$x = \pi/2$

$x = \pi/4$

7

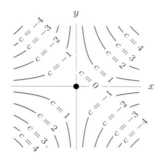

21 100, 150, 200, 250

23 Answers in °C:

(a)

(b)

(c)

(d)

(b) $f = 0$; ends of string don't move

33 Cross-sections are lines parallel to y-axis

35 $f(x, y) = x^2 - 1$

37 False

39 True

41 True

43 True

45 True

47 True

49 False

51 (c)

9

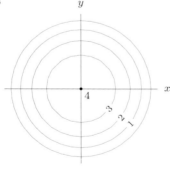

Section 12.3

1 (a) (III)
  (b) (I)
  (c) (V)
  (d) (II)
  (e) (IV)

3 Contours evenly spaced

13

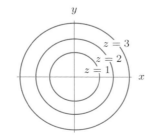

25 (a) A
  (b) B
  (c) A

27

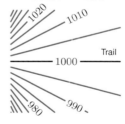
Elevation in meters

29 (a) I
  (b) IV
  (c) II
  (d) III

(a)

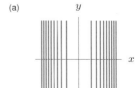

15

1156

(b)

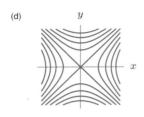

(c)

(d)

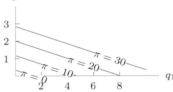

31 (a) $\pi = 3q_1 + 12q_2 - 4$ (thousands)
   (b) q_2

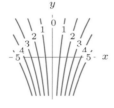

33 $\alpha + \beta > 1$: increasing
   $\alpha + \beta = 1$: constant
   $\alpha + \beta < 1$: decreasing

35 (I) h
   (II) j
   (III) f
   (IV) g

37 (a)

(b)

43 Contour diagram is curves in xy-plane
45 $f(x, y) = x^2$
47 Could not be true

49 Might be true
51 True
53 True
55 False
57 False
59 True

Section 12.4

1 Linear
3 Linear function
5 Linear function
7 -1.0
9 $z = \frac{4}{3}x - \frac{1}{2}y$
11 $z = -2y + 2$
13 $\Delta z = 0.4; \quad z = 2.4$
15 CDs cost $8
   DVDs cost $12
17 (I) h
   (II) n
   (III) f
   (IV) m
19 (a) Linear
   (b) Linear
   (c) Not linear
21 85 kg person at 14 km/h
   55 kg person at 16 km/h
23 $f(x, y) = 2x - 0.5y + 1$
25 $g(x, y) = 3x + y$
27

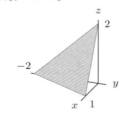

29

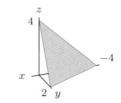

33 (a) $7/\sqrt{29}$
   (b) $-5/\sqrt{104}$
35 $f(x, y) = xy$ has linear cross-sections
37 $z = -2x + y$
39 False
41 True
43 True
45 False
47 False
49 False

Section 12.5

1 (a) I
   (b) II
3 $f(x, y) = \frac{1}{3}(5 - x - 2y)$

5 $f(x, y) = (1 - x^2 - y^2)^2$
7 Elliptical and hyperbolic paraboloid, plane
9 Hyperboloid of two sheets
11 Ellipsoid
13 Yes, $f(x, y) = (2x + 3y - 10)/5$
15 No
17 $f(x, y) = \sqrt{10 - x^2 - y^2}$
   $g(x, y, z) = x^2 + y^2 + z^2 = 10$
19 (a) Parallel planes: $2x - 3y + z = c + 20$
   (b) $f_z(0, 0, 0) = 1$
   (c) $\vec{n} = 2\vec{i} - 3\vec{j} + \vec{k}$
   (d) Yes; $-3°$C
21 $f(x, y) = 3\sqrt{1 - x^2 - y^2/4}$;
   $g(x, y) = -3\sqrt{1 - x^2 - y^2/4}$
23 (a) Graph of f is graph of
      $x^2 + y^2 + z^2 = 1, z \geq 0$
   (b) $\sqrt{1 - x^2 - y^2} - z = 0$
25 $g(x, y, z) = 1 - \sqrt{x^2 + z^2} - y = 0$
27 Ellipsoid
29 Parallel planes
33

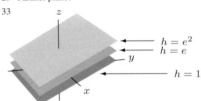

35

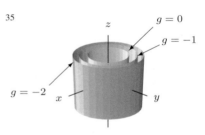

37 Level surfaces hyperbolic cylinders
39 $f(x, y, z) = y + z$
41 $f(x, y, z) = (x + y + z)^2$
43 True
45 True
47 True
49 False
51 False

Section 12.6

1 Not continuous
3 Continuous
5 Not continuous
7 1
9 0
11 1
19 No
21 $c = 1$
23 (c) No

25 For quotient, need $g(a, b) \neq 0$
27 $f(x, y) = (x^2 + 2y^2)/(x^2 + y^2)$
29 $f(x, y) = 1/((x - 2)^2 + y^2)$

Chapter 12 Review

1

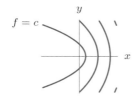

3 Yes, $f(x, y) = 5 - 3x + 2y$
5 Yes, $f(x, y) = 2 + x/5 + y/5 - 3x^2/5 + y^2$
7 (a) (IV)
  (b) (II)
  (c) (I)
  (d) (III)
9 Lines with slope $3/5$, evenly spaced

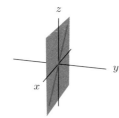

11 Ellipses

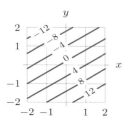

13 Vertical line through $(2, 1, 0)$
15 $(x/5) + (y/3) + (z/2) = 1$
17 Not linear
19 (a)

		y	
	2.5	3.0	3.50
	-1	6	7	8
x	1	0	1	2
	3	-6	-5	-4

  (b) $f(x, y) = -2 - 3x + 2y$
21 $f(x, y) = x^2 + y^2 + 5$
   $g(x, y, z) = x^2 + y^2 + 5 - z = 0$

23 $f(x, y) = \sqrt{1 - x^2 - y^2}$
   $g(x, y, z) = x^2 + y^2 + z^2 = 1$
25 Cylinder
27 $g(x, y) = 200x + 100y$
31 (a)

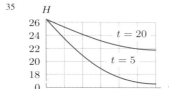

  (b) $c \geq 0$
35

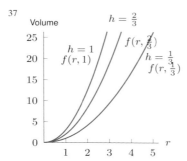

37

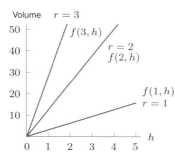

39 $f(x, y) = 2x - y + 2$
43 (a) For $t = 0$:

For $t = \pi/4$:

For $t = \pi/2$:

For $t = 3\pi/4$:

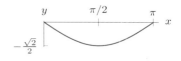

For $t = \pi$:

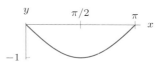

45 (a) $(1, \sqrt{3}, 0)$
  (b) $(1, 1/\sqrt{3}, 2\sqrt{2}/\sqrt{3})$
  (c) Tetrahedron
47 (a) $f(1, 1, 3) \approx 26$, $f(1, 2, 1) \approx 12$
  (b) $f(1, 1, 3)$ exact, $f(1, 2, 1)$ not exact
  (c) $5x^2 + 2yz + 3zx^3 + 6 \cdot 2^{x-y}$
  (d) $f(1, 1, 3) = 26$, $f(1, 2, 1) = 15$

Section 13.1

1 $2\vec{i} + \vec{j}$,
  $2\vec{i}$,
  $-2\vec{i}$,
  $-2\vec{i} + 2\vec{j}$,
  $-2\vec{i} - \vec{j}$
3 $3\vec{i} + 4\vec{j}$
5 $\vec{u} = \vec{i} + \vec{j} + 2\vec{k}$
  $\vec{v} = -\vec{i} + 2\vec{k}$
7 $-\vec{i} + \vec{j}$
9 $-5\vec{i} - \vec{j}$
11 $0.3\vec{i} - 1.8\vec{j} + 0.03\vec{k}$
13 $-6\vec{i} - 5\vec{j} + 11\vec{k}$
15 $4\vec{i} - \vec{j} + 3\vec{k}$
17 $\sqrt{11}$
19 7.6
21 $6\vec{i} + 3\vec{j} - 3\vec{k}$
23 $4\vec{i} + 3\vec{j}$
25 $-10\vec{i} + 6\vec{j} + 7\vec{k}$
27 (a)

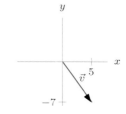

(b) $\|\vec{v}\| = 8.602$
(c) $54.46°$

29 $(-\vec{i} + \vec{j} - \vec{k})/\sqrt{3}$

31 $(2/\sqrt{6})\vec{i} - (2/\sqrt{6})\vec{j}$
   $+ (4/\sqrt{6})\vec{k}$

33 (a) $(3/5)\vec{i} + (4/5)\vec{j}$
   (b) $6\vec{i} + 8\vec{j}$

35 (a) $\sqrt{2}\vec{i} + \sqrt{2}\vec{j}$
   (b) $(\sqrt{3}/2)\vec{i} + \vec{k}/2$

37 $\vec{p} = -\frac{4\sqrt{5}}{5}\vec{i} - \frac{2\sqrt{5}}{5}\vec{j}$

39 $-\vec{u}, \vec{v}, \vec{v} - \vec{u}, \vec{u} - \vec{v}$

41 $\vec{v} = 3\vec{i} + 4\vec{j}$ and $\vec{w} = 3\vec{i} - 4\vec{j}$

45 $\|\vec{u} + \vec{v}\|$ could be less than 1

47 Longer diagonal if angle between $\vec{u}$ and $\vec{v}$ more than $90°$

49 $\vec{v} = \vec{j} + \sqrt{3}\,\vec{k}$

51 $\vec{u} = \vec{i}, \vec{v} = 3\vec{i} + 3\vec{j}$

53 False

55 False

57 False

59 False

61 False

Section 13.2

1 Scalar

3 Vector

5 Scalar

7 (a) $50\vec{i}$
   (b) $-50\vec{j}$
   (c) $25\sqrt{2}\vec{i} - 25\sqrt{2}\vec{j}$
   (d) $-25\sqrt{2}\vec{i} + 25\sqrt{2}\vec{j}$

9 $-37.59\vec{i}, -13.68\vec{j}$

11 $50.194°$

13 6.986 km/hr

15 548.6 km/hr

17 $\vec{E} = 27.908\vec{i} + 1.823\vec{j}$
   Speed 27.97 km/hr in direction $3.74°$ north of east

19 $0.4v\vec{i} + 0.7v\vec{j}$

21 $0.0006\vec{i} + 0.0004\vec{j} + 0.001\vec{k}$

23 $\vec{F}_3 = 4.45\vec{i} - 37.23\vec{j}$
   37.50 newtons $83.20°$ south of east

25 $48.3°$ east of north
   744 km/hr

27 (a) 0.11 meters/second
   (b) CW: $\vec{v} = 2\pi\vec{j}$; CCW: $\vec{v} = -2\pi\vec{j}$ for speed in meters/minute

37 (a) $384\vec{i}, 280\vec{i} + 90\vec{j}, -104\vec{i} + 90\vec{j}$
   (b) $294.109, 137.535$ thousand km
   (c) $419.225\vec{i} - 158.084\vec{j}$

39 $\vec{i}$ or $\vec{j}$ components could be different

41 $\vec{F} = \vec{i} - 2\vec{j}$

43 Yes

45 Yes

47 Yes

Section 13.3

1 12

3 -38

5 -7

7 -2

9 238

11 $2\vec{i} + \vec{j} - \vec{k}$
   (multiples of)

13 $3\vec{i} + 4\vec{j} - \vec{k}$ (multiples of)

15 $\pi\vec{i} + (\pi - 1)\vec{j} + (1 - \pi)\vec{k}$
   (multiples of)

17 $2\pi/3$ radians $(120°)$

19 $\pi/6$ radians $(30°)$

21 $3x - y + 4z = 6$

23 $x - y + z = 3$

25 $2x + 4y - 3z = 5$

27 $2x - 3y + 5z = -17$

29 (a) $\lambda = -2.5$
   (b) $a = -6.5$

31 (a) $\vec{n} = 3\vec{i} - \vec{j} - \vec{k}$
   (b) $\vec{i} - \vec{j} + 4\vec{k}$ (answers may differ)

33 (a) is (I); (b) is (III), (IV); (c) is (II), (III); (d) is (II)

35 $\vec{v}_1, \vec{v}_4, \vec{v}_8$ all parallel
   $\vec{v}_3, \vec{v}_5, \vec{v}_7$ all parallel
   $\vec{v}_1, \vec{v}_4, \vec{v}_8$ perpendicular to $\vec{v}_3, \vec{v}_5, \vec{v}_7$
   $\vec{v}_2$ and $\vec{v}_9$ perpendicular

37 $\vec{u} \perp \vec{v}$ for $t = 2$ or -1.
   No values of t make $\vec{u}$ parallel to $\vec{v}$

39 $\vec{a} = -\frac{8}{21}\vec{d} + (\frac{79}{21}\vec{i} + \frac{10}{21}\vec{j} - \frac{118}{21}\vec{k})$

41 Lengths: $\sqrt{34}, \sqrt{29}, \sqrt{13}$
   Angles: $37.235°, 64.654°, 78.111°$

43 (a) $\vec{F}$ parallel $= -0.168\vec{i} - 0.224\vec{j}$
   (b) $\vec{F}$ perp $= 0.368\vec{i} - 0.276\vec{j}$
   (c) $W = -1.4$

45 (a) $\vec{F}$ parallel $= \vec{0}$
   (b) $\vec{F}$ perp $= \vec{F}$
   (c) $W = 0$

47 (a) $\vec{F}$ parallel $= -3.48\vec{i} - 4.64\vec{j}$
   (b) $\vec{F}$ perp $= 0.48\vec{i} - 0.36\vec{j}$
   (c) $W = -29$

49 (a) $\vec{F}$ parallel $= 3.846\vec{i} - 0.769\vec{j}$
   (b) $\vec{F}$ perp $= -3.846\vec{i} - 19.231\vec{j}$
   (c) $W = 20$

51 (a) $\vec{F}$ parallel $= \vec{0}$
   (b) $\vec{F}$ perp $= \vec{F}$
   (c) $W = 0$

53 0.727804

55 $\vec{w}_4$ increases most
   $\vec{w}_3$ decreases most

57 $8\vec{i} - \vec{j}$

59 \$710 revenue

73 Need $\vec{u}$ a unit vector

75 $(-3, 3)$

77 False

79 True

81 True

83 False

85 True

Section 13.4

1 $\vec{k}$

3 $\vec{j} - \vec{k}$

5 $-2\vec{i} + 2\vec{j}$

7 $\vec{0}$

9 $2\vec{k}$

11 $\vec{i}$

13 $x + y + z = 1$

15 4

17 6

19 $\vec{a} \times \vec{b} = -2\vec{i} - 7\vec{j} - 13\vec{k}$
   $\vec{a} \cdot (\vec{a} \times \vec{b}) = 0$
   $\vec{b} \cdot (\vec{a} \times \vec{b}) = 0$

21 $4\vec{i} + 26\vec{j} + 14\vec{k}$
   (multiples of)

23 $4x + 26y + 14z = 230$

25 $-7\vec{i} + 6\vec{j} + 23\vec{k}$

27 $4x + 26y + 14z = 0$

29 $x + 4y + 3z = 0$

31 (a) $11x - 12y - 14z = -47$
   (b) 10.74

33 (a) 0.6
   (b) 0.540

35 (a) 1.625
   (b) 1.019

37 (b) $(-y, x)$

39 48

45 (b) ABC equilateral

53 $4\pi\vec{i}$

55 (a) $((u_2v_3 - u_3v_2)^2 + (u_3v_1 - u_1v_3)^2 + (u_1v_2 - u_2v_1)^2)^{1/2}$
   (b) $|u_1v_2 - u_2v_1|$
   (c) $m = (u_2v_3 - u_3v_2)/(u_2v_1 - u_1v_2)$,
       $n = (u_3v_1 - u_1v_3)/(u_2v_1 - u_1v_2)$

57 Parallel, not perpendicular

59 $\vec{v} = (8\vec{i} - 6\vec{j})/5$

61 False

63 True

65 True

67 True

69 False

Chapter 13 Review

1 Scalar; -1

3 -1

5 $\vec{a} = -2\vec{j}, \vec{b} = 3\vec{i}, \vec{c} = \vec{i} + \vec{j},$
   $\vec{d} = 2\vec{j}, \vec{e} = \vec{i} - 2\vec{j}, \vec{f} = -3\vec{i} - \vec{j}$

7 $5\vec{i} + 30\vec{j}$

9 $3\sqrt{2}$

11 $3\vec{i} + 7\vec{j} - 4\vec{k}$

13 -3

15 $\vec{0}$

17 0

19 $\vec{0}$

21 (a) 4
   (b) $-4\vec{i} - 11\vec{j} - 17\vec{k}$
   (c) $3.64\vec{i} + 2.43\vec{j} - 2.43\vec{k}$
   (d) $79.0°$
   (e) 0.784.
   (f) $2\vec{i} - 2\vec{j} + \vec{k}$ (many answers possible)
   (g) $-4\vec{i} - 11\vec{j} - 17\vec{k}$.

23 $\pm(-\vec{i} + \vec{j} - 2\vec{k})/\sqrt{6}$

25 $\vec{n} = 4\vec{i} + 6\vec{k}$

27 $-3\vec{i} + 4\vec{j}$

29 $\vec{u}$ and $\vec{w}$; $\vec{v}$ and $\vec{q}$.

31 $\vec{F}_{\text{parallel}} = \vec{F}$
$\vec{F}_{\text{perp}} = \vec{0}$
$W = -10$

33 $\vec{F}_{\text{parallel}} = -(6/5)\vec{i} + (8/5)\vec{j}$
$\vec{F}_{\text{perp}} = (16/5)\vec{i} + (12/5)\vec{j}$
$W = -10$

35 $\vec{F}_{\text{parallel}} = 2\vec{j}$
$\vec{F}_{\text{perp}} = 5\vec{i}$
$W = 6$

37 (a) True
(b) False
(c) False
(d) True
(e) True
(f) False

39 (a) $17.93\vec{i} - 7.07\vec{j}$
(b) 19.27 km/hr
(c) $21.52°$ south of east

41

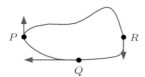

43 Parallel:
$3\vec{i} + \sqrt{3}\vec{j}$ and $\sqrt{3}\vec{i} + \vec{j}$
Perpendicular:
$\sqrt{3}\vec{i} + \vec{j}$ and $\vec{i} - \sqrt{3}\vec{j}$
$3\vec{i} + \sqrt{3}\vec{j}$ and $\vec{i} - \sqrt{3}\vec{j}$

45 2, 8

47 $2x - 3y + 7z = 19$

49 $\sqrt{6}/2$

51 (a) $\vec{u}$ and $-\vec{u}$ where
$\vec{u} = \frac{12}{13}\vec{i} - \frac{4}{13}\vec{j} - \frac{3}{13}\vec{k}$
(b) $\theta \approx 49.76°$
(c) 13/2
(d) $13/\sqrt{29}$

53 $9x - 16y + 12z = 5$
0.23

55 $7.0710\vec{i} + 2.5882\vec{j} + 6.580\vec{k}$

57 228.43 newtons
$85.5°$ south of east

59 $|ywr - vzr + zus - xws + xvt - yut|$

61 (b) $(1, 1/\sqrt{3}, \sqrt{6}/6)$
(c) $109.471°$

Section 14.1

1 $f_x(3,2) \approx -2/5$; $f_y(3,2) \approx 3/5$

3 $-3.15, 2.244$
$-3.29, 2.21$

5 (a) Dollars/Year
(b) Negative
(c) Dollars/Dollar
(d) Positive

7 $\partial P/\partial t$:
dollars/month
Rate of change in payments with time
negative
$\partial P/\partial r$:
dollars/percentage point
Rate of change in payments with interest rate
positive

9 (a) Negative

11 $f_x > 0, f_y > 0$

13 $f_x < 0, f_y < 0$

15 $f_w(16, -4) \approx -\frac{1}{4}$

17 $f_w(8, -10) \approx -0.25$

19 About -0.7

21 $\partial Q/\partial b < 0$
$\partial Q/\partial c > 0$

23 -1.5 and -1.22

25 (a)

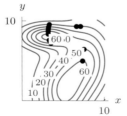

(b)

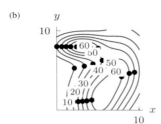

27 (a) Negative
(b) Positive

29 (a) 2.5, 0.02
(b) 3.33, 0.02
(c) 3.33, 0.02

31 (a) $-3/8°$C/m, $2/5°$C/min
(b) $-1°$C/m, $3/28°$C/min

35 (b)

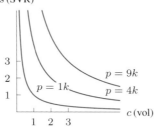

(c)

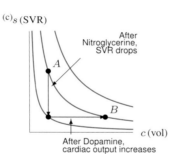

(d)

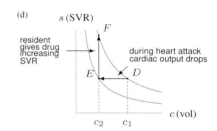

37 Same only if x and y have same units

39

		y		
		0	5	10
	0	50	52	54
x	10	40	42	44
	20	30	32	34

41 True

43 True

45 True

47 True

49 True

Section 14.2

1 (a) 7.01
(b) 7

3 $f_x = 2xy + 3x^2 - 7y^6$
$f_y = x^2 - 42xy^5$

5 $\partial z/\partial x = \frac{14x+7}{(x^2+x-y)^{-6}}$
$\partial z/\partial y = -7(x^2 + x - y)^6$

7 $f_x = (\alpha + \beta)A^\alpha x^{\alpha+\beta-1} y^{1-\alpha-\beta}$
$f_y = (1 - \alpha - \beta)A^\alpha x^{\alpha+\beta} y^{-\alpha-\beta}$

9 $(15a^2bcx^7 - 1)/(ax^3 y)$

11 $V_r = \frac{2}{3}\pi rh$

13 $-2\pi r/T^2$

15 $e^{\sqrt{xy}}(1 + \sqrt{xy}/2)$

17 g

19 $(a + b)/2$

21 $2B/u_0$

23 $2mv/r$

25 Gm_1/r^2

27 $-e^{-x^2/a^2}(a^2 - 2x^2)/a^4$

29 $v_0 + at$

31 $-\pi r^{3/2}/(M\sqrt{GM})$

33 $(15x^2y - 3y^2)\cos(5x^3y - 3xy^2)$

35 $\partial F/\partial L = \frac{3}{2}\sqrt{K/L}$

37 $\epsilon_0 E$

39 $c\gamma a_1 b_1 K^{b_1 - 1}(a_1 K^{b_1} + a_2 L^{b_2})^{\gamma - 1}$

41 13.6

43 (a) 3.3, 2.5
(b) 4.1, 2.1
(c) 4, 2

45 (a) $a < 3$
(b) Up

47 (a) $\partial g/\partial m = G/r^2$
$\partial g/\partial r = -2Gm/r^3$

(b)

g

m

g

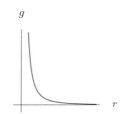

r

49 (a) $c^2((1 - v^2/c^2)^{-1/2} - 1)$; positive
   (b) $mv(1 - v^2/c^2)^{-3/2}$; positive

51 (a)

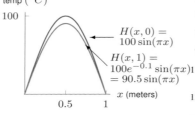

temp (°C)

100

$H(x,0) = 100\sin(\pi x)$

$H(x,1) = 100e^{-0.1}\sin(\pi x) = 90.5\sin(\pi x)$

0.5   1 x (meters)

   (b) $254.2e^{-0.1t}$ °C/m
   (c) $-254.2e^{-0.1t}$ °C/m
   (d) $-10e^{-0.1t}\sin(\pi x)$ °C/sec

53 $f(x,y) = x^4y^2 - 3xy^4 + C$

55 Derivative is limit, not a difference quotient
    Counterexample: $f(x,y) = x^2 + y^2$

57 $f(x,y) = xy$, $g(x,y) = xy + y^2$

59 True

61 True

63 False

65 False

67 (d)

Section 14.3

1 $z = -4 + 2x + 4y$

3 $z = 6 + 3x + y$

5 $z = -1$

7 $z = -36x - 24y + 148$

9 $dg = (2u + v)\,du + u\,dv$

11 $dz = (-y\sin(xy) + 2x)dx - x\sin(xy)dy.$

13 $dh = e^{-3t}\cos(x + 5t)\,dx$
    $+ (-3e^{-3t}\sin(x + 5t)$
    $+ 5e^{-3t}\cos(x + 5t))\,dt$

15 $dg = 4\,dx$

17 $dP \approx 2.395\,dK + 0.008\,dL$

19 (c) $z = 12x - 6y - 7$

21 $z = 9 + 6(x - 3) + 9(y - 1)$

23 (a) $z = 1 + 2(x - 1) + y$

   (b) $1 + 2(x - 1) + y$
   (c) $2dx + dy$

25 (a) $dg = (2u + v)\,du + u\,dv$
   (b) 0.9

27 39270 cm³

29 $V \approx 1.0235 + 0.00173T - 0.00478P$

31 $+2.2$°C

33 (a) $T = 304.9$
    $dT = 12.16\,dV + 304.29\,dP$
   (b) ≈ 2.5 dm³

35 8246.68

37 (b) $-f_x(a,b)/f_y(a,b)$
   (c) $10(x - 3) + 3(y - 4) = 0$

39 Tangent planes may be parallel, not identical

41 $f(x,y) = xy^2$, $g(x,y) = xy^2 + 10$

43 True

45 False

47 False

49 True

Section 14.4

1 $(\frac{15}{2}x^4)\vec{i} - (\frac{24}{7}y^5)\vec{j}$

3 $50\vec{i} + 100\vec{j}$

5 $\nabla z = e^y\vec{i} + e^y(1 + x + y)\vec{j}$

7 $(x\vec{i} + y\vec{j})/\sqrt{x^2 + y^2}$

9 $\sin\theta\vec{i} + r\cos\theta\vec{j}$

11 $\nabla z = \frac{1}{y}\cos(\frac{x}{y})\vec{i} - \frac{x}{y^2}\cos(\frac{x}{y})\vec{j}$

13 $\left(\frac{-12\beta}{(2\alpha - 3\beta)^2}\right)\vec{i} + \left(\frac{12\alpha}{(2\alpha - 3\beta)^2}\right)\vec{j}$

15 $60\vec{i} + 85\vec{j}$

17 $10\pi\vec{i} + 4\pi\vec{j}$

19 $\frac{1}{100}(2\vec{i} - 6\vec{j})$

21 $0.45\vec{i} + 0.2\vec{j}$

23 5

25 $-46/5$

27 84/5

29 $ydx + xdy$

31 $2x\vec{i} + 10y\vec{j}$

33 Negative

35 Approximately zero

37 Negative

39 $\vec{i}$

41 $-\vec{i}$

43 $\vec{i} + \vec{j}$

45 $-\vec{i} + \vec{j}$

47 (a) Should be number
   (b) 11/5

49 (a) $(3.96838, 5.09487)$
   (b) 0.1052
   (c) $1/(3\sqrt{10})$

51 4.4

53 No

55 Yes

57 -0.5; better estimate is -1.35

59 -1.1; better estimate is -1.8

61 $4\vec{i} + 6\vec{j}$; $4(x - 2) + 6(y - 3) = 0$

63 $-4\vec{i} + \vec{j}$; $-4(x - 2) + (y - 3) = 0$

65 (a) $(1, 1)$ or $(3, 1)$
   (b) $(1, 2)$ or $(1, -1/2)$
   (c) $(1, 2)$ or $(-1, -2)$

67 Fourth quadrant

69 P

71 (a) $-16\vec{i} + 12\vec{j}$
   (b) $16\vec{i} - 12\vec{j}$
   (c) $12\vec{i} + 16\vec{j}$; answers may vary

73 (a) -3.268
   (b) -4.919

75 (a)

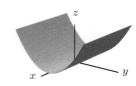

(b)

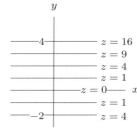

y

$z = 16$
$z = 9$
$z = 4$
$z = 1$
$z = 0$ x
$z = 1$
$z = 4$

(c) $\vec{j}$

77 (a) Circles centered at P
   (b) away from P
   (c) 1

79 $(3\sqrt{5} - 2\sqrt{2})\vec{i} + (4\sqrt{2} - 3\sqrt{5})\vec{j}$

81 $4\sqrt{2}$,
   $6\vec{i} + 2\vec{j}$

83 $5/\sqrt{2}$

85 (a) P, Q
   (b)

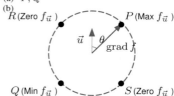

R (Zero $f_{\vec{u}}$) P (Max $f_{\vec{u}}$)

$\vec{u}$ θ

grad f

Q (Min $f_{\vec{u}}$) S (Zero $f_{\vec{u}}$)

   (c) $\|\operatorname{grad} f\|$
    $f_{\vec{u}} = \|\operatorname{grad} f\|\cos\theta$

91 $f_{\vec{u}}(0,0)$ is scalar, not vector

93 Closer contours give longer gradients

95

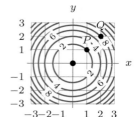

y

3
2
1

Q

P

x

-1
-2
-3

-3 -2 -1 1 2 3

97 False

99 False

101 True

103 False

105 True

107 True

109 True

111 True

113 False

Section 14.5

1 $2x\vec{i}$

3 $2x\vec{i} + 3y^2\vec{j} - 4z^3\vec{k}$

5 $-\sin(x+y)\vec{i}$
$+(\cos(y+z) - \sin(x+y))\vec{j}$
$+\cos(y+z)\vec{k}$

7 $-2(x\vec{i} + y\vec{j} + z\vec{k})/(x^2+y^2+z^2)^2$

9 $e^y \sin z\vec{i} + xe^y \sin z\vec{j} + xe^y \cos z\vec{k}$

11 $(2x_1 x_2^3 x_3^4)\vec{i} + (3x_1^2 x_2^2 x_3^4)\vec{j} + (4x_1^2 x_2^3 x_3^3)\vec{k}$

13 $(2xy/(x^2+2))\vec{i} + \ln(x^2+5)\vec{j} + 2ze^{z^2}\vec{k}$

15 $2\vec{i} + 3\vec{j} + 4\vec{k}$

17 $6\vec{i} + 3\vec{j} + 2\vec{k}$

19 $\vec{i} + 2\vec{j} + (2/e)\vec{k}$

21 $9/\sqrt{14}$

23 $1/\sqrt{2}$

25 $\sqrt{77}$

27 $2\vec{i} - 2\vec{j} + \vec{k}$;
$2(x+1) - 2(y-1) + (z-2) = 0$

29 $-4\vec{i} + 2\vec{j} + \vec{k}$;
$-4(x+1) + 2(y-1) + (z-2) = 0$

31 $2\vec{i} + 4\vec{j} + 3\vec{k}$;
$2(x+1) + 4(y-1) + 3(z-2) = 0$

33 $10/3$

35 $2z + 3x + 2y = 17$

37 $x + 3y + 7z = -9$
$\vec{i} + 3\vec{j} + 7\vec{k}$

39 $20(x-10) + 900(y+10) - 600(z-30) = 0$

41 $x = 1$

43 $z = 2x + y + 3$

45 grad $g(-1,-1)$ lies directly under path of steepest descent

47 (a) $(x-2) + 4(y-3) - 6(z-1) = 0$
(b) $z = 1 + (1/6)(x-2) + (2/3)(y-3)$

49 (a) $\vec{k}$
(b) $z = 1$

51 (a) (I)-(E), (II)-(F), (III)-(G), (IV)-(H)
(b) (I)-(L), (II)-(J), (III)-(M), (IV)-(K)

53 (a) $6.33\vec{i} + 0.76\vec{j}$
(b) -34.69

55 (a) 23
(b) -9.2
(c) $-16\vec{i} + 6\vec{j}$
(d) $16x - 6y - z = 23$

57 (a) $(0,0,1)$
(b) $(3,-5,35)$

59 $3x + 10y - 5z + 19 = 0$

61 (a) $-25/\sqrt{21}$
(b) $-8\vec{i} + 7\vec{j} + 4\vec{k}$
(c) $\sqrt{129}$

63 (a) is (V); (b) is (IV); (c) is (V)

71 $f_x(0,0,0)x + f_y(0,0,0)y$
$\quad + f_z(0,0,0)z = 0$

73 $f(x,y,z) = 2x + 3y + 4z + 100$

75 False

77 False

79 (a) $^\circ$C per meter
(b) $^\circ$C per second
(c) $^\circ$C per second

Section 14.6

1 $\dfrac{dz}{dt} = e^{-t} \sin(t)(2\cos t - \sin t)$

3 $2e^{1-t^2}(1 - 2t^2)$

5 $2t\sin(\ln t) + 2t\ln(t)\cos(t^2)$
$\quad + t\cos(\ln t) + \dfrac{\sin t^2}{t}$

7 $\dfrac{\partial z}{\partial u} = \dfrac{e^v}{u}$
$\dfrac{\partial z}{\partial v} = e^v \ln u$

9 $\dfrac{\partial z}{\partial u} = 2ue^{(u^2-v^2)}(1 + u^2 + v^2)$
$\dfrac{\partial z}{\partial v} = 2ve^{(u^2-v^2)}(1 - u^2 - v^2)$

11 $\dfrac{\partial z}{\partial u} = \dfrac{1}{vu}\cos\left(\dfrac{\ln u}{v}\right)$
$\dfrac{\partial z}{\partial v} = -\dfrac{\ln u}{v^2}\cos\left(\dfrac{\ln u}{v}\right)$

13 $\dfrac{\partial z}{\partial u} = \dfrac{-2uv^2}{u^4 + v^4}$
$\dfrac{\partial z}{\partial v} = \dfrac{2vu^2}{u^4 + v^4}$

15 $\dfrac{\partial z}{\partial u} = -2u\sin u^2$
$\dfrac{\partial z}{\partial v} = 0$

17 $-2\rho\cos 2\phi, 0$

19 401.1

21 -5 pascal/hour

25 (a) Spheres centered at the origin
(b) $(\vec{i} + 2\vec{j} + 2\vec{k})/3$
(c) About $1/3$
(d) About $(\vec{i} + 2\vec{j} + 2\vec{k})/9$
(e) About $\vec{i}/3$
(f) (i) Same
(ii) Parallel to $\vec{r}$ at each point

27 $\dfrac{\partial w}{\partial u} = \dfrac{\partial w}{\partial x}\dfrac{\partial x}{\partial u} + \dfrac{\partial w}{\partial y}\dfrac{\partial y}{\partial u} + \dfrac{\partial w}{\partial z}\dfrac{\partial z}{\partial u}$
$\dfrac{\partial w}{\partial v} = \dfrac{\partial w}{\partial x}\dfrac{\partial x}{\partial v} + \dfrac{\partial w}{\partial y}\dfrac{\partial y}{\partial v} + \dfrac{\partial w}{\partial z}\dfrac{\partial z}{\partial v}$

29 (a) $F_u(x,3)$
(b) $F_v(3,x)$
(c) $F_u(x,x) + F_v(x,x)$
(d) $F_u(5x,x^2)(5) + F_v(5x,x^2)(2x)$

31 $b \cdot k + d \cdot q$

37 $b \cdot k + d \cdot q$

41 $\dfrac{\partial U_1}{\partial P}$

47 $F(b,x)$

49 $dz/dt = f_x(g(t),h(t))g'(t) + f_y(g(t),h(t))h'(t)$

51 $dz/dt|_{t=0} = f_x(2,3)g'(0) + f_y(2,3)h'(0)$

53 $f(x,y) = 4x + 2y$

55 $w = uv, u = 2s^2 + t$ and $v = e^{st}$, many other answers are possible

57 (c)

Section 14.7

1 $f_{xx} = 6y$
$f_{xy} = 6x + 15y^2$
$f_{yx} = 6x + 15y^2$
$f_{yy} = 30xy$

3 $f_{xx} = 6(x+y)$
$f_{yy} = 6(x+y)$
$f_{yx} = 6(x+y)$
$f_{xy} = 6(x+y)$

5 $f_{xx} = 4y^2 e^{2xy}$
$f_{xy} = 4xye^{2xy} + 2e^{2xy}$
$f_{yx} = 4xye^{2xy} + 2e^{2xy}$
$f_{yy} = 4x^2 e^{2xy}$

7 $f_{xx} = \dfrac{y^2}{(x^2+y^2)^{3/2}}$
$f_{xy} = \dfrac{-xy}{(x^2+y^2)^{3/2}} = f_{yx}$
$f_{yy} = \dfrac{x^2}{(x^2+y^2)^{3/2}}$

9 $f_{xx} = 30xy^2 + 18$
$f_{xy} = 30x^2 y - 21y^2$
$f_{yx} = 30x^2 y - 21y^2$
$f_{yy} = 10x^3 - 42xy$

11 $f_{xx} = -12\sin 2x \cos 5y$
$f_{xy} = -30\cos 2x \sin 5y$
$f_{yx} = -30\cos 2x \sin 5y$
$f_{yy} = -75\sin 2x \cos 5y$

13 (a) Negative
(b) Zero
(c) Negative
(d) Zero
(e) Zero

15 (a) Positive
(b) Zero
(c) Positive
(d) Zero
(e) Zero

17 (a) Zero
(b) Negative
(c) Zero
(d) Negative
(e) Zero

19 (a) Positive
(b) Positive
(c) Zero
(d) Zero
(e) Zero

21 (a) Positive
(b) Negative
(c) Negative
(d) Negative
(e) Positive

23 $Q(x,y) = 1 + 2x - 2y + x^2 - 2xy + y^2$

25 $Q(x,y) = 1 + x + x^2/2 - y^2/2$

27 $Q(x,y) = 1 - x^2/2 - 3xy - (9/2)y^2$

29 $Q(x,y) = -y + x^2 - y^2/2$

31 $1 + x - y/2 - x^2/2 + xy/2 - y^2/8$

33 $L(x,y) = 1 + (1/2)(x-1) + y$
$Q(x,y) = 1 + (1/2)(x-1) + y$
$\quad -(1/8)(x-1)^2 - (1/2)(x-1)y - (1/2)y^2$
$L(0.9, 0.2) = 1.15$
$Q(0.9, 0.2) = 1.13875$
$f(0.9, 0.2) \approx 1.14018$

35 $L(x,y) = Q(x,y) = x - 1$
$L(0.9, 0.2) = Q(0.9, 0.2) = -0.1$
$f(0.9, 0.2) \approx -0.098$

39 $a = -b^2$

41 Positive, negative

43 (a) $z_{yx} = 4y$

(b) $z_{xyx} = 0$

(c) $z_{xyy} = 4$

45 (a) Increasing, decreasing

(b) $f_x > 0, f_y < 0$

(c) $f_{xx} > 0, f_{yy} < 0$

(d) y

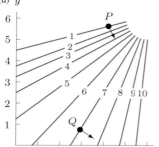

(e) P

47 $d > 0, e = f = 0$

49 $d > 0, e < 0, f > 0$

51 (a) $\partial Y/\partial L > 0$

(b) $\partial^2 Y/(\partial K \partial L) > 0$

53 (a) $a = k$

(b) (x_2, y_2)

(c) Minimum

(d) $m < 0$
$n > 0$

55 (a) (i) Dollars/Year, negative, (ii) Dollars/Dollar, positive

(b) $\partial^2 P/(\partial A \partial C) < 0$

(c) eCA

57 (a) $1 + x/2 + y$

(b) $1 + x/2 + y - x^2/8 - xy/2 - y^2/2$

61 Counterexample: $f(x, y) = x^3 + y^4$

63 $f(x, y) = x^2 + y^2$

65

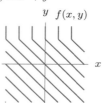

Section 14.8

1 $(0, 0)$

3 None

5 y-axis

7 None

9 $(1, 2)$

11 (a)

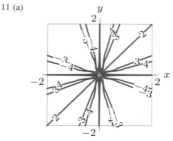

(b) No

(c) No

(d) No

(e) Exist, not continuous

13 (a)

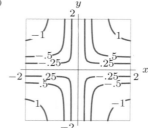

(b) Yes

(c) Yes

(d) No

(e) Exist, not continuous

15 (a)

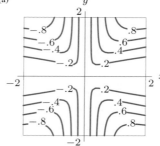

(b) Yes

(d) No

(f) No

17 (a)

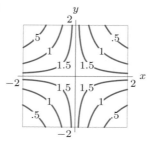

(b)

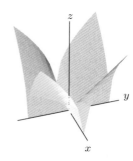

(c) No

(e) No

19 (a) No

21 Counterexample: $\sqrt{x^2 + y^2}$

23 $f(x, y) = \sqrt{x^2 + y^2}$

25 (a) Differentiable

(b) Not differentiable

(c) Not differentiable

(d) Differentiable

Chapter 14 Review

1 Vector; $3e^{-1}\vec{i} - \frac{1}{2}e^{-1}\vec{j}$

3 Vector; $-(\sin x)e^y\vec{i} + (\cos x)e^y\vec{j} + \vec{k}$

5 $f_x = 10xy^3 + 8y^2 - 6x$,
$f_y = 15x^2y^2 + 16xy$

7 $\pi/\sqrt{lg}$

9 $f_x = \frac{2xy^3}{(x^2+y^2)^2}$, $f_y = \frac{x^4-x^2y^2}{(x^2+y^2)^2}$

11 $\partial f/\partial p = (1/q)e^{p/q}$
$\partial f/\partial q = -(p/q^2)e^{p/q}$

13 $f_N = c\alpha N^{\alpha-1}V^\beta$

15 $x/\left(2\sqrt{\omega x}\cos^2\left(\sqrt{\omega x}\right)\right)$

17 $(5x - 2y + 2xy\beta)e^{x\beta-3}/(2y\beta + 5)^2$

19 $[x^2y(-3\lambda + 10) - 3\lambda^4(8\lambda^2 - 27\lambda + 50)]/2(\lambda^2 - 3\lambda + 5)^{3/2}$

21 $e^x\cos(xy) + e^x \cdot (-\sin(xy)) \cdot y$
$e^x \cdot (-\sin(xy)) \cdot x + 2ay$
y^2

23 $f_{xx} = (2x^2 - y^2)(x^2 + y^2)^{-5/2}$
$f_{xy} = 3xy(x^2 + y^2)^{-5/2}$

25 $V_{rr} = 2\pi h$, $V_{rh} = 2\pi r$

27 $(a^2 + b^2)e^{ax-bt}$

29 $(3x^2 - yz)\vec{i} - xz\vec{j} + (3z^2 - xy)\vec{k}$

31 $\cos(y^2 - xy)(-y\vec{i} + (2y - x)\vec{j})$

33 $(e^y + 1/x)\vec{i} + xe^y\vec{j} + (1/z)\vec{k}$

35 $\sin\phi\cos\theta\vec{i} + \rho\cos\phi\cos\theta\vec{j} - \rho\sin\phi\sin\theta\vec{k}$

37 $\left(\frac{5\alpha}{\sqrt{5\alpha^2+\beta}}\right)\vec{i} + \left(\frac{1}{2\sqrt{5\alpha^2+\beta}}\right)\vec{j}$

39 $\vec{0}$

41 $15/\sqrt{2}$

43 -4

45 $24/\sqrt{19}$

47 $4\vec{i} - 2\vec{j}$

49 $2\vec{i} - 2\vec{j} - 6\vec{k}$

51 $3x + z = 2$

53 $f_{xx} = 2y^2$, $f_{xy} = f_{yx} = 4xy - 15y^2$, $f_{yy} = 2x^2 - 30xy$

55 $(8t + 4t^3)\cos(4t^2 + t^4)$

57 $2te^{t^2}$

59 $\cos(\sin t \cos t^2)\left(\cos t^2 \cos t - 2t \sin t \sin(t^2)\right)$

61 $Q(x, y) = 1 - (1/2)x^2 - (9/2)y^2$

63 (a) is (IV); (b) is (V); (c) is (I); (d) is (II)

65 (c) Positive
    (d) Negative

67 (A) 0.06, −0.06
    (B) 0, −0.05
    (C) 0, 0

69 (a) 41.1
    (b) 41.11, yes

73 (a) Increases
    (b) Increases
    (c) 55 joules

75 0.3

77 0.7

79 −0.4

81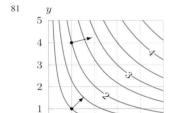

83 (a) Incorrect
    (b) Correct
    (c) Incorrect

85 (a) $y = (2x + 13)/5$
    (b) $2x - 5y - z + 20 = 0$

87 $2x - y - z = 4$

89 Yes

91 (a) −1
    (b) 23

95 −0.008544

97 (a) $z = 7 - 3(x - 2) + 4(y - 1)$
    (b) $-3(x - 2) + 4(y - 1) = 0$

99 0.06

101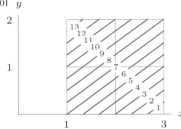

103 $f_r(2, 1) = -2/\sqrt{5}$, $f_\theta(2, 1) = 11$

105 (a) −14, 2.5
     (b) −2.055

(c) 14.221 in direction of $-14\vec{i} + 2.5\vec{j}$
(d) $f(x, y) = f(2, 3) = 7.56$
(e) For example, $\vec{v} = 2.5\vec{i} + 14\vec{j}$
(f) −0.32

107 (a) $e^{10} - 2e^{10}x - 6e^{10}y$
     (b) $1 + (x - 1)^2 + (y - 3)^2$
     (c) $-2\vec{i} - 6\vec{j}$
     (d) $-2e^{10}\vec{i} - 6e^{10}\vec{j} - \vec{k}$

109 (a) $(1/6) + (1/9)x + y - (1/108)x^2 + (2/3)xy + (3/2)y^2$
     (b) $(1/6) + (1/9)x - (1/108)x^2$,
          $1 + 6y + 9y^2$,
          $(1/6) + (1/9)x + y - (1/108)x^2 + (2/3)xy + (3/2)y^2$

111 (a) $(A_0 + A_1 + 2A_2 + A_3 + 2A_4 + 4A5) + (A_1 + 2A_3 + 2A_4)(x - 1) + (A_2 + A_4 + 4A_5)(y - 2) + A_3(x - 1)^2 + A_4(x - 1)(y - 2) + A_5(y - 2)^2$
     (b) Expands to f
     (c) $A_0 - A_3 + A_4 - 4A_5 + (A_1 + 2A_3 + 2A_4)x + (A_2 + A_4 + 4A_5)y$, does not expand to f

(b)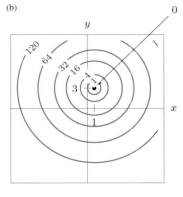

33 (a) Local maximum
    (b) Saddle point
    (c) Local minimum
    (d) None of these

35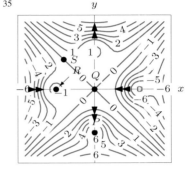

37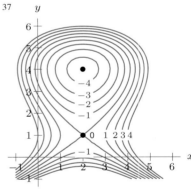

Section 15.1

1 A: no
   B: yes, max
   C: yes, saddle

3 (a) None
   (b) E, G
   (c) D, F

5 Saddle point

7 Saddle point

9 Local min: $(2, 2)$

11 Saddle point: $(-2, 1)$

13 Saddle point: $(3, 2)$

15 Saddle pt: $(0, -5)$
    Local min: $(2, -5)$

17 Local max: $(0, -1)$
    Saddle pts: $(0, 1)$, $(2, -1)$
    Local min: $(2, 1)$

19 Saddle pts: $(1, -1)$, $(-1, 1)$

21 Local min: $(0, 0)$

23 $a = -9, b = -12, c = 50$

25 (a) $f_x(x, y) = (a - 2ax^2 - 2bxy)e^{-(x^2 + y^2)}$,
         $f_y(x, y) = (b - 2by^2 - 2axy)e^{-(x^2 + y^2)}$
    (b) and (c) $a/\sqrt{2(a^2 + b^2)}, b/\sqrt{2(a^2 + b^2)}$
         local max;
         $(-a/\sqrt{2(a^2 + b^2)}, -b/\sqrt{2(a^2 + b^2)})$
         local min

27 Saddle point: $(0, 0)$.

29 Critical points: $(0, 0)$, $(\pm\pi, 0)$,
    $(\pm 2\pi, 0)$, $(\pm 3\pi, 0)$, $\cdots$
    Local minima: $(0, 0)$,
    $(\pm 2\pi, 0)$, $(\pm 4\pi, 0)$, $\cdots$
    Saddle points: $(\pm\pi, 0)$,
    $(\pm 3\pi, 0)$, $(\pm 5\pi, 0)$, $\cdots$

31 (a) $(1, 3)$ is a minimum

39 (a) $(0, 0)$
    (b) $D = -24x^2$
    (c) Saddle point

41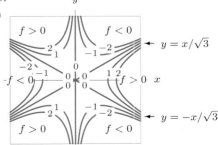

43 $(1, 3)$ could be saddle point

45 Can be saddle if f_{xy} large

47 $f(x, y) = 4 - (x - 2)^2 - (y + 3)^2$

49 False

51 True

53 False

55 True

57 False

Section 15.2

1 Mississippi:
$37 - 38$ (max), $23 - 27$ (min)
Alabama:
$28 - 29$ (max), $23 - 27$ (min)
Pennsylvania:
30 (max), 20 (min)
New York:
$21 - 24$ (max), $14 - 16$ (min)
California:
$41 - 42$ (max), $5 - 8$ (min)
Arizona:
$42 - 47$ (max), $25 - 27$ (min)
Massachusetts:
$21 - 24$ (max), 10 (min)

3 Max: 11 at $(5.1, 4.9)$
Min: -1 at $(1, 3.9)$

5 Neither

7 Neither

9 Global max $= 0$
No global min

11 Min $= 0$ at $(0, 0)$
(not on boundary)
Max $= 2$ at $(1, 1), (1, -1)$,
$(-1, -1)$ and $(-1, 1)$
(on boundary)

13 max$= 1$ at $(1, 0)$ and $(-1, 0)$
(on boundary)
min$= -1$ at $(0, 1), (0, -1)$
(on boundary)

15 $h = 25\%, t = 25°C$

17 $1/162$

19 $w = 6.35$ cm
$l = 6.35$ cm
$h = 12.7$ cm

21 $\sqrt{6}$

23 (a) $P = p_1^2/8 + p_2^2/8 - 10$
(b) $\partial(\max P)/\partial p_1 = p_1/4$

25 (b) $p_1 = p_2 = 25$
Max revenue is 4375

29 $y = \frac{25}{6} - \frac{3}{2}x$

31 (b) $f(\sqrt{1/2}, \sqrt{3/5}) =$
$4\sqrt{2} + 2\sqrt{15} \approx 13.403$

33 (a) Decrease; increase
(d) Both zero

35 Some do, like $f(x, y) = x^2 + y^2$; some don't

37 $f(x, y) = x + y$

39 True

41 True

43 True

45 False

47 True

Section 15.3

1 (a) Min
(b) Neither

(c) Neither
(d) Max

3 Max: 12 at $(1, 3)$;
Min: -8 at $(-1, -3)$

5 Max: 2 at $(1, 1)$ and $(0, 2)$;
Min: -2 at $(-1, -1)$ and $(0, -2)$

7 Maximum $f(10, 12.5) = 250$;
No minimum

9 Min $= \frac{3}{4}$, no max

11 Max $= 0$, no min

13 Max:$f(0, 2) = f(0, -2) = 8$
Min:$f(0, 0) = 0$

15 Max $= \frac{\sqrt{2}}{4}$, min $= -\frac{\sqrt{2}}{4}$

17 Max: 32 at $(1, -1)$;
Min:8 at $(-1, 1)$

19 Maximum of 400 at $(6, 6)$,
minimum of less than 100 at approx $(10.5, 0)$
or $(0, 13.5)$

21 (a)–(d)

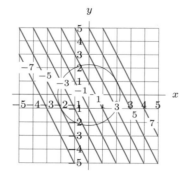

(e) Negative

23 (a) and (b)

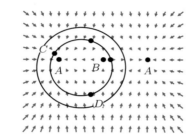

(c) Max of 5 at $(2, 1)$
Min of -5 at $(-2, -1)$
(d) Max of 5 at $(2, 1)$
Min of -5 at $(-2, -1)$

25 $\Delta c/4; -\Delta c/4$

29 (a) $C = \$4349$
(b) $\$182$

31 (a) $W = 225$
$K = 37.5$
(c) $W = 225$
$K = 37.5$
$\lambda = 0.29$

33 (a) No
(b) Yes
(c) $a + b = 1$

35 $x_1 = ((v_1)^{1/2} + (v_2)^{1/2})/(m(v_1)^{1/2})$
$x_2 = ((v_1)^{1/2} + (v_2)^{1/2})/(m(v_2)^{1/2})$

37 (a) $f_1 = \frac{k_1}{k_1+k_2} mg, f_2 = \frac{k_2}{k_1+k_2} mg$
(b) Distance the mass stretches the top spring and compresses the lower spring

41 (a) Cost of producing quantity u when prices are p, q
(b) $2\sqrt{pqu}$

43 (a) $-5\lambda^2 + 15\lambda$
(b) 1.5, 11.25
(c) 11.25, 1.5
(d) same

45 (a) $S = \ln(a^a(1 - a)^{(1-a)}) + \ln b - a\ln p_1 - (1 - a)\ln p_2$
(b) $b = e^c p_1^a p_2^{(1-a)}/(a^a(1 - a)^{(1-a)})$

47 Maximum value is 1

49 $f(x, y) = 3x + 4y$

51 $f(x, y) = x^2 + y^2$

53

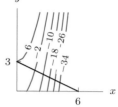

57 False

59 False

61 False

63 False

65 True

67 True

Chapter 15 Review

1 $(0, 0)$: Local Max
$(1, 1), (1, -1)$: Saddle point

3 $(0, 0), (2, 0)$: Saddle point
$(1, 1/4)$: Local min

5 Local maximum: $(\pi/3, \pi/3)$

7 $(0, 0)$ saddle point
$(\frac{9}{4}, \frac{3}{2})$ local minimum

9 $(\sqrt{2}, -\sqrt{2}/2)$ saddle point

11 Min $= \sqrt{2}$, max $= 2$

13 No max, min $= 0.866$

15 Min $= -2$, max $= 2$

17 Max $= 5/2$, min $= -2$

19 Maxima: $(-1, 1)$ and $(1, -1)$
Minimum: $(0, 0)$

21 Max $= 1$
Min $= -1$

23 Minimum

25 Neither

27 (a) $\sqrt{(x - 3)^2 + (y - 4)^2}$
(b) 4 at $(0.6, 0.8)$
(c) 6 at $(-0.6, -0.8)$

29 $y = 2/3 - x/2$

31 $A = 10, B = 4, C = -2$

33 $q_1 = 50$ units
$q_2 = 150$ units

35 6340

37 (a) Reduce K by 1/2 unit,
increase L by 1 unit.

39 Along line $x = 2y$

41 $p_1 = 110, p_2 = 115$.

43 $x \approx 23.47, y \approx 23.47, z \approx 75.1$

45 (a) $i_1 = R_2 I/(R_1 + R_2)$,
$i_2 = R_1 I/(R_1 + R_2)$
(b) $\lambda = 2 \cdot$ Voltage

49 $d \approx 5.37$ m, $w \approx 6.21$ m,
$\theta = \pi/3$ radians

51 Student B correct. Local max, not global

Section 16.1

1 Lower sum : 0.34
Upper sum : 0.62

3 Over: Approx 137
Under: Approx 60

5 Upper sum $= 46.63$
Lower sum $= 8$
Average ≈ 27.3

7 Positive

9 Zero

11 Negative

13 Negative

15 294 mg

17 210

19 Depends on sign of f on R

21 $f(x, y) = -1$

23 True

25 False

27 False

29 False

31 True

Section 16.2

1
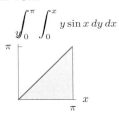
$$\int_0^\pi \int_0^x y \sin x \, dy \, dx$$

3
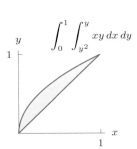
$$\int_0^1 \int_{y^2}^y xy \, dx \, dy$$

5 150

7 54

9 $e - 2$

11 $3 - \sin 3$

13 $\int_1^4 \int_1^2 f \, dy \, dx$ or $\int_1^2 \int_1^4 f \, dx \, dy$

15 $\int_{-1}^3 \int_{-2}^{(1-3x)/4} f \, dy \, dx$
or $\int_{-2}^1 \int_{-1}^{(1-4y)/3} f \, dx \, dy$

17 $\int_1^3 \int_{\frac{1}{2}(y-1)}^{-\frac{1}{2}(y-5)} f \, dx \, dy$

19 $(e^4 - 1)(e^2 - 1)e$

21 -2.678

23 94.5

25 14

27 2.38176

29 $1/24$

31 $1/2$

33 $1/3$

35 (a) 180
(b) 144

37 $\frac{1}{2}(1 - \cos 1) = 0.23$

39 $\frac{1}{4}(1 - \cos 81) = 0.056$

41 2203.2

43 $44/3$

45 (a) $\int_0^3 \int_x^{2x} f(x, y) \, dy \, dx$
$\int_0^3 \int_{y/2}^y f(x, y) \, dx \, dy$
$+$ $\int_3^6 \int_{y/2}^3 f(x, y) \, dx \, dy$
(b) $1/2 + 4e^9$

47 6

49 $\int_{-3}^3 \int_{-\sqrt{9-y^2}}^{\sqrt{9-y^2}} (9 - x^2 - y^2) \, dx \, dy$

51 4

53 117.45

55 Volume $= 6$

57 $a^4/12$

59 $1/10$

61 (a) $a + (3/2)b = 20$
(b) $f(x, y) = x + \frac{38}{3} y$:

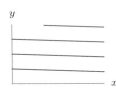

$f(x, y) = -3x + \frac{46}{3} y$:

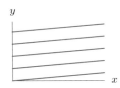

65 Integrals not over same region

67 $\int_{-1}^1 \int_{-\sqrt{1-x^2}}^{\sqrt{1-x^2}} 2 \, dx \, dy$

69 $\int_0^2 \int_0^{6-3y} 1 \, dx \, dy$

71 False

73 False

75 False

77 False

Section 16.3

1
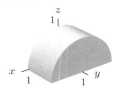

3

5
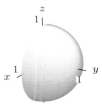

7

9
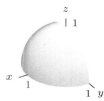

11 $a + b + 2c$

13 $(1 - e^{-a})(1 - e^{-b})(1 - e^{-c})$

15 $V = \int_{-1}^1 \int_{-\sqrt{1-x^2}}^{\sqrt{1-x^2}} \int_{x^2+y^2}^{\sqrt{4-x^2-y^2}} 1 \, dz \, dy \, dx$
Can reverse order x, y

17 $V = \int_0^2 \int_y^{(y+2)/2} \int_0^{\sqrt{9-x^2-y^2}} 1 \, dz \, dx \, dy$

19 $V = \int_0^1 \int_0^{\sqrt{4-x^2}} \int_0^{\sqrt{4-x^2-y^2}} 1 \, dz \, dy \, dx$

21 $\int_0^1 \int_{-2}^2 \int_0^{\sqrt{4-z^2}} f(x, y, z) \, dy \, dz \, dx$

23 $\int_0^r \int_{-\sqrt{r^2-x^2}}^{\sqrt{r^2-x^2}} \int_0^{\sqrt{r^2-x^2-y^2}} f(x, y, z) \, dz \, dy \, dx$

25 Positive

27 Positive

29 Zero

31 Positive

33 Zero

35 Positive

37 Positive

39 125

41 (a) $2 \int_0^{10} \int_0^1 \int_y^1 e^{-3x} \, dz \, dy \, dx$
Other answers are possible
(b) $(1 - e^{-30})/3$

43 1/2

45 162

47 1

49 29 gm

51 243

53 (a) $x + y + z = 1$
(b) 1/6

55 (a) $z = \sqrt{1 - x^2}, 0 \le y \le 10$
(b) $\int_0^{10} \int_{-1}^1 \int_0^{\sqrt{1-x^2}} f(x, y, z) \, dz \, dx \, dy$

57 $\int_0^2 \int_0^{\sqrt{12-3y^2}} \int_0^{6y^2} f(x, y, z) \, dz \, dx \, dy$

59 $\int_0^{\sqrt{12}} \int_0^{24-2x^2} \int_{\sqrt{\frac{x}{6}}}^{\sqrt{\frac{12-x^2}{3}}} f(x, y, z) \, dy \, dz \, dx$

61 (a) $\int_0^2 \int_{\sqrt{\frac{x}{2}}}^{\frac{4-x}{2}} \int_0^{4-x-2y} f(x, y, z) \, dz \, dy \, dx$

(b) $\int_0^2 \int_0^{4-x-\sqrt{2x}} \int_{\sqrt{\frac{x}{2}}}^{\frac{4-x-z}{2}} f(x, y, z) \, dy \, dz \, dx$

63 $m = 1/36; (\bar{x}, \bar{y}, \bar{z}) = (1/4, 1/8, 1/12)$

65 $m(b^2 + c^2)/3$

67 Not true for $f(x, y, z) = z$

69 $f(x, y, z) = 7/(12\pi)$

71 False

73 False

75 True

77 False

79 False

Section 16.4

1 $\int_0^{2\pi} \int_0^{\sqrt{2}} f \, r \, dr \, d\theta$

3 $\int_{\pi/2}^{3\pi/2} \int_1^2 f \, r \, dr \, d\theta$

5 $\int_0^{2\pi} \int_0^5 f(r \cos \theta, r \sin \theta) \, r \, dr \, d\theta$

7 $\int_1^5 \int_2^4 f(x, y) \, dy \, dx$

9 $r = 1$

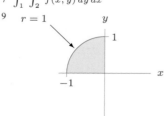

11

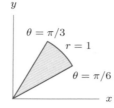

13

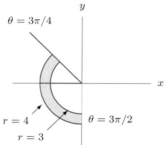

15

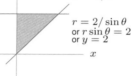
$\theta = \pi/4$
$r = 2/\sin \theta$
or $r \sin \theta = 2$
or $y = 2$

17 0

19 (a)

$y = x/3$

(b) $\int_0^1 \int_0^{3y} f(x, y) \, dx \, dy$

(c) $\int_{\tan^{-1}(1/3)}^{\pi/2} \int_0^{1/\sin \theta} f(r \cos \theta, r \sin \theta) r \, dr \, d\theta$

21 6

23 $625\pi/2$

25 (a) $\int_{\pi/2}^{3\pi/2} \int_1^4 \delta(r, \theta) \, r \, dr \, d\theta$
(b) (i)
(c) About 39,000

27 $250\pi/3$ grams

29 $2/\sqrt{3}$

31 (a) $\int_0^{2\pi} \int_0^3 r/(r^2 + 1) \, dr \, d\theta$
(b) $\pi \ln 10$

33 (a)

(b) $(4\pi - 3\sqrt{3})/12$

35 (a) $\int_{-\sqrt{3}/2}^{\sqrt{3}/2} \int_{1-\sqrt{1-y^2}}^{\sqrt{1-y^2}} \, dx \, dy$
(b) $\int_0^1 \int_{-\arccos(r/2)}^{\arccos(r/2)} r \, d\theta \, dr$

37 Upper limit for inner integral $1/\cos \theta$

39 Quarter disk $0 \le x \le 1, 0 \le y \le \sqrt{1 - x^2}$

41 (a), (c), (e)

Section 16.5

1 (a) is (IV); (b) is (II); (c) is (VII); (d) is (VI); (e) is (III); (f) is (V)

3 $\phi = \pi/4$

5 $\theta = \pi/4$

7 $\rho = 10/\cos \phi$

9 $4\pi(1 - \cos 1) = 5.78$

11 $7\pi(\pi - 2)/12$

13 $\int_0^3 \int_0^1 \int_0^5 f \, dz \, dy \, dx$

15 $\int_0^4 \int_0^{\pi/2} \int_0^2 f \cdot r \, dr \, d\theta \, dz$

17 $\int_0^{2\pi} \int_0^{\pi/6} \int_0^3 f \cdot \rho^2 \sin \phi \, d\rho \, d\phi \, d\theta$

19 $\int_0^{\pi} \int_0^2 \int_0^r f r \, dz \, dr \, d\theta$

21 $\int_0^{\pi} \int_0^K \int_0^{2\pi} \rho^2 \sin \phi \, d\theta \, d\rho \, d\phi$

23 $\int_0^{2\pi} \int_0^{\arctan(1/2)} \int_0^{4/\cos \phi} g(\rho, \phi, \theta) \rho^2 \sin \theta \, d\rho \, d\phi \, d\theta$

25 (a) $\int_0^1 \int_0^{\sqrt{1-x^2}} \int_{-\sqrt{1-x^2-y^2}}^0 dz \, dy \, dx$
(b) $\int_0^{\pi/2} \int_0^1 \int_{-\sqrt{1-r^2}}^0 r \, dz \, dr \, d\theta$
(c) $\int_0^{\pi/2} \int_{\pi/2}^{\pi} \int_0^1 \rho^2 \sin \phi \, d\rho \, d\phi \, d\theta$

27 (a) $\int_{-1/\sqrt{2}}^{1/\sqrt{2}} \int_{-\sqrt{(1/2)-x^2}}^{\sqrt{(1/2)-x^2}} \int_{\sqrt{x^2+y^2}}^{1/\sqrt{2}} dz \, dy \, dx$
(b) $\int_0^{2\pi} \int_0^{1/\sqrt{2}} \int_r^{1/\sqrt{2}} r \, dz \, dr \, d\theta$
(c) $\int_0^{2\pi} \int_0^{\pi/4} \int_0^{1/(\sqrt{2} \cos \phi)} \rho^2 \sin \phi \, d\rho \, d\phi \, d\theta$

29 (a) $\int_0^{2\pi} \int_0^{\pi} \int_1^2 \rho^2 \sin \phi \, d\rho \, d\phi \, d\theta$
(b) $\int_0^{2\pi} \int_0^2 \int_{-\sqrt{4-r^2}}^{\sqrt{4-r^2}} r \, dz \, dr \, d\theta$
$- \int_0^{2\pi} \int_0^1 \int_{-\sqrt{1-r^2}}^{\sqrt{1-r^2}} r \, dz \, dr \, d\theta$

31 $V = \int_0^{2\pi} \int_0^{3/\sqrt{2}} \int_r^{\sqrt{9-r^2}} r \, dz \, dr \, d\theta$
or
$V = \int_0^{2\pi} \int_0^{\pi/4} \int_0^3 \rho^2 \sin \phi \, d\rho \, d\phi \, d\theta$
Order of integration can be altered; other coordinates can be used

33 $V = \int_0^{\pi/2} \int_0^{\sqrt{7}} \int_0^r r \, dz \, dr \, d\theta$
Order of integration can be altered; other coordinates can be used

35 $V = \int_0^{2\pi} \int_{\pi/4}^{\pi/2} \int_0^{\sqrt{8}} \rho^2 \sin \phi \, d\rho \, d\phi \, d\theta$
Order of integration can be altered; other coordinates can be used

37 $28\pi/3$

39 $5\pi/3$

41 $25\pi/6$

43 (a) Positive
(b) Zero

45 $\int_0^{2\pi} \int_0^l \int_a^{a+h} r \, dr \, dz \, d\theta = \pi l((a + h)^2 - a^2)$

47 $\int_0^{2\pi} \int_0^a \int_{hr/a}^h r\,dz\,dr\,d\theta = \pi h a^2/3$

49 (a) $\int_0^{2\pi} \int_1^5 \int_{-\sqrt{25-r^2}}^{\sqrt{25-r^2}} r\,dz\,dr\,d\theta$

   (b) $64\sqrt{6}\pi = 492.5$ mm^3

51 $81\pi(-\sqrt{2}+2)/4$

53 $324\pi/5$ gm

55 1702π gm

57 $3/\sqrt{2}$

59 (a) $\pi/5$
   (b) $5/6$

61 $3a/8b$ above center of base

63 $a^2/4$

65 Stored energy $= (q^2/4\pi\epsilon)\ln(b/a)$

67 28π gm

69 $W = \int_0^1 \int_0^{2\pi} \int_{\sqrt{1-r^2}}^{(\sqrt{9-r^2})-1} r\,dz\,d\theta\,dr +$
   $\int_1^{2\sqrt{2}} \int_0^{2\pi} \int_0^{(\sqrt{9-r^2})-1} r\,dz\,d\theta\,dr$

71 Total charge $= \frac{k\pi}{2} h^2 R^2$

73 $r^2 \sin\theta\,dr\,d\theta\,d\phi$

75 Need factor of $\rho^2 \sin\phi$

77 $\int_0^{2\pi} \int_0^{\pi/2} \int_0^5 \rho^2 \sin\phi\,d\rho\,d\phi\,d\theta$

Section 16.6

1 0

3 1

5 7/8

7 1/16

9 Not a joint density function

11 Not a joint density function

13 Is joint density function

15 (a) 20/27
   (b) 199/243

17 (a) $k = 3/8$
   (b) 15/32
   (c) 1/16

19 $\int_{65}^{100} \int_{0.8}^1 f(x,y)\,dx\,dy$

21 (a) 0 if $t \le 0$, $2t^2$ if $0 < t \le 1/2$,
       $1 - 2(1-t)^2$ if $1/2 < t \le 1$,
       1 if $1 < t$
   (b) 0 if $t \le 0$, $4t$ if $0 < t \le 1/2$,
       $4 - 4t$ if $1/2 < t \le 1$,
       0 if $1 < t$

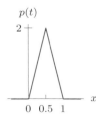

   (c) x, y: All equally likely
       z: Near $1/2$

23 $\int_{-\infty}^{\infty} \int_{-\infty}^{\infty} (p_1(x,y) + p_2(x,y))\,dx\,dy = 2$

25 $a = 0$, $b = 1$, $c = 0$, $d = 1$

27 True

29 False

Chapter 16 Review

1

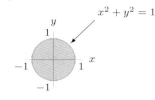

3

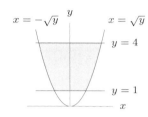

5

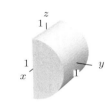

7 $\int_0^2 \int_0^3 \int_0^5 f\,dx\,dy\,dz$

9 $\int_0^{2\pi} \int_{\pi/2}^{\pi} \int_2^5 f\rho^2 \sin\phi\,d\rho\,d\phi\,d\theta$

11 $\int_0^4 \int_{(y/2)-2}^{-y+4} f(x,y)\,dx\,dy$ or
   $\int_{-2}^0 \int_0^{2x+4} f(x,y)\,dy\,dx +$
   $\int_0^4 \int_0^{-x+4} f(x,y)\,dy\,dx$

13 $7\pi/3$

15 $\frac{1}{3}(\sin 5 - \sin 2)(\cos 1 - \cos 2)$

17 $-4\cos 4 + 2\sin 4$
   $+ 3\cos 3 - 2\sin 3 - 1$

19 $85/12$

21 $\int_0^{2\pi} \int_{3\pi/4}^{\pi} \int_0^{\sqrt{2}} f(\rho,\phi,\theta)\rho^2 \sin\phi\,d\rho\,d\phi\,d\theta$

23 $\int_{-1}^1 \int_{-\sqrt{1-x^2}}^{\sqrt{1-x^2}} \int_{-\sqrt{2-x^2-y^2}}^{-\sqrt{x^2+y^2}} h(x,y,z)\,dz\,dy\,dx$

25 Positive

27 Positive

29 Negative

31 Zero

33 Positive

35 $\int_0^2 \int_0^{1-\frac{z}{4}} \int_0^{2-2y-\frac{z}{2}} f(x,y,z)\,dx\,dy\,dz$

37 (a) $\int_{-2}^0 \int_0^{2+x} \int_0^{x-y+2} dz\,dy\,dx$
       Other orders possible
   (b) $4/3$

39

41

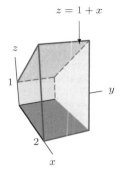

$\int_0^2 \int_0^1 \int_0^{1+x} f(x,y,z)\,dz\,dy\,dx$;
Order of integration can be altered

43

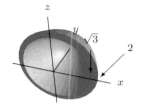

$\int_0^{\pi} \int_0^{\pi} \int_{\sqrt{3}}^2 \rho^2 \sin\phi\,d\rho\,d\phi\,d\theta$
Order of integration can be altered

45 Positive

47 Positive

49 Can't tell

51 Zero

53 Zero

55 Positive

57 Positive

59 $3\pi/16$

61 8π

63 20.1 km^3

65 $125/3$

67 (a) $\int_0^{2\pi} \int_0^{12} (1+2r)r\,dr\,d\theta$
   (b) 2448π gm

69 $\int_0^{2\pi} \int_0^{\arccos(3/5)} \int_{3/\cos\phi}^5 \rho^2 \sin\phi\,d\rho\,d\phi\,d\theta$
   $= 52\pi/3 = 54.45$ cm^3

71 ≈ 183

73 $4\pi(a^2 - R^2)^{3/2}/3$

75 $I_x = 25/2$
   $I_y = 65/2$
   $I_z = 85/2$

77 $2\pi Gm(r_2 - r_1 - \sqrt{r_2^2 + h^2} + \sqrt{r_1^2 + h^2})$

79 (a) $3/2000$
   (b) $343/1000$

81 $6\pi/7$

83 $a, b+c, 0$

Section 17.1

1 $x = \cos t$, $y = 1 + \sin t$, $\pi \le t \le 2\pi$

3 $x = 2\cos t$, $y = 2\sin t$, $0 \le t \le \pi/2$

5 $x = 0$, $y = t$, $-2 \le t \le 1$

7 $x = 3 + t, y = 2t, z = -4 - t$

9 $x = -3 + 2t, y = 4 + 2t, z = -2 - 3t$

11 $x = t, y = 1, z = -t$

13 $x = 1 + 4t, y = 5 - 5t, z = 2 - 3t$

15 $x = -3 + 2t, y = -2 - t, z = 1 - 2t$

17 $x = 3 - 3t, y = 0, z = -5t$

19 $x = 3\cos t, y = 3\sin t, z = 5, 0 \le t < 2\pi$

21 $x = 2\cos t, y = -2\sin t, z = 0$

23 $x = 2\cos t, y = 0, z = 2\sin t$

25 $x = 0, y = 3\cos t, z = 2 + 3\sin t$

27 $x = t, y = t^3, z = 0$

29 $x = -3t^2, y = 0, z = t$

31 $x = t, y = 4 - 5t^4, z = 4$

33 $x = 0, y = 1 + \frac{5}{2}\cos t, z = -2 + \sin t$

35 $\vec{r}(t) = -\vec{i} - 3\vec{j} + t(6\vec{i} + 5\vec{j}), 0 \le t \le 1$,
$x = -1 + 6t, y = -3 + 5t, 0 \le t \le 1$

37 $x = -1 + 3t, y = 2, z = -3 + 5t$

39 $x = 0, y = 5\cos t, z = -5\sin t$,
$-\pi/2 \le t \le \pi/2$

41 Two arcs:
$\vec{r}(t) = 5\vec{i} + 5(-\cos t\vec{i} + \sin t\vec{j})$,
$0 \le t \le \pi$ or
$\vec{r}(t) = 5\vec{i} + 5(\cos t\vec{i} + \sin t\vec{j})$,
$\pi \le t \le 2\pi$

43 Different lines

45 Different lines

47 $\vec{r}(t) = (2 + (t/5)10)\vec{i} + (5 + (t/5)4)\vec{j}$

49 $\vec{r}(t) = (2 + (t-10)10)\vec{i} + (5 + (t-10)4)\vec{j}$

51 $\vec{r}(t) = (12 - (10/21)(t + 11))\vec{i} + (9 - (4/21)(t + 11))\vec{j}$

53 No

55 (b) $-\vec{i} - 10\vec{j} - 7\vec{k}$
(c) $\vec{r} = (1 - t)\vec{i} + (3 - 10t)\vec{j} - 7t\vec{k}$

57 (a) $\vec{r} = (\vec{i} + 3\vec{j} + 7\vec{k}) + t(2\vec{i} - 3\vec{j} - \vec{k})$
(b) $(3, 0, 6)$
(c) $\sqrt{14}$

59 (a) $2\vec{i} - 5\vec{j} + 3\vec{k}$
(c) $x = 1 + 2t, y = -1 - 5t, z = 1 + 3t$

61 $c = 2$

63 $x = \frac{8}{3}, \; y = t, \; z = \frac{1}{3} + t$

65 $x = 1 + 2t, \; y = 2 + 4t, \; z = 5 - t$

67 Yes

69 (a) Circle, center (a, a), rad b, per $2\pi/k$
(b) (i) Increases radius
(ii) Center moves outward on $y = x$
(iii) Speeds up
(iv) Touches axes

71 (a) $-2e^{-1}/3 \, \mu g/m^3/m$
(b) $t = \pm\sqrt{3}/2$ sec

73 (a) II, $y = x$
(b) IV, $x + y = a$
(c) V, $x^2 - y^2 = a^2$
(d) I, $x^2 + y^2 = a^2$
(e) III, $x^2 + y^2 = a^2$

75 Many possible answers
(a) $a = -2, b = 7, c = 4, d = 0$
(b) $a = -2, b = 7, c = 4, d = 11$
(c) $a = 7, b = 2, c = 0, d = 41$

77 Line Equation:
$x = 1 + 2t$
$y = 2 + 3t$
$z = 3 + 4t$
Shortest distance: $\sqrt{6/29}$

79 (a) $x = 1 + 2t, y = 5 + 3t, z = 2 - t$
(b) $(-8/7, 25/14, 43/14)$

81 (a) Yellow
(b) Direction $(-0.4 + t)\vec{i} + (-3.8 + t)\vec{j} + (t - 1)\vec{k}$

83 (a) (i) is (C); (ii) is (A); (iii) is (D); (iv) is (G)
(b) (iii)

85 Distance $|R|$ from z-axis
Distance $\sqrt{R^2 + t^2}$ from origin

87 $\vec{i} + 2\vec{j} + 3\vec{k} + t(\vec{i} + 2\vec{j})$
$\vec{i} + 2\vec{j} + 3\vec{k} + t(\vec{i} - \vec{k})$

89 False

91 False

93 False

95 True

97 True

99 True

Section 17.2

1 $\vec{v} = \vec{i} + 2t\vec{j} + 3t^2\vec{k}$,
Speed $= \sqrt{1 + 4t^2 + 9t^4}$,
Particle never stops

3 $\vec{v} = 6t\vec{i} + 3t^2\vec{j}$,
$\|\vec{v}\| = 3|t| \cdot \sqrt{4 + t^2}$,
Stops when $t = 0$

5 $\vec{v} = 6t\cos(t^2)\vec{i} - 6t\sin(t^2)\vec{j}$,
$\|\vec{v}\| = 6|t|$,
Stops when $t = 0$

7 $\vec{v} = 3\vec{i} + \vec{j} - \vec{k}$, $\vec{a} = \vec{0}$

9 $\vec{v} = \vec{i} + 2t\vec{j} + 3t^2\vec{k}$, $\vec{a} = 2\vec{j} + 6t\vec{k}$

11 $\vec{v} = -3\sin t\vec{i} + 4\cos t\vec{j}$,
$\vec{a} = -3\cos t\vec{i} - 4\sin t\vec{j}$

13 $x = 1 + 2(t - 2), y = 2, z = 4 + 12(t - 2)$

15 Length ≈ 24.6

17 $3 + \ln 2$

19 $\vec{v} = 6\cos(3t)\vec{j} - 6\sin(3t)\vec{k}$,
$\vec{a} = -18\sin(3t)\vec{j} - 18\cos(3t)\vec{k}$,
$\vec{v} \cdot \vec{a} = 0, \|\vec{v}\| = 6, \|\vec{a}\| = 18$

21 Line through $(1, -5, -2)$ in direction of $-\vec{i} + 2\vec{j} + 3\vec{k}$,
$\vec{v} = (6t^2 + 3)(-\vec{i} + 2\vec{j} + 3\vec{k})$,
$\vec{a} = 12t(-\vec{i} + 2\vec{j} + 3\vec{k})$

23 Vertical: $t = 3$
Horizontal: $t = \pm 1$
As $t \to \infty, x \to \infty, y \to \infty$
As $t \to -\infty, x \to \infty, y \to -\infty$

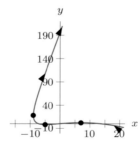

25 $x = 5 + 2(t - 4), y = 4 - 3(t - 4), z = -2 + (t - 4)$.

27 (a) $x = 2 + 0.6t, y = -1 + 0.8t, z = 5 - 1.2t, 0 \le t \le 5$

(b) $x = 2 + 1.92t, y = -1 + 2.56t, z = 5 - 3.84t, 0 \le t \le 1.56$

29 (a) 6.4 meters
(b) 1.14 sec
(c) 15.81 m/sec
(d) $(11.4, -5.7, 0)$
(e) -9.8 m/sec^2

31 (a) 5 secs; $(10, 15, 100)$
(b) $t = 0, 10$ secs, $\sqrt{113}$ cm/sec
(c) 5 secs, $\sqrt{13}$ cm/sec

33 (a) $t = 5.181$ sec
(b) $x = 103.616$ meters
(c) 2 meters
(d) 9.8 meters/sec^2
(e) $\theta = 0.896; v = 32.016$ meters/sec

35 (a) (IV); 4.5 sec; $(0, 8.9$ m, $0)$
(b) (II); 3.2 sec; base of tower
(c) (V); 10 sec; halfway up

37 $\vec{r}(t) = 22.1t\vec{i} + 66.4t\vec{j} + (442.7t - 4.9t^2)\vec{k}$

39 At $t = -\sqrt{6}$

43 SHORT ANSWER NOT WRITTEN

45 (a) R, counterclockwise, $2\pi/\omega$
(b) $\vec{v} = -\omega R\sin(\omega t)\vec{i} + \omega R\cos(\omega t)\vec{j}$
(c) $\vec{a} = -\omega^2\vec{r}$

47 Same path, B moves twice as fast

49 Counterclockwise

51 (a) $x - \sqrt{6}y + z = 3 - 7\sqrt{6}i$
(b) $\pi/3$
(c) 4 ppm/sec

55 Orthogonal only if speed is constant

57 Length $= \int_A^B \|\vec{v}(t)\| \, dt$

59 $0 \le t \le 10/\sqrt{2}$

61 True

63 False

65 True

67 False

69 False

71 False

Section 17.3

1 (a) y-axis
(b) Increasing
(c) Neither

3 (a) x-axis
(b) Increases
(c) Decreases

5 $\vec{V} = x\vec{i}$

7 $\vec{V} = x\vec{i} + y\vec{j} = \vec{r}$

9 $\vec{V} = -x\vec{i} - y\vec{j} = -\vec{r}$

11

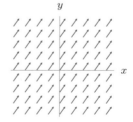

13

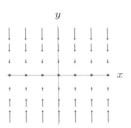

15

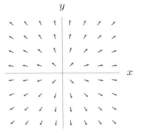

17

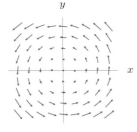

19

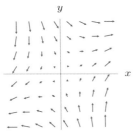

21 (a) III
   (b) II
   (c) IV
   (d) VI

23 $3\vec{i} - 4\vec{j}$, other answers possible

25 $(1/\sqrt{1 + x^2})(\vec{i} - x\vec{j})$, other answers possible

27 $\vec{F}(x, y) = (y + \cos x)((1 + y^2)\vec{i} - (x + y)\vec{j})$, other answers possible

29 I, II, III

31 $\vec{F}(x, y) = x\vec{i}$ (for example)

33 $\vec{F}(x, y) = \frac{y\vec{i} - x\vec{j}}{\sqrt{x^2 + y^2}}$ (for example)

35 (a) Circles origin counterclockwise

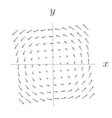

(b) Spirals outward counterclockwise around origin

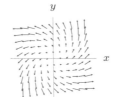

(c) Spirals inward counterclockwise around origin

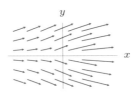

37 (c) $A = 1, K = 1$:

$A = 2, K = 1$:

$A = 0.2, K = 2$:

41 $x^2 - yz$ is not a vector

43 $\left((x^2 + 1)\vec{i} + z\vec{j} + y\vec{k}\right) / \sqrt{(x^2 + 1)^2 + z^2 + y^2}$

Section 17.4

1 Field:

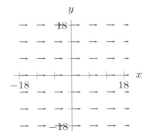

Flow, $y = $ constant:

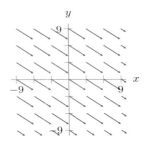

3 Field:

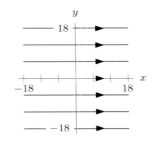

Flow, $y = -(2/3)x + c$:

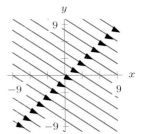

5 Field:

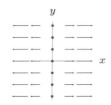

Flow:

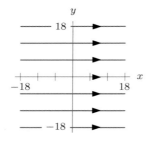

7 Field:

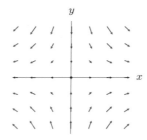

Flow:

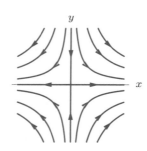

9 Field:

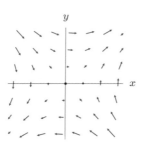

Flow:

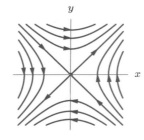

15 (a) Same directions, different magnitudes
(b) Same curves, different parameterizations

17 (a) $\vec{v} \cdot \text{grad } H = 0$

19 (a) $\vec{v} = \pi(-y\vec{i} + x\vec{j})/12$
(b) Horizontal circles

23 (a) $h'(t) = 0$

25 Counterexample: $\vec{F}(x, y, z) = (x^2 + 1)\vec{i}$

27 $\vec{F}(x, y, z) = x\vec{i} + y\vec{j} + z\vec{k}$

29 False

31 False.

33 False

35 False

37 False

Chapter 17 Review

1 $\vec{r} = 2\vec{i} - \vec{j} + 3\vec{k} + t(5\vec{i} + 4\vec{j} - \vec{k})$

3 $x = t, y = 5$

5 $x = 4 + 4\sin t, y = 4 - 4\cos t$

7 $x = 2 - t, y = -1 + 3t, z = 4 + t.$

9 $x = 1 + 2t, y = 1 - 3t, z = 1 + 5t.$

11 $x = 3\cos t$
$y = 5$
$z = -3\sin t$

13 $\vec{r} = 10\cos(2\pi t/30)\vec{i} - 10\sin(2\pi t/30)\vec{j} + 7\vec{k}$

15 $\vec{v} = \vec{i} + (3t^2 - 1)\vec{j}$

17 $\vec{v} = 6t\vec{i} + 2t\vec{j} - 2t\vec{k}$

19 Vector; $\left((3\cos\sqrt{2t+1})\vec{i} - (3\sin\sqrt{2t+1})\vec{j} + \vec{k}\right)/\sqrt{2t+1}$

21 Vector; $-(\cos t/(2\sqrt{3+\sin t}))\vec{i} - (\sin t/(2\sqrt{3+\cos t}))\vec{j}$

23 No

25 Same direction $-\vec{i} + 4\vec{j} - 2\vec{k}$, point $(3, 3, -1)$ in common

27

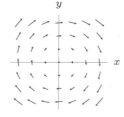

29 When $t = 0, (1, -2, 3)$
When $t = \frac{4}{7}, \left(\frac{15}{7}, -\frac{2}{7}, \frac{17}{7}\right)$

31 (a) $(I) = C_4, (II) = C_1,$
$(III) = C_2, (IV) = C_6$
(b) $C_3 : 0.5\cos t\vec{i} - 0.5\sin t\vec{j},$
$C_5 : -2\cos(\frac{t}{2})\vec{i} - 2\sin(\frac{t}{2})\vec{j}$

33 (A) = (II); (B) = (III); (C) = (IV); (D) = (I)

35 $x = 5 + 3(t - 7), y = 4 + 1(t - 7),$
$z = 3 + 2(t - 7).$

37 $x = (3/\sqrt{5})t, y = 5 - (6/\sqrt{5})t$
$x = -(3/\sqrt{5})t, y = 5 + (6/\sqrt{5})t$

39 $x = -2t, y = 7 - 6t$

41 (a) Vel: $(-4\sin 4t\vec{i} + 4\cos 4t\vec{j} + 3\vec{k})$
Acc: $(-16\cos 4t\vec{i} - 16\sin 4t\vec{j})$
(b) 5
(d) $180°$

43 (a) Posn $= (0, 0, 15)$
Vel $= 5\vec{i} + 3\vec{j} + 2\vec{k}$
Acc $= -2\vec{k}$
(b) $t = 5$
Speed $= \sqrt{98}$

45 (a) $(2t, 0)$ when $0 \leq t \leq \frac{1}{2}$
$(\cos\frac{3\pi}{2}(t - \frac{1}{2}), \sin\frac{3\pi}{2}(t - \frac{1}{2}))$
when $\frac{1}{2} < t \leq \frac{5}{6}$
$(0, -2(t - \frac{4}{3}))$
when $\frac{5}{6} < t \leq \frac{4}{3}$

(b) $(0, 2t)$ when $0 \leq t \leq \frac{1}{2}$
$(\sin\frac{3\pi}{2}(t - \frac{1}{2}), \cos\frac{3\pi}{2}(t - \frac{1}{2}))$
when $\frac{1}{2} < t \leq \frac{5}{6}$
$(-2(t - \frac{4}{3}), 0)$
when $\frac{5}{6} < t \leq \frac{4}{3}$

47 $x = -2 + 9(t + 2)$
$y = 8 - 6(t + 2)$
Other answers possible

49 Line Equation:
$x = 1 + 2t$
$y = 2 + 3t$
$z = 3 + 4t$
Shortest distance: $\sqrt{174}/29$

51 (a) $x = 8t,$
$y = -\frac{1}{2}gt^2 + 10t + 1.5$
(b) $x = 0.2\cos(4\pi t),$
$y = 0.2\sin(4\pi t)$
(c) $x = 8t + 0.2\cos(4\pi t),$
$y = -\frac{1}{2}gt^2 + 10t + 1.5 + 0.2\sin(4\pi t)$
(d)

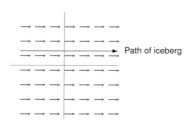

53 (a) Yes
(b) Yes

55 Location at $t = 7$ is:
(a) $x = 8, y = 3$

(b) $x = e^{14}, \ y = 3e^7$

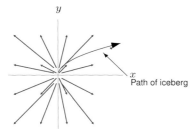

(c) $x = \cos 7 - 3\sin 7$
$y = \sin 7 + 3\cos 7$

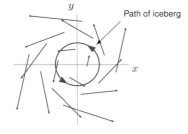

59 (a) Circles centered at origin
(c) Parameterized at different speeds

61 (a) $x = \frac{20}{13} - \frac{6t}{13}, y = \frac{-1}{13} - \frac{t}{13}, z = t$
(b) $x = t, y = \frac{1}{6}(-2 - 2t + 3t^2),$
$z = \frac{1}{6}(20 - 10t - 3t^2).$
(c) $x = \sqrt{2 - t^2}, y = t,$
$z = 5 + 5t - 3\sqrt{2 - t^2}$ and
$x = -\sqrt{2 - t^2}, y = t,$
$z = 5 + 5t + 3\sqrt{2 - t^2}$

Section 18.1

1 Positive

3 Positive

5 Negative

7 (a) Zero
(b) C_1, C_3: Zero
C_2: Negative
C_4: Positive
(c) Zero

9 0

11 0

13 28

15 16

17 -48

19 $19/3$

21 20

23 28

25 C_1 is pos; C_2, C_3 are zero

27 C_1 is neg; C_2 is neg; C_3 is zero

29 $\int_{C_3} \vec{F} \cdot d\vec{r} < \int_{C_1} \vec{F} \cdot d\vec{r} < \int_{C_2} \vec{F} \cdot d\vec{r}$

31 (b)

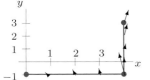

(c) 60

33 $a > 0$

35 $b < 0$

37 $c > 3$

39 0

41 Negative

43 0

45 Yes

51 $-GMm/8000$

53 Sphere of radius a centered at the origin

55 (a) $\phi(\vec{r}) = -\frac{Q}{4\pi\epsilon}\frac{1}{a} + \frac{Q}{4\pi\epsilon}\frac{1}{||\vec{r}||}$
(b) Because then $\phi(\vec{r}) = \frac{Q}{4\pi\epsilon}\frac{1}{||\vec{r}||}$

57 Value of a line integral is not a vector

59

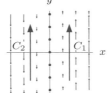

61 True

63 True

65 False

67 False

69 False

Section 18.2

1 $\int_0^\pi (\cos^2 t - \sin^2 t)\, dt$
Other answers are possible

3 $\int_0^{2\pi} (-\sin t \cos(\cos t)$
$\cos t \cos(\sin t))dt$

5 8

7 π

9 $4\cos 2$

11 -2π

13 12

15 85.32

17 24π

19 $39/2$

21 $-7/2$

23 $-157/12$

25 $\int_C y^2\, dx + z^2\, dy + (x^2 - 5)dz$

27 $e^{-3y}\vec{i} - yz(\sin x)\vec{j} + (y + z)\vec{k}$

29 $18\pi^2$

31 -18π

33 (a) $11/6$
(b) $7/6$

35 $560g$

37 (a) $3/2$
(b) $3/2$

43 Sign depends on C

45 $y = \pi/2, x = t, 0 \le t \le 3, \int_C \vec{F} \cdot d\vec{r} = 3$

47 True

49 True

51 False

53 False

Section 18.3

1 Path-independent

3 Path-independent

5 Path-independent

7 12

9 Negative, not path-independent

11 Negative, not path-independent

13 $f(x, y) = x^2 y + K$

15 $f(x, y, z) = e^{xyz} + \sin(xz^2) + C$
$C = $ constant

17 -2

19 -7

21 $\cos(2) - \cos(54)$

23 $(e^7 - 1)/7 - 3/5$

25 0

27 PQ

29 Each integral is 1

31 Yes

33 No

35 $e^9 - 1$

37 (a) $e - 1$: total change in temperature;
no need to find ∇T
(b) No

39 $9/2$

41 $\frac{3}{\sqrt{2}} \ln(\frac{3}{\sqrt{2}} + 1)$

43 $e^{(1.25\pi)^2/2} - 1$

45 (a) $7e^3 - 2e$
(b) $7e^3 - 2e$

47 (a) 9
(b) 0

49 (a)

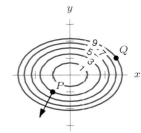

(b) Shorter
(c) 6

51 (a) Positive
(b) Not gradient
(c) $\vec{F_2}$

53 (a) $(8, 9)$
(b) 50

55 (a) $2\pi mg$
(b) Yes

59 (a) $\vec{F} - \text{grad } \phi = -y \, \text{grad } h$
(b) 30

61 (a) $\vec{F} - \text{grad } \phi = -(x + 2y) \, \text{grad } h$
(b) -50

63 (a) Increases

65 $f(Q) - f(P)$ where $\vec{F} = \text{grad } f$

67 Methods other than Theorem 18.1 can be used

69 Gradient of any function

75 True

77 True

79 False

81 True

83 False

Section 18.4

1 No

3 No

5 $f(x, y) = x^3/3 + xy^2 + C$

7 Yes, $f = \ln A|xyz|$ where $A > 0$

9 No

11 -2π

13 -6

15 $-3\pi m^2$

17 -12

19 $e - \cos 1$

21 $-9\pi/8$

23 (a) 0
    (b) 0
    (c) 0
    (d) -6π
    (e) -6π
    (f) 0
    (g) -6π

25 πab

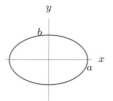

27 3/2

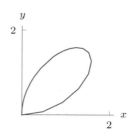

29 (b) 0
    (c) $\vec{G} = \nabla(xyz + zy + z)$
    (d) $\vec{H}_1 = \nabla(yx^2), \vec{H}_2 = \nabla(y(x + z))$

31

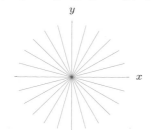

33 (a) Possible answers are:
$$\vec{F} = \text{grad}(xy)$$
$$\vec{G} = \text{grad}(\arctan(x/y)),\ y \neq 0$$
$$\vec{H} = \text{grad}\left((x^2 + y^2)^{1/2}\right),\ (x, y) \neq$$
$$(0, 0)$$
    (b) $0, -2\pi, 0$
    (c) Does not apply to $\vec{G}, \vec{H}$; holes in domain

35 $L_1 < L_2 < L_3$

37 (a) $21\pi/2$
    (b) 2

41 Must have $\int_C \vec{F} \cdot d\vec{r} = 0$ for all closed paths

43 $Q(x, y) = x^2/2$

45 $x\vec{i} + x^2\vec{j}$, scalar curl nonzero

47 False

49 False

51 True

53 False

Chapter 18 Review

1 Negative

3 Scalar; 12

5 -58

7 50

9 18

11 1372

13 Not path-independent

15 Path-independent

17 Path-independent

19 Path-independent

21 350

23 $27\pi/2$

25 45

27 0, 100

29 (ii), (iv)

31 (a) 24
    (b) 12
    (c) -12

35 18

37 -36

39 (a) 0
    (b) 24

41 (a) 0
    (b) 6
    (c) $75\pi/2$
    (d) 14

43 (a) 9/2
    (b) $-9/2$

45

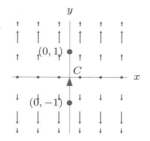

47 (a) Closed curve oriented counterclockwise
    (b) Closed curve oriented clockwise
         with $y > 0$ or
         Closed curve oriented counterclockwise
         if $y < 0$

(Other answers are possible)

49 (a)

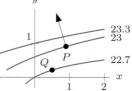

    (b) Longer
    (c) -0.3

51 (a) $\pi/2$
    (b) 0

55 (a) $\omega = 3000 \text{ rad/hr}$
         $K = 3 \cdot 10^7 \text{ m}^2 \cdot \text{rad/hr}$
    (b) Inside tornado:

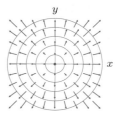

View from great distance:

    (c) $r < 100$ m, circulation is $2\omega\pi r^2$
         $r \geq 100$ m, circulation is $2K\pi$

57 (a) $-(\pi/2)(-6a^2 + a^4), a = \sqrt{3}$
    (b) Integrand $3 - x^2 - y^2$ positive inside disk
         of radius $\sqrt{3}$, negative outside

59 $18a + 18b + 36c + (81d/2)$,
    $-18a - 18b - 36c - (81d/2)$,
    curves go in opposite directions

Section 19.1

1 Scalar; -100π

3 $25\pi\vec{k}$

5 $-15\vec{j}$

7 (a) 45
    (b) -45

9 (a) Positive
    (b) Negative

(c) Zero
(d) Zero
(e) Zero
11 (a) Zero
(b) Zero
(c) Zero
(d) Negative
(e) Zero
13 4
15 3/2
17 2π
19 0
21 3
23 108π
25 $9\pi\sqrt{75}$
27 2π
29 28
31 28π
33 0
35 114
37 64/3
39 756π
41 18
43 $81\pi/2$
45 $\pi(e^4 - 1)$
47 8π
49 (a)   (i) Positive
          (ii) Zero
         (iii) Zero
          (iv) Negative
      (b) Integral over B
51 6π cm^3/sec
53 $\vec{j}$ components parallel to S, other components same
55 $4\pi R$
57 (a) Zero
   (b) Zero
59 (a) π
   (b) $\vec{n} = \frac{1}{\sqrt{x^2+y^2+z^2}}\left(x\vec{i} + y\vec{j} - z\vec{k}\right)$
   (c) 0
63 (a)

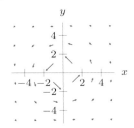

   (b) 0
   (c) $Ih \ln|b/a|/2\pi$
65 Value of a flux integral is not a vector
67 $\vec{F} = -\vec{i} + \vec{j}$
69 True
71 False
73 True
75 True
77 True

Section 19.2

1 $\left(-8\vec{i} - 7\vec{j} + \vec{k}\right) dx\, dy$
3 $\left(-y\vec{i} + (-x - 2y)\vec{j} + \vec{k}\right) dx\, dy$
5 $\int_0^8 \int_0^4 (4z - 10x + y)\, dx\, dy$
7 $\int_0^5 \int_0^{\sqrt{25-x^2}} -3xe^{3y}\cos(x + 2y)\, dy\, dx$
9 8
11 -500
13 $-5/3 - \sin 1 = -2.508$
15 $\int_0^{\pi/2} \int_0^5 10\,(\cos\theta + 2\sin\theta)\, dz\, d\theta$
17 $\int_0^{2\pi} \int_{-8}^8 \left(6z^2\cos\theta + 6\sin\theta e^{6\cos\theta}\right) dz\, d\theta$
19 $\int_0^{2\pi} \int_0^{\pi/2} 100\,(\sin\phi\cos\theta + 2\sin\phi\sin\theta + 3\cos\phi)\sin\phi\, d\phi\, d\theta$
21 $\int_{-\pi/2}^{\pi/2} \int_0^{\pi} 16\cos^2\phi\sin^2\phi\cos\theta\, d\phi\, d\theta$
23 2000
25 $100\sqrt{2}/3$
27 8000/3
29 $(8 - 5\sqrt{2})\pi/6 = 0.486$
31 36π
33 1/6
35 3/2
37 $e + \sin 1 - 1$
39 7/3
41 625π
43 $81\pi/4$
45 1296π
47 1/4
49 $2\pi/3$
51 $4\pi a^3$
53 -1
55 $11\pi/2$
57 (a) Constant inside sphere of radius a
   (b) $\vec{E} = \begin{cases} k\dfrac{\delta_0}{3}\rho\vec{e}_\rho & \rho \le a \\ k\dfrac{\delta_0 a^3}{3r^3}\vec{e}_\rho & \rho > a \end{cases}$
59 Cone not on a cylinder
61 $f(x, y) = -x - y$
63 True
65 False

Section 19.3

1 Scalar; 0
3 0
5 -1
7 $6x - (x + z)\cos(xz)$
9 0
11 $2/\|\vec{r} - \vec{r}_0\|$
13

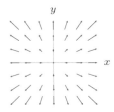

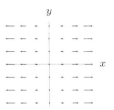

15

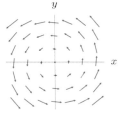

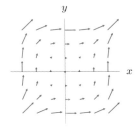

17 $0.00002\pi/3$
19 (a) 2.984
   (b) 0.100
21 div $\vec{F} = 3$
23 (a) $2ay + 6y$
   (b) $a = -3$
25 Undefined
27 (a) 1
29 Positive charges concentrated near line $x = -1$ and negative ones near $x = 1$
31 (a)

   (b) $\vec{v}(x) = (80 - x/10)\vec{i}$ km/hr
        if $0 \le x < 600$
        $\vec{v}(x) = 20\vec{i}$ km/hr
        if $600 \le x < 2400$
        $\vec{v}(x) = (20 + (x - 2400)/5)\vec{i}$ km/hr
        if $2400 \le x < 2700$
        $\vec{v}(x) = 80\vec{i}$ km/hr
        if $x \ge 2700$
   (c) div $\vec{v}(300) = -1/10$, div $\vec{v}(1800) = 0$
        div $\vec{v}(2500) = 1/5$, div $\vec{v}(3000) = 0$
        km/hr/m
33 (a) 0
   (b) 0
37 0

39 $\vec{b} \cdot (\vec{a} \times \vec{r})$

43 (d)

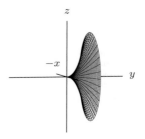

45 div $\vec{F} = 2x + 2 - 2z$

47 $\vec{F}(x, y, z) = 2x\vec{i} + 3y\vec{j} + 4z\vec{k}$

49 $\vec{F}(x, y) = 2x\vec{i}$

51 False

53 False

55 False

57 False

59 False

61 False

Section 19.4

1 24

3 3π

5 2π

9 24

11 72

13 288

15 36π

17 620π

19 48π

21 1.571

23 Yes; -3.22

25 $\int_S \vec{F} \cdot d\vec{A} = \int_W \text{div } \vec{F} \, dV = 0$

27 (a) 0
   (b) No
   (c) 4π
   (d) 4π

29 (a) $cb(12a - a^2)$
   (b) 6, 10, 10; 3600

31 4π

33 (a) $24\pi/5$
   (b) $\int_{S_2} \vec{F} \cdot d\vec{A} < \int_{S_3} \vec{F} \cdot d\vec{A} < \int_{S_4} \vec{F} \cdot d\vec{A}$

35 S closed, oriented inward

37 (a) 30 watts/km³
   (b) $\alpha = 10$ watts/km³
   (d) $6847°$C

39 (a) 0
   (c) No

41 S not the boundary of a solid region

43 Any sphere

45 False

47 False.

49 True.

51 False

53 True

55 False

57 False

Chapter 19 Review

1 Scalar; 54π

3 8

5 -6

7 $2x + xe^z$

9 0

11 0

13 75π

15 80

17 -12

19 0

21 Zero

23 $3(8 + \sin 4)$

25 20

27 2π

29 $b > 0$

31 $a > 0$

33 (a) 3/2
   (b) 3/2

35 32π

37 162

39 120π

41 Flux through S_1 = Flux through S_2 < Flux through S_3 < Flux through S_4

43 $a = 6$
   Cannot say anything about b and c

45 (a) Negative
   (b) Positive
   (c) Zero

47 (a) Flux = c^3
   (b) div $\vec{F} = 1$
   (c) div $\vec{F} = 1$

49 (a) $2c^3$
   (b) 2
   (c) 2

51 -2π

53 42π

55 $-7(\vec{i} + \vec{k})/(16\pi)$

57 (a) div $\vec{F} = 0$
   (b) 12

59 $(20,000\pi/3) - 128$

61 (a) 35π
   (b) 105π
   (c) 70π

63 (a) 0, except at origin
   (b) 4π
   (c) 0
   (d) 4π
   (e) 4π if origin inside
       0 if origin outside

67 (a)   (i) Total charge inside W

        (ii) Total current out of S

69 (b) $\|\vec{v}\| = K/r^2$
   (c) Flux = $4\pi K/3$
   (d) Zero

71 (a)

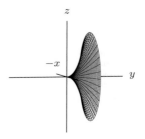

(b) $32\pi/5$

Section 20.1

1 Vector: $\vec{i} + \vec{j} - \vec{k}$

3 $4y\vec{k}$

5 $4x\vec{i} - 5y\vec{j} + z\vec{k}$

7 $\vec{0}$

9 $\vec{0}$

11 Zero curl

13 Nonzero curl

15 0

17 $50\vec{i} + 300\vec{j} + 2\vec{k}$

19 (a) $(f - c)\vec{i} + (be^z - e\cos x)\vec{j} + (2dx - 3ay^2)\vec{k}$
   (b) $f = c$
   (c) $f = c, b = e = 0$

21 (a) Horizontal
   (b) Vertical
   (c) Parallel to the yz-plane, making angle t with horizontal

23 (a) $w = 1$

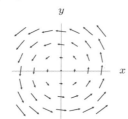

$w = -1$

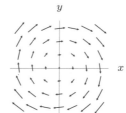

(b) $|\omega| \cdot \sqrt{x^2 + y^2}$
(c) div $\vec{v} = 0$
    curl $\vec{v} = 2\omega\vec{k}$
(d) $2\pi\omega R^2$

33 curl $\vec{F}$ is always a vector field

35 $\vec{F} = x^2\vec{i} + y^2\vec{j} + z^2\vec{k}$

37 True

39 False

41 False

43 True

45 True

Section 20.2

1 25

3 -8π

5 $-\pi$

7 (a) -200π
  (b) 0

9 π

11 $-6\sqrt{3}\pi$

13 0

15 0

17 $8\pi/\sqrt{3}$

19 (a) All 3-space
  (b) $\frac{2ax\vec{i} + 2by\vec{j} + 2cz\vec{k}}{1 + ax^2 + by^2 + cz^2}$
  (c) 0
  (d) $\ln(3 + 507\pi^2/4) - \ln(2)$

21 Negative

23 (a) $-\vec{i} - \vec{j} - \vec{k}$
  (b) -6

25 (a) $\vec{i} + \vec{j} + \vec{k}$
  (b) $-4\sqrt{3}$

27 72

31 120π

33 (a) Parallel to xy-plane; same in all horizontal planes
  (b) $(\partial F_2/\partial x - \partial F_1/\partial y)\vec{k}$
  (d) Green's Theorem

35 C not the boundary of a surface

37 Any oriented circle

39 True

41 False

43 True

45 True

47 False

Section 20.3

1 Yes

3 Yes

5 Yes

7 Yes

9 Yes

11 Yes

13 Curl no; Divergence yes

15 Curl yes; Divergence no

19 $\vec{F} = (-\frac{3}{2}z - 2y)\vec{i} + (2x - z)\vec{j} + (y + \frac{3}{2}x)\vec{k}$

23 No

25 (a) Yes
  (b) Yes
  (c) Yes

27 (a) curl $\vec{E} = \vec{0}$
  (b) 3-space minus a point if $p > 0$
       3-space if $p \leq 0$.
  (c) Satisfies test for all p.
       $\phi(r) = r^{2-p}$ if $p \neq 2$.
       $\phi(r) = \ln r$ if $p = 2$.

29 (b) $\nabla^2\psi = -\,\mathrm{div}\,\vec{A}$

31 Curl of scalar function not defined

33 $f(x, y, z) = x^2$

35 False

37 True

Chapter 20 Review

1 $\vec{i} - \vec{j} - \vec{k}$

3 (c), (d), (f)

5 C_2, C_3, C_4, C_6

7 Defined; scalar

9 Defined; vector

11 Nonzero curl

13 div $\vec{F} = 2x + 3y^2 + 4z^3$
    curl $\vec{F} = \vec{0}$
    $\vec{F}$ not solenoidal; is irrotational

15 div $\vec{F} = -\sin x + e^y + 1$
    curl $\vec{F} = \vec{i} - \vec{j}$
    $\vec{F}$ not solenoidal; not irrotational

17 (a) 30π
  (b) 30π

19 (a) π
  (b) π

21 -5π

23 6

25 (a) $\vec{a}$
  (b) Not defined
  (c) Not defined
  (d) Not defined
  (e) 0
  (f) $-2\vec{a}$

27 (a) Defined; (e), (f) by Stokes'
  (b) Not defined
  (c) Not defined
  (d) Defined
  (e) Defined; (a), (f) by Stokes'
  (f) Defined; (a), (e) by Stokes'
  (g) Defined

29 $600\vec{i} - 500\vec{j} - 2\vec{k}$

31 (a) $-10\sqrt{5}\pi$
  (b) $2\sqrt{3}$
  (c) $2\sqrt{3}(\pi - 1)$

33 $27\pi/2$

35 6π

37 $16\pi/3$

39 -147

41 768π

43 4π

45 (a) $128\pi/3$
  (b) Cannot
  (c) 0

47 (a) $\vec{0}$; all except z-axis
  (b) 2π
  (c) 0
  (d) 2π
  (e) $\pm 2\pi$ if encircles z-axis
       0 otherwise

49 (a) (II), (III), (V)
  (b) (I), (II), (III), (IV), (V)
  (c) (I),(III)

51 (a) $-6\vec{i} - 8\vec{j} + \vec{k}$
  (b) $-6\pi a^2$; limit -6
  (c) $-8\pi a^2$; limit -8
  (d) $\pi(a^2 - (3/4)a^4)$; limit 1
  (e) Limits give $\vec{i}$, $\vec{j}$, $\vec{k}$ components of curl

53 (a) 5
  (b) 0.02094...
  (c) $(20\pi a^3/3) + (4\pi a^5/5)$, limit 5

Section 21.1

1 Horizontal disk of radius 5 in plane $z = 7$

3 Helix radius 5 about z-axis

5 Top hemisphere

7 Vertical segment

9 Curve

11 Surface

13 $s = s_0$: lines parallel to y-axis with $z = 1$
    $t = t_0$: lines parallel to x-axis with $z = 1$

15 $s = s_0$: parabolas in planes parallel to yz-plane
    $t = t_0$: parabolas in planes parallel to xz-plane

17 $\vec{r}(s, t) = (s + 2t)\vec{i} + (2s + t)\vec{j} + 3s\vec{k}$,
    other answers possible

19 $\vec{r}(s, t) = (3 + s + t)\vec{i} + (5 - s)\vec{j} + (7 - t)\vec{k}$,
    other answers possible

21 (a) Yes
  (b) No

23 $s = 4, t = 2$
    $(x, y, z) = (x_0 + 10, y_0 - 4, z_0 + 18)$

25 Horizontal circle

27 (a) $x = \left(\cos\left(\frac{\pi}{3}t\right) + 3\right)\cos\theta$
       $y = \left(\cos\left(\frac{\pi}{3}t\right) + 3\right)\sin\theta$
       $z = t \qquad 0 \leq \theta \leq 2\pi, \, 0 \leq t \leq 48$
  (b) 456π cm^3

29 If $\theta < \pi$, then $(\theta + \pi, \pi/4)$
    If $\theta \geq \pi$, then $(\theta - \pi, \pi/4)$

31 $x = r\cos\theta, \qquad 0 \leq r \leq a$
    $y = r\sin\theta, \qquad 0 \leq \theta \leq 2\pi$
    $z = (1 - r/a)\, h$

33 (a) $-x + y + z = 1$,
       $0 \leq x \leq 2$,
       $-1 \leq y - z \leq 1$
  (b)

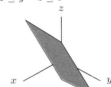

35 (a) $z = (x^2/2) + (y^2/2)$
       $0 \leq x + y \leq 2$
       $0 \leq x - y \leq 2$
  (b)

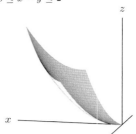

37 Radius: $R\sin\phi$

39 $x + y - z - 3 = 0$

41 True

43 True

45 True

47 False

Section 21.2

1 1

3 e^{2s}

5 -13

7 $a = 1/10, b = 1$

9 $a = 1/50, b = 1/10$

11 3

13 $\rho^2 \sin\phi$

15 13.5

17 72

19 (a) $(1/(2\pi\sigma^2)) \int_{-\infty}^{\infty} \int_{-\infty}^{2t-x} e^{-(x^2+y^2)/(2\sigma^2)}\, dy\, dx$

   (b) $(1/(\sqrt{\pi}\sigma)) \int_{-\infty}^{t} e^{-u^2/\sigma^2}\, du$

   (c) $(1/(\sqrt{\pi}\sigma)) e^{-t^2/\sigma^2}$

   (d) Normal, mean 0, standard deviation $\sigma/\sqrt{2}$

21 R does not correspond to T

23 $x = 2s, y = 3t$

25 False

Section 21.3

1 $((s+t)\vec{i} - (s-t)\vec{j} - 2\vec{k})\, ds\, dt$

3 $-e^s(\cos t\, \vec{j} + \sin t\, \vec{k})\, ds\, dt$

5 0

7 0

9 $2\pi a L$

11 $3\pi/16$

15 (a) $x = (1-v)\cos u + av$, $y = (1-v)\sin u + bv$, $z = cv$, $0 \le u \le 2\pi$, $0 \le v \le 1$

   (b) $\frac{1}{2}\int_0^{2\pi} \sqrt{c^2 + (1 - a\cos u - b\sin u)^2}\, du$

   (c) 5.805

21 Area element, dA is not $ds\, dt$

23 $\vec{F} = -\vec{j}$.

25 True

27 (f)

Chapter 21 Review

1 $a = 1/15, b = 1/15$

3 Cone, height 7 and radius 14

5 $-1/2$

7 0

9 $x = 2 + 5\sin\phi\cos\theta$
  $y = -1 + 5\sin\phi\sin\theta$
  $z = 3 + 5\cos\phi$

11 (a) Cylinder

(b) Helices

13 $x = a\sin\phi\cos\theta \quad 0 \le \phi \le \pi$
   $y = b\sin\phi\sin\theta \quad 0 \le \theta \le 2\pi$
   $z = c\cos\phi$

15 (a) $x^2 + y^2 = 9, x \ge 0, 1 \le z \le 2$.

   (b)

17 0

19 $2\pi c(a^2 + b^2)$

Appendix A

1 (a) $y \le 30$

   (b) two zeros

3 -1.05

5 2.5

7 $x = -1.1$

9 0.45

11 1.3

13 (a) $x = -1.15$

   (b) $x = 1, x = 1.41$, and $x = -1.41$

15 (a) $x \approx 0.7$

   (b) $x \approx 0.4$

17 (a) 4 zeros

   (b) $[0.65, 0.66], [0.72, 0.73],$ $[1.43, 1.44], [1.7, 1.71]$

19 (b) $x \approx 5.573$

21 Bounded $-5 \le f(x) \le 4$

23 Not bounded

Appendix B

3 $\sqrt{2}e^{i\pi/4}$

5 $0e^{i\theta}$, for any θ.

7 $\sqrt{10}e^{i(\arctan(-3)+\pi)}$

9 $-3 - 4i$

11 $-5 + 12i$

13 $1/4 - 9i/8$

15 $-1/2 + i\sqrt{3}/2$

17 $-125i$

19 $\sqrt{2}/2 + i\sqrt{2}/2$

21 $\sqrt{3}/2 + i/2$

23 -2^{50}

25 $2i\sqrt[3]{4}$

27 $(1/\sqrt{2})\cos(-\pi/12) + (i/\sqrt{2})\sin(-\pi/12)$

29 $-i, -1, i, 1$
   $i^{-36} = 1, i^{-41} = -i$

31 $A_1 = 1 + i$
   $A_2 = 1 - i$

37 True

39 False

41 True

Appendix C

1 (a) $f'(x) = 3x^2 + 6x + 3$

   (b) At most one

   (c) $[0, 1]$

   (d) $x \approx 0.913$

3 $\sqrt[4]{100} \approx 3.162$

5 $x \approx 0.511$

7 $x \approx 1.310$

9 $x \approx 1.763$

11 $x \approx 0.682328$

Appendix D

1 3, 0 radians

3 2, $3\pi/4$ radians

5 $7\vec{j}$

7 $\|3\vec{i} + 4\vec{j}\| = \|-5\vec{i}\| = \|5\vec{j}\|, \|\vec{i} + \vec{j}\| = \|\sqrt{2}\vec{j}\|$

9 $5\vec{j}$ and $-6\vec{j}$; $\sqrt{2}\vec{j}$ and $-6\vec{j}$

11 (a) $(-3/5)\vec{i} + (4/5)\vec{j}$

    (b) $(3/5)\vec{i} + (-4/5)\vec{j}$

13 $8\vec{i} - 6\vec{j}$

15 $\vec{i} + 2\vec{j}$

17 Equal

19 Equal

21 $\vec{i} + \vec{j}, \sqrt{2}, \vec{i} - \vec{j}$

23 Pos: $(1/\sqrt{2})\vec{i} + (1/\sqrt{2})\vec{j}$
   Vel: $(-1/\sqrt{2})\vec{i} + (1/\sqrt{2})\vec{j}$
   Speed: 1

INDEX